METHODS FOR DETECTING DNA DAMAGING AGENTS IN HUMANS: APPLICATIONS IN CANCER EPIDEMIOLOGY AND PREVENTION

The International Agency for Research on Cancer (IARC) was established in 1965 by the World Health Assembly, as an independently financed organization within the framework of the World Health Organization. The headquarters of the Agency are at Lyon, France.

The Agency conducts a programme of research concentrating particularly on the epidemiology of cancer and study of potential carcinogens in the human environment. Its field studies are supplemented by biological and chemical research carried out in the Agency's laboratories in Lyon and, through collaborative research agreements, in national research institutions in many countries. The Agency also conducts a programme for the education and training of personnel for cancer research.

The publications of the Agency are intended to contribute to the dissemination of authoritative information on different aspects of cancer research. A complete list is printed at the end of this book.

This volume comprises the proceedings of a symposium on Detection of DNA-damaging Agents in Man, held in Espoo, Finland, on 2–4 September 1987. The symposium was co-sponsored by the Finnish Work Environment Fund; the US National Cancer Institute; the US National Center for Toxicological Research, Food and Drug Administration; the Swedish Work Environment Fund; and Shell Research, Sittingbourne, UK.

INTERNATIONAL AGENCY FOR RESEARCH ON CANCER
INSTITUTE OF OCCUPATIONAL HEALTH, FINLAND
COMMISSION OF THE EUROPEAN COMMUNITIES

METHODS FOR DETECTING DNA DAMAGING AGENTS IN HUMANS: APPLICATIONS IN CANCER EPIDEMIOLOGY AND PREVENTION

EDITORS

H. BARTSCH, K. HEMMINKI & I.K. O'NEILL

IARC SCIENTIFIC PUBLICATIONS NO. 89

INTERNATIONAL AGENCY FOR RESEARCH ON CANCER

LYON 1988

Published by the International Agency for Research on Cancer, 150 cours Albert Thomas, 69372 Lyon Cedex 08, France

CEC-EUR 11381

Distributed for the International Agency for Research on Cancer
by Oxford University Press, Walton Street, Oxford OX2 6DP, UK

Distributed in the USA
by Oxford University Press, New York

ISBN 92 832 1189 8
ISSN 0300 5085

PRINTED IN THE UNITED KINGDOM

CONTENTS

Editorial Board and Programme Committee xi

Forewords

L. Tomatis 1

J. Rantanen 3

P. Bourdeau 5

Introduction 7

Summaries

G.N. Wogan 9

P.H.M. Lohman 13

L.C. Hogstedt 21

KEYNOTE ADDRESSES

Dose monitoring and cancer risk
L. Ehrenberg 23

Detection of DNA damage in studies on cancer etiology and prevention
G.N. Wogan 32

DIETARY EXPOSURES

Do aflatoxin-DNA adduct measurements in humans provide accurate data for cancer risk assessment?
J.D. Groopman 55

Detection of exposure to aflatoxin in an African population
H. Autrup & J. Wakhisi 63

Application of antibody methods to the detection of aflatoxin in human body fluids
C.P. Wild, B. Chapot, E. Scherer, L. Den Engelse & R. Montesano 67

Detection in human cells of alkylated macromolecules attributable to exposure to nitrosamines
R. Montesano, H. Brésil, P. Degan, G. Martel-Planche, M. Serres & C.P. Wild 75

Urinary *N*-nitrosamino acids as an index of exposure to *N*-nitroso compounds
H. Ohshima & H. Bartsch 83

Urinary excretion of 3-methyladenine in humans as a marker of nucleic acid methylation
D.E.G. Shuker & P.B. Farmer 92

N-Nitrosoproline excretion in patients with gastric lesions and in a control population
T.M. Knight, D. Forman, S. Leach, P. Packer, C. Vindigni, C. Minacci, L. Lorenzini, P. Tosi, G. Frosini, M. Marini & N. Carnicelli 97

Immunocytochemical localization of DNA adducts in rat tissues following treatment with *N*-nitrosomethylbenzylamine
J. Van Benthem, C.P. Wild, E. Vermeulen, H.H.K. Winterwerp, L. Den Engelse & E. Scherer 102

Recoverable, semipermeable, microencapsulated DNA surrogates for monitoring the colorectal cavity; in-situ effects of fibre and meat in human diets on benzo[*a*]pyrene and possible endogenous cross-linking agents
I.K. O'Neill, A.C. Povey, S. Bingham, I. Brouet, J.-C. Béréziat & A. Ellul . 107

A model system for studying covalent binding of food carcinogens MeIQx, MeIQ and IQ to DNA and protein
H. Wallin & J. Alexander 113

TOBACCO USE

Approaches to the development of assays for interaction of tobacco-specific nitrosamines with haemoglobin and DNA
S.S. Hecht, S.G. Carmella, N. Trushin, T.E. Spratt, P.G. Foiles & D. Hoffmann 121

Assessment of passive and transplacental exposure to tobacco smoke
M. Sorsa & K. Husgafvel-Pursiainen 129

Haemoglobin adducts of aromatic amines in people exposed to cigarette smoke
M.S. Bryant, P. Vineis, P.L. Skipper & S.R. Tannenbaum 133

DNA adducts, micronuclei and leukoplakias as intermediate endpoints in intervention trials
H.F. Stich & B.P. Dunn 137

Detection of benzo[*a*]pyrene-DNA addutts in cultured cells treated with benzo[*a*lpyrene diol epoxide by quantitative immunofluorescence microscopy and ^{32}P-postlabelling; immunofluorescence analysis of benzo[*a*]pyrene-DNA adducts in bronchial cells from smoking individuals
R.A. Baan, P.T.M. van den Berg, M.-J.S.T. Steenwinkel & C.J.M. van der Wulp 146

OCCUPATIONAL EXPOSURES

Haemoglobin binding in control of exposure to and risk assessment of aromatic amines
H.-G. Neumann 157

Assessment of exposure and susceptibility to aromatic amine carcinogens
F.F. Kadlubar, G. Talaska, N.P. Lang, R.W. Benson & D.W. Roberts 166

DNA adduct formation during continuous feeding of 2-acetylaminofluorene at multiple concentrations
F.A. Beland, N.F. Fullerton, T. Kinouchi & M.C. Poirier 175

Dosimeters of human exposure to carcinogens: polycyclic aromatic hydrocarbon-macromolecular adducts
A. Weston, J.C. Willey, D.K. Manchester, V.L. Wilson, B.R. Brooks, J.-S. Choi, M.C. Poirier, G.E. Trivers, M.J. Newman, D.L. Mann & C.C. Harris 181

Aromatic DNA adducts in white blood cells of foundry workers
K. Hemminki, F.P. Perera, D.H. Phillips, K. Randerath, M.V. Reddy & R.M. Santella 190

A comparison of ^{32}P-postlabelling and immunological methods to examine human lung DNA for benzo[*a*]pyrene adducts
R.C. Garner, B. Tierney & D.H. Phillips 196

Binding efficiency of antibodies to DNA modified with aromatic amines and polycyclic aromatic hydrocarbons: implications for quantification of carcinogen-DNA adducts *in vivo*
E. Kriek, F.J. Van Schooten, M.J.X. Hillebrand & M.C. Welling 201

Synchronous fluorescence spectrophotometry of benzo[*a*]pyrene diol epoxide-DNA adducts: a tool for detection of in-vitro and in-vivo DNA damage by exposure to benzo[*a*]pyrene
K. Vähäkangas, O. Pelkonen & C.C. Harris 208

Micronuclei in cytokinesis-blocked lymphocytes as an index of occupational exposure to alkylating cytostatic drugs
J.W. Yager, M. Sorsa & S. Selvin 213

Detection of styrene oxide-DNA adducts in lymphocytes of a worker exposed to styrene
S.F. Liu, S.M. Rappaport, K. Pongracz & W.J. Bodell 217

Single-strand breaks in DNA of peripheral lymphocytes of styrene-exposed workers
S.A.S. Walles, H. Norppa, S. Osterman-Golkar & J. Mäki-Paakkanen 223

Induction of single-strand breaks in liver DNA of mice after inhalation of vinyl chloride
S.A.S. Walles, B. Holmberg, K. Svensson, S. Osterman-Golkar, K. Sigvardsson & K. Lindblom 227

Determination of specific mercapturic acids in human urine after experimental exposure to toluene or *o*-xylene
Å. Norström, B. Andersson, L. Aringer, J.-O. Levin, A. Löf, P. Näslund & M. Wallèn 232

ALKYLATING EXPOSURES

Prospective detection and assessment of genotoxic hazards: a critical appreciation of the contribution of L.G. Ehrenberg
A.S. Wright, T.K. Bradshaw & W.P. Watson 237

Dosimetry of ethylene oxide
S. Osterman-Golkar 249

Estimation of the cancer risk of genotoxic chemicals by the rad-equivalence approach
A. Kolman, D. Segerbäck & S. Osterman-Golkar 258

Epidemiological studies on ethylene oxide and cancer: an updating
L.C. Hogstedt 265

An immunoassay for monitoring human exposure to ethylene oxide
M.J. Wraith, W.P. Watson, C.V. Eadsforth, N.J. van Sittert & A.S. Wright. 271

Determination of specific urinary thioethers derived from acrylonitrile and ethylene oxide
M. Gérin, R. Tardif & J. Brodeur 275

2-Hydroxyethylation of haemoglobin in man
B.J. Passingham, P.B. Farmer, E. Bailey, A.G.F. Brooks & D.W. Yates ... 279

Immunocytochemical analysis of DNA adducts in single cells: a new tool for experimental carcinogenesis, chemotherapy and molecular epidemiology
E. Scherer, J. Van Benthem, P.M.A.B. Terheggen, E. Vermeulen, H.H.K. Winterwerp & L. Den Engelse 286

Detection of O^4-ethylthymine in human liver DNA
N. Huh, M.S. Satoh, J. Shiga & T. Kuroki 292

Determination of *N*7-methylguanine by immunoassay
D.E.G. Shuker 296

Detection of O^6-methylguanine in human DNA
R. Saffhill, A.F. Badawi & C.N. Hall 301

Detection of DNA adducts by postlabelling with ^{3}H-acetic anhydride
K. Hemminki, K. Savela, E. Linkola & A. Hesso 306

MEDICINAL EXPOSURES

DNA adducts of cisplatin and carboplatin in tissues of cancer patients
M.C. Poirier, M.J. Egorin, A.M.J. Fichtinger-Schepman, S.H. Yuspa & E. Reed 313

Induction and removal of cisplatin-DNA adducts in human cells *in vivo* and *in vitro* as measured by immunochemical techniques
A.M.J. Fichtinger-Schepman, F.J. Dijt, P. Bedford, A.T. van Oosterom, B.T. Hill & F. Berends 321

Determination of cisplatin in blood compartments of cancer patients
R. Mustonen, K. Hemminki, A. Alhonen, P. Hietanen & M. Kiilunen 329

Detection and quantification of 8-methoxypsoralen-DNA adducts
R.M. Santella, X.Y. Yang, V.A. DeLeo & F.P. Gasparro 333

Immunoassay of dithymidine cyclobutane dimers in nanogram quantities of DNA
P.T. Strickland & J.S. Creasey 341

UNIDENTIFIED DNA DAMAGING AGENTS

Novel uses of mass spectrometry in studies of adducts of alkylating agents with nucleic acids and proteins
P.B. Farmer, J. Lamb & P.D. Lawley 347

Pentafluorobenzylation of alkyl and related DNA base adducts facilitates their determination by electrophore detection
M. Saha, O. Minnetian, D. Fisher, E. Rogers, R. Annan, G. Kresbach, P. Vouros & R. Giese 356

Monitoring human exposure to carcinogens by ultrasensitive postlabelling assays: application to unidentified genotoxicants
K. Randerath, R.H. Miller, D. Mittal & E. Randerath 361

An aromatic DNA adduct in colonic mucosa from patients with colorectal cancer
D.H. Phillips, A. Hewer, P.L. Grover & J.R. Jass 368

Enhancement of sensitivity of fluorescence line narrowing spectrometry for detection of carcinogen-DNA adducts
R. Jankowiak, R.S. Cooper, D. Zamzow, G.J. Small, G. Doskocil & A.M. Jeffrey 372

Search for unknown adducts: increase of sensitivity through preselection by biochemical parameters
M. Törnqvist 378

Human organ culture techniques for the detection and evaluation of genotoxic agents
W.P. Watson, R.J. Smith, K.R. Huckle & A.S. Wright 384

Nonselective and selective methods for biological monitoring of exposure to coal–tar products
R.P. Bos & F.J. Jongeneelen 389

An improved standardized procedure for urine mutagenicity testing
A. Rannug, M. Olsson, L. Aringer & G. Brunius 396

Problems in monitoring mutagenicity of human urine
H. Hayatsu, T. Hayatsu, Q.L. Zheng, Y. Ohara & S. Arimoto 401

OXIDATIVE DAMAGE

Measuring oxidative damage in humans: relation to cancer and ageing
B.N. Ames 407

Formation of reactive oxygen species and of 8-hydroxy-2′-deoxyguanosine in DNA *in vitro* with betel-quid ingredients
M. Friesen, G. Maru, V. Bussachini & H. Bartsch 417

Formation of the DNA adduct 8-hydroxy-2′-deoxyguanosine induced by man-made mineral fibres
P. Leanderson, P. Söderqvist, C. Tagesson & O. Axelson 422

APPLICATIONS IN MOLECULAR EPIDEMIOLOGY AND CANCER ETIOLOGY

Prospects for epidemiological studies on hepatocellular cancer as a model for assessing viral and chemical interactions
F.X. Bosch & N. Muñoz 427

DNA restriction fragment length polymorphism analysis of human bronchogenic carcinoma
J.C. Willey, A. Weston, A. Haugen, T. Krontiris, J. Resau, E. McDowell, B. Trump & C.C. Harris 439

Application of biological markers to the study of lung cancer causation and prevention
F.P. Perera, R.M. Santella, D. Brenner, T.-L. Young & I.B. Weinstein 451

Epidemiological studies of the relationship between carcinogenicity and DNA damage
J. Kaldor & N.E. Day 460

Karyotypes of human T-lymphocyte clones
B. Lambert, K. Holmberg, S.-H. He & N. Einhorn 469

Role of oncogenes in chemical carcinogenesis: extrapolation from rodents to humans
M.W. Anderson, R.R. Maronpot & S.H. Reynolds 477

Looking ahead: algebraic thinking about genetics, cell kinetics and cancer
W.G. Thilly 486

LIST OF PARTICIPANTS 493

INDEX OF AUTHORS 502

SUBJECT INDEX 505

FOREWORD

L. Tomatis

International Agency for Research on Cancer, Lyon, France

As a follow-up to the symposium held in 1983 in Espoo, the present conference, organized jointly by the Institute of Occupational Health (Finland) and the Agency, had as its primary goal discussion of the potential of and priorities for conducting integrated laboratory and epidemiological investigations. The wide interest aroused by the subject of the meeting is witnessed not only by the list of very distinguished participants, but also by the impressive list of sponsors, whom I should like to thank for their support:

the Commission of the European Communities, Environment Research Programme; the Finnish Work Environment Fund; the National Cancer Institute, USA; the National Center for Toxicological Research/Food and Drug Administration, USA; the Swedish Work Environment Fund; and Shell Research, UK.

It is one of the most encouraging signs of the present period of cancer research that understanding of certain stages of the carcinogenic process is actually progressing side by side with the development of much more refined methods than have ever existed before for monitoring low doses of exposure at the individual level. The possibilities that now exist for detecting and quantifying the interactions of minute doses of carcinogens with critical cellular macromolecules provide means for new, more efficient epidemiology aimed at identifying carcinogenic hazards. It may also provide means for more accurate evaluation and quantification of cancer risks.

There is little doubt that such possibilities represent a considerable step forward towards improving the prevention of human cancer. Hopefully, several of the most promising dosimetric methods that were discussed will receive priority for further validation and application.

It is, however, important not to forget that the primary role of everyone concerned with public health is to protect people against harmful exposures and that in no case, therefore, can these advances in the development of dosimetric methods be used as an excuse for the deliberate or careless exposure of individuals to carcinogens, or for disregarding the long-term health risks that they may entail.

It is a pleasure for us all to recognize publicly Professor Lars Ehrenberg on this occasion as a pioneer in the line of research that has permitted the developments discussed at this meeting.

I should like to thank the Programme Committee for their work and the Institute of Occupational Health for sponsoring and supporting this meeting.

FOREWORD

J. Rantanen

Institute of Occupational Health, Helsinki, Finland

In 1983, more than 200 scientists convened in this same conference centre to attend an international meeting on monitoring human exposure to carcinogenic and mutagenic agents. The proceedings were published in the IARC Scientific Publications series in the form of a useful compendium. The purpose of the present meeting was to review progress in and practical applications of research on DNA damage.

The focus of the present symposium is more specific than that of the 1983 meeting. This reflects a new trend in cancer research in which the importance of DNA interactions of chemicals is recognized as the primary step in cancer formation. That finding was as fundamental as were later discoveries of the mechanisms of regulation of cellular growth, highlighted by the characterization of oncogenes and their products in cellular systems. How carcinogen-induced DNA damage activates oncogenes is today a most fascinating field for research. The answer to this question will bridge the gap in our knowledge of how lesions in the DNA molecule trigger uncontrolled cellular growth. We also need to elucidate the relationships between DNA damage and structural changes of chromosomes, such as point mutations, sister chromatid exchanges and chromosomal aberrations, and other genetic endpoints measured in humans. The four years intervening between the two meetings have witnessed an extraordinary vigour in the rapidly developing field of research on chemical carcinogenesis, and we had here an interesting opportunity to make an inventory of the present status of our knowledge.

Several industrialized countries have adopted premarketing requirements for new chemicals. These safety requirements also include testing for genotoxic activity and sometimes for carcinogenicity. In the early 1980s, much effort was put into establishing testing protocols, and test systems were energetically worked out. More recently, the question has been raised as to how such test results, obtained mostly in lower organisms and laboratory animals, should be interpreted in relation to human risk assessment. We would probably gain little by horizontal expansion of test systems; we need vertical expansion in order to evaluate human risk. Therefore, measurement of DNA damage in humans is an essential topic of research, and I hope that this meeting has provided a much clearer understanding of the possibilities for using DNA damage as an indicator of human risk.

Short-term testing and animal bioassays have been given an important role in premarketing safety screens and as warning signals for adverse effects in humans. The value of animal models is sometimes disputed, but there is a bulk of scientific evidence on their relevance in predictive testing. In addition, an enormous range of human biochemical reactions was first worked out in experimental animals. These substitutes, however, must not dilute our efforts to obtain direct information on risk on the human organism.

On this occasion, we have the special privilege of honouring Professor Lars Ehrenberg, a pioneer of research on DNA damage. Lars Ehrenberg the scientist has gone through the same stages of choosing models for risk assessment as has the whole scientific community — his thinking has, however, gone ten years ahead of that of the rest of us. In the 1960s, he was developing short-term test systems, in which he compared the mutagenicity of chemicals and of radiation. In the 1970s, he started to measure adducts of carcinogens in humans, using protein adducts as surrogates for DNA damage. And, what is most exciting, he developed and tested methods for risk assessment based on measurements of adducts with human macromolecules. Many of us were surprised to hear how long ago Professor Ehrenberg generated these ideas. The organizers would thus like to dedicate this symposium to Professor Ehrenberg to honour his most remarkable contribution to research on DNA damage.

We should also like to express our gratitude to our co-organizer, the International Agency for Research on Cancer, for smooth collaboration. We are also grateful to the Commission of the European Communities, the National Cancer Institute, USA, the Food and Drug Administration of the USA, Shell Research and the Work Environment Funds of Sweden and of Finland for the financial support that made this meeting economically possible.

I hope that the exciting scientific results presented at this meeting will also benefit people outside the small island of Hanasaari, as well as stimulating those of us who attended the meeting.

FOREWORD

P. Bourdeau

Environment and Non-nuclear Energy Research, Directorate-General for Science, Research and Development, Commission of the European Communities, Brussels, Belgium

The European Communities have important regulatory responsibilities for environmental protection. These include the protection of workers, of the general population and of the environment from risks due to environmental chemicals — both 'new' chemicals and those already present in the environment. The scientific basis for such regulations is, to a significant degree, provided by environmental research at the level of the Community.

The emphasis of research in this area has been the development of methods and techniques that support the implementation of chemical notification systems, in particular the Directive concerning dangerous substances, and other related Community regulations. The work has focused on genetic effects (mutagenesis and carcinogenesis), and some 20 laboratories in the Member States have participated in coordinated research work.

The European Communities are committed to preventive and anticipatory environment policies, and notification of chemicals is a good example of this. However, useful as it is, notification schemes have limitations in terms of predictive ability, applicability to existing chemicals and assessment of multiple exposures. A second line of defence is necessary — one that keeps a watchful eye on people actually exposed or at risk of exposure. Until recently, suitable methods were not readily available for this task, and much of the effort was directed toward estimations of external dose and recognition of clinical effects. However, during the last ten years or so, great progress has been made in developing highly sensitive, rapid methods for detecting and estimating internal dose, biologically effective dose and early, reversible effects of exposure to chemicals.

In no other field have such advances been more exciting and important, or the need for and advantages of international collaboration been more clear, than in the area of genetic effects of environmental chemicals.

The Commission of the European Communities is very pleased to continue its participation in collaborative work on detection methods for assessing exposure to and biological effects of genotoxic chemicals and to be associated with this particular international conference. As well as providing an excellent opportunity to review progress in the field and discuss research directions for the future, the conference is a fitting tribute to the major scientific achievements of Professor Lars Ehrenberg.

INTRODUCTION

The organizers of this meeting would like to express their satisfaction with its outcome. Since 1983, when a similar conference was held, a wealth of new data has been presented, many obtained directly from humans exposed to carcinogens; in addition, many of the participants have initiated collaborative studies with others.

As much is expected of the developing field of metabolic and molecular epidemiology, it was somewhat surprising that epidemiologists and clinicians were under-represented at this meeting. Whatever the reasons were, it would seem that laboratory scientists and epidemiologists still have some difficulty in understanding each other's language. Therefore, special acknowledgement should be made of those workers in epidemiology and related fields who attended the conference, where many of the presentations were packed with technical terms. It could well be that these few will be the first to incorporate into their studies the new tools and methods that have been presented and which will be developed over the next few years.

Clearly, using these powerful tools, not only individuals in the general population who are exposed to carcinogens but also individuals who are more vulnerable to carcinogenic insults, due to predisposition, could be identified, before clinical manifestation of malignancies. Indeed, exploitation of molecular markers for genetic predisposition to cancer, although not discussed to a great extent at this meeting, will be an obligatory adjunct in such studies! Additionally, chemotherapeutic treatment schedules with alkylating anticancer drugs could be optimized for individual cancer patients.

The request that IARC try to standardize some of the new methods will be given serious consideration. Indeed, quality assurance and a guarantee of interlaboratory reproducibility for many of the new methods will be essential.

In this volume, special homage is paid to Lars Ehrenberg (see p. 23), who stimulated research in this area so much, despite the fact that his pioneering contributions, made more than two decades ago, were recognized only after some delay. As a humble gesture from the organizers, the proceedings of this meeting are dedicated to his life's work.

The Editors

SUMMARY: METHODS

G. N. Wogan

Various aspects of methods for detecting DNA damaging agents were discussed in oral presentations as well as in posters presented at the meeting. The main points that emerged can be grouped for convenience into the following categories with respect to their subject matter: existing methodology, including methods that have been used in published reports as well as new applications; modifications of existing methods; new methods currently under development; and studies concerning the validation and characterization of existing methods. The following summarizes the major findings that were presented and avenues of current investigation.

Existing methods

The results of previously published applications of existing methods for detecting exposure to DNA damaging agents were summarized in many presentations. The successful use of urinary markers of genotoxic exposures was reported, as in the detection of *N*-nitrosoproline as an indicator of exposure to *N*-nitroso compounds. The same approach has been used to detect aflatoxin B_1 and aflatoxin B_1-7-guanine as markers of exposure to aflatoxin B_1; 3-methyladenine produced as a result of exposure to methylating agents; and thymine glycol as an indicator of exposure to agents that cause oxidative damage to DNA. Detection of adducts formed between genotoxic agents and haemoglobin has been reported in studies of populations occupationally exposed to ethylene oxide, in which 3-hydroxyhistidine and 3-hydroxyvaline have been measured, and in smokers, whose haemoglobin has been found to contain levels of 4-aminobiphenyl and 3-hydroxyvaline that are correlated with the frequency of cigarette smoking.

Detection of DNA adducts of genotoxic agents in the cells and tissues of exposed individuals has also been accomplished through the use of existing analytical methodology. In several studies, exposure to the ubiquitous polycyclic aromatic hydrocarbon benzo[*a*]pyrene has been detected by the determination of derivatives covalently bound to DNA. Immunoassays and physicochemical methods have been used to detect adducts formed *via* the major intermediate in the activation pathway, the benzo[*a*]pyrene-7,8-diol-9,10-epoxide. This adduct has been identified in the DNA of peripheral leucocytes of workers in foundries, aluminium manufacturing plants, roofers, coke oven workers, and cigarette smokers, by synchronous scanning fluorescence as well as by immunoassays conducted in the enzyme-linked immunosorbent assay or ultrasensitive enzyme radioimmunoassay modes. The successful application of immunoassays to detect DNA adducts of cisplatinum in leucocytes of ovarian cancer patients receiving chemotherapy, and O^6-methylguanine in the blood of populations at high risk for oesophageal cancer was also discussed.

The method of ^{32}P-postlabelling for the detection of DNA adducts has also been used in analysing the DNA of cells and tissues of individuals exposed to environmental carcinogenic insults. The postlabelling technique has been used to detect adducts in placentas, peripheral leucocytes and oral mucosal cells of tobacco smokers as well as of coke oven and foundry workers. Increased total levels of adducts were in general reflective of elevated levels of exposure.

The use of other biomarkers of genotoxic exposure was discussed in the context of the measurement of urinary mutagens as markers for smoking and the ingestion of mutagens formed through the cooking of beef. Measurement of micronuclei in oral mucosal cells was compared with ^{32}P-postlabelling as an indicator of genotoxic damage resulting from betel chewing and inverted smoking. The method in which sister chromatid exchange is measured was found to be of inadequate sensitivity and specificity for the detection of DNA damage resulting from passive smoking.

New applications of existing methods: recent and ongoing studies

Many studies recently completed or currently in progress, in which existing methods were being used to detect DNA damage in exposed populations, were discussed. Essentially, all the methods mentioned above are in current use, including physico-chemical methods (e.g., synchronous scanning fluorescence spectrophotometry), as well as immunoassays and postlabelling. The enzyme-linked immunosorbent assay is being successfully applied to analyse the following: 4-aminobiphenyl adducts in DNA of urinary bladder cells and of haemoglobin; aflatoxin B_1 in urine, blood, tissues and breast milk; O^6-ethylthymine in liver DNA; 7-methylguanine and 8-methoxypsoralen in cells and tissues; and thymine-thymine cyclobutane dimers induced in DNA of cells exposed to ultraviolet light. Radioimmunoassay is being applied in the analysis of hydroxyvaline in haemoglobin.

The method of ^{32}P-postlabelling is being applied in studies of DNA adduct formation in colonic mucosal cells of persons at elevated risk for colorectal cancer, as well as in the detection of adducts with styrene oxide and with fluoranthene in populations occupationally exposed to these agents.

Reports were also made concerning measurement of excretion of urinary markers related to specific exposure or risk situations. These include: *N*-nitrosoproline excretion in relation to stomach cancer risk; mercapturic acid excretion as a measure of exposure to toluene and *n*-xylene; thioethers derived from occupational exposure to ethylene oxide and acrylonitrile; and thioethers together with metabolites such as 1-hydroxybenzo[*a*]pyrene derived from coal-tar products.

New methodological approaches

Detection of DNA adducts

Several new methodological approaches to the detection of DNA adducts are under development and were discussed in both oral and poster presentations. Among the methods based on physicochemical properties, several approaches to improving the sensitivity of detection are being investigated. These include the technique of fluorescence line-narrowing spectrophotometry for the detection and quantification of fluorescent adducts, such as those formed with polycyclic aromatic hydrocarbons. Derivatization with pentafluorobenzene followed by analysis by gas chromatography-mass spectrometry was also discussed as a sensitive method for the detection and quantification of O^4-methylthymine. The property of chemiluminescence forms the basis of a method being developed for quantification of levels of 8-hydroxyguanine in DNA hydrolysates separated by high-performance liquid chromatography and electro-chemical detection.

Additional applications of antibodies were discussed, such as the use of immobilized antibodies for immunopurification of DNA adducts followed by analysis using established procedures, such as the enzyme-linked immunosorbent assay and postlabelling. Such a method is being developed for the detection of benzo[*a*]pyrene-diol epoxide adducts. The possibility for further extension of immunological detection of DNA adducts was afforded by the observation that anti-DNA adduct antibodies are present in the serum of many individuals exposed to polycyclic aromatic hydrocarbons, and whose peripheral leucocytes also contain adducts with these compounds. The development of methods capable of detecting these autoimmune antibodies may therefore provide a sensitive means of quantifying past exposures to genotoxins.

An extension of the conceptual approach of postlabelling was discussed in the context of the utilization of ^{3}H-acetic anhydride for postlabelling adducts such as 7-modified guanines as an alternative to the existing ^{32}P-labelling technique.

Detection of haemoglobin adducts

Several new methodological approaches to the detection of haemoglobin adducts were discussed. Increased applications of the techniques of mass spectrometric analysis were proposed as a means of identification as well as quantification of derivatives of genotoxic agents covalently bound to a haemoglobin and possibly to other blood proteins. Methods that maximally utilize such advances as fast-atom bombardment, selective-ion monitoring and tandem mass spectrometry would provide powerful tools for detecting adducts of a wide range of chemical types and molecular weights. It could therefore substantially expand the existing information base concerning the usefulness of haemoglobin adducts as indicators of genotoxic exposure. Proposals were also made to develop analytical methods for specific haemoglobin adducts for which detection methods do not currently exist. These include a method for detecting 4-(3-pyridyl)-4-oxybutylation of globin (or of DNA) as an indicator of damage by tobacco-specific nitrosamines, and measurement of terminal valine modifications by gas chromatography-mass spectrometry as a means of detecting damage by alkylating agents of a wider range of molecular weights (C_1 to C_n) than is currently possible.

Biomarkers of genotoxic exposure

Development of new biomarkers of genotoxic exposure was also proposed, using several lines of investigation. Detection and quantification of micronuclei in cytokinesis-blocked lymphocytes was described as a method deserving further development on the basis of improved sensitivity in comparison to measurements of sister chromatid exchange or chromosomal aberrations for detecting chromosomal damage. Applications of molecular biological techniques for the detection of restriction fragment length polymorphisms and oncogene activation were put forward as areas of promise for further development in the detection of individuals at high risk or susceptibility. Immunocytochemical analysis of DNA adducts *in situ* in cellular DNA can be accomplished through histochemical methods such as peroxidase-antiperoxidase staining. This detection method combined with quantification by computer-assisted microdensitometry would make possible the characterization of DNA adduct levels in single cells, with the possibility of eventual semi-automation of analysis.

Characterization/validation of existing and new methods

Information presented at the meeting demonstrated the usefulness of existing methods for detecting various indicators of DNA damage in samples collected from human subjects exposed to genotoxic agents. Studies conducted to date have essentially consisted of feasibility trials, designed to determine the adequacy of the analytical methods for detecting the consequences of known or predictable exposures.

In order to provide statistically valid measures of DNA damage useful for assessing risk, much further validation of the methodology will be required. General methodological criteria to be validated include: *sensitivity,* adequate to detect ambient exposures and to yield measurements that accurately reflect exposure levels; *specificity,* established by the use of authentic internal standards of known identity; *accuracy,* defined by recovery of authentic standards; *repeatability/precision,* established through collaborative studies designed to determine intra- and interlaboratory variation in results of the application of a standardized methodology. The need for appropriately designed validation studies was emphasized by many participants, as was the importance of a coordinated effort to achieve these objectives with minimal delay.

Several specific issues were discussed relating to evaluation of the validity of existing methods as well as methods currently under development. It was noted that there has been no systematic effort to determine the level of agreement among results obtained from independent methods of analysis of the same endpoint, such as analysis of a single sample of DNA for adducts by postlabelling, immunoassay and physicochemical methods. Such multiple analyses will be essential to determine the adequacy of each method to fulfil its intended purpose. The importance of complete characterization of antibodies used as reagants in the analytical methods was also emphasized, in particular with respect to their cross-reactivity and the dependence of antigen-antibody binding on levels of DNA adduction. The accuracy of observed measurements may be greatly influenced by such factors. Interpretation of the possible health significance of observed levels of indicators of DNA damage will ultimately depend upon determinations of interrelationships among different indices of exposure, e.g., DNA adduct levels *versus* haemoglobin adduct levels and *versus* biomarkers of damage. At present, information concerning these interrelationships is very limited, and studies designed to provide such data will be very valuable. Finally, an issue of central importance was identified as the interpretation of observed levels of markers of DNA damage in human populations. This concerns the nature and significance of so-called 'background' levels of adducts that have been observed in DNA of individuals not known to have been exposed to DNA damaging agents. The existence of 'background' levels of 3-hydroxyhistidine, 3-methyladenine and 8-hydroxyguanine has been demonstrated by the use of methods capable of their detection, and unidentified DNA adducts have also been detected by postlabelling. It is currently unknown whether these observations are real or artefactual, and what their sources are, if they are real. Future studies must be designed to address these important issues.

SUMMARY: ADDUCTS

P.H.M. Lohman

Introduction

In the 1983 meeting held in the Hanasaari Centre, Finland, one of the general conclusions was that 'the field of quantitative risk estimation is in its infancy... However, newer methods, especially at the molecular level, are being developed that may make possible a beginning of risk estimation, even on an individual basis' (Lohman *et al.*, 1984a).

Now, four years later, it can be concluded that new molecular methods have been developed faster than was expected in 1983, and even the application of such methods for monitoring populations has in some instances become both technically and economically feasible (Garner, 1985; Farmer *et al.*, 1987). However, as will be discussed later in this summary, the field of quantitative risk estimation has not yet escaped its childhood. In other words, it is like playing Russian roulette with our genes: we know to what extent and how genes can be damaged but we do not know which type of damage is critical.

The reasons that the new molecular methods — and especially the measurement of so-called protein and DNA adducts — have become popular, are that (i) they meet the requirements of sufficient intrinsic sensitivity and specificity that make measurements possible in occupational and environmental settings, (ii) they are quantitative, (iii) most fulfil the practical requirements of being relatively cheap, fast and reproducible, and (iv) they can be applied to body fluids like blood and urine or small samples of cells, such as those from buccal mucosa and skin.

Although the study of the interaction of genotoxicants with cellular macromolecules is relatively new, the pyrimidine dimer was recognized as the first radiation-induced DNA adduct in the early 1960s by the group of Berends in The Netherlands (Beukers & Berends, 1960), and DNA damage by chemical agents was demonstrated by the groups of Lawley in the UK (Brookes & Lawley, 1964) and of Ehrenberg in Sweden (see review by Ehrenberg, this volume). In other words, the improvements made in recent years have been technical, not conceptional.

Sensitivity and specificity of new assay systems

The technical improvements with regard to the sensitivity and specificity of the new assay systems are striking, because we are approaching the level of sensitivity that will allow us to measure low, but extensive, exposure of humans under environmental and occupational circumstances. On the basis of molecular dosimetric assays and mutation analysis in mammalian cells in culture, the upper range of subtoxic levels of a genotoxicant can be expected to differ between, e.g., 2000 adducts/cell for exposure to aromatic amines and 100 000 adducts/cell for exposure to ultraviolet light (Lohman *et al.*, 1985). These numbers can also be expressed as number of lesions per unmodified DNA base; exposure to aromatic amines would lead to 1×10^{-7} adducts/base, while in normal sunlight at least 1×10^{-4} adducts/base per day can be expected in the upper layer of the skin as a result of exposure to ultraviolet light.

In Table 1, the intrinsic sensitivity of the available assay systems for measuring quantitatively the interaction between genotoxic agents and cellular macromolecules is estimated. Intrinsic sensitivity is defined as the expected power of each test to detect the same, hypothetical chemical. Of course, the values are derived from incidental observations described during this conference. It is important to realize, however, that the limit of detection per assay may differ by orders of magnitude depending on the physicochemical nature of the chemical or adduct.

The sensitivity of an assay is usually expressed as fmol adduct/μg DNA. This value does not, however, indicate directly whether reliable measurements can be performed in the human situation. Often, for most of the existing physicochemical methods, a sensitivity expressed as number of adducts/base cannot be reached, because a minimal amount of DNA must be available in order to make a measurement.

It will be clear from Table 1 that the immunochemical assays and the postlabelling assay have an excellent record with regard to sensitivity and minimal practical problems for obtaining samples. The tandem mass spectrometry method is the only physicochemical method that meets the same criteria, but the expensive and sophisticated equipment required still reduces its widespread practical application. Practical applications are still also limited for the spectacular, but sophisticated, immunochemical methods for the detection of DNA adducts at the single-cell level as described at this conference (Benthem *et al.*,Perera *et al.* and Baan *et al.*). This holds true especially for the recently introduced laser-scan immunofluorescence microscopy (Baan *et al.*, 1986).

In order to measure occupational exposure to known or suspected genotoxic chemicals, most of the assays for detecting adducts are sufficiently sensitive. However, the tandem mass spectrometry described by Farmer *et al.* (this volume) and the ^{32}P-postlabelling method developed by Randerath *et al.* (this volume) have an additional advantage that interaction of low levels of unknown genotoxic agents can be detected under both occupational and environmental circumstances.

New developments in the detection of protein adducts, especially those involving haemoglobin as the target molecule (Neumann, 1984 and this volume; Osterman-Golkar, this volume), have shown remarkable sensitivity and specificity for detecting exposure of humans to genotoxic agents. In a number of studies, a direct correlation can be made between levels of haemoglobin adducts and of DNA adducts in the same exposed individual. The levels of sensitivity reached with the protein adduct methods are similar to those found with the immunochemical methods for detecting DNA adducts. An additional practical advantage of the protein adduct method is that better 'signal-to-noise' ratios are often obtained than with the DNA adduct assays, because large amounts of protein (especially haemoglobin) can be obtained from individuals.

Methods for the detection of adducts or metabolites of genotoxic agents in urine have also reached high levels of sensitivity and specificity, including both physicochemical and immunochemical methods (Ohshima & Bartsch, this volume; Shuker & Farmer, this volume; Vanderlaan *et al.*, 1987). However, unless they can be clearly validated, such assays carry the same problems as other nonspecific urine assays, such as those for detecting mutagenic activity, as discussed extensively at the previous meeting on this topic (Lohman *et al.*, 1984b; van Sittert, 1984).

It is interesting to note that in some publications dealing with assays of human urine, high responses are assumed to indicate a high internal exposure. However, if a proper balance between intake of the genotoxic agent and excretion of the compound or its metabolite in the urine is not reached, a high response in a urine assay might just as well indicate a very low internal exposure.

Table 1. Assays for the detection of interaction between genotoxic agents and cellular macromolecules

DNA adducts

Method[a]	Estimated lower limit of detection[b]		Amount needed for assay (μg DNA)	Amount usually available in human biopsies (μg DNA)	
	fmol/μg DNA	Adducts/base		Tissue	Blood
UV/HPLC	100 000	1×10^{-5}	1000	1	100
AAS	100	1×10^{-6}	10	1	100
FL/HPLC	50	1×10^{-7}	100	1	100
SSFS	5	1×10^{-8}	100	1	100
MS/MS	0.5	1×10^{-9}	10	1	100
Immunochemical					
competitive	5	5×10^{-8}	0.01–1[c]	1	100
direct	1	1×10^{-8}	0.01–1[c]	1	100
single cell				100 cells	
ES		1×10^{-6}	100 cells	or tissue	
FL		1×10^{-7}	or tissue	section	
FL/Laser		1×10^{-8}	section	Surgical/autopsy sample	
with HPLC	1	1×10^{-8}	100–1000		
^{32}P-Postlabelling	0.1	1×10^{-9}	0.01[c]	0.1	100

Protein adducts[d]

Method	Estimated lower limit of detection[b]		Amount needed for assay (mg protein)	Amount available in human samples (mg blood protein)
	fmol/mg protein	DNA adducts[e]		
GC/HPLC	10	1×10^{-7}	50	150[f]

Adducts/metabolites in urine

Method	Estimated lower limit of detection[b,g] (fmol/ml urine)	Amount needed for assay (ml urine)	Amount available in human samples (ml urine)
Immunochemical	10	5–500	1000
HPLC	10	5–500	1000
GC-MS	10 000	5–10	1000

[a]Abbreviations: UV/HPLC, ultraviolet light/high-performance liquid chromatography; AAS, atomic absorption spectrometry; FL/HPLC, immunofluorescence microscopy/HPLC; SSFS, synchronous scanning fluorescence spectrophotometry; MS/MS, tandem mass spectrometry; ES, enzyme staining; GC/HPLC, gas chromatography/HPLC

[b]The estimated lower limit of detection is expected to reflect the intrinsic sensitivity of a test, i.e., the estimated sensitivity for measuring the same, hypothetical genotoxic agent in each of the individual assays. The values are given only for comparison and may differ by several orders of magnitude depending on the physicochemical properties of the agent or adduct.

[c]For these assays, the DNA must be isolated and purified; current purification methods usually require at least 1 μg DNA

[d]Estimations from Farmer *et al.* (1987)

[e]Calculated as equivalent DNA adduct according to the method of Neumann (1984), using alkylating compound as a model agent

[f]Corresponds to 1–2 ml blood; in practice, much more can be obtained

[g]Often hampered by high background levels

Confounding factors

In using the new molecular assays, a number of confounding factors must be addressed before the methods can be used as quantitative tools for measuring exposure. A major problem seems to be the organ and tissue specificity of the induction of DNA damage. Baan *et al.* (this volume) show, for instance, that the amount of benzo[*a*]pyrene adducts in different cell types in the buccal mucosa of smokers varies significantly, and Wild *et al.* (1986; this volume) demonstrate that, in comparison to liver cells, blood lymphocytes contain extremely low levels of DNA adducts after exposure to aflatoxins *in vivo*. In addition, the interspecies and interindividual differences reported by Poirier *et al.* (this volume) and by Fichtinger-Schepman *et al.* (this volume) in mammals exposed to the cytostatic agent cisplatin are of considerable magnitude. Organ, tissue and species differences may be due to variations in DNA repair processes in different cells. However, the experimental evidence presented here shows clearly that most of the variations occur before DNA repair takes place.

Attention should also be paid to the chemical instability of certain DNA adducts, as described, for instance, by Montesano *et al.* (this volume) for 7-alkylguanine. This and the repair of DNA adducts after induction make accurate measurement of exposure heavily dependent on the time at which a sample is taken. For accurate determinations of exposure, but not necessarily—as discussed later—for risk estimation, measurements of stable, persistent adducts (with both protein and DNA) are therefore more reliable.

Several authors (Groopman, Poirier *et al.*, Fichtinger *et al.* and Randerath *et al.*) indicated that in using the new methods careful attention must be paid to interlaboratory variations. For instance, in immunochemical assays, the source of variation can be attributed to differences in the properties of the antibodies or the presentation of the antigens. It may also be appropriate to signal another source of nonreproducible results, which was not discussed extensively during the meeting, namely, the statistical analysis, or often the lack of statistical analysis, of the data obtained with the new, supersensitive methods.

A confounding factor that was discussed only indirectly during the meeting is the possible occurrence of secondary lesions in DNA. Such 'secondary lesions' may be caused by endogenous genotoxic agents, such as reactive oxygen species, in the adducts themselves which may be formed as a result of the interaction of the genotoxic agent with drug-metabolizing enzyme systems. Oxygen species may induce such lesions. As indicated by Neumann (this volume), at least liver cells can cope with large amounts of such endogenous reactive products. However, the formation of endogenous, secondary lesions would not be related to the dose of the primary, exogenous genotoxic agent; and, therefore, the total amount of adduct formed by both exogenous and endogenous lesions would also not be dose-related.

Another interfering factor I should like to mention is the often appreciable amount of background DNA adducts, the significance of which remains to be elucidated. Their existence is clearly demonstrated in studies of both protein and DNA alkyl adducts (Ehrenberg, this volume), and the ^{32}P-postlabelling method reveals the presence mainly of unknown DNA adducts (Randerath *et al.*, this volume).

Of course, the existence of confounding factors in measuring exposure to genotoxic agents is not unique to the new methods described but is also true for other tests. At the previous meeting, a concise table was prepared summarizing the 'in's and 'out's of the application of the available tests (Lohman *et al.*, 1984a). It is remarkable that in the four years that have passed hardly any of the uncertainties indicated in the table

Table 2. Evaluation of the applicability of new molecular methods for measuring human exposure to genotoxic agents[a]

Criterion	Method of biological monitoring			
	DNA adducts			Protein adducts
	Physical methods	Immunochemical methods	^{32}P-Post-labelling	
Appropriateness for measuring exposure				
qualitative	(+)	+	+	+
recent (one week) internal dose	?	+	+	+
long-term body burden	?	(+)	(+)	(+)
dose at target site	?	+	+	−
Appropriateness for assessing health effects				
nonadverse (reversible)	?	(−)	(−)	(−)
adverse		(−)	(−)	(−)
Interpretation of results				
on individual basis	+	+	+	+
on group basis	+	+	+	+
Precision of method				
technical reproducibility	?	(+)	(+)	+
stability of parameter over time	(+)	(+)	+	(+)
interlaboratory reproducibility	?	(+)	(+)	+
Sensitivity				
for certain environmental exposures	?	+	+	+
for occupational exposures	(+)	+	+	+
for acute exposures	+	+	+	+
Chemical specificity	+	±	−	±
Absence of interference by confounding factors	?	(+)	(+)	(+)
Absence of background levels	?	(−)	(−)	(−)
Simplicity of analysis	−	±	±	±
Ease of sample storage	+	+	+	+
Current applicability				
research level	(+)	+	+	+
routine use	(−)	(+)	(+)	(+)

[a]+, applicable/true; (+) probably applicable/probably true; −, not applicable/not true; (−), not presently applicable/not presently true; ±, cannot be generalized; ?, unknown

have been resolved. The only major change has been the production of new methods for measuring human exposure to genotoxic agents (Table 2).

Risk estimation

In spite of the fact that technological improvements have made sensitive measurement of exposure of humans to genotoxic agents feasible, no new method has been developed for risk estimation (Wogan, this volume). With regard to tumour formation, the rad-equivalent approach of Ehrenberg's group (for reviews, see Ehrenberg, and Wright *et al.*, this volume) still stands alone as a heroic attempt to relate target dose to adverse biological effects of *low levels of exposure* of humans to genotoxic agents. The results obtained so far with the new molecular assay systems have confirmed the validity of the rad-equivalent approach, at least for a limited set of alkylating agents, such as ethylene oxide. The general applicability of the approach must still be proven, especially for agents that form bulky or bifunctional lesions in DNA.

Fig. 1. Factors involved in tumour initiation

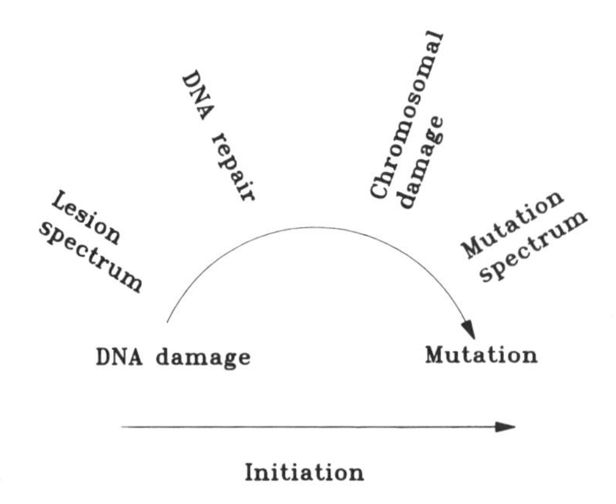

The concept of tumour initiation, promotion and progression was developed from experimental carcinogenesis studies. In this model, carcinogen-DNA adducts are considered to be of prime importance during early initiation of tumour formation, although in the later stages of tumour formation DNA lesions may also play a role in the conversion of benign into malignant cells (for review, see Harris, 1985).

Tumour initiation is often considered to be a single first step in the process of tumour formation; however, as depicted in Figure 1, even initiation is a complex, multistep process. It would be more surprising than logical, on the basis of current scientific knowledge, if the electrophilic reactivity of chemicals and their metabolites, with the induction of primary chemical damage to DNA (including DNA adducts) led stoichiometrically to mutation and/or cancer. With regard to the initiation process, it should be investigated whether, other than for some monoalkylating agents, a direct quantitative relation exists between the induction of DNA damage and ultimate mutation in target cells (Fig. 1). The spectrum of lesions in DNA is often very complex, and lesions are not repaired equally. It has often been questioned whether persistent DNA lesions should be considered 'key' events in the induction of mutations; however, this generalization is not justifiable scientifically. Persistent DNA lesions should be considered good markers for measuring exposure; whether they can also be considered 'key' lesions for initiation is heavily dependent on the properties of the genotoxic agent in question.

DNA repair has also been shown to depend on chromosomal structure. Considerable evidence is accumulating that many agents induce mutations not randomly over the chromosome but at preferential spots (see, e.g., Thilly, 1985; Drobetsky *et al.*, 1987; Vrieling *et al.*, 1988).

As concluded by Wright *et al.* (this volume), the determination of rad-equivalent values for genotoxic action appears to provide the only currently practical operational approach to evaluating both heritable and cancer risks posed by genotoxic chemicals.

Future needs

The new molecular methods for measuring DNA damage *in vivo* after environmental or occupational exposure to genotoxicants show vast interlaboratory variation. Calibration procedures, especially in the human situation, are needed. Therefore, it is recommended that an international collaboration be established to exchange cells from people exposed *in vivo* to known concentrations of genotoxicants.

Because human protein samples can be obtained in sufficient quantities and because most compounds bind at higher levels to proteins than to DNA, determination of protein adducts seems preferable for measuring exposure to monofunctional alkylating agents. However, for risk estimation, measurement of DNA adduct formation in target-cell populations appears to be essential.

The new methods for molecular dosimetry after in-vivo exposure to genotoxic agents have shown considerable interindividual variation in the induction of DNA damage and mutations (Baan *et al.*, Fichtinger-Schepman *et al.*, Poirer *et al.*, this volume). The nature of such variations (whether genetically predetermined or not) should be further investigated.

The new molecular methods (especially the ^{32}P-postlabelling method) have demonstrated the presence of considerable levels of persistent (mostly unknown) DNA lesions in various cells of the human body (Randerath *et al.*, this volume). The importance of the presence of such 'background' lesions for risk estimation and for the analysis of a possible threshold *in vivo* for the adverse effects of genotoxic agents should be investigated further.

In order to develop new approaches for risk estimation, the relation between ('key') DNA lesions and the induction of mutations in target cells in humans and experimental animals should be studied in detail. Current techniques allow quantitative detection of DNA lesions *in vivo* after *low levels of exposure* to genotoxicants. However, mutations occur at much lower levels than DNA adducts in cells of organs and tissues, and the measurement of mutations *in vivo* is still a challenge for the future. A breakthrough in the analysis of mutation induction *in vivo* at the molecular level is expected and should be stimulated (Thilly, 1985; Lohman *et al.*, 1987; Vrieling *et al.*, 1988; Harris, this volume).

Approaches to risk estimation on the basis of measurements of DNA adducts are usually discussed in relation to tumour formation. However, the heritable effects of genotoxic agents should be considered an equally heavy burden on the human population. Therefore, it is recommended that attention be focused on studying the differences between DNA adduct formation and repair *in vivo* in somatic and germ cells.

The role of the genomic instability caused by genotoxic agents *in vivo* should also be studied in view of current understanding of the mechanisms of ageing and evidence of its involvement in the promotion of age-associated diseases other than cancer (Committee on Chemical Toxicity and Aging, 1987).

References

Baan, R.A., Lohman, P.H.M., Fichtinger-Schepman, A.M.J., Muysken-Schoen, M.A. & Ploem, J.S. (1986) *Immunochemical approach to detection and quantitation of DNA adducts resulting from exposure to genotoxic agents.* In: Sorsa, M. & Norppa, H., eds, *Monitoring of Occupational Genotoxicants,* New York, Alan R. Liss, pp. 135–146

Beukers, R. & Berends, W. (1960) Isolation and identification of the irradiation product of thymine. *Biochim. biophys. Acta, 41*, 550

Brookes, P. & Lawley, P.D. (1964) Evidence for the binding of polynuclear aromatic hydrocarbons to the nucleic acid of mouse skin: relation between carcinogenic power of hydrocarbons and their binding to DNA. *Nature, 202*, 781–784

Committee on Chemical Toxicity and Aging (1987) *Aging in Today's Environment (Recommendations),* Washington DC, National Academy Press, pp. 165–168

Drobetsky, E.A., Grosovsky, A.J. & Glickman, B.W. (1987) The specificity of UV-induced mutations at an endogenous locus in mammalian cells. *Proc. natl Acad. Sci. USA, 84,* 9103–9107

Farmer, P.B., Neumann, H.-G. & Henschler, D. (1987) Estimation of exposure of man to substances reacting covalently with macromolecules. *Arch. Toxicol., 60,* 251–260

Garner, R.C. (1985) Assessment of carcinogen exposure in man. *Carcinogenesis, 6,* 1071–1078

Harris, C. C. (1985) Future directions in the use of DNA adducts as internal dosimeters for monitoring human exposure to environmental mutagens and carcinogens. *Environ. Health Perspect., 623,* 185–191

Lohman, P.H.M., Lauwerys, R. & Sorsa, M. (1984a) *Methods of monitoring human exposure to carcinogenic and mutagenic agents.* In: Berlin, A., Draper, M., Hemminki, K. & Vainio, H., eds, *Monitoring Human Exposure to Carcinogenic and Mutagenic Agents* (*IARC Scientific Publications No. 59*), Lyon, International Agency for Research on Cancer, pp. 423–427

Lohman, P.H.M., Jansen, J.D. & Baan, R.A. (1984b) *Comparisons of various methodologies with respect to specificity and sensitivity in biomonitoring occupational exposure to mutagens and carcinogens.* In: Berlin, A., Draper, M., Hemminki, K. & Vainio, H., eds, *Monitoring Human Exposure to Carcinogenic and Mutagenic Agents* (*IARC Scientific Publications No. 59*), Lyon, International Agency for Research on Cancer, pp. 259–277

Lohman, P.H.M., Baan, R.A., Fichtinger-Schepman, A.M.J., Muysken-Schoen, M.A., Lansbergen, M.J. & Berends, F. (1985) *Molecular dosimetry of genotoxic damage: biochemical and immunochemical methods to detect DNA-damage in vitro and in vivo.* In: *TIPS-FEST Supplement,* Amsterdam Elsevier, pp. 1–7

Lohman, P.H.M., Vijg, J., Uitterlinden, A., Gossen, J., Slagboom, P. & Berends, F. (1987) DNA methods for detecting and analyzing mutations *in vivo. Mutat. Res., 181,* 227–234

Neumann, H.-G. (1984) Review: analysis of hemoglobin as a dose monitor for alkylating and arylating agents. *Arch. Toxicol., 56,* 1–6

van Sittert, N.J. (1984) *Biomonitoring of chemicals and their metabolites.* In: Berlin, A., Draper, M., Hemminki, K. & Vainio, H., eds, *Monitoring Human Exposure to Carcinogenic and Mutagenic Agents* (*IARC Scientific Publications No. 59*), Lyon, International Agency for Research on Cancer, pp. 153–172

Thilly, W.G. (1985) The potential use of gradient denaturing gel electrophoresis to obtain mutation spectra in human cells. *Carcinogenesis, 10,* 511–528

Vanderlaan, M., Watkins, B.E., Hwang, M., Knize, M.G. & Felton, J.S. (1987) Monoclonal antibodies for the immunoassay of mutagenic compounds produced by cooking beef. *Carcinogenesis, 9*, 153–160

Vrieling, H., Simons, J.W.I.M. & van Zeeland, A.A. (1988) Nucleotide sequence determination of point mutations at the mouse HPRT locus using *in vitro* amplification of HPRT mRNA sequences. *Mutat. Res., 198,* 107–113.

Wild, C.A., Garner, R.C., Montesano, R. & Tursi, F. (1986) Aflatoxin B_1 binding to plasma albumin and liver DNA upon chronic administration to rats. *Carcinogenesis, 7,* 853–858

SUMMARY: EPIDEMIOLOGICAL APPLICATIONS

L.C. Hogstedt

Although the title of this conference was 'Detection methods for DNA damaging agents in man: applications in cancer epidemiology and prevention', relatively few papers have addressed 'applications in cancer epidemiology and prevention' — and it is not surprising. The remarkable number of methods developed during the 1980s could not yet have been applied in many epidemiological studies or as the basis for preventive action. These methods must be tested for reproducibility and validity in the experimental setting before they can be applied in field research, and they must be economically and practically feasible.

Epidemiology is usually defined as the 'science of the occurrence of diseases', implying the study of determinants of health parameters in populations. The two major components of epidemiological analysis are the outcome, usually a disease, and the determinants of exposure, as reflected in the subdisciplines of epidemiology, such as radiation, occupational and smoking epidemiology, defined on the basis of the exposure, and cancer, cardiovascular and skin epidemiology, defined on the basis of the outcome (disease).

It might be helpful to consider whether the approaches discussed at this conference are exposure- or disease-oriented. The term 'molecular epidemiology' has come into fashion, but 'molecules' are not easily classified as either exposures or disorders. Dr Frederica Perera recently reviewed 'molecular cancer epidemiology' (1987) and gave the following explanation of 'molecular epidemiology':

> '. . . seeks to combine the precision of laboratory methods to quantify carcinogenic dose (exposure) or preclinical response in humans (outcome) with the relevance and rigor of analytic epidemiology'.

It is easy to agree with the author that 'just as infectious disease epidemiology was greatly advanced by methods to identify viruses and to elucidate the mechanisms involved, biological markers of dose and response spring from greater understanding of basic biologic processes and have significant potential in identifying and assessing carcinogenic risks to humans from exposures to man-made environmental chemicals'.

Exposure assessment

Epidemiological studies of cancer usually involve the use of occupational titles, questionnaires or industrial hygiene data for assessing exposure. The new techniques discussed during this conference provide intriguing possibilities for more refined, accurate, relevant exposure-time estimates. It would be an epidemiologist's dream to be able to use an estimate of the biologically effective dose in the target tissue or a (validated) surrogate for exposure assessment. However, a long time lag from the start of exposure to the diagnosed tumour will be a considerable problem, as will be the enormous amount of work and expense involved in monitoring a large number of people for relatively rare cancers.

A cost-effective approach might be to store samples of blood, urine or tissues for decades and to analyse them only for the case entities of concern and for an appropriate number of controls. In a similar way, food and other environmental compounds could be stored. Such comprehensive, rational data banks would not be cheap, but their future value cannot be overestimated. Some samples have been stored already, and plans for more extensive banks are under discussion in several countries.

In one of the very few epidemiological studies presented at this conference, Autrup and Wakhisi correlated aflatoxin B_1-guanine concentrations in urine with liver cancer incidence in a 'geographical epidemiology study'. 'Geographical epidemiology' correlates the occurrence of a disease and the intensity of exposure in a geographical area, without individual data; it can involve, for instance, 'mapping' of cancer incidence and food consumption in a country or state. Warnings against drawing the wrong conclusions from such studies (the 'ecological fallacy') can be found in the epidemiological literature. In studies of diseases that develop a long time after the start of exposure, it is particularly difficult to draw any meaningful inference from such correlations. However, food consumption has probably been fairly stable in some countries, and the approach might be useful if this is substantiated.

Some papers reported on relationships between external dose and DNA adducts in field studies: for instance, Groopman on aflatoxin in the diet compared with total aflatoxin B_1 excretion in urine; and Hemminki *et al.* on aromatic adducts in DNA of white blood cells and polycyclic aromatic hydrocarbon categories among foundry workers. However, large epidemiological studies remain to be performed. In this regard, two very interesting protocols were presented — one by Dr J. Kaldor on cytostatic drugs and second tumours and another by Dr F.X. Bosch and Dr N. Muñoz on liver cancer among 3000 male carriers of hepatitis B surface antigen with measurement of hepatitis B virus markers, aflatoxin adducts to albumin and haemoglobin adducts to other chemicals.

The preventive potential

The preventive potential of validated markers of biologically effective dose and of preclinical response is impressive. They will facilitate epidemiological studies in that they are specific and more numerous than the clinical disease, thus allowing detection of relationships using fewer subjects. If decreasing exposure reduces the number of cancer cases, this action would also have direct preventive importance for the group under study, and not only for all other people that might have been exposed had the demonstrated risk not been observed — albeit that preventive actions are based on scientific results. Future conferences on these new markers of exposure and outcome will probably have more presentations on epidemiological applications, and more applied epidemiologists will be present.

Reference

Perera, F. (1987) Molecular cancer epidemiology: a new tool in cancer prevention. *J. natl Cancer Inst.*, *78*, 887–898

DOSE MONITORING AND CANCER RISK

L. Ehrenberg

Department of Radiobiology, Stockholm University, Stockholm, Sweden

When I was honoured by the dedication of this conference, and when the organizers of the conference asked for a summary of the development of ideas and procedures for dose monitoring in humans, my immediate reaction was that the honour should be distributed to all the collaborators who, through hard work and independent initiative, rendered this development possible.

This introduction also appears to be the right place to express the humble recognition that a scientific achievement is practically never independent of contributions from other schools and institutions. In the present context, the achievements of Elizabeth and James Miller and of Peter Brookes and Phil Lawley thus played a fundamental role. The Millers demonstrated in the 1940s (1947), at the time we initiated our work on induced mutation in agricultural plants, that exposure to chemical carcinogens led to the formation *in vivo* of intermediates that form reaction products ('adducts') with cellular macromolecules; their work also led to the important recognition (1966) that most chemical carcinogens (as it turned out later, those with initiating capability) are, or are metabolized to, electrophilic reagents. Of comparable importance was the basic work of Brookes and Lawley which led them, in 1964 (Brookes & Lawley, 1964a,b), to suggest that DNA is the key target in chemical mutagenesis. Of great importance also was the work at the Chester Beatty Research Institute, London, which involved the introduction of reaction-kinetic thinking into this field of biology, partly with the aim of optimizing the properties of chemotherapeutic drugs (Ross, 1962). In this atmosphere, Loveless (1969) could show that the mutagenic and carcinogenic potency of alkylating agents was associated with reactivity towards certain centres such as O^6-guanine.

Induced mutagenesis in agricultural plants

On the initiative of the plant geneticist, Åke Gustafsson, studies on induced mutagenesis began in the Department of Organic Chemistry and Biochemistry of the University of Stockholm in 1948. The material used was barley (*Hordeum vulgare*) and other agricultural crop species. As stated by the 'father of Swedish genetics', Herman Nilsson-Ehle, these organisms offer the opportunity to carry out basic and applied scientific work in parallel, and the studies on mechanisms of mutagenesis were carried out in close collaboration with plant breeders, who selected agronomically interesting variants from the hereditary variation created. In some cases, this work led to new, economically valuable varieties.

In the beginning, ionizing radiation of different qualities was the mutagenic agent studied. Various ontogenic stages, especially resting and germinating seeds, were exposed under different conditions, and the main endpoints observed were lethal action, growth inhibition (which could have genetic or epigenetic causes), sterility (often due to translocation) and mutation (forward mutation in the approximately 1000 genes that control the chlorophyll apparatus; Gustafsson, 1940).

In order to render the studies of mechanisms meaningful, it was necessary to refer effects to dose; for this purpose, methods for dosimetry of, especially, fast and thermal neutrons were developed (Ehrenberg & Saeland, 1954). In collaboration with A. Ehrenberg, K.G. Zimmer and B. Sparrman, it was possible to demonstrate that free radicals play a role in the biological effects of radiation (Zimmer *et al.*, 1957; Sparrman *et al.*, 1959).

Plant breeders were interested in obtaining as high a mutation frequency and as broad a variation as possible and, of course, also to bring about mutations that induced specific, desirable changes. Although in a few instances certain mutations were induced preferentially by high-linear energy transfer (LET) radiation (neutrons) or by low-LET radiation (γ-radiation, X-rays), evidently due to viable or lethal, multilocus deletions (which are preferentially induced by neutrons), respectively, it soon became clear that such goals were not achievable through variations in irradiation conditions.

With inspiration from publications on the mutagenicity of certain reactive chemicals, especially by Oehlkers (1943), Auerbach (1948) and Rapoport (1948), chemical mutagenesis was introduced into the work around 1955 (Ehrenberg *et al.*, 1956), primarily with the purpose of investigating the potential value for mutation breeding of plants. Certain chemicals, like ethylenimine, indeed gave rise to a number of mutant phenotypes that had never been seen before (Ehrenberg *et al.*, 1959). Although some chemical mutagens offered possibilities for directing mutagenesis towards point mutation, and others towards chromosomal aberrations (multilocus deletions; see Ong & de Serres, 1975) and permitted higher mutation frequencies to be obtained than was possible with ionizing radiation, the general impression was that the locus specificity was low (see Hagberg *et al.*, 1958; Persson & Hagberg, 1969). Such specificity had instead to be achieved by the development of methods for fast screening of desirable hereditary deviations. Physical methods for this purpose were suggested and developed (Zupančič *et al.*, 1967; Johansson *et al.*, 1969).

It was soon realized that mutagenicity should be considered a general property of alkylating agents. As a first step in systematizing mutagenic potency, partly as a tool for optimizing plant breeding, the concepts of mutagenic *efficiency* and mutagenic *effectiveness* were introduced, the former reflecting the ratio of frequencies of mutagenic *versus* lethal and other events that tend to eliminate mutations or reflecting limitations due to factors such as low solubility, and the latter expressing the frequency of mutations per unit dose (Ehrenberg, 1960). In the incipient nuclear age, characterized by a fear of genetic hazards from ionizing radiation, the observation that certain chemicals, some of which were in use, are far more efficient mutagens than radiation indicated the necessity for considering the risks (of hereditary diseases or cancer) to humans from exposure to such agents. This concern was expressed in a lecture in 1958, and in the following year in a letter to the Swedish National Board of Health (Ehrenberg & Gustafsson, 1959). In view of this new aspect of the work with mutagens, the usefulness of higher plants for characterizing the patterns of action of mutagens was explored (for review, see Ehrenberg, 1971).

Early efforts to monitor effects in humans

The lecture mentioned above led to a lot of sensational journalism. In consequence, leaders of an industry producing and using ethylene oxide asked for our assistance in estimating the extent to which their employees were subjected to health hazards due to exposure to this alkylating agent. At that time (1959), it was too early to score chromosomal aberrations in lectin-stimulated lymphocytes, since the conditions for reproducible cell cultivation and assay were still under development. We therefore chose to study the effects on the numbers of peripheral lymphocytes of the kinds found earlier in radiological workers. In addition to one case of leukaemia, about one-half of 35 workers in an ethylene oxide department exhibited a lymphocytosis that could be considered a 'radiomimetic' effect (Ehrenberg & Hällström, 1967). Following an accident in 1961 in which eight people had considerable acute exposure, increased frequencies of chromosomal aberrations were demonstrated (Ehrenberg & Hällström, 1967; see also Ehrenberg *et al.*, 1981).

Struggle to increase the sensitivity of methods for detecting mutagens and carcinogens

Indications that dose-response curves for the mutagenic action of both radiation and chemicals could be extrapolated linearly to the origin, i.e., a no-effect threshold could not be defined, and the possibility that weak 'genotoxic' agents (a concept suggested by H. Druckrey, see Ehrenberg *et al.*, 1973) with a wide distribution could be dangerous to human populations, led to speculation about the sensitivity, or resolving power, of test systems needed to protect humans against health hazards. By characterizing the sensitivity of test systems with respect to dose of γ-radiation that had a detectable mutagenic or carcinogenic effect, most test systems were found to be too insensitive by several orders of magnitude: the detection levels were of the order of 100 rad, whereas doses in the mrad range are considered to be acceptable. Thus, a struggle to increase the sensitivity of test systems, including the problem of interpretation of negative tests (see Ehrenberg, 1977, 1984), became an important component of our work.

Efforts to measure low (but possibly unacceptable) genetic risks in humans had in fact been initiated in about 1957, for a different although related reason: in a radio interview on genetic risks, a colleague and I had expressed the opinion that there were factors more dangerous than radiation. Forced by the mass media to disclose which these factors were led us to suggest, as an example, the rise in testicular temperature due to wearing trousers and, especially, tight underwear (Ehrenberg *et al.*, 1957). Early studies had indicated that the activation energy for spontaneous mutagenesis is high, and we considered it possible that the extracorporal position of the testes during the mating seasons of some mammalian species could function as a protection against spontaneous mutation by lowering the gonadal temperature.

As we felt a responsibility to demonstrate and measure this mutagenic action, we studied mutation in germ cells of human males and, simultaneously, of plants, in order to obtain a greater resolving power owing to the increased numbers of observed individuals. The work with human sperm, using radiolabelled antibodies for surface antigens, and exploratory studies with erythrocytes were premature (especially due to difficulties caused by nonspecific adsorption). The work with plant pollen (including that of barley) led, however, to a highly sensitive system which permitted detection of the mutagenic action of doses comparable to the annual background dose (100 mrad; see Ehrenberg & Eriksson, 1966). This system was highly sensitive to volatile chemical mutagens (Lindgren, 1971).

Reaction kinetic parameters and mutagenicity

In the mid-1960s, Siv Osterman entered the laboratory and undertook a reaction-kinetic characterization of alkylating agents, with the purpose of elucidating the properties that determine mutagenic efficiency and effectiveness. One starting point was the linear free energy relationship of Swain and Scott (1953):

$$\log(k_n/k_{H_2O}) = s \cdot n, \tag{1}$$

where k_n and k_{H_2O} are the second-order rate constants for reaction at nucleophilicity n, s is a substrate constant describing the selectivity of the alkylator, and n is the nucleophilic strength (on an arbitrary scale, in which, by definition, $s = 1$ for methyl bromide and $n = 0$ for H_2O; on this scale, oxygens have n in the range 0–4, nitrogens 3.5–5 and sulphurs 5–7).

These studies exhibited a proportionality between the mutagenic effectiveness in plants and in bacteria of simple monofunctional alkylating agents and their reaction rates ($k_{n=2}$) at $n = \sim 2$ (Turtóczky & Ehrenberg, 1969; Osterman-Golkar *et al.*, 1970), in agreement with accumulating evidence that alkylations at DNA base oxygens are premutagenic events (Loveless, 1969). In contrast, the effectiveness for immediate killing was found to be proportional to reaction rates at $n = \sim 5$, compatible with the idea that killing is caused by inactivation of proteins (key enzymes) through alkylation of —SH and —NH_2 groups. A relatively low s-value is therefore associated with high mutagenic efficiency; in fact, 'supermutagenic' (Rapoport *et al.*, 1966) nitrosamides have very low s-values (Velemínský *et al.*, 1970). For compounds with low s, i.e., effective mutagens, 'immediate' killing did not occur but there was a delayed killing due to genetic damage.

Bifunctional agents were found to be one to two orders of magnitude more effective than monofunctional agents (Ehrenberg & Hussain, 1981) but less efficient due to delayed killing (Ehrenberg & Gustafsson, 1957).

Dose monitoring and risk estimation

The realization that the mutagenic effectiveness of monofunctional alkylating agents is proportional to the rate of reaction towards nucleophiles of a certain strength would have the following implications:

(a) Any low-molecular weight alkylating agent (or metabolite) gives rise to genotoxic effects, provided it can penetrate the DNA space of target cells. This conclusion is valid to the extent that there is no no-effect threshold, and it can most probably be extended to other classes of electrophiles (Ehrenberg & Osterman-Golkar, 1980).
(b) Since demonstration of chemical reactivity is easier than demonstration of genotoxicity, it could be used primarily to judge whether negative results in biological tests, i.e., cases where the null hypothesis of nongenotoxicity was accepted, are false or true (see Ehrenberg, 1977/1984). Demonstration of chemical reactivity could thus be a sensitive means of detecting weak mutagens/carcinogens, which, if widespread in populations, might be causes of considerable genotoxic risks.

The only environmental carcinogenic factor which at the onset of our work (and probably still now) was subjected to regulation on the basis of reasonably realistic risk estimation, is ionizing radiation, particularly γ- and X-radiation. It appeared, therefore, logical to compare chemicals and radiation with respect to current environmental doses and genotoxic effectiveness. In studies of induced forward mutation in barley and *Escherichia coli,* it was then somewhat surprisingly found that, for a number of monofunctional alkylating agents, the 'critical' degree of alkylation at $n = 2$ that gave rise (in the linear low-dose region) to the same response as 1 rad was approximately 1×10^{-7} in both species (after readjustment of the data for barley to 37°C), and this relationship was also found to be approximately valid in other species, including mammalian systems. This indicated that the relationship

$$1\ \text{rad} \leftrightarrow \frac{[RY_{n=2}]}{[Y_{n=2}]} = 1 \times 10^{-7} \qquad (2)$$

might apply also to man (see Ehrenberg, 1979; Ehrenberg & Osterman-Golkar, 1980).

The (cumulative) degree of alkylation at $n = 2$, $[R_iY_{n=2}]^0$, is equal to the product of the second-order rate constant, k_i, for the formation of R_iY, the dose (D) being defined as the time integral of concentration ($[R_iX]$) of the (ultimate) carcinogen (R_iX) under consideration (Ehrenberg *et al.*, 1974):

$$D = \int_t [R_iX](t)\, dt. \tag{3}$$

Measurement of the dose would then allow an estimate of risks in terms of the risk from 1 rad γ-radiation.

When a one-compartment model applies (as is the case for several simple alkylating agents), a solution of (3) is

$$D = \frac{[R_iX]^0}{\lambda_i}, \tag{4}$$

where $[R_iX]^0$ is the cumulative concentration and λ_i the first-order rate constant for disappearance, e.g., through detoxification. The mean life span ($1/\lambda_i$) of R_iX molecules in the body is usually in the order of seconds or minutes, i.e., the concentrations of R_iX during exposure rarely reach the detection levels of analytical methods. Doses can therefore preferably be measured through products of reaction ('adducts') with suitable monitors (Y_{mon}):

$$R_iX + Y_{mon} \rightarrow R_iY_{mon} + X. \tag{5}$$

Exploratory experiments with various tissue proteins showed that such dosimetry was feasible. Ideas of implanting permeable disks containing a nucleophilic gel (polyethylenimine) were abandoned when it was realized that the haemoglobin of the red cells constitutes a suitable natural monitor (long and mostly well-defined life span) which is available in large amounts in both animals and humans (Osterman-Golkar *et al.*, 1976). Haemoglobin was preferred to DNA partly because the methods for adduct identification by mass spectrometric methods are easier. In principle, measurement of adducts to the DNA of white blood cells is no more informative than measurement of haemoglobin adducts. In both cases, if the dose varies between organs, the target:blood dose ratio has to be determined in animal models, e.g., by measuring DNA adduct levels in target cells (see Ehrenberg & Osterman-Golkar, 1980; Segerbäck, 1983).

On the basis of estimates of target dose (D_{targ}), cancer risk can be estimated from expressions of the type

$$\text{Risk} = P_{can}(D) = k_\gamma \cdot Q \cdot D_{targ}, \tag{6}$$

where Q is a quality factor for monofunctional alkylators containing the inverse of thc factor 1×10^{-7} in (2), and k_γ is the risk coefficient for cancer induced by low-LET radiation. For bifunctional alkylating agents, the values of Q are appreciably higher (for diepoxybutane by one to two orders of magnitude), probably due to intra- or interstrand cross-linking of 7-guanine nitrogens, which become premutagenic events (see Ehrenberg, 1979; Ehrenberg & Hussain, 1981).

Throughout our work, it has been clear that precise measurement of the dose as defined in (3) is by no means an obligatory step in risk estimation; its main import lies in its conceptual role in the pharmacokinetic-reaction kinetic model of the in-vivo fate of carcinogens and mutagens (see Ehrenberg *et al.*, 1983). When the ultimate carcinogen is a very short-lived electrophile, it may even be impracticable to determine rate constants for specific reactions with haemoglobin and DNA. It is, however, always possible to determine the *ratios* of reaction rates *in vivo,* and risks could then be referred to the rad equivalent of the (cumulative) level of DNA adducts, this level being calculated from the observed level of the corresponding haemoglobin adducts (see Osterman-Golkar *et al.*, 1976). Viewpoints on the determination of rad equivalence are given by Kolman *et al.* (this volume), and a critical appraisal of this approach is presented by Wright *et al.* (this volume).

Throughout our work, ethylene oxide (and, later, its metabolic precursor, ethene) has been used as a model compound. This is due not only to the initial challenge, in 1959, of a Swedish industry, but also because ethylene oxide is the single or clearly dominant genotoxic agent in certain work environments and thus offers a possibility — otherwise relatively rare — for checking the reasonableness of estimated risks epidemiologically. Osterman-Golkar (this volume) summarizes our work on in-vivo dosimetry of the compound in humans, and Hogstedt (this volume) presents epidemiological observations of cancer risks. By and large, estimated risks were confirmed (Calleman *et al.*, 1978; Ehrenberg & Hussain, 1981).

Risk identification

Macroepidemiological studies indicate that a large proportion of current cancer incidence is due to environmental factors, taken in the broad sense that includes life style and dietary habits (Higginson & Muir, 1979). The precise nature of these factors is, however, largely unknown, particularly with regard to the genotoxic factors (mutagens) that are supposed to act with promoting and cocarcinogenic conditions as initiators and to contribute to malignancy in later stages ('progression') of the development of a tumour.

Since electrophiles react indiscriminately with cellular nucleophiles, at reaction rates determined by kinetic laws, any electrophile can react to some extent with 'critical sites' of DNA. For this reason, any electrophile that can appear in cell nuclei (by permeating membranes or by intranuclear formation) should be considered a mutagen or a cancer initiator (Ehrenberg & Osterman-Golkar, 1980).

Against this background, it was clear that monitoring of in-vivo doses of electrophilically reactive compounds through their adducts with haemoglobin or other macromolecules could have implications far beyond the problem of risk estimation of a-priori defined chemicals or exposures. An unprejudiced screening of members of the general public for unknown adducts would disclose at least the most important mutagens and cancer initiators that occur and permit one to identify their chemical structure and whether they were of exogenous or endogenous origin. Methods could further be suggested to measure the total load, and its variations, of mutagens and cancer initiators, leading to elucidation of the relative role of mutation in current cancer incidence.

Törnqvist (this volume) illustrates — using ethylene oxide and ethene as a model — problems encountered in the identification of causative reactive chemicals and their origin on the basis of adducts observed in unexposed persons.

Such monitoring requires (i) that the sensitivity of the analytical methods be sufficiently high to permit detection of the main contributors to risk; and (ii), as a basis for risk estimation, knowledge of the shape of dose-response curves at the very low doses and levels that occur. These two problems constitute the main components of current research.

With respect to analytical sensitivity, we set as a goal in the development of a method to determine adducts to the *N*-terminals of haemoglobin that adducts be detectable at levels at which the corresponding risks are acceptably low (Törnqvist *et al.*, 1986). A risk of cancer death due to radiation from man-made sources in the range of 10^{-6}–10^{-5} per year might be considered acceptable by members of the public (International Commission on Radiological Protection, 1977). Levels of adducts with ethylene oxide and with chemicals with similar *s*-values (see equation 1 above) associated with these risks (corresponding to 0.05–0.5% of the overall risk of cancer death) are in the range 1–10 pmol/g haemoglobin, and this detection level could be achieved (Törnqvist *et al.*, 1986). It should be stressed, however, that the setting of this goal was provisional. For more effective chemicals (e.g., polyfunctional alkylating and certain bulky agents) and for the components of mixed exposures, work continues to be directed toward still greater sensitivity, in conjunction with the development of methods for analysing other classes of electrophiles.

Acknowledgements

The work summarized in this paper was supported financially by many sources; at early stages, particularly, by the Swedish Agricultural Research Council and the Swedish Atomic Research Council. The quantitative work on in-vivo dose monitoring and risk assessment has, however, met with little sympathy and has left Swedish organizations that support basic research, or research aiming at cancer prevention, indifferent. Since the mid-1970s, the staff has therefore had to fight continuously for its economic existence. Of decisive importance for the achievement of anything at all during this period is the support of the Swedish Work Environment Fund, not the least through the untiring interest and efforts of Nils Boman. In later years, the National Swedish Environment Protection Board and Shell Internationale Research Maatschappij BV have also given important support.

References

Auerbach, C. (1948) Chemical induction of mutations. *Hereditas, Suppl.*, 128–147

Brookes, P. & Lawley, P.D. (1964a) Evidence for the binding of polynuclear aromatic hydrocarbons to the nucleic acids of mouse skin: relation between carcinogenic power of hydrocarbons and their binding to deoxyribonucleic acid. *Nature, 202,* 781–784

Brookes, P. & Lawley, P.D. (1964b) Alkylating agents. *Br. Med. Bull., 20,* 91–95; see also Brookes, P. (1966) Quantitative aspects of the reaction of some carcinogens with nucleic acids and the possible significance of such reactions in the process of carcinogenesis. *Cancer Res., 26,* 1994–2003

Calleman, C.-J., Ehrenberg, L., Jansson, B., Osterman-Golkar, S., Segerbäck, D., Svensson, K. & Wachtmeiser, C.A. (1978) Monitoring and risk assessment by means of alkyl groups in hemoglobin in persons occupationally exposed to ethylene oxide. *J. environ. Pathol. Toxicol., 2,* 427–442

Ehrenberg, L. (1960) Induced mutation in plants: mechanisms and principles. *Genet. Agr., 12,* 364–389

Ehrenberg, L. (1971) *Higher plants.* In: Hollaender, A., ed., *Chemical Mutagens, Principles and Methods for Their Detection,* New York, Plenum, pp. 365–386

Ehrenberg, L. (1977/1984) *Aspects of statistical inference in testing for genetic toxicity.* In: Kilbey, B.J., Legator, M., Nichols, W. & Ramel, C., eds, *Handbook of Mutagenicity Test Procedures,* Amsterdam, Elsevier, 1st ed., pp. 419–459; 2nd ed., pp. 775–822

Ehrenberg, L. (1979) *Risk assessment of ethylene oxide and other compounds.* In: McElheny, V.K. & Abrahamson, S., eds, *Assessing Chemical Mutagens: The Risk to Humans* (Banbury Report 1), Cold Spring Harbor, NY, CSH Press, pp. 157–190

Ehrenberg, L. & Eriksson, G. (1966) The dose dependence of mutation rates in the rad range, in the light of experiments with higher plants. *Acta radiol., 254* (Suppl.), 73–81

Ehrenberg, L. & Gustafsson, Å. (1957) On the mutagenic action of ethylene oxide and diepoxybutane in barley. *Hereditas, 43,* 595–602

Ehrenberg, L. & Gustafsson, Å. (1959) *Chemical Mutagens: Their Uses and Hazards in Medicine and Technology. A Report of February 1959 to the National Board of Health* (English translation published by Bloms Boktryckeri, Lund, 1970)

Ehrenberg, L. & Hällström, T. (1967) *Hematologic studies on persons occupationally exposed to*

Ehrenberg, L. & Hällström, T. (1967) *Hematologic studies on persons occupationally exposed to ethylene oxide.* In: *Radiosterilization of Medical Products* (*Report SM 92/96*), Vienna, International Atomic Energy Agency, pp. 327–334

Ehrenberg, L. & Hussain, S. (1981) Genetic toxicity of some important epoxides. *Mutat. Res., 86,* 1–113

Ehrenberg, L. & Osterman-Golkar, S. (1980) Alkylation of macromolecules for detecting mutagenic agents. *Teratog. Mutag. Carcinog., 1,* 105–127

Ehrenberg, L. & Saeland, E. (1954) Chemical dosimetry of radiations giving different ion densities. An experimental determination of G values for Fe^{2+} oxidation. *J. nucl. Energy, 1,* 150–159

Ehrenberg, L., Gustafsson, Å. & Lundqvist, U. (1956) Chemically induced mutation and sterility in barley. *Acta chem. scand., 10,* 492–494

Ehrenberg, L., von Ehrenstein, G. & Hedgran, A. (1957) Gonad temperature and spontaneous mutation-rate in man. *Nature, 180,* 1433–1434

Ehrenberg, L., Gustafsson, Å. & Lundqvist, U. (1959) The mutagenic effects of ionizing radiations and reactive ethylene derivatives in barley. *Hereditas, 45,* 351–368

Ehrenberg, L., Brookes, P., Druckrey, H., Lagerlöf, B., Litwin, J. & Williams, G. (1973) The relation of cancer induction and genetic damage. *Ambio, Special Report, 3,* 15–16

Ehrenberg, L., Hiesche, K.D., Osterman-Gokar, S. & Wennberg, I. (1974) Evaluation of genetic risks of alkyating agents: tissue doses in the mouse from air contaminated with ethylene oxide. *Mutat. Res., 24,* 83–103

Ehrenberg, L., Hällström, T. & Osterman-Golkar, S. (1981) Etylenoxid. Kriteriedokument för Gränsvärden. *Arbete Hälsa, 6*

Ehrenberg, L., Moustacchi, E. & Osterman-Golkar, S. (1983) Dosimetry of genotoxic agents and dose-response relationships of their effects. *Mutat. Res., 123,* 121–182

Gustafsson, Å. (1940). The mutation system of the chlorophyll apparatus. *Lunds Univ. Årsskr. N.F. Avd. 2, 36,* 1–40

Hagberg, A., Gustafsson, Å. & Ehrenberg, L. (1958) Sparsely contra densely ionizing radiations and the origin of erectoid mutations in barley. *Hereditas, 44,* 523–530

Higginson, J. & Muir, C.S. (1979) Environmental carcinogenesis: misconceptions and limitations to cancer control. *J. natl Cancer Inst., 63,* 1291–1298

International Commission on Radiological Protection (1977) *Recommendations* (Publ. 26), Oxford, Pergamon Press

Johansson, A., Larsson, B., Tibell, G. & Ehrenberg, L. (1969) *On the possibilities of nitrogen determination in seeds by direct nuclear reactions.* In: *New Approaches to Breeding for Improved Plant Protein,* Vienna, International Atomic Energy Agency, pp. 169–171

Lindgren, K. (1971) *The Mutagenic Effects of Ethylene Oxide in Air,* Thesis, Stockholm University

Loveless, A. (1969) Possible relevance of O–6 alkylation of deoxyguanosine to the mutagenicity and carcinogenicity of nitrosamines and nitrosamides. *Nature, 223,* 206–207

Miller, E.C. & Miller, J.A. (1947) Presence and significance of bound aminoazo dyes in livers of rats fed *p*-dimethylaminoazobenzene. *Cancer Res., 7,* 468–480

Miller, E.C. & Miller, J.A. (1966) Mechanisms of chemical carcinogenesis: nature of proximate carcinogens and interactions with macromolecules. *Pharmacol. Rev., 18,* 805–838

Oehlkers, F. (1943) Die Auslösung von Chromosomenmutationen in der Meiosis durch Einwirkung von Chemikalien. *Z. indukt. Abstamm. Vererbungsl.,* 81, 313–341

Ong, T. & de Serres, F.J. (1975) Mutation induction by difunctional alkylating agents in *Neurospora crassa. Genetics, 80,* 475–482

Osterman-Golkar, S., Ehrenberg, L. & Wachtmeister, C.-A. (1970) Reaction kinetics and biological action in barley of monofunctional methanesulfonic esters. *Radiat. Bot., 10,* 303–327

Osterman-Golkar, S., Ehrenberg. L., Segerbäck, D. & Hällström, I. (1976) Evaluation of genetic risks of alkylating agents. II. Haemoglobin as a dose monitor. *Mutat. Res., 34,* 1–10

Persson, G. & Hagberg, A. (1969) Induced variation in a quantitative character in barley. Morphology and cytogenetics of erectoides mutants, *Hereditas, 61,* 115–178

Rapoport, I.A. (1948) Action of ethylene oxide, glycidol and glycols on gene mutation (Russ.). *Dokl. Akad. Nauk SSSR, 60,* 469–472

Rapoport, I.A. *et al.*, eds (1966) *Supermutageny,* Moscow, Izd. Nauka
Ross, W.C.J. (1962) *Biological Alkylating Agents,* London, Butterworths
Segerbäck, D. (1983) Alkylation of DNA and hemoglobin in the mouse following exposure to ethene and ethene oxide. *Chem.-biol. Interact., 45,* 139–151
Sparrman, B., Ehrenberg, L. & Ehrenberg, A. (1959) Scavenging of free radicals and radiation protection by nitric oxide in plant seeds. *Acta chem. scand., 13,* 199–200
Swain, C.G. & Scott, C.B. (1953) Quantitative correlation of relative rates. Comparison of hydroxide ion with other nucleophilic reagents toward alkyl halides, esters, epoxides and acyl halides. *J. Am. chem. Soc., 75,* 141–147
Törnqvist, M., Mowrer, J., Jensen, S. & Ehrenberg, L. (1986) Monitoring of environmental cancer initiators through hemoglobin adducts by a modified Edman degradation method. *Anal. Biochem., 154,* 255–266
Turtóczky, I. & Ehrenberg, L. (1969) Reaction rates and biological action of alkylating agents. Preliminary report on bactericidal and mutagenic action in *E. coli. Mutat. Res., 8,* 229–238
Velemínský, J., Osterman-Golkar, S. & Ehrenberg, L. (1970) Reaction rates and biological action of *N*-methyl- and *N*-ethyl-*N*-nitrosourea. *Mutat. Res., 10,* 169–174
Zimmer, K.G., Ehrenberg, L. & Ehrenberg, A. (1957) Nachweis langlebiger magnetischer Zentren in bestrahlten biologischen Medien und deren Bedeutung für die Strahlenbiologie. *Strahlentherapie, 103,* 3–15
Zupančič, I., Vrščaj, S., Porok, J., Levstek, I., Eržen, V., Blinc, R., Paušak, S., Ehrenberg, L. & Dunmanović, J. (1967) Transient NMR selection method in plant breeding. *Acta chem. scand., 21,* 1664–1665

DETECTION OF DNA DAMAGE IN STUDIES ON CANCER ETIOLOGY AND PREVENTION

G.N. Wogan

Department of Applied Biological Sciences, Massachusetts Institute of Technology, Cambridge, MA, USA

This international conference on methods for detecting DNA damage and their application in studies on cancer etiology and prevention concerns a very timely and important subject. The process of risk assessment is being used increasingly by regulatory decision-making bodies in formulating policies intended to minimize health risks resulting from exposure to hazardous substances. The process of risk assessment requires the use of factual data to define the health effects of exposure of individuals or of populations to such substances. In one current definition (National Academy of Sciences/National Research Council, 1983), the process of risk assessment includes three elements: hazard identification, dose-response assessment and risk characterization. Exposure assessment and epidemiology data are key components of the dose-response assessment, the objective of which is to define dose-incidence relationships for adverse health effects (such as cancer) in human populations. These components, together with animal bioassay data, appropriate extrapolations of information concerning dose-effects in animals and interspecies differences in response effectively determine the quantitative features of the risk estimation, and are therefore of critical importance.

Epidemiological studies designed to evaluate the health significance of environmental chemicals, including carcinogens, are seriously compromised by a lack of quantitative data on individuals in exposed populations. Data on levels of compounds in environmental media often represent the only information available, and average population exposure is therefore the only quantitative parameter that can be calculated. Biological monitoring, i.e., measurements that can be made on cells, tissues or body fluids of exposed people, has the objective of defining 'internal dose', or 'biologically effective dose', on an individual basis. Information gained from such measurements can be utilized to detect potentially hazardous exposures before adverse health effects appear, and also to establish exposure limits that minimize the likelihood of significant health risks. Because measurements are made on individual bases, they provide for that person an indication not only of external exposure to a given substance, but also of the amount absorbed, metabolically transformed to inactive and activated derivatives, and the fraction bound to functionally important cellular sites. Thus, biological monitoring data are complementary to information derived from analysis of environmental media, inasmuch as they can be interpreted in the context of known mechanisms of action, and are therefore more directly relevant to assessment of health risks. Because the monitoring strategy can be designed to take into account exposure through multiple routes and to integrate the consequences of intermittent as well as continuous exposures, it can also provide evidence of total risk from multiple sources.

Certain attributes are required of a method adequate to provide accurate, quantitative measurements of exposure to environmental carcinogens and to fulfil the objective of providing early indication of long-term risk of cancer. Some of those indispensable to attainment of these objectives include the following: (1) the analytical methodology should be adequate to detect and quantify exposure to carcinogens and mutagens at ambient levels in the environment; (2) the methods should be applicable to cells or body fluids that are readily accessible; (3) measured values should be related quantitatively to exposure levels over a wide range; and (4) the methods should integrate consequences of intermittent or continuous exposures to multiple agents. All of these attributes are applicable to accurate dosimetry of *exposure* alone. In order to be applicable to the assessment of *health risk,* the method should also make it possible to detect early biological effects predictive of long-term adverse health risks (e.g., cancer). All of the methods discussed in this volume are potentially applicable to detection of exposure; at present, none can be considered adequately validated to provide direct evidence of cancer risk. Strategies for using exposure information in the prospective assessment of genetic risk represent important future applications of these methodological developments, the validation of which will require a great deal of additional research.

Biological monitoring and its applications in the field that has come to be known as 'molecular epidemiology' have been the subjects of several recent, comprehensive reviews. This volume, and therefore this discussion, deals specifically with detection methods. Broader perspectives can be found in the reviews of Berlin *et al.* (1979, 1984), Bridges (1980), Sorsa and Norppa (1986), Wogan and Gorelick (1985), Garner (1985), Perera (1987) and Farmer *et al.* (1987).

Indicators of genotoxic exposures

Currently available methods for detecting DNA damaging agents fall into two categories on the basis of the character of the experimental systems used and the endpoints detected: measurements of levels of genotoxic chemicals, their metabolites and derivatives in cells, tissues, body fluids and excreta; and measurements of biological responses such as cytogenetic changes in exposed individuals. This discussion emphasizes recent developments in the former category, and only a brief summary of biological indicators of genotoxic exposure is included, consistent with the emphasis reflected in this volume.

Biological indicators

Cytogenetic changes in chromosomes are among the earliest manifestations of genetic injury in persons exposed to ionizing radiation. The linear interdependence of dose of ionizing radiation with the number of structural chromosomal aberrations in circulating lymphocytes has made it possible for the method to be used for biological dosimetry, even at low radiation levels (Evans *et al.*, 1979). Cytogenetic monitoring based on the detection of chromosomal damage in somatic cells is finding increasing application, especially in the occupational setting (Vainio *et al.*, 1983).

Damage to chromosomes can take various forms: structural aberrations, sister chromatid exchanges (SCE) and numerical abnormalities (Bloom, 1981; Evans, 1983). The type of alteration produced depends on the lesions induced in the chromosomes and, therefore, on the nature of the genotoxic injury in question. Structural chromosomal aberrations result from the breakage and rearrangement of whole chromosomes and are most efficiently induced by those substances that directly break the backbone of DNA (e.g., ionizing radiation and radiomimetic chemicals) or significantly distort the DNA helix (such as intercalating agents). SCE result from the breakage and rejoining of DNA strands, without observable morphological distortion of chromosomal structure. They are efficiently induced by genotoxic agents that form covalent adducts with DNA or interfere with DNA synthesis or repair. Numerical alterations (aneuploidy) represent gains or losses of whole chromosomes or parts thereof. It is believed that aneuploidy arises following exposure of the cell to substances that interfere with the apparatus of cell division. Because the three types of cytogenetic endpoints respond to different cellular lesions, information about them is complementary, and all should be taken into account in assessing the possible genotoxicity of environmental chemicals.

The system used most frequently in monitoring exposure to clastogenic agents involves the study of mitogen-stimulated lymphocytes in short-term cultures of blood samples. Because of the long life span of lymphocytes, detection of aberrations in mitogen-stimulated cells offers some possibility for detecting both short-term and accumulated damage, although the kinetics of damage of this type are poorly understood. Structural aberrations can be classified as either unstable or stable, depending on whether they persist in dividing cell populations. Unstable aberrations (dicentrics, rings, deletions and other asymmetrical rearrangements) lead to cell death. Stable alterations consist of balanced translocations, inversions and other symmetrical rearrangements, and are transmitted to progeny at cell division. The biological relevance of these stable rearrangements in somatic cells is not well defined, but recent evidence suggests that certain of them, such as stable translocations, can persist and may confer a growth advantage in dividing cell populations. For example, translocation of an oncogene, the *c-myc* gene, to a site at which its transcription is placed under the control of regulatory elements of the immunoglobulin genes has been observed in patients with Burkitt's lymphoma, and the abnormal expression of these genes in such patients may be causally related to the rearrangement. An oncogene has also been mapped to the site of the rearrangement in chronic myelogenous leukaemia, and other rearrangements in malignancy may be related to the translocation of other oncogenes.

Recent developments in chromatin staining methods have made possible the detection of intrachromosomal SCE. Although the molecular mechanisms underlying these changes have not been fully characterized, it is clear that measurements of SCE offer a sensitive indicator of DNA damaging agents. While the biological consequences of SCE formation are poorly understood, formation of these lesions in a chromosome represents, at the very least, the breakage and rejoining of four DNA strands. It has been amply demonstrated that the frequency of SCE formation is dramatically increased when cells, animals or people are exposed to known carcinogens (Latt, 1981).

The assay methods are quantitative and sensitive, and the occurrence of SCE has been linearly and positively correlated with specific locus mutations *in vitro* and with lung tumour induction in mice. Experiments in rats, mice and rabbits have demonstrated that it is possible to measure an increase in SCE in peripheral blood lymphocytes for several days following exposure *in vivo*. Increased SCE frequencies have been observed in the lymphocytes of cigarette smokers, workers exposed occupationally to ethylene oxide, individuals undergoing cancer chemotherapy and those exposed to certain drugs. As a result of these findings, SCE techniques are considered a valuable adjunct to measurement of chromosomal aberrations in cytogenetic monitoring. The significance of cytogenetic endpoints in the assessment of genotoxic effects of ethylene oxide and their relevance to human cancer has been reviewed recently by Kolman *et al.* (1986).

Recent advances in recombinant DNA procedures are being applied successfully to the identification of molecular defects in man that account for certain heritable diseases, and also to somatic mutations associated with neoplasia. Recombinant DNA methods are highly sensitive and discriminating in the detection of DNA mutations and sequence alterations. As noted above, oncogenes can be activated by point mutations or chromosomal rearrangements, as in the case of Burkitt's lymphoma. Recently, T-cell leukaemias have been found to be associated with translocations that involve the *c-myc* gene and the α chain of the T-cell receptor. The precise site of joining for the translocation has been identified and sequenced. Each junction was initially identified by DNA hybridization analysis with *c-myc* and T-cell receptor probes. If the translocation occurs reproducibly, through aberrant homologous recombination involving specific nucleotide sequences, precise and simple methods for detecting the translocation should be possible. This might in turn make possible the detection of early-stage effects of environmental chemicals capable of inducing leukaemias through this mechanism. Other leukaemias associated with chromosomal translocations include chronic myelogenous leukaemia and acute promyelocytic leukaemia, indicating the possibility of applying a similar approach to their early detection. Restriction fragment length polymorphism (RFLP) analysis is based on the combined use of restriction endonucleases which cleave DNA at specific recognition sites together with oligonucleotide probes to identify changes in fragment length induced by point mutations or insertion or deletion alterations in DNA sequences. The use of RFLP analysis to identify changes in DNA sequence associated with environmental exposures and their possible relationships to cancer risk are discussed elsewhere in this volume.

Biochemical markers

The majority of chemical carcinogens and mutagens exert their effects only after metabolic conversion to chemically reactive forms, which bind covalently to cellular macromolecules, including nucleic acids and proteins, to form addition products (adducts). Certain low-molecular-weight chemicals important in cellular functions, such as water and glutathione, are also readily attacked. In some cases, formation of these adducts may be catalysed by enzymes such as glutathione-*S*-transferase. The overall process whereby electrophiles react with nucleophiles forms a central theorem of carcinogenesis, and particular emphasis has been placed on DNA adducts, since these are thought to represent the initiating events leading ultimately to mutation or malignant transformation. It has been established empirically that the carcinogenic potency of a large number of chemicals is proportional to their ability to bind to DNA—the so-called covalent binding index—when reacted *in vivo* with DNA (Lutz, 1979). Covalent adducts formed in RNA and proteins have no putative mechanistic role in carcinogenesis but are expected to be related quantitatively to total exposure and activation and therefore to represent dosimeters for both exposure and activating capability. The proportionality of response between different target molecules in different tissues and cells forms the basis of an approach to biological monitoring in which the goal is to determine the accumulated dose of the ultimate carcinogenic form of a chemical at the critical target which leads to the unwanted biological response.

A practical approach to the problem of monitoring the critical target dose must be based on the complex relations among the stages of exposure, metabolic and physiological processing, and the ultimate biological effects of genotoxic chemicals. Any measure of exposure or target dose should reflect the capacity of the individual for absorption, metabolism and excretion of the particular carcinogen to give a more accurate and relevant index than simple measurement of the concentration of the compound in air, water or food. Several types of chemical and biochemical methods have been developed for detecting DNA damage in human populations exposed to genotoxic substances. These include measurement of chemicals and their metabolites in blood and urine by chemical analysis and immunoassay, detection of mutagens in urine and measurement of covalent adducts of haemoglobin and DNA. Several analytical methods have been devised to detect covalent adducts, including gas chromatography-mass spectrometry (GC-MS) for determining haemoglobin adducts and physicochemical, immunological and postlabelling for detection and quantification of DNA adducts. For the purposes of this discussion, it is necessary only to summarize briefly the analytical strategies, inasmuch as the reviews cited earlier provide detailed discussion of the various methods involved.

The use of DNA adducts to determine the critical tissue dose may be approached in two ways. First, measurements can be made of the levels of DNA adducts derived from a chemical of interest in cells of an accessible tissue, such as white blood cells or biopsy or autopsy material. If the chemical nature and stability of the DNA adducts for the compound of interest have been fully characterized, qualitative as well as quantitative identification of adduct levels can, in principle, provide an indication of exposure history as well as individual ability to activate the carcinogen to DNA-binding forms. A second approach takes advantage of the fact that some adducts are known to be removed from cellular DNA (and also from RNA) and excreted in urine. Detection and measurement of their excretion rates can provide information on recent exposure of the subject, and possibly also indications of that individual's capability for DNA repair. Thus, studies of urinary excretion of adducts may provide data complementary to measurement of adduct levels in cellular DNA in the same individual.

Although DNA adducts offer the most direct biological monitor for a carcinogen in which DNA is clearly the ultimate target, interpretation of data derived from DNA adduct measurements is in fact highly complex. It is well established that carcinogens — varying in structural complexity from simple alkylating agents to more complex molecules that require multiple steps of activation — react to form covalent bonds at a variety of nucleophilic sites on all four DNA bases as well as on the phosphate backbone of DNA. Thus, from a qualitative viewpoint, detection of all DNA adducts derived from even a single carcinogen is a very complex analytical problem. From a quantitative point of view, the problem is complicated even further by the fact that adducts are removed from DNA by chemical and enzymatic processes at different rates. The rates may vary for each type of adduct, from one tissue to another, or even for the same adduct in different types of cells in the same tissue. Singer and Grunberger (1983) gave a comprehensive review of the types of DNA adducts formed by different carcinogens and mutagens, as well as their repair characteristcs.

In the past few years, analytical methods of several types have been devised for the detection and quantitative analysis of carcinogen-DNA adducts at levels reflecting exposure to ambient levels of the agents. Most of the available information concerning DNA adducts in experimental systems has been obtained through the use of physicochemical or radiochemical detection. The usefulness of these methods of detection in human monitoring is limited by their relative insensitivity and inapplicability, respectively. In a few instances (e.g., benzo[*a*]pyrene and aflatoxins), ultrasensitive physicochemical methods based on the inherent property of fluorescence have been successfully applied to the detection of carcinogen-DNA adducts in human material. A postlabelling method, in which modified DNA bases are detected by ^{32}P-labelling of mononucleotides produced by enzymatic hydrolysis of DNA, has been devised and shown to be effective in detecting adducts of a large number of carcinogens of differing chemical structures (Randerath *et al.*, 1985). Immunoassays have been developed with sensitivities that approach the levels required for the detection of modified DNA in exposed populations (Poirier, 1981; IARC/IPCS Working Group, 1982). These methods are currently being evaluated in studies of workers and others known to be exposed to polycyclic aromatic hydrocarbons (PAH) and other environmental carcinogens, and representative results are summarized in the following section.

The chemical modification of proteins is a well-established phenomenon. Indeed, clinical application of a haemoglobin dosimeter has already been used successfully in monitoring excessive exposure of diabetics to blood glucose. Circulating glucose reacts *via* a carbonylamine condensation to form a Schiff's base, predominantly with the *N*-terminal amino groups of the globin chains, to yield a glycosylated haemoglobin named haemoglobin A_{1c}. Levels of this modified haemoglobin are significantly elevated in diabetics, and its level in blood in a given patient reflects the degree of diabetic control. This approach has been regarded as more useful than measurement of glucose levels in blood or urine because it gives an integrated picture of exposure. The projected use of haemoglobin alkylation products formed as a result of exposure to environmental carcinogens is based on the same principles.

The main nucleophilic centres in proteins are the sulphur atoms of cysteine and methionine, nitrogens of amino groups, guanido groups, ring systems and oxygen atoms. The chemical reactivity of the nucleophilic sites of protein is determined by several factors, such as polarizability, protonation state and pK_a, among others. Experience has shown that the simple concept of electrophile reacting with nucleophile is not sufficient to predict the nature of the site participating in protein-carcinogen adduct formation. The type of adduct formed is frequently unpredictable and may depend on the affinity of the carcinogen for induced binding sites on the protein which steer the proximal reactive form of the carcinogen to a particular nucleophilic site on the protein.

Ehrenberg and his colleagues conducted the pioneering work on the use of protein adducts as dosimeters and have made many contributions to the field, only the main points of which can be mentioned here. Ehrenberg and Osterman-Golkar (1980) reviewed the rationale and technical requirements for the use of protein alkylation for detecting mutagenic agents. Important among the requirements is that exposure must result in the formation of stable covalent derivatives of amino acids for which assay methods of adequate sensitivity and specificity can be devised. Further, the target protein should be found in easily accessible fluids (e.g., blood) and should be present in concentrations adequate to provide sufficient material for analysis. Although any protein could, in principle, be used for monitoring alkylated derivatives, haemoglobin was suggested by Osterman-Golkar *et al.* (1976) as a suitable dose-monitoring protein, and the majority of the relevant literature concerns studies of haemoglobin alkylation. More recently, albumin has been recognized as a potentially useful dose monitor, since it is abundant and reactive, has a long half-life and is synthesized in the hepatocyte — a site where many carcinogens are metabolized to their most reactive forms. Also, since albumin is a component of the interstitial fluid which bathes all cells, it may capture carcinogen metabolites from any tissue.

In an extensive series of studies, Ehrenberg and his colleagues characterized the essential attributes of haemoglobin as a dosimeter, which can be summarized as follows. The stability of alkylated residues in haemoglobin modified by ethylene oxide or *N*-nitrosodimethylamine was established, and the half-life of alkylation levels produced by a single dose of either agent was found to be equivalent to the life span of erythrocytes in mice. The validity of the steady-state level of alkyl residues in haemoglobin as a measure of chronic, repeated exposure was subsequently established in mice dosed repeatedly with methylmethane sulphonate. Osterman-Golkar *et al.* (1983) reviewed studies of haemoglobin alkylation in people occupationally exposed to ethylene oxide. Blood samples were obtained from workers exposed to known levels of ethylene oxide (established by analysis of air samples), and haemoglobin was analysed for the presence of *N*-3-(2-hydroxyethyl)histidine by mass spectrometry and by ion-exchange amino acid analysis. The authors concluded that the haemoglobin alkylation values accurately reflected exposure and were in good agreement with earlier data derived for ethylene oxide in mice.

Pereira and Chang (1981) surveyed the ability of carcinogens and mutagens representing a broad spectrum of chemical classes to bind covalently to haemoglobin in rats. Animals were dosed with ^{14}C-labelled carcinogens and blood was collected 24 h later. Covalent binding was determined by analysis of purified haemoglobin for ^{14}C bound to the protein. All the carcinogens studied were found to form covalent haemoglobin adducts in a dose-related manner, but the absolute binding level was *not* related to known carcinogenic potency.

In order for protein alkylation to be useful as a monitoring procedure, reliable dose-response relationships between exposure dose and production of alkylated amino acids must be established. This requirement has been satisfied for the exposures studied to date, all of the observations having been made in experimental animals dosed with various known carcinogens. Thus, GC-MS determination of the levels of *S*-methylcysteine in haemoglobin of rats following injection of methylmethane sulphonate showed that the level of alkylated amino acid was linearly related to dose. For ethylene oxide, a virtually linear relationship was observed between dose and production of *N*-3-(2-hydroxyethyl)histidine in haemoglobin, in experiments in which the alkylating agent was administered by inhalation at doses of 0–100 ppm in air, 30 h per week for two years. Other agents for which dose-response relationships have been established include *trans*-4-dimethylaminostilbene, chloroform, *N*-nitrosodimethylamine and 4-aminobiphenyl, which are discussed in more detail below.

On this basis, it can be concluded that the amount of alkylated haemoglobin is directly related to erythrocyte dose, and that erythrocyte dose and exposure dose are almost always linearly related to each other. Thus, many of the requirements for validating haemoglobin adducts as exposure dosimeters have been met. It is important to consider whether this parameter could also be used to estimate carcinogenic risk. The observation of haemoglobin alkylation *per se* can be taken as an indicator of genotoxic risk only when it has been shown that such alkylation correlates with reactions at the target DNA site, i.e., that the erythrocyte dose is directly related to the target dose. This relationship has been studied in only a few experimental systems, in which the amounts of DNA and haemoglobin binding products have been compared following dosing with carcinogens. Observed levels of alkylation of guanine in DNA of liver and testis induced by ethylene oxide deviated by no more than two-fold from the amount expected on the basis of haemoglobin alkylation. Thus, the degree of alkylation of DNA could be estimated approximately from the dose of this compound in erythrocytes. Similar relationships have been shown for *trans*-4-dimethylaminostilbene and 2-acetylaminofluorene. Thus, in at least *some* instances, it seems possible to predict DNA binding by measuring protein binding, and therefore the latter may in some cases be taken as an indication of genotoxic risk.

Detection of DNA damaging agents in man

The development of these methods has made possible their evaluation for detecting DNA damaging agents in man. Results of several such studies have been published,

Table 1. Indicators of genotoxic exposure: urinary excretion of markers

Compound analysed	Exposure source	Method of analysis[a]	Principal findings[b]		Reference
Nitrosamino acids	Unknown	GC-TEA	High-risk area (44)	21.2 μg/day	Lu *et al.* (1986)
			Low-risk area (40)	5.6 μg/day	
N-Nitrosoproline	Cigarette smoke	GC-TEA	Smokers (13)	5.9 μg/day	Hoffman *et al.* (1986)
			Nonsmokers (13)	3.6 μg/day	
N-Nitrosoproline	Unexposed	GC-MS	Nonsmokers (24)	3.3 μg/day	Garland *et al.* (1986)
Aflatoxin B_1, aflatoxin B_1-7-guanine	Diet	IA-HPLC	Exposed subjects (20)	0.1–10 μg/day aflatoxin B_1 equivalent	Groopman *et al.* (1985)
Aflatoxin B_1-7-guanine	Diet	HPLC-SSFS	Low/high-risk areas (983)	12% positive	Autrup *et al.* (1987)
3-Methyladenine	Unexposed	GC-MS (SIM)	Excretion rate (9)	4.5–16.1 μg/day	Shuker *et al.* (1987)
Thymine glycol, thymidine glycol	Unexposed	HPLC	Excretion rate (9)	0.39 nmol/kg/day 0.10 nmol/kg/day	Cathcart *et al.* (1984)

[a] GC, gas chromatography; TEA, thermal energy analysis; MS, mass spectrometry; IA, infra-red analysis; HPLC, high-performance liquid chromatography; SSFS, synchronous scanning fluorescence spectrophotometry; SIM, single-ion monitoring

[b] In parentheses, no. of subjects

and it is useful to summarize the nature of the findings in the context of the objectives of the present volume. Tables 1–5 contain data extracted from published reports dealing with various aspects of biomonitoring of exposure of humans to genotoxic agents of a variety of types and sources. In preparing these tables, an attempt has been made to select data that are representative of the results obtained, with particular emphasis on those aspects of the study that are pertinent to method validation.

Urinary excretion of markers

Table 1 summarizes findings from a series of studies in which urinary excretion of markers of the interactions of carcinogens with proteins or nucleic acids were measured as indices of carcinogenic exposures. Lu *et al.* (1986) determined the daily excretion of nitrosamino acids in population groups residing in areas of high and low risk with respect to development of cancer of the oesophagus. This study was designed to gain evidence about the possible etiological role of *N*-nitroso compounds, some of which are effective oesophageal carcinogens in animals, as risk factors in populations at differing levels of risk. The substantially higher values for persons in the high-risk area indicate a higher level of exposure to these carcinogens during the study period. A similar approach was used by Hoffmann and Brunnemann (1983) to assay the potential of inhaled cigarette smoke for endogenous *N*-nitrosation of amines. The data revealed not only the existence of a substantial background of endogenous nitrosation in nonsmoking controls, but also a significant increase in smokers. Garland *et al.* (1986) conducted an extensive investigation of individual and interindividual differences in the excretion of *N*-nitrosoproline in healthy volunteer subjects, all but two of whom were nonsmokers. As indicated in Table 1, the average value for these subjects is very close to that for the nonsmoking controls and for residents of the low-risk areas in the earlier studies. In each of these studies, intake of ascorbic acid was shown to lower the excretion rates, as expected. The general interlaboratory agreement with respect to the values reported indicates the potential value of this measurement in detecting exposure to nitrosating agents in the environment.

Assessment of human exposure to aflatoxins through measurement of urinary excretion of the major DNA adduct has been reported in two studies. Groopman *et al.* (1985) used the technique of immunoaffinity purification coupled with high-performance liquid chromatography (HPLC) detection to identify the guanine adduct of aflatoxin B_1 in the urine of persons residing in a commune in Guangxi Province, China, where the dietary content of the carcinogen was known to be high. The analytical method was adequately sensitive to detect the presence of the adduct in persons exposed to high levels and also to quantify the excretion of aflatoxin M_1, a metabolite, as well as of unmetabolized carcinogen. Aflatoxin exposure was also measured by urinary excretion of the guanine adduct in populations living in areas with different liver cancer incidence in Kenya (Autrup *et al.*, 1987). The adduct, detected by HPLC used in combination with synchronous scanning fluorescence spectrophotometry (SSFS) was found in 12% of a large series (983) of samples collected over a period of years in different areas of the country. Both analytical methods require sophisticated analytical instrumentation and are not yet suitable for routine monitoring.

Two additional approaches based on analysis of urinary components are being developed as potential monitors for exposure to genotoxic exposures. Shuker *et al.* (1987) have explored the measurement of 3-methyladenine in urine as an indicator of exposure to methylating agents. As indicated in Table 1, they have developed a GC-MS method using single-ion monitoring that is capable of detecting the methylated base in nominally unexposed individuals. Further validation of the method will be required to determine its usefulness for monitoring exposure. Cathcart *et al.* (1984) have developed an HPLC assay for free thymine glycol and for thymidine glycol in urine, in order to provide a noninvasive assay for oxidative DNA damage, since these compounds are products of DNA damage caused by ionizing radiation and other oxidative mutagens. While they were able to determine excretion rates for both compounds in healthy individuals, the method in its present form is not applicable for routine application but will require further development.

Haemoglobin adducts

A summary of representative data on measurement of haemoglobin adducts derived from exposure to carcinogens is presented in Table 2. Most of the work done to date has concerned measurement of *N*-3-(2-hydroxyethyl)histidine and *N*-(2-hydroxyethyl)-valine as monitors of exposure to ethylene oxide. This methodology, in which the alkylated derivative was measured by GC-MS, was developed by Professor Ehrenberg and his colleagues and was used in a study of occupationally exposed workers (Calleman *et al.*, 1978). Average data shown in Table 2 for exposed subjects and unexposed controls indicate the sensitivity of the method for detecting exposure. Other findings in this study indicate the general agreement between data obtained in man and those predicted from earlier studies in mice and also demonstrate the superiority of the method for providing a cumulative measure of exposure as compared to point monitoring of air levels. The same method was subsequently applied by Van Sittert *et al.* (1985) in a study of workers in an ethylene oxide manufacturing plant, in which cytogenetic and immunological endpoints were measured simultaneously in the same individuals. These investigators reported a substantially higher background of *N*-3-(2-hydroxyethyl)histidine in their control subjects and reported no significant difference between exposed and unexposed persons. Farmer *et al.* (1986) compared levels of this compound as determined by GC-MS on protein hydrolysates in which the adduct was concentrated by ion-exchange chromatography, with levels of *N*-(2-hydroxyethyl)valine determined by GC-MS analysis of haemoglobin subjected to Edman degradation

Table 2. Indicators of genotoxic exposure: haemoglobin adducts

Compound analysed	Exposure source	Method of analysis[a]	Principal findings[b]		Reference
N-3-(2-Hydroxy-ethyl)histidine	Ethylene oxide (occupational)	GC-MS	Exposed subjects (5) Control subjects (2)	0.5–13.5 nmol/g Hb 0.05 nmol/g Hb	Calleman *et al.*. (1978)
N-3-(2-Hydroxy-ethyl)histidine	Ethylene oxide (occupational)	GC-MS	Exposed subjects (32) Control subjects (31)	2.08 nmol/g Hb 1.59 nmol/g Hb	Van Sittert *et al.* (1985)
N-3-(2-Hydroxy-ethyl)histidine	Ethylene oxide (occupational)	Ion exchange + GC-MS	Exposed subjects (7) Control subjects (3)	0.68–8.0 nmol/g Hb 0.53–1.6 nmol/g Hb	Farmer *et al.* (1986)
N-(2-Hydroxy-ethyl)valine	Ethylene oxide (occupational)	Edman degradation + GC-MS	Exposed subjects (7) Control subjects (3)	0.02–7.7 nmol/g Hb 0.03–0.93 nmol/g Hb	Farmer *et al.* (1986)
N-(2-Hydroxy-ethyl)valine	Cigarette smoke	Edman degradation + GC-MS	Smokers (11) Nonsmokers (14)	389 pmol/g Hb 58 pmol/g Hb	Törnqvist *et al.* (1986)
4-Amino-biphenyl	Cigarette smoke	GC-MS (NCI)	Smokers (15) Nonsmokers (26)	154 pg/g Hb 28 pg/g Hb	Bryant *et al.* (1987)

[a] GC-MS, gas chromatography-mass spectrometry; NCI, negative-ion chemical ionization
[b] In parentheses, no. of subjects

before analysis. They found that the two methods of analysis gave consistent results, especially at high levels, and also that higher levels of background alkylation (of unknown origin) were obtained in measurements of *N*-3-(2-hydroxyethyl)histidine than with *N*-(2-hydroxyethyl)valine, suggesting that the latter assay would show greater sensitivity in monitoring exposure to ethylene oxide. This method was subsequently applied in a study of cigarette smokers and nonsmokers (Törnqvist *et al.*, 1986). The results demonstrated an elevation of *N*-(2-hydroxyethyl)valine levels in smokers that was quantitatively compatible with measured levels of ethene in the smoke to which they were exposed.

Bryant *et al.* (1987) have developed a method for the analysis of 4-aminobiphenyl covalently bound as the sulphinic acid amide to the 93-β cysteine of human haemoglobin. The method involves hydrolysis of the haemoglobin followed by GC-MS determination of the parent amine after derivatization. Application of the method to smokers and nonsmokers revealed consistently elevated levels in smokers and a detectable background of undetermined origin of the adduct in nonsmokers. Collectively, these results indicate the applicability of analysis of haemoglobin adducts as monitors of exposure to carcinogens of different structural types and modes of action.

Detection of benzo[a]pyrene-7,8-diol-9,10-epoxide (BPDE) adducts in DNA by immunoassay and chemical analysis

Several studies have been conducted that were designed to determine exposure to the ubiquitous PAH benzo[*a*]pyrene by the detection of derivatives covalently bound to DNA (Table 3). Immunoassays and physicochemical methods have been applied to detect derivatives formed through the major intermediate in the activation pathway, BPDE. Immunoassays have been applied in two modes, enzyme-linked immunosorbent assay (ELISA) and ultrasensitive enzyme radioimmunoassay (USERIA), both employing polyclonal antisera that recognize BPDE-DNA adducts, with various levels of cross-reactivity with structually related congeners. In an early pilot study in humans,

Table 3. Detection of BPDE adducts in DNA by immunoassay or chemical analysis[a]

Type of assay	Principal findings			Reference
ELISA (PC)	Lung DNA (tumorous/non-tumorous)	(5/27)	0.14–0.18 fmol/μg DNA	Perera *et al.* (1982)
ELISA (PC)	WBC Roofers	(7/28)	2–120 fmol/50 μg DNA	Shamsuddin *et al.* (1985)
	WBC Foundry workers	(7/20)		
	WBC Controls (smokers)	(2/9)	37–47 fmol/50 μg DNA	
SSFS	WBC Aluminium workers	(1/30)	Positive	Vähäkangas *et al.* (1985)
	WBC Controls	(0/10)		
USERIA (PC)	WBC Coke-oven workers	(18/27)	0.4–34.3 fmol/μg DNA	Harris *et al.*. (1985)
SSFS	WBC Coke-oven workers	(31/41)	Positive	
		(11/41)	Positive for serum Ab	
USERIA (PC)	WBC Coke-oven workers	(13/38)	0.1–13.7 fmol/μg DNA	Haugen *et al.* (1986)
SSFS	WBC Coke-oven workers	(4/38)	0.38–2.2 fmol/μg DNA	
		(13/38)	Positive for serum Ab	
ELISA (PC)	WBC Foundry workers (heavy, medium, light exposure)	(22)	1.2 fmol/μg DNA 0.53 fmol/μg DNA 0.32 fmol/μg DNA	Perera *et al.* (1987)
	Controls	(10)	0.06 fmol—μg DNA	

[a] PC, polyclonal; WBC, white blood cells

Perera *et al.* (1982) demonstrated that the ELISA is sufficiently sensitive to detect adducts in DNA extracted from lung tumours as well as from nontumour tissue of lung cancer patients; but the small number of subjects studied precluded conclusions concerning exposure history. Shamsuddin *et al.* (1985) employed ELISA and USERIA to investigate the levels of BPDE-DNA adducts in the white blood cells of roofers and foundry workers, in view of their exposure to high levels of benzo[*a*]pyrene. Adducts were detected in a significant proportion of exposed individuals and also in two of nine control subjects, both of whom were cigarette smokers. The SSFS technique was applied in the analysis of DNA in white blood cells collected from workers in an aluminium plant for the presence of BPDE-DNA adducts (Vähäkangas *et al.*, 1985). The limit of detection of this method as applied is about one adduct in 10^7 nucleotides. One sample of white blood cell DNA from the series of 30 exposed subjects showed the presence of a detectable level of BPDE. Harris *et al.* (1985) analysed the DNA of white blood cells from coke-oven workers by USERIA and by SSFS to determine the frequency and levels of BPDE adducts as markers of exposure, since these workers are known to be exposed to high levels of benzo[*a*]pyrene and are also at an elevated risk of lung cancer. Approximately two-thirds of the workers had detectable levels of BPDE-DNA adducts as determined by immunoassay, and an even larger proportion showed evidence of the presence of BPDE by the SSFS assay. Antibodies to the DNA adducts were also detected in the serum of 27% of the workers. Coke-oven workers were also the subjects of a study by Haugen *et al.* (1986), who sought to evaluate the genotoxic effects of their occupational exposure through a study of BPDE-DNA adducts, with simultaneous measurements of urinary excretion of PAH metabolites and air monitoring. As in the earlier study, DNA adducts were measured by USERIA and SSFS, and the sera were examined for the presence of anti-DNA adduct antibodies. The results of the study were in close agreement with those obtained earlier by Harris *et al.* (1985) in all respects. Perera *et al.* (1987) have recently applied the ELISA in the study of PAH adducts in the DNA of white blood cells collected from Finnish foundry workers classified as having high, medium or low exposure, as well as from unexposed control subjects. Mean levels of DNA adducts increased with exposure, and there was a highly significant difference between the control and pooled exposure group means.

Immunoassay techniques have also been used to detect adducts of DNA damaging agents other than BPDE (Table 4). Radioimmunoassay of O^6-methyldeoxyguanine was performed by Wild *et al.* (1986) on human oesophageal and cardiac stomach mucosal DNA from tissue samples obtained during surgery in Linxian County, China, an area of high risk for both oesophageal and stomach cancer. Similar analyses were conducted on samples collected from hospitals in Europe. Using this methodology, O^6-methyldeoxyguanine was detected in about two-thirds of samples from the high-risk area, as compared to five out of 12 from the control area. Adduct levels were also higher in samples from people living in the high-risk areas. DNA adducts induced by the anticancer drug cisplatin were detected in the white blood cells of cancer patients treated with the drug through application of ELISA methodology (Fichtinger-Schepman *et al.*, 1987). The immunoassay procedure involved four antisera capable of detecting different adducts, including: intrastrand cross-links on pGpG sequences (which proved to be the major adduct); intrastrand cross-links on pApG sequences; inter- or intrastrand cross-links on two guanines separated by one or more bases; and a monofunctional adduct bound to guanine. Results indicated that the susceptibility of white blood cells to adduct formation can show strong individually determined differences. Reed *et al.* (1987) also used ELISA methodology to study adducts derived from cisplatin in white blood cells of ovarian cancer patients being treated with the drug. Values for median adduct levels were grouped by complete response, partial

Table 4. Detection of O^6-methylguanine and DNA adducts of cisplatin by immunoassay[a]

Compound analysed	Type of assay	Principal findings			Reference
O^6-Methyldeoxy-guanine	RIA	Oesophagus/Stomach (tumorous/nontumorous)	(27/37)	25–160 fmol/mg DNA	Wild *et al.* (1986)
		European controls	(5/12)	25–45 fmol/mg DNA	
Cisplatin-DNA	ELISA (4 PC)	WBC Infused patient		pGpG 5.6 fmol/μg DNA pApG 1.9 fmol/μg DNA $(GMP)_2$ 1.1 fmol/μg DNA GMP 0.06 fmol/μg DNA	Fichtinger-Schepman *et al.* (1987)
Cisplatin-DNA	ELISA (PC)	WBC Ovarian cancer patients	(55) (peak)	Complete response 212 amol/μg DNA Partial response 193 amol/μg DNA No response 62 amol/μg DNA	Reed *et al.* (1987)

[a] RIA, radioimmunoassay; PC, polyclonal; WBC, white blood cells.

response and no response, and statistical analysis of the data showed that higher level of adduct formation correlated with clinical responsiveness to the drug.

Taken together, these data show that adducts formed in DNA of accessible cells of people exposed to DNA damaging agents of a variety of types can be detected, and in many instances quantified, by currently available immunological and chemical methods. Application of these methods in well-designed longitudinal studies in man will permit further evalutions of their validity and of their limitations in actual use.

Detection of aromatic DNA adducts by ^{32}P-postlabelling

The postlabelling procedure developed by Randerath and his collaborators has been extensively applied to studies of DNA adduct formation in a variety of experimental systems, and the capacity of the method to detect adducts of a large number of carcinogens (about 50) has been reported. The procedure has recently been extended to studies in humans (Table 5). Everson *et al.* (1986) investigated the presence of DNA adducts in human term placentas by both ELISA and the ^{32}P-postlabelling assay. The immunoassay revealed a small but insignificant increase in BPDE adduct levels in placentas from smokers compared to nonsmokers; however, the postlabelling assay detected a number of adduct types, the major one of which was strongly related to maternal smoking during pregnancy. Subsequently, Randerath *et al.* (1986) compared adduct levels in bronchus and larynx from smokers with those produced in mouse skin treated with cigarette tar. The human tissues were found to contain detectable levels of adducts, one of which (designated adduct 1) was identical to a major and persistent adduct formed in mouse skin. Dunn and Stich (1986) used the postlabelling assay to investigate DNA adducts in exfoliated mucosal cells collected from the oral cavity of three groups at high risk for oral cancer. Five chromatographically distinct adducts were found in both the high-risk groups and nonsmoking controls. Individual adducts were detected in 30–90% of samples, and no adduct was found in high-risk groups that did not also appear in control groups. Thus, although the method was useful for detecting adducts derived from unknown sources, it did not differentiate between exposed and unexposed populations. Using the same approach, Chacko and Gupta (1987) analysed the DNA from oral mucosal cells of cigarette smokers and nonsmokers to determine whether smoking-related adducts could be identified. Two chromatographically distinct adducts of unknown identity were found in smokers but not in nonsmokers. In addition, the levels of these major and several minor adducts were substantially higher in smokers. Phillips *et al.* (1986) analysed DNA from normal human bone-marrow mononuclear and non-mononuclear cells for the presence of aromatic adducts. Ten out of ten individuals showed the presence of adducts that were not present in fetal bone marrow but were present at lower levels in the DNA of peripheral white blood cells. Their data suggest that the adducts result from environmental exposure to unidentified genotoxic agents. Phillips *et al.* (1987) also employed the postlabelling assay to analyse DNA from white blood cells of foundry workers for the presence of adducts that might reflect differing levels of exposure to benzo[*a*]pyrene. Adducts were found in ten of ten workers at levels detectable by the analytical method, but none of these had the chromatographic properties characteristic of adducts derived from benzo[*a*]pyrene. Adducts were also identified in unexposed control subjects. The results indicated significant interindividual differences in DNA binding among people exposed to similar levels. Reddy *et al.* (1987) investigated white blood cell and placental DNA for the possible presence of adducts derived from exposure of pregnant women to residential wood combustion smoke. Detectable levels of unidentified adducts were found in all placentas of exposed and

Table 5. Detection of aromatic DNA adducts by ^{32}P-postlabelling[a]

Source of exposure	DNA analysed	Principal findings		References
Cigarette smoke	Placenta	Smokers (16/17)	Adduct 1: 1.4 per 10^8 nucleotides (postlabelling) 2.0 per 10^6 nucleotides (ELISA)	Everson *et al.* (1986)
		Nonsmokers (3/14)	Adduct 1	
Cigarette smoke	Bronchus, larynx	Smokers (2)	Total adducts: 1 per ($1.7–2.9 \times 10^7$ nucleotides) [0.10–0.18 fmol/μg DNA] Adduct 1: 8–14% of total	Dunn & Stich (1986)
Betel chewing, tobacco chewing, inverted smoking	Oral mucosa	Exposed (59)	Adducts found in 30–95% [1 per 10^9 nucleotides to 1 per 10^7 nucleotides]	Chacko & Gupta (1987)
Cigarette smoke	Oral mucosa	Smokers (11/14)	Total adducts: 0.1–210 amol/μg DNA	Phillips *et al.* (1986)
		Nonsmokers (2/8)	Total adducts: 0.4–1.7 amol/μg DNA	
Foundry workers	WBC		Total adducts:	Phillips *et al.* (1987)
		Exposed (10/10)	0.2–11.6 per 10^8 nucleotides	
		Controls (5/10)	0.4 per 10^8 nucleotides	
Wood smoke	Placenta, WBC		Total adducts:	Reddy *et al.* (1987)
		Exposed: Placenta (4)	12 per 10^9 nucleotides	
		WBC (8)	ND	
		Control: Placenta (5)	12 per 10^9 nucleotides	
		WBC (8)	ND	

[a] WBC, white blood cells; ND, not determined

unexposed women; none of the nine adducts found was present in DNA of white blood cells from any of the subjects. The results suggest that residential wood smoke does not elicit aromatic DNA adducts at detectable levels, but that placental DNA contains detectable levels of adducts of unknown identity and origin. Collectively, these results demonstrate the capacity of the postlabelling method to identify DNA damage arising from a variety of environmental exposures. However, additional experience with its applications in human studies will be required to establish fully the important analytical characteristics of the method.

DNA adducts and haemoglobin adducts as measures of exposure and susceptibility

The information summarized above provides valuable evidence concerning some features of currently available methods for detecting and quantifying covalent adducts to DNA and haemoglobin in human material. Most of the studies constitute feasibility trials designed to test the use of the methods for certain specific purposes. Taken together, the available data show that detection of exposure to DNA damaging agents of a variety of chemical types will be feasible using the methods currently in hand or being developed. Substantial additional technological validation will be required for routine application of any of the current methods, but it seems likely that many of them will eventually come into broader application in epidemiological surveys. In interpreting the data generated by their application, it is important to take into account all of the available information concerning the parameters being measured. Some of the important points in this context can be summarized as follows.

Concerning the use of DNA adduct levels as measures of exposure and susceptibility, virtually all of the available evidence comes from experimental systems, especially experiments in which adduct formation in carcinogen-treated animals was studied in relation to tumour formation in response to the treatment. Most carcinogens have been shown to form complex spectra of DNA adducts, and qualitatively similar adduct profiles can be formed in sensitive and resistant species, strains and tissues. Adduct persistence may or may not be related to susceptibility and target tissue specificity. In target tissues, maximal total adduct levels usually reflect carcinogen potency and dose, and maximal total adduct levels are linearly related to dose over a wide range. Many factors have been identified that can affect the levels of DNA adducts in cells of treated animals at any given time. The exposure-sampling interval can markedly affect observed levels, as determined by the pharmacokinetic properties of carcinogens of specific chemical types. An important component of this process is the capacity for metabolic activation, which can be affected by a variety of genetic and environmental factors (e.g., inducers and inhibitors). Exposure to protective agents such as antioxidants and other dietary ingredients can also exert marked effects. DNA repair capacity, as determined by genetic factors, kinetics (e.g., saturability) and modulating factors, is also an important determinant of adduct stability. Collectively, these factors indicate the complexity of the problem of accurately interpreting DNA adduct levels as quantitative measures of exposure.

The use of DNA adduct levels as indicators of long-term risk is complicated by additional factors. Generally, the target tissue at risk to a given genotoxic exposure is unknown, and reliance must be placed upon measurements of adduct levels in surrogate cells such as peripheral white blood cells. The validity of analysing DNA in surrogate cells for target cell DNA is difficult to assess. Extrapolation of DNA adduct levels as indicators of cancer risk necessarily entails simplifying assumptions, which cannot be evaluated on the basis of current information, about the multistage nature of the cancer process. In addition, point measurements of DNA adduct levels may or may not provide evidence of multiple and variable exposures.

As discussed earlier, protein adducts have no putative mechanistic role in carcinogenesis and are primarily regarded as measures of exposure alone. Evidence from experimental animals and from humans supports the interpretation that protein adduct levels are exposure monitors. Measurement of haemoglobin adducts has certain obvious technical advantages, inasmuch as the protein can readily be obtained in abundant quantities. Sensitive and specific methods have been developed for two classes of carcinogens, and additional ones are in the process of development. Carcinogens of diverse chemical structures have been shown to bind haemoglobin *in vivo*. Adducts are stable over the erythrocyte life span, thereby giving an integrated measure of multiple exposures over a substantial time period. Levels of haemoglobin adducts have been shown to be linearly related to dose for at least seven carcinogens of different types. Importantly, haemoglobin adduct levels have been shown to be related, within a factor of two, to DNA adduct levels in target tissues for three agents, ethylene oxide, *trans*-4-dimethylaminostilbene and 2-acetylaminofluorene. Haemoglobin adducts therefore provide very useful complementary data to DNA adduct levels as dosimeters of carcinogen exposure.

The information summarized above concerning the application of current methods for detecting and quantifying levels of DNA and haemoglobin adducts in human samples indicates that adequate levels of sensitivity have been attained for several classes of DNA damaging agents to make possible the detection of adducts resulting from ambient levels of exposure. Specificity for the detection of compounds of known structure is also a common feature of most of the methods, and some are suitable for detecting total DNA damage from multiple sources. With respect to practicability, certain of the methods, e.g., immunoassays, are applicable to large numbers of samples, whereas others, e.g., postlabelling and thymine glycol analysis, are technically more complex and time-consuming. Validation will be required to avoid systematic errors in measurement before the methods are applicable to large-scale epidemiological studies. Methodological attributes such as accuracy, reproducibility (intra- and interlaboratory) and applicability to stored samples must be determined by appropriately designed collaborative studies among qualified laboratories. Additional factors such as variations in observed values due to age, sex and race, as well as effects of possible interfering factors (e.g., diet, smoking, alcohol) must be defined in properly designed human studies of limited scope in order to validate the approach for epidemiological surveys.

References

Autrup, H., Seremet, T., Wakhisi, J. & Wasunna, A. (1987) Aflatoxin exposure measured by urinary excretion of aflatoxin B_1-guanine adduct and hepatitis B virus infection in areas with different liver cancer incidence in Kenya. *Cancer Res.*, *47*, 3430–3433

Berlin, A., Wolff, A.H. & Hasegawa, Y. (1979) *The Use of Biological Specimens for the Assessment of Human Exposure to Environmental Pollutants*, London, Martinus Nijhoff

Berlin, A., Draper, M., Hemminki, K. & Vainio, H., eds (1984) *Monitoring of Human Exposure to Carcinogenic and Mutagenic Agents* (*IARC Scientific Publications No. 59*), Lyon, International Agency for Research on Cancer

Bloom, A.D. (1981) *Guidelines for Studies of Human Populations Exposed to Mutagenic and Reproductive Hazards*, New York, March of Dimes Birth Defects Foundation

Bridges, B.A. (1980) An approach to the assessment of the risk to man from DNA damaging agents. *Arch. Toxicol., Suppl. 3*, 271–281

Bryant, M.S., Skipper, P.L., Tannenbaum, S.R. & Maclure, M. (1987) Hemoglobin adducts of 4-aminobiphenyl in smokers and nonsmokers. *Cancer Res., 47,* 602–608

Calleman, C.-J., Ehrenberg, L., Jansson, B, Osterman-Golkar, S., Segerbäck, D., Svensson, K. & Wachtmeister, C.A. (1978) Monitoring and risk assessment by means of alkyl groups in hemoglobin in persons occupationally exposed to ethylene oxide. *J. Environ. Pathol. Toxicol., 2,* 427–442

Cathcart, R., Schweirs, E., Saul, R.L. & Ames, B.N. (1984) Thymine glycol and thymidine glycol in human and rat urine: a possible assay for oxidative DNA damage. *Proc. natl Acad. Sci. USA, 81,* 5633–5637

Chacko, M. & Gupta, R.C. (1987) Evaluation of DNA damage in the oral mucosa of cigarette smokers and nonsmokers by ^{32}P-adduct assay. *Proc. Am. Assoc. Cancer Res., 28,* 101

Dunn, B.P. & Stich, H.F. (1986) ^{32}P-Postlabelling analysis of aromatic DNA adducts in human oral mucosal cells. *Carcinogenesis, 7,* 1115–1120

Ehrenberg, L. & Osterman-Golkar, S. (1980) Alkylation of macromolecules for detecting mutagenic agents. *Teratog. Carcinog. Mutag., 1,* 105–127

Evans, H.J. (1983) Cytogenetic methods for detecting effects of chemical mutagens. *Ann. N.Y. Acad. Sci., 407,* 131–141

Evans, H.J., Buckton, K.E., Hamilton, G.E. & Carothers, A. (1979) Radiation induced chromosome aberrations in nuclear dockyard workers. *Nature, 277,* 531–534

Everson, R.B., Randerath, E., Santella, R.M., Cefalo, R.C., Avitts, T.A. & Randerath, K. (1986) Detection of smoking-related covalent DNA adducts in human placenta. *Science, 231,* 54–57

Farmer, P.B., Bailey, E., Gorf, S.M., Törnqvist, M., Osterman-Golkar, S., Kautiainen, A. & Lewis-Enright, D.P. (1986) Monitoring human exposure to ethylene oxide by the determination of haemoglobin adducts using gas chromatography-mass spectrometry. *Carcinogenesis, 7,* 637–640

Farmer, P.B., Neumann, H.-G. & Henschler, D. (1987) Estimation of exposure of man to substances reacting covalently with macromolecules. *Arch. Toxicol., 60,* 251–260

Fichtinger-Schepman, A.-M.J., van Oosterom, A.T., Lohman, P.H.M. & Berends, F. (1987) *cis*-Diamminedichloroplatinum(II)-induced DNA adducts in peripheral leukocytes from seven cancer patients: quantitative immunochemical detection of the adduct induction and removal after a single dose of *cis*-diamminedichloroplatinum(II). *Cancer Res., 47,* 3000–3004

Garland, W.A., Kuenzig, W., Rubio, F., Kornychuk, H., Norkus, E.P. & Conney, A.H. (1986) Urinary excretion of nitrosodimethylamine and nitrosoproline in humans: interindividual differences and the effect of administered ascorbic acid and α-tocopherol. *Cancer Res., 46,* 5392–5400

Garner, R.C. (1985) Assessment of carcinogen exposure in man. *Carcinogenesis, 6,* 1071–1078

Groopman, J.D., Donahue, P.R., Zhu, J., Chen, J. & Wogan, G.N. (1985) Aflatoxin metabolism in humans: detection of metabolites and nucleic acid adducts in urine by affinity chromatography. *Proc. natl Acad. Sci. USA, 82,* 6492–6496

Harris, C.C., Vähäkangas, K., Newman, M.J., Trivers, G.E., Shamsuddin, A., Sinopoli, N., Mann, D.L. & Wright, W.E. (1985) Detection of benzo[a]pyrene diol epoxide-DNA adducts in peripheral blood lymphocytes and antibodies to the adducts in serum from coke oven workers. *Proc. natl Acad. Sci. USA, 82,* 6672–6676

Haugen, A., Becher, G., Benestad, C., Vähäkangas, K., Trivers, G.E., Newman, M.J. & Harris, C.C. (1986) Determination of polycyclic aromatic hydrocarbons in the urine, benzo[a]pyrene diol epoxide-DNA adducts in lymphocyte DNA, and antibodies to the adducts in sera from coke oven workers exposed to measured amounts of polycyclic aromatic hydrocarbons in the work atmosphere. *Cancer Res., 46,* 4178–4183

Hoffmann, D. & Brunnemann, K.D. (1983) Endogenous formation of *N*-nitrosoproline in cigarette smokers. *Cancer Res., 43,* 5570–5574

IARC/IPCS Working Group (1982) Development and possible use of immunological techniques to detect individual exposure to carcinogens. *Cancer Res., 42,* 5236–5239

Kolman, A., Naslund, M. & Calleman, C.J. (1986) Genotoxic effects of ethylene oxide and their relevance to human cancer. *Carcinogenesis, 7,* 1245–1250

Latt, S.A. (1981) Sister chromatid exchange formation. *Ann. Rev. human Genet., 15,* 11–55

Lu, S.-H., Ohshima, H., Fu, H.-M., Tian, Y., Li, F.-M., Blettner, M., Wahrendorf, J. & Bartsch, H. (1986) Urinary excretion of *N*-nitrosamino acids and nitrate by inhabitants of high- and low-risk areas for esophageal cancer in northern China: endogenous formation of nitrosoproline and its inhibition by vitamin C. *Cancer Res., 46,* 1485–1491

Lutz, W.K. (1979) *In vivo* covalent binding of organic chemicals to DNA as a quantitative indication in the process of chemical carcinogenesis. *Mutat. Res., 65,* 289–356

National Academy of Sciences/National Research Council (1983) *Risk Assessment in the Federal Government: Managing the Process,* Washington DC, National Academy Press

Osterman-Golkar, S., Ehrenberg, L., Segerbäck, D. & Hällström, I. (1976) Evaluation of genetic risks of alkylating agents. II. Haemoglobin as a dose monitor. *Mutat. Res., 34,* 1–10

Osterman-Golkar, S., Farmer, P.B., Segerbäck, D., Bailey, E., Calleman, C.J., Svensson, K. & Ehrenberg, L. (1983) Dosimetry of ethylene oxide in the rat by quantification of alkylated histidine in hemoglobin. *Teratog. Carcinog. Mutag., 3,* 395–405

Pereira, M.A. & Chang, L.W. (1981) Binding of chemical carcinogens and mutagens to rat haemoglobin. *Chem.-biol. Interactions, 33,* 301–305

Perera, F.P. (1987) Molecular cancer epidemiology: a new tool in cancer prevention. *J. natl Cancer Inst., 78,* 887–898

Perera, F.P., Poirier, M.C., Yuspa, S.H., Nakayama, J., Jaretzki, A., Curnen, M.M., Knowles, D.M. & Weinstein, I.B. (1982) A pilot project in molecular cancer epidemiology: determination of benzo[a]pyrene-DNA adducts in animal and human tissues by immunoassays. *Carcinogenesis, 3,* 1405–1410

Perera, F.P., Hemminki, K.H., Santella, R.M., Brenner, D. & Kelly, G. (1987) DNA adducts in white blood cells of foundry workers. *Proc. Am. Assoc. Cancer Res., 28,* 94

Phillips, D.H., Hewer, A. & Grover, P.L. (1986) Aromatic DNA adducts in human bone marrow and peripheral blood leukocytes. *Carcinogenesis, 7,* 2071–2075

Phillips, D.H., Hemminki, K., Hewer, A. & Grover, P.L. (1987) ^{32}P-Postlabelling of white blood cell DNA from foundry workers. *Proc. Am. Assoc. Cancer Res., 28,* 102

Poirier, M.C. (1981) Antibodies to carcinogen-DNA adducts. *J. natl Cancer Inst., 67,* 515–519

Randerath, E., Avitts, T.A., Reddy, M.V., Miller, R.H., Everson, R.B. & Randerath, K. (1986) Comparative ^{32}P-analysis of cigarette smoke-induced DNA damage in human tissues and mouse skin. *Cancer Res., 46,* 5869–5877

Randerath, K., Randerath, E., Agrawal, H.P., Gupta, R.C., Schurdak, M.E. & Reddy, M.V. (1985) Postlabeling methods for carcinogen-DNA adduct analysis. *Environ. Health Perspectives, 62,* 57–65

Reddy, M.V., Kenny, P.C. & Randerath, K. (1987) ^{32}P-Assay of DNA adducts in white blood cells (WBC) and placentas of pregnant women exposed to residential wood combustion (RWC) smoke. *Proc. Am. Assoc. Cancer Res., 28,* 97

Reed, E., Ozols, R., Tarone, R., Yuspa, S.H. & Poirier, M.C. (1987) Platinum-DNA adducts in leukocyte DNA correlate with disease response in ovarian cancer patients receiving platinum-based chemotherapy. *Proc. natl Acad. Sci. USA, 84,* 5024–5028

Shamsuddin, A.K.M., Sinopoli, N.T., Hemminki, K., Boesch, R.R. & Harris, C.C. (1985) Detection of benzo[*a*]pyrene:DNA adducts in human white blood cells. *Cancer Res., 45,* 66–68

Shuker, D.E.G., Bailey, E., Parry, A., Lamb, J. & Farmer, P.B. (1987) The determination of urinary 3-methyladenine in humans as a potential monitor of exposure to methylating agents. *Carcinogenesis, 8,* 959–962

Singer, B. & Grunberger, D. (1983) *Molecular Biology of Mutagens and Carcinogens,* New York, Plenum Press

Sorsa, M. & Norppa, H., eds (1986) *Monitoring of Occupational Genotoxicants,* New York, Alan R. Liss

Törnqvist, M., Osterman-Golkar, S., Kautiainen, A., Jensen, S., Farmer, P.B. & Ehrenberg, L. (1986) Tissue doses of ethylene oxide in cigarette smokers determined adduct levels in haemoglobin. *Carcinogenesis, 7,* 1519–1521

Vähäkangas, K., Haugen, A. & Harris, C.C. (1985) An applied synchronous fluorescence spectrophotometric assay to study benzo[a]pyrene-diol epoxide-DNA adducts. *Carcinogenesis, 6,* 1109–1116

Vainio, H., Sorsa, M. & Hemminki, K. (1983) Biological monitoring in surveillance of exposure to genotoxicants. *Am. J. ind. Med., 4,* 87–103

Van Sittert, N.J., De Jong, G., Slare, M.G., Davies, R., Dean, B.J., Wren, L.J. & Wright, A.S. (1985) Cytogenetic, immunological and haematological effects in workers in an ethylene oxide manufacturing plant. *Br. J. ind. Med., 42,* 19–26

Wild, C.P., Umbenhauer, D., Chapot, B. & Montesano, R. (1986) Monitoring of individual human exposure to aflatoxins (AF) and N-nitrosamines (NNO) by immunoassays. *J. cell. Biol., 30,* 171–179

Wogan, G.N. & Gorelick, N.J. (1985) Chemical and biochemical dosimetry of exposure to genotoxic chemicals. *Environ. Health Perspectives, 62,* 5–18

DIETARY EXPOSURES

DO AFLATOXIN-DNA ADDUCT MEASUREMENTS IN HUMANS PROVIDE ACCURATE DATA FOR CANCER RISK ASSESSMENT?

J.D. Groopman

Environmental Health Section, Boston University School of Public Health, Boston, MA, USA

Primary hepatocellular carcinoma is one of the most lethal and most common cancers in Africa and Asia and is associated with exposure to aflatoxin (AF) B_1. To date, many human studies have relied upon presumptive intake data, rather than on quantitative analyses of AF-DNA adduct and metabolite content obtained by monitoring biological fluids from exposed people. Information obtained by monitoring exposed individuals for specific DNA adducts and metabolites will define the pharmacokinetics of AFB_1, thereby facilitating risk assessment. In addition, using an animal model based on the differential effects of ethoxyquin on the kinetics of AF-DNA adduct and γ-glutamyl transpeptidase-positive foci formation, we have data to support the concept that measurement of the major, rapidly excised AFB-7-guanine (Gua) adduct in tissues and fluids is an appropriate dosimeter for estimating exposure status and risk in individuals consuming this mycotoxin.

Liver cancer is one of the most lethal and common cancers in Africa and Asia and is associated with dietary exposure to AFB_1. In areas of high AF exposure, daily ingestion has been found to vary from 200 to 17 500 ng per day, with corresponding liver cancer incidence rates extended from a minimum of 2.0 to a maximum of 50.0 cases/100 000 population per year (Van Rensburg *et al.*, 1985; Groopman *et al.*, 1986; Peers *et al.*, 1987). In order to develop individual monitoring procedures for exposed people, monoclonal antibodies specific for AF metabolites, especially the DNA adducts, have been produced for use in the preparative purification of AF from biological fluids. These methods, in conjunction with other analytical procedures, have resulted in the development of rapid protocols to measure quantitatively aflatoxins in urine obtained from people exposed to this carcinogen (Groopman *et al.*, 1985, 1986, 1987). Despite the development of these new technologies, we must continue to define animal model systems that can be used to help interpret human data. Preliminary evidence from a rat model based on the differential effects of antioxidants on the kinetics of AF-DNA adduct and tumour formation (Kensler *et al.*, 1986) indicates that the direct linear extrapolation of total adduct content in target tissues to dose may be inappropriate for assigning risk to people. However, our findings do support the concept that measurement of the major, rapidly excised AFB-7-Gua adduct in tissues and fluids is an appropriate dosimeter for estimating exposure status and risk in individuals consuming this mycotoxin.

The objective of this article is to review some of the relevant data on the development of monoclonal antibodies that recognize AF as they have been applied to the production of preparative affinity columns. These columns, in turn, have been extremely useful for the rapid isolation of AF-DNA adducts from complex fluids, such as urine. The most important analytical advantage of this approach is the one-step removal of interfering materials from urine or other samples, thereby permitting the use of standard reversed-phase high-performance liquid chromatography (HPLC) analysis with ultraviolet monitoring to resolve and quantify AF metabolites without the need for radiolabelled material. This method permits the rapid, easy analysis of human samples from environmentally exposed people. However, it has become clear that, while the measurement of adduct levels is an important indicator of exposure, the presence of adducts does not itself provide the requisite perspective for assigning risk to an exposed individual. Therefore, we have sought to validate the use of DNA adduct measurements for risk assessment by developing animal models in which the carcinogenic outcome may be predictably altered through the use of graded doses of chemoprotective agents, such as dietary antioxidants. In this manner, the kinetics of DNA adduct formation and removal may be examined relative to a predetermined carcinogenic endpoint. Taken together, these systems will be useful for evaluating the hypothesis that the measurement of carcinogen-DNA adducts in biological fluids provides accurate estimates of exposure and subsequent risk for neoplasia.

Monoclonal antibodies for aflatoxins and preparative affinity chromatography

We have produced various monoclonal antibodies that recognize AF, using antigens ranging from AF-modified DNA to AF-adducted proteins (Groopman *et al.*, 1982, 1984). We have combined procedures of analytical chemistry, such as HPLC, with a preparative monoclonal antibody-affinity chromatography column, both to purify the aflatoxin adducts and metabolites from urine from interfering compounds and to confirm their identities.

The development of a reusable monoclonal antibody-affinity chromatographic column for rapid isolation of AF metabolites and their adducts from human urine and serum samples required the production of high-affinity monoclonal antibodies. One of the antibodies isolated had an affinity constant for AFB_1 of 1×10^9 l/mol (Groopman *et al.*, 1984). This antibody also had significant recognition and cross-reactivity for AFB_2, AFM_1 and the two major AF-DNA adducts, AFB_1-formamidopyrimidine (FAPyr) and AFB_1-7-Gua.

To test the validity of the affinity chromatography technique and the applicability of the column to biological samples, AFB_1 added to human urine, serum and milk was analysed. Freshly collected human urine was spiked with ^{3}H-AFB_1, and we were able to obtain quantitative binding of the ^{3}H-AFB_1 to the column and subsequent 95% recovery in 50% dimethyl sulphoxide buffer. Furthermore, human serum or human milk spiked with ^{3}H-AFB_1 was applied directly to the antibody column without prior treatment, and quantitative binding and recovery into 50% dimethyl sulphoxide were observed in both instances. These data indicate that a rapid preparative tool could be devised to isolate AF from biological fluids.

AFB_1 dietary antioxidant animal model

Experimental AFB_1 hepatocarcinogenesis is amenable to modification by dietary manipulation. We (Kensler *et al.*, 1986) have developed a more refined exposure protocol for assessing the mechanisms of chemoprotection by the antioxidant, ethoxyquin, to validate the possible use of adduct dosimetry in risk assessment. In our standard protocol, male Fischer 344 rats are placed on a purified diet of the AIN-76A formulation supplemented with 0.4% ethoxyquin. Beginning one week later, animals are dosed orally with 250 μg AFB_1/kg body weight on five days a week for two weeks.

Fig. 1. Effect of ethoxyquin on AF-DNA adduct formation and removal in rat liver

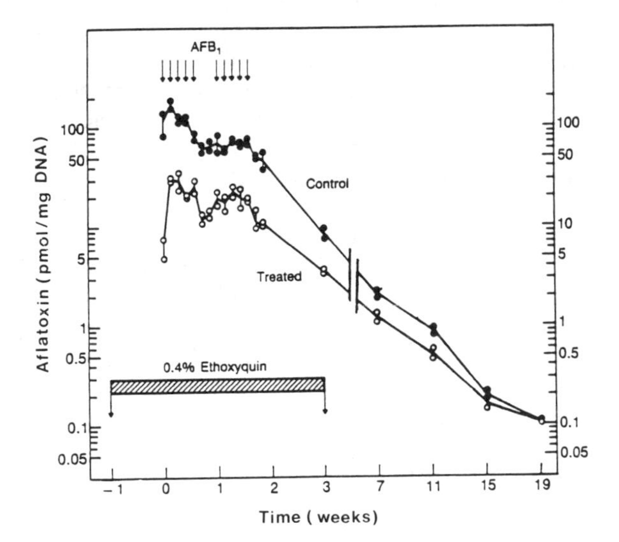

One week after cessation of AF dosing, rats are restored to the basal diet. With this protocol, ethoxyquin supplementation reduces by >95% the number and volume of presumptive preneoplastic hepatic lesions (γ-glutamyl transpeptidase-positive foci) observed at four months when compared to rats maintained on the basal (unsupplemented) diet throughout the experimental period.

It is well established that a single dose of AFB_1 is not an efficient carcinogenic regimen in rats; however, a dosing regimen of small repeated doses can induce a high incidence of hepatocellular carcinoma. Therefore, the effect of ethoxyquin on the kinetics of AF-DNA adduct formation and removal was examined in rats treated in the multiple-dosing protocol described for the γ-glutamyl transpeptidase foci studies. The time course for the formation and removal of total AF-DNA adducts in liver in rats receiving oral intubations of 250 μg AFB_1/kg on each of days 8–12 and 15–19 is shown in Figure 1. Maximal binding levels were achieved following the second dose, and binding following the next three doses remained at a plateau level of about 140 pmol AF equivalents bound per mg DNA. Overall binding declined after cessation of the first dosing period; however, resumption of AFB_1 dosing produced only minor elevation of binding levels as the cycle of adduct formation and removal was renewed. This 50% diminution of AF-DNA binding during the second cycle presumably results from AFB_1-induced alterations in cytochrome P450-mediated AFB_1 activation. Total DNA adduct levels dropped five-fold in the first week following cessation of dosing and continued to decline at a comparable rate over the next four months to a level of 100 fmol AF equivalents bound per mg DNA at 133 days. Inclusion of ethoxyquin in the diet, beginning one week prior to and extending to one week beyond dosing with AFB_1, produced a dissimilar pattern of effects and yielded substantially lower binding levels during the early period. At 2 h after the first AFB_1 dose, approximately 18-fold less binding was observed in the ethoxyquin-treated animals. By day 2, the difference had declined to six-fold and was about 3.5-fold throughout the second dosing cycle. Remarkably, the difference in binding levels diminished during the period after dosing such that binding levels in control and ethoxyquin-treated rats were indistinguishable at days 106 and 133.

Liquid chromatographic analysis of hydrolysed DNA from the livers of these animals revealed no remarkable qualitative difference in the adduct profile induced by ethoxyquin treatment at any time; however, ethoxyquin treatment reduced the amount of the AFB-7-Gua adduct by more than 95%. The relative concentrations of the ring-opened formamidopyrimidine adducts were also decreased to a comparable degree. When integrated across the four-month time frame of the experiment, ethoxyquin treatment reduced the accumulation of AFB-7-Gua, AFB-7-FAPyr (major) and AFB-7-FAPyr (minor) adducts by 77, 71 and 76%, respectively. However, the temporal patterns for the different adducts were quite distinct: the levels of the two formamidopyrimidine adducts remained constant over the two-week dosing period, at approximately 40 and 10 pmol bound/mg DNA for the major and minor derivatives, respectively. AFB-7-FAPyr (major) was the only adduct detectable after day 49, and ethoxyquin treatment had no effect on levels of this adduct at these late time points (days 106 and 133). By contrast, although it was the dominant species on the first day of dosing, levels of the AFB-7-Gua adduct decreased rapidly after day 1, such that levels during the second dosing cycle were only one-fifth to one-third those observed during the first cycle. No AFB-7-Gua adduct was detectable after day 21, indicating that this adduct is rapidly removed from DNA by chemical and/or enzymatic processes.

When these data are considered in the context of the quantitative two- and three-dimensional analyses of γ-glutamyl transpeptidase lesions in livers of rats treated with an identical antioxidant-aflatoxin exposure protocol, it is apparent that a strong relationship exists between the initial amount of DNA modification (AFB-7-Gua) in target tissue by AF and its pathological effect. However, levels of AF-DNA adducts (i.e., the formamidopyrimidine derivatives) at later times did not appear to be related to the greatly diminished neoplastic outcome in the ethoxyquin-treated animals. These experiments indicate that dramatic alterations in the formation of specific adducts can result in a change in the carcinogenic outcome, but also serve to underscore the difficulties associated with exposure dosimetry assigned by simplistic approaches of monitoring total adduct levels by the affinity chromatographic or other methods.

Monoclonal antibody affinity chromatography of rat urine

Preliminary experiments (Groopman *et al.*, 1985) demonstrated the efficacy of using the monoclonal antibody affinity technique as a preparative tool to isolate AF from exposed animals. The next concern addressed in these studies was whether the levels of DNA adduct excreted into the urine corresponded to the levels of initial DNA adduct formation within the liver. Six adult male Fischer 344 rats were each intubated orally with 0.25 mg ^{14}C-AFB$_1$ per kg body weight. Three of the rats were maintained for one week prior to dosing on a diet containing 0.4% ethoxyquin, while the other three were maintained on an AIN-76A diet. After 24 h, the rats were sacrificed and the DNA was isolated from the livers. The urine and faeces excreted by the rats over the 24-h period were also collected and analysed. The rats maintained on the ethoxyquin diet had a greater than 90% reduction in AFB-DNA adduct formation. The urine from each of the rats in the treated and control dietary groups contained about 20% of the administered dose, while the faeces accounted for about 60% of the administered ^{14}C-AFB$_1$. Therefore, there was no apparent difference between ethoxyquin-treated and control rats in the total amount of AF excreted into urine and faeces.

Excretion and DNA binding in this experiment are summarized in Table 1. Excretion of the major AF-DNA adduct into the urine of rats maintained on the ethoxyquin diet was reduced by 66% compared to the controls. This corresponds

Table 1. Effect of ethoxyquin on aflatoxin-DNA adduct formation and metabolite removal into urine and faeces after 24 h: mean (range)

Dietary treatment	% Dose in urine	% Dose in faeces	nmol AFB-DNA adducts in urine
Control	13.0 (9.5–15.3)	43.0 (36.3–50.3)	0.57 (0.41–0.74)
Ethoxyquin	13.7 (7.9–19.8)	57.7 (56.3–59.0)	0.19 (0.14–0.23)

Each rat was intubated with ^{14}C-AFB_1 at a total dose of 80.128 nmol. The initial levels of DNA adduct formation can be calculated from the dosing data in a previous publication (Kensler *et al.*, 1986). The mean (range) AFB-DNA adduct levels in the liver DNA of control and ethoxyquin rats after 24 h were 61.0 (52.8–70.5) and 6.2 (5.5–7.0) pmol/mg DNA, respectively.

qualitatively with the reduction in DNA adduct levels in the livers of the treated animals; however, the reduction in DNA adduct excretion is not the same quantitatively, owing perhaps to the contribution of DNA adducts formed in other organs which are not as dramatically reduced by ethoxyquin as in the liver. Another factor in this difference may be the contribution of excised (or turned over) AF-RNA adducts.

These data indicate that a general correlation does exist between levels of AF-DNA adducts excreted into urine and initial levels of binding to DNA in a target organ. While further and more extensive investigations must be performed to obtain more details about the kinetics of the excretion patterns in multiply or chronically dosed animals, our findings strongly suggest that measurement of AFB-7-Gua adducts in urine is a valid and quantitative indicator of recent exposure to AF.

Analysis of human urine using the monoclonal antibody affinity column

People exposed to AFB_1 from dietary sources were identified for pilot studies by collaborators at the Institute of Health in Beijing, China (Groopman *et al.*, 1985, 1987). These urine samples were used to gain preliminary evidence of the applicability of the monoclonal antibody affinity column technique and of HPLC analysis procedures for monitoring individuals for exposure to aflatoxins. For the initial study, 20 individuals were selected and two 25-ml aliquots of urine were obtained from a morning voiding for each individual. The intake of AFB_1 from the diet, primarily corn, over the previous day (24 h) was calculated: exposures ranged from 13.4–87.5 μg AFB_1. Competitive radioimmunoassay of the samples eluted from the monoclonal antibody column demonstrated that the AF concentration in the collected urine was 0.1–10 ng/ml, calculated from a linear extrapolation of the radioimmunoassay standard curves generated using AFB_1.

Urine samples from four individuals who had been exposed to the highest level (87.5 μg) the previous day were prepared with the antibody affinity column and then measured by analytical HPLC. HPLC analysis demonstrated the presence of the major AFB_1-DNA adduct, AFB_1-7-Gua, at levels representing between 7–10 ng of the adduct. Thus, the monoclonal antibody columns, coupled with HPLC, can be used to quantify AF-DNA adducts in human urine samples obtained from environmentally exposed people.

These initial findings stimulated a more extensive study the following year. Since the relationship between dose and excretion of AFB_1 and its adducts had not yet been determined in chronically exposed people, the following protocol was developed. The

Fig. 2. Scatterplot of total male and female intake of AF compared with excretion

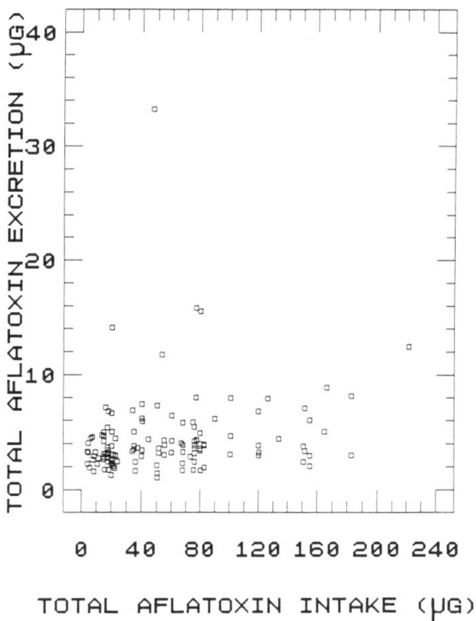

diets of 30 men and 12 women aged 25 to 64 years were monitored for one week and total AF intake determined for each day. Urine was obtained in two 12-h fractions on three consecutive days during the one-week period. These urine samples were obtained only after dietary AF levels had been measured for at least three consecutive days; therefore, the urine collections were initiated on the fourth day of the protocol. These samples have also been analysed by another analytical methodology (Zhu *et al.*, 1987). The average male intake of AFB_1 was 48.4 μg per day, for a total exposure over the seven-day period of 276.8 μg. The average female daily intake was 92.4 μg/day. Immunoassays were performed on aliquots of the 12-h urines following clean-up of the samples by C18 Sep-Pak and monoclonal antibody affinity chromatography.

Total AFB_1 excretion for each 12-h sample period was calculated by multiplying the urine volume by the concentration of AFB_1 determined in the aliquot of urine. Figure 2 depicts a scatterplot comparison of AF intake with AF metabolite excretion for men and women combined. The AF intake represents the total integrated ingestion by an individual on the day prior to urine collection and during the three days of urine collection. The excretion data are the composite of all AF metabolites excreted into the urine during the three days of urine sampling. Despite a 20-fold range of AFB_1 intake, the amount of AF excreted generally varied over only a three-fold range, indicating that urinary excretion of AFB_1 is a saturable process.

Figures 3 and 4 depict box-and-whisker plot analyses of individual male intake and excretion. In Figure 3, it is seen that the day-to-day variability in intake is low, but on any given day usually one or two people fall outside the general population. However, in Figure 4, which shows the excretion characteristics for each of the 12-h urine collections, many more outlying individuals are seen. The importance of identifying these outliers is that they often skew standard regression analyses of data, since these

Fig. 3. Box-and-whisker plot analysis of male intake of AF

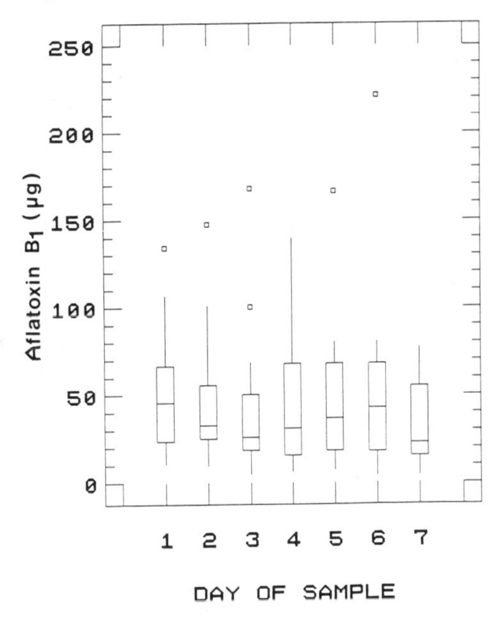

The box-and-whisker plot represents an analysis of the data to provide the 10th and 90th percentiles (ends of the whiskers), the 25th and 75th percentiles (ends of the box) and a line for the 50th percentile (within the box). Any outlying points are depicted as individual squares. This type of data analysis reveals spread and outlier characteristics of the data which can be lost or suppressed if the data are calculated using standard mean and standard error analysis.

Fig. 4. Box-and-whisker plot analysis of male 12 h excretion of AF

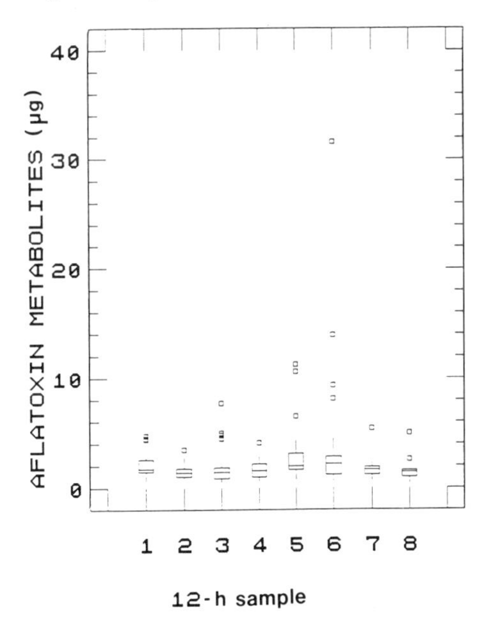

points do not represent members of the population, leading to a potentially inappropriate interpretation of data.

We also performed HPLC analysis of the urine samples for AFM_1, AFP_1 and the major AFB-DNA adducts. Taken together, it appears that urine is a valid compartment in which to sample people for exposure to AF, but more data must be collected for developing a risk model for people.

Acknowledgements

This work was supported by grants PO1ES00597 and CA39416 from the US Public Health Service.

References

Groopman, J.D., Haugen, A., Goodrich, G.R. & Harris, C.C. (1982) Quantitation of aflatoxin B1 modified DNA using monoclonal antibodies. *Cancer Res., 42,* 3120–3124.

Groopman, J.D., Trudel, L.J., Donahue, P.R., Rothstein, A. & Wogan, G.N. (1984) High affinity monoclonal antibodies for aflatoxins and their application to solid phase immunoassay. *Proc. natl Acad. Sci. USA, 81,* 7728–7731

Groopman, J.D., Donahue, P.R., Zhu, J., Chen, J. & Wogan, G.N. (1985) Aflatoxin metabolism in humans: detection of metabolites and nucleic acid adducts in urine by affinity chromatography. *Proc. natl Acad. Sci. USA, 82,* 6492–6497

Groopman, J.D., Busby, W.F., Donahue, P.R. & Wogan, G.N. (1986) *Aflatoxins as risk factors for liver cancer: an application of monoclonal antibodies to monitor human exposure.* In: Harris, C.C., ed., *Biochemical and Molecular Epidemiology of Cancer,* New York, Alan R. Liss, pp. 233–256

Groopman, J.D., Donahue, P.R., Zhu, J., Chen, J. & Wogan, G.N. (1987) Temporal patterns of aflatoxin metabolites in urine of people living in Guangxi Province, P.R.C. *Proc. Am. Assoc. Cancer Res., 28,* 36

Kensler, T.W., Egner, P.A., Davidson, N.E., Roebuck, B.D., Pikul, A. & Groopman, J.D. (1986) Modulation of aflatoxin metabolism, aflatoxin N7-guanine formation and hepatic tumorigenesis in rats fed ethoxyquin: role of induction of glutathione S-transferases. *Cancer Res., 46,* 3924–3931

Peers, F., Bosch, X., Kaldor, J., Linsell, A. & Pluijmen, M. (1987) Aflatoxin exposure, hepatitis B virus infection and liver cancer in Swaziland. *Int. J. Cancer, 39,* 545–553

Van Rensburg, S.J., Cook-Mozaffari, P., Van Schalkwyk, D.J., Van der Watt, J.J., Vincent, T.J. & Purchase, I.F. (1985) Hepatocellular carcinoma and dietary aflatoxin in Mozambique and Transkei. *Br. J. Cancer, 51,* 713–726

Zhu, J.-Q., Zhang, L.-S., Hu, X., Xiao, Y., Chen, J.-S., Xu, Y.-C., Fremy, J. & Chu, F.S. (1987) Correlation of dietary aflatoxin B_1 levels with excretion of aflatoxin M_1 in human urine. *Cancer Res., 47,* 1848–1852

DETECTION OF EXPOSURE TO AFLATOXIN IN AN AFRICAN POPULATION

H. Autrup[1] & J. Wakhisi[2]

[1] *Laboratory of Environmental Carcinogenesis, The Fibiger Institute, Copenhagen, Denmark; and*
[2] *Department of Surgery, University of Nairobi, Medical School, Nairobi, Kenya*

Urinary excretion of 8,9-dihydro-8-(7′-guanyl)-9-hydroxyaflatoxin (AFB-Gua) was studied in areas of different liver cancer incidence in Kenya. Of 983 urine samples analysed for AFB-Gua by high-performance liquid chromatography, 12.6% gave positive results. The chemical identity of AFB-Gua was verified by synchronous scanning fluorescence spectrophotometry. A moderate degree of correlation between prevalence of exposure to aflatoxin B_1 and liver cancer incidence could be established in Bantu. People living in areas with high exposure to aflatoxin B, form antibodies that recognize an aflatoxin B_1 epitope.

Consumption of an aflatoxin-contaminated diet is considered to be an important factor in the etiology of liver cancer (Linsell & Peers, 1977). Exposure to aflatoxins has been established by food analysis (Peers & Linsell, 1973; Peers *et al.*, 1976), by excretion of aflatoxins and their metabolites in urine (Campbell *et al.*, 1970), and by excretion of aflatoxin B_1 (AFB)-Gua (Autrup *et al.*, 1985; Groopman *et al.*, 1985). The latter product is formed by depurination of the major DNA adduct, which is formed when metabolically activated AFB interacts with DNA (Essigmann *et al.*, 1977). In this report, we present the results of a cross-sectional study in Kenya to correlate excretion of AFB-Gua with liver cancer incidence. Determination of urinary carcinogen-DNA adducts offers a viable alternative for assessing exposure to environmental carcinogens.

Materials and methods

Biological samples

Morning urine samples (minimum volume, 25 ml) were collected at outpatient clinics of selected district hospitals using sterile, disposable containers. Patients with stomach and intestinal complaints were excluded from the study, since it could be expected that they had changed their regular eating habits. The donors represented the normal sex and age distribution of the Kenyan population; however, only people over the age of ten years were included. Blood samples (maximum, 7 ml) were collected from the urine donors in Vacutainer blood collection tubes containing EDTA, and the plasma was prepared by centrifugation. A total of 983 urine samples were collected over a four-year period (January 1981–June 1984), while blood collection started in April 1983. No individual was sampled more than once. Urine samples were processed in C18 Sep-Pak cartridges and analysed for AFB-Gua as described previously (Autrup *et al.*, 1985).

A sample was considered positive for AFB-Gua if it cochromatographed with authentic AFB-Gua in two different high-performance liquid chromatography systems and had the characteristic synchronous fluorescent spectrum (Autrup *et al.*, 1983). About 85% of the samples positive in both systems were also positive by the third criterion, giving 15% false-positives if the chemical verification was excluded. The lower level of detectability for AFB-Gua by this assay is 0.3 pmol/25 ml urine. This level of AFB-Gua corresponds to an initial formation of 0.3 adduct per 10^6 bases.

Liver cancer incidence

Records for more than 2000 patients attending the clinic for liver disease at Kenyatta National Hospital in Nairobi during the period 1978–1982 were analysed. A diagnosis of hepatocellular carcinoma (HCC) was confirmed on the basis of two criteria: (1) presence of α-protein and (2) histological examination of needle or surgical biopsies. The 1981 census (Kenyan Government report) was used to calculate the incidence rates of HCC in various districts.

Formation of antibodies against AFB

An enzyme-linked immunosorbent assay has been developed to detect AFB antibodies. A 96-well NUNC immunoplate (NUNC, Roskilde, Denmark) is coated with AFB-bovine serum albumin (BSA) (1 μg/ml; 25 mol AFB/mol BSA; 100 μl/well; Sigma) dissolved in phosphate-buffered saline (PBS). Control wells are treated with BSA only (1 μg/ml; 100 μl/well). The plates are incubated at 37°C overnight; coated plates are stored at −20°C. Human sera are diluted serially (1:50–1:10 000) with PBS and added to the wells (100 μl; six wells for each dilution) and incubated overnight at 4°C. The plates are washed eight times with PBS-Tween 20 (NUNC-Immuno SERA Washer model 596, Nippon Intermed, Japan) prior to addition of the secondary antibody. Alkaline phosphatase-conjugated rabbit antihuman immunoglobulin G (Dakopatt, Copenhagen, Denmark) diluted 1:500 in 1% BSA is used as the secondary antibody. The plates are incubated for 1.5 h at room temperature, washed and incubated with the substrate (4-nitrophenylphosphate; 1 mg/ml) at pH 9 for 1 h at room temperature. The optical density is read at 405 nm (EIA Reader, Model 307; Bioteck Instruments Inc.).

Results and discussion

A total of 122 samples (12.4%) were positive for AFB-Gua. A large variation in presence of AFB-Gua was seen between districts and between males and females (Autrup *et al.*, 1987). The association between prevalence of AFB exposure and liver cancer incidence is shown in Table 1. The seasonal variation in AFB-Gua excretion corresponded to the seasonal variation in the amount of AFB in prepared food (Peers & Linsell, 1973).

On a national basis, no correlation could be established between exposure to AFB and liver cancer incidence; however, if the analysis is limited to the Bantu people, a moderate association is found using a Spearman nonparametric rank correlation analysis ($r = 0.75$). The Kenyan population consists of many different ethnic groups, which may have different capacities to convert AFB into its active metabolite. By limiting the study to the Bantu people, the genetic difference in metabolizing AFB has been eliminated.

Antibodies that recognize an aflatoxin epitope were detected in the sera. The specific activity was defined as the greatest dilution of serum at which the optical density in the AFB-BSA-coated wells was still twice that in the BSA-coated wells

Table 1. Prevalence per 100 000 of exposure to AFB and liver cancer: all ethnic groups

District	AFB exposure		Liver cancer incidence	
	Males	Females	Males	Females
Machakos	0.118	0.071	2.32	0.30
Makueni	0.129	0.073	2.69	0.56
Meru/Embu	0.417	—	1.71	—
Murang'a	0.123	0.078	1.42	0.78
Kerichio	0.197	0.150	1.37	0.45
Kiambu	0.214	0.048	2.38	0.88
S. Nyanza	0.314	0.147	0.56	0.24
Kitale	0.194	0.221	1.06	0.48
Busia	—	0.050	—	0.30

Table 2. Antigenic activity against an AFB epitope — provincial level

	No. of cases	
Titre	Central Province	Western Province
500–5000	8	3
5000–10 000	3	9
≃10 000	5	9

(Table 2). The highest activity was detected in sera from western Kenya, an area with a high prevalence of AFB exposure. An association between recent exposure to AFB, as measured by urinary AFB-Gua, and high specific activity of the antibody was established.

This study gives additional support to the hypothesis that AFB plays a major role in the induction of human liver cancer. Hepatitis B infection may play a minor role in East Africa, since no correlation was established between the prevalence of infection, as measured by surface antigen, and liver cancer incidence (Autrup *et al.*, 1987).

Acknowledgements

This project has been supported by Yamagiwa-Yoshida Memorial International Cancer Study Awards, by an award from the Friedman Foundation, and by grants from the Neye Foundation and the Danish Cancer Society. The assistance of K.A. Bradley, T. Seremet, Dr K. Vähäkangas and Dr A.K.M. Shamsuddin, and the encouragement of Dr C.C. Harris are highly appreciated.

References

Autrup, H., Bradley, K.A., Shamsuddin, A.K.M., Wakhisi, J. & Wasunna, A. (1983) Detection of putative adduct with fluorescence characteristics identical to 2,3-dihydro-(7′-guanyl)-3-hydroxyaflatoxin B_1 in human urine collected in Muranga district, Kenya. *Carcinogenesis, 4,* 1193–1195

Autrup, H., Wakhisi, J., Vähäkangas, K., Wasunna, A. & Harris, C.C. (1985) Detection of 8,9-dihydro-8-(7′-guanyl)-9-hydroxy aflatoxin B_1 in human urine. *Environ. Health Perspectives, 62,* 105–108

Autrup, H., Seremet, T., Wakhisi, J. & Wasunna, A. (1987) Aflatoxin exposure measured by urinary excretion of aflatoxin B_1-guanine adduct and hepatitis B virus infection in areas with different liver cancer incidence in Kenya. *Cancer Res., 47,* 3430–3433

Campbell, T.C., Caedo, J.P. & Bulatao-Jayme, J. (1970) Aflatoxin M_1 in human urine. *Nature, 227,* 403–404

Essigmann, J.M., Croy, R.G., Dzan, A.M., Busby, W.F., Jr., Reinhold, V.N., Buchu, G. & Wogan, G.N. (1977) Structural identification of the major DNA adduct formed by aflatoxin B_1 *in vitro. Proc. natl Acad. Sci. USA, 74,* 1870–1874

Groopman, J.D., Donahue, P.R., Zhu, J., Chen, J. & Wogan, G.N. (1985) Aflatoxin metabolism in humans: detection of metabolism and nucleic acid adducts in urine by affinity chromatography. *Proc. natl Acad. Sci. USA, 82,* 6492–6496

Linsell, C.A. & Peers, F.G. (1977) *Field studies on liver cell cancer.* In: Hiatt, H.H., Watson, J.D. & Winsten, J.A., eds, *Origins of Human Cancer,* Cold Spring Harbor, NY, CSH Press, pp. 549–556

Peers, F.G. & Linsell, C.A. (1973) Dietary aflatoxins and liver cancer: a population based study in Kenya. *Br. J. Cancer, 27,* 473–484

Peers, F.G., Gilman, G.A. & Linsell, C.A. (1976) Dietary aflatoxins and human liver cancer: a study in Swaziland. *Int. J. Cancer, 17,* 167–175

APPLICATION OF ANTIBODY METHODS TO THE DETECTION OF AFLATOXIN IN HUMAN BODY FLUIDS

C.P. Wild,[1,3] B. Chapot,[1] E. Scherer,[2] L. Den Engelse[2] & R. Montesano[1]

[1]International Agency for Research on Cancer, Lyon, France; and [2]Netherlands Cancer Institute, Amsterdam, The Netherlands

Four different approaches to the quantification of human exposure to aflatoxins (AF) are presented: (i) analysis of urinary AF metabolites and DNA adducts, (ii) assay of AF bound to blood proteins and to lymphocyte DNA, (iii) immunocytochemical localization of AF in individual cells, and (iv) detection of AF in human breast milk. The potential applications of these approaches for assessing the role of both AF and hepatitis B virus (HBV) in primary hepatocellular carcinoma (HCC) are presented. The advantages and limitations of the methods for use in large-scale epidemiological studies are discussed, with particular attention to sensitivity.

Primary hepatocellular carcinoma (HCC) is the seventh most common cancer in the world and is particularly prevalent in south-east Asia, China and sub-Saharan Africa (Parkin *et al.*, 1984). Following experimental observations in animals of the potent hepatocarcinogenicity of AF (Busby & Wogan, 1984), epidemiological studies have implicated these widespread food contaminants in the etiology of HCC, either independently or in combination with HBV (for reviews see Busby & Wogan, 1984; Harris & Sun, 1984; Muñoz & Bosch, 1987). Data on exposure to AF in these studies are based exclusively upon analysis of food samples. This paper reports developments in methods for assessing exposure at an individual level by providing an integrated value of exposure resulting from intake, distribution, metabolic (in)activation and excretion of AF. A summary of the approaches used in our laboratory and others (see also Groopman, this volume) is given in Figure 1. To date, our work has involved pilot studies on small sample sizes in order to evaluate these experimental approaches for future, larger-scale biochemical epidemiological studies. The results obtained, together with a critique of the various approaches, are given below.

AF in urine

Urine samples have been analysed from four populations, in The Gambia, the Philippines, Singapore and Lyon, France. The staple food source, groundnuts, in The Gambia is known to be contaminated with AF, and exposure to AF has been associated with an increased risk of HCC in the Philippines (Bulatao-Jayme *et al.*, 1982), with individual daily consumption of AFB_1 estimated at a mean of 22 μg for high-risk groups. Urine samples from the Philippines analysed in this study were obtained from outpatients in hospitals in Manila, Roxar and Cebu. The samples from

[3] To whom correspondence should be addressed

Fig. 1. Assessment of human exposure to aflatoxins at the individual level

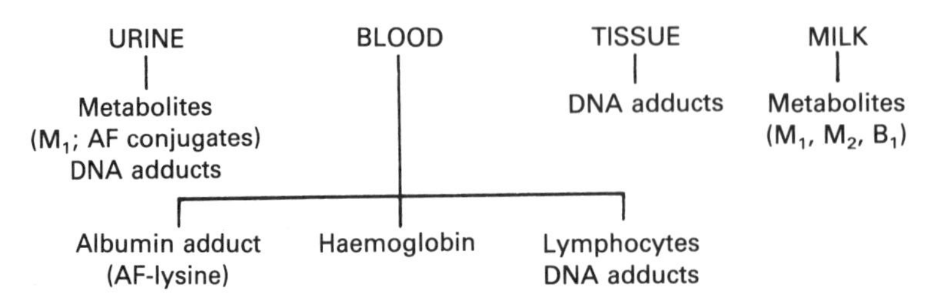

Singapore, where there is some evidence of AF exposure, were collected from hospital staff in Singapore General Hospital, while the urines from France were from laboratory volunteers or hospital patients who were involved in an unrelated study. It should be emphasized that the main purpose of these initial analyses was to assess the value of this approach in terms of its sensitivity and specificity rather than to examine AF exposure as a risk factor for HCC in these populations. Immunoassay of urinary AF was performed following an immunoaffinity column extraction, as described previously (Wild *et al.,* 1986a, 1987), except that for urine samples from The Gambia and some from France AF was eluted from the affinity column with 60% methanol in phosphate-buffered saline adjusted to pH 3.0 (Fig. 2). The data obtained are presented in Figure 3.

The assay proved to be (1) sufficiently sensitive to detect urinary AF levels resulting from environmental exposure and (2) specific enough to allow differences in high (The Gambia and Philippines), moderate (Singapore) and low (Lyon, France) exposure groups to be detected. In addition, the fact that the affinity columns are reusable and the assays quite simple and rapid to perform, using only 5-ml samples, suggests that this approach is valid for examining exposure to AF. Most samples contained 0.1–1.0 ng AFB_1 equivalent per ml urine, which corresponds to an exposure of the order of 15–150 ng AFB_1/kg body wt per day (assuming 15% dietary AF excreted in 1500 ml urine daily for a 70-kg man). These figures agree well with estimates of exposure from data on dietary intake (see Van Rensburg *et al.*, 1985). It is interesting to note, on the basis of figures from Ames *et al.* 1987), that exposure to the level of AF permitted in a western diet (up to 20 ppb) would be expected to result in 0.04 ng AF/ml urine using the same assumptions as above. Thus, the urine assay is sufficiently sensitive to measure contamination levels in populations at low-level exposure and low risk of HCC.

As discussed previously (Wild *et al.,* 1986a), the enzyme-linked immunosorbent assay (ELISA) can be used to quantify AF on the basis of a composite inhibition resulting from any AF metabolite present in the urine and bound by the antibody employed in the immunopurification. As the antibody has a different affinity for each metabolite, it remains important to establish how well the inhibition value in ELISA reflects dietary intake of AF at the individual level. In order to do this, comprehensive knowledge of antibody specificity is required. A particular problem is the water-soluble AF conjugates likely to be present in human urine. These compounds are as yet uncharacterized, and their reaction with antibodies being used to immunopurify AF from urine has not been reported.

It is possible to chromatograph immunopurified material further, allowing identification and quantification of several metabolites in one sample (Groopman *et al.,* 1986). This may give a metabolic 'fingerprint' for an individual, allowing some assessment of the effects of, for example, HBV status, parasite infestation or protein

Fig. 2. Analysis of human urine sample for aflatoxins

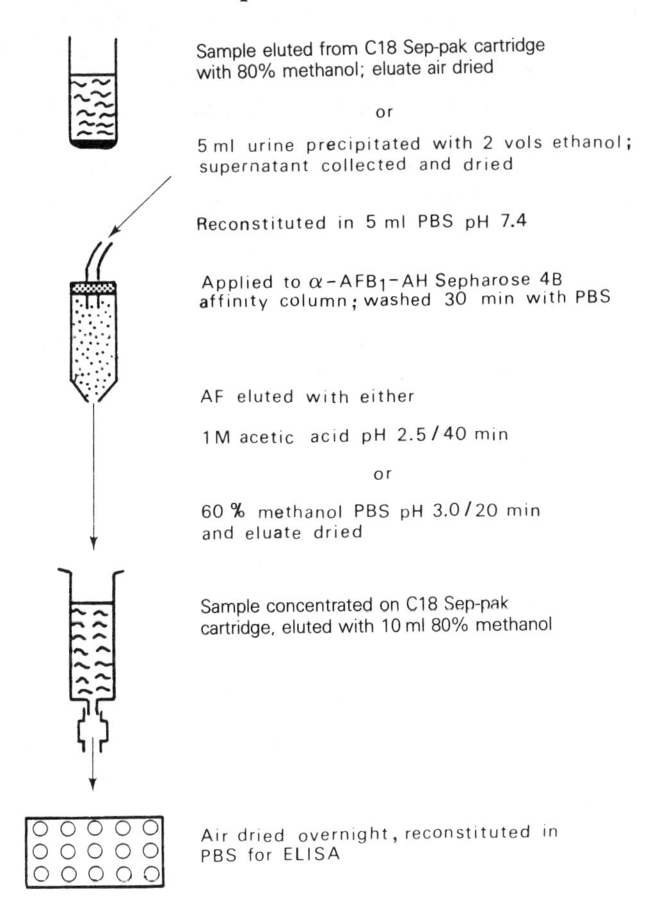

Fig. 3. Aflatoxin levels in human urine

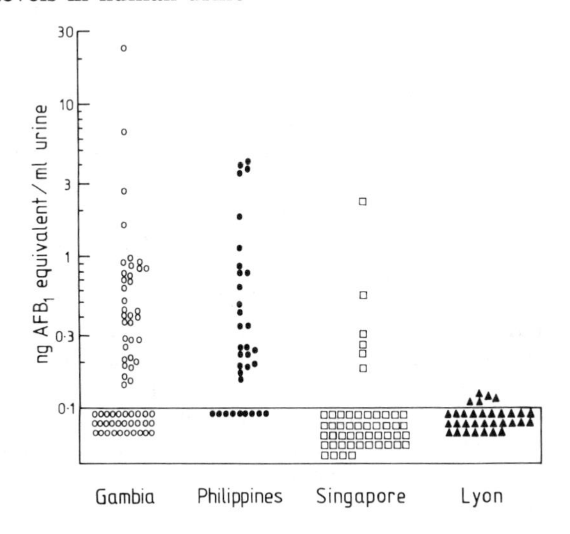

intake on AF metabolism. An alternative approach is to use a purification system and an antibody which give a high specificity for only one metabolite (Zhu *et al.*, 1987). The latter group have shown a correlation between dietary intake of AFB_1 and urinary AFM_1 (correlation coefficient, 0.66) in Chinese subjects. It should be stressed that in this case the metabolite is a detoxification product and may not necessarily reflect the reaction pathway for AF which leads to DNA modification in the target cells.

Finally, an important consideration is that urine analysis may reflect only relatively recent exposure to AF (some days), and thus the suitability of these measurements for epidemiological studies may be somewhat limited.

AF in peripheral blood

In order to measure AF exposure over a longer period, we have used a rat model system to look for the formation and persistence of AF adducts in peripheral blood proteins and white blood cells.

A constant relationship between AF bound to plasma albumin and liver DNA has been observed in Wistar rats following single (3.5–200 μg/kg) and multiple doses (3.5 μg/kg) of AFB_1 (Wild *et al.*, 1986b). The major AF-albumin adduct in rats has been demonstrated by Sabbioni *et al.* (1987) to be an AF-lysine residue. We have extended our experiments to look for AF binding to total white blood cells and plasma albumin in a strain of rat (Sprague Dawley (SD)) which is less sensitive to the hepatocarcinogenicity of AFB_1 than the Wistar (WI) strain used previously. Rats were given single oral doses of ^3H-AFB_1 (Moravek, Brea, CA, USA), and the levels of AF bound to plasma protein and liver DNA were quantified 24 h later, as reported (Wild *et al.*, 1986b); white blood cells were obtained by centrifugation after lysis of red blood cells. Levels of AF bound to plasma protein were similar in the two strains (4.9 and 5.9 pg AF/mg plasma protein per unit dose (1 μg/kg) for SD and WI, respectively) over a dose range of 3.5–200 μg AFB_1 per kg body weight. In contrast, binding to DNA was three-fold lower in SD rats than in WI. Thus, while for each strain a constant ratio of plasma protein:DNA-bound AF was found, the mean ratios were 3.3 (WI) and 1.1 (SD).

In the same animals, no binding to white blood cell DNA could be detected, and DNA extracted from the spleens of these rats contained no detectable AF adduct (limit of sensitivity, 10 pg AF/mg DNA). In addition, upon immunocytochemical analysis (see below) of spleen tissue from rats exposed to 14 consecutive daily doses of 50 μg AFB_1, no AF-DNA adduct could be visualized (limit of sensitivity, 180 pg AF/mg DNA).These results are in agreement with our previous observations (Wild *et al.*, 1986b) and imply that the use of lymphocyte DNA for assessing human exposure to AF is not very promising. Binding of AF to haemoglobin was also very low, in agreement with observations from other laboratories (<0.05% of administered dose; Tannenbaum & Skipper, 1984). In contrast, (i) the constant relationship between liver DNA and albumin-bound AF, (ii) the relatively high proportion of AF bound to albumin (1–3% of a single dose), (iii) its accumulation to a steady-state level upon chronic dosing and (iv) the stability of the circulating albumin adduct in rats (Skipper *et al.*, 1985; Wild *et al.*, 1986b; Sabbioni *et al.*, 1987) are promising indications of the use of this adduct as a biological dosimeter for human exposure to AF. Additional factors which suggest that development of assays to measure this adduct may be useful are that adduct formation probably occurs in the liver, the putative target organ in man, and albumin has a half-life of ~20 days in man, thus allowing accumulation upon chronic exposure to levels 30-fold greater than after a single exposure (see Sabbioni *et al.*, 1987).

AF in tissues

Quantification of DNA adducts in the target organ for tumour induction would in theory be the most direct measure of the biologically effective dose of an environmental carcinogen. This has been achieved for some adducts (Perera *et al.*, 1982; Umbenhauer *et al.*, 1985) but necessitates invasive sampling and a relatively large amount of tissue, precluding measurements on small cell numbers such as those obtained by biopsy. An additional disadvantage when using isolated bulk DNA is the loss of information regarding the cellular heterogeneity of the distribution of adducts.

Visualizing adducts in single cells is a potential way of overcoming the above problems, and immunocytochemical methods have allowed the detection of DNA adducts in tissue sections from carcinogen-treated animals (Heyting *et al.*, 1983), including those treated with AFB_1 (Pestka *et al.*, 1983; Shamsuddin *et al.*, 1987).

For diagnoses of HCC, a needle biopsy is sometimes performed, providing a potential source of human material for immunocytochemical analysis of AF adducts. We have applied the immunostaining technique developed by Heyting *et al.* (1983) to detect AF adducts in rat tissues after single or multiple doses of AFB_1 and AFG_1 (C.P. Wild, L. Den Engelse and E. Scherer, in preparation). In brief, male SD rats (180–200 g) were treated with AF, and, 24 h after treatment, tissues were removed and frozen on tissue blocks for immunocytochemistry. The antibody specific for AF was the polyclonal antiserum used in ELISA (see above). The only major modification of the technique of Heyting *et al.* (1983) was that the alkali treatment used to make the DNA accessible to the specific antibody was performed for 10 min with 0.05 N NaOH in 40% ethanol.

Positive nuclear staining, after single or multiple doses of AF, was observed in liver, kidney and lung tissue but not in oesophagus, forestomach, colon, spleen, pancreas or testis. At the lower doses of AFB_1 (10, 30 and 80 μg/kg), the tritiated compound (5–6 μCi per rat) was administered, allowing the absolute modification level to be determined in DNA extracted from liver. Using this comparison, we measured a lower limit of detection by immunocytochemistry of one adduct in 4×10^6 nucleotides, or about 2000 adducts per mammalian diploid genome.

The advantage of this approach is illustrated in particular by observations on lung, where staining was found exclusively in the nuclei of cells lining the small bronchi. Adduct levels in these cells were easily detected by immunostaining; in DNA isolated from whole lung, the levels would be diluted by at least two orders of magnitude. Consequently, adducts concentrated in what may be the target cell population of a particular tissue could be diluted to nondetectable levels during isolation of bulk DNA. This result would be particularly misleading in tissues such as lung, which have a great heterogeneity of cell type.

AF in breast milk

AFM_1, a hydroxylated metabolite of AFB_1, has been found in milk from a variety of species and is hepatocarcinogenic in rats and trout (for review, see Busby & Wogan, 1984). In countries with a high rate of HCC, peak incidence is seen at a relatively early age, e.g., an age-specific rate for males (20–30 years old) of 43.5 per 100 000 in Inhambane Province, Mozambique (Van Rensburg *et al.*, 1985), with a similar trend in other high-incidence countries, including Zimbabwe (Muñoz & Bosch, 1987). It is possible that this incidence pattern is influenced by exposure to environmental carcinogens early in life. We have been investigating the occurrence of AFM_1 in human breast milk using an ELISA after extraction of AF on C18 Sep-pak Cartridges (Waters

Assoc., MO, USA). This procedure allows the quantification of 2 pg AFM_1 per ml milk, using less than 10 ml of sample (Wild *et al.*, 1987) and has been used to analyse milks collected from rural areas of Zimbabwe, Thailand and The Gambia. AF was present in breast milk from mothers in Zimbabwe (6/54 samples positive) and Thailand (3/12 samples positive), confirming one previous report on breast milk from mothers in Sudan (Coulter *et al.*, 1984). However, surprisingly, in preliminary analyses in The Gambia, where AF levels were high in the urine (see above), no positive sample was detected. This may be due to the fact that the samples were collected in January and February, when dietary contamination may be lower than in the rainy season. Further milk collection is under way in order to examine this hypothesis. The levels of AFM_1 found in Zimbabwe and Thailand (up to 50 pg AF/ml milk) correspond to exposure levels of around 10 ng AFM_1 per kg body weight for an infant, based on average milk consumption (Laupus, 1969). Assuming that 1% of ingested AFB_1 is excreted in milk, the mothers would have been exposed to 10–100 ng AFB_1/kg body weight. This level in adults has been associated with an elevated risk of developing HCC (see Van Rensburg *et al.*, 1985), and it is of interest that in experimental studies young animals may be more sensitive to the carcinogenic action of AF than adults (Vesselinovitch *et al.*, 1972). In addition, vertical transmission of HBV can occur soon after birth (Steven *et al.*, 1975; Whittle *et al.*, 1983); thus, the two environmental factors that have been most strongly linked both epidemiologically and experimentally with HCC are present at or soon after birth. Our method should facilitate studies examining the importance of the interaction between these two risk factors for HCC.

Conclusions

Considerable progress has been made in recent years towards developing assays to measure human exposure to AF. Table 1 presents a summary of the levels of AF expected in human tissues or body fluids on the basis of animal and human exposure data. It is clear that the assays in urine and milk using samples of 5–10 ml are sensitive enough to determine the levels occurring after daily exposures to as little as 1 μg AFB_1

Table 1. Aflatoxin adduct levels found in rats or estimated in humans in relation to assay sensitivity

Dose AFB_1	Urine (ng/ml)	Plasma albumin (pg/mg)	Liver DNA (pg/mg)	Milk (pg/ml)
Wistar rat[a]				
3.5 μg/kg single	5.5	72	108	ND
3.5 μg/kg for 24 days	8.6	258	250	ND
Human				
10 μg single	1.0[b]	0.5[c]	8[d]	200[e]
10 μg chronic	1.6[b]	15.0[c,f]	18	—
Assay sensitivity	0.1	—	180	2.0

ND, not determined
[a]Data derived from Wild *et al.* (1986b)
Assumptions:
[b]15% AF intake excreted into 1500 ml urine per day, 70-kg man
[c]1% AFB_1 intake bound to albumin, 40 mg albumin per ml blood, 70 ml blood per kg body weight
[d]Comparison of adduct formation in human and rat liver based on dose/surface area; activation of AFB_1 to reactive epoxide in man is one-tenth the level in rats (Booth *et al.*, 1981); comparable rate of adduct loss or removal in the two species
[e]1% excretion of AFB_1 as AFM_1 in 500 ml of milk per day, 50-kg woman
[f]Accumulation, ~30 times over a single dose (see Sabbioni *et al.*, 1987)

(14 ng AFB_1/kg body weight). The low level of binding of AF to haemoglobin seems to discourage its use as a dosimeter for exposure to AF, despite the fact that the long half-life of red blood cells (120 days) means that the level of carcinogen-modified haemoglobin could reflect exposure over a period of months. No assay is yet available for the AF plasma albumin adduct, although a method for 4-aminobiphenyl-haemoglobin adduct can measure down to 10 fg adduct/g haemoglobin (Bryant *et al.*, 1987), suggesting that measurement of the levels estimated in Table 1 of 15 pg AF/mg protein should be feasible (taking into consideration that about 0.5 g albumin is available in a 10-ml blood sample). Immunocytochemistry in its present form appears to be an order of magnitude too insensitive; however, in this case, the adducts may be concentrated in some cell types (see discussion above), making the levels in these cells much higher than the estimates in Table 1.

The techniques for assessing exposure to AF have moved rapidly from the stage of methodological development to application in pilot biochemical epidemiological studies. Experience gained in the type of studies discussed above serves to refine the methods for future applications. The choice of assay depends naturally on the exact nature of the information being sought. For example, AF-plasma protein adducts may be particularly useful as an index of exposure over a period of months ($t\frac{1}{2}$ albumin in man, around 20 days), whereas urinary levels of AF are likely to represent exposure over one or two days prior to sampling. In many cases, complementary data may be obtained by using more than one technique, and one can envisage that the same assay may give not only a measure of exposure but also information concerning factors that contribute to individual susceptibility to HCC, e.g., the effect of chronic active hepatitis on AF metabolism using analysis of metabolites in the urine, or the interaction of HBV infection and AF exposure during the first months of life. These studies can now be complemented by experiments in animal models, e.g., duck, with the duck hepatitis virus, where the same methods can be used for analyses. Thus, advances in assay technology can provide opportunities to examine the mechanisms of AF carcinogenesis, in addition to measuring human exposure to these compounds.

Acknowledgements

The authors wish to express their thanks to the people involved in the collection and analysis of the samples used in these studies, specifically to Dr A. Hall, Dr R. Ryder, Dr H. Whittle, Dr F.X. Bosch and F. Loik in the Gambian studies, Dr N. Muñoz in the Philippines study, Dr Oon in Singapore, Dr S. Petcharin in Thailand and Professor R. Sohier for organizing the supply of human milk samples from Lyon, France. The important contribution of Dr C. Garner and Dr F. Tursi to the work on plasma albumin adducts and of Dr C. Chetsanga, C. Mutiro and F. Pionneau to that on breast milk is fully acknowledged. C.P. Wild was the recipient of an IARC research training grant and a Royal Society European Exchange Programme fellowship during the period of this work.

References

Ames, B.N., Magaw, R. & Gold, L.S. (1987) Ranking possible carcinogenic hazards. *Science, 236,* 271–280

Booth, S.C., Bösenburg, H., Garner, R.C., Hertzog, P.J. & Norpoth, K. (1981) The activation of aflatoxin B_1 in liver slices and in bacterial mutagenicity assays using livers from different species including man. *Carcinogenesis, 2,* 1063–1068

Bryant, M.S., Skipper, P.L., Tannenbaum, S.R. & Maclure, M. (1987) Hemoglobin adducts of 4-aminobiphenyl in smokers and non-smokers. *Cancer Res., 47*, 602–608

Busby, W.F. & Wogan, G.M. (1984) *Aflatoxins.* In Searle, C.D., ed., *Chemical Carcinogens* (*ACS Monograph* 182), Washington DC, American Chemical Society, pp. 945–1136

Bulatao-Jayme, J., Almero, E.M., Castro, M.A.C.A., Jardeleza, M.A.T.R. & Salamat, L.A.

(1982) A case-control dietary study of primary liver cancer risk from aflatoxin exposure. *Int. J. Epidemiol.*, *11*, 113–119

Coulter, J.B.S., Lamplugh, S.M., Suliman, G.I., Omer, M.I.A. & Hendrickse, R.G. (1984) Aflatoxins in human breast milk *Ann. trop. Paed.*, *4*, 61–66

Groopman, J.D., Donahue, P.R., Zhu, J.Q., Chen, J.S. & Wogan, G.N. (1986) Aflatoxin metabolism in human: detection of metabolites and nucleic acid adduct in urine by affinity chromatography. *Proc. natl Acad. Sci. USA*, *82*, 6492–6496

Harris, C.C. & Sun, T-T. (1984) Multifactoral etiology of human liver cancer. *Carcinogenesis*, *5*, 697–701

Heyting, C., Van der Laken, C.J., Van Raamsdonk, W. & Pool C.W. (1983) Immunohistochemical detection of O^6-ethyldeoxyguanosine in the rat brain after *in vivo* applications of N-ethyl-N-nitrosourea. *Cancer Res.*, *43*, 2935–2941

Laupus, W.E. (1969) *Feeding of infants*. In: Nelson, W.E., Vaughan, V.C. & McKay, R.J., eds, *Textbook of Paediatrics*, Philadelphia, Saunders, pp. 143–157

Muñoz, N. & Bosch, F.X. (1987) *Epidemiology of hepatocellular carcinoma*. In: Okuda, K. & Ishak, K.G., eds, *Neoplasms of the Liver*, Tokyo, Springer (in press)

Parkin, D.M., Stjernsward, J. & Muir, C.S. (1984) Estimates of the worldwide frequency of twelve major cancers. *Bull. World Health Organ.*, *62*, 163–182

Perera, F., Poirier, M.C., Yuspa, S.H., Nakayama, J., Jaretzki, A., Curnen, M.M., Knowles, D.M. & Weinstein, I.B. (1982) A pilot project in molecular cancer epidemiology: determination of benzo(a)pyrene DNA adducts in animal and human tissues by immunoassays. *Carcinogenesis*, *3*, 1405–1410

Pestka, J.J., Beery, J.T. & Chu, F.S. (1983) Indirect immunoperoxidase localisation of aflatoxin B_1 in rat liver. *Food chem. Toxicol.*, *21*, 41–48

Sabbioni, G., Skipper, P.L., Buchi, G. & Tannenbaum, S.R. (1987) Isolation and characterisation of the major serum albumin adduct formed by aflatoxin B_1 in rats. *Carcinogenesis*, *8*, 819–824

Shamsuddin, A.M., Harris C.C. & Hinzman, M.J. (1987) Localization of aflatoxin B_1–nucleic acid adducts in mitochondria and nuclei. *Carcinogenesis*, *8*, 109–114

Skipper, P.L., Hutchins, D.H., Turesky, R.J.K., Sabbioni, G. & Tannenbaum, S.R. (1985) Carcinogen binding to serum albumin. *Proc. Am. Assoc. Cancer Res.*, *26*, 356

Steven, C.E., Beasley, R.P., Tsui, J. & Lee, W.C. (1975) Vertical transmission of hepatitis B antigen in Taiwan. *New Engl. J. Med.*, *293*, 771–774

Tannenbaum, S.R. & Skipper, P.L. (1984) Biological aspects to the evaluation of risk: dosimetry of carcinogens in man. *Fundam. appl. Toxicol.*, *4*, 5367–5373

Umbenhauer, D., Wild, C.P., Montesano, R., Saffhill, R., Boyle, J.M., Huh, N., Kirstein, U., Thomale, J., Rajewsky, M.F. & Lu, S.H. (1985) O^6-Methyldeoxyguanosine in oesophageal DNA among individuals at high risk of oesophageal cancer. *Int. J. Cancer*, *36*, 661–665

Van Rensburg, S.J., Cook-Mozaffari, P., Van Schalkwyk, D.J., Van Der Watt, J.J., Vincent, T.J. & Purchase, I.F. (1985) Hepatocellular carcinoma and dietary aflatoxin in Mozambique and Transkei. *Br. J. Cancer*, *51*, 713–726

Vesselinovitch, S.D., Mihailovich, N., Wogan, G.N., Lombard, L.S. & Rao, K.V.N. (1972) Aflatoxin B_1, a hepatocarcinogen in the infant mouse. *Cancer Res.*, *32*, 2289–2291

Whittle, H.C., Bradley, A.K., McLauchlan, K., Ajdukiewicz, A.B., Howard, C.R., Zuckerman, A.J. & McGregor, I.A. (1983) Hepatitis B virus infection in two Gambian villages. *Lancet*, *i*, 1203–1206

Wild, C.P., Umbenhauer, D., Chapot, B. & Montesano, R. (1986a) Monitoring of individual human exposure to aflatoxin (AF) and *N*-nitrosamines (NNO) by immunoassays. *J. cell. Biochem.*, *24*, 171–179

Wild, C.P., Garner, R.C., Montesano, R. & Tursi, F. (1986b) Aflatoxin B_1 binding to plasma albumin and liver DNA upon chronic administration to rats. *Carcinogenesis*, *7*, 853–858

Wild, C.P., Pionneau, F.A., Montesano, R., Mutiro, C.F. & Chetsanga, C.J. (1987) Aflatoxin detected in human breast milk. *Int. J. Cancer*, *40*, 328–333

Zhu, J-Q., Zhang, L-S., Hu, X., Xiao, Y., Chen, J-S., Xu, Y-C., Fremy, J. & Chu, F.S. (1987) Correlation of dietary aflatoxin B_1 levels with excretion of aflatoxin M_1 in human urine. *Cancer Res.*, *47*, 1848–1852

DETECTION IN HUMAN CELLS OF ALKYLATED MACROMOLECULES ATTRIBUTABLE TO EXPOSURE TO NITROSAMINES

R. Montesano, H. Brésil, P. Degan, G. Martel-Planche, M. Serres & C.P. Wild

International Agency for Research on Cancer, Lyon, France

Various methods for detecting DNA alkylation adducts are described briefly, with emphasis on immunoassays using antibodies against O^6-methyldeoxyguanosine (O^6-medGua), O^4-methylthymidine (O^4-meThy) and 7-methyldeoxyguanosine (7-medGua). The application of these methods to epidemiological studies is discussed, and results obtained so far on the presence of DNA alkylation adducts in human tissues are presented.

Epidemiological studies and, to a lesser extent, experimental carcinogenicity studies have resulted in the identification of various agents (factors) that are causally associated with human cancer (IARC, 1987). These factors can account for only one-third of the cancer mortality in countries like the USA and the UK and much remains to be done in identification of causes of human cancer (Peto, 1985). It should be noted that even for well-established exposures associated with cancer (e.g., tobacco, alcohol, schistosomal infection), the specific agent(s) directly responsible for tumour induction is not well defined. In addition, for some cancers an interaction between various risk factors is important, and the nature of the interaction is not at present understood. This lack of information represents a considerable limitation on the implementation of measures for primary prevention and on intervention studies aimed at reducing cancer morbidity and mortality.

Epidemiology has provided useful and accurate descriptions of some of the risk factors associated with cancer; however, it is difficult to foresee the extent to which it could contribute more to the identification of still unknown causes of human cancer, particularly when the increased associated risk is small but nevertheless significant in terms of number of cancers. The integration of highly sensitive laboratory methods into future epidemiological studies promises to contribute much to this task. In particular, the assessment of individual exposures to carcinogen-induced adducts in cellular macromolecules could result in a significant improvement in the sensitivity and specificity of epidemiological studies. This point has been discussed recently by Saracci (1984). Such considerations are particularly pertinent to the role of nitrosamines in the induction of human cancer and, specifically, to the development of markers in human cells attributable to nitrosamine exposure, which is the topic of this contribution. The sensitivity and specificity of such measurements as indicators of past exposure to nitrosamines as well as their applicability to epidemiological studies are addressed briefly.

It is well established that humans are exposed to nitrosamines from various sources and that nitrosamines are carcinogenic to experimental animals at dose levels that approach human exposures. Table 1 shows that large portions of the population are

Table 1. Estimated exposures of humans to nitrosamines[a]

Source of exposure	Daily intake (μg/person)
Cigarette smoking	17
Endogenous (dietary)	0.32–20
Occupational	10–180

[a] From National Research Council/National Academy of Sciences (1981)

exposed to these carcinogens at a level of tens of micrograms. In addition, in snuff users and in workers in occupations such as leather tanning, exposures could reach 180–200 μg/day per person (see Bartsch & Montesano, 1984). The evidence that nitrosamines are the cause of some human cancers is, however, limited or nonexistent: Table 2 indicates the strength of the evidence linking nitrosamine exposure to cancer at some sites. However, as already stated, evidence for a causal association is nonexistent, with the possible exception of limited data on a link between oesophageal and stomach cancer in Linxian county, China, and cancer of the nasopharynx in southern China, and intake of food containing nitrosamines (Yang, 1980; Yu *et al.*, 1986).

Table 2. Evidence for nitrosamine (NNO) exposure in the etiology of some human cancers

Cancer site(s)	Some identified or suggested risk factor(s)	Evidence for NNO exposure	Modified cellular macromolecules
Oral cavity	Tobacco products, snuff, *nass, khaini*	Strong	
Oral cavity	Betel quid with:		
	tobacco	Strong	
	lime	Moderate	
Lung, bladder, oral cavity, larynx, pharynx, oesophagus, pancreas, renal pelvis	Tobacco smoke	Strong	
Oesophagus (Normandy, France)	Tobacco-alcohol	Strong	
Oesophagus, stomach (Linxian, China)	Pickled vegetables, NO_3^-/NO_2^-, fungi contaminated cornbread and NNO precursors	Moderate	O^6-Methyldeoxyguanosine
Nasopharynx	Cantonese salted fish	Moderate	
Stomach	Endogenous carcinogens derived from NO_3^-/NO_2^--rich diet, salt, carbohydrate	Weak	
Urinary bladder	Bilharzia, bacterial infection	Weak	
Brain	Prenatal exposure to NNO or precursors (drugs, comestics, food items)	Weak	

Markers, such as cellular macromolecular adducts, attributable to nitrosamine exposure are available in one study only (Umbenhauer *et al.*, 1985).

Development of antibodies against DNA alkylation adducts

Nitrosamines like *N*-nitrosodimethylamine (NDMA) and *N*-nitrosodiethylamine, following metabolic activation, alkylate both DNA and proteins. Some 12 sites of alkylation in DNA have been identified, the major methylation products being 7-medGua (~70% of total alkylation), total phosphotriester (12%), O^6-medGua (~7%), 3-methyladenine (3-meAde) (~8%) and O^4-meThy (~0.1%) (see Singer & Grunberger, 1983). Although there are variations at the cellular, tissue and species levels in the capacity to repair such DNA adducts, 7-medGua and the phosphotriesters are generally repaired with very low efficiency as compared to O^6-medGua and O^4-meThy; 3-meAde is repaired rapidly and undergoes spontaneous depurination, and this modified base appears in the urine (see Lindahl, 1982).

Various antibodies (polyclonal and monoclonal) are now available against many DNA adducts, and these have been used in sensitive immunoassays to detect the adducts at low levels in isolated DNA (see Müller & Rajewsky, 1981; Strickland & Boyle, 1984). Table 3 lists the methods developed and used in our laboratory to detect the presence of DNA methylation adducts in human tissues. The antibodies have very high affinity and specificity, and, following chromatographic purification of DNA adducts, the immunoassay sensitivity is such that they can be applied to the analysis of human tissues (see below). In comparison, the detection of methylation adducts of haemoglobin and of 3-meAde in urine appears, at present, to be much less sensitive, since relatively high background levels (of unknown origin) of these adducts are found in human blood and urine (see, e.g., Table 3). A similar problem exists for 7-meGua, which is found in human urine at a level of approximately 10 mg/24 h (Gombar *et al.*, 1983).

In developing a rationale as to which DNA adducts to measure as markers of human exposure, it is important to note that the levels of such adducts depend not only on the initial degree of interaction between the methylating species and the various nucleophilic sites in DNA, but also on the degree of repair. For example, O^6-medGua is formed at levels about 100 times higher than O^4-meThy after a single dose of a methylating agent. However, the more rapid repair of the former adduct in rat liver can result in accumulation of O^4-meThy after multiple exposures to levels similar to those of O^6-medGua (Richardson *et al.*, 1985). This observation and the fact that human exposure is likely to be chronic indicated the importance of developing

Table 3. Detection of DNA and protein adducts attributable to nitrosamine exposure

Adduct	Method[a]	Sensitivity (adduct/normal nucleoside)	Reference
O^6-medGua	Mab-RIA	$0.4/10^7$ deoxyguanosine	Wild *et al.* (1983)
7-medGua (imidazole ring open)	Pab-ELISA	$3.2/10^7$ deoxyguanosine	Degan *et al.* (unpublished data)
O^4-meThy	Pab-RIA	$1.0/10^7$ thymine	Wild *et al.* (1987)
Methylcysteine-haemoglobin	GC-MS	>100 nmol/g haemoglobin (>10 mg *N*-nitrosodimethylamine/kg)	Bailey *et al.* (1981)

[a]Mab, monoclonal antibody; RIA, radioimmunoassay; Pab, polyclonal antibody; ELISA, enzyme-linked immunosorbent assay; GC-MS, gas chromatography-mass spectrometry

antibodies to this relatively minor DNA adduct. 7-medGua, unlike O^6-medGua and O^4-meThy, does not appear to be directly mutagenic but is the major methylation adduct (see above) and persists in DNA longer than the *O*-alkylated nucleosides. We have thus also developed antibodies for use in a highly sensitive enzyme-linked immunosorbent assay to detect 7-medGua in human DNA (Table 3). Detection of 7-medGua may be more informative about past exposure to methylating agents, although it is noteworthy that O^6-medGua has been detected in brain DNA of rats and gerbils six months after a single exposure to *N*-methyl-*N*-nitrosourea (Kleihues & Bucheler, 1977; Kleihues *et al.*, 1980). Thus, the detection of these DNA adducts in human tissues could reflect exposure to nitrosamines that occurred several months previously. In addition, determination of these adducts in a given tissue provides, at an individual level, an integrated value that is the result of a number of variables: endogenous and exogenous exposures to nitrosamines, metabolism, repair of DNA damage and other factors, for example, diet and genetic predisposition, that can affect these variables.

DNA alkylation adducts in human tissues

The availability of these antibodies permits the examination of human tissues for the presence of DNA alkylation adducts attributable to nitrosamine exposure, with the aim of determining whether there is a significant difference in levels among individuals at different risks of developing certain cancers. So far, such studies are few and suffer some limitations. In one study (Umbenhauer *et al.*, 1985), O^6-medGua was detected in oesophageal muscosal DNA from individuals in populations at high risk of oesophageal cancer and for which there is some evidence of exposure to nitrosamines. The prevalence of positive samples, as well as the level of O^6-medGua, was lower among individual samples in populations at lower risk of oesophageal cancer. These findings indicate that O^6-medGua can be detected in human tissue DNA and support the applicability of such methods to epidemiological studies. The determinations were made, however, in surgical specimens of oesophageal tissues; this approach does not permit a random selection of samples to be analysed and also limits the number of individuals who can be assayed. In addition, the type of epidemiological study is restricted: for example, while retrospective analyses can be performed, prospective studies or intervention studies cannot easily be done using this type of material.

The possibility of determining the presence of DNA alkylation adducts in DNA of peripheral blood cells could overcome some of these difficulties, and the antibodies against 7-medGua appear to be particularly promising in this respect. Table 4 shows that 7-medGua can be detected in blood lymphocytes of rats treated with 1 mg/kg NDMA and that the level of this DNA adduct is similar to that in liver. In 20 ml of

Table 4. DNA adducts in rats treated with *N*-nitrosodimethylamine (1 mg/kg, orally) and measured 6 h afterwards[a]

Tissue	μmol adduct/mol deoxyguanosine[b]	
	O^6-medGua	7-medGua
Liver	19 (*82*)	253 (*1087*)
Lymphocytes	27 (*6.3*)	263 (*60*)

[a]Degan *et al.* (unpublished data)
[b]Numbers in parentheses represent total pmol detected

Table 5. Comparison of O^6-medGua levels in human and animal tissues

Species	Organ	Alkylating agent	Dose	Level of modification (fmol) O^6-medGua/μg DNA	Reference
Rat	Liver	NDMA	1 μg/kg	0.064	Pegg & Perry (1981)
Hamster	Liver	NDMA	10 μg/kg	0.213	Stumpf *et al.* (1979)
Human	Liver	NDMA	~20 mg/kg	300.0	Herron & Shank (1980)
Human (FRG)	Liver	NDMA	1 μg/day	0.015	Calculated from Preussmann *et al.* (1979)
Human (China)	Oesophagus, stomach	Unknown	Unknown	0-0.161	Umbenhauer *et al.* (1985)
Human (Europe)	Oesophagus, stomach, colon	Unknown	Unknown	0-0.044	Umbenhauer *et al.* (1985)

human blood (4×10^7 lymphocytes, ~200 μg DNA), one can detect 1 pmol 7-medGua by enzyme-linked immunosorbent assay, that is, 5 fmol adduct/μg DNA. Pegg and Hui (1978) have shown that a single dose to rats of 1 μg/kg NDMA results in 0.31 fmol 7-medGua/μg DNA in liver. If the relationship between DNA alkylation in liver and lymphocytes is similar in humans, one could try to extrapolate these data. On the basis of a dose per surface area comparison and assuming a similar activation of NDMA in rat and human liver (Montesano & Magee, 1969), a level of 7-medGua ranging from 0.5 to 5 fmol/μg DNA may occur in human blood-cell DNA as a result of a single exposure to 10–100 μg NDMA. These levels of exposures are consistent with the data presented in Table 1.

It is also consistent that the levels of O^6-medGua found in tissue sample DNA from Linxian county, China, and from Europe are of the same order of magnitude as that expected on the basis of environmental levels of nitrosamine exposure and of the results of experimental studies in rodents (see Table 5). It is evident, however, that these conclusions must be supported by more data on human tissues (see Saffhill *et al.*, this volume), and it is important that more information become available on the 'background' levels of these DNA adducts in humans and on their variation within the same individual. It is necessary, for example, to establish the levels of adducts present in samples from populations at low risk for specific cancers before assessing the significance of increased adduct levels in high-risk populations. To facilitate the accumulation of this type of 'background' data, a close interaction between research groups is called for, so that the most information can be obtained from human samples when these become available. Not only will this make best use of valuable samples, but it should also allow an effective interlaboratory validation of methods.

Future considerations

The actual DNA alkylation damage in target tissues and cells of individuals exposed to nitrosamines is the result of a number of factors that modulate and interact with each other, such as tissue distribution, metabolism and DNA repair. This implies that the level of DNA adducts in a given tissue, e.g., oesophagus, could be quite different in individuals of two populations exposed to the same amount of nitrosamines. Thus, environmental measurements of nitrosamines or their precursors are not necessarily related to the level of risk for cancer in a population, as shown by epidemiological and experimental studies.

Fig. 1. Mechanism of carcinogenicity of consumption of alcohol in conjunction with other factors. NNO, *N*-nitrosamines

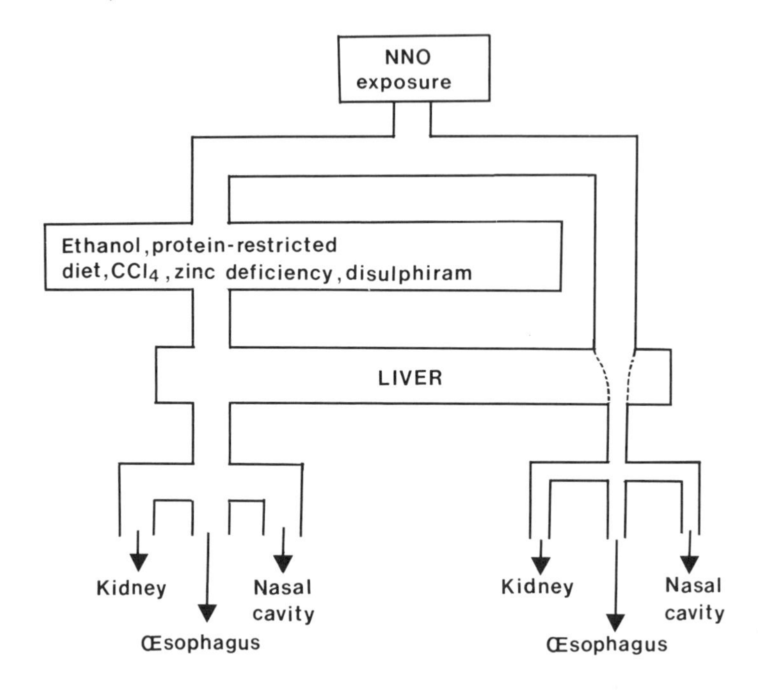

For instance, Tuyns *et al.* (1987) have clearly shown that tobacco smoke [containing a high level of nitrosamines (Hoffman & Hecht, 1985)] and alcohol are the main risk factors in the induction of oesophageal cancer in Normandy, France, and, in addition, that a diet low in fresh meat, citrus fruit and oils entails a nine-fold higher risk of oesophageal cancer than diets rich in these components. The important role of dietary components has been recognized in other populations at high risk of oesophageal cancer (see Day, 1984). These observations are consistent with the hypothesis that consumption of alcohol in conjunction with a diet that is restricted in certain components alter the metabolism in the liver and the distribution to extrahepatic tissues of nitrosamines derived from the diet, from tobacco smoke and from endogenous formation (Swann, 1982). This hypothesis is supported by some data from humans (see Swann, 1984) and by a considerable amount of data from experimental animals (see Bartsch & Montesano, 1984, and Fig. 1).

Recent epidemiological studies in the UK have shown an inverse relationship between nitrate/nitrite (nitrosamine precursors) intake and risk of stomach cancer (Forman *et al.*, 1985; Knight *et al.*, 1987). However, as mentioned above, it would be important to determine whether such intake is related to a biologically effective dose (DNA adducts) in the target tissue.

Another relevant situation for clarification is the role of nitrosamines in tobacco-associated cancers in humans (see also Hecht *et al.*, this volume). It now appears possible to determine alkylation adducts in DNA and other cellular macromolecules in smokers and nonsmokers.

This paper addresses the value of DNA alkylation adducts as an index of exposure to nitrosamines for use in epidemiological studies, but it does not necessarily imply that the presence of such adducts will result in the appearance of cancer many years later.

Tumour development is the result of a number of stages, which are affected by various factors, including genetic predisposition, and it is important that for each type of cancer there be markers of the impact of additional factors. For example, preliminary studies in Lyon (Hollstein *et al.*, 1986) revealed the presence of a restriction fragment length polymorphism at *c-mos* in two of 12 oesophageal cancer patients examined, and studies are in progress to ascertain if this association is spurious or indicates a predisposition to oesophageal cancer.

Acknowledgements

This study was partially supported by US National Cancer Institute Grant No. 1 UO1 ESO4281-01.

References

Bailey, E., Connors, T.A., Farmer, P.B., Gorf, S.M. & Rickard, J. (1981) Methylation of cysteine in hemoglobin following exposure to methylating agents. *Cancer Res.*, *41*, 2514–2517

Bartsch, H. & Montesano R. (1984) Relevance of nitrosamines to human cancer. *Carcinogenesis*, *5*, 1381–1393

Day, N.E. (1984) The geographic pathology of cancer of the oesophagus. *Br. med. Bull.*, *40*, 329–334

Forman, D., Al-Dabbagh, S. & Doll, R. (1985) Nitrates, nitrites and gastric cancer in Great Britain. *Nature*, *313*, 620–625

Gombar, C.T., Zubroff, J., Strahan, G.D. & Magee, P.N (1983) Measurement of 7-methylguanine as an estimate of the amount of dimethylnitrosamine formed following administration of aminopyrine and nitrite to rats. *Cancer Res.*, *43*, 5077–5080

Herron, D.C. & Shank, R.C. (1980) Methylated purines in human liver DNA after probable dimethylnitrosamine poisoning. *Cancer Res.*, *40*, 3116–3117

Hoffmann, D. & Hecht, S. (1985) Nicotine-derived N-nitrosamines and tobacco-related cancer; current status and future directions. *Cancer Res.*, *45*, 935–944

Hollstein, M., Montesano, R. & Yamasaki, H. (1986) Presence of an EcoRI RFLP of the c-*mos* locus in normal and tumor tissue of esophageal cancer patients. *Nucleic Acids Res.*, *14*, 8695

IARC (1987) *IARC Monographs on the Evaluation of Carcinogenic Risks of Chemicals to Humans*, Suppl. 7, *Overall Evaluations of Carcinogenicity*: *An Updating of Selected IARC Monographs from Volumes 1 to 42*, Lyon

Kleihues, P. & Bucheler, J. (1977) Long-term persistence of O^6-methylguanine in rat brain DNA. *Nature*, *269*, 625–626

Kleihues, P., Bamborschke, S. & Doerjer, G. (1980) Persistence of alkylated DNA bases in the Mongolian gerbil (*Meriones unguiculatus*) following a single dose of methylnitrosourea. *Carcinogenesis*, *1*, 111–113

Knight, T.M., Forman, D., Al-Dabbagh, S.A. & Doll, R. (1987) Estimation of dietary intake of nitrate and nitrite in Great Britain. *Food chem. Toxicol.*, *25*, 277–285

Lindahl, T. (1982) DNA repair enzymes. *Ann. Rev. Biochem.*, *51*, 61–87

Montesano, R. & Magee, P.N. (1970) Metabolism of dimethylnitrosamine by human liver slices *in vitro*. *Nature*, *228*, 173–174

Müller, R. & Rajewsky, M.F. (1981) Antibodies specific for DNA components structurally modified by chemical carcinogens. *J. Cancer Res. clin. Oncol.*, *102*, 99–113

National Research Council/National Academy of Sciences (1981) *The Health Effects of Nitrate, Nitrite, and* N-*Nitroso Compounds*, Washington DC, National Academy Press

Pegg, A.E. & Hui, G. (1978) Formation and subsequent removal of O^6-methylguanine from deoxyribonucleic acid in rat liver and kidney after small doses of dimethylnitrosamine. *Biochem. J.*, *173*, 739–748

Pegg, A.E. & Perry, W. (1981) Alkylation of nucleic acids and metabolism of small doses of dimethylnitrosamine in the rat, *Cancer Res.*, *41*, 3128–3132

Peto, R. (1985) *The preventibility of cancer.* In: Vessey, M.P. & Gray, M., eds, *Cancer Risks and Prevention*, Oxford, Oxford University Press, pp. 1–14

Preussmann, R., Spiegelhalder, B., Eisenbrand, G. & Janzowiski, C., (1979), N-*Nitroso compounds in foods*. In: Miller, E.C., Miller, J.A., Hirono, I., Sugimura, T. & Takayama, S., eds, *Naturally Occurring Carcinogens, Mutagens and Modulators of Carcinogenesis,* Tokyo, Japanese Scientific Societies Press, pp. 185–194

Richardson, F.C., Dryoff, M.C., Boucheron, J.A. & Swenberg, J.A. (1985) Differential repair of O^4-alkylthymidine following exposure to methylating and ethylating hepatocarcinogens. *Carcinogenesis, 6*, 625–629

Saracci, R. (1984) *Assessing exposure of individuals in the identification of disease determinants.* In: Berlin, A., Draper, M., Hemminki, K. & Vainio, H., eds, *Monitoring Human Exposure to Carcinogenic and Mutagenic Agents* (*IARC Scientific Publications No. 59, IPCS Joint Symposia No. 7*), Lyon, International Agency for Research on Cancer, pp. 135–142

Singer, B. & Grunberger, D., eds (1983) *Molecular Biology of Mutagens and Carcinogens,* New York, Plenum Press

Strickland, P.T. & Boyle, J.M. (1984) Immunoassay of carcinogen-modified DNA. *Prog. Nucleic Acids Res. mol. Biol., 31*, 1–58

Stumpf, R., Margison, G.P., Montesano, R. & Pegg, A.E. (1979) Formation and loss of alkylated purines from DNA of hamster liver after administration of dimethylnitrosamine. *Cancer Res., 39*, 50–54

Swann, P.F. (1982) *Metabolism of nitrosamines: observations on the effect of alcohol on nitrosamine metabolism and on human cancer.* In: Magee, P.N., ed., *Nitrosamines and Human Cancer* (*Banbury Report 12*), Cold Spring Harbor, NY, CSH Press, pp. 53–68

Swann, P.F. (1984) *Effect of ethanol on nitrosamine metabolism and distribution. Implications for the role of nitrosamines in human cancer and for the influence of alcohol consumption on cancer incidence.* In: O'Neill, I.K., von Borstel, R.C., Miller, C.T., Long, J. & Bartsch, H., eds, N-*Nitroso Compounds: Occurrence, Biological Effects and Relevance to Human Cancer* (*IARC Scientific Publications No. 57*), Lyon, International Agency for Research on Cancer, pp. 501–512

Tuyns, A.J., Riboli, E., Doornbof, G. & Pequignot, G. (1987) Diet and esophageal cancer in Calvados (France). *Nutr. Cancer, 9*, 81–92

Umbenhauer, D., Wild, C.P., Montesano, R., Saffhill, R., Boyle, J.M., Huh, N., Kirstein, U., Thomale, J., Rajewsky, M.F. & Lu, S.H. (1985) O^6-Methyldeoxyguanosine in oesophageal DNA among individuals at high risk of oesophageal cancer. *Int. J. Cancer, 36*, 661–665

Wild, C.P. Smart, G., Saffhill, R. & Boyle, J.M. (1983) Radioimmunoassay of O^6-methyldeoxyguanosine in DNA of cells alkylated *in vitro* and *in vivo. Carcinogenesis, 4,* 1605–1609

Wild, C.P., Lu, S.H. & Montesano, R. (1987) *Radioimmunoassay used to detect DNA alkylation adducts in tissues from populations at high risk for oesophageal and stomach cancer.* In: Bartsch, H., O'Neill, I.K. & Schulte-Hermann, R., eds, *The Relevance of* N-*Nitroso Compounds to Human Cancer*: *Exposures and Mechanisms* (*IARC Scientific Publications No. 84*), Lyon, International Agency for Research on Cancer, pp. 534–537

Yang, C.S. (1980) Research on esophageal cancer in China: a review. *Cancer Res., 40,* 2633–2644

Yu, M.C., Ho, J.H.C., Lai, S.H. & Henderson, B.E. (1986) Cantonese-style salted fish as a cause of nasopharyngeal carcinoma: report of a case-control study in Hong Kong. *Cancer Res., 46,* 956–961

URINARY *N*-NITROSAMINO ACIDS AS AN INDEX OF EXPOSURE TO *N*-NITROSO COMPOUNDS

H. Ohshima & H. Bartsch

International Agency for Research on Cancer, Lyon, France

On the basis of results from animal experiments and studies in human subjects, the amount of nitrosoproline (NPRO) excreted in 24-hr urine following ingestion of precursors (proline, nitrate) has been measured as an index of endogenous nitrosation. Several protocols of the NPRO test have been applied to human subjects, in order to study the kinetics and dietary modifiers of endogenous nitrosation, and in clinical and epidemiological studies. These studies have demonstrated that endogenous nitrosation in humans is highly complex and is influenced by factors such as gastric pH and amounts of precursors, catalysts and inhibitors. Thus, individual monitoring for nitrosation potential, rather than analyses of precursors in saliva, urine and gastric juice, is necessary in order to establish a causal relationship between endogenous nitrosation and human cancer. Results obtained after application of the NPRO test to subjects at high risk for cancers of the stomach, oesophagus, oral cavity and urinary bladder are summarized.

Humans are exposed to a wide range of nitrogen-containing compounds which can react with nitrosating agents to form *N*-nitroso compounds (NOC), a versatile class of carcinogen (National Research Council, 1981; Shephard *et al.*, 1987). In addition, nitrosation of certain polyaromatic hydrocarbons and phenolic compounds results in the formation of *C*-nitroso or *C*-nitro compounds, some of which have been reported to be mutagenic and carcinogenic (IARC, 1984). Humans are also exposed to various types of nitrosating agents, such as nitrous acid and nitrogen oxides, in the diet, tobacco smoke, air and water. Nitrite, nitrate and nitrosating agents can be synthesized endogenously in reactions mediated by bacteria and activated macrophages (Stuehr & Marletta, 1985; Calmels *et al.*, 1987; Miwa *et al.*, 1987). In this way, endogenous formation of NOC can occur in various ways at various sites in the body, including nitrosation in the oral cavity, stomach and intestine, reaction of nitrogen oxides in the lung, and reactions mediated by bacteria and macrophages in infected or inflamed organs.

Exposure to endogenously formed NOC has been associated with increased risks of cancer of the stomach, oesophagus and bladder, but convincing epidemiological evidence is still lacking. One of the reasons has been the dearth of reliable methods to estimate the extent of in-vivo formation of NOC. We have developed a simple and sensitive method for the quantitative estimation of endogenous nitrosation in humans (Ohshima & Bartsch, 1981; Bartsch *et al.*, 1983). This method can be used to measure urinary *N*-nitrosamino acids (NAA) as indices of exposure to NOC. In the following, we briefly summarize the method and results obtained from its application to clinical and epidemiological studies.

Urinary NAA as indices of human exposure to NOC

Human urine contains several NAA, NPRO, *N*-nitrosothiazolidine 4-carboxylic acid (NTCA) and *N*-nitroso 2-methylthiazolidine 4-carboxylic acid (NMTCA) being the major ones (Ohshima *et al.*, 1983; Tsuda *et al.*, 1983; Ohshima *et al.*, 1984a; Tsuda *et al.*, 1984). Some new NAA have recently been identified, including *N*-nitrosomethylaminopropionic acid and *N*-nitrosoazetidine carboxylic acid (Nair *et al.*, 1986). In view of the carcinogenicity of most members of this class, detection of new NOC in human urine may reveal hitherto unknown sources of human exposure; particularly as many NOC are metabolized *in vivo* and subsequently excreted as NAA in the urine. All of the NAA mentioned above are currently analysed as indicators of human exposure to exogenous and endogenously formed NOC, although their origin, except for NPRO, is mostly unknown.

Procedure for the NPRO test

Several forms of the method have been applied in clinical and epidemiological studies. L-Proline is utilized as a probe for nitrosatable amines, and NPRO excreted in the urine is determined as a marker for endogenous nitrosation. The rationale for applying this test in human studies is based on the following: (i) NPRO has been reported to be neither carcinogenic nor mutagenic (IARC, 1978; Mirvish *et al.*, 1980); (ii) after gavage of rats with ^{14}C-NPRO, $^{14}CO_2$ production and DNA alkylation were negligible (Chu & Magee, 1981), but urinary excretion of NPRO (as the unchanged compound) was rapid and almost complete (Dailey *et al.*, 1975; Chu & Magee, 1981; Ohshima *et al.*, 1982a); (iii) in humans, preformed NPRO ingested in food extracts was also eliminated rapidly and almost quantitatively in the urine within 24 h after ingestion (Ohshima *et al.*, 1982b). The difference between the amount of NPRO excreted in the 24-hr urine and that ingested in foods can therefore be used as an indicator of daily endogenous nitrosation (Ohshima & Bartsch, 1981). Thus, application of the NPRO test does not entail risk to the health of study subjects, and it was cleared by the IARC ethical committee.

Method A: Loading test with nitrate and proline

Human subjects are given either vegetable juice rich in nitrate (e.g., 200 ml beetroot juice containing 260 mg nitrate) or sodium ^{14}N- or ^{15}N-nitrate and, 30 min later, L-proline (500 mg). In order to minimize the confounding effect of modifiers of nitrosation and dietary (preformed) NPRO, subjects fast for a further 2 h and avoid consuming cured meats or smoked fish during the 24-h urine collection. Whenever possible, information is collected from each subject on demographic data, smoking, drinking and dietary habits and clinical findings. Dietary records are also obtained on the day of urine collection.

Method B: Loading with proline alone or with proline and ascorbic acid

Three 24-h urine samples are collected from each subject according to the following protocols: (i) undosed, (ii) after intake of 100 mg proline three times a day after each meal, and (iii) after intake of 100 mg proline and 100 mg vitamin C three times a day. Alternatively, 12-h overnight urine samples are collected after intake of a single dose of proline with or without vitamin C after dinner. During a day of urine collection, diets are not strictly controlled, but detailed records are obtained from each subject.

Sample collection and analysis of NAA

Samples of 24-h urine are collected in plastic bottles containing sodium hydroxide or ammonium sulphamate in dilute sulphuric acid to avoid artefactual formation of NPRO during collection and storage of the sample. Samples of 12-h urine collected without preservatives are divided into two aliquots: either sodium hydroxide or ammonium sulphamate is added and used for analyses of nitrite, nitrate and NAA; the other is stored without alkali or acid to be used for analyses of creatinine and other compounds, if necessary.

Urine samples are spiked with *N*-nitrosopipecolic acid as internal standard and analysed for NPRO and other NAA after conversion to their methyl esters by diazomethane in a gas chromatograph with a thermal energy analyser, a nitrosamine-specific detector. Other derivatizing agents, such as boron trifluoride-methanol (Ladd *et al.*, 1984; Leaf *et al.*, 1987) and pentafluorobenzyl bromide (Garland *et al.*, 1986), have been used; however, the sulphur-containing NAA, NTCA and NMTCA, are acid-labile and decompose during derivatization with boron trifluoride or hydrochloric acid in methanol (Ohshima *et al.*, 1984a). A mass spectrometer equipped with a gas chromatograph has also been used to quantify ^{14}N- and ^{15}N-NPRO (Wagner *et al.*, 1985; Garland *et al.*, 1986); this, however, requires purification of NPRO by high-performance liquid chromatography or thin-layer chromatography prior to analysis.

Application of Method A to experimental and clinical studies on endogenous nitrosation in human subjects

Table 1 summarizes experimental studies on endogenous nitrosation in humans. Method A was originally designed for study of kinetics and factors affecting endogenous nitrosation in humans (Ohshima & Bartsch, 1981). Proline can be given with or without a dietary nitrosation modifier (Stich *et al.*, 1983, 1984a); ^{15}N-nitrate has also been used to study incorporation of ^{15}NO into proline (Wagner *et al.*, 1985). Further, nitrosation of proline has been investigated in human volunteers consuming a standard diet consisting of a high-nitrate salad meal with and without foods rich in vitamin C (Knight & Forman, 1987).

As shown in Table 2, Method A has also been used in clinical studies to test the hypothesis that subjects with precancerous conditions of the stomach have an elevated potential for endogenous nitrosation (Bartsch *et al.*, 1984; Crespi *et al.*, 1987; Hall *et al.*, 1987a,b). Consistent with the nitrosamine hypothesis (Correa *et al.*, 1975), intragastric levels of bacteria and nitrite were positively related to intragastric pH; but, contrary to the hypothesis, no increased level of NPRO and other NAA was detected in subjects with more advanced lesions, compared to those with a normal stomach. This discrepancy is the subject of further studies; in particular, the adequacy of proline and other amino acids as substrates for bacteria-mediated nitrosation is being investigated.

Application of Method B to epidemiological field studies

Method B was designed to test whether the capacity of an individual to nitrosate proline is different for those living in high- and in low-risk areas for particular cancers. The findings are as follows. (i) Background levels of NAA in the urine of undosed subjects (no loading with nitrate or proline) may be a marker for exposure to these compounds either as dietary components or by endogenous synthesis. However, it should be noted that there is considerable person-to-person and day-to-day variation in background urinary levels (Ohshima *et al.*, 1984a; Garland *et al.*, 1986). It has also

Table 1. Human studies on endogenous nitrosation using urinary NAA as an indicator

Observations	References
NPRO formation in a volunteer; effect of doses of nitrate and proline, inhibition by vitamins C and E; rapid and complete excretion of preformed NPRO ingested in food extracts; NPRO formation from pickled vegetables plus proline	Ohshima & Bartsch (1981, 1982); Ohshima *et al.* (1982a,b)
NPRO formation by $NaNO_3$ plus proline in six subjects on a controlled diet; incorporation of ^{15}N-nitrate into NPRO and its inhibition by vitamins C and E	Wagner *et al.* (1982, 1985)
Inter- and intraindividual differences in background NPRO in 24 subjects; effect of vitamin C	Garland *et al.* (1986)
Relation between vitamin C dose and NPRO excretion	Leaf *et al.* (1987)
Inhibition of NPRO formation by betel-nut extract, caffeine, ferulic acid, tea and coffee in two subjects; ingestion of nitrite-preserved meats and NPRO excretion	Stich *et al.* (1983, 1984a,b)
NPRO formation after high-nitrate meal plus proline and its inhibition by vitamin C from dietary source	Knight & Forman (1987)
Increased NPRO excretion by 12 volunteers with high doses (1.8–8.1 g) of nitrate	Ellen & Schuller (1984)
Percutaneous excretion of NPRO in humans	Bogovski & Rooma (1984)
Increased NPRO formation in cigarette smokers	Hoffmann & Brunnemann (1983); Ladd *et al.* (1984); Bartsch *et al.* (1984)
Nitrosation in the oral cavity of betel-quid chewers	Nair *et al.* (1986, 1987)
Identification of NTCA and NMTCA in human urine	Ohshima *et al.* (1983); Tsuda *et al.* (1983)
Levels of NPRO, NTCA, NMTCA and NSAR in 15 subjects; day-to-day variation in one subject; effect of vitamin C	Ohshima *et al.* (1984b)
Increased NTCA and NMTCA formation after cigarette smoking; dietary effect and sex difference on their levels	Tsuda *et al.* (1986, 1987)
Effect of nitrate, cysteine and nitrate + cysteine on NPRO, NTCA, NMTCA and NHPRO levels	Tricker & Preussmann (1987)
Effect of interval between doses of nitrate and proline; limitations and current perspective of NAA as indices of exposure to NOC	Tannenbaum (1987)

been shown that diet is a significant contributor to background NPRO (Stich *et al.*, 1984b). (ii) On the basis of previous observations that endogenous nitrosation of proline can be blocked to a large extent by ingested ascorbic acid, the difference in NPRO levels in urine collected after intake of proline and after proline and ascorbic acid intake may be used as an indicator of endogenous nitrosation.

Method B is currently being used in several field studies, as shown in Table 2. Kamiyama *et al.* (1987) observed endogenous nitrosation of ingested proline in the subjects living in a high-risk area for stomach cancer in northern Japan but not in those living in a low-risk area. This endogenous nitrosation was effectively inhibited by intake of ascorbic acid. Lu *et al.* (1986, 1987) compared the nitrosation potential in populations at different risks for oesophageal cancer in China and showed that the amounts of NPRO and other NAA excreted in the urine of eight different populations were positively correlated with the mortality rates for oesophageal cancer. Similarly, Chen *et al.* (1987) collected samples of 12-h overnight urine from 1035 subjects, representing approximately 40 male adults in each of 26 counties with a wide range of mortality rates for oesophageal, gastric and liver cancers in China. Two urine specimens — one after a loading dose of proline and ascorbic acid and the other after a dose of proline — were collected from each subject, and 5% of the volume of each sample was pooled, to give one sample for each commune; these were analysed for

Table 2. Clinical and epidemiological studies using urinary NAA as indices of exposure to NOC

Observations	References
Clinical studies	
Studies on intragastric nitrosation and precancerous lesions of the stomach: no increase in NPRO in subjects with chronic atrophic gastritis or after cimetidine dose	Bartsch *et al.* (1984)
No increase in NPRO formation in subjects with pernicious anaemia and gastrectomy	Hall *et al.* (1987a,b)
No increase in NPRO formation in subjects with high bacteria count and high gastric pH	Crespi *et al.* (1987)
Increased background NPRO, NTCA and NMTCA in cirrhosis patients	Habib *et al.*[a]
Epidemiological studies	
Field study in high- and low-risk areas for oesophageal cancer in China; higher levels of NAA in a high-risk population; endogenous nitrosation of proline and its inhibition by vitamin C; NAA levels correlated positively with mortality rate from oesophageal cancer in eight areas of China	Lu *et al.* (1986, 1987)
Field study in 26 counties in China: moderate correlation between oesophageal cancer mortality rates and NPRO formation after intake of proline	Chen *et al.* (1987)
Field study in high- and low-risk areas for stomach cancer in Japan: increased nitrosation potential in a high-risk population and protective factors in the diet of a low-risk population	Kamiyama *et al.* (1987)
High urinary excretion of NPRO and nitrate in subjects with liver fluke, a high-risk factor for cholangiocarcinoma	Srianujata *et al.* (1987)

[a]IARC (1985b)

nitrate, NPRO and other NAA. There was a moderate tendency for oesophageal cancer mortality rates to be associated positively with nitrosation potential and negatively with background ascorbate levels in plasma. Recently, Umbenhauer *et al.* (1985) reported that the levels of O^6-methyldeoxyguanosine determined by radioimmunoassay are elevated in the DNA of specimens of oesophageal and stomach mucosa removed surgically from cancer patients in Linxian, China, a high-risk area for oesophageal cancer. Such DNA lesions may arise from exposure to NOC. These observations further support the notion that NOC are an important factor in the etiology of oesophageal cancer in certain areas of China.

Application of the method to study endogenous nitrosation in the oral cavity of betel-quid chewers

Chewing of betel quid with tobacco is causally associated with human cancer (IARC, 1985a), and tobacco-specific nitrosamines are suspected to play a major role in the etiology of oral cancer (Hoffmann & Hecht, 1985). In order to evaluate endogenous nitrosation in the oral cavity of chewers of betel quid with tobacco, saliva samples collected from such subjects given a quid supplemented with proline were analysed for NPRO and other NOC. It was demonstrated that NPRO is formed during chewing, implying that other NOC may also be formed endogenously in the oral cavity.

Conclusion

The following conclusions can be drawn. (i) It has been demonstrated unequivocally that endogenous nitrosation of proline occurs in the human body after ingestion of amounts of precursors (nitrate, amine) that are considered to be the normal dietary intake. Intake of nitrate above a dose of 260 mg/day per person led to a sharp increase in the amount of NPRO formed *in vivo*. (ii) Some sulphur-containing NAA (NTCA and NMTCA) are formed in the human body, possibly through a two-step synthesis from cysteine and aldehydes, followed by nitrosation; the urinary levels of these NAA increase significantly upon administration of nitrate (Ellen & Schuller, 1984; Wagner *et al.*, 1985) or of nitrate and cysteine (Tricker & Preussmann, 1987). (iii) Nitrosation inhibitors, like vitamins C and E and polyphenolic compounds, significantly reduce the amounts of NPRO and other NAA formed in healthy human subjects (Ohshima & Bartsch, 1981; Stich *et al.*, 1983; Ohshima *et al.*, 1984b; Stich *et al.*, 1984a; Wagner *et al.*, 1985; Leaf *et al.*, 1987). (iv) Excretion of NPRO and other NAA is increased in cigarette-smoking subjects, perhaps due to the high level of thiocyanate (a catalyst of nitrosation) in the saliva of smokers and to higher exposure to aldehydes and nitrosating agents like nitrogen oxides present in cigarette smoke (Hoffmann & Brunnemann, 1983; Ladd *et al.*, 1984; Tsuda *et al.*, 1986). (v) The method can satisfactorily be applied to human subjects in clinical and field studies. When formation of endogenous NOC was assessed in subjects living in high- and low-incidence areas for stomach cancer and for oesophageal cancer in northern Japan and in China, generally higher exposures to endogenous NOC were found in high-risk populations. (vi) The process of endogenous nitrosation in humans is highly complex and is influenced by many factors, such as the pH of stomach contents and the occurrence of bacteria, catalysts and inhibitors. Therefore, determination only of nitrate and nitrite in saliva, urine or gastric juice is insufficient to assess the in-vivo nitrosation process in humans.

The method has, however, certain limitations, and further studies are needed: (i) the adequacy of proline and other amino compounds as substrates for nitrosation mediated by bacteria and macrophages, and for nitrosation by nitrogen oxides in the lung, must be investigated; (ii) more reliable markers for exposure to nitrosating agents, e.g., nitrosated or deaminated protein and DNA bases, should be developed and validated; (iii) relevant, biologically active NOC formed *in vivo* in man must be identified and methods to monitor exposure to them be developed.

References

Bartsch, H., Ohshima, H. Muñoz, N., Crespi, M. & Lu, S.H. (1983) *Measurement of endogenous nitrosation in humans: potential applications of a new method and initial results.* In: Harris, C.C. & Autrup, H.N., eds, *Human Carcinogenesis,* New York, Academic Press, pp. 833–855

Bartsch, H., Ohshima, H., Muñoz, N., Crespi, M., Casale, V., Ramazzotti, V., Lambert, R., Minaire, Y., Forichon, J. & Walters, C.L. (1984) *In-vivo nitrosation, precancerous lesions and cancers of the gastrointestinal tract: on-going studies and preliminary results.* In: O'Neill, I.K., von Borstel, R.C., Long, J.E., Miller, C.T. & Bartsch, H., eds, N-*Nitroso Compounds: Occurrence, Biological Effects and Relevance to Human Cancer* (*IARC Scientific Publications No. 57*), Lyon, International Agency for Research on Cancer, pp. 955–962

Bogovski, P.A. & Rooma, M.A. (1984) *Studies on the excretion of endogenously formed* N-*nitrosoproline. I. Percutaneous excretion of* N-*nitrosoproline in humans.* In: O'Neill, I.K., von Borstel, R.C., Long, J.E, Miller, C.T. & Bartsch, H., eds, N-*Nitroso Compounds: Occurrence, Biological Effects and Relevance to Human Cancer* (*IARC Scientific Publications No. 57*), Lyon, International Agency for Research on Cancer, pp. 199–204

Calmels, S., Ohshima, H., Rosenkranz, H., McCoy, E. & Bartsch, H. (1987) Biochemical studies on the catalysis of nitrosation by bacteria. *Carcinogenesis, 8*, 1085–1088

Chen, J., Ohshima, H., Yang, H., Li, J., Campbell, T.C., Peto, R. & Bartsch, H. (1987) *A correlation study on urinary excretion of* N-*nitroso compounds and cancer mortality in*

China: interim results. In: Bartsch, H., O'Neill, I.K. & Schulte-Hermann, R., eds, *The Relevance of* N-*Nitroso Compounds to Human Cancer: Exposures and Mechanisms* (*IARC Scientific Publications No. 84*), Lyon, International Agency for Research on Cancer, pp. 503–506

Chu, C. & Magee, P.N. (1981) Metabolic fate of nitrosoproline in the rat. *Cancer Res., 41,* 3653–3657

Correa, P., Haenszel, W., Cuello, C., Tannenbaum, S. & Archer, M. (1975) A model for gastric cancer epidemiology. *Lancet, ii,* 58–60

Crespi, M., Ohshima, H., Ramazzotti, V., Muñoz, N., Grassi, A., Casale, V., Leclerc, H., Calmels, S., Cattoen, C., Kaldor, J. & Bartsch, H. (1987) *Intragastric nitrosation and precancerous lesions of the gastrointestinal tract: testing of an etiological hypothesis.* In: Bartsch, H., O'Neill, I.K. & Schulte-Hermann, R., eds, *The Relevance of* N-*Nitroso Compounds to Human Cancer: Exposures and Mechanisms* (*IARC Scientific Publications No. 84*), Lyon, International Agency for Research on Cancer, pp. 511–517

Dailey, R.E., Braunberg, R.C. & Blaschka, A.M. (1975) The absorption, distribution and excretion of [^{14}C]-nitrosoproline by rats. *Toxicology, 3,* 23–28

Ellen, G. & Schuller, P.L. (1984) N-*Nitrosoproline in urine from patients and healthy volunteers after administration of large amounts of nitrate*: In: O'Neill, I.K., von Borstel, R.C., Long, J.E., Miller, C.T. & Bartsch, H., eds, N-*Nitroso Compounds: Occurrence, Biological Effects and Relevance to Human Cancer* (*IARC Scientific Publications No. 57*), Lyon, International Agency for Research on Cancer, pp. 193–198

Garland, W.A., Kuenzig, W., Rubio, F., Kornychuk, H., Norkus, E.P. & Conney, A.H. (1986) Studies on the urinary excretion of nitrosodimethylamine and nitrosoproline in humans: interindividual and intraindividual differences and the effect of administered ascorbic acid and α-tocopherol. *Cancer Res., 46,* 5392-5400

Hall, C.N., Kirkham, J.S. & Northfield, T.C. (1987a) Urinary *N*-nitrosoproline excretion: a further evaluation of the nitrosamine hypothesis of gastric carcinogenesis in precancerous conditions. *Gut, 28,* 216–220

Hall, C.N., Darkin, D., Viney, N., Cook, A., Kirkham, J.S. & Northfield, T.C. (1987b) *Evaluation of the nitrosamine hypothesis of gastric carcinogenesis in man.* In: Bartsch, H., O'Neill, I.K. & Schulte-Hermann, R., eds, *The Relevance of* N-*Nitroso Compounds to Human Cancer: Exposures and Mechanisms* (*IARC Scientific Publications No. 84*), Lyon, International Agency for Research on Cancer, pp. 527–530

Hoffman, D. & Brunnemann, K. (1983) Endogenous formation of *N*-nitrosoprolinc in cigarette smokers. *Cancer Res., 43,* 5570–5574

Hoffman, D. & Hecht, S.S. (1985) Nicotine-derived *N*-nitrosamines and tobacco related cancer: current status and future directions. *Cancer Res., 45,* 935–942

IARC (1978) *IARC Monographs on the Evaluation of the Carcinogenic Risk of Chemicals to Humans,* Vol. 17, *Some* N-*Nitroso Compounds,* Lyon, pp. 303–308

IARC (1984) *IARC Monographs on the Evaluation of the Carcinogenic Risk of Chemicals to Humans,* Vol. 33, *Polynuclear Aromatic Compounds, Part 2, Carbon Blacks, Mineral Oils and Some Nitroarenes,* Lyon, pp. 167–222

IARC (1985a) *IARC Monographs on the Evaluation of the Carcinogenic Risk of Chemicals to Humans,* Vol. 37, *Tobacco Habits other than Smoking: Betel-quid and Areca-nut Chewing: and some Related Nitrosamines,* Lyon

IARC (1985b) *Annual Report 1985,* Lyon, p. 36

Kamiyama, S., Ohshima, H., Shimada, A., Saito, N., Bourgade, M.-C., Ziegler, P. & Bartsch, H. (1987) *Urinary excretion of* N-*nitrosamino acids and nitrate by inhabitants in high- and low-risk areas for stomach cancer in northern Japan.* In: Bartsch, H., O'Neill, I.K. & Schulte-Hermann, R., eds, *The Relevance of* N-*Nitroso Compounds to Human Cancer: Exposures and Mechanisms* (*IARC Scientific Publications No. 84*), Lyon, International Agency for Research on Cancer, pp. 497–502

Knight, T.M. & Forman, D. (1987) *The availability of dietary nitrate for the endogenous nitrosation of L-proline.* In: Bartsch, H., O'Neill, I.K. & Schulte-Hermann, R., eds, *The Relevance of* N-*Nitroso Compounds to Human Cancer: Exposures and Mechanisms* (*IARC Scientific Publications No. 84*), Lyon, International Agency for Research on Cancer, pp. 518–523

Ladd, K.F., Newmark, H.L. & Archer, M.C. (1984) *N*-Nitrosation of proline in smokers and nonsmokers. *J. natl Cancer Inst., 73,* 83–87

Leaf, C.D., Vecchio, A.J., Roe, D.A. & Hotchkiss, J.H. (1987) Influence of ascorbic acid dose on *N*-nitrosoproline formation in humans. *Carcinogenesis, 8,* 791–795

Lu, S.H., Ohshima, H., Fu, H.-M., Tian, Y., Li, F.-M., Blettner, M., Wahrendorf, J. & Bartsch, H. (1986) Urinary excretion of *N*-nitrosamino acids and nitrate by inhabitants of high- and low-risk areas for esophageal cancer in northern China: endogenous formation of nitrosoproline and its inhibition by vitamin C. *Cancer Res., 46,* 1485–1491

Lu, S.H., Yang, W.X., Guo, L.P., Li, F.M., Wang, G.J., Zhang, J.S. & Li, P.Z. (1987) *Determination of* N-*nitrosamines in gastric juice and urine and a comparison of endogenous formation of* N-*nitrosoproline and its inhibition in subjects from high- and low-risk areas for oesophageal cancer.* In: Bartsch, H., O'Neill, I.K. & Schulte-Hermann, R., eds, *The Relevance of* N-*Nitroso Compounds to Human Cancer: Exposures and Mechanisms* (*IARC Scientific Publications No. 84*), Lyon, International Agency for Research on Cancer, pp. 538–543

Mirvish, S., Bulay, O., Runge, R.G. & Patil, K. (1980) Study of the carcinogenicity of large doses of dimethylnitramine, *N*-nitroso-L-proline, and sodium nitrite administered in drinking water to rat. *J. natl Cancer Inst., 64,* 1435–1442

Miwa, M., Stuehr, D.J., Marletta, M.A., Wishnok, J.S & Tannenbaum, S.R. (1987) Nitrosation of amines by stimulated macrophages. *Carcinogenesis, 8,* 955–958

Nair, J., Ohshima, H., Pignatelli, B., Friesen, M., Malaveille, C., Calmels, S. & Bartsch, H. (1986) *Modifiers of endogenous carcinogen formation*: *studies on in-vivo nitrosation in tobacco users.* In: Hoffmann, D. & Harris, C.C., eds, *New Aspects of Tobacco Carcinogenesis* (*Banbury Report No. 23),* Cold Spring Harbor, NY, CSH Press, pp. 45–61

Nair, J., Nair, U.J., Ohshima, H., Bhide, S.V. & Bartsch, H. (1987) *Endogenous nitrosation in the oral cavity of chewers while chewing betel quid with or without tobacco.* In: Bartsch, H., O'Neill, I.K. & Schulte-Hermann, R., eds, *The Relevance of* N-*Nitroso Compounds to Human Cancer: Exposures and Mechanisms* (*IARC Scientific Publications No. 84*), Lyon, International Agency for Research on Cancer, pp. 465–469

National Research Council (1981) *The Health Effects of Nitrate, Nitrite and* N-*Nitroso Compounds, Part I of a 2-part Study,* Washington DC, National Academy Press

Ohshima, H. & Bartsch, H. (1981) Quantitative estimation of endogenous nitrosation in humans by monitoring *N*-nitrosoproline excreted in the urine. *Cancer Res., 41,* 3658–3662

Ohshima, H. & Bartsch, H. (1982) *Quantitative estimation of endogenous nitrosation in humans by measuring excretion of* N-*nitrosoproline in the urine.* In: Sugimura, T. & Kondo, S., eds, *Environmental Mutagens and Carcinogens,* New York, Alan R. Liss, pp. 577–585

Ohshima, H., Béréziat, J.-C. & Bartsch, H. (1982a) Monitoring *N*-nitrosamino acids excreted in the urine and faeces of rats as an index for endogenous nitrosation. *Carcinogenesis, 3,* 115–120

Ohshima, H., Béréziat, J.-C. & Bartsch, H. (1982b) *Measurement of endogenous* N-*nitrosation in rats and humans by monitoring urinary and faecal excretion of* N-*nitrosamino acids.* In: Bartsch, H., O'Neill, I.K., Castegnaro, M. & Okada, M., eds, N-*Nitroso Compounds: Occurrence and Biological Effects* (*IARC Scientific Publications No. 41*), Lyon, International Agency for Research on Cancer, pp. 397–411

Ohshima, H., Friesen, M., O'Neill, I.K. & Bartsch, H. (1983) Presence in human urine of a new *N*-nitroso compound, *N*-nitrosothiazolidine 4-carboxylic acid. *Cancer Lett., 20,* 183–190

Ohshima, H., O'Neill, I.K., Friesen, M., Pignatelli, B. & Bartsch, H. (1984a) *Presence in human urine of new sulfur-containing* N-*nitrosamino acids:* N-*nitrosothiazolidine 4-carboxylic acid and* N-*nitroso 2-methylthiazolidine 4-carboxylic acid.* In: O'Neill, I.K., von Borstel, R.C., Long, J.E., Miller, C.T. & Bartsch, H., eds, N-*Nitroso Compounds: Occurrence, Biological Effects and Relevance to Human Cancer* (*IARC Scientific Publications No. 57),* Lyon, International Agency for Research on Cancer, pp. 77–85

Ohshima, H., O'Neill, I.K., Friesen, M., Béréziat, J.-C. & Bartsch, H. (1984b) Occurrence in human urine of new sulphur-containing *N*-nitrosamino acids *N*-nitrosothiazolidine 4-carboxylic acid and its 2-methyl derivative and their formation. *J. Cancer Res. clin. Oncol., 108,* 121–128

Shephard, S.E., Schlatter, C. & Lutz, W.K. (1987) Assessment of the risk of formation of carcinogenic *N*-nitroso compounds from dietary precursors in the stomach. *Food chem. Toxicol., 25,* 91–108

Srianujata, S., Tonbuth, S., Bunyaratvej, S., Valyasevi, A., Promvanit, N. & Chaivatsagul, W. (1987) *High urinary excretion of nitrate and* N-*nitrosoproline in opisthorchiasis subjects.* In: Bartsch, H., O'Neill, I.K. & Schulte-Hermann, R., eds, *The Relevance of* N-*Nitroso Compounds to Human Cancer: Exposures and Mechanisms* (*IARC Scientific Publications No. 84*), Lyon, International Agency for Research on Cancer, pp. 544–546

Stich, H.F., Ohshima, H., Pignatelli, B., Michelon, J. & Bartsch, H. (1983) Inhibitory effect of betel nut extracts on endogenous nitrosation in humans. *J. natl Cancer Inst., 70,* 1047–1050

Stich, H.F., Dunn, B.P., Pignatelli, B., Ohshima, H. & Bartsch, H. (1984a) *Dietary phenolics and betel nut extracts as modifiers of* N-*nitrosation in rat and man.* In: O'Neill, I.K., von Borstel, R.C., Long, J.E., Miller, C.T. & Bartsch, H., eds, N-*Nitroso Compounds: Occurrence, Biological Effects and Relevance to Human Cancer* (*IARC Scientific Publications No. 57*), Lyon, International Agency for Research on Cancer, pp. 213–222

Stich, H.F., Hornby, A.P. & Dunn, B.P. (1984b) The effects of dietary factors on nitrosoproline levels in human urine. *Int. J. Cancer, 33,* 625–628

Stuehr, D.J. & Marletta, M.A. (1985) Mammalian nitrate biosynthesis: mouse macrophages produce nitrite and nitrate in response to *Escherichia coli* lipopolysaccharide. *Proc. natl Acad. Sci. USA, 82,* 7738–7742

Tannenbaum, S.R. (1987) *Endogenous formation of* N-*nitroso compounds: a current perspective.* In: Bartsch, H., O'Neill, I.K. & Schulte-Hermann, R., eds, *The Relevance of* N-*Nitroso Compounds to Human Cancer: Exposures and Mechanisms* (*IARC Scientific Publications No. 84*), Lyon, International Agency for Research on Cancer, pp. 292–296

Tricker, A.R. & Preussmann, R. (1987) Influence of cysteine and nitrate on the endogenous formation of *N*-nitrosamino acids. *Cancer Lett., 34,* 39–47

Tsuda, M., Hirayama, T. & Sugimura, T. (1983) Presence of *N*-nitroso-L-thioproline and *N*-nitroso-L-methylthioproline in human urine as major *N*-nitroso compounds. *Gann, 74,* 331–333

Tsuda, M., Kakizoe, T., Hirayama, T. & Sugimura, T. (1984) *New type of* N-*nitrosamino acids,* N-*nitroso-L-thioproline and* N-*nitroso-L-methylthioprolines, found in human urine as major* N-*nitroso compounds.* In: O'Neill, I.K., von Borstel, R.C., Long, J.E., Miller, C.T. & Bartsch, H., eds, N-*Nitroso Compounds: Occurrence, Biological Effects and Relevance to Human Cancer* (*IARC Scientific Publications No. 57*), Lyon, International Agency for Research on Cancer, pp. 223–230

Tsuda, M., Niitsuma, J., Sato, S., Hirayama, T., Kakizoe, T. & Sugimura, T. (1986) Increase in the levels of *N*-nitrosoproline, *N*-nitrosothioproline and *N*-nitroso-2-methylthioproline in human urine by cigarette smoking. *Cancer Lett., 30,* 117–124

Tsuda, M., Nagai, A., Suzuki, H., Hayashi, T., Ikeda, M., Kuratsune, M., Sato, S. & Sugimura, T. (1987) *Effect of cigarette smoking and dietary factors on the amounts of* N-*nitrosothiazolidine 4-carboxylic acid and* N-*nitroso-2-methylthiazolidine 4-carboxylic acid in human urine.* In: Bartsch, H., O'Neill, I.K. & Schulte-Hermann, R., eds, *The Relevance of* N-*Nitroso Compounds to Human Cancer: Exposures and Mechanisms* (*IARC Scientific Publications No. 84*), Lyon, International Agency for Research on Cancer, pp. 446–450

Umbenhauer, D., Wild, C.P., Montesano, R., Saffhill, R., Boyle, J.M., Huh, N., Kirstein, U., Thomale, J., Rajewsky, M.F. & Lu, S.H. (1985) O^6-Methyldeoxyguanosine in oesophageal DNA among individuals at high risk of oesophageal cancer. *Int. J. Cancer, 36,* 661–665

Wagner, D.A., Shuker, D.E.G., Hasic, G. & Tannenbaum, S.R. (1982) *Endogenous nitrosoproline in humans.* In: Magee, P.N., ed., *Nitrosamines and Human Cancer* (*Banbury Report 12*), Cold Spring Harbor, NY, CSH Press, pp. 319–323

Wagner, D.A., Shuker, D.E.G., Bilmazes, C., Obiedzinski, M., Baker, I., Young, V.R. & Tannenbaum, S.R. (1985) Effect of vitamins C and E on endogenous synthesis of *N*-nitrosamino acids in humans: precursor-product studies with ^{15}N-nitrate. *Cancer Res., 45,* 6519–6522

URINARY EXCRETION OF 3-METHYLADENINE IN HUMANS AS A MARKER OF NUCLEIC ACID METHYLATION

D.E.G. Shuker[1] & P.B. Farmer[2]

[1]*International Agency for Research on Cancer, Lyon, France; and* [2]*MRC Toxicology Unit, MRC Laboratories, Carshalton, Surrey, UK*

Antisera to 3-methyladenine (3-meAde) were obtained using a novel analogue of 3-meAde which was bound covalently to methylated bovine serum albumin and used as an antigen. 3-meAde keyhole limpet haemocyanin (3-meAde-KLH) was detected at high dilutions of the antisera (1 in 10^4), and 3-meAde itself inhibited the recognition. In an enzyme-linked immunosorbent assay (ELISA), at room temperature, 3-meAde was detected at 1 pmol/well; however, in a non-equilibrium assay at 4°C, a considerable enhancement of sensitivity was obtained, and 3-meAde was detected at 70 fmol/well. Due to the presence of other purines, the direct determination of urinary 3-meAde was not possible, and a high-performance liquid chromatography (HPLC) clean-up step, followed by ELISA, has been developed. Human urine samples have been analysed by this method and the results compared to those obtained by a gas chromatography-mass spectrometric (GC-MS) method.

3-meAde is one of the major DNA adducts produced by methylating agents (Lawley, 1976). In experimental animals, it is excised intact from DNA and excreted unchanged in urine (Shuker *et al.*, 1987a). Unlike other methylated purines (e.g., 7-methylguanine), 3-meAde has not been reported to occur naturally in either DNA or RNA, indicating that the determination of urinary levels of 3-meAde could be a useful marker of exposure to methylating agents.

It has been shown that 3-meAde can be quantified in human urine by a GC-MS method and that it is normally present at low levels (5–15 μg/24 h; Shuker *et al.*, 1987b).

In order to develop a method suitable for the determination of 3-meAde in large numbers of samples, it was considered that immunological methods would be desirable and that an antiserum to 3-meAde would be required. This paper describes preliminary studies on the immunochemical detection of 3-meAde and its applications to human samples.

Preparation of antisera to 3-meAde

A novel analogue of 3-meAde, *N*6-carboxymethyl-3-methyladenine (CM-3-meAde; Fig. 1; D. Shuker, manuscript in preparation), was used to prepare an antigen for immunization. The synthesis of 3-meAde-methylated bovine serum albumin is summarized in Figure 2.

Two rabbits were immunized with 3-meAde-methylated bovine serum albumin (0.5 mg in phosphate-buffered saline (PBS)-Freund's complete adjuvant (0.5 ml)) by subcutaneous injection in the hindquarters. Booster injections of the same amount of

Fig. 1. Structures of 3-meAde and its analogue, CM-3-meAde

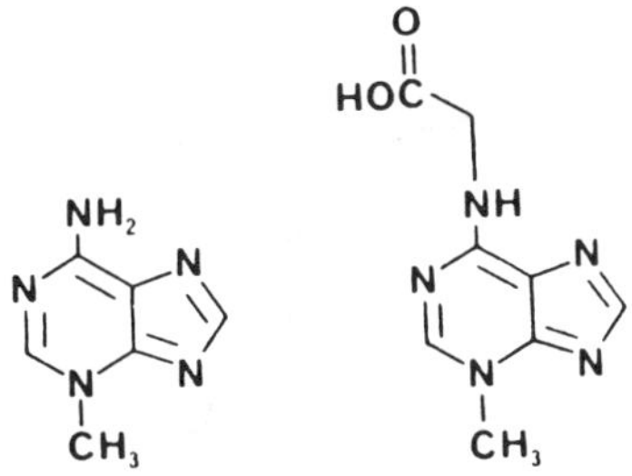

Fig. 2. Synthesis of 3-meAde-protein conjugates using CM-3-meAde. Protein was either methylated bovine serum albumin (for immunogen) or KLH (for coating antigen).

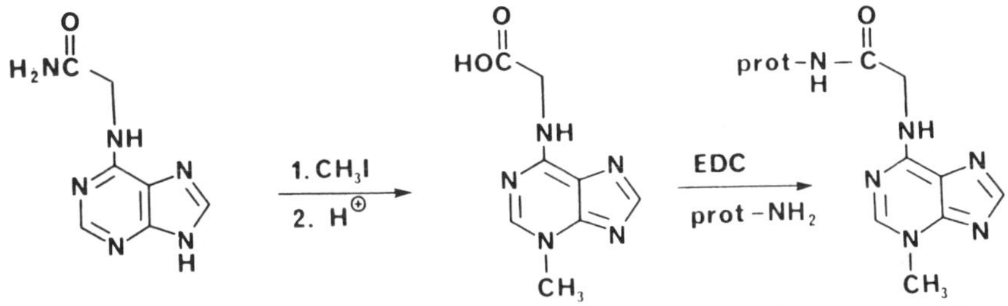

antigen in Freund's incomplete adjuvant were given four and eight weeks after the initial immunization. Two weeks after the second booster dose, both rabbits were bled from the lateral ear vein (local anaesthesia) and serum was prepared. Sera were assayed for anti-3-meAde activity by ELISA using a checkerboard procedure: 96-well microtitre plates were coated with 3-meAde-KLH (1 ng to 10 μg per well), and sera were diluted with PBS from 1 in 10 to 1 in 10^6. Binding of rabbit immunoglobulin G was detected using horseradish peroxidase-linked goat anti-rabbit immunoglobulin with 3,3′,5,5′-tetramethylbenzidine as a substrate. Both rabbits produced active antisera of approximately equal titre.

ELISA for 3-meAde

The checkerboard procedure established conditions for optimal colour development (with a limiting dilution of antiserum) of 10 ng coating antigen per well and a working serum dilution of 1 in 10^4. Unmodified KLH did not show appreciable binding. 3-meAde itself was detected using a competitive ELISA procedure and gave a characteristic inhibition curve (Fig. 3). Preliminary studies on the cross-reactivity of related purines (Table 1) show that the antiserum is quite selective for 3-meAde and CM-3-meAde. The parent purine, adenine, is recognized well, but, for example, 3-methylxanthine is detectable only at high concentrations.

It has been found that considerable enhancement of sensitivity can be obtained if the first steps of the ELISA are carried out at 4°C. This is due to nonequilibrium phenomena, in which the rapid association of antibody and hapten occurs at only a slightly reduced rate at 4°C, whereas the disassociation is significantly inhibited

Fig. 3. Inhibition curve for 3-meAde using antiserum A (1 in 10^4 dilution) and 3-meAde-KLH (10 ng/well), with 90-min antibody-antigen incubation

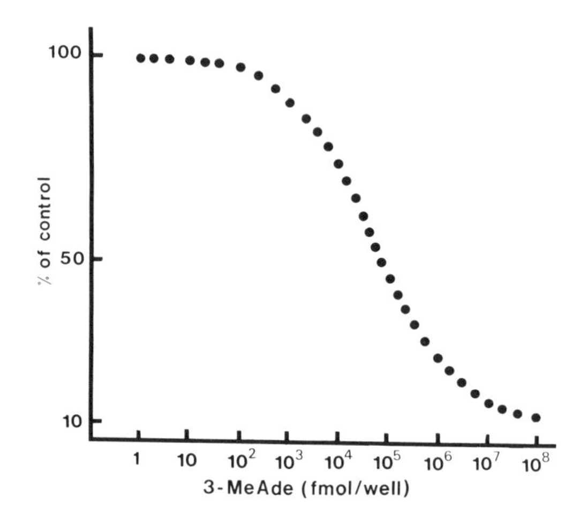

(Zettner & Duly, 1974). Thus, usable standard curves were obtained at concentrations essentially undetectable at room temperature (Fig. 4). These assay conditions were used in the determination of 3-meAde in human urine.

3-meAde in human urine samples

The direct determination of 3-meAde in urine samples by ELISA was not possible due to the cross-reactivity of normal urinary metabolites present at relatively high concentrations (>1 mg/l; e.g., adenine). Thus, HPLC fractionation was chosen as a preliminary clean-up step. Small aliquots of filtered urine (<100 μl) could be injected onto the HPLC without affecting the chromatographic properties of 3-meAde. Collection of the 3-meAde fraction was optimized using 3-[^{3}H]meAde, and good recoveries were obtained.

We initially attempted to validate the ELISA using samples that had been analysed previously by GC-MS. Ten human urine samples were analysed in the following way: filtered human urine (100 μl) was injected onto a HPLC column (Beckmann ultrasphere — ODS (5 μm) 4.6 mm × 150 mm plus a guard column (4.6 × 45 mm) of

Table 1. Cross-reactivity of purines with 3-meAde antiserum A

Purine	Concentration for 50% inhibition (pmol/well)	Concentration for 20% inhibition (pmol/well)
CM-3-meAde	20	1.6
3-meAde	100	1.6
Adenine	2500	130
Triacanthine	10 000	650
Caffeine	100 000	4000
3-Carboxymethyladenine	>100 000	30 000
3-Methylxanthine	>100 000	40 000
Theophylline	>100 000	40 000

Fig. 4. Standard ELISA curve for 3-meAde determination at 4°C with a 30-min antibody-antigen incubation, and at about 22°C with a 90-min antibody-antigen incubation

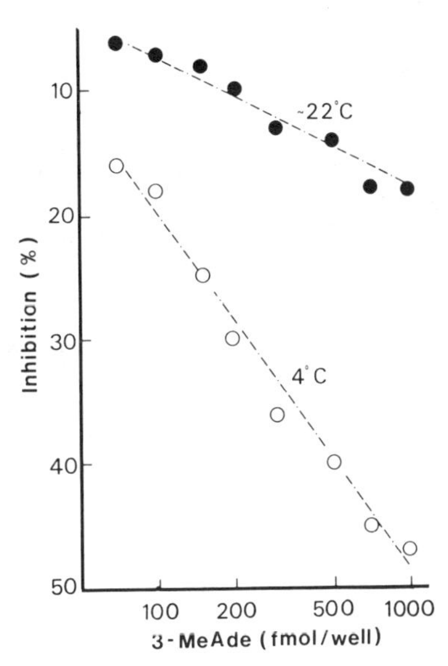

the same material) and fractionated using a gradient of 0.1% aqueous heptafluorobutyric acid and methanol. With the particular gradient used, a fraction eluting between 14.5 min and 16.25 min was collected into 1.5-ml Eppendorf tubes and reduced to dryness in a Speed-Vac concentrater. For ELISA, the samples were reconstituted in PBS (250 μl). Blank samples, with HPLC solvent only, did not give a response.

Using a GC-MS procedure (Shuker *et al.*, 1987b), the level of 3-meAde in the ten samples in this preliminary study ranged between 3.14 and 20.30 ng/ml. The HPLC-ELISA method gave results that were consistently much higher (five to ten fold) than those obtained with GC-MS. Results obtained after dilution of the samples (1:2 and 1:10) indicated that the response was due to a specific substrate for the antiserum, since the slope of the dilution curve was the same as that of the standard curve. Fractionation of the '3-meAde' HPLC peak followed by ELISA showed that the interfering compound eluted at almost the same retention time as 3-meAde. A large number of likely urinary purines were screened for interference; the only one detected so far, which has almost the same HPLC retention time and some cross-reactivity, is theophylline. As theophylline is a metabolite of caffeine and is also present in tea, its presence in urine is not surprising, since the subjects consumed a diet of their own choice.

Conclusions

3-meAde can be detected at low levels by ELISA using a rabbit antiserum. Initial attempts to use the assay to determine 3-meAde levels in human samples by HPLC-ELISA have been unsuccessful, due to an interfering compound — probably theophylline — which has similar chromatographic properties. At present, no suitable separation of 3-meAde from this interference in human urine samples has been found.

These preliminary results illustrate that, despite the ease and speed of immunological assays, some caution must be exercised in their application, particularly in the case of a complex body fluid like urine.

Acknowledgements

One of us (D.E.G.S.) gratefully acknowledges the award of a Royal Society European Science Exchange Programme Fellowship and the continuing support and encouragement of Dr H. Bartsch, Chief, Unit of Environmental Carcinogens and Host Factors, IARC. It is also a pleasure to acknowledge the assistance of J. Green in the preparation of antisera, A. Parry and J. Lamb in the GC-MS analyses and Dr C.P. Wild for stimulating discussions.

References

Lawley, P.D. (1976) *Methylation of DNA by carcinogens; some applications of chemical analytical methods.* In: Montesano, R., Bartsch, H., Tomatis, L. & Davis, W., eds, *Screening Tests in Chemical Carcinogenesis* (*IARC Scientific Publications No. 12*), Lyon, International Agency for Research on Cancer, pp. 181–208

Shuker, D.E.G., Bailey, E. & Farmer, P.B. (1987a) *The excretion of methylated nucleic acid bases as an indicator of exposure to nitrosatable drugs.* In: Bartsch, H., O'Neill, I.K. & Schulte-Hermann, R., eds, *The Relevance of* N-*Nitroso Compounds to Human Cancer: Exposures and Mechanisms* (*IARC Scientific Publications No. 84*), Lyon, International Agency for Research on Cancer, pp. 407–410

Shuker, D.E.G., Bailey, E., Parry, A., Lamb, J. & Farmer, P.B. (1987b) The determination of urinary 3-methyladenine in humans as a potential monitor of exposure to methylating agents. *Carcinogenesis, 8,* 959–962

Zettner, A. & Duly, P.E. (1974) Principles of competitive binding assays (saturation analyses). II. Sequential saturation. *Clin. Chem., 20,* 5–14

N-NITROSOPROLINE EXCRETION IN PATIENTS WITH GASTRIC LESIONS AND IN A CONTROL POPULATION

T.M. Knight,[1] D. Forman,[1,5] S. Leach,[2] P. Packer,[2] C. Vindigni,[3] C. Minacci,[3] L. Lorenzini,[3] P. Tosi,[3] G. Frosini,[4] M. Marini[4] & N. Carnicelli[4]

[1]*Imperial Cancer Research Fund, Cancer Epidemiology & Clinical Trials Unit, Radcliffe Infirmary, Oxford, UK;* [2]*Bacterial Metabolism Research Laboratory, CPHL, Porton Down, Salisbury, UK;* [3]*Institute of Pathological Anatomy, University of Siena, Italy; and* [4]*Gastroenterology and Digestive Endoscopy Unit, Siena, Italy*

***N*-Nitrosoproline (NPRO) excretion was measured in a group of hospital in-patients who were identified as either bearing gastric lesions or having apparently healthy stomachs. NPRO was assayed in background 24-h urine samples and then in urines collected after loading doses of nitrate and proline. The presence of gastric lesions was associated with altered gastric juice pH and nitrite concentration, but not with NPRO excretion. The significance of NPRO excretion as a marker of endogenous nitrosation is dependent on the interpretation of this result.**

This study was designed to address the question of whether individuals with chronic gastric disease, who are at a higher risk of developing gastric cancer, synthesize more NPRO *in vivo* than individuals without such disease. NPRO synthesis has been used as an indicator of nitrosating ability and is potentially a powerful marker of *N*-nitroso compound formation (Ohshima & Bartsch, 1981; Bartsch *et al.*, 1983; Ohshima *et al.*, 1985). It is, therefore, hoped that this test could be used as a surrogate measure for exposure to an important class of DNA damaging agents and could be adapted to epidemiological studies. The interrelationships between NPRO synthesis and gastric juice pH, nitrate and nitrite concentration, and mutagenicity have also been considered.

A total of 81 patients, aged between 32 and 69 years, were recruited from the Central Hospital in Siena, Italy. Some were attending the gastroenterology department with gastric symptoms, while others had been admitted to other departments with a wide range of nongastric illnesses. All subjects underwent a gastric endoscopic examination, at which biopsy samples were obtained. The results of the endoscopies and histological examination of the biopsies were all reviewed by one pathologist (C.V.), and, on the basis of this review, subjects were categorized into one of a number of groups (see Table 1) which represent different degrees of gastric disease. Also included were a few subjects diagnosed with gastric carcinoma.

The subjects consumed similar diets during the test period and for the two days prior to endoscopy. These diets excluded foods known to contain preformed NPRO, and the intake of foods with high levels of nitrate, nitrite or ascorbic acid was restricted. Subjects fasted for 12 h prior to endoscopy; at endoscopy, gastric juice samples were

[5] To whom correspondence should be sent

Table 1. Proportion of subjects in different gastric pH groups and mean pH by diagnostic category

Diagnosis[a]	No. of subjects	Gastric pH				
		<3.0	3.1–6.0	>6.0	Mean	95% confidence interval
		(% in each group)				
Control	30	50	27	23	4.0	3.1–4.9
SG	8	50	25	25	4.1	2.4–5.9
CAG	33	27	15	58	5.5	4.7–6.4
GC	10	10	20	70	6.2	4.7–7.7

Comparison of pH (control *versus* CAG), $p = 0.011$ (*t*-test)
[a]Control, nothing of pathological significance; SG, superficial gastritis; CAG, chronic atrophic gastritis; GC, gastric carcinoma

aspirated, measured for pH, and kept under appropriate storage conditions for nitrate and nitrite analysis and mutagenicity testing (Bartholomew, 1984; Venitt *et al.*, 1984). Two consecutive 24-h urine collections were made, commencing on the morning of endoscopy: the first was to assess background excretion of NPRO, and the second followed consumption of beetroot juice, containing approximately 300 mg nitrate, and 30 min later a 500-mg dose of L-proline dissolved in drinking-water. After the urine volume had been recorded, aliquots were stored for subsequent analysis of NPRO using standard procedures (Ohshima & Bartsch, 1981).

Five subjects did not consume all the beetroot juice, and a further 29 provided urine collections of less than 23.5 or more than 24.5 h on either study day. Data for these subjects were excluded from the appropriate sections of the current analysis. One subject was excluded as he had taken daily tablets of ascorbic acid; no other drug or medication had any effect on the parameters under investigation.

Table 1 shows the proportion of subjects within each diagnosis group with low (≤3.0), moderate (3.1–6.0) and high (>6.0) gastric juice pH values, together with the mean pH for each group, indicating the expected relationship between increased pH and disease severity.

Table 2 shows the mean gastric juice nitrate and nitrite and urinary nitrate levels together with NPRO excretion for the different diagnostic groups and for the different gastric pH categories. There was a significant relationship between decreased level of gastric juice nitrate and increased severity of gastric symptoms and increased gastric pH. The relationship with gastric juice nitrite was similar but in the inverse direction, i.e., increased nitrite was associated with increased severity and pH, but this was statistically significant only in the case of pH.

Table 2 also shows a lack of relationship between gastric disease or pH and NPRO excretion, either for background levels of NPRO or for levels after loading doses of nitrate and proline or for the difference between test and background. All disease and pH groups showed approximately an order of magnitude difference between the test and background levels of NPRO, the differences all being highly significant. In comparison with other groups, both the normally acidic and normal pathology groups had an apparently increased synthesis of NPRO after the loading doses, but in neither case was this increase statistically significant. When a correction was carried out for background NPRO, by looking at the difference between test and background, the effect was no longer of even borderline significance. The results show clearly that neither hypochlorhydric individuals nor those with atrophic gastritis produce more NPRO than subjects with normal gastric conditions. It is also noteworthy that not one

Table 2. Mean gastric juice nitrate and nitrite concentrations and urinary NPRO excretion by diagnostic category and by pH group (geometric means with 95% confidence intervals in parentheses)

Diagnosis[a]	Gastric juice			Urinary NPRO (μg/24 h)			
	No.	Nitrate (μg/ml)	Nitrite (ng/ml)	No.	Background	Test	Difference[b]
Control	28	11.9 (9.0–15.7)	14.0 (2.5–77.0)	18	0.5 (0.1–2.7)	4.1 (2.4–7.1)	0.8 (0.1–4.6)
SG	8	8.9 (6.2–12.8)	13.2 (0.2–1071.5)	6	0.1 (0.0–9.6)	0.6 (0.0–57.0)	0.3 (0.0–23.4)
CAG	30	5.7 (4.3–5.7)	65.8 (0.1–294.4)	18	0.3 (0.1–2.2)	2.5 (0.6–10.1)	1.1 (0.2–6.5)
GC	10	6.3 (3.8–10.4)	62.2 (2.5–1563.1)	5	0.4 (0.0–119.9)	2.9 (0.5–15.2)	1.3 (0.4–3.9)
p value[c]		<0.001	0.17		0.69	0.49	0.78
Gastric juice pH							
<3.0	27	11.3 (8.5–15.1)	9.3 (1.6–54.6)	17	0.7 (0.1–4.2)	5.0 (2.8–8.8)	1.4 (0.3–6.1)
3.1–6.0	16	7.7 (5.2–11.5)	15.3 (1.3–176.2)	11	0.2 (0.0–3.3)	1.5 (0.2–14.7)	0.4 (0.0–6.7)
>6.0	33	6.0 (4.6–7.7)	119.0 (30.8–461.3)	19	0.2 (0.0–1.3)	1.9 (0.5–6.9)	0.8 (0.2–4.2)
p value[d]		<0.001	0.04		0.63	0.13	0.57
All	76	7.9 (6.6–9.4)	31.2 (11.6–84.1)	47	0.3 (0.1–1.0)	2.6 (1.3–5.2)	0.8 (0.3–2.3)

[a]See footnote to Table 1
[b]Difference between background excretion and test (following loading doses) excretion calculated for each individual and then as mean
[c]Comparison of control group with CAG (t-test)
[d]Significance of regression, pH on variable

of the samples of gastric juice was identified as being mutagenic in a bacterial test system (Venitt *et al.*, 1984).

There was no significant effect of smoking on gastric pH, gastric juice nitrite concentration, or background NPRO excretion. However gastric juice nitrate was significantly increased in current nonsmokers (mean, 9.0 μg/ml *versus* 6.1 in smokers; $p = 0.040$), while current smokers showed a moderate significant increase in NPRO excretion under the test conditions (mean, 5.7 μg/24 h *versus* 1.6 in nonsmokers; $p = 0.03$). Adjusting the test NPRO for background excretion reduced the size of the effect and made it of borderline significance (mean, 1.9 μg/24 h *versus* 0.5; $p = 0.18$). These results therefore support the suggestion (Ladd *et al.*, 1984) that the higher intakes of thiocyanate received by smokers in comparison with nonsmokers catalyse NPRO formation.

In general, this study is in agreement with others which show that a decrease in stomach acidity is associated with a rise in gastric juice nitrite concentration (e.g., Ruddell *et al.*, 1976; Gledhill *et al.*, 1985). This effect is due almost certainly to colonization of the stomach by bacteria, some species of which are nitrate reducers (Reed *et al.*, 1981; Kyrtopoulos *et al.*, 1985; Crespi *et al.*, 1987). However, whereas some studies have reported an increase in the concentration of *N*-nitroso compounds in gastric juice under conditions of hypochlorhydria (Correa *et al.*, 1975; Schlag *et al.*, 1980; Reed *et al.*, 1981; Schlag *et al.*, 1982; Pignatelli *et al.*, 1987), the results of this study are consistent with those of the two others in which NPRO excretion in urine was examined (Hall *et al.*, 1986; Crespi *et al.*, 1987): it does not increase in patients with precursor lesions for gastric cancer.

Normally, formation of *N*-nitroso compounds (including NPRO) proceeds optimally at an acid pH, as occurs in the normal stomach. The formation of these compounds in hypochlorhydric gastric juice is thought to be a result of an alternative pathway of bacterial catalysis (Sander, 1968; Ruddell *et al.*, 1976; Bartsch *et al.*, 1985). The kinetics and mechanism of the bacterially mediated reaction differ significantly from those of the acid-catalysed reaction (Leach *et al.*, 1987). At present, it is not possible to be certain that proline is a suitably representative nitrosatable substrate for the bacterial reaction. If NPRO synthesis is indicative of nitrosation potential only in relation to acid-catalysed *N*-nitrosation, then its use as a test system would be most appropriate in assessing the level of nitrosation in healthy populations with normal stomachs (Lu *et al.*, 1986; Kamiyama *et al.*, 1987).

Another problem with this particular configuration of the NPRO test is that the intragastric nitrite concentration in these subjects is deliberately and considerably elevated as a consequence of the large oral dose of nitrate (equivalent to three times the average per-caput daily intake). This artefactual increase in intragastric nitrite is likely to be sufficiently large to mask the approximately ten-fold difference ordinarily observed between normal and achlorhydric individuals. It is conceivable that the test would be made more meaningful by exclusion of the nitrate loading dose, so that nitrosation potential was dependent *inter alia* upon normal endogenous levels of nitrite. This latter form of the test is that now utilized in studies of healthy populations.

Further work is needed to clarify both the role of bacterial catalysis and the importance of the nitrate dose in NPRO formation, in order to explain the inconsistency of the results in gastritis patients between studies in which *N*-nitroso compounds have been measured in gastric juice and those investigating urinary NPRO excretion.

Acknowledgements

We thank Dr S. Venitt and Miss J. Kidd for performing the mutagenicity assays and Miss C. Bates for typing the manuscript.

References

Bartholomew, B. (1984) A rapid method for the assay of nitrate in urine using the nitrate reductase enzyme of *Escherichia coli. Food chem. Toxicol., 22,* 541–543

Bartsch, H., Ohshima, H., Muñoz, N., Crespi, M. & Lu, S.-H. (1983) *Measurement of endogenous nitrosation in humans: potential applications of a new method and initial results.* In: Harris, C.C. & Autrup, H.N., eds, *Human Carcinogenesis,* New York, Academic Press, pp. 833–855

Bartsch, H., Ohshima, H., Muñoz, N., Calmels, S., Crespi, M., Cassale, V. & Ramazotti, V. (1985) Intragastric formation of *N*-nitroso compounds (NOC): testing of an etiological hypothesis. *Eur. J. Cancer, 21,* 1370

Correa, P., Haenszel, W., Cuello, C., Tannenbaum, S. & Archer, M. (1975) A model for gastric cancer epidemiology. *Lancet, ii,* 58–59

Crespi, M., Ohshima, H., Ramazzotti, V., Muñoz, N., Grassi, A., Cassale, V., Leclerc, H., Calmels, S., Cattoen, C., Kaldor, J. & Bartsch, H. (1987) *Intragastric nitrosation and precancerous lesions of the gastrointestinal tract: testing of an etiological hypothesis.* In: Bartsch, H., O'Neill, I.K. & Schulte-Hermann, R., eds, *The Relevance of* N-*Nitroso Compounds to Human Cancer: Exposures and Mechanisms* (*IARC Scientific Publications No. 84*), Lyon, International Agency for Research on Cancer, pp. 511–517

Gledhill, T., Leicester, R.… rd, J., Viney, N., Darkin, D. & Hunt, R.H. (1985) Epi… *290,* 1383–1386

Hall, C.N., Darkin, D., Brim… J.S. & Northfield, T.E. (1986) An evaluation of the n… arcinogenesis in precancerous conditions. *Gut, 27,* 491–…

Kamiyama, S., Ohshima, H., S… M.C., Ziegler, P. & Bartsch, H. (1987) *Urinary excretion … by inhabitants in high- and low-risk areas for stomach* … tsch, H., O'Neill, I.K. & Schulte-Hermann, R., eds, … *ounds to Human Cancer: Exposures and Mechanisms* (… *84*), Lyon, International Agency for Research on Cance…

Kyrtopoulos, S.A., Dasalakis, G., … rou, E., Bonatsos, G., Golematis, B., Lakiotis, G., … 85) Studies in gastric carcinogenesis. II. Absence of ele… mpounds in the gastric juice of Greek hypochlorhydric in… 140

Ladd, K.F., Newark, H.L. & Archer, … oline in smokers and non-smokers. *J. natl Cancer Inst., 7…*

Leach, S., Cook, A., Challis, B., Hil… mpson, M. (1987) *Bacterially mediated* N-*nitrosation reactions and endogenou… ation of* N-*nitroso compounds.* In: Bartsch, H., O'Neill, I.K. & Schulte-Hermann, R., eds, *The Relevance of* N-*Nitroso Compounds to Human Cancer: Exposures and Mechanisms* (*IARC Scientific Publications No. 84*), Lyon, International Agency for Research on Cancer, pp. 396–399

Lu, S.-H., Ohshima, H., Fu, H.M., Tian, Y., Li, F.-M., Blettner, M., Wahrendorf, J. & Bartsch, H. (1986) Urinary excretion of *N*-nitrosamino acids and nitrate by inhabitants of high- and low-risk areas for esophageal cancer in northern China: endogenous formation of nitrosoproline and its inhibition by vitamin C. *Cancer Res., 46,* 1485–1491

Ohshima, H. & Bartsch, H. (1981) Quantitative estimation of endogenous nitrosation in humans by monitoring *N*-nitrosoproline excreted in the urine. *Cancer Res., 41,* 3658–3662

Ohshima, H., Muñoz, N., Nair, J., Calmels, S., Pignatelli, B., Crespi, M., Leclerc, H., Lu, S.-H., Bhide, S.V., Vincent, P., Gounot, A.M. & Bartsch, H. (1985) Monitoring for endogenous formation of carcinogenic *N*-nitroso compounds by measurement of urinary *N*-nitrosamino acids. Its application to epidemiological and clinical studies. *Ann. biol. Clin., 43,* 463–474

Pignatelli, B., Richard, I. & Bourgade, M.C. (1987) Improved group determination of total *N*-nitroso compounds in human gastric juice by chemical denitrosation and thermal energy analysis. *Analyst, 112,* 945–949

Reed, P.I., Smith, P.L.R., Haines, K., House, F.R. & Walters, C. (1981) Gastric juice *N*-nitrosamines in health and gastroduodenal disease. *Lancet, ii,* 550–555

Ruddell, W.S.J., Bone, E.S., Hill, M.J., Blendis, L.M. & Walters, C.L. (1976) Gastric juice nitrite: a risk factor for cancer in the hypochlorhydric stomach? *Lancet, ii,* 1037–1039

Sander, J. (1968) Nitrosaminsynthese durch bakterien. *Hoppe-Seyler's Z. Physiol. Chem., 349,* 429–432

Schlag, P., Bockler, R., Ulrich, H., Peter, M., Merkle, P. & Herfarth, C. (1980) Are nitrite and *N*-nitroso compounds in gastric juice risk factors for carcinoma in the operated stomach? *Lancet, i,* 727–729

Schlag, P., Bockler, R. & Peter, M. (1982) Nitrite and nitrosamines in gastric juice: risk factors for gastric cancer? *Scand. J. Gastroenterol., 17,* 145–150

Venitt, S., Crofton-Sleigh, C. & Forster, R. (1984) *Bacterial mutation assays using reverse mutation.* In: Venitt, S. & Parry, J.M., eds, *Mutagenicity Testing — A Practical Approach,* Oxford, IRL Press, pp. 45–98

IMMUNOCYTOCHEMICAL LOCALIZATION OF DNA ADDUCTS IN RAT TISSUES FOLLOWING TREATMENT WITH *N*-NITROSOMETHYLBENZYLAMINE

J. Van Benthem, C.P. Wild[1], E. Vermeulen, H.H.K. Winterwerp, L. Den Engelse & E. Scherer

Division of Chemical Carcinogenesis, The Netherlands Cancer Institute, Amsterdam, The Netherlands

Immunocytochemical visualization of O^6-methylguanosine (meGua) and 7-meGua shows that DNA methylation by *N*-nitrosomethylbenzylamine(NMBzA) occurs not only in the target organs for tumour induction by this nitrosamine, the oesophagus and (occasionally) the tongue, but also in other tissues (liver, lung, trachea, tracheal glands and nasal cavity) for which no tumour induction by NMBzA has been reported. Thus, the organotropic carcinogenic action of NMBzA cannot be exclusively ascribed to differences in levels of DNA methylation. Additional determinants of the cancer risk in extra-oesophageal tissues could be the small size of the NMBzA-activating target cell population and a low proliferative activity.

It has been suggested that *N*-nitrosamines may contribute to the high incidence of oesophageal cancer in northern China (Umbenhauer *et al.*, 1985). Experimental studies have indicated that many *N*-nitrosamines are tissue-specific carcinogens. An example is the methylating agent NMBzA, which induces squamous-cell carcinomas in the upper gastrointestinal tract of rats, from the pharyngeal region including the basal tongue down to the basal oesophagus and occasionally the forestomach (Stinson *et al.*, 1978). Methylation of DNA by NMBzA has been reported in oesophagus, to a lesser extent in lung, liver and forestomach, and at very low levels in kidney (Hodgson *et al.*, 1980; Kleihues *et al.*, 1981; Kouros *et al.*, 1983). We have used an immunoperoxidase staining technique (Heyting *et al.*, 1983; Scherer *et al.*, this volume) to establish the cellular heterogeneity of NMBzA-induced methylation in a variety of rat tissues. The results are discussed in the light of the hypothesis that organotropism of NMBzA in rats may be related to differences in DNA methylation (compare Hodgson *et al.*, 1980; Kleihues *et al.*, 1981; Kouros *et al.*, 1983).

In the first experiment, male Sprague-Dawley rats (180–200 g) received a single intraperitoneal injection of NMBzA in 0.14 M NaCl (0.5, 1.0 or 2.5 mg/kg body weight) and were killed by exsanguination 6 or 72 h later. Oesophagus and liver were dissected, mounted onto blocks for cryostat sectioning and stored at −80°C. In all experiments, tissues from vehicle-treated rats were placed onto the same block as tissues from two NMBzA-treated rats, so that immunostaining of treated and control tissues was on the same slide. In the second experiment, only one dose-time combination (2.5 mg/kg NMBzA; killed after 6 h) was used, but a larger variety of tissues was collected.

[1] *Present address*: International Agency for Research on Cancer, Lyon, France

Fig. 1. Immunocytochemical DAB staining of O^6-meGua (A,B,C,D) and 7-meGua (E, F) in oesophageal epithelium 6 h after treatment with various doses of NMBzA; A, 0; B, 0.5; C, 1.0; D, 2.5; E, 0; and F, 2.5 mg/kg

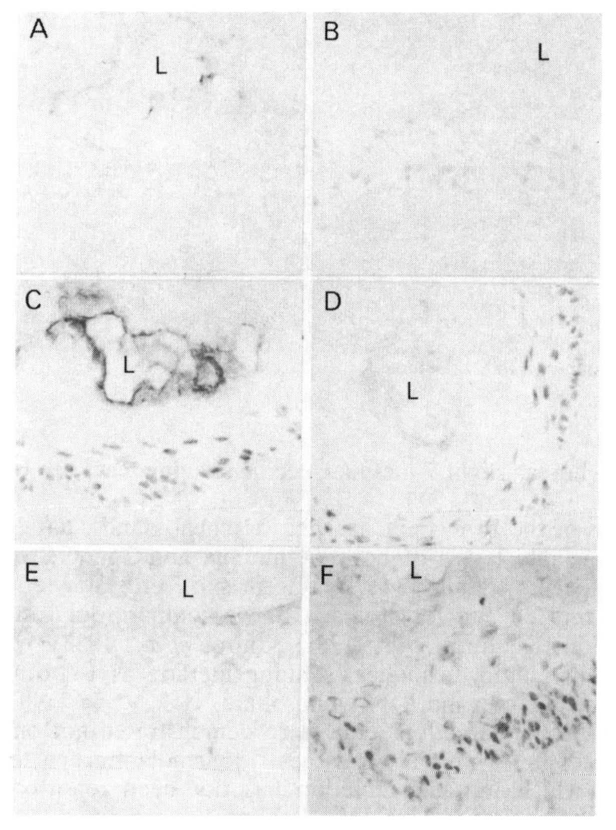

First antibodies were diluted 1:6000. L, lumen

Cryostat sections from all tissues were stained immunocytochemically (Scherer *et al.*, this volume) for O^6-meGua and 7-meGua in DNA, using rabbit antisera raised against an O^6-methylguanosine-bovine serum albumin conjugate and an imidazole ring-opened 7-methylguanosine-haemocyanin conjugate, respectively. The properties of the anti-O^6-meGua antibody have been reported previously (Wild *et al.*, 1983). The anti-7-meGua antibody has its highest affinity for imidazole-ring opened 7-meGua but recognizes to a lesser extent the ring-closed form of this adduct (Degan *et al.*, unpublished data).

Six hours after administration of each of the NMBzA doses, O^6- and 7-meGua-specific staining was observed in the oesophagus (Fig. 1). Staining intensity was dose dependent and restricted to the nuclei of mucosal epithelial cells (Fig. 1). After 72 h, both adducts were still visible, but the average staining intensity had declined, and the stained cells were found closer to the lumen of the oesophagus (Fig. 2). 7-meGua staining intensity was still dose dependent after 72 h, while in contrast O^6-meGua could be detected only after the highest dose of NMBzA, indicating active repair of the latter adduct by the oesophageal epithelium. In the liver, no O^6-meGua could be detected at

Fig. 2. Immunocytochemical staining of O^6-meGua (A) and 7-meGua (B) in oesophageal epithelium 72 h after treatment with 2.5 mg/kg (A) or 0.5 mg/kg (B) NMBzA

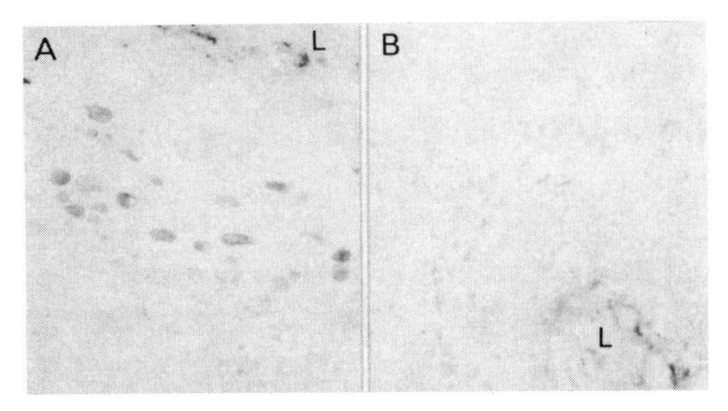

First antibodies were diluted 1:6000. L, lumen

either 6 or 72 h, whereas slight 7-meGua-specific staining was seen 6 h after the highest dose of NMBzA.

Epithelial cells of the bronchioli, trachea, tracheal glands, nasal cavity and tongue all contained detectable levels of both O^6-meGua and 7-meGua at 6 h (Fig. 3). In contrast, adduct-specific staining was lacking in skin, prostate, testis, small intestine, pancreas, kidney, spleen and forestomach. If we compare our results with published data on methylation (Kleihues *et al.*, 1981; Kouros *et al.*, 1983), we can estimate the sensitivity of the immunocytochemical staining method as 1–2 μmol O^6-meGua and about 25 μmol 7-meGua per mol DNA-phosphate.

Strikingly, O^6-MeGua and 7-MeGua were demonstrated not only in target tissues (oesophagus and tongue), but also in tissues (bronchioli, trachea, tracheal glands and nasal cavity) for which tumour induction has not been reported. Apparently, the carcinogenic organotropism of NMBzA cannot be explained by differences in its methylation pattern. Explanations for the low cancer risk in extra-oesophageal tissues might be either the small size of cell populations capable of NMBzA activation, or a low proliferative activity. Efficient repair of relevant DNA adducts is a rather unlikely explanation, since the promutagenic base O^6-meGua accumulates particularly in non-target tissues after repeated applications of NMBzA (Van Benthem *et al.*, manuscript in preparation). A rather trivial explanation may be that microscopic tumours or precancerous lesions in non-target tissues have not been reported in previous studies owing to a slower progression of these lesions as compared to those in the oesophagus. Another factor of possible importance may be the intrinsic sensitivity of a cell type to neoplastic transformation, which is probably related to the possibility of activating specific proto-oncogenes, and to the number of steps (Scherer, 1987) required before a neoplastic endpoint is reached in a tissue.

DNA adducts due to NMBzA treatment were demonstrated exclusively in epithelial cells. Apparently, these cells contain the enzymes necessary for the activation of NMBzA. It is striking that predominantly the nuclei of those epithelial cells which are derived from the foregut of the endoderm showed DNA adducts. Staining in epithelial cells derived from other parts of the endoderm was either absent (intestine, kidney and pancreas) or weak (liver). It should be noted, however, that other groups have found alkylation at low levels in forestomach, kidney and spleen (Hodgson *et al.*, 1980;

Fig. 3. Immunocytochemical staining of O^6-meGua (6 h after 2.5 mg/kg NMBzA) in epithelial cells of trachea and tracheal glands (A), oesophagus (B), bronchioli (C), nasal turbinate (D), maxillary turbinate (E) and serous glands (D, E) in the nasal cavity

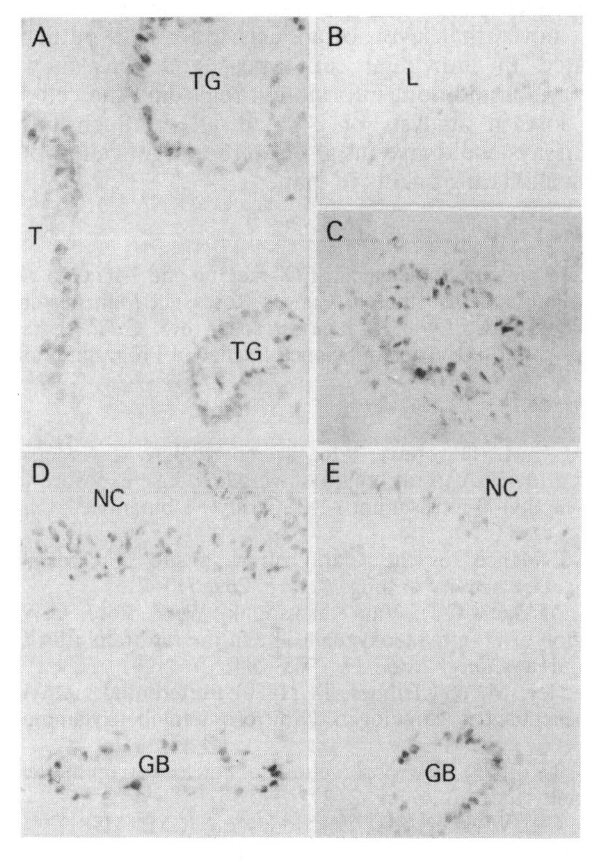

First antibody was diluted 1:20 000. GB, serous glands; L, lumen; NC, nasal cavity; T, trachea; TG, tracheal gland

Kouros *et al.*, 1983). Such levels may have been below the present limit of detection in our assay.

In the bronchioli, several epithelial cell types can be distinguished (Jeffery & Reid, 1975). Our experiments show that the DNA in at least some bronchiolar cells becomes strongly methylated, indicating their ability to activate NMBzA. In this respect, it is of interest that Boyd (1977) and Belinsky *et al.* (1987) have reported that Clara cells, a bronchiolar epithelial cell type, can enzymatically activate the tobacco-derived nitroso compound 4-(*N*-nitrosomethylamino)-1-(3-pyridyl)-1-butanone (NNK).

The advantages of visualizing adducts in single cells are illustrated by the present experiments. For example, methylation levels of isolated, bulk DNA reported in the literature are always several times higher for oesophagus than for lung (Hodgson *et al.*, 1980; Kouros *et al.*, 1983). The present experiments show that the low level of lung DNA methylation is not due to a general paucity of metabolic activation of NMBzA in all lung cells, but results from a striking concentration of the relevant enzymes in only one (or a few) cell type(s).

Studies of human tissues, including oesophagus, to detect alkylation adducts have involved analysis of isolated DNA (Umbenhauer *et al.*, 1985). Although this appears to be a promising approach to detecting human exposure to environmental alkylating agents, it should be noted that levels of adducts too low to be detected in bulk DNA may be concentrated in individual cell types, and thus may be detectable by immunocytochemistry. In addition, information regarding the cellular heterogeneity of alkylation will be lost in analysis of isolated DNA. Such information from immunocytochemical assays could give insight into the target cell population for initiation of carcinogenesis by alkylating agents in man.

Acknowledgements

We highly appreciate that antibodies against O^6-meGua and 7-meGua were made available to us by Dr R. Saffhill (Paterson Institute for Cancer Research, Manchester, UK), Dr P. Degan and Dr R. Montesano (IARC, Lyon, France), respectively. C.P.W. gratefully acknowledges receipt of a fellowship from the European Science Exchange Programme of the Royal Society.

References

Belinsky, S.A., White, C.M., Devereux, T.R., Swenberg, J.A. & Anderson, M.W. (1987) Cell selective alkylation of DNA in rat lung following low dose exposure to the tobacco specific carcinogen 4-(N-methyl-N-nitrosamino)-1-(3-pyridyl)-1-butanone. *Cancer Res., 46,* 1143–1148

Boyd, M.R. (1977) Evidence for the Clara cell as a site of cytochrome P450-dependent mixed-function oxidase activity in lung. *Nature, 269,* 713–715

Heyting, C., Van Der Laken, C.J., Van Raamsdonk, W. & Pool, C.W. (1983) Immunohistochemical detection of O^6-ethyldeoxyguanosine in the rat brain after *in vivo* applications of N-ethyl-N-nitrosourea. *Cancer Res., 43,* 2935–2941

Hodgson, R.M., Wiessler, M. & Kleihues, P. (1980) Preferential methylation of target organ DNA by the oesophageal carcinogen N-nitrosomethylbenzylamine. *Carcinogenesis, 1,* 861–866

Jeffery, P.K. & Reid, L. (1975) New observations of rat airway epithelium: a quantitative and electron microscopic study. *J. Anat., 2,* 295–320

Kleihues, P., Veit, C., Wiessler, M. & Hodgson, R.M. (1981) DNA methylation by N-nitrosomethylbenzylamine in target and non-target tissues of NMRI mice. *Carcinogenesis, 2,* 897–899

Kouros, M., Mönch, W., Reiffer, F. J. & Dehnen, W. (1983) The influence of various factors on the methylation of DNA by the oesophageal carcinogen N-nitrosomethylbenzylamine. I. The importance of alcohol. *Carcinogenesis, 4,* 1081–1084

Scherer, E. (1987) Relationship among histochemically distinguishable early lesions in multistep-multistage hepatocarcinogenesis. *Arch. Toxicol., Suppl., 10,* 81–94

Stinson, S.F., Squire, R.A. & Sporn, M.B. (1978) Pathology of esophageal neoplasms and associated proliferative lesions induced in rats by N-methyl-N-benzylnitrosamine. *J. natl Cancer Inst., 61,* 1471–1475

Umbenhauer, D., Wild, C.P., Montesano, R., Saffhill, R., Boyle, J.M., Kirstein, U., Thomale, J., Rajewsky, M.F. & Lu, S.H. (1985) O^6-Methyldeoxyguanosine in oesophageal DNA among individuals at high risk of oesophageal cancer. *Int. J. Cancer., 36,* 661–665

Wild, C.P., Smart, G., Saffhill, R. & Boyle, J.M. (1983) Radioimmunoassay of O^6-methyldeoxyguanosine in DNA of cells alkylated *in vitro* and *in vivo*. *Carcinogenesis, 4,* 1605–1609

RECOVERABLE, SEMIPERMEABLE, MICROENCAPSULATED DNA SURROGATES FOR MONITORING THE COLORECTAL CAVITY; IN-SITU EFFECTS OF FIBRE AND MEAT IN HUMAN DIETS ON BENZO[*a*]PYRENE AND POSSIBLE ENDOGENOUS CROSS-LINKING AGENTS

I.K. O'Neill,[1] A.C. Povey,[1] S. Bingham,[2] I. Brouet,[1] J.-C. Béréziat[1] & A. Ellul[1]

[1]*International Agency for Research on Cancer, Lyon, France; and* [2]*Dunn Clinical Nutrition Centre, Cambridge, UK*

Semipermeable magnetic microcapsules containing polyethyleneimine (PEI) as a DNA surrogate are shown to trap ^{14}C-benzo[*a*]pyrene and hitherto unknown, endogenous, putative cross-linking agent(s) within the gut of male Fischer rats. Trapping is substantially modulated by complete, cooked human diets fed isocalorically and varied three-fold in either beef, fat or bran fibre nonstarch polysaccharide within the normal human intake levels. Preliminary results indicate that the cross-linking agent(s) are derived from microflora. Using metabolized benzo[*a*]pyrene as a model DNA damaging agent within the gut, beef and decreased bran fibre were found to increase its availability, paralleling risk alterations found in nutritional epidemiology. These novel microcapsules are capable of intercepting a range of substances relevant to DNA damage.

Both epidemiological and laboratory studies suggest that dietary intake and intraluminal DNA damaging agents are major risk factors for gastrointestinal (GI) cancers. We are developing magnetic, recoverable, semipermeable, electrophile-trapping microcapsules (Povey *et al.*, 1986) for GI transit in humans. These microcapsules contain dilute aqueous PEI as a DNA surrogate which cannot escape, although carcinogens can enter; they are 10–50 μm in diameter and are administered (Povey *et al.*, 1987a) in several millions to present a high surface area dispersed through a large volume of the gut. The microcapsules contain PEI both in free solution in the core and covalently bound within the membrane. Reactive substances of <1000 molecular weight enter the microcapsules (Povey *et al.*, 1987b) and react with both states of PEI throughout the gut (Povey *et al.*, 1987c,d; O'Neill *et al.*, 1987). As part of this study, we are evaluating their use with human diets in rodents and investigating the effects of both unknown and administered DNA-damaging agents. We present here results obtained using these novel microcapsules to explore effects of human diets and of apparently endogenous cross-linking substances as DNA damaging agents.

Preparation of human diets and magnetic microcapsules

Six nutritionally adequate, low- and high-fat diets (fat providing 15% or 45% of total energy) designed for isocaloric feeding (I.K. O'Neill, A.C. Povey, S.A. Bingham, I. Brouet, J.-C. Béréziat, submitted for publication) were prepared from large batches

Table 1. Weight percentage of macro-components in diets[a]

Component	Low-fat, low-NSP[b] high-beef	High-fat, low-NSP, high-beef	Low-fat, low-NSP, low-beef	High-fat, low-NSP, low-beef	Low-fat, high-NSP, high-beef	Low-fat, high-NSP, low-beef	Rat chow
Protein[c]	15.5 (8.4)	19.0 (10.3)	16.6 (2.7)	19.9 (9.4)	15.6 (8.0)	15.9 (2.6)	24 (11)
Fat[d]	7.1	25.7	7.1	25.5	5.7	6.7	5
Sugars	41	12.7	40	11.6	39	40	60.5
Starch	30	40	30	40	27	27	
NSP	2.6	3.2	2.5	2.8	6.4	6.6	4.5[e]

[a]For low-fat human diets, protein, fat, sugars and starch supplied 16, 15, 40 and 29%, respectively, of total energy; for high-fat diets, the values were 15, 45, 10 and 30%. Fed isocalorically, diets provide the same weights of nutrients and the same intakes of fat, nonstarch polysaccharide and beef
[b]NSP, Nonstarch polysaccharide
[c]Animal protein given in parentheses; for rat chow, protein was from fish meal
[d]Average polyunsaturated: saturated ratio, 0.46; range, 0.36–0.59
[e]Crude fibre (Weende method)

of cooked foods widely consumed in the UK (Table 1). Beef, fat and bran-fibre nonstarch polysaccharide were independently varied three-fold; starch, total protein and calcium were kept constant throughout by the use of powdered potato, gluten and calcium carbonate.

Microcapsules (Povey *et al.*, 1986) with cross-linked polyhexamethyleneterephthalamide membranes and containing PEI and magnetite were prepared by interfacial polymerization; they had a geometric mean diameter of 23 μm (range, 14–32 μm) with a PEI core: membrane ratio of 1. A batch of microcapsules was treated with [^{14}C] methyl iodide in ethanol to produce microcapsules (0.1 μCi/million) with ^{14}C-labelled PEI in core and membrane; 29% of total radioactivity was found to be attached to the membrane following treatment with simulated gastric juice and then ultrasonication to release core.

Protocol for in-vivo benzo[*a*]pyrene trapping

Male Fischer rats (six weeks old) were fed isocalorically on the prepared low-fat diets; others were fed on standard rat chow as comparison. Administration of intragastric doses of microcapsules and of ^{14}C-benzo[*a*]pyrene was started on either day 15 (period I; group I) or day 22 (period II; group II), and microcapsules were given by gavage (3.9 million/dose) at 0, 24 and 48 h and ^{14}C-benzo[*a*]pyrene (7 μCi/rat) at 2 h. Microcapsules were extracted magnetically from faeces collected at 24, 48 and 72 h and pooled; residual faecal solid and aqueous extract, and collected urine, were measured for radioactivity. Microcapsules were extracted successively (Povey *et al.*, 1987d) and extracts analysed by high-performance liquid chromatography for benzo[*a*]pyrene metabolites.

Protocol for endogenous cross-linking agents

Male Fischer rats (six weeks old), three per subgroup, were fed isocalorically on either a low-fat diet or two high-fat diets for four weeks, with benzo[*a*]pyrene treatment and PEI microcapsules given in the fourth week. After two weeks, another group on the low-fat diet received an antibiotic cocktail in drinking-water; lack of faecal enzyme activity (β-glucuronidase and β-galactosidase) confirmed them to be germ-free. The radiolabelled microcapsules (0.33 μCi/rat) were given by gavage during the fifth week after benzo[*a*]pyrene excretion had finished, then faeces were collected

Table 2. Diet-dependent excretion of ^{14}C-benzo[*a*]pyrene[a]

Diet	% Administered benzo[*a*]pyrene excreted				Faecal metabolite distribution (water : solids)		
	Faeces		Urine		Overall	10^3 × normalized[b]	
	I	II	I	II	II	I	II
Human diet, low-fat							
Low-NSP[c], high-beef	78.8 ± 2.7	71.3 ± 5.0	4.0 ± 0.3	5.5 ± 1.0	5.2 ± 1.0	14.0 ± 1.5	12.3 ± 2.0
Low-NSP, low-beef	77.2 ± 6.8	71.9 ± 5.5	3.1 ± 0.2	5.6 ± 0.8	5.2 ± 0.2	12.8 ± 2.2	7.2 ± 0.5
High-NSP, high-beef	74.4 ± 0.8	79.5 ± 3.7	3.9 ± 0.7	5.1 ± 0.6	5.0 ± 1.2	13.7 ± 2.7	17.3 ± 4.0
High-NSP, low-beef	82.2 ± 1.1	80.3 ± 2.8	3.7 ± 1.1	5.2 ± 0.3	3.2 ± 0.4	12.8 ± 1.7	12.8 ± 2.7
Rat chow	78.5 ± 10.7	87.6 ± 10.8[d]	3.4 ± 0.1	4.4 ± 0.4	0.8 ± 0.3	11.0 ± 3.7	11.5 ± 3.3
Trend association (significance)	NS	NSP ($p < 0.001$)	NS	NSP ($p < 0.05$)	NSP ($p < 0.001$)	NS	Beef/NSP ($p < 0.01$)

[a]See Table 1 for diets. Results expressed as mean ± SD (n = 3); data for excretion in periods I (third week) and II (fourth week) by groups I and II, respectively; NS, no statistically significant trend. Animals show incomplete adaptation to diets in period I.
[b]Normalized for weight of faeces (g) and volume of water (ml) used for microcapsule extraction
[c]NSP, nonstarch polysaccharide
[d]A third rat excreted <19% of the benzo[*a*]pyrene dose.

for three days and microcapsules extracted magnetically; microcapsule aliquots were measured for radioactivity or sonicated prior to counting of membrane radioactivity.

Weight gain was similar for animals fed on the six human diets or on rat chow over the five-week experimental period; no obvious ill effect of the diets was apparent.

Benzo[a]pyrene metabolism

Animals excreted 70–99% of administered ^{14}C-benzo[*a*]pyrene within 70 h *via* the faeces and urine (Table 2). In period II, animals fed on high-fibre diets excreted more radioactivity *via* the faeces and less in the urine than those fed low-fibre diets; these and the following trends were statistically significantly related to nonstarch polysaccharide content.

The level of ^{14}C-benzo[*a*]pyrene metabolites bound to microcapsules was 1–5% that excreted in the faeces; thus, the microcapsules sampled but did not interfere with the distribution of the metabolites (Table 3). High-nonstarch polysaccharide and no or low-beef diets increased the overall binding of luminal ^{14}C-benzo[*a*]pyrene metabolites to faecal solids, as indicated by the ratio of soluble to bound radioactivity, and decreased the specific metabolite binding per million recovered microcapsules (Table 3). Microcapsules bound 15–78 times more benzo[*a*]pyrene metabolites than did faecal solids on a weight basis; this was dependent on beef content and independent of the nonstarch polysaccharide content of the diet (Table 3).

Extraction of microcapsules with ammoniacal methanol and high-performance liquid chromatography analysis revealed an influence of diet on the metabolites bound to microcapsules — in particular, that dietary beef may increase the formation of benzo[*a*]pyrene diones and in particular the 1,6-dione (Table 3). Taking microcapsule activity as an indicator of available metabolites within the intestine (trend related to beef : nonstarch polysaccharide ratio in the same way as the faecal water : solid ratio), and dividing by faecal bulk suggests a 40-fold variation in availability to gut mucosa (Table 3).

Table 3. Effects of nonstarch polysaccharide (NSP) and beef on binding of ^{14}C-benzo[*a*]pyrene metabolites in microcapsules extracted from faeces[a]

Diet	Millions of microcapsules recovered[b]		Specific activity of bound benzo[*a*]pyrene metabolites (nmol/million)	Metabolite binding ratio w/w microcapsules/faecal solids[c]	Relative availability to gut mucosa[d]	Benzo[*a*]pyrene 1,6-dione as % of total[e]
	I	II	(Group II)	(Group II)	(Group II)	(Group II)
Human diet, low-fat						
Low-NSP, high-beef	4.1 ± 0.5	3.7 ± 1.1	1.04 ± 0.10	74 ± 10	0.80 ± 0.08	1.3
Low-NSP, low-beef	4.5 ± 1.2	1.8 ± 0.3	0.93 ± 0.41	44 ± 21	1.2 ± 0.5	0.4
High-NSP, high-beef	6.2 ± 0.8	3.1 ± 0.4[f]	0.97 ± 0.02	78 ± 11	0.57 ± 0.01	1.8
High-NSP, low-beef	7.7 ± 5.4	4.5 ± 0.9	0.59 ± 0.11	41 ± 11	0.30 ± 0.06	0.6
Rat chow	7.6 ± 0.5	12.2 ± 1.7[g]	0.24	29 ± 6[g]	0.032	1.0
Trend association (significance)[h]	NSP $p < 0.01$	Interaction $p < 0.001$	Beef/NSP[i] $p < 0.001$	Beef $p < 0.001$	Not tested	Beef $p < 0.001$

[a] Faeces collected in period I (third week; group I) or period II (fourth week; group II), pooled, homogenized and microcapsules extracted magnetically. Results expressed as mean ± SD (n = 3). Animals showed incomplete adaptation to diets in period I.
[b] Subject to variable efficiency in recovery of microcapsules
[c] Normalized on weight/weight basis for radioactivity per milligram of faecal solids and microcapsule total PEI
[d] Specific microcapsule activity divided by faecal mass
[e] By comparison, benzo[*a*]pyrene 3,6-dione was constant (2.1–2.3%) for human diets
[f] Mean of two rats; a third animal excreted 0.6 million microcapsules, which bound 0.31 nmol benzo[*a*]pyrene.
[g] Mean of two rats; a third animal excreted 7.5 million microcapsules binding 3.55 nmol benzo[*a*]pyrene.
[h] Statistical analysis for linear effects of beef, NSP, their ratios or mutual interaction
[i] Beef/NSP ratio, 5:5:5:1.5:0

Table 4. Effects of diets and germ-free conditions on cross-linking of *methyl*-^{14}C-labelled microcapsules passed through Fischer rats and ultrasonicated in 1 M hydrochloric acid[a]

Human diet	Membrane/total[b] (%)	Daily faecal weight (g)
High-NSP, low-beef, low-fat	80.6 ± 12.0	3.5 ± 0.5
High-NSP, low-beef, low-fat (germ-free)	55.9 ± 15.0	4.5 ± 2.1
Low-NSP, high-beef, high-fat	73.7 ± 8.2	2.8 ± 0.5
Low-NSP, low-beef, high-fat	64.7 ± 8.7	2.2 ± 0.2
Before gut transit	29	—

[a]Average of three rats in each group
[b]Radioactivity of membrane and its covalently-attached PEI in relation to total recovered microcapsules

Endogenous cross-linking agents

Transit through the gut greatly diminished release of core PEI (Table 4). Germ-free animals showed substantially less cross-linking than normal animals consuming the same diet, suggesting the involvement of microflora. The extent of cross-linking is related to faecal bulk, and possibly to beef content (Table 4), suggesting that the high-nonstarch polysaccharide, low-fat, low-beef diet may protect against colorectal carcinogenesis by diminishing the amount of reactive cross-linking molecules reaching the gut mucosa.

Discussion

We demonstrated previously that microcapsules can collect administered carcinogens and their metabolites or precursors during transit through the gut. The present results reveal effects within the intestinal cavity that (i) can be attributed to food components known to modulate colorectal cancer risk and (ii) are consistent with faecapentaene cross-linking characteristics (Plummer *et al.*, 1986) and that microflora are the source of these agents. A recent case-control study of colorectal cancer in Melbourne, Australia (Kune *et al.*, 1987) showed a 3.2-fold risk difference between humans consuming high-beef/low-fibre and low-beef/high-fibre diets. The microcapsule trapping range, normalized for faecal bulk (Table 3), suggests that UK diets can alter exposure of mucosa to reactive benzo[*a*]pyrene metabolites to a similar extent and that there might be a very much smaller exposure with high-residue diets. Benzo[*a*]pyrene was used as a model xenobiotic because of the high proportion of faecal excretion of this compound and its DNA damaging action on both rat and human colonic tissues (Autrup *et al.*, 1980). Although we still do not know which DNA damaging agents are responsible for human colorectal carcinogenesis, the present results show there is no justification for continuing to use unrepresentative, non-human diets in investigating xenobiotic metabolism in experimental animals.

Microcapsules appear to offer the first opportunity for sampling *in situ* noninvasively, and we found previously that they are neither damaged nor blocked (Povey *et al.*, 1987c) during gut transit. The present results on cross-linking accord with our earlier data (Povey *et al.*, 1987a), although diet-dependent precipitation of the PEI, resistant to acid treatment, cannot be completely excluded but seems unlikely. Two experiments (unpublished) reveal no adverse effect of microcapsules in animals that could preclude their use in humans. The microcapsules are presently being developed to permit interception and characterization of unknown DNA damaging agents within the gut.

These results show that diet has a considerable effect on DNA damaging agents arising from biological availability; this factor may represent as large a variation as exogenous exposures.

Acknowledgements

We thank Ms E. Cardis for statistical analysis, Mlle F. Elghissassi for technical assistance and Mrs M. Wrisez and Mrs A. Zitouni for preparing the manuscript. This work was supported in part by NIH grant R01-CA 39417-01.

References

Autrup, H.N., Swartz, R.D., Essigman, J.M., Smith, L., Trump, B.F. & Harris, C.C. (1980) Metabolism of aflatoxin B_1, benzo(*a*)pyrene and 1,2-dimethylhydrazine by cultured rat and human colon. *Teratog. Carcinog. Mutagenesis, 1,* 3–13

Kune, S., Kune, G.A. & Watson, L.F. (1987) Case-control study of dietary etiological factors: the Melbourne colorectal cancer study. *Nutr. Cancer, 9,* 21–42

O'Neill, I.K., Castegnaro M., Brouet, I. & Povey A.C. (1987) Magnetic semi-permeable polyethyleneimine microcapsules for monitoring of *N*-nitrosation in the gastrointestinal tract. *Carcinogenesis, 8,* 1469–1474

Plummer, S.M., Grafstrom, R.C., Yang, L.L., Curren, R.D., Linnainmaa, K. & Harris, C.C. (1986) Fecapentaene-12 causes DNA damage and mutations in human cells. *Carcinogenesis, 7,* 1607–1609

Povey, A.C., Bartsch, H., Nixon, J.R. & O'Neill, I.K. (1986) Trapping of chemical carcinogens with magnetic polyethyleneimine microcapsules. I. Microcapsule preparation and *in vitro* reactivity of encapsulated nucleophiles. *J. pharm. Sci., 75,* 831–837

Povey, A.C., Brouet, I., Nixon, J.R. & O'Neill, I.K. (1987a) Trapping of chemical carcinogens with magnetic polyethyleneimine (PEI) microcapsules. III. In-vivo trapping of electrophiles from *N*-nitroso *N*-methylurea and recovery from faeces. *J. pharm. Sci., 76,* 201–207

Povey, A.C., Nixon, J.R. & O'Neill, I.K. (1987b) Trapping of chemical carcinogens with magnetic polyethyleneimine (PEI) microcapsules. II. Effect of membrane and reactant structures. *J. pharm. Sci., 76,* 194–200

Povey, A.C., Bartsch, H. & O'Neill, I.K. (1987c) Magnetic polyethyleneimine (PEI) microcapsules as retrievable traps for carcinogen electrophiles formed in the gastrointestinal tract. *Cancer Lett., 36,* 45–53

Povey, A.C., Brouet, I., Bartsch, H. & O'Neill, I.K. (1987d) Binding of benzo[*a*]pyrene (BaP) metabolites in the rat intestinal lumen by magnetic polyethyleneimine (PEI) microcapsules following an intragastric dose of [^{14}C]-benzo[*a*]pyrene. *Carcinogenesis, 8,* 825–831

A MODEL SYSTEM FOR STUDYING COVALENT BINDING OF FOOD CARCINOGENS MeIQx, MeIQ AND IQ TO DNA AND PROTEIN

H. Wallin & J. Alexander

Department of Toxicology, National Institute of Public Health, Oslo, Norway

We have studied the covalent binding of carcinogenic aminoimidazoazaarene compounds to macromolecules in a microsomal model system. The ^{14}C-labelled compounds were incubated with rat-liver microsomes, and binding to macromolecules was measured after their precipitation on glass filters, which were washed several times in organic solvents. The amount of radioactivity was determined by liquid scintillation counting. Covalent binding was dependent on the addition of NADPH, with an optimal concentration of about 1 mM. The binding appeared to follow saturation kinetics when carcinogen concentrations were lower than 200 µM, with K_m values of less than 20 µM. At 50 µM, 2-amino-3,4-dimethylimidazo[4,5-*f*]-quinoline (MeIQ) and 2-amino-3-methylimidazo[4,5-*f*]quinoline (IQ) bound more effectively than 2-amino-3,8-dimethylimidazo[4,5-*f*]quinoxaline (MeIQx). When DNA was included in the incubations, binding to this macromolecule was ten-fold less per milligram than binding to proteins. In comparison with microsomes from untreated animals, those from rats treated with β-naphthoflavone caused up to nine-fold more binding of MeIQx, six-fold more of IQ and three times as much of MeIQ. Induction by Aroclor 1254 caused up to 17-fold more binding, whereas induction by phenobarbital caused up to three-fold more binding. The effects of the inducers were greatest for MeIQx and IQ, while smaller effects were seen for MeIQ. The results are most consistent with cytochrome P450-dependent metabolic activation of the carcinogens to hydroxylamine metabolites, for which an isoenzyme(s) inducible by polyaromatic and polychlorinated hydrocarbons is most effective. To our knowledge, this is the first report that MeIQx is metabolized to reactive species capable of covalent binding to macromolecules.

Over a dozen heterocyclic amines isolated from cooked food may cause cancer in humans (Felton *et al.*, 1986; Sugimura *et al.*, 1986). We are developing methods for measuring macromolecular adducts to these carcinogens in human tissues, but, since little is known about their chemical structures, we developed an in-vitro model system in which to study the covalent binding of IQ, MeIQ and MeIQx to proteins and DNA.

In the model system, 1 mg rat liver microsomal protein/ml was incubated with 50 μM ^{14}C-labelled substrates synthesized according to Adolfson and Olsson (1983): IQ, 0.971 Ci/mol; MeIQx, 2.51 Ci/mol; MeIQ, 1.19 Ci/mol), 1 mM NADPH, 5 mM $MgCl_2$, 50 mM Tris-HCl, pH 7.4 at 37°C. After 10 min, 50-μl samples were transferred to 2.5-cm Whatman GF/C glass filters, which were then immersed in 95% ethanol. The filters were washed several times with organic solvents and counted for radioactivity, as described previously (Wallin *et al.*, 1981).

Covalent binding increased when microsomes from rats treated with well known

Table 1. Covalent binding of MeIQx, IQ and MeIQ to liver microsomal proteins: effect of enzyme induction and substrate concentration

Source of microsomes	Substrate concentration (μM)	Covalent binding (nmol/mg)		
		MeIQx	IQ	MeIQ
Control	50	0.14 ± 0.33	0.32 ± 0.08	0.42 ± 0.05
β-Naphthoflavone	4	0.59 ± 0.02	0.82 ± 0.04	0.79 ± 0.04
β-Naphthoflavone	10	0.91 ± 0.06	1.10 ± 0.10	1.06 ± 0.10
β-Naphthoflavone	30	1.18 ± 0.18	1.55 ± 0.06	1.22 ± 0.04
β-Naphthoflavone	50	1.28 ± 0.04	1.78 ± 0.10	1.41 ± 0.11
β-Naphthoflavone	100	1.40 ± 0.37	1.86 ± 0.05	1.80 ± 0.12
Phenobarbital	50	0.27 ± 0.02	0.84 ± 0.10	0.38 ± 0.25
Aroclor 1254	50	2.38 ± 0.39	4.66 ± 0.18	1.68 ± 0.03

Male Wistar rats weighing 150 g were treated by single intraperitoneal injection of 75 mg Aroclor 1254, four daily intraperitoneal injections of 7.5 mg β-naphthoflavone or 0.1% phenobarbital in drinking-water. Animals were starved 24 h before being killed on day 5. The procedure for preparation of microsomes is described elsewhere (Morgenstern *et al.*, 1987). Background binding without NADPH was subtracted from the values [mean ($n = 3$) ± SD].

inducers (Conney, 1986) were used (Table 1). The induction pattern was similar for all three substances: Aroclor 1254 was more effective as an inducer than β-naphthoflavone, while phenobarbital treatment caused little or no increase in covalent binding compared to microsomes from untreated rats. The effect of the inducers was greater for MeIQx and IQ than for MeIQ (up to 17-, 15- and four-fold, respectively). The radioactivity incorporated initially was too low to obtain accurate determinations of initial velocities. However, a double reciprocal plot of 10-min incubations indicated that the K_m values for all three substances were less than 20 μM. At substrate concentrations greater than 200 μM, binding did not follow simple saturation kinetics (data not shown), owing perhaps to the participation of other mechanisms for carcinogen activation than that characterized at lower concentrations.

The time-dependent covalent binding of MeIQx to microsomal protein is shown in Table 2. The binding was nonlinear with time but increased over 60 min, at which time about 4% of the radiolabel had bound. The covalent binding was dependent on NADPH, with a near to optimal concentration at 1 mM. Binding was not proportional to the amount of microsomal protein included in the incubation; thus, there was less specific binding, expressed as binding per milligram of protein, with higher amounts of protein.

The effect of inducers and the dependence on NADPH points to a cytochrome P450-mediated metabolic activation of the carcinogens. Particularly effective were enzymes induced by Aroclor 1254 and β-naphthoflavone. These results corroborate the findings of Kato (1986), who showed that a high-spin cytochrome P448 enzyme induced by polychlorinated biphenyls was highly effective in transforming IQ to the proximate mutagen *N*-hydroxy-IQ.

Binding to nucleic acids

Covalent binding to nucleic acids was studied in microsomal incubations scaled up to 1 ml. Nucleic acids were prepared as described elsewhere (Wallin *et al.*, 1987),

Table 2. Effects of different experimental parameters on the covalent binding of MeIQx to liver microsomal proteins from rats treated with β-naphthoflavone

Parameter	Covalent binding (nmol/mg protein)
Time (min)	
0	0
5	0.67 ± 0.03
10	1.06 ± 0.04
20	1.40 ± 0.03
30	1.52 ± 0.12
60	2.12 ± 0.21
Concentration of NADPH (mM)	
0.2	0.64 ± 0.06
0.6	1.38 ± 0.01
1.0	1.58 ± 0.09
1.5	1.72 ± 0.04
Concentration of microsomal protein (mg/ml)	
0.5	1.99 ± 0.14
1.0	1.34 ± 0.07
2.0	0.79 ± 0.12

Unless indicated differently, 1 mg/ml microsomal protein was incubated with 50 μM MeIQx and 1 mM NADPH in Tris buffer at 37°C for 10 min. Background binding without NADPH was subtracted from the values [mean (n = 3) ± SD]

except that RNase treatment was excluded when RNA was prepared. Aliquots of 70 μg RNA were recovered from incubations containing microsomes from Aroclor 1254-treated animals and MeIQx. This RNA contained about 300 pmol MeIQx/mg RNA. Assuming that the ratio between the amounts of RNA and protein is one to seven in microsomes, about 2% of the adducts measured by the filter method as protein adducts are RNA adducts. Covalent binding was also measured after including 1 mg/ml calf thymus DNA in the incubation medium. When DNA was included, total macromolecular binding was reduced by approximately one-half, perhaps because the carcinogens became intercalated in the DNA and there were thus lower concentrations of carcinogen available for metabolic activation (Watanabe *et al.*, 1982). When microsomes from β-naphthoflavone-treated rats were used, covalent binding of 50-μM concentrations of the carcinogens were 100 ± 24 pmol MeIQx/mg DNA (mean of four experiments ± SD), 180/170 pmol MeIQ/mg DNA and 264/260 pmol IQ/mg DNA (results from two independent experiments). Using microsomes from Aroclor 1254-treated rats, DNA bound 258 ± 16 pmol MeIQx/mg DNA (mean of three experiments ± SD), 226/205 pmol MeIQ/mg DNA and 330/329 pmol IQ/mg DNA (results from two independent experiments). Covalent binding to DNA using microsomes from untreated or phenobarbital-treated rats was less than 50 pmol/mg DNA. The metabolites bound firmly to DNA, as no radiolabelled compound could be extracted from DNA by further washes with organic solvents. Furthermore, the radiolabel comigrated with DNA during gel filtration chromatography and was retained on Amicon PM-10 filters after extensive washing with buffer. The ratios between covalent binding to proteins (see Table 1) and DNA were about ten for all three carcinogens in all instances, and there was no significant change in the ratios when different preparations of microsomes were used.

Conclusions

Recent research interest has been focused on a number of highly potent bacterial mutagens that are formed upon heating protein-rich food. Aminoimidazoazaarene compounds are found in particularly large amounts in food consumed by humans (Felton *et al.*, 1986; Sugimura *et al.*, 1986). Although these compounds have proved to be less genotoxic to mammalian cells than to *Salmonella typhimurium* strains, the compounds tested so far are capable of causing cancer in several tissues of rodents (Sugimura *et al.*, 1986; Wild *et al.*, 1986; Holme *et al.*, 1987a). We have initiated studies on the metabolism, genotoxic effects and occurrence of these compounds, for investigations including biological effects, human exposure and measurement of damage in human macromolecules (Brunborg *et al.*, 1988; Holme *et al.*, 1987a,b,c, 1988; Størmer *et al.*, 1987; Becher *et al.*, 1988).

Using the model system presented here, it was possible to study the metabolic activation of aminoimidazoazaarene compounds to reactive species which bind to proteins and nucleic acids, and it should be possible to generate adducts in amounts suitable for structural determinations. It should also be useful for producing standard macromolecules containing well-defined amounts of radiolabelled adducts, which can be used for the development of methods for measuring nonradioactive adducts. Then, it will be possible to include other macromolecules, such as haemoglobin and serum albumin, in the incubation. However, it should be noted that deviation from the conditions described here, such as using carcinogen concentrations greater than 200 μmol, may lead to other mechanisms of activation of the carcinogens and therefore other chemical identities of the adducts.

The reactive species formed during metabolic activation of IQ is probably derived from a hydroxylamine metabolite (Okamoto *et al.*, 1981). Since the kinetic features of the metabolic activation, and the reactivity of the binding species towards DNA and protein, were remarkably similar for all three carcinogens, MeIQ and MeIQx are probably activated by the same mechanism as IQ.

In the model system, covalent binding to DNA was greater than that to proteins, indicating that DNA from blood cells may be a suitable target molecule for analysing adducts in human materials.

Acknowledgement

This study was supported in part by a grant from the Royal Norwegian Council for Scientific and Industrial Research. Microsomes were kindly prepared by A. Mikalsen.

References

Adolfsson, L. & Olsson, K. (1983) A convenient synthesis of mutagenic ^{3}H-imidazo[4,5-f]quinoline-2-amines and their 2-^{14}C-labelled analogues. *Acta chem. scand. Ser. B, 37,* 157–159

Becher, G., Knize, M.G., Nes, I.F. & Felton, J.S. (1988) Isolation and identification of mutagens from a fried Norwegian meat product. *Carcinogenesis, 9,* 247–253

Brunborg, G., Holme, J.A., Alexander, J. & Becher, G. (1988) Genotoxic effects of MeIQ and IQ and their N-acetylated metabolites. *Mutagenesis* (in press)

Conney, A.H. (1986) Induction of microsomal cytochrome P-450 enzymes: the first Bernard B. Brodie lecture at Pennsylvania State University. *Life Sci., 39,* 2493–2518

Felton, J.S., Knize, M.G., Shen, N.H., Anderson, B.D., Bjeldanes, L.F. & Hatch, F.T. (1986) Identification of mutagens in cooked beef. *Environ. Health Perspectives, 67,* 17–24

Holme, J.A., Hongslo, J., Søderlund, E., Brunborg, G., Christensen, T., Alexander, J. & Dybing, E. (1987a) Comparative genotoxic effects of IQ and MeIQ in Salmonella typhimurium and cultured mammalian cells. *Mutat. Res.*, *187*, 181–190

Holme, J.A., Alexander, J. & Dybing, E. (1987b) Mutagenic activation of IQ and MeIQ by 9000 g-fraction and cells isolated from small intestine, kidney and liver. *Cell. Biol. Toxicol.*, *3*, 51–61

Holme, J.A., Brunborg, G., Alexander, J., Trygg, B. & Bjørnstad, C. (1987c) Modulation of the mutagenic effects of IQ and MeIQ in bacteria with rat liver 9000 g supernatant or monolayers of rat hepatocytes as an activation system. *Mutat. Res.*, *197*, 39–49

Holme, J.A., Alexander, J. & Becher, G. (1988) Metabolism of IQ and MeIQ in suspensions of isolated rat hepatocytes. *Toxicol. In Vitro*, *1*, 175–181

Kato, R. (1986) Metabolic activation of mutagenic heterocyclic aromatic amines from protein pyrolysates. *CRC crit. Rev. Toxicol.*, *16*, 307–349

Morgenstern, R., Wallin, H. & De Pierre, J.W. (1987) *Mechanisms of activation of the microsomal glutathione transferase*. In: Mantle, T.J., Pickett, C.B. & Hayes, J.D. eds, *Glutathione S-Transferases and Carcinogenesis*, London, Taylor & Francis

Okamoto, T., Shudo, K., Hashimoto, Y., Kosuge, T., Sugimura, T. & Nishimura, S. (1981) Identification of a reactive metabolite of the mutagen 2-amino-3-methylimidazolo[4,5-f]quinoline. *Chem. Pharm. Bull.*, *29*, 590–593

Størmer, F.C., Alexander, J. & Becher, G. (1987) Formation and fluorometric detection of N-acetylated metabolites of IQ and MeIQ in the rat. *Carcinogenesis*, *8*, 1277–1280

Sugimura, T., Sato, S., Ohgaki, H., Takayama, S., Nagao, M. & Wakabayashi, K. (1986) *Mutagens and carcinogens in cooked food*. In: Knudsen, I., ed., *Genetic Toxicology of the Diet*, New York, Alan R. Liss, pp. 85–107

Wallin, H., Schelin, C., Tunek, A. & Jergil, B. (1981) A rapid and sensitive method for determination of covalent binding of benzo[*a*]pyrene to proteins. *Chem.-biol. Interactions*, *38*, 109–118

Wallin, H., Jeffrey, A.M. & Santella, R.M. (1987) Investigation of benzo(*a*)pyrene globin adducts. *Cancer Lett.*, *35*, 139–146

Watanabe, T., Yokoyama, S., Hayashi, K., Kasai, H., Mishimura, S. & Miyazawa, T. (1982) DNA-binding of IQ, MeIQ and MeIQx, strong mutagens found in broiled foods. *FEBS Lett.*, *150*, 434–438

Wild, D., Kaiser, G., King, M.-T. & Harnasch, D. (1986) *Genetic activity of IQ* (2-*amino*-3-*methylimidazo*[4,5-*f*]*quinoline*) *and structural analogs*. In: Knudsen, I., ed., *Genetic Toxicology of the Diet*, New York, Alan R. Liss, pp. 145–154

TOBACCO USE

APPROACHES TO THE DEVELOPMENT OF ASSAYS FOR INTERACTION OF TOBACCO-SPECIFIC NITROSAMINES WITH HAEMOGLOBIN AND DNA

S.S. Hecht, S.G. Carmella, N. Trushin, T.E. Spratt, P.G. Foiles & D. Hoffmann

American Health Foundation, Valhalla, NY, USA

The tobacco-specific, nicotine-derived nitrosamines 4-(*N*-nitrosomethylamino)-1-(3-pyridyl)-1-butanone (NNK) and *N'*-nitrosonornicotine (NNN) are among the most important carcinogens in tobacco and tobacco smoke. Treatment of Fischer 344 rats with these carcinogens resulted in alkylation of haemoglobin and DNA by the 4-(3-pyridyl)-4-oxobutyl group formed during their metabolism. This alkyl group can be detached from globin or DNA under mild hydrolytic conditions as 4-hydroxy-1-(3-pyridyl)-1-butanone, which appears to be a potentially useful dosimeter for human exposure to, and activation of, tobacco-specific nitrosamines.

Among the many carcinogens present in tobacco smoke, we have focused our recent efforts on the tobacco-specific nitrosamines NNK and NNN (Fig. 1). There are several reasons why we believe that NNK and NNN are particularly relevant to cancer induction by tobacco products. First, they are found in mainstream and sidestream tobacco smoke and in unburnt tobacco in quantities sufficiently high that the estimated exposure of a long-term smoker or snuff-dipper would be similar to the total doses of these compounds required to produce tumours in experimental animals (Hoffmann & Hecht, 1985). Second, they are strong carcinogens in mice, rats and hamsters; the high tumorigenicity of NNK in the lung, and the ability of NNK and NNN to induce local tumours in the oral cavity of rats are of particular relevance to tobacco-induced cancer in man (Hoffmann & Hecht, 1985; Hecht *et al.*, 1986a,b). Third, they are formed from nicotine and thus represent a specific link to tobacco and other nicotine-containing products, in contrast to other tobacco-smoke carcinogens, such as benzo[*a*]pyrene, which are ubiquitous in the environment. This point is particularly important in considering the development of dosimeters for exposure to carcinogens specifically from tobacco smoke.

In extensive analytical studies, the levels of NNK and NNN in tobacco products have been documented thoroughly. Using these data, one can estimate individual exposure; however, there are many uncertainties in such estimates which could be misleading. Investigations of nicotine uptake and metabolism in smokers have clearly shown that inhalation patterns can affect dose (IARC, 1986). Exposure of smokers and nonsmokers to environmental tobacco smoke is also difficult to quantify. In addition, the dose of nitrosamines would be affected by their endogenous formation in smokers, which has been clearly demonstrated using *N*-nitrosoproline as a marker (Hoffmann & Brunnemann, 1983; Bartsch & Montesano, 1984; Ladd *et al.*, 1984). Finally, metabolic studies have shown that human tissues vary widely in their capacity to metabolize NNK and NNN (Castonguay *et al.*, 1983). This is important because their metabolic

Fig. 1. Intermediates and products involved in the binding of NNK and NNN to DNA and globin

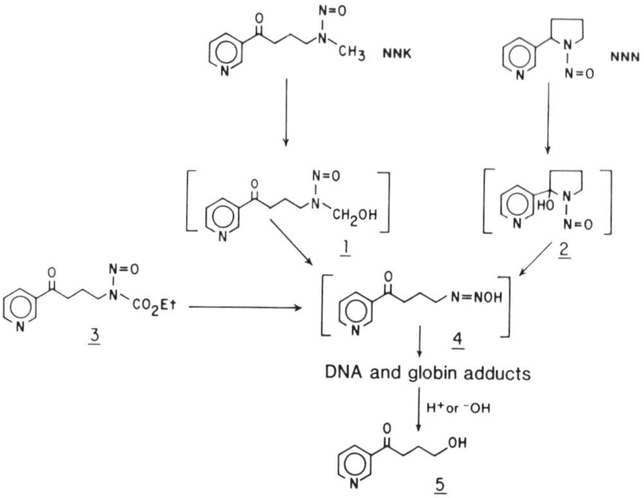

pathways are linked to their abilities to initiate carcinogenesis. Thus, our interest is to develop a practical dosimeter for the determination of the internal dose of biologically active alkylating agents formed from NNK and NNN. For this purpose, we have focused on the interactions of NNK and NNN with haemoglobin and DNA.

Interactions with haemoglobin

Figure 2 illustrates the total binding of tritium to globin 24 h after intraperitoneal injection of Fischer 344 rats with various doses of [5-^{3}H]NNK, which has the label in the pyridine ring (Carmella & Hecht, 1987). Binding was linear over a 100-fold range. Binding of tritium to globin was also observed after treatment with [C^{3}H$_3$]NNK, but the dose-response characteristics have not yet been examined in detail.

Figure 3 summarizes the binding of [5-^{3}H]NNK to globin during chronic dosing and the disappearance of tritium after cessation of treatment. The characteristics of this curve are in line with those predicted for binding to rat haemoglobin, on the basis of the lifetime of the erythrocyte and of the half-life of NNK-globin adducts, as discussed below (Ehrenberg & Osterman-Golkar, 1976).

When the globin from rats treated with [5-^{3}H]NNK was incubated with dilute sodium hydroxide at room temperature, about 15% of the radioactivity was released and could be extracted into chloroform. When the resulting chloroform extracts were analysed by high-performance liquid chromatography (HPLC), chromatograms such as those illustrated in Figure 4 were obtained. Most of the radioactivity released coeluted with 4-hydroxy-1-(3-pyridyl)-1-butanone (structure 5, Fig. 1); the structure of the released material was confirmed by gas chromatography-mass spectrometry. Although this compound accounted for only 15% of the radioactivity bound to globin, we have focused our attention on it because it may be suitable as a dosimeter for human studies.

The persistence of total tritium, and of the precursor to structure 5, was studied in Fischer 344 rats treated with a single intraperitoneal dose of [5-^{3}H]NNK, as illustrated

Fig. 2. Presence of tritium in globin 24 h after treatment of Fischer 344 rats with various doses of [5-^{3}H]NNK

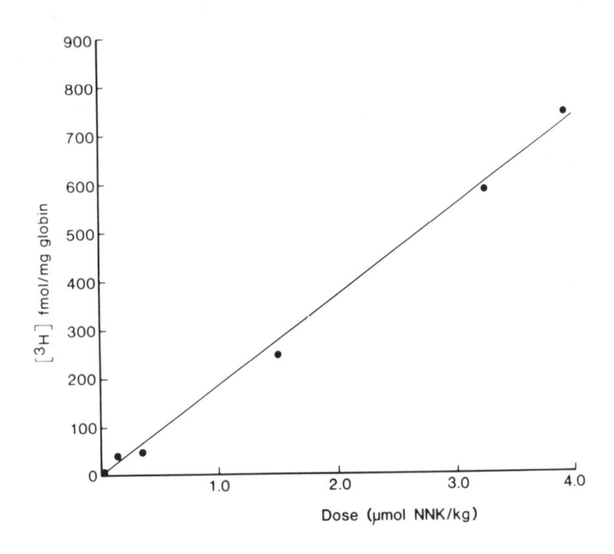

in Figure 5. The half-lives of total adducts and of the precursor adduct to structure 5 were 12 days and 9.1 days, respectively, indicating that these adducts are somewhat unstable since they disappear more rapidly than the turnover of erythrocytes, which have a lifetime of 60 days in rats.

As illustrated in Figure 1, 2′-hydroxylation of NNN, a known metabolic process, is expected to give the same intermediate (structure 4) as produced from methyl

Fig. 3. Presence of tritium in globin during chronic treatment of Fischer 344 rats with [5-^{3}H]NNK

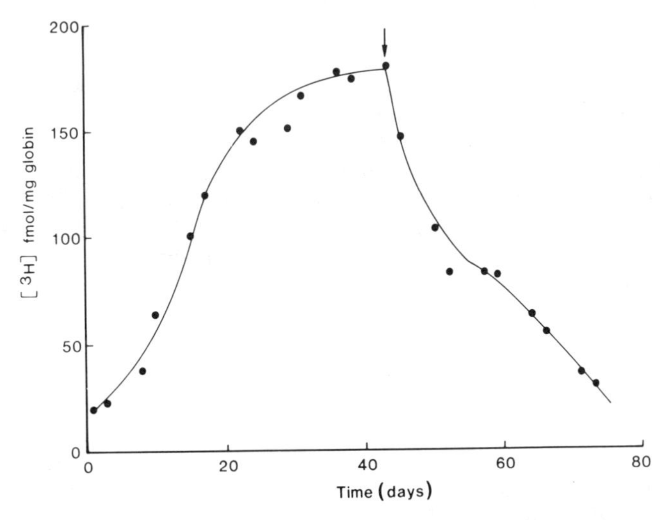

The arrow represents termination of [5-^{3}H]NNK treatment.

Fig. 4. Chromatograms obtained upon HPLC analysis of (A) standard 4-hydroxy-1-(3-pyridyl)-1-butanone (structure 5, Fig. 1), and chloroform extracts of base-treated globin 24 h (B) and 2 weeks (C) after intraperitoneal injection of Fischer 344 rats with [5-^{3}H]NNK

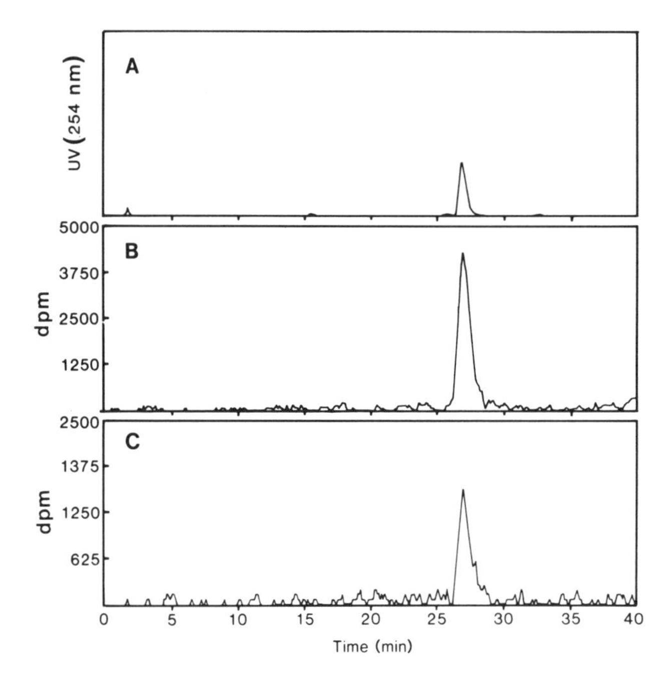

hydroxylation of NNK. This expectation was confirmed by analysis of globin from rats treated with [5-^{3}H]NNN, in which the tritium is in the pyridine ring. The presence of structure 5 in the chloroform extracts of base-treated globin was confirmed by HPLC and gas chromatography-mass spectrometry, as for NNK. The persistence of total tritium and precursor to structure 5 is illustrated in Figure 6 for rats treated with a single dose of [5-^{3}H]NNN.

These results are consistent with the mechanism summarized in Figure 1 and provide evidence for 4-(3-pyridyl)-4-oxobutylation of haemoglobin by NNK and NNN. This alkylation mechanism had been predicted on the basis of our metabolic studies of NNK and NNN and of the chemistry of the model compound, structure 3 (Chen *et al.*, 1978; Hecht & Chen, 1979; Hecht *et al.*, 1980). Although the structure of the globin adduct precursor to structure 5 is not known, it appears likely, because of its release by mild base hydrolysis, that the adduct is an aspartate or glutamate ester. Studies are in progress to obtain further structural information on the globin adducts of NNK and NNN.

Interactions with DNA

In earlier studies, we showed that NNK is a methylating agent and that 7-methylguanine and O^6-methylguanine can be readily detected in the target tissues for NNK tumorigenicity in Fischer 344 rats: liver, lung and nasal mucosa (Castonguay *et al.*, 1984; Hecht *et al.*, 1986a). Experiments by Belinsky and co-workers (1986) have also shown that O^4-methylthymidine is produced *in vivo* from NNK and that

Fig. 5. Levels of total tritium in globin (○) and of structure 5 released upon base treatment of globin (●) at intervals after injection of Fischer 344 rats with [5-^{3}H]NNK

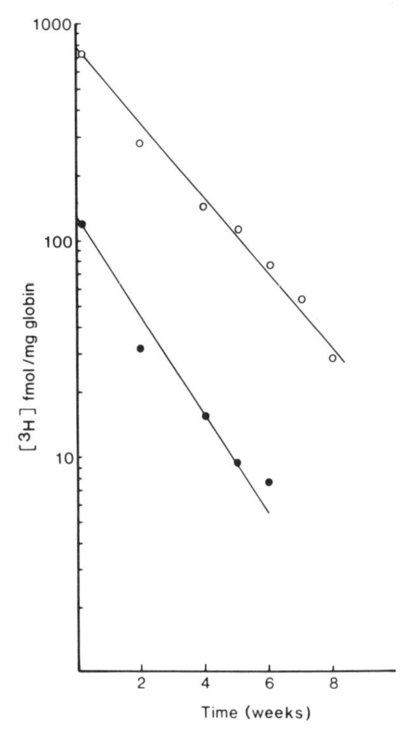

The 24-h point corresponds to approximately 0.1% of dose for total tritium and 0.02% for structure 5.

O^6-methylguanine persists in the lungs of rats treated chronically with NNK. Methylation of DNA by NNK results at least partially from metabolic hydroxylation of the methylene group α to the nitroso functionality, with generation of the methylating agent, methyl diazohydroxide. As illustrated in Figure 1, hydroxylation of the methyl group yields structure 4, which should cause 4-(3-pyridyl)-4-oxobutylation of DNA by analogy to its interaction with globin. Whereas methylation of DNA and globin could be caused by a number of environmental or endogenous agents, 4-(3-pyridyl)-4-oxobutylation should result only from nicotine-related compounds. For dosimetry studies, the latter adducts will probably be more useful than the former in specifically delineating exposure to tobacco-related carcinogens.

In order to investigate 4-(3-pyridyl)-4-oxobutylation of DNA by NNK, Fischer 344 rats were each given a subcutaneous injection of 2 mCi (2 μmol) [5-^{3}H]NNK and sacrificed 24 h later. DNA was isolated from liver, kidney, oesophagus, lung and nasal mucosa, hydrolysed by treatment with 0.8 N HCl at 80°C for 6 h and analysed by HPLC; typical results for liver DNA are illustrated in Figure 7. The peak at 50 min coeluted with structure 5 of Figure 1. Further evidence for its identity was obtained by rechromatography of the collected peak by normal-phase HPLC; the retention time was identical to that of a standard. When the collected peak was treated with sodium

Fig. 6. Levels of total tritium in globin (○) and of structure 5 released upon base treatment of globin (●) at intervals after injection of Fischer 344 rats with [5-^{3}H]NNN

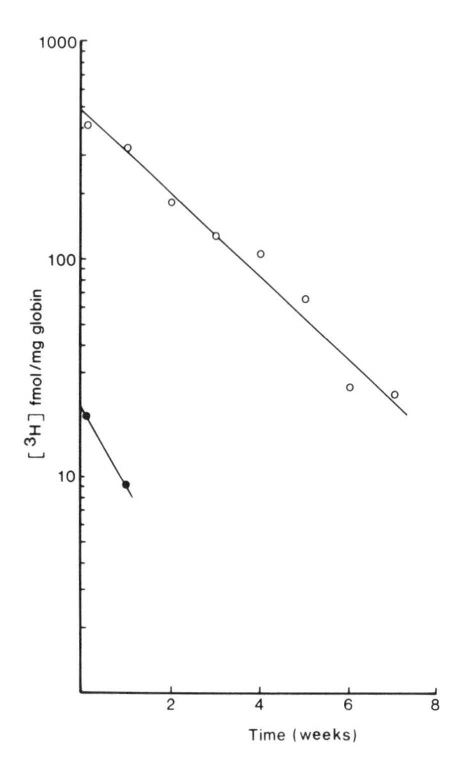

Fig. 7. Chromatogram obtained upon HPLC analysis of (A) standard 4-hydroxy-1-(3-pyridyl)-1-butanone (structure 5, Fig. 1) and (B) acid hydrolysate of hepatic DNA isolated from the liver of a Fischer 344 rat treated with [5-^{3}H]NNK

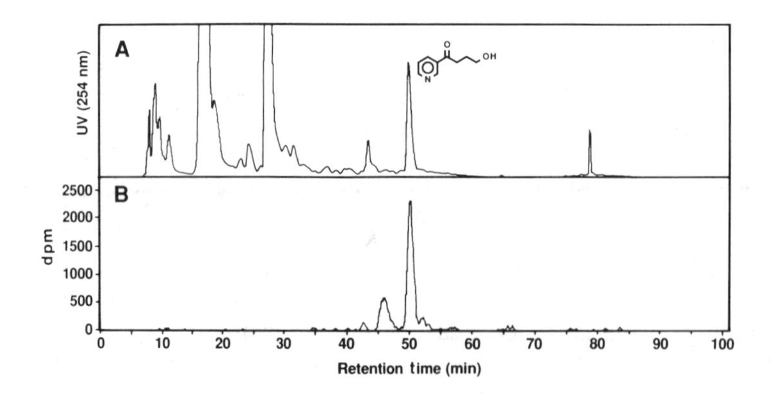

borohydride, material that was chromatographically indistinguishable from 4-hydroxy-1-(3-pyridyl)-1-butanol was formed. These results demonstrate that the peak eluting at 50 min in Figure 7 is 4-hydroxy-1-(3-pyridyl)-1-butanone (structure 5). Its extent of formation in liver DNA was approximately 0.67 ± 0.1 pmol/mg DNA ($n = 3$). Smaller amounts were detected in lung DNA, and no significant radioactivity above background was found in DNA isolated from the other tissues. 4-Hydroxy-1-(3-pyridyl)-1-butanone was also released from hepatic DNA upon hydrolysis at 100°C for 35 min in 10 mM sodium cacodylate buffer, pH 7. The amount detected was similar to that found upon acid hydrolysis. The amount of structure 5 released upon acid hydrolysis of the DNA was approximately 50% of the total radioactivity present in DNA. An unidentified peak that eluted just prior to structure 5 accounted for approximately 20% of the radioactivity. These results demonstrate that 4-(3-pyridyl)-4-oxobutylation of DNA by NNK does occur *in vivo* in rats.

On the basis of the pathways illustrated in Figure 1, 4-(3-pyridyl)-4-oxobutylation of DNA should also occur upon administration of NNN. Accordingly, three male Fischer 344 rats were each treated with 1.7 mCi (0.35 μmol) [5-^{3}H]NNN by subcutaneous injection and sacrificed 24 h later. DNA was isolated and analysed as described above. The results were similar to those obtained with NNK, in that 4-hydroxy-1-(3-pyridyl)-1-butanone was released from hepatic DNA upon either acid or neutral thermal hydrolysis. The amount released upon acid hydrolysis was 0.08 ± 0.01 pmol/mg DNA. These results are consistent with those obtained in experiments on alkylation of globin by NNK and NNN.

Approaches to dosimetry

The hydrolytic release of 4-hydroxy-1-(3-pyridyl)-1-butanone (structure 5) from globin or DNA would appear to provide a potentially valuable approach to assessing adduct formation from NNK or NNN in individuals exposed to tobacco products or other nicotine-containing products. The carbonyl or hydroxy group of structure 5 could be derivatized to facilitate its detection. We are exploring derivatization to the corresponding pentafluorobenzoate, which should be readily detected by gas chromatography coupled with negative-ion chemical ionization mass spectrometry. Experiments to date indicate that the detection limit is approximately 10 fmol when specific-ion monitoring is used. The sensitivity of this detection method would appear to be potentially adequate for analysis of structure 5 in human globin, if sufficient preliminary purification of the sample can be accomplished.

On the basis of the results obtained after treating rats with tritium-labelled NNK or NNN, it would appear that the measurement of globin adducts will provide a more practical approach to human dosimetry than the measurement of DNA adducts. Although the levels of 4-(3-pyridyl)-4-oxobutylation of globin and DNA are similar when expressed on a per milligram basis, larger amounts of globin than of DNA are readily obtained. DNA from potential target tissues for tobacco-induced cancers is rarely available from humans. It will be necessary to carry out studies in experimental animals to relate the levels of globin adducts to those of DNA adducts under various chronic dosing conditions. The data from such studies will provide a base for interpreting the human data.

Acknowledgements

This study was supported by Grants 29580, 32391 and 44377 from the US National Cancer Institute.

References

Bartsch, H. & Montesano, R. (1984) Relevance of nitrosamines to human cancer. *Carcinogenesis, 6*, 1381–1393

Belinsky, S.A., White, C.M., Boucheron, J.A., Richardson, F.C., Swenberg, J.A. & Anderson, M. (1986) Accumulation and persistence of DNA adducts in respiratory tissue of rats following multiple administrations of the tobacco specific carcinogen 4-(N-methyl-N-nitrosamino)-1-(3-pyridyl)-1-butanone. *Cancer Res., 46*, 1280–1284

Carmella, S.G. & Hecht, S.S. (1987) Formation of hemoglobin adducts upon treatment of F344 rats with the tobacco-specific nitrosamines 4-(methylnitrosamino)-1-(3-pyridyl)-1-butanone and N′-nitrosonornicotine. *Cancer Res., 47*, 2626–2630

Castonguay, A., Stoner, G.D., Schut, H.A.J. & Hecht, S.S. (1983) Metabolism of tobacco-specific N-nitrosamines by cultured human tissues. *Proc. natl Acad. Sci. USA, 80,* 6694–6697

Castonguay, A., Tharp, R. & Hecht, S.S. (1984) *Kinetics of DNA methylation by the tobacco-specific carcinogen* 4-(*methylnitrosamino*)-1-(3-*pyridyl*)-1-*butanone in the F*344 *rat.* In: O'Neill, I.K., von Borstel, R.C., Miller, C.T., Long, J. & Bartsch, H., eds, N-*Nitroso Compounds*: *Occurrence, Biological Effects and Relevance to Human Cancer* (*IARC Scientific Publications No.* 57), Lyon, International Agency for Research on Cancer, pp. 787–796

Chen, C.B., Hecht, S.S. & Hoffmann, D. (1978) Metabolic α-hydroxylation of the tobacco specific carcinogen N′-nitrosonornicotine. *Cancer Res., 38,* 3639–3645

Ehrenberg, L. & Osterman-Golkar, S. (1976) Alkylation of macromolecules for detecting mutagenic agents. *Teratog. Carcinog. Mutag., 1,* 105–127

Hecht, S.S. & Chen, C.B. (1979) Hydrolysis of model compounds for α-hydroxylation of the carcinogens, N-nitrosopyrrolidine and N′-nitrosonornicotine. *J. org. Chem., 44,* 1563–1566

Hecht, S.S., Young, R. & Chen, C.B. (1980) Metabolism in the F344 rat of 4-(N-methyl-N-nitrosamino)-1-(3-pyridyl)-1-butanone, a tobacco specific carcinogen. *Cancer Res., 40,* 4144–4150

Hecht, S.S., Trushin, N., Castonguay, A. & Rivenson, A. (1986a) Comparative carcinogenicity and DNA methylation in F344 rats by 4-(methylnitrosamino)-1-(3-pyridyl)-1-butanone and N-nitrosodimethylamine. *Cancer Res., 46,* 498–502

Hecht, S.S., Rivenson, A., Braley, J., DiBello, J., Adams, J.D. & Hoffmann, D. (1986b) Induction of oral cavity tumors in F344 rats by tobacco-specific nitrosamines and snuff. *Cancer Res., 46,* 4162–4166

Hoffmann, D. & Brunnemann, K. (1983) Endogenous formation of N-nitrosoproline in cigarette smokers. *Cancer Res., 43,* 5570–5574

Hoffmann, D. and Hecht, S.S. (1985) Nicotine-derived N-nitrosamines and tobacco related cancer: current status and future directions. *Cancer Res., 45,* 935–944

IARC (1986) *IARC Monographs on the Evaluation of the Carcinogenic Risk of Chemicals to Humans,* Vol. 38, Tobacco Smoking, Lyon, pp. 163–179

Ladd, K. F., Newmark, H.L. & Archer, M.C. (1984) N-Nitrosation of proline in smokers and nonsmokers. *J. natl Cancer Inst., 73,* 83–87

ASSESSMENT OF PASSIVE AND TRANSPLACENTAL EXPOSURE TO TOBACCO SMOKE

M. Sorsa & K. Husgafvel-Pursiainen

Institute of Occupational Health, Helsinki, Finland

Although tobacco smoke has been shown to be highly genotoxic in various experimental systems, most nonmolecular methods designed to assess exposure to mutagens are too insensitive to detect passive exposure to tobacco smoke. Biochemical markers of intake — cotinine and thiocyanates in body fluids — were shown to be elevated after occupational, passive or transplacental exposure to tobacco smoke, while no response was seen in the frequency of sister chromatid exchanges (SCE) in cultured blood lymphocytes. After occupational exposure to environmental tobacco smoke, the intake marker levels are generally less than 5% of the levels found in active smokers, while cord blood levels (representing fetal exposure) are at about the same level as in the mothers at the time of delivery.

Involuntary exposure to an established human carcinogen, tobacco smoke (IARC, 1986), may occur in indoor spaces or transplacentally due to maternal smoking. In the first case, the agent is environmental tobacco smoke, which is composed of aged, atmospherically transformed sidestream smoke and exhaled mainstream smoke; in the latter case, the fetus is exposed by active or passive maternal exposure. All forms of tobacco smoke — mainstream smoke, sidestream smoke and environmental tobacco smoke — have been shown convincingly to be genotoxic in various short-term assay systems (IARC, 1986). Therefore, the use of biomonitoring methods designed to detect exposure to genotoxic agents is justified. We report here the results of biomonitoring two different situations involving exposure to tobacco smoke, using both specific biochemical markers and nonspecific genotoxicity markers.

Occupational exposure to tobacco smoke

Nonsmoking personnel working in indoor restaurants without restrictions on smoking by the public represent one of the groups probably most heavily exposed to environmental tobacco smoke at work. Environmental monitoring data, including analysis of polyaromatic compounds, total particulate matter and genotoxic activity of particulate samples from typical night restaurants, were published previously (Husgafvel-Pursiainen *et al.*, 1986), and details of the results of biological monitoring of smoking and nonsmoking waiters and controls are given in a separate report (Husgafvel-Pursiainen *et al.*, 1987).

The group results of biological monitoring (Table 1) clearly show that passive exposure can be detected only by single compound intake markers — cotinine in plasma and thiocyanate in plasma — and the indicators of genotoxic exposure — urinary mutagenicity (as measured by the indicator strain *Salmonella typhimurium* TA98 with exogenous metabolic activation) and SCE frequency — do not significantly respond to passive smoking.

Table 1. Assessment of active and passive exposure of restaurant personnel to tobacco smoke using different exposure parameters

Group	Positive in urine mutagenicity assay (%)	Mean sister chromatid exchanges/cell ± SD	Plasma cotinine (ng/ml)	Plasma thiocyanate (μmol/l)
Smokers $n = 22$	12/19 (63)	9.1 ± 1.1**	246 ± 91***	144 ± 45***
Nonsmokers				
Exposed to ETS $n = 27$	4/26 (15)	7.9 ± 0.7	10 ± 4***	58 ± 18*
Unexposed $n = 20$	1/17 (6)	7.9 ± 0.7	5.2 ± 1.5	46 ± 16

*** $p < 0.001$, ** $p < 0.01$, * $p < 0.05$ compared with the unexposed group; two-tailed Student's *t*-test

Transplacental exposure to tobacco smoke

The risks of maternal smoking to the health and normal development of the fetus have long been known (see, e.g., Johnston, 1981; National Research Council, 1986). The risks of spontaneous abortion, premature births, prenatal deaths and low full-term birth weights have been associated with maternal smoking during pregnancy. A recent case-control study pointed to an increased risk of cancer in the offspring of smoking mothers (Stjernfeldt *et al.*, 1986).

Far fewer studies have been made on the effects of maternal passive smoking on pregnancy outcome. Lowering of the birth weight has been associated with passive smoking (Martin & Bracken, 1986; Rubin *et al.*, 1986), and exposure of the fetus to maternal passive smoking has been documented by measurements of cotinine in amniotic fluid from passive smokers (Andresen *et al.*, 1982).

In the present study we compared biochemical (cotinine, thiocyanate) and biological (SCE) intake markers in groups of mothers and newborns to estimate fetal exposure to tobacco smoke.

Altogether, 50 pairs of mothers and newborns from one delivery unit participated in the study. Maternal blood samples were collected just before delivery and cord blood samples immediately after delivery. All mothers were interviewed two to three days after delivery about their smoking habits and possible exposure to environmental tobacco smoke. Among the active smokers, all the intake markers measured were significantly increased as compared with nonsmoking mothers (Table 2). The cotinine and thiocyanate levels in mothers and their newborns correlated significantly, while the mean SCE frequency was significantly lower in all newborns as compared with that of mothers (see Table 2) and of adults in general. No effect of maternal passive smoking could be detected in any of the cord blood parameters; however, according to interview data none of the mothers had had substantial passive exposure at their homes, and all of them had been on maternity leave from work for at least three to four weeks before the delivery. A slightly increasing trend in exposure parameters is seen for the maternal values when unexposed nonsmokers are compared with exposed nonsmokers.

Conclusions

Several other reports on passive smoking have concluded that the presence of cotinine in plasma, urine or saliva is a good marker for intake of tobacco smoke (Jarvis

Table 2. Assessment of tobacco smoke exposure in groups of smoking and nonsmoking mothers and their newborns

Parameter[a]	Smoking mothers		Nonsmoking mothers		
	All	Heavy smokers[b]	All	Unexposed	Passively exposed
P-COT (ng/ml)					
Mothers	89.4 ± 72.7*** $n = 17$	99.2 ± 75.2*** $n = 14$	2.6 ± 5.7 $n = 29$	1.7 ± 2.3 $n = 22$	5.4 ± 10.9 $n = 7$
Cord blood	61.8 ± 44.2** $n = 11$	68.2 ± 46.8** $n = 9$	1.4 ± 2.0 $n = 17$	1.5 ± 2.2 $n = 14$	< detection limit $n = 3$
P-SCN (μmol/l)					
Mothers	109.1 ± 28.7*** $n = 17$	109.2 ± 28.6*** $n = 15$	43.8 ± 12.7 $n = 26$	45.1 ± 11.8 $n = 19$	40.1 ± 15.2 $n = 7$
Cord blood	90.8 ± 24.7*** $n = 18$	90.6 ± 23.8*** $n = 16$	38.3 ± 10.4 $n = 25$	38.2 ± 10.7 $n = 19$	38.5 ± 10.4 $n = 6$
$\bar{x}$ SCE/cell ± SD					
Mothers	9.0 ± 0.9** $n = 17$	9.0 ± 1.0** $n = 15$	8.1 ± 0.9 $n = 25$	7.9 ± 0.9 $n = 18$	8.4 ± 1.0 $n = 7$
Cord blood	6.1 ± 0.5 $n = 17$	6.1 ± 0.6 $n = 15$	5.9 ± 0.5 $n = 26$	5.9 ± 0.4 $n = 19$	5.9 ± 0.8 $n = 7$

[a] P-COT, plasma cotinine; P-SCN, plasma thiocyanate
[b] Smokers of ≥10 cigarette day throughout pregnancy *** $p < 0.001$, ** $p < 0.01$ compared to the total nonsmoking group

et al., 1984, 1985; IARC, 1986). Since cotinine is the main metabolite of nicotine, it is specific to tobacco; further, its half-life in the body is long enough (~30 h; Matsukura *et al.*, 1984) for measurements to be made of exposures over the previous few days. Cotinine is not vulnerable to dietary or other confounders, as is thiocyanate. Thiocyanate reflects exposure to hydrogen cyanide in tobacco smoke or other sources, but it and its precursors are also found in vegetables especially of the genus Brassica. However, thiocyanate has a half-life of up to 14 days (Hauth *et al.*, 1984) and can thus be used as a measure of long-term exposure, especially when dietary confounders can be controlled. The personal interviews showed that the subjects in the two studies described here did not have special dietary habits, e.g., vegetarians, which may explain the clear differences in plasma thiocyanate values between the smoking and nonsmoking groups. Plasma thiocyanate was also significantly increased in the nonsmoking group of restaurant personnel with long-term exposure to environmental tobacco smoke.

The measures of genotoxic exposure used — urinary mutagenicity and SCE frequency — suffer from nonspecificity in relation to the exposure. Even if these parameters revealed intake of genotoxic compounds in active smoking, they are too insensitive, nonspecific and vulnerable to other confounding exposures to be used as indicators of passive or transplacental smoking.

Acknowledgements

The authors are grateful to colleagues at the Institute of Occupational Health, Dr K. Koskimies, H. Salo, S. Valkonen, Dr K. Engström. and H. Järventaus for help with the analyses. Grants from the Research Council for the Environmental Sciences (30/064) and the National Board of Health (3.1.3./86) are gratefully acknowledged.

References

Andresen, B.D., Ng, K.J., Iams, J.D. & Bianchine, J.R. (1982) Cotinine in amniotic fluid from passive smokers. *Lancet, i,* 791–792

Hauth, J.C., Hauth, J., Drawbaugh, R.B., Gilstrap, L.C. & Pierson, W.P. (1984) Passive smoking and thiocyanate concentrations in pregnant women and newborns. *Obstet. Gynecol., 63,* 519–522

Husgafvel-Pursiainen, K., Sorsa, M., Møller, M. & Benestad, C. (1986) Genotoxicity and polynuclear aromatic hydrocarbon analysis of environmental tobacco smoke samples from restaurants. *Mutagenesis, 1,* 287–292

Husgafvel-Pursiainen, K., Sorsa, M., Engström, K. & Einistö, P. (1987) Passive smoking at work: biochemical and biological measures of exposure to environmental tobacco smoke. *Int. Arch. occup. environ. Health, 59,* 337–345

IARC (1986) *IARC Monographs on the Evaluation of the Carcinogenic Risks of Chemicals to Humans,* Vol. 38, *Tobacco Smoking,* Lyon

Jarvis, M.J., Tunstall-Pedoe, H., Feyerabend, C., Vesey, C. & Saloojee, Y. (1984) Biochemical markers of smoke absorption and self reported exposure to passive smoking. *J. Epidemiol. Commun. Health, 38,* 335–339

Jarvis, M.J., Russell, M.A.H., Feyerabend, C., Eiser, J.R., Morgan, M., Gammage, P. & Gray, E.M. (1985) Passive exposure to tobacco smoke: saliva cotinine concentrations in a representative population sample of non-smoking school-children. *Br. med. J., 291,* 927–929

Johnston, C. (1981) Cigarette-smoking and the outcome of human pregnancies: a status report on the consequences. *Clin. Toxicol., 18,* 189–209

Martin, T.R. & Bracken, M.B. (1986) Association of low birth weight with passive smoke exposure in pregnancy. *Am. J. Epidemiol., 124,* 633–642

Matsukura, S., Taminato, T., Kitano, N., Seino, Y., Hamada, H., Uchihashi, M., Nakajima, H. & Hirata, Y. (1984) Effects of environmental tobacco smoke on urinary cotinine excretion in non-smokers. *New Engl. J. Med., 311,* 828–832

National Research Council (1986) *Environmental Tobacco Smoke. Measuring Exposures and Assessing Health Effects,* Washington DC, National Academy Press

Rubin, D.H., Krasilnikoff, P.A., Leventhal, J.M., Weile, B. & Berget, A. (1986) Effect of passive smoking on birth-weight. *Lancet, ii,* 415–417

Stjernfeldt, M., Berglund, K., Lindsten, J. & Ludvigsson, J. (1986) Maternal smoking during pregnancy and risk of childhood cancer. *Lancet, i,* 1350–1352

HAEMOGLOBIN ADDUCTS OF AROMATIC AMINES IN PEOPLE EXPOSED TO CIGARETTE SMOKE

M.S. Bryant,[1] P. Vineis,[2] P.L. Skipper[1] & S.R. Tannenbaum[1]

[1] *Department of Applied Biological Sciences, Massachusetts Institute of Technology, Cambridge, MA, USA; and* [2] *Unit of Cancer Epidemiology, Department of Biomedical Science and Human Oncology, Turin, Italy*

In a population-based study in Turin, Italy, smokers of blond tobacco showed 4-aminobiphenyl (4-ABP) adduct levels some three times higher than nonsmoking subjects, and smokers of black tobacco showed levels about five times greater than nonsmokers. A dose-response relationship between the number of cigarettes smoked per day and 4-ABP adduct level was observed, but did not account for the higher adduct levels observed in smokers of black tobacco. Smoking-related increases in haemoglobin adducts were also observed for *o*-toluidine, *p*-toluidine, 2,4-dimethylaniline and 2-ethylaniline. Smoking subjects showed 3-aminobiphenyl adduct levels about 12 times greater than those of nonsmokers, who rarely showed a detectable level. This may indicate that there are fewer sources of 3-aminobiphenyl exposure not related to tobacco smoke. Smokers of black tobacco showed higher adduct levels than smokers of blond tobacco for 4-ABP, *p*-toluidine and 2,4-dimethylaniline.

Epidemiological studies have shown an increased relative risk for urinary bladder cancer among cigarette smokers as compared to suitable nonsmoking control populations. These relative risks range from 1.5 to 3.0 (Vineis *et al.*, 1984). A recent Danish study has also implicated other forms of tobacco consumption (cigars/cigarillos, chewing tobacco and pipe tobacco) as factors for increased risk for bladder cancer (Mommsen & Aagard, 1983).

The work of Patrianakos and Hoffmann (1979) showed that the level of 4-ABP in the mainstream smoke of a US nonfilter cigarette, consisting of blond tobacco, was about 2.4 ng per cigarette. The level in the mainstream smoke of a French cigarette, containing black tobacco, was 4.6 ng. The work of Vineis *et al.* (1984) has shown that the risk for bladder cancer among smokers of black tobacco is higher than that of smokers of blond tobacco cigarettes. It is possible that the increased level of aromatic amines in the black tobacco cigarettes plays a role in the increased risk for bladder cancer in this population.

In order to determine whether the level of 4-ABP-haemoglobin adducts was greater in the blood of smokers of black tobacco, who were likely to have greater exposure to 4-ABP than smokers of blond tobacco, a study was conducted on the population of Turin, which was the site of the earlier epidemiological study (Vineis *et al.*, 1984).

Methods

Male volunteers were recruited at a blood donor centre to donate an additional 10-ml blood sample and to fill out a questionnaire on their smoking history and other parameters, such as occupation. During two collection periods, 87 samples were collected, from 25 nonsmoking controls, 40 smokers of blond tobacco, 18 smokers of black tobacco, three smokers of mixed blond/black tobacco and one cigar smoker. The samples were recoded so that the identity of the samples would be unknown at the time of the analysis. The washed red blood cells were sent to the USA in a coded, blind fashion for analysis of the adduct levels.

The method of analysis of 4-ABP has been described (Bryant *et al.*, 1987) and involves gas chromatography-mass spectrometry. Analysis for other amines was achieved by scanning chromatograms at suitable single ions. Additional internal standards included d_5-aniline and 5-F-2-naphthylamine.

Results

4-Aminobiphenyl

Comparison of the 4-ABP-haemoglobin adduct levels in the three groups (non-smokers, smokers of blond and of black tobacco) shows that both smoking groups have elevated adduct levels compared to nonsmokers, and that smokers of black tobacco have somewhat higher adduct levels than smokers of blond tobacco. The difference between these three groups was found to be statistically significant ($p = 0.0001$) by analysis of variance. The unadjusted means and standard errors of the 4-ABP-haemoglobin adduct levels are 51 ± 18 for nonsmokers, 176 ± 14 for smokers of blond tobacco and 288 ± 21 for smokers of black tobacco.

The age distribution by type of tobacco showed that a higher proportion of older subjects smoked black tobacco (90% were over 35, compared with 50% of smokers of blond tobacco). In addition, the smokers of black tobacco consumed more cigarettes per day than the smokers of blond tobacco (black: 67% smoked 20 + cigarettes/day; blond: 44% smoked 20 + cigarettes/day). The data were thus analysed as to whether 4-ABP adduct levels were correlated with age or amount smoked per day. The age of the subjects showed a trend that was correlated with higher adduct levels but was not statistically significant ($p = 0.099$). The amount smoked per day was highly correlated ($p = 0.0015$) with adduct level. The dose-response relationship was examined within each tobacco type and found to be statistically significant for blond tobacco ($p = 0.0074$). A trend among the smokers of black tobacco was evident, but was not statistically significant ($p = 0.183$). These dose-response relationships are given in Table 1. Within each category of amount smoked, the smokers of black tobacco showed adduct levels that were 40–50% higher than those of smokers of blond tobacco.

Table 1. 4-ABP-haemoglobin adduct levels by amount smoked (least squares means and standard errors)

Amount smoked (no. of cigarettes/day)	Blond tobacco	Black tobacco
< 10	124.4 (24.9)	—
10 – 19	154.5 (19.3)	222.7 (57.8)
20 +	216.4 (17.1)	321.2 (40.8)

Table 2. Adduct levels observed (least squares means and standard errors)[a]

Aromatic amine	Nonsmokers (25)		Tobacco type			
			Blond (43)		Black (18)	
2-Naphthylamine	11.6	(1.1)	17.2	(1.8)	21.3	(4.9)
o-Toluidine	188	(19)	290	(19)	329	(22)
m-Toluidine	1141	(138)	1097	(108)	1140	(138)
p-Toluidine	209	(24)	306	(35)	415	(73)
2-Ethylaniline	38	(3)	70	(10)	80	(12)
3-Ethylaniline	102	(16)	115	(22)	129	(34)
2,5-Dimethylaniline	50	(6)	67	(10)	70	(14)
2,4-Dimethylaniline	40	(5)	73	(10)	114	(17)
2,6-Dimethylaniline	264	(90)	86	(15)	98	(30)
2,3-Dimethylaniline	52	(8)	56	(11)	64	(19)
3,5-Dimethylaniline	93	(12)	112	(31)	135	(31)
3,4-Dimethylaniline	47	(11)	46	(11)	65	(19)
3-Aminobiphenyl	1.2	(0.4)	13.8	(1.9)	12.9	(3.0)
4-Aminobiphenyl	51	(18)	176	(14)	288	(21)

[a] Values expressed in pg amine/g haemoglobin

Substituted anilines, 2-naphthylamine and 3-aminobiphenyl

The adduct levels for the remaining amines observed in the three groups are given in Table 2. Methods for determination of adducts with other aromatic amines may not have the same degree of accuracy as those for 4-ABP and 3-aminobiphenyl. Thus, the levels were examined in a semiquantitative manner to determine if they correlated with smoking status, or perhaps with type of tobacco. Adducts with seven amines appeared to be associated with smoking status: 4-ABP, 3-aminobiphenyl, 2-naphthylamine, *o*-toluidine, *p*-toluidine, 2-ethylaniline and 2,4-dimethylaniline. Only five of these amines, however, showed an association with tobacco type: 4-ABP, *o*-toluidine, *p*-toluidine, 2-ethylaniline and 2,4-dimethylaniline. In the case of 3-aminobiphenyl, a dose-response relationship was observed among smokers of blond tobacco ($p = 0.02$), but not for smokers of black tobacco. Thus, for smokers of both tobacco types, there was a dose-response relationship of borderline significance ($p = 0.06$).

Conclusions

Due to the complex nature of tobacco smoke, it may ultimately be difficult to prove that it is the aromatic amines contained in the smoke that are responsible for the increased bladder cancer risk in smoking populations. Other types of compounds present in the smoke may also be human bladder carcinogens: some *N*-nitroso compounds are bladder carcinogens in various species, and these may also be produced during tobacco smoking. *N*-Nitrosodi-*n*-butylamine is a bladder carcinogen in rats, mice, hamsters and guinea-pigs (IARC, 1974) and has been detected in tobacco smoke (Schmeltz & Hoffmann, 1977). Nonetheless, it is the aromatic amines which appear most likely to be the cause of the increased risk for bladder cancer among cigarette smokers.

It is interesting to note that the adduct ratios for smokers of black and blond tobacco *versus* nonsmokers closely resemble the relative risks for bladder cancer seen historically in these populations (Vineis *et al.*, 1984). One conclusion to be drawn from this finding is that the aromatic amines, including 4-ABP, are involved not only in the increased bladder cancer risk of cigarette smokers but also in the increased potency of black tobacco. Thus, the adduct levels of 4-ABP would be two to three times higher in the blood of smokers of blond tobacco and five to six times higher in the blood of smokers of black tobacco as compared to nonsmokers. The bladder cancer risk of any individual is a function of a great number of factors: the amount and duration of smoking, type of tobacco, exposure to promoting agents, interaction of occupation, and biochemical parameters such as urine pH and metabolic status (e.g., acetylator phenotype). Nonetheless, when comparing groups of individuals, among whom differences may be normally distributed or may offset one another, the overall risk for the group may be correlated with one of these factors. Although the exact relationship between 4-ABP-haemoglobin adduct level and bladder cancer risk is not understood, it appears that measurement of this parameter may be a useful step in examining the biochemical basis for the increased risk in smokers of blond and black tobacco.

Acknowledgements

We wish to thank Professor B. Terracini for encouragement and advice, and Dr R. Falcetta and Dr G. Mazzucco for technical assistance. The project was supported by funds partly provided by the International Cancer Research Data Bank Program of the National Cancer Institute, National Institutes of Health (USA) under contract No. NO1-CO-65341 (International Cancer Research Technology Transfer—ICRETT) and partly by the International Union Against Cancer; by Grant No. SIG-10-I from the American Cancer Society; Grant No. PO1-ES00597, from the National Institute of Environmental Health Sciences; and PHS Toxicology Training Grant No. T32-ES07020, awarded by the National Institutes of Health.

References

Bryant, M.S., Skipper, P.L., Tannenbaum, S.R. & Maclure, M. (1987) Hemoglobin adducts of 4-aminobiphenyl in smokers and nonsmokers. *Cancer Res.*, *47*, 602–608

IARC (1974) *IARC Monographs on the Evaluation of Carcinogenic Risk of Chemicals to Man*, Vol. 4, *Some Aromatic Amines, Hydrazine and Related Substances*, N-*Nitroso Compounds and Miscellaneous Alkylating Agents*, Lyon, pp. 197–210

Mommsen, S. & Aagard, J. (1983). Tobacco as a risk factor in bladder cancer. *Carcinogenesis*, *4*, 335–338

Patrianakos, C. & Hoffmann, D. (1979) Chemical studies on tobacco smoke. LXIV. On the analysis of aromatic amines in cigarette smoke. *J. anal. Chem.*, *3*, 150–154

Schmeltz, I. & Hoffmann, D. (1977) Nitrogen-containing compounds in tobacco and tobacco smoke. *Chem. Rev.*, *77*, 295–311

Vineis, P., Estève, J. & Terracini, B. (1984) Bladder cancer and smoking in males: types of cigarettes, age at start, effect of stopping and interactions with occupation. *Int. J. Cancer*, *34*, 165–170

DNA ADDUCTS, MICRONUCLEI AND LEUKOPLAKIAS AS INTERMEDIATE ENDPOINTS IN INTERVENTION TRIALS

H.F. Stich & B.P. Dunn

Environmental Carcinogenesis Unit, British Columbia Cancer Research Centre, Vancouver, BC, Canada

Internal dosimeters that can provide information about responses to chemopreventive agents in a short time would be invaluable for planning treatment protocols for large-scale intervention trials. Micronuclei meet many of the prerequisites of a good intermediate endpoint. They can be quantified in cultured cells, animal tissues and human exfoliated cells and biopsies. With image scanning, up to 10^5 cells can be screened for micronuclei within a few minutes. The predictive value of micronuclei has been demonstrated using cultured cells exposed to carcinogens and chemopreventive agents and using oral mucosa of betel-quid chewers. DNA adducts, as detected by ^{32}P-postlabelling techniques, could conceivably be another potentially useful marker. However, prior to their use in intervention trials, interindividual variations in their levels in primary, secondary and nontarget tissues and the relationship with doses of carcinogens must be established. The wide scatter of DNA adduct levels in the bronchial mucosa of smokers and of nonsmokers reveals one difficulty that can be encountered using this marker in intervention trials.

Intervention trials that use cancer as an endpoint require a large number of participants; they also last for a long time, are costly and are difficult to control. The application of 'intermediate endpoints', which may reveal responses to chemopreventive regimes within a short time and which may act as surrogates for cancer, is an attractive idea worthy of study. The use of markers in small-scale, pretrial studies could prove invaluable in planning treatment protocols for large-scale, long-term intervention trials. However, a more profound understanding of the factors measured by short-term tests is required before they can be accepted on a routine basis. In this paper, we point out some of the resolved and unresolved issues, and discuss the pros and cons of using three markers — DNA adducts, micronuclei and leukoplakia — as intermediate endpoints in intervention trials.

Quantifying chemopreventive efficacy with in-vitro model systems

Many intervention trials have been initiated without knowledge of the most effective dose of a chemopreventive agent. In-vitro tests, which have been successfully used to screen compounds for mutagenic, clastogenic and carcinogenic activity, should prove helpful in revealing the efficacy of chemopreventive agents. In order to transfer results between studies on cultured cells, animal models and human tissues, endpoints should be used that are applicable to all three systems. Micronuclei are one of the few markers that can be applied to virtually all mammalian cells, regardless of whether they are maintained in culture, obtained as exfoliated cells or collected from biopsies. They

Fig. 1. Contour plot of nuclei (N) and micronuclei (M) in a Chinese hamster ovary cell culture exposed for 3 h to methyl methanesulphonate

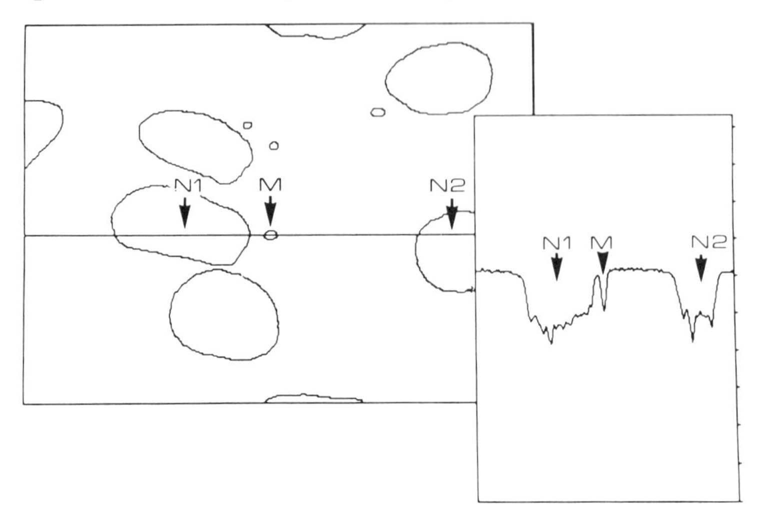

The insert shows the optical density along the line shown in the main square

can be analysed on smear or imprint preparations of cells or on sections used routinely for histopathological diagnosis (Stich, 1987). The possibility of automating the scoring of micronucleated cells makes this assay highly suitable for analysing entire population groups and for following the response to chemopreventive agents in large-scale intervention trials. A newly developed, dynamic microscope image processing scanner (Jaggi *et al.*, 1986; Palcic & Jaggi, 1986) has been used successfully to automate the scoring of the percentage of micronucleated cells, the number of micronuclei per cell and the amount of DNA per micronucleus. Slides can be screened at a typical rate of 1 cm^2/min, which permits the analysis of approximately 10^5 cells (nuclei)/min. An example is given in Figure 1.

The potential for obtaining relevant predictive values from in-vitro studies should not be underestimated, considering that many conditions prevailing in man can be simulated readily. Cultured human cells that retain a high degree of tissue specificity are becoming available for in-vitro experiments. In addition, a high degree of relevance can be achieved by using neoplastic transformation as an endpoint and viruses or oncogenes believed to be involved in the development of human cancers as etiological factors. The two following examples show how in-vitro models can be used to yield information about the doses of chemopreventive agents that can inhibit the pathogenic action of chemical carcinogens, and about the transforming effect of viral DNA. Once protective cellular levels have become known from in-vitro experiments, they should be aimed for in human target tissues during intervention trials on population groups at elevated risk for cancer. For example, intracellular levels of β-carotene at 3–4.5 ng/10^6 cultured cells protect against the genotoxic effects (including the induction of micronuclei) of methyl methanesulphonate (Fig. 2) and 4-nitroquinoline-1-oxide ($1–4 \times 10^{-6}$ M) (Stich & Dunn, 1986). Comparable levels were found in exfoliated oral mucosal cells following the administration of β-carotene at a dose (180 mg/week) that reduced the frequency of micronucleated cells in tobacco and areca-nut chewers (Stich *et al.*, 1984) and in snuff dippers (Stich *et al.*, 1985). The

Fig. 2. Inhibitory effect of β-carotene on the genotoxic activity of methyl methanesulphonate (MMS) in Chinese hamster ovary cells

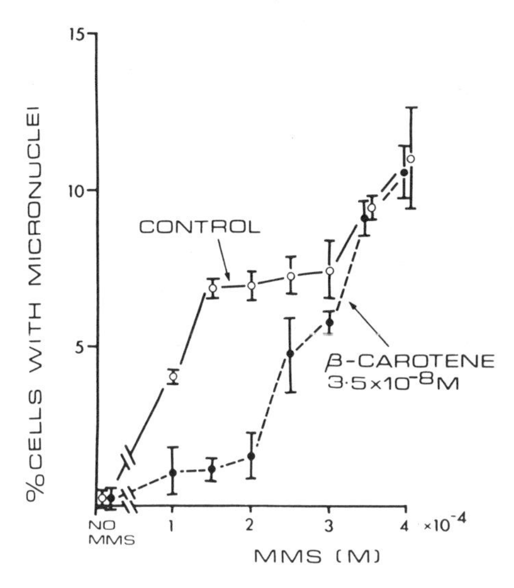

Cells were exposed to MMS for 3 h and harvested 17 h later. All experiments were performed in triplicate.

second example deals with the effect of all-*trans*-retinoic acid on the formation of neoplastic colonies in C127 cells transfected with bovine papilloma virus DNA (plasmid pd BPV-1), which does not become integrated into the genome of the host cell (Lancaster, 1981). In the presence of retinoic acid at levels of 5×10^{-5} M and higher, neoplastic transformation is completely inhibited (Fig. 3). However, transformed colonies develop after the removal of the retinoic acid. Such results must be considered when intervention trials of retinoids are conducted on patients with human papilloma virus DNA-containing lesions. The BPV–DNA transformation system could conceivably act as a model for cervical dysplasias, in which the human papilloma virus DNA seems to persist in a nonintegrated manner (Dürst *et al.*, 1985).

Variations of intermediate endpoints in target tissues of human 'control' groups

An intermediate endpoint must meet several prerequisites. It should be amenable to quantification and be scorable in tissues that can be obtained by noninvasive procedures or during routine diagnostic procedures; it should respond rapidly to cessation of exposure to carcinogens; and it should show little variation in the tissues of individuals who do not belong to a high cancer risk group. The sensitivity of the original ^{32}P-postlabelling method, which has been greatly improved by the application of several prepurification steps (Dunn & Stich, 1986; Reddy & Randerath, 1986), permits the analysis of DNA adducts in small biopsies of human tissue that can be taken, for example, during bronchoscopic examination or in approximately 2×10^6 exfoliated cells easily obtained by brushing the oral mucosa (Dunn & Stich, 1986). One basic unresolved issue is the extent to which types and levels of DNA adducts vary in tissues of human subjects who are not exposed to excessive amounts of carcinogens. As shown in Figure 4, relatively high levels of DNA adducts were detected in the bronchial epithelium of some nonsmokers by the ^{32}P-postlabelling technique. The observed variation may not be unexpected, considering the large interindividual

Fig. 3. Transformation of C127 cells induced by BPV DNA as an endpoint for examining the inhibitory effect of retinoic acid

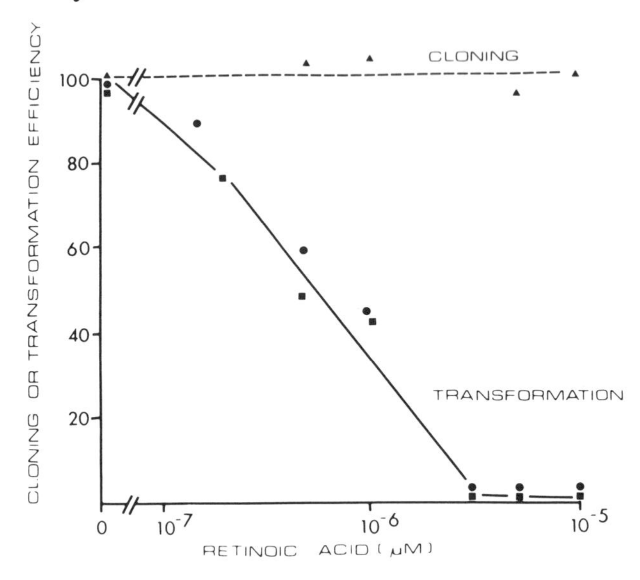

Mouse (C127) cells were transfected with BPV DNA (plasmid pd BPV-1). Two days after transfection, the cultures were subdivided and incubated in media containing either all-*trans*-retinoic acid dissolved in dimethylsulphoxide or dimethylsulphoxide only (0.02%). Transformed foci were scored 18 days after subculturing (Tsang *et al.*, 1988)

Fig. 4. Levels of DNA adducts in bronchial biopsies from current cigarette smokers and nonsmokers

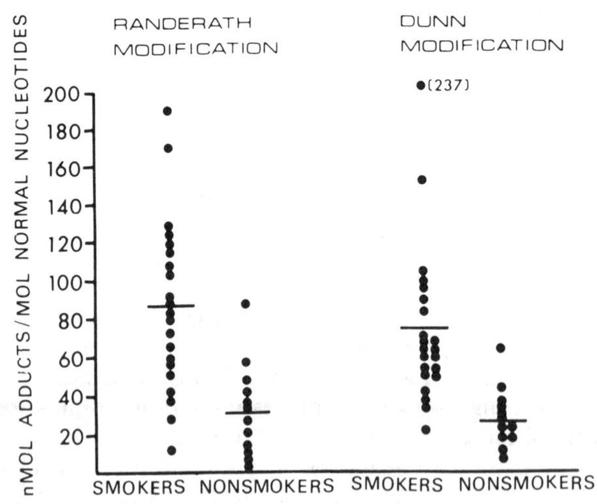

Two bronchial biopsies were required to extract enough DNA for the ^{32}P-postlabelling procedure. Isolation of the DNA adducts was enhanced using the nuclease P1 procedure (Randerath modification) and high-performance liquid chromatography (Dunn modification).

Fig. 5. Frequencies of micronucleated cells in the buccal mucosa of newborns (BUI), adults (BUII) and betel-quid chewers (BUC), in the bronchial epithelium of nonsmokers (BR), in the urinary bladder of nonsmokers (UR) and smokers (URS) and in mild dysplasia of the cervix (CVD)

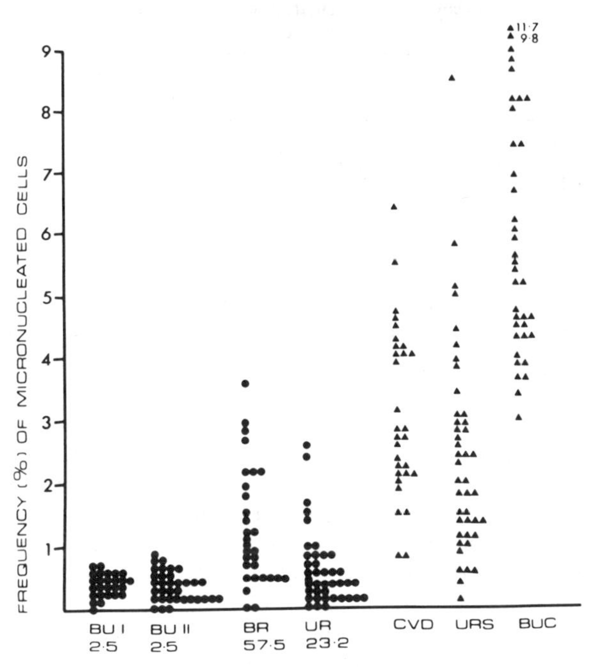

The figures at the bottom represent the cancer incidence for each tissue (Waterhouse *et al.*, 1982).

differences in aryl hydrocarbon hydroxylase activity in normal lung tissue (Sabadie *et al.*, 1981). Similarly, the frequencies of micronucleated bronchial cells from non-smokers with no known occupational exposure to polycyclic aromatic hydrocarbons are scattered over a relatively wide range (Fig. 5). This variation in the bronchial epithelium contrasts greatly with the low frequencies of micronucleated cells found in the oral and nasal mucosa of nonsmokers and nondrinkers of alcoholic beverages. Whether background frequencies of micronucleated cells in particular tissues are associated with their cancer incidences is an intriguing question; however, additional information is required on a larger series of different tissues and organs (Fig. 5).

Intermediate endpoints in tissues at elevated risk for cancer

Intermediate events that are directly involved in carcinogenesis should occur at highest levels in those tissues that are the targets of carcinogens and in which cancer will develop. However, there is not a great deal of evidence either for or against such an idea. Using ^{32}P-postlabelling techniques, the average level of DNA adducts was found to be higher in bronchial biopsies from a group of current smokers (Table 1). The 2.8-fold increase in DNA adduct levels in the bronchi of smokers over those in the group of nonsmokers is in contrast to the 9.6- to 20.6-fold increase in relative risk for the development of cancer in individuals with apparently comparable smoking habits

Table 1. DNA adducts in bronchial biopsies of current smokers and nonsmokers (from Stich *et al.*, 1987)

Subjects	Number	Age (years)	Number of cigarettes per day	DNA adducts[a]	Relative risk for lung cancer[b]
Smokers[c]	25	63.5 ± 10.5	27.3 ± 12.2	99.8 ± 47.9	9.6 – 20.6
Nonsmokers	38	65.0 ± 11.4	0	35.2 ± 29.4	1

[a]In nmol adducts per mol nucleotides
[b]Data from cohort studies in Canada and the USA (IARC, 1986)
[c]Smoking continued up to day of bronchoscopy

(IARC, 1986). The simplest explanation is that factors other than polycyclic aromatic hydrocarbon-induced DNA adducts must be involved in the etiology of lung cancer. Considering the difficulty of obtaining even one sample from many human organs, and the virtual impossibility of taking repeated samples, the idea was expressed that levels of DNA adducts from easily accessible tissues such as exfoliated or peripheral blood cells could act as predictors for events in, for example, the bronchial epithelium. We explored this issue by comparing levels of DNA adducts in the bronchial epithelium, which is the primary target tissue, with those in peripheral white blood cells. In the group of smokers examined, no correlation between DNA adduct levels in bronchi and those in white peripheral blood cells was evident (Table 2). It thus appears unlikely that the levels of DNA adducts in one tissue can be used to predict the behaviour of DNA adducts in other organs.

We investigated not only whether different human tissues respond differently, but also whether different regions of an organ or tissue can vary in their behaviour towards carcinogens. To yield information on this issue, biopsies were taken from three areas of the bronchial tree of two cigarette smokers. Considerable variation was found in the DNA adduct levels at different bronchial sites (Table 3). Micronucleus frequencies, which can be analysed in a few hundred cells, also appear to be ideal for examining the responses of subpopulations within a narrow range of a tissue. Indeed, unequal distributions of micronucleated cells have recently been reported in several human tissues. For example, the site in the oral mucosa at which either snuff or betel quid is kept for long periods showed higher frequencies of micronucleated cells than other regions of the oral cavity (Stich *et al.*, 1982, 1983, 1985). Preneoplastic lesions of the cervical mucosa or vulva can be composed of several small regions that differ considerably in their frequencies of micronucleated cells (Stich, 1987). The results show that, even within a tissue, the frequency of micronucleated cells in one area (e.g., oral mucosa, which is in close contact with betel quid or snuff) cannot be used to

Table 2. Levels of DNA adducts in bronchial biopsies and white blood cells of the same smokers

	nmol adducts per mol normal nucleotides				
Bronchial epithelium	237	68	66	61	21
White blood cells	5	10	3	3	6
Number of cigarettes smoked per day	20	10	30	16	20

For technique, see legend to Figure 4

Table 3. Variations in DNA adduct levels in biopsies taken from three different sites of the bronchial tree of cigarette smokers

Subject code	Site of biopsy[b]	Number of cigarettes per day	DNA adducts[a] (nmol per mol normal nucleotides)
B90	RLL	40	234
	RUL		228
	RML		130
B92	RLL	20	42
	RUL		154
	RML		76

[a] ^{32}P-postlabelling technique (high-performance liquid chromatographic variation)
[b] RLL, right lower lobe; RUL, right upper lobe; RML, right middle lobe

predict their frequency in a region distant from the location of the carcinogenic mixture. This observation carries several implications that cannot be ignored. In intervention trials using intermediate endpoints, cell samples taken prior to, during and after administration of chemopreventive agents must be taken from the same area in order to avoid erroneous conclusions. The relatively large quantities of tissue necessary for chemical or immunological measurement of DNA adducts can yield only average values for many millions of cells. The method of choice seems to be the detection of DNA adducts in single nuclei by immunofluorescence microscopy (Baan *et al.*, 1986). This approach combines the advantages of detecting molecular changes with that of a pathological diagnosis.

Responses of intermediate endpoints to the administration of chemopreventive agents

Intervention trials using precancerous lesions or cancer as an endpoint last for five or more years. The objective of applying intermediate endpoints is to gain statistically significant results within a much shorter time. At present, little is known about the survival of DNA adducts in carcinogen-exposed human tissues. This issue is confounded by the possibility that different DNA adducts formed by a single carcinogen and its metabolites may be repaired at different rates. Without knowledge of these basic issues, it would seem premature to consider DNA adducts as markers for intervention trials. Somewhat more data are available on the behaviour of micronucleated cells. For example, upward shifts in the frequency of exfoliated buccal mucosal cells containing micronuclei are readily detectable within seven to ten days after the onset of radiotherapy to the head and neck region (Stich *et al.*, 1983). This is the time required for the genotoxic anomalies that occur in the dividing cells of the basal layer to reach the surface of the mucosa, from which exfoliated cells are collected. Significant downward trends in the frequency of micronucleated cells in the mucosa of individuals were observed within four weeks following the cessation of radiotherapy, and in betel-quid chewers after they stopped this habit. A chemopreventive treatment that is as effective as the total removal of carcinogens should become detectable within a comparable time. Table 4 shows the responses of different endpoints to the administration of β-carotene (180 mg/week) and vitamin A (100 000 IU/week) for one, three and six months. Remission of oral leukoplakias that were formed prior to the onset of the trial, and inhibition of the development of new leukoplakias within the six-month treatment period, follow a different course from the

Table 4. Responses of various endpoints to cessation of carcinogenic injury or chemopreventive treatment

Endpoints in oral mucosa[a]	Number of individuals	Treatment[b]	Average changes (%)[c]		
			1 month	3 months	6 months
Reduction in MNC	5	Cessation of radiotherapy	985	—	—
Reduction in MNC	17	β-carotene + vitamin A	57.9	89.5	—
Reduction in NC	51	β-carotene + vitamin A	0	0	0
Remission of leukoplakias	51	β-carotene + vitamin A	0	9.5	24.5
Inhibition of new leukoplakias	51	β-carotene + vitamin A	0	45.4	63.2

[a]MNC, frequency of exfoliated buccal mucosal cells with at least one micronucleus; NC, frequency of exfoliated buccal mucosal cells with severely clumped nuclear chromatin (De Campos Vidal *et al.*, 1973)
[b]Oral administration of β-carotene (180 mg/week) plus vitamin A (100 000 IU/week)
[c]Percentage of change in treated individuals as compared to those receiving a placebo

frequency of micronuclei. An understanding of the time course of responses appears to be essential when different intermediate endpoints are considered for intervention trials.

Outlook

The recently awakened interest in chemoprevention as a means of controlling cancer has led to the initiation of many intervention trials on human population groups at elevated risk for cancer. In the course of these studies, the necessity of using intermediate endpoints and short-term tests to gain the basic information required in the design of long-term, large-scale, costly clinical trials has become evident. In the search for such intermediate endpoints, various DNA adducts and cytological markers (e.g., micronuclei), which can be applied to tissue samples obtainable by noninvasive procedures, are being explored. The adaptation of new, highly sensitive techniques for measuring DNA alterations and genetic changes to human biopsies taken routinely, and the development of automated or semiautomated scoring procedures could become tools applicable to what may be called 'molecular epidemiology'.

Acknowledgements

These studies were supported by the National Cancer Institute of Canada, and by the Natural Sciences and Engineering Research Council of Canada. H.F.S. is a Terry Fox Cancer Research Scientist of the National Cancer Institute of Canada.

References

Baan, R.A., Lohman, P.H.M., Fichtinger-Schepman, A.M.J., Muysken-Schoen, M.A. & Ploem, J.S. (1986) *Immunochemical approach to detection and quantitation of DNA adducts resulting from exposure to genotoxic agents.* In: Sorsa, M. & Norppa, H., eds, *Monitoring of Occupational Genotoxicants,* New York, Alan R. Liss, pp. 135–146

De Campos Vidal, B., Schlüter, G. & Moore, G.W. (1973) Cell nucleus pattern recognition: influence of staining. *Acta cytol., 17,* 510–521

Dunn, B.P. & Stich, H.F. (1986) ^{32}P-Postlabelling analysis of aromatic DNA adducts in human oral mucosal cells. *Carcinogenesis, 7,* 1115–1120

Dürst, M., Kleinheinz, A., Hotz, M. & Gissmann, L. (1985) The physical state of human papillomavirus type 16 DNA in benign and malignant genital tumors. *J. gen. Virol., 66,* 1515–1522

IARC (1986) *IARC Monographs on the Evaluation of the Carcinogenic Risk of Chemicals to Humans,* Vol. 38, *Tobacco Smoking,* Lyon

Jaggi, B.W., Poon, S. & Palcic, B. (1986) *Implementation and evaluation of the DMIPS cell analyzer.* In: *IEEE Proc. Engineering in Medicine and Biology,* Vol. 3, Chicago, pp. 906–911

Lancaster, W.D. (1981) Apparent lack of integration of bovine papillomavirus DNA in virus-induced equine and bovine tumor cells and virus-transformed mouse cells. *Virology, 108,* 251–255

Palcic, B. & Jaggi, B. (1986) The use of solid state image sensor technology to detect and characterize live mammalian cells growing in tissue culture. *Int. J. Radiat. Biol., 50,* 345–352

Reddy, M.V. & Randerath, K. (1986) Nuclease P1-mediated enhancement of sensitivity of ^{32}P-postlabelling test for structurally diverse DNA adducts. *Carcinogenesis, 7,* 1543–1551

Sabadie, N., Richter-Reichhelm, H.B., Saracci, R., Mohr, U. & Bartsch, H. (1981) Inter-individual differences in oxidative benzo(a)pyrene metabolism by normal and tumorous surgical lung specimens from 105 lung cancer patients. *Int. J. Cancer, 27,* 417–425

Stich, H.F. (1987) Micronucleated exfoliated cells as indicators for genotoxic damage and as markers in chemoprevention trials. *J. Nutr. Growth Cancer, 4,* 9–18

Stich, H.F. & Dunn, B.P. (1986) Relationship between cellular levels of beta-carotene and sensitivity to genotoxic agents. *Int. J. Cancer, 38,* 713–717

Stich, H.F., Stich, W. & Parida, B.B. (1982) Elevated frequency of micronucleated cells in the buccal mucosa of individuals at high risk for oral cancer: betel quid chewers. *Cancer Lett., 17,* 125–134

Stich, H.F., San, R.H.C. & Rosin, M.P. (1983) Adaptation of the DNA repair and micronucleus tests to human cell suspensions and exfoliated cells. *Ann. N.Y. Acad. Sci., 407,* 93–105

Stich, H.F., Stich, W., Rosin, M.P. & Vallejera, M.O. (1984) Use of the micronucleus test to monitor the effect of vitamin A, beta-carotene and canthaxanthin on the buccal mucosa of betel nut/tobacco chewers. *Int. J. Cancer, 34,* 745–750

Stich, H.F., Hornby, A.P. & Dunn, B.P. (1985) A pilot beta-carotene intervention trial with Inuits using smokeless tobacco. *Int. J. Cancer, 36,* 321–327

Stich, H.F., Dunn, B.P., Enarson, D.A., Nelems, B., Miller, R.R., Ostrow, D. & Champion, P. (1987) *Quantitating cytological and biochemical markers for carcinogen exposures in the bronchi of smokers, ex-smokers and non-smokers* (Abstract). In *6th World Conference on Smoking and Health, Tokyo, 9–12 November*

Tsang, S.S., Li, G. & Stich, H.F. (1988) Effect of retinoic acid on bovine papillomavirus (BPV) DNA-induced transformation and number of BPV DNA copies. *Int. J. Cancer* (in press)

Waterhouse, J., Muir, C., Shanmugaratnam, K. & Powell, J., eds (1982) *Cancer Incidence in Five Continents,* Vol. IV (*IARC Scientific Publications No. 42*), Lyon, International Agency for Research on Cancer

DETECTION OF BENZO[*a*]PYRENE-DNA ADDUCTS IN CULTURED CELLS TREATED WITH BENZO[*a*]PYRENE DIOL-EPOXIDE BY QUANTITATIVE IMMUNOFLUORESCENCE MICROSCOPY AND ^{32}P-POSTLABELLING; IMMUNOFLUORESCENCE ANALYSIS OF BENZO[*a*]PYRENE-DNA ADDUCTS IN BRONCHIAL CELLS FROM SMOKING INDIVIDUALS

R.A. Baan[1], P.T.M. van den Berg, M.-J.S.T. Steenwinkel & C.J.M. van der Wulp

Department of Genetic Toxicology, TNO Medical Biological Laboratory, Rijswijk, The Netherlands

Monoclonal antibodies were raised against the reaction product of benzo[*a*]pyrene diol-epoxide (BPDE) and deoxyguanosine-5′-monophosphate. The antibodies were used for detection of DNA adducts *in situ* in BPDE-treated cultured human fibroblasts by immunofluorescence microscopy. Analogue-digital conversion of the fluorescence signal and further image processing allowed measurement of the immunospecific fluorescence in the nuclei of these cells. The results are compared with the adduct levels measured in isolated DNA by ^{32}P-postlabelling. Preliminary results are shown of the application of the immunofluorescence method to the analysis of DNA adducts in bronchial cells obtained from smoking individuals.

The determination of DNA adducts is considered to be a promising approach for monitoring humans with respect to exposure to carcinogenic or mutagenic chemicals. This method can be used to estimate directly a relevant effect of the exposure, i.e., the amount of the chemicals bound to the target molecule. Promutagenic lesions may play an important role during the onset of carcinogenesis. Recent studies on proto-oncogene activation have shown that point mutations at specific sites within the genome can give rise to cell transformation (Zarbl *et al.*, 1985; Vousden *et al.*, 1986).

Several biochemical, biophysical and immunochemical methods to detect and measure primary damage in DNA have been developed over the past years (this volume). Methods for biomonitoring must be suitable for assessing exposure to low levels of nonradioactive genotoxicants at the workplace or in the general environment. Immunochemical techniques are often applied for the detection of DNA lesions present as such in isolated DNA, as nucleotide adducts in an enzymic DNA digest or as base adducts after depurination of DNA. The highly sensitive postlabelling methods and gas chromatographic-mass spectrometry procedures to detect DNA adducts also yield data on overall adduct levels in DNA. Information on cell type-specific induction of DNA damage is not provided by these approaches. Immunochemical detection of

[1] To whom correspondence should be addressed

DNA damage at the single cell level, as illustrated in this paper, offers the possibility for analysing the distribution of DNA damage in different types of cells.

We describe the development of an immunochemical method to monitor (human) exposure to benzo[*a*]pyrene (BP), a widely studied member of the group of polycyclic aromatic hydrocarbons. A major BP-DNA adduct, the reaction product of the reactive BPDE and deoxyguanosine-5′-monophosphate, was used as a hapten to raise antibodies in mice. The monoclonal antibodies obtained were used to detect BP-DNA damage at the level of the single cell by immunofluorescence microscopy. Quantification of the fluorescence signal was achieved by analogue-digital conversion and image processing programmes. The method was applied to study the presence of adducts in the nuclei of cultured human fibroblasts treated with various concentrations of BPDE. To determine the adduct levels in these cells, DNA was isolated and analysed by ^{32}P-postlabelling.

The immunofluorescence method was applied to investigate the presence of BP-DNA adducts in specimens of bronchial cells obtained from smoking individuals. Preliminary results show an increased level of fluorescence in these samples when compared to that in specimens from nonsmokers. Furthermore, an apparent cell type-specific distribution of immunospecific fluorescence was noticed in the samples from smokers. This approach may offer possibilities for investigating the correlation between DNA damage and cell type-specific neoplastic transformation.

Materials and methods

Antigen and antibodies

Preparation of the BP-DNA adduct, (±)-N^2-(7*R*,8*S*,9*R*-trihydroxy-7,8,9,10-tetrahydrobenzo[*a*]pyren-10*S*-yl)-2′-deoxyguanosine-5′-monophosphate (BP-dGMP), and isolation of adduct-specific monoclonal antibodies were described in detail earlier (Baan *et al.*, 1988). Briefly, deoxyguanosine-5′-monophosphate was incubated with the reactive BP metabolite BPDE. The adduct was purified by chromatography on Sephadex LH20. Its identity was confirmed by fast atom bombardment-mass spectrometry (courtesy of Dr W.P. Watson, Shell Research Centre, Sittingbourne, UK). The adduct was coupled to chicken γ-globulin and injected into mice. After fusion of spleen cells from immunized mice with SP2/0 plasmacytoma cells, hybrid cultures were screened for antibody production in an enzyme-linked immunosorbent assay, with control DNA and BP-modified DNA as immobilized antigens. Antibodies from clone II.E4 were used in the experiments described in this paper (see Baan *et al.*, 1988, for further details).

In-situ detection of BP-DNA adduct formation

Human fibroblasts (79Rd172) were cultured on 10 × 10-mm glass cover slips and treated with various concentrations of BPDE in F10 medium without serum for 30 min at 37°C. The cells were washed, fixed and processed for immunofluorescence microscopy as described earlier (Muysken-Schoen *et al.*, 1985), with some modifications (Baan *et al.*, 1988). After incubation with BP-deoxyguanosine adduct-specific antibodies from clone II.E4 and fluorescein(FITC)-labelled second antibodies, the preparations were counterstained with propidium iodide, a DNA-specific dye.

Smears of human bronchial cells were obtained from Dr H.F. Stich, British Columbia Cancer Research Centre, Vancouver, BC, Canada. These samples had been taken from current smokers and nonsmokers by brushing the bronchi. The cell smears had been fixed in 85% ethanol and were then sent to our laboratory. The slides were processed as indicated above. To detect fluorescence, a Leitz fluorescence microscope was used, equipped with a Ploemopak filter block, containing the appropriate filter sets to observe FITC and propidium fluorescence.

Quantification of immunospecific fluorescence

The fluorescence signal, as observed in the fluorescence microscope, was enhanced with an image amplifier and recorded with a very sensitive television camera. The data points in 256 lines on the image were digitalized into a 256 × 256 matrix, each picture element (pixel) of which was assigned a grey-value between 0 and 256. The digitalized data were then processed using the TIPS-tcl image-analysis programme, developed at TNO. In this analysis, the red fluorescence pattern of the propidium iodide is used as a mask to define the nuclear area. This is especially important when background values are measured, e.g., in untreated control cells, when it is often hard to distinguish the fluorescence in the cell nuclei from that of the surrounding cytoplasm. The details of this quantitative immunofluorescence microscopy technique will be described elsewhere (C.J.M. Van der Wulp *et al.*, in preparation).

Analysis of BP-DNA adduct formation with the ^{32}P-postlabelling method

Postlabelling analysis of DNA samples isolated from fibroblasts treated with BPDE in parallel to those described above was carried out according to a published procedure (Gupta, 1985), as summarized earlier (Baan *et al.*, 1988). A schematic outline of the steps in the postlabelling procedure is given in Figure 1.

Results

The monoclonal antibodies raised against the BP-deoxyguanosine adduct are highly specific for the BP modification (see Baan *et al.*, 1988). The antibodies appear to be suitable for in-situ detection, by immunofluorescence microscopy, of BP-DNA adducts in cultured cells treated with BPDE. To visualize the antibody molecules attached to the adducts, an FITC-labelled second antibody was used. Figure 2 shows the FITC images of treated and control cells, as observed with the fluorescence microscope. Immunospecific fluorescence is present in the nuclei of cells treated with various concentrations of BPDE, while the nuclei of untreated cells show a low background signal. The FITC fluorescence signal is quantified by analogue-digital conversion and

Fig. 1. Schematic outline of steps in the ^{32}P-postlabelling assay (according to Gupta, 1985)

DNA (containing adducts X, Y,...)

|

micrococcal endonuclease, spleen exonuclease

|

Ap,Cp,Gp,Tp,mCp,Xp,Yp,...

|

extraction with butanol: Xp,Yp,...

|

T4 polynucleotide kinase, ^{32}P-ATP

|

^{32}pXp, ^{32}pYp,...

|

PEI-cellulose chromatography

|

autoradiography

Fig. 2. Detection of BP-DNA adducts in cultured human fibroblasts by immunofluorescence microscopy

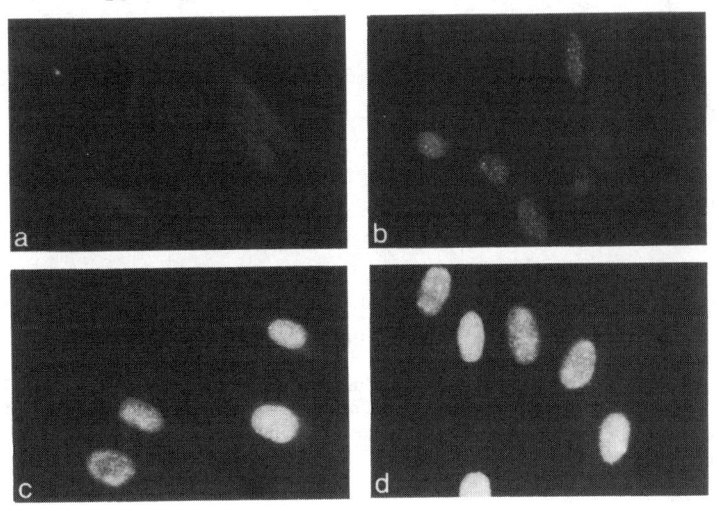

The cells were treated with 0 (panel a), 1 (panel b), 5 (panel c) and 10 (panel d) μM BPDE for 30 min at 37°C. The cells were washed, fixed and processed for immunofluorescence microscopy (see Baan *et al.*, 1988). The adduct-specific antibody was from clone II.E4; the second antibody was a fluorescein-conjugated goat antiserum directed against mouse immunoglobulins.

image processing, as described in Materials and Methods. The results, expressed as fluorescence per pixel, averaged over 50–100 nuclei at each dose of BPDE, are given in Figure 3. A dose-dependent increase in fluorescence is observed.

Results of the ^{32}P-postlabelling analysis of adduct levels in DNA samples isolated from cells treated in parallel to those used in the immunofluorescence experiment are shown in Figure 4. In this case, a dose-dependent increase in DNA adduct level is seen (see also Fig. 3).

Fig. 3. Measurement of BP-DNA adducts in cultured human fibroblasts by quantitative immunofluorescence microscopy

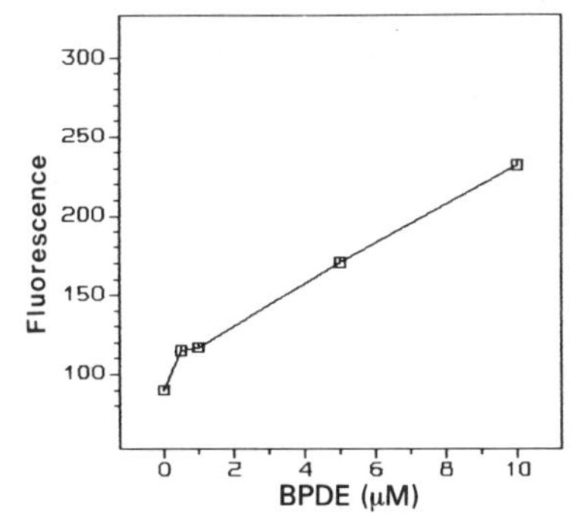

The immunofluorescence signals in the nuclear areas of BPDE-treated cells (see Fig. 2) were quantified by analogue-digital and image processing (C.J.M. Van der Wulp *et al.*, in preparation). The results are given as averaged fluorescence (arbitrary units) per pixel in 50–100 cell nuclei.

Fig. 4. Measurement of BP-DNA adducts in cultured human fibroblasts by ^{32}P-postlabelling

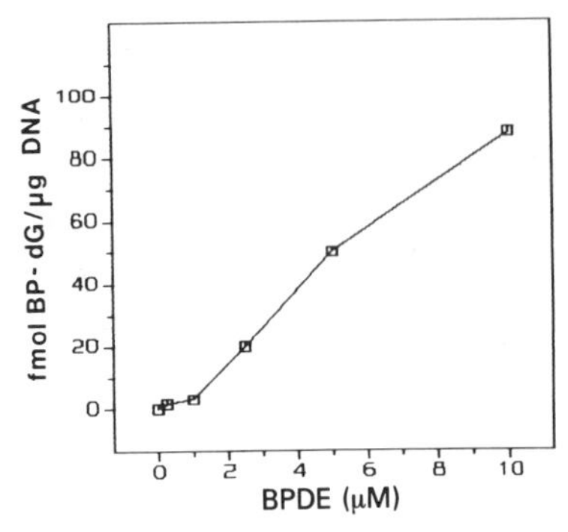

Cells were treated with BPDE in parallel to those used in the immunofluorescence experiment (Figs 2 and 3). DNA was isolated and assayed according to the schematic outline of Figure 1 (see Gupta, 1985; Baan *et al.*, 1988). Results are given as fmol BP-deoxyguanosine adduct/μg DNA, which corresponds to the number of BP-deoxyguanosine adducts/3×10^6 nucleotides.

The immunofluorescence microscopy technique with BP adduct-specific antibodies was recently used to investigate the possible presence of immunospecific fluorescence in the nuclei of bronchial cells obtained from current smokers. The first qualitative results are shown in Figure 5. Immunospecific fluorescence is observed in these specimens. Furthermore, these bronchial smears contain various types of cells which appear to be differentially stained with FITC (Fig. 5, right). Figure 6 shows the result of immunofluorescence analysis of cells from the oral cavity of a nonsmoking individual. No fluorescence is seen in the cell nuclei in this preparation.

Discussion

This paper describes the development of an immunochemical method for the detection of DNA damage induced in cultured cells by exposure to a reactive derivative of BP. The monoclonal antibodies used had been raised against the mononucleotide adduct BP-dGMP, and appear suitable for staining BP-modified DNA in treated cells. After exposure, a clear nuclear fluorescence is observed, while the cytoplasm of the cells and the nuclei of untreated cells show a low background fluorescence. The same background is seen when the adduct-specific antibodies are omitted (not shown). The immunofluorescence has been quantified by, first, recording the propidium iodide staining pattern, the coordinates of which are stored in the computer. The FITC fluorescence present in the same area is then recorded with the appropriate filter setting of the microscope. The data are digitalized, processed and finally expressed as the averaged fluorescence per pixel of 50–100 cell nuclei. The results show a dose-dependent increase in the fluorescence (Fig. 3). The background value observed in the untreated cells is due partly to fluorescence generated in the optical system of the conventional microscope (Baan *et al.*, 1986). Preliminary experiments have shown that with the laser-scan microscope measurements can be performed with higher sensitivity, mainly because of a reduced background signal (C.J.M. Van der Wulp *et al.*, in preparation).

The analysis of DNA adduct levels by ^{32}P-postlabelling was carried out according to the method published by Gupta (1985), which involves enrichment of the adduct

Fig. 5. Analysis of DNA adducts in bronchial cells from current smokers

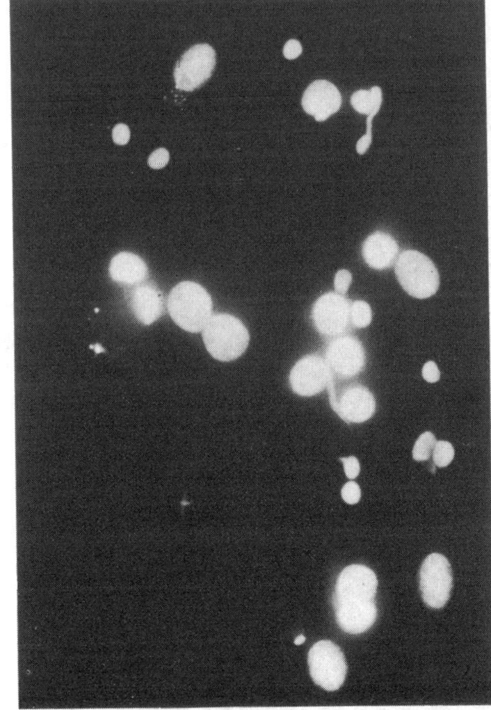

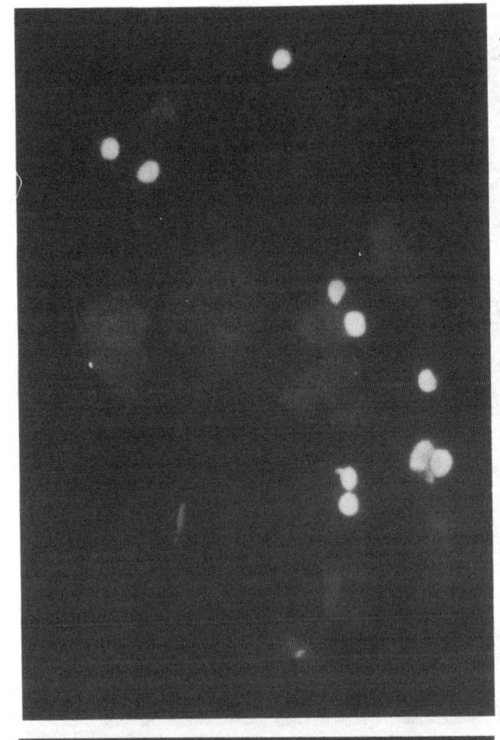

Samples were obtained by brushing bronchi. The slides were processed as described (Baan *et al.*, 1988). Left, propidium iodide staining; right, FITC-fluorescence.

Fig. 6. Analysis of DNA adducts in cells from the oral cavity of current nonsmokers

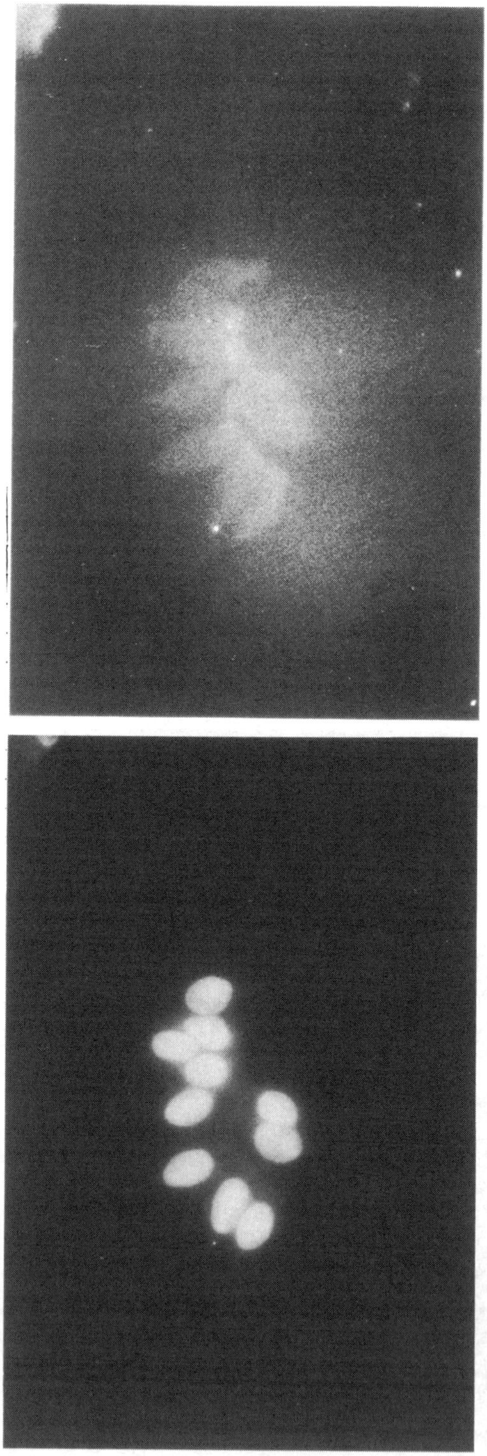

Samples were obtained by brushing the buccal mucosa. The slides were processed as described (Baan *et al.*, 1988). Left, propidium iodide staining; right, FITC-fluorescence

fraction by extraction of the DNA digest with butanol. This modification of the basic assay enhances the sensitivity by lowering background levels of radioactivity on the chromatogram (see Baan *et al.*, 1988). The postlabelling data show a dose-dependence that is similar to that observed in the immunofluorescence experiment. Comparison of the results of the two assays allows calibration of the fluorescence in terms of adduct content. In this case, a fluorescence value of 170 units would correspond to 4×10^5 adducts/cell (see also Figs 3 and 4).

The first qualitative results of adduct analysis in smears of human bronchial cells clearly show the presence of fluorescence in samples from current smokers. Although the number of specimens from nonsmokers analysed so far is still limited, we have observed background fluorescence levels in most of these (not shown). An interesting phenomenon in the bronchial cells from smokers is the clear cell type-specific fluorescence (see also the propidium pattern with the FITC image in Fig. 5). Because the cell morphology is preserved, identification of different types of cells by histochemical staining is still feasible. This offers the opportunity to investigate a possible relation between induction of DNA damage and, e.g., neoplastic transformation in distinct cell types. Other investigators have also used adduct-specific antibodies in combination with fluorescence techniques (Adamkiewicz *et al.*, 1983; Huitfeldt *et al.*, 1987) or histochemical staining methods (Menkveld *et al.*, 1985) for analysis of site- or cell type-specific distribution of DNA damage.

The antibodies used in these experiments were raised against a BP-DNA adduct; however, cross-reactivity of similar antibodies with DNA damage induced by other polycyclic aromatic hydrocarbons has been reported (Haugen *et al.*, 1986). The immunofluorescence observed in the human samples, as shown in this paper, may therefore represent more than just BP-DNA adducts. Experiments to characterize our antibodies with respect to such cross-reactivity are in progress.

For obvious reasons, bronchial brushings are not suitable for routine application in biomonitoring. Therefore, specimens from white blood cells, buccal mucosa cells and bronchial cells, all from the same individual, are being analysed for the presence of immunospecific fluorescence with the method described above. Comparison of the results of such experiments can provide information on the relevance of monitoring studies on readily accessible material (blood, buccal samples) to exposure of the target tissue. Investigations on site-specific induction and removal of DNA damage in target tissues in relation to biological effects will provide relevant information for assessment of the health risk involved.

Acknowledgements

We thank Dr H.F. Stich, British Columbia Cancer Research Centre, Vancouver, BC, Canada, for providing the specimens of human bronchial and buccal mucosa cells, and for his interest in this work. This research was partly sponsored by Shell and by the Directorate of Labour, Dutch Ministry of Social Affairs and Employment.

References

Adamkiewicz, J., Ahrens, O., Huh, N. & Rajewsky, M.F. (1983) Quantitation of alkyldeoxynucleosides in the DNA of individual cells by high-affinity monoclonal antibodies and electronically intensified, direct immunofluorescence. *J. Cancer Res. clin. Oncol., 105,* A15

Baan, R.A., Lohman, P.H.M., Fichtinger-Schepman, A.M.J., Muysken-Schoen, M.A. & Ploem, J. S. (1986) *Immunochemical approach to detection and quantitation of DNA adducts resulting from exposure to genotoxic agents.* In: Sorsa, M. & Norppa, H., eds, *Monitoring of Occupational Genotoxicants,* New York, Alan R. Liss, pp. 135–146

Baan, R.A., Van den Berg, P.T.M., Watson, W.P. & Smith, R.J. (1988) *In situ* detection of

DNA adducts formed in cultured cells by benzo(*a*)pyrene diolepoxide (BPDE), with monoclonal antibodies specific for the BP-deoxyguanosine adduct. *Toxicol. environ. Chem.*, *16*, 325–339

Gupta, R.C. (1985) Enhanced sensitivity of ^{32}P-postlabeling analysis of aromatic carcinogen: DNA adducts. *Cancer Res.*, *45*, 5656–5662

Haugen, A., Becher, G., Benestad, C., Vähäkangas, K., Trivers, G.E., Newman, M.J. & Harris, C.C. (1986) Determination of polycyclic aromatic hydrocarbons in the urine, benzo(*a*)pyrene diolepoxide-DNA adducts in lymphocyte DNA and antibodies to the adducts in the sera from coke oven workers exposed to measured amounts of polycyclic aromatic hydrocarbons in the work atmosphere. *Cancer Res.*, *46*, 4178–4183

Huitfeldt, H.S., Spangler, E.F., Baron, J. & Poirier, M.C. (1987) Microfluorimetric determination of DNA adducts in immunofluorescent-stained liver tissue from rats fed 2-acetylaminofluorene. *Cancer Res.*, *47*, 2098–2102

Menkveld, G.J., Van der Laken, C.J., Hermsen, T., Kriek, E., Scherer, E. & Den Engelse, L. (1985) Immunohistochemical localization of O^6-ethyldeoxyguanosine and deoxyguanosin-8-yl-(acetyl)aminofluorene in liver sections of rats, treated with diethylnitrosamine, ethylnitrosourea or N-acetylaminofluorene. *Carcinogenesis*, *6*, 263–270

Muysken-Schoen, M.A., Baan, R.A. & Lohman, P.H.M. (1985) Detection of DNA adducts in N-acetoxy-2-acetylaminofluorene-treated human fibroblasts by means of immunofluorescence microscopy and quantitative immunoautoradiography. *Carcinogenesis*, *6*, 999–1004

Vousden, K.H., Bos, J.L., Marshall, C.J. & Phillips, D.H. (1986) Mutations activating human c-Ha-ras1 protooncogene (HRAS1) induced by chemical carcinogens and depurination. *Proc. natl Acad. Sci. USA*, *83*, 1222–1226

Zarbl, H., Sukumar, S., Arthur, A.V., Martin-Zanca, D. & Barbacid, M. (1985) Direct mutagenesis of H-ras-1 oncogenes by nitroso-methyl-urea during initiation of mammary carcinogenesis in rats. *Nature*, *315*, 382–385

OCCUPATIONAL EXPOSURES

HAEMOGLOBIN BINDING IN CONTROL OF EXPOSURE TO AND RISK ASSESSMENT OF AROMATIC AMINES

H.-G. Neumann

Institute of Pharmacology and Toxicology, University of Würzburg, Würzburg, Federal Republic of Germany

Haemoglobin is proposed as a dose monitor for aromatic amines. Metabolically formed nitrosoarenes react with sulphydryl groups of haemoglobin and, after intramolecular rearrangement, yield sulphinic acid amides. This type of adduct is stable *in vivo* but can readily be hydrolysed after haemoglobin is isolated from blood samples, usually yielding the parent amine, which is quantified by gas chromatography or high-performance liquid chromatography. The haemoglobin binding index was determined in rats for a series of monocyclic aromatic amines, benzidine and some benzidine congeners. The following relationships are discussed: between binding of metabolites to DNA and to proteins; between haemoglobin binding and biological endpoints such as carcinogenesis and methaemoglobin formation; and between haemoglobin binding and molecular endpoints such as DNA binding and protein binding in liver and kidney. Haemoglobin binding correlates with a biologically active dose of aromatic amines and is thus well suited for monitoring exposure. The relationship between haemoglobin binding and the dose at critical targets is more complex, and, at present, carcinogenic risk cannot be assessed from biological monitoring data.

The large-scale production of aromatic amines as intermediates for synthetic dyestuffs coincided with the beginning of the industrial production of chemicals. It is also an important example of a health hazard caused by an uncontrolled exposure in the work environment. Many workers developed cyanosis and some developed bladder tumours, after long exposure. Thus, the problem of controlling exposure to these compounds is an old one, and a number of methods have been used. Proceeding from monitoring of environmental concentrations to measuring the actual uptake, by analysing blood concentrations or amounts of the chemicals or their metabolites excreted in urine, was a decisive step forward in this endeavour. This concept has been developed recently into the dosimetry of genotoxic compounds that react with macromolecular components of the organism. The use of haemoglobin as a dosimeter for alkylating agents was introduced by Ehrenberg and his associates (1974). Our proposal to use haemoglobin adducts with aromatic amines as a measure of the bioavailability of reactive metabolites (Groth & Neumann, 1971; Wieland & Neumann, 1978) led to the development of practical methods for biological monitoring. Several reviews on this subject have been published (Ehrenberg *et al.*, 1983; Neumann, 1984a, 1986, 1987; Farmer *et al.*, 1987). Although this method has already been successfully applied in monitoring humans for exposure to 4-aminobiphenyl (Green *et al.*, 1984), aniline and *p*-chloroaniline (Lewalter & Korallus, 1985), the experimental basis requires broadening. Before the method can be used to monitor a particular arylamine, the haemoglobin adduct must be verified and the analytical method worked out.

Fig. 1. Metabolic activation of aromatic amines

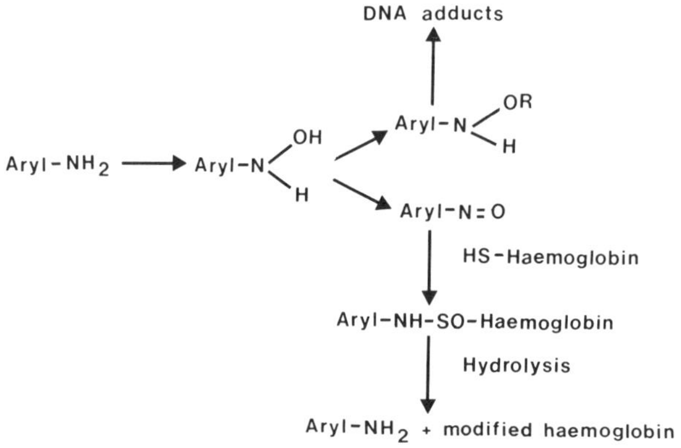

DNA adducts

Aryl-N(OR)(H)

Aryl-NH_2 → Aryl-N(OH)(H)

Aryl-N=O

HS-Haemoglobin

Aryl-NH-SO-Haemoglobin

Hydrolysis

Aryl-NH_2 + modified haemoglobin

Arylhydroxylamines are common precursors of both DNA adducts and haemoglobin binding.

Further, before risk estimations can be made, we need to know more about the role of macromolecular binding in carcinogenesis. The importance of developing biological monitoring procedures and of using them in humans is that we will ultimately be able to bridge the gap between experimental animals and man and to assess human risk.

Relationship between binding to DNA and to haemoglobin

Acute and chronic toxic effects of aromatic amines are attributed to the generation of reactive metabolites that interact with tissue components. The metabolism of aromatic amines is very complex, and several pathways can lead to reactive metabolites (Miller & Miller, 1969), *N*-oxidation being the most important step. Arylhydroxylamines (or aryl-*N*-hydroxyacetamides) are the common precursors of both the ultimate genotoxic metabolites that react with DNA and nitrosoarenes, which readily react with sulphydryl groups of proteins (Fig. 1). Some of the primary reaction product of nitrosoarenes with sulphydryl groups rearranges to form a stable sulphinic acid amide. The level of this protein adduct is particularly high in haemoglobin because nitrosoarenes are formed within erythrocytes in the course of methaemoglobin production, a cyclic process in which one molecule of arylhydroxylamine may be oxidized to the nitrosoarene several times, thereby increasing the chances of adduct formation. The haemoglobin adduct therefore provides an amplified measure of the bioavailability of *N*-oxidation products and is in this way correlated with the generation of genotoxic metabolites. This notion is strongly supported by the observation that binding of metabolites to DNA, RNA and proteins in several tissues and to haemoglobin and albumin in blood was directly proportional to dose over a wide range of doses (Neumann 1980, 1984b), indicating that the pharmacokinetics apparently follow first order. Haemoglobin can be obtained from blood samples more easily and in greater amounts than DNA, and hydrolysis of the sulphinic acid amide usually releases the parent amine, which can be extracted and quantified.

Haemoglobin binding of some aromatic amines in rats

We have recently studied the haemoglobin binding of a number of monocyclic aromatic amines, most of which are listed as suspected or established animal

Table 1. Haemoglobin binding indices (HBI) of monocyclic aromatic amines[a]

Arylamine	HBI
Aniline	22
o-Toluidine	4.0
m-Toluidine	4.9
p-Toluidine	4.2
2,4-Dimethylaniline	2.3
2,4,5-Trimethylaniline	0.7
4-Chloroaniline	569
4-Chloro-*o*-toluidine	28
5-Chloro-*o*-toluidine	28
6-Chloro-*o*-toluidine	0.6

[a]Haemoglobin was isolated 24 h after oral administration of 0.6 mmol/kg of the arylamine to female Wistar rats
[b]Bound (mmol): dose (mmol/kg) Hb

carcinogens in the MAK list of the Federal Republic of Germany (Henschler, 1986). The compounds were administered orally to female Wistar rats, blood samples were taken after 24 h, haemoglobin isolated and hydrolysed, and the extracted amines analysed by capillary gas chromatography with a flame-ionization and/or nitrogen-sensitive detector (Albrecht & Neumann, 1985; Birner & Neumann, 1987). The calculated haemoglobin binding indices are listed in Table 1. The values represent only the hydrolysable fraction, and not total binding of metabolites, and may therefore be at variance with data for total binding obtained with labelled compounds.

The parent amine was identified as the only cleavage product in all cases. The level of binding varied by a factor of 1000 (0.6–569), the value for 4-chloroaniline being particularly high. It appears that substitution of aniline by methyl groups decreases and substitution by chloro atoms increases haemoglobin binding. Steric hindrance by one or two substituents in the *ortho* positon to the amino group may be responsible for decreasing haemoglobin binding. However, it should be kept in mind that adduct formation occurs at the end of many competing distribution and reaction steps, which may be influenced differently by molecular modifications. Much more information will be necessary to interpret these results and to establish general rules for the comprehensive group of monocyclic aromatic amines, if they exist. Nevertheless, exposure to all of these arylamines can be monitored by analysing the haemoglobin adduct.

Similarly, we studied benzidine and some congeners. In this case, the extracted cleavage products were analysed by high-performance liquid chromatography using an electrochemical detector (G. Birner & H.-G. Neumann, unpublished). It can be seen from Table 2 that the human carcinogen benzidine forms three hydrolysable adducts with haemoglobin, and the animal carcinogens 3,3′-dichlorobenzidine and 3,3′-dimethoxybenzidine only two adducts. The major product of hydrolysis with benzidine has been identified as the monoacetylated derivative, the parent diamine accounting for only about 10%. The result indicates that these diamines are partly *N*-oxidized in their monoacetylated form to give *N*-hydroxy-*N*′-acetylbenzidines as the proximate genotoxic metabolites. This conforms with the finding that the major DNA adduct of benzidine is a guanine-C8-monoacetylbenzidine derivative (Kennelly *et al.*, 1984) and supports the existence of a correlation between metabolites binding to haemoglobin and to DNA (see Fig. 1).

Table 2. Haemoglobin binding indices (HBI) of benzidine and some congeners[a]

Arylamine	No. of adducts	HBI of cleavage products		
		Monoacetyl derivative	Diamine	Unknown
Benzidine	3	19 ± 3.5	2.4 ± 0.1	3.0 ± 0.1
3,3′-Dichlorobenzidine	2	1.5 ± 0.6	2.0 ± 0.7	
3,3′-Dimethoxybenzidine	2	2.5	1.0	
3,5,3′,5′-Tetramethylbenzidine	0	not detectable		

[a]Haemoglobin was isolated 24 h after oral administration of 0.5 mmol/kg of the arylamine to female Wistar rats (mean ± SD; $n = 3$).

No haemoglobin adduct could be detected with 3,5,3′,5′-tetramethylbenzidine. This does not prove that a *N*-hydroxy derivative is not formed metabolically: steric hindrance could prevent the reaction of the nitroso derivative with haemoglobin; but, in contrast to several other benzidine congeners, this compound is also nonmutagenic (McCann *et al.*, 1975), indicating that DNA-binding derivatives are either not formed under the conditions of the test system or do not react with DNA. The compound is oxidized by peroxidases, a reaction which generates a blue charge-transfer complex (Josephy *et al.*, 1982) and forms the basis for the use of 3,5,3′,5′-tetramethylbenzidine as a reagent for the detection of occult blood (Holland *et al.* 1974). Although the relevance of this reaction for toxic effects *in vivo* cannot be ascertained, the results support the use of this compound in blood tests as a less hazardous replacement for other benzidine congeners.

Correlation with biological endpoints

The target dose concept (Ehrenberg *et al.*, 1983) implies that haemoglobin binding of carcinogenic chemicals is correlated with modification of critical DNA targets, and possibly with carcinogenesis. In order to substantiate this concept further, we are looking for correlations between haemoglobin binding and biological endpoints. Table 3 gives a list of aromatic amines with their classification by the IARC and by the MAK Commission. The list comprises established human carcinogens, as well as established and doubtful animal carcinogens. The new listing of 4-chloro-*o*-toluidine as a human

Table 3. Haemoglobin binding indices (HBI) in the rats of selected aromatic amines, and their classifications as carcinogens

Arylamine	HBI[a]	Genotoxicity	Classification	
			IARC	MAK Commission
4-Chloroaniline	569	+	—	—
4-Aminobiphenyl	344	+	1	III A1
4-Chloro-*o*-toluidine	28	+	—	III A1
Benzidine	24	+	1	III A1
Aniline	22	+	3	III B
o-Toluidine	4	+	2A	III A2
3,3′-Dichlorobenzidine	3.5	+	2B	III A2
3,3′-Dimethoxybenzidine	3.5	+	2B	III A2
2,4-Dimethylaniline	2.3	+	—	III B

[a]Calculated from hydrolysable adducts

Table 4. Correlation between haemoglobin binding index (HBI) and methaemoglobin (MetHb) formation in rats

Arylamine	HBI	% MetHb[a]	Time to maxima (h)
4-Chloroaniline	569	49 ± 4.5	1
4-Aminobiphenyl	344	52 ± 7.2	1.5
5-Chloro-*o*-toluidine	28	2.0 ± 0.4	1.5
Aniline	22	1.7 ± 0.7	0.5
2,4-Dimethylaniline	2.3	1.0 ± 0.5	1.25

[a]0.6 mmol/kg of the arylamines was administered orally to female Wistar rats and serial blood samples were taken for up to 6 h (mean ± SD, $n = 3$)

carcinogen should be emphasized, because this is the first of the industrial monocyclic aromatic amines to be placed in this category. All of the listed compounds are genotoxic in more than one short-term test; therefore, it may be concluded that all of them yield DNA-binding and protein-binding metabolites *in vivo* and that they have a certain carcinogenic potential. DNA binding has been measured in only a few cases, but haemoglobin binding proves the generation of *N*-oxidized metabolites in all cases. However, it is impossible to classify carcinogenic potential on the basis of the available information.

Another endpoint that demonstrates the availability of *N*-oxidized arylamine metabolites is methaemoglobin formation. We have measured the generation of methaemoglobin in female Wistar rats after oral administration of representative arylamines (G. Birner & H.-G. Neumann, unpublished). The maximal levels of methaemoglobin correlate remarkably well with the haemoglobin binding index (Table 4), showing that the two parameters indicate in parallel the availability of *N*-oxidation products within erythrocytes. This correlation could show that the haemoglobin binding index indicates erythrocyte stress, resulting in premature, excessive degradation of erythrocytes in the spleen, an effect that has been related to the generation of haemangiosarcoma in this tissue (Gralla *et al.*, 1979; Bus, 1983; Neumann, 1985). However, the extent of methaemoglobin formation is not the only determinant of ageing or haemolysis of erythrocytes, and more information about the toxic effects of arylamines must be obtained in order to interpret these complex relationships.

Carcinogenic aromatic amines exhibit notable species- and tissue-specific effects. Usually, these are explained, at least in part, by differences in metabolism. We have investigated whether haemoglobin binding reflects differences in target dose due to pharmacokinetic parameters. Haemangiosarcoma is produced by aniline in male rats, by *o*-toluidine in rats and mice, and by 4-chloro-*o*-toluidine and 5-chloro-*o*-toluidine only in mice. The last of these compounds also produces liver tumours in mice. We have measured haemoglobin binding of these compounds in mice (G. Birner & H.-G. Neumann, unpublished) and found the levels to be generally lower in mice than in rats (Table 5) and not to correlate consistently with tumour formation. With 5-chloro-toluidine, the haemoglobin binding index would be expected to be higher; however, 4-chloroaniline is the most active, even in mice.

Correlation with molecular endpoints

The level of DNA binding in livers of animals treated with 4-chloro-*o*-toluidine was greater in mice than in rats (8 *versus* 4.8 pmol/mg), but binding to liver proteins was

Table 5. Haemoglobin binding index (HBI) and carcinogenicity in different species

Arylamine	Haemangiosarcoma		HBI	
	Rat (Fischer)	Mouse ($B6C3F_1$)	Rat (Wistar)	Mouse (B6C3F)
Aniline	+	–	22	2.2
o-Toluidine	+	+	4	2.1
4-Chloro-*o*-toluidine	–	+	28	2.5
5-Chloro-*o*-toluidine	–	+	28	1.0
4-Chloroaniline	(±)	(±)	569	132

more extensive in rats than in mice (200 *versus* 60 pmol/mg; Bentley *et al.*, 1986). These authors concluded that different metabolites are responsible for binding to DNA and protein and that the pattern of metabolites differs between the species. Liver is not a target tissue for this compound in mice, but neither could DNA damage be demonstrated in target capillary endothelial cells. The relationships between tissue dose or haemoglobin binding (see Table 5) and the haemangiosarcomatogenic effect are obviously not simple.

In experiments designed to correlate macromolecular damage in rat liver with the initiating activity of 2-acetylaminofluorene, a complete carcinogen, and of *trans*-4-acetylaminostilbene, a putative pure initiator for this tissue, we have measured macromolecular binding in liver, kidney and blood at the end of the initiation period (Table 6; Ruthsatz *et al.*, 1986; M. Ruthsatz & H.-G. Neumann, unpublished). One variable is the time that elapsed between the last treatment and the day of analysis. The results show that both haemoglobin binding and plasma protein binding reflect the tissue dose semiquantitatively. Plasma protein binding compares quite closely to liver protein binding. Haemoglobin binding is considerably greater and decreases more slowly than plasma protein binding, as expected. Haemoglobin binding correctly predicts the greater DNA binding and initiating potential of *trans*-4-acetylaminostilbene as compared to 2-acetylaminofluorene.

Macromolecular binding in liver, kidney and blood was also measured in an experiment in which rats received a toxic dose of *trans*-4-acetylaminostilbene either with no other treatment or after pretreatment with methylcholanthrene. Whereas in the first case most of the animals die after a latent period of 12 days, from stomach

Table 6. Binding of *trans*-4-acetylaminostilbene (AAS)[a] and 2-acetylaminofluorene (AAF)[b] metabolites to different macromolecular targets on day 28 after oral administration to female Wistar rats (pmol/mg)

Oral administration (day)		Liver		Kidney	Blood	
0, 3, 6, 9	14, 17, 20, 23	DNA	Protein	DNA	Haemoglobin	Plasma protein
AAS		31	11	13	89	6
	AAS	54	52	30	285	46
AAF		58	11	47	84	12
	AAF	115	25	74	165	53

[a] 4 × 20 μmol/kg
[b] 4 × 100 μmol/kg

Table 7. Binding of *trans*-4-acetylaminostilbene metabolities (pmol/mg) to different macromolecular targets after oral administration of an acutely toxic dose (211 μmol/kg) to female Wistar rats

Treatment[a]	Liver		Kidney		Blood	
	DNA	Protein	DNA	Protein	Haemoglobin	Plasma protein
24 h						
Untreated	80	219	20	53	674	151
MC-pretreated	53	301	6	27	29	60
7 days						
Untreated	139	184	122	172	634	181
MC-pretreated	29	86	28	16	13	5

[a]MC, methylcholanthrene

bleeding, methylcholanthrene pretreatment prevents toxicity (Marquardt *et al.*, 1985; Pfeifer & Neumann, 1986). Methylcholanthrene acts by increasing the rate of metabolism, thereby reducing the liver first-pass effect, and by shifting metabolism towards more efficient inactivation (A. Pfeifer & H.-G. Neumann, unpublished). The drastic reduction in macromolecular binding by methylcholanthrene pretreatment in extrahepatic tissues, such as kidney (Table 7), is directly reflected by binding to blood proteins. The extent and time course of macromolecular binding in liver parallels neither those in extrahepatic tissues nor those in blood.

Conclusions

Binding to haemoglobin correlates with biologically active doses of aromatic amines and would therefore be expected to correlate with the target dose for genotoxic affects. However, most targets are ill-defined, and the relationships between DNA binding and protein binding *in vivo* are not simple. Knowing the rates of reaction of putative ultimate aromatic amine metabolites with DNA and proteins *in vitro* is not sufficient for calculating satisfactorily the genotoxic target dose on the basis of haemoglobin binding. The exact relationship between target dose and genotoxic effect is unknown for all the compounds considered, nor is it clear how genotoxic effects are quantitatively related to initiation. Finally, many data indicate that the extent of DNA damage may correlate with initiation but not necessarily with tumour formation. Our understanding of the mechanism of action of carcinogenic aromatic amines must advance considerably before we can assess the tumour risk from data obtainable by biological monitoring. This should not inhibit us from using the new methods or from monitoring exposure. Only if we know more about the biologically active doses of carcinogens required to produce cancer in man will we be able to define acceptable environmental exposures.

Acknowledgements

Work in this laboratory was supported by the Deutsche Forschungsgemeinschaft.

References

Albrecht, W. & Neumann, H.-G. (1985) Biomonitoring of aniline and nitrobenzene. Hemoglobin binding in rats and analysis of adducts. *Arch. Toxicol.*, *57*, 1–5

Bentley, P., Bieri, F., Muecke, W., Waechter, F. & Stäubli, W. (1985) Species differences in the toxicity of *p*-chloro-*o*-toluidine to rats and mice. Covalent binding to hepatic macromolecules and hepatic non-parenchymal cell DNA and an investigation of effects upon the incorporation of (^{3}H) thymidine into capillary endothelial cells. *Chem.-biol. Interactions, 57,* 27–40

Birner, G. & Neumann, H.-G. (1987) Biomonitoring of aromatic amines. Binding of substituted anilines to rat hemoglobin. *Naunyn Schmiedeberg's Arch. Pharmacol., Suppl. 335,* 73

Bus, J.S. (1983) Aniline and nitrobenzene: erythrocyte and spleen toxicity. *Chemical Industry Institute of Technology (CIIT) Activities 3, No. 12*

Ehrenberg, L., Hiesche, K.D., Osterman-Golkar, S. & Wenneberg, I. (1974) Evaluation of genetic risk of alkylating agents: tissue dose in the mouse from air contaminated with ethylene oxide. *Mutat. Res., 24,* 83–103

Ehrenberg, L., Moustacchi, E. & Osterman-Golkar, S. (1983) Dosimetry of genotoxic agents and dose-response relationships of their effects. *Mutat. Res., 123,* 121–182

Farmer, P.B., Neumann, H.-G. & Henschler, D. (1987) Estimation of exposure of man to substances reacting covalently with macromolecules. *Arch. Toxicol., 60,* 251–260

Gralla, E.J., Bus, J.S., Reno, F., Cushman, J.R. & Ulland, B.N. (1979) Studies of aniline HCl in rats. *Toxicol. appl. Pharmacol., 48,* A97

Green, L.C., Skipper, P.L., Turesky, R.J., Bryant, M.S. & Tannenbaum, S.R. (1984) *In vivo* dosimetry of 4-aminobiphenyl in rats via a cysteine adduct in hemoglobin. *Cancer Res., 44,* 4254–4259

Groth, U. & Neumann, H.-G. (1971) The relevance of chemico-biological interactions for the toxic and carcinogenic effects of aromatic amines. V, The pharmacokinetics of related aromatic amines in blood. *Chem.-biol. Interactions, 4,* 409–419

Henschler, D., ed. (1986) *Deutsche Forschungsgemeinschaft, Maximum Concentrations at the Workplace and Biological Tolerance Values for Working Materials,* Weinheim, VCH Verlagsgesellschaft mbH

Holland, V.R., Saunders, B.C., Rose, F.L. & Walpole, A.L. (1974) A safer substitute for benzidine in the detection of blood. *Tetrahedron, 30,* 3299–3302

Josephy, P.D., Eling, T. & Mason, R.P. (1982) The horseradish peroxidase-catalyzed oxidation of 3,5,3′,5′-tetramethylbenzidine. *J. biol. Chem., 257,* 3669–3675

Kennelly, J.C., Beland, F.A., Kadlubar, F.F. & Martin, C.N. (1984) Binding of *N*-acetylbenzidine and *N,N′*-diacetylbenzidine to hepatic DNA of rat and hamster *in vivo* and *in vitro*. *Carcinogenesis, 5,* 407–412

Lewalter, J. & Korallus, U. (1985) Blood protein conjugates and acetylation of aromatic amines. New findings on biological monitoring. *Int. Arch. occup. environ. Health, 56,* 179–196

Marquardt, P., Romen, W. & Neumann, H.-G. (1985) Tissue specific, acute toxic effects of the carcinogen trans-4-dimethylaminostilbene. *Arch. Toxicol., 56,* 151–157

McCann, J., Choi, E., Yamasaki, E. & Ames, B.N. (1975) Detection of carcinogens as mutagens in the Salmonella/microsome test: assay of 300 chemicals. *Proc. natl Acad. Sci. USA, 72,* 5135–5139

Miller, E.C. & Miller, J.A. (1969) The metabolic activation of carcinogenic aromatic amines and amides. *Prog. exp. Tumor Res., 11,* 273–301

Neumann, H.-G. (1980) Dose-response relationships in the primary lesion of strong electrophilic carcinogens. *Arch. Toxicol., Suppl. 3,* 69–77

Neumann, H.-G. (1984a) Analysis of hemoglobin as a dose monitor for alkylating and arylating agents. *Arch. Toxicol., 56,* 1–6

Neumann, H.-G. (1984b) *Dosimetry and dose-response relationships.* In: Berlin, A., Draper, M., Hemminki, K. & Vainio, H., eds, *Monitoring Human Exposure to Carcinogenic and Mutagenic Agents (IARC Scientific Publications No. 59),* Lyon, International Agency for Research on Cancer, pp. 115–126

Neumann, H.-G. (1985) *Die Rolle der Promotion bei der Einwirkung krebserzeugender Stoffe.* In: Appel, K.E. & Hildebrandt, A.G., eds, *Tumorpromotoren, bga-Schriften,* Munich, MMV Medizin Verlag, pp. 98–107

Neumann, H.-G. (1986) Macromolecules as dose monitors for exposure to environmental chemicals. *Trends pharmacol. Sci., 7,* 173–174

Neumann, H.-G. (1987) *Concepts for assessing the internal dose of chemicals in vivo.* In: Fowler,

B.A., ed., *Mechanisms of Cell Injury: Implications for Human Risk,* New York, John Wiley & Sons, pp. 241–254

Pfeifer, A. & Neumann, H.-G. (1986) Organ specific acute toxicity of the carcinogen trans-4-acetylaminostilbene is not correlated with macromolecular binding. *Chem.-biol. Interactions, 59,* 185–201

Ruthsatz, M., Franz, R. & Neumann, H.-G. (1986) *DNA-damage, initiation and promotion by aromatic amines.* In: Friedberg, T. & Oesch, F., eds, *Primary Changes and Control Factors in Carcinogenesis,* Wiesbaden, Deutscher Fachschriften-Verlag, pp. 88–91

Wieland, E. & Neumann, H.-G. (1978) Methemoglobin formation and binding to blood constituents as indicators for the formation, availability and reactivity of activated metabolites derived from trans-4-aminostilbene and related aromatic amines. *Arch. Toxicol., 40,* 17–35

ASSESSMENT OF EXPOSURE AND SUSCEPTIBILITY TO AROMATIC AMINE CARCINOGENS

F.F. Kadlubar,[1] G. Talaska,[2] N.P. Lang,[2] R.W. Benson[1] & D.W. Roberts[1]

[1]*Division of Biochemical Toxicology, National Center for Toxicological Research, Jefferson, AR; and* [2]*the John L. McClellan Memorial Veterans Administration Medical Center, Little Rock, AR, USA*

As a consequence of human exposure to carcinogenic aromatic amines, biochemical approaches to risk assessment have emphasized metabolic determinants of individual susceptibility and quantification of arylamine-macromolecular adducts. A known genetic polymorphism in humans, hepatic arylamine acetyltransferase activity, has been associated with differences in individual susceptibility to urinary bladder (slow acetylators) and colorectal (rapid acetylators) cancers. Similarly, the high specificity of an inducible human cytochrome P450 towards the *N*-oxidation of 4-aminobiphenyl and other aromatic amines is consistent with metabolic differences that can be used to predict relative human risk. Exposure to aromatic amines has also been documented, primarily by quantification of adducts with protein or DNA. Using ^{32}P-postlabelling methods and a competitive avidin/biotin-amplified enzyme-linked immunoassay, we have estimated 4-aminobiphenyl-DNA adduct levels in surgical samples of human peripheral lung and urinary bladder epithelium and report values ranging from 2 to 97 adducts per 10^8 nucleotides.

Aromatic amines, such as 4-aminobiphenyl (ABP) and 2-naphthylamine, have been widely recognized as human occupational carcinogens (Parkes & Evans, 1984). Although industrial exposure to these chemicals has been severely restricted, their presence in cigarette smoke and in synthetic fuels has been reported (Patrianakos & Hoffmann, 1979; Haugen *et al.*, 1982). Recently, Bryant *et al.* (1987) have shown the presence of ABP-haemoglobin adducts in human blood samples. While the levels found in cigarette smokers were five to six times higher than in nonsmokers, the fact that adducts were detected consistently in individuals that were not overtly exposed to tobacco or synthetic fuels strongly suggests the existence of unknown and perhaps ubiquitous sources of ABP contamination in the environment. Furthermore, the presence of heterocyclic aromatic amines in cooked foods (Kato, 1986) and of polycyclic nitroaromatic hydrocarbons in diesel exhaust and ambient air (Rosenkranz & Mermelstein, 1986) provides additional sources of aromatic amine exposure to humans. Thus, the development of biomarkers for exposure and susceptibility to aromatic amines represents an important goal for human risk assessment.

Role of acetylation polymorphism in human urinary bladder and colorectal cancers

Hepatic aromatic amine *N*-acetyltransferase has been studied extensively as a genetic polymorphism in humans and has been shown to predispose individuals to several drug toxicities (reviewed by Weber & Hein, 1985). More recently, evidence has emerged that phenotypically slow acetylators are at elevated risk for urinary bladder cancer, while rapid acetylators are predominant among individuals with a history of colorectal cancer (Cartwright, 1984; Lang *et al.*, 1986; Ilett *et al.*, 1987). As discussed below, these observations are consistent with the known metabolic activation pathways for aromatic amine carcinogens and indicate that acetylator phenotype should be a useful predictive tool for determining individual susceptibility to aromatic amine carcinogenesis.

For those primary aromatic amines that are carcinogenic for the urinary bladder, metabolic activation is generally thought to involve hepatic *N*-oxidation to form an *N*-hydroxy arylamine, which is then transported through the circulation to the lumen of the urinary bladder where it reacts with urothelial DNA under slightly acidic conditions. Although *N*-glucuronidation has long been regarded as an important transport mechanism, recent studies with ABP in dogs have shown that the unconjugated *N*-hydroxy metabolite is the predominant form that enters the blood and is filtered into the urinary bladder lumen (reviewed by Kadlubar *et al.*, 1987). Consequently, hepatic *N*-acetylation represents a direct competing pathway for arylamine *N*-oxidation and for formation of the ultimate carcinogenic metabolite. In a summary of eight pharmacogenetic studies of acetylators in Denmark, Sweden, the UK and the USA, Mommsen *et al.* (1985) noted that 65% of bladder cancer patients were slow acetylators as compared with 59% of the controls. While this difference was statistically significant ($p < 0.025$), the observed association between increased risk for bladder cancer and slow acetylator phenotype in a general population was not particularly impressive. However, when exposure to aromatic amine carcinogens has been documented, the excess of individuals with the slow acetylator phenotype has been striking. For example, 22 of 23 bladder cancer patients who worked in the synthetic dye industry were slow acetylators compared to 52 of 88 in the control group (Cartwright *et al.*, 1982). Furthermore, this correlation has been consistently observed in instances of known occupational exposure to aromatic amines and in highly industrialized areas (Cartwright, 1984).

For those aromatic amines that are carcinogenic for the intestinal tract, metabolic activation is also believed to involve hepatic *N*-oxidation. However, effective *N*-glucuronidation of these *N*-hydroxy metabolites appears to result in their biliary excretion, subsequent hydrolysis by intestinal bacterial β-glucuronidases, and their reabsorption into the colonic mucosa. Here, colonic *O*-acetyltransferase further activates the *N*-hydroxy arylamine by forming a reactive *N*-acetoxy derivative that binds covalently to DNA. Since *O*-acetyltransferase activity appears to be catalysed by the arylamine *N*-acetyltransferase (Flammang *et al.*, 1987a,b), it is likely that metabolic activation of *N*-hydroxy arylamines in human colon will exhibit the same genetic polymorphism. Thus, the recent findings that phenotypically rapid acetylators are more prevalent among patients with a history of colorectal cancer are consistent with this route of metabolic activation. These data, which are summarized in Table 1, suggest that aromatic amines may play an important role in the etiology of this cancer in a general population. Potential sources of human exposure to this class of carcinogens include food-borne heterocyclic amines, aromatic amine herbicides and airborne nitroaromatic hydrocarbons.

Thus, available evidence indicates that acetylator polymorphism should be a useful biomarker of individual susceptibility to aromatic amine carcinogenesis. When exposure to urinary bladder carcinogens is suspected, the slow acetylator individual can be regarded as being at greater relative risk. In a general population, however, rapid acetylators seem to be predisposed towards colorectal cancer. In this case, identification of an etiological agent and its quantification would serve to improve further the utility of acetylator phenotyping as a predictive tool in epidemiological studies.

Table 1. Distribution of *N*-acetyltransferase phenotypes in controls and in patients with a prior history of colorectal cancer

Group	No. of fast acetylators/total no. of individuals		
	Lang *et al.* (1986)	Ilett *et al.* (1987)	Combined
Controls	11/41	10/41	21/82
Cancer patients	20/43[a]	27/49[b]	47/92[c]

[a]Compared to controls, $\chi^2 = 3.49$, $p = 0.0100$, odds ratio = 2.36
[b]Compared to controls, $\chi^2 = 8.70$, $p = 0.0015$, odds ratio = 3.80
[c]Compared to controls, $\chi^2 = 11.82$, $p = 0.0001$, odds ratio = 3.03

Role of cytochromes P450 in susceptibility to aromatic amine carcinogenesis

Cytochromes P450 are a family of monooxygenases that catalyse the oxidative metabolism of a wide variety of drugs and carcinogens. Depending on the species, there are some 10–15 different isozymes the sequence homology of which may vary between 20 and 75%. The majority of these are inducible by environmental and endogenous factors, and several appear to exhibit genetic polymorphisms in humans (Distlerath & Guengerich, 1987). Since the *N*-oxidation of carcinogenic arylamines represents a critical initial step in their metabolic activation (*vide supra*), the role of specific hepatic cytochromes P450 in this process has received considerable research emphasis. In the last few years, studies with purified rodent liver P450s have indicated that hydrocarbon-inducible forms corresponding to rat P450c (P450/BNF-B) and P450d (P450/ISF-G) exhibit the highest catalytic activity toward arylamines. For example, 2-aminofluorene and 3-amino-1-methyl-5*H*-pyrido[4,3-*b*]indole (Trp-P-2) were found to be *N*-oxidized effectively by both of these P450s, while *N*-oxidation of ABP, 2-naphthylamine, 2-amino-6-methyldipyrido-[1,2-*a*:3′,2′-*d*]imidazole (Glu-P-1), and 2-amino-3-methylimidazo[4,5-*f*]quinoline (IQ) were catalysed selectively by cytochrome P450d (reviewed by Kadlubar & Hammons, 1987; Kadlubar *et al.*, 1988). In addition, the human orthologue of rat cytochrome P450d has been purified by Distlerath *et al.* (1985) and by Wrighton *et al.* (1986), who have shown that the human and rodent isozymes possess similar immunochemistry and amino-terminal sequences, as well as catalytic activity for phenacetin *O*-deethylase. We have recently compared the ability of 19 preparations of human liver microsomes to catalyse phenacetin *O*-deethylation and the *N*-oxidation of ABP. Although a wide range of activities was observed (Fig. 1), a strong correlation between the rates of ABP *N*-oxidation and phenacetin *O*-deethylation was seen ($p < 0.001$), suggesting their common catalysis by the human orthologue of P450d. This conclusion has been further strengthened by recent findings (Davies *et al.*, 1988; McManus & Burgess, 1988; Sesardic *et al.*, 1988) which indicate that phenacetin *O*-deethylase activity is well correlated with immunoquantified levels of human P450d and with the mutagenic activation of Trp-P-2, IQ and *N*-acetyl-2-aminofluorene by human liver microsomes. In addition, this cytochrome P450 was highly variable in different individuals (60-fold range), appeared to be inducible in the liver by cigarette smoking (about fourfold), and exhibited a genetic polymorphism in humans, with 5–10% of the Caucasian population being deficient in phenacetin *O*-deethylase activity. Thus, individual variation in the capacity for metabolic activation of several aromatic amines is apparent and indicates the need for a noninvasive method for determining the phenotype of those at risk for tumour induction by this class of chemical carcinogens.

Fig. 1. Rates of ABP *N*-oxidation and phenacetin *O*-deethylation by human liver microsomes ($n = 19$; from Kadlubar *et al.*, 1987)

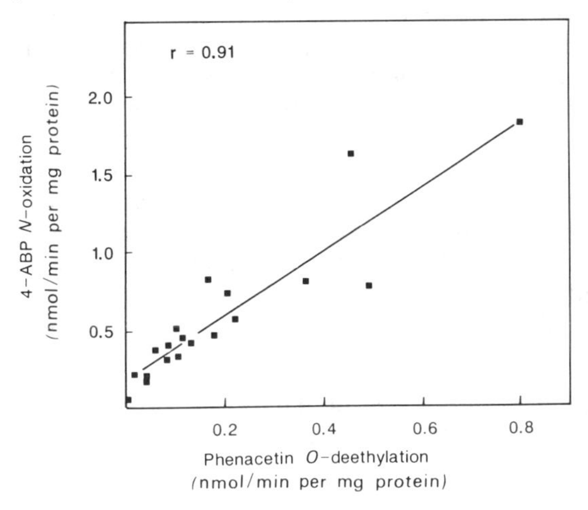

^{32}P-Postlabelling and immunochemical methods for determining human exposure to ABP

Recent studies on assessment of exposure to aromatic amines have focused on the estimation of arylamine-macromolecular adducts. Thus far, these efforts have primarily involved the measurement of adducts with blood proteins, namely haemoglobin (Neumann, 1984; Bryant *et al.*, 1987) and serum albumin (Skipper *et al.*, 1985). With regard to ABP, haemoglobin adducts appear to be formed as a direct consequence of hepatic *N*-oxidation and the subsequent entry of free *N*-hydroxy-ABP into the circulation. A secondary oxidation of *N*-hydroxy-ABP then occurs within erythrocytes to form 4-nitrosobiphenyl, which undergoes an addition reaction with a single cysteine residue on the haemoglobin (Bryant *et al.*, 1987). Additional studies have indicated that ABP-haemoglobin adducts are an accurate measure of the amounts of unconjugated *N*-hydroxy-ABP entering the urinary bladder lumen (Kadlubar *et al.*, 1987); however, the levels of ABP-DNA adducts formed within the urothelium are not always correlated with ABP-haemoglobin adduct levels and, instead, are directly dependent on urine voiding intervals (i.e., retention of *N*-hydroxy-ABP in the bladder lumen). Thus, the need exists to determine both ABP-haemoglobin and ABP-DNA adducts in order to assess both individual exposure and relative risk.

With the development of highly sensitive ^{32}P-postlabelling methods to detect aromatic DNA adducts (Gupta *et al.*, 1982), it became possible to assess genotoxic exposure of humans to carcinogenic arylamines. We initially attempted to apply this technique to the detection in human tissues of *N*-(deoxyguanosin-8-yl)-ABP, the major ABP-DNA adduct likely to be formed in the urinary bladder (Beland *et al.*, 1983). Enzymatic hydrolysis of DNA samples (10 μg) to 3′-nucleotides and postlabelling with polynucleotide kinase and ^{32}P-ATP (4500 Ci/mmol; 20 mCi) were carried out by the published method, except that higher amounts of radioactivity were employed and *n*-butanol extraction was used to obtain a ten-fold enrichment of the ABP-adducted 3′-nucleotide. To determine recovery of the postlabelled ABP adduct, we synthesized the unlabelled *N*-(deoxyguanosin-8-yl)-ABP-3′,5′-bis(phosphate) (C8-pdGp-ABP; Lasko *et al.*, 1987) and added it to the incubation mixture just after postlabelling and apyrase addition. Instead of using thin-layer chromatography to separate the 3′,[5′-^{32}P]-bis(phosphates), we performed a second *n*-butanol extraction, which also provided a ten-fold enrichment over the unadducted ^{32}P-labelled bis(phosphates). Reconstituted samples were applied to a Waters high-performance liquid chromatography (HPLC) system equipped with a μBondpack C_{18} Semi-prep column and then developed with a gradient system of 0.1 M ammonium acetate/0.001 M ammonium phosphate, pH 5.7 (solvent A) and acetonitrile (solvent B), as described previously (Lasko *et al.*, 1987). Fractions that eluted at the retention time (11–13 min) of the synthetic C8-pdGp-ABP were collected, acetonitrile was removed under a stream of argon, and the adducted sample then extracted into butanol, evaporated, reconstituted in column buffer, and reapplied to a second identical HPLC system. Fractions (30-sec) were again collected and ^{32}P levels were determined by liquid scintillation counting (Fig. 2). The overall recovery of the unlabelled C8-pdGp-ABP was $60 \pm 5\%$.

Fig. 2. Rechromatography of ABP-modified calf thymus DNA (A) and human bladder epithelial DNA (B) after ^{32}P-postlabelling

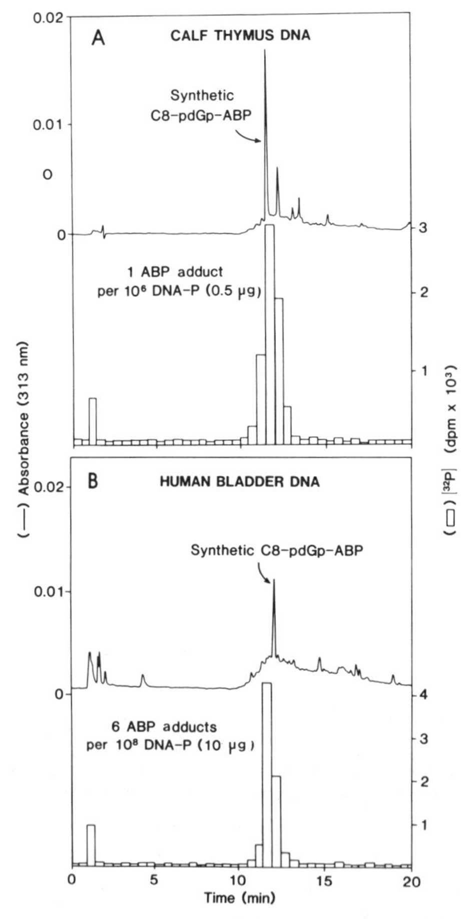

Sample preparation, ^{32}P-postlabelling, addition of synthetic C8-pdGP-ABP and HPLC were carried out as described in the text. The solid line represents ultraviolet-absorbing components in the eluent, and the histogram shows the elution of the radiolabelled bis(phosphate).

ABP-adducted DNA standards were prepared by reaction of calf thymus DNA with ^{3}H-*N*-hydroxy-ABP (55 mCi/mmol) and serial dilution with unmodified DNA to give C8-pdGp-ABP levels of one adduct per 10^6, one per 10^7, one per 10^8 and one per 10^9 nucleotides (DNA-P). As shown in Figure 2A, using this method we readily detected 1.7 fmol or one ABP adduct per 10^6 DNA-P upon injection of an aliquot corresponding to 0.5 μg of the ABP-DNA standard. The other ABP-DNA standards yielded similar results, with a limit of detection of about 6 attomol or 0.2 adducts per 10^9 DNA-P (<100 dpm; 10 μg DNA sample).

Human bladder epithelial cells were then collected by centrifugation of bladder lavages, and DNA was isolated by the method of Cox and Irving (1977). Results obtained by the ^{32}P-postlabelling method (Fig. 2B) indicate the presence of post-labelled C8-pdGp-ABP in human urothelium at a level of six ABP adducts per 10^8 DNA-P. In order to extend the applicability of this method, we then prepared a series of adducted DNA standards, including DNA modified by reactions with *N*-hydroxy-2-naphthylamine, *N*-hydroxy-1-aminopyrene, *N*-hydroxy-*N*′-acetylbenzidine and aflatoxin 2,3-dichloride. Unfortunately, we found that each of these DNA samples yielded postlabelled adduct standards that chromatographed with retention times ranging from 10–16 min in the HPLC system used. Thus, the proximity of these and perhaps other unknown adducts to the C8pdGp-ABP adduct rendered actual detection of an ABP-DNA adduct in human bladder by postlabelling/HPLC subject to question. As further attempts to achieve greater separation of different postlabelled adducts by HPLC were unsuccessful, we turned our attention to the development of an equally sensitive but more selective immunochemical method.

To obtain the desired antibody specificity, the immunogen was prepared by covalently attaching *N*-(guanosin-8-yl)-ABP through its ribose moiety to keyhole limpet haemocyanin using a periodate coupling procedure (Roberts *et al.*, 1985). High-affinity polyclonal rabbit antiserum was obtained, and a sensitive competitive avidin/biotin-amplified enzyme-linked immunoassay (ELISA) was developed, the solid-phase antigen consisting of the same adduct covalently attached to bovine serum albumin (Roberts *et al.*, 1985). This antiserum was characterized extensively and shown to have high specificity for the purine and biphenyl rings of the adduct, namely *N*-(guan-8-yl)-ABP and its corresponding nucleoside and 5′-nucleotide. Accordingly, the antibody did not appreciably recognize *N*-(deoxyadenosin-8-yl)-ABP, guanine, deoxyguanosine, deoxyadenosine, unhydrolysed ABP-modified DNA, ABP-modified serum albumin or the C8-deoxyguanosine adducts of 2-naphthylamine, 1-aminopyrene, 3,2′-dimethyl-4-aminobiphenyl or 2-aminofluorene. Although there was significant cross-reactivity toward *N*-(deoxyguanosin-N^2-yl)-ABP, and the C8-deoxyguanosine adducts of benzidine and *N*-acetylbenzidine, their potential interference in the assay was eliminated by selective hydrolysis of DNA samples with trifluoroacetic acid at 70°C for 1 h. Under these conditions, modified DNA standards released only *N*-(guan-8-yl)-ABP. Furthermore, acid hydrolysis resulted in greater assay sensitivity by allowing the use of 10–100 μg DNA (compared to only 1–10 μg with the postlabelling method). Using a limiting amount of antibody in the presence of an excess of solid-phase coating antigen, the limit of detection was 0.5–5.0 *N*-(guan-8-yl)-ABP adducts per 10^8 DNA-P. This ELISA was then validated by administration of ABP to dogs and analysis of liver and urothelial DNA for the C8-guanine adduct (Kadlubar *et al.*, 1987, 1988).

For our initial studies on the detection of *N*-(guan-8-yl)-ABP in human tissues, we obtained surgical samples of histologically normal peripheral lung or urinary bladder epithelial tissues from 11 individuals with a current history of cigarette smoking and thus of known exposure to ABP (Patrianakos & Hoffmann, 1979). As shown in Figure 3, the values obtained ranged from two to 97 adducts per 10^8 DNA-P, which is within

Fig. 3. Immunochemical determination of *N*-(guan-8-yl)-ABP in human lung and bladder epithelial DNA

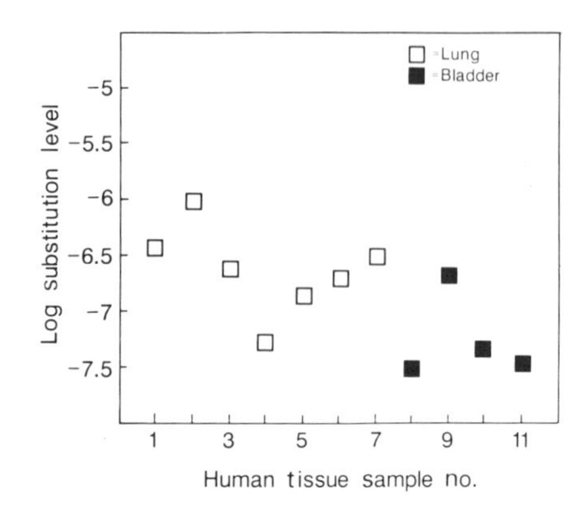

the same range as that determined from the ^{32}P-postlabelling assay. These putative adduct levels are comparable with those reported for other carcinogen-DNA adducts in human tissues from individuals who were not known to be exposed to occupational carcinogens or to chemotherapeutic drugs. In this regard, O^6-methylguanine and polycyclic aromatic hydrocarbon-DNA adduct levels have been detected at levels of one adduct per 10^6 to 10^8 DNA-P in human oral mucosa, oesophagus, bronchus, lung, placenta and white blood cells (reviewed by Dunn & Stich, 1986; Perera, 1987). In future studies to confirm the presence of ABP-DNA adducts in human tissues, we will quantify them by at least one other method, such as ^{32}P-postlabelling or electrophore-labelling/mass spectrometry. In addition, we will examine the relationship between ABP-DNA adducts in human tissues and smoking status, occupation and pharmacogenetic and pharmacodynamic variables.

References

Beland, F.A., Beranek, D.T., Dooley, K.L., Heflich, R.H. & Kadlubar, F.F. (1983) Arylamine-DNA adducts *in vitro* and *in vivo*: their role in bacterial mutagenesis and urinary bladder carcinogenesis. *Environ. Health Perspect., 49,* 125–134

Bryant, M.S., Skipper, P.S., Tannenbaum, S.R. & Maclure, M. (1987) Hemoglobin adducts of 4-aminobiphenyl in smokers and non-smokers. *Cancer Res., 47,* 602–608

Cartwright, R.A. (1984) *Epidemiological studies on N-acetylation and C-center oxidation in neoplasia.* In: Omenn, G.S. & Gelboin, H.V., eds, *Genetic Variability in Responses to Chemical Exposure* (*Banbury Report* 16), Cold Spring Harbor, NY, CSH Press, pp. 359–368

Cartwright, R.A., Glashan, R.W., Rogers, H.J., Ahmad, R.A., Barham-Hall, D., Higgins, E. & Kahn, M.A. (1982) Role of *N*-acetyltransferase phenotypes in bladder carcinogenesis: a pharmacogenetic approach to bladder cancer. *Lancet, ii,* 842–846

Cox, R. & Irving, C.C. (1977) Selective accumulation of O^6-methylguanine in DNA of rat bladder epithelium after intravesical administration of *N*-methyl-*N*-nitrosourea. *Cancer Lett., 3,* 265–270

Davies, D.S., Boobis, A.R., Sesardic, D., Murray, S., Rice, J. & Gooderham, N. (1988) *Isozyme specificity of drug oxidation by animal and human tissues.* In: Abstracts, VIIth International Symposium on Microsomes and Drug Oxidations, 18–21 August 1987, Adelaide, Australia, Abstract S31

Distlerath, L.M. & Guengerich, F.P. (1987) *Enzymology of human liver cytochromes P-450.* In: Guengerich, F.P., ed., *Mammalian Cytochromes P-450,* Vol. I, Boca Raton, FL, CRC Press, pp. 133–198

Distlerath, L.M., Reilly, P.E.B., Martin, M.V., Davis, G.G., Wilkinson, G.R. & Guengerich, F.P. (1985) Purification and characterization of the human liver cytochromes P-450 involved in debrisoquine 4-hydroxylation and phenacetin *O*-deethylation, two prototypes for genetic polymorphism in oxidative drug metabolism. *J. biol. Chem., 260,* 9057–9067

Dunn, B.P. & Stich, H.F. (1986) ^{32}P-Postlabelling analysis of aromatic DNA adducts in human oral mucosal cells. *Carcinogenesis, 7,* 1115–1120

Flammang, T.J., Hein, D.W., Talaska, G. & Kadlubar, F.F. (1987a) *N-Hydroxyarylamine O-acetyltransferase and its relationship to aromatic amine N-acetyltransferase polymorphism in the inbred hamster and in human tissue cytosol.* In: King, C.M., Romano, L.J. & Schuetzle, D., eds, *Carcinogenic and Mutagenic Responses to Aromatic Amines and Nitroarenes,* New York, Elsevier, pp. 137–148

Flammang, T.J., Yamazoe, Y., Guengerich, F.P. & Kadlubar, F.F. (1987b) The S-acetyl coenzyme A-dependent metabolic activation of the carcinogen *N*-hydroxy-2-aminofluorene by human liver cytosol and its relationship to the aromatic amine *N*-acetyltransferase phenotype. *Carcinogenesis, 8,* 1967–1970

Gupta, R.C., Reddy, M.V. & Randerath, K. (1982) ^{32}P-Postlabeling analysis of non-radioactive aromatic carcinogen-DNA adducts. *Carcinogenesis, 3,* 1081–1092

Haugen, D., Peak, M.J., Suhrbler, K.M. & Stamoudis, V.C. (1982) Isolation of mutagenic aromatic amines from a coal conversion oil by cation exchange chromatography. *Anal. Chem., 54,* 32–37

Ilett, K.F., David, B.M., Detchon, P., Castleden, W.M. & Kwa, R. (1987) Acetylation phenotype in colorectal carcinoma. *Cancer Res., 47,* 1466–1469

Kadlubar, F.F. & Hammons, G.J. (1987) *The role of cytochromes P*-450 *in the metabolism of chemical carcinogens.* In: Guengerich, F.P., ed., *Mammalian Cytochromes P*-450, Vol. II, Boca Raton, FL, CRC Press, pp. 81–130

Kadlubar, F.F., Dooley, K.L., Benson, W.R., Roberts, D.W., Butler, M.A., Teitel, C.H. & Young, J.F. (1987) *Pharmacokinetic model of aromatic amine-induced urinary bladder carcinogenesis in beagle dogs administered* 4*-aminobiphenyl.* In: King, C.M., Romano, L.J. & Schuetzle, D., eds, *Carcinogenic and Mutagenic Responses to Aromatic Amines and Nitroarenes,* New York, Elsevier, pp. 173–180

Kadlubar, F.F., Butler, M.A., Hayes, B.W., Beland, F.A. & Guengerich, F.P. (1988) *Role of microsomal cytochrome P*-450 *and prostaglandin H synthase in* 4*-aminobiphenyl-DNA adduct formation.* In: Miners, J.O., Birkett, D.J., Drew, R., May, B. & McManus, M. eds, *Microsomes and Drug Oxidations,* London, Taylor & Francis (in press)

Kato, R. (1986) Metabolic activation of mutagenic heterocyclic aromatic amines from protein pyrolysates. *CRC crit. Rev. toxicol., 16,* 307–348

Lang, N.P., Chu, D.Z.J., Hunter, C.F., Kendall, D.C., Flammang, T.J. & Kadlubar, F.F. (1986) Role of aromatic amine acetyltransferase in human colo-rectal cancer. *Arch. Surg., 121,* 1259–1261

Lasko, D.D., Basu, A.K., Kadlubar, F.F., Evans, F.E., Lay, J.O., Jr & Essigmann, J.M. (1987) A probe for the mutagenic activity of the carcinogen 4-aminobiphenyl: synthesis and characterization of an M13mp10 genome containing the major carcinogen-DNA adduct at a unique site. *Biochemistry, 26,* 3072–3082

McManus, M.E. & Burgess, W.M. (1987) *Mutagenic activation of the food derived heterocyclic amine* 2*-amino-*3*-methylimidazo*[4,5-*f*]*quinoline by rabbit and human liver microsomes.* In: Abstracts, VIIth International Symposium on Microsomes and Drug Oxidations, 18–21 August 1987, Adelaide, Australia, Abstract P161

Mommsen, S., Barfod, N.M. & Aagard, J. (1985) *N*-Acetyltransferase phenotypes in the urinary bladder carcinogenesis of a low-risk population. *Carcinogenesis, 6,* 199–201

Neumann, H.-G. (1984) Analysis of hemoglobin as a dose monitor for alkylating and arylating agents. *Arch. Toxicol., 56,* 1–6

Parkes, H.G. & Evans, A.E.J. (1984) *Epidemiology of aromatic amine cancers.* In: Searle, C.E., ed., *Chemical Carcinogens,* 2nd Ed., Washington DC, American Chemical Society, pp. 277–301

Patrianakos, C. & Hoffmann, D. (1979) Chemical studies on tobacco smoke. LXIV. On the analysis of aromatic amines in cigarette smoke. *J. anal. Toxicol., 3,* 150–154

Perera, F.P. (1987) Molecular cancer epidemiology: a new tool in cancer prevention. *J. natl Cancer Inst., 78,* 887–898

Roberts, D.W., Benson, R.W., Flammang, T.J. & Kadlubar, F.F. (1985) *Development of an avidin-biotin amplified enzyme-linked immunoassay for detection of DNA adducts of the human bladder carcinogen* 4-*aminobiphenyl.* In: Simic, M., Grossman, L. & Upton, A.C., eds, *Mechanisms of DNA Damage and Repair: Implications for Carcinogenesis and Risk Assessment,* New York, Plenum Press, pp. 479–488

Rosenkranz, H.S. & Mermelstein, R. (1986) The genotoxicity, metabolism and carcinogenicity of nitrated polycyclic aromatic hydrocarbons. *J. environ. Sci. Health, 2,* 221–272

Sesardic, D., Boobis, A.R., Harries, G.C., Edwards, R.J. & Davies, D.S. (1987) *Inducibility of the form of cytochrome P*-450 *catalyzing aromatic amine oxidation in man.* In: Abstracts, VIIth International Symposium on Microsomes and Drug Oxidations, 18–21 August 1987, Adelaide, Australia, Abstract P160

Skipper, P.L., Obiedzinski, M.W., Tannenbaum, S.R., Miller, D.W., Mitchum, R.K. & Kadlubar, F.F. (1985) Identification of the serum albumin adduct formed by 4-aminobiphenyl *in vivo* in rats. *Cancer Res., 45,* 5122–5127

Weber, W.W. & Hein, D.W. (1985) *N*-Acetylation pharmacogenetics. *Pharmacol. Rev., 37,* 25–79

Wrighton, S.A., Campanile, C., Thomas, P.E., Maines, S.L., Watkins, P.B., Parker, G., Mendez-Picon, G., Haniu, M., Shively, J.E., Levin, W. & Guzelian, P.S. (1986) Identification of a human liver cytochrome P-450 homologous to the major isosafrole-inducible cytochrome P-450 in the rat. *Mol. Pharmacol, 29,* 405–410

DNA ADDUCT FORMATION DURING CONTINUOUS FEEDING OF 2-ACETYLAMINOFLUORENE AT MULTIPLE CONCENTRATIONS

F.A. Beland,[1,2] N.F. Fullerton,[1] T. Kinouchi[1,2] & M.C. Poirier[3]

[1]*National Center for Toxicological Research, Jefferson, AR;* [2]*University of Arkansas for Medical Sciences, Little Rock, AR; and* [3]*National Cancer Institute, Bethesda, MD, USA*

A linear relationship was observed between the administered dose and DNA adduct levels in the livers and bladders of BALB/c mice fed the carcinogen, 2-acetylaminofluorene (2-AAF), continuously for one month. A similar linear correlation was found between the probit of the liver tumour incidence and the log of the liver DNA adduct levels; however, because there is a no-observable-effect level for bladder tumour induction, the relationship between the probit of the bladder tumour incidence and bladder DNA adduct levels was not linear. These data suggest that the relationship between DNA adduct formation and tumour incidence may be tissue specific.

Carcinogen-DNA adducts have been proposed as biomarkers to monitor the exposure of humans to carcinogens (Wogan & Gorelick, 1985); however, accurate estimates of risk will depend upon determining the relationship between adduct levels and eventual tumour incidence. Animal models in which a large range of concentrations are administered chronically can be very useful in elucidating this relationship. Studies in rats (Poirier *et al.*, 1984) have shown that adduct levels in the liver increase during the first two weeks of feeding and then maintain steady-state conditions in which the rate of adduct formation equals the rate of adduct removal. Similar studies of DNA adducts have not been conducted in mice, but Jackson *et al.* (1980) reported that tissue levels of 2-AAF and/or its metabolites in mice reach steady-state concentrations after approximately two weeks of feeding. In this study, we determined the carcinogen-DNA adduct levels in the livers and bladders of BALB/c mice fed 0–150 mg 2-AAF per kg diet continuously. In addition to correlating adduct levels with the administered dose, these levels were compared with the tumour incidence in mice fed the same diets for up to 33 months (Farmer *et al.*, 1979).

Carcinogen administration and adduct analyses

Weanling female BALB/c mice (four to five per group; obtained from the breeding colony at the National Center for Toxicological Research) were fed 0, 5, 10, 15, 30, 45, 60, 75, 100 or 150 mg 2-AAF per kg diet for one month. Upon sacrifice, the livers and urinary bladders were excised, and, following preparation of hepatic nuclei (Basler *et al.*, 1981), DNA was isolated from each tissue (Beland *et al.*, 1984). Typical DNA yields were 940 ± 180 μg (mean $\pm$ SD) for liver DNA and 32 ± 8 μg for bladder DNA. Adduct levels from each sample were determined by radioimmunoassay, using a polyclonal antibody specific for *N*-(deoxyguanosin-8-yl)-2-aminofluorene (dG-C8-AF),

Fig. 1. DNA adduct levels in the livers (□) and bladders (○) of female BALB/c mice fed 2-AAF continuously for one month

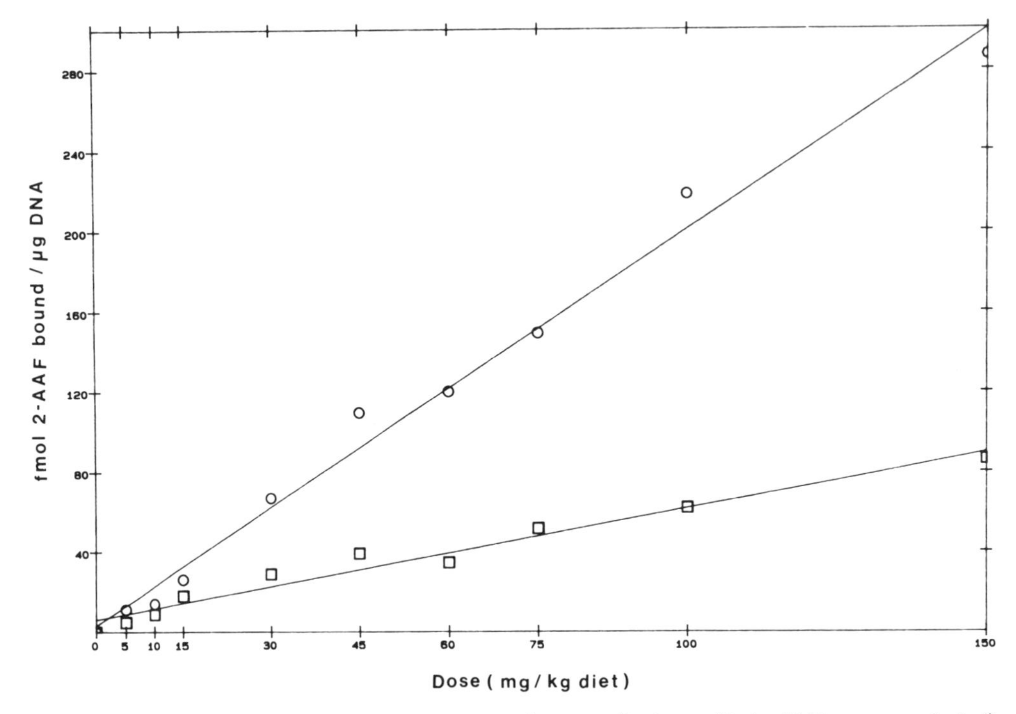

Adduct levels were determined by radioimmunoassay using an antibody specific for *N*-(deoxyguanosin-8-yl)-2-aminofluorene. See Table 1 for additional details. The solid lines were obtained from linear regression analyses by the method of least squares.

with ^{3}H-G-C8-AF as the tracer (Poirier *et al.*, 1984). This adduct has been shown to be the major product bound to DNA in mouse liver after administration of the proximate carcinogen, *N*-hydroxy-2-AAF (Lai *et al.*, 1985).

DNA adducts *versus* dose

The relationship between administered dose and adduct levels is shown in Figure 1. In liver DNA, there was a linear correlation between the log of the administered dose and the level of dG-C8-AF ($r = 0.98$). A similar relationship was observed for bladder DNA ($r = 0.99$), except that at each dose level there was four times more dG-C8-AF bound to bladder DNA than to hepatic DNA.

DNA adducts *versus* tumours

2-AAF is carcinogenic for both the liver and bladder of BALB/c mice. The kinetics of tumour formation in these tissues has been studied by administering 0–150 mg 2-AAF per kg diet continuously for up to 33 months with sacrifices at 18, 24 and 33 months (Farmer *et al.*, 1979). Interestingly, while the hepatic tumour incidence was linear throughout the entire dose range, a no-effect level was observed for bladder tumour induction (Table 1). Farmer *et al.* (1979) suggested that the no-observable-

Table 1. Liver and bladder DNA adduct levels and tumour incidence in mice fed 2-AAF continuously

Dose (mg/kg)	DNA binding[a] (fmol/μg DNA)	Tumour incidence[b] (%)		
		18 months	24 months	33 months
Liver				
0	0 ± 0	1.1	2.6	17.0
5	4.89 ± 0.95	—	—	—
10	8.61 ± 1.53	—	—	—
15	17.95 ± 4.38	—	—	—
30	29.05 ± 7.71	1.7	7.8	30.3
35	—	1.8	9.4	72.0
45	39.30 ± 8.77	2.7	11.0	40.8
60	34.60 ± 2.47	3.2	15.6	44.8
75	51.73 ± 12.40	4.5	20.1	49.3
100	61.85 ± 7.83	8.9	25.6	71.0
150	86.13 ± 10.62	11.4	40.1	81.8
Bladder				
0	0 ± 0	0.4	0.3	1.0
5	11.3 ± 3.9	—	—	—
10	14.2 ± 5.4	—	—	—
15	26.1 ± 9.8	—	—	—
30	67.0 ± 10.5	0.7	0.4	1.1
35	—	0.4	0.4	0.0
45	109.6 ± 59.3	1.1	0.5	1.9
60	120.2 ± 31.5	1.2	0.8	3.0
75	149.6 ± 35.8	1.0	2.2	16.0
100	218.6 ± 12.1	4.7	17.2	67.7
150	286.9 ± 13.3	51.7	75.4	100.0

[a]Weanling female BALB/c mice were fed 2-AAF at the concentration indicated for one month. Adduct levels were determined by radioimmunoassay and represent the mean ± SD of three to five animals
[b]Tumour data are from Farmer *et al.* (1979). Dead or moribund mice were grouped with those sacrificed at the time indicated. The mice were grouped as follows: 18 months = 533 to 637 days; 24 months = 638 to 864 days; 33 months = 865 to 1001 days

effect level resulted from efficient hepatic detoxification of reactive metabolites at low doses of carcinogen, which protects the bladder from tumorigenesis. However, in a subsequent study (Jackson *et al.*, 1980), a linear correlation was found between bladder tissue levels of 2-AAF and/or its metabolites and the administered dose throughout the entire dose range. Relating total tissue levels of a carcinogen and/or its metabolites to tumour incidence may be misleading because DNA adducts represent only a small fraction of this material. We therefore compared the DNA adduct levels in mice fed 2-AAF for one month to the tumour incidence reported by Farmer *et al.* (1979). As shown in Figure 2, there was a linear relationship between the probit of the liver tumour incidence and the log of the liver adduct levels. This was not the case for bladder tumours (Fig. 3), because, as had been observed when dose was compared with tumour incidence, a no-observable-effect level exists, below which the tumour incidence does not exceed that of the untreated controls.

Conclusions

In this study we found a linear correlation between the administered dose of 2-AAF and the level of dG-C8-AF in both liver and bladder DNA, the adduct levels in the

Fig. 2. Relationship between liver DNA adduct levels and tumour incidence

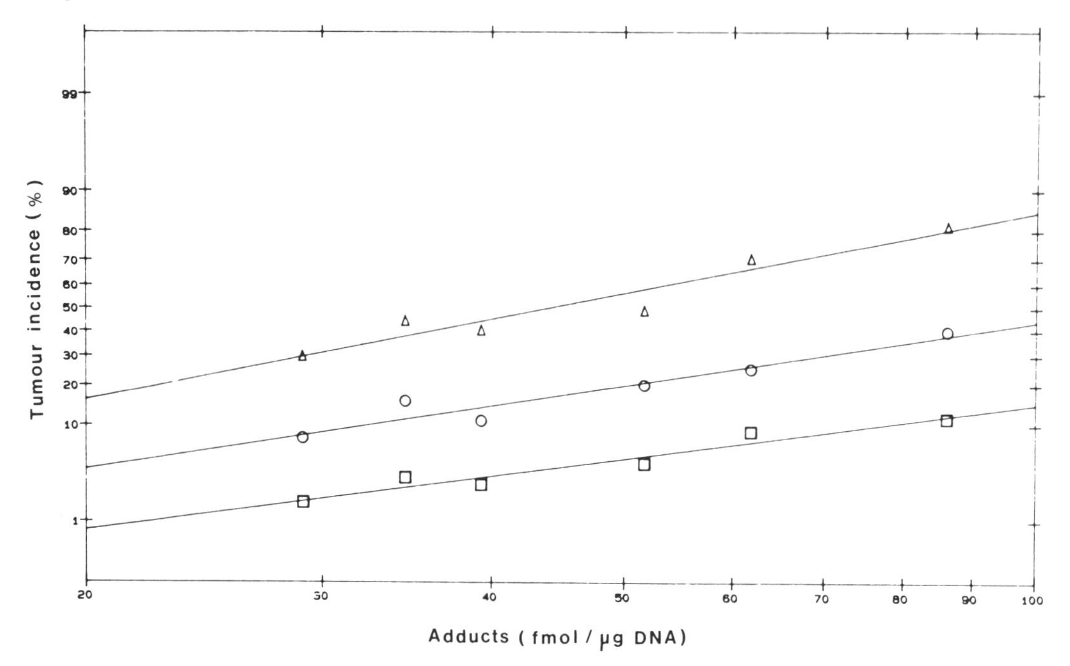

Tumour data are from Farmer *et al.* (1979) for female BALB/c mice fed 2-AAF continuously for 18 (□), 24 (○) or 33 (△) months at 30, 45, 60, 75, 100 or 150 mg 2-AAF per kg diet. Dead or moribund mice were grouped with those sacrificed at the time indicated. See Table 1 for additional details. Probit analysis of tumour data was conducted by the method of Finney (1952). The solid lines were obtained from linear regression analyses by the method of least squares. The correlation coefficients (r) were 0.97, 0.96 and 0.96 for 18, 24 and 33 months, respectively.

bladder being four times higher than those found in the liver (Fig. 1). Regression analyses of the data for liver, using a probit log adduct model, yielded a linear relationship (Fig. 2), indicating that hepatic adduct levels are predictive of tumour incidence for this tissue. Such a relationship was not observed with bladder DNA adducts; instead, there appeared to be an adduct level below which the tumour incidence did not exceed that observed in the control mice. These adduct levels were approximately 120 fmol dG-C8-AF per μg DNA for mice fed 2-AAF for 33 months and 240 fmol dG-C8-AF per μg DNA in mice fed for 18 months (Table 1). Interestingly, a substantial tumour incidence would be expected to occur in liver containing these adduct levels. (Using these adduct values, the probit log adduct model predicts a hepatic tumour incidence of ~40% at 18 months and ~90% at 33 months.) Beyond these threshold adduct values, the slope of the probit log adduct curve for the bladder (Fig. 3) becomes considerably steeper than that for the liver. This change in slope for the adduct response curve in bladder suggests that there may be a sufficient number of initiated cells in this tissue at low doses of 2-AAF but that there is an insufficient promotional stimulus to allow for their progression.

In summary, the results of this experiment indicate that the relationship between DNA adduct formation and tumour incidence may be tissue specific. In this particular instance, the limiting factor for liver tumour induction appears to be the formation of DNA adducts, while bladder tumour induction may be limited by the amount of promoter present.

Fig. 3. Relationship between bladder DNA adduct levels and tumour incidence

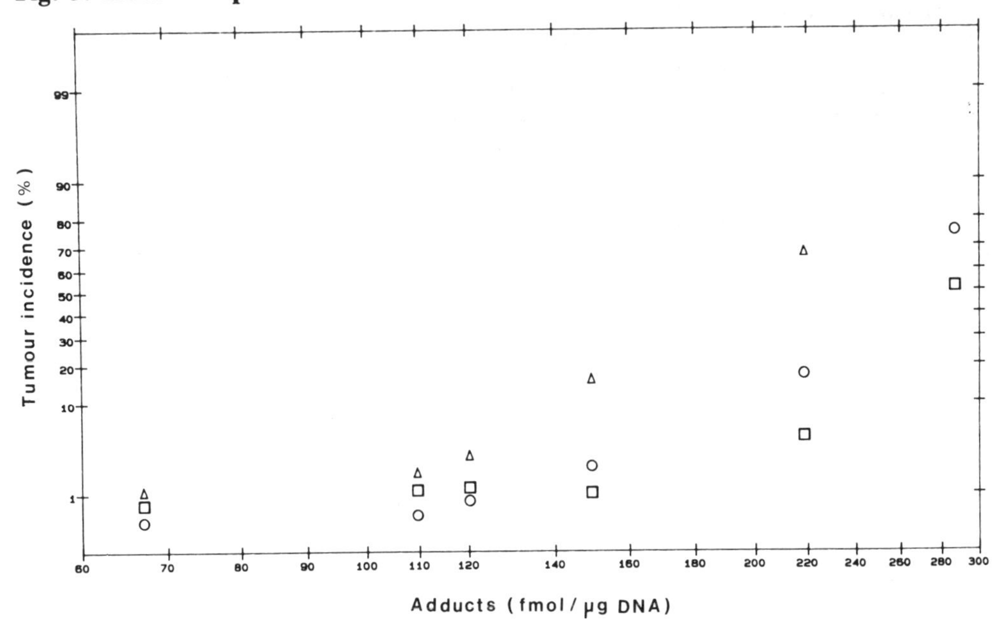

Tumour data are from Farmer *et al.* (1979) for female BALB/c mice fed 2-AAF continuously for 18 (□), 24 (○) or 33 (△) months at 30, 45, 60, 75, 100 or 150 mg 2-AAF per kg diet. Dead or moribund mice were grouped with those sacrificed at the time indicated. See Table 1 for additional details. Probit analysis of tumour data was conducted by the method of Finney (1952).

Acknowledgement

We thank C. Hartwick for helping to prepare this manuscript. This work was supported in part by American Cancer Society grant NP-593.

References

Basler, J., Hastie, N.D., Pietras, D., Matsui, S.-I., Sandberg, A.A. & Berezney, R. (1981) Hybridization of nuclear matrix attached deoxyribonucleic acid fragments. *Biochemistry, 20,* 6921–6929

Beland, F.A., Fullerton, N.F. & Heflich, R.H. (1984) Rapid isolation, hydrolysis and chromatography of formaldehyde-modified DNA. *J. Chromatogr., 308,* 121–131

Farmer, J.H., Kodell, R.L., Greenman, D.L. & Shaw, G.W. (1979) Dose and time response models for the incidence of bladder and liver neoplasms in mice fed 2-acetylaminofluorene continuously. *J. environ. Pathol. Toxicol., 3,* 55–68

Finney, D.J. (1952) *Probit Analysis. A Statistical Treatment of the Sigmoid Response Curve,* Cambridge, University Press

Jackson, C.D., Weis, C. & Shellenberger, T.E. (1980) Tissue binding of 2-acetylaminofluorene in BALB/c and C57Bl/6 mice during chronic oral administration. *Chem.-biol. Interactions, 32,* 63–81

Lai, C.-C., Miller, J.A., Miller, E.C. & Liem, A. (1985) N-Sulfoöxy-2-aminofluorene is the major ultimate electrophilic and carcinogenic metabolite of N-hydroxy-2-acetylaminofluorene in the livers of infant male C57BL/6J × C3H/HeJ F_1(B6C3F_1) mice. *Carcinogenesis, 6,* 1037–1045

Poirier, M.C., Hunt, J.M., True, B., Laishes, B.A., Young, J.F. & Beland, F.A. (1984) DNA adduct formation, removal and persistence in rat liver during one month of feeding 2-acetylaminofluorene. *Carcinogenesis, 5,* 1591–1596

Wogan, G.N. & Gorelick, N.J. (1985) Chemical and biochemical dosimetry of exposure to genotoxic chemicals. *Environ. Health Perspectives, 62,* 5–18

DOSIMETERS OF HUMAN EXPOSURE TO CARCINOGENS: POLYCYCLIC AROMATIC HYDROCARBON-MACROMOLECULAR ADDUCTS

A. Weston,[1] J.C. Willey,[1] D.K. Manchester,[1] V.L. Wilson,[1] B.R. Brooks,[2] J.-S. Choi,[1] M.C. Poirier,[3] G.E. Trivers,[1] M.J. Newman,[4] D.L. Mann[1] & C.C. Harris[1]

[1]*Laboratory of Human Carcinogenesis, National Cancer Institute, National Institutes of Health, Bethesda, MD;* [2]*Division of Computer Research and Technology, National Institutes of Health, Bethesda, MD;* [3]*Laboratory of Cellular Carcinogenesis, National Cancer Institute, National Institutes of Health, Bethesda, MD; and* [4]*School of Veterinary Medicine, Louisiana State University, Baton Rouge, LA, USA*

The metabolic activation of polycyclic aromatic hydrocarbons (PAH), for example benzo[*a*]pyrene, leads to the formation of carcinogen-macromolecular adducts. Methods that make it possible to detect low levels of these adducts in human peripheral blood samples should be useful in the dosimetry of human exposure to carcinogens. We demonstrated previously the usefulness of enzyme immunoassays and of synchronous fluorescence spectroscopy (SFS) for detecting and characterizing low levels of PAH-macromolecular adducts present in synthetic adduct mixtures. These methods have now been refined and applied to the analysis of samples of peripheral blood collected from occupationally exposed individuals (coke-oven workers) and from people attending smoking cessation clinics. The results of both immunoassays and SFS show the presence of benzo[*a*]pyrene diol epoxide (BPDE)-DNA, BPDE-haemoglobin and other putative PAH-macromolecular adducts in peripheral blood samples from certain individuals.

Most carcinogens, or their electrophilic metabolites, have mutagenic activity (Tennant *et al.*, 1987). Studies with human tissues and cells *in vitro* have shown that individual components of these mixtures, such as PAH, undergo metabolic activation and covalent binding to DNA in processes that are qualitatively similar to those observed in experimental animals (Autrup & Harris, 1983; Weston *et al.*, 1986). Thus, it may be concluded that carcinogen-DNA adducts are formed in humans as a result of environmental exposures to chemical mixtures *in vivo*. Chemical DNA damage resulting in the activation of oncogenes (Sukumar *et al.*, 1983; Balmain *et al.*, 1984) has been shown to occur through the formation of covalent adducts in DNA (Marshall *et al.*, 1984; Vousden *et al.*, 1986), and a similar mechanism might be responsible for the possible deletion or inactivation of regulatory genes (Knudsen, 1985) or tumour suppressor genes. Therefore, the concept of carcinogen-DNA adduct formation is central to theories of chemical carcinogenesis (Theall-Arce *et al.*, 1987). Detection, identification and quantification of carcinogen-DNA adducts in humans is extremely difficult (Harris, 1987). Most, if not all, humans are continually exposed to a plethora of complex carcinogenic chemical mixtures, largely in the atmosphere, food and water. The levels of carcinogens bound to human DNA and other macromolecules also challenge the detection limits of conventional assay systems, and complex mixtures of adducted materials confound simple assay systems. Laboratory investigations have centred on the development of techniques that are suitable for the detection and measurement of carcinogen residues in human DNA (Garner, 1985; Harris *et al.*, 1987). In our laboratory, emphasis has been placed on the simultaneous development of immunoassays, high-performance liquid chromatography (HPLC)/fluorescence spectroscopy and ^{32}P-postlabelling as corroborative indicators of human exposure to chemical carcinogens. Since drugs and chemicals that modify DNA can elicit an immune response and the production of antibodies in humans and laboratory animals (Reidenberg & Drayer, 1978; Dubroff & Reid, 1980; Utrecht *et al.*, 1981; Leitmen *et al.*, 1986), we have used immunoassays (ultrasensitive enzymatic radioimmunoassay (USERIA) and enzyme-linked immunosorbent assay (ELISA)) both as direct monitors of the presence of adducts in human DNA samples and as indirect monitors to determine the presence of anti-PAH-DNA adduct antibodies in human serum samples.

All of the methods currently available for human biomonitoring have their own peculiar advantages and disadvantages, and the corroborative approach that we have used has proved to be of considerable value. The aim of our initial epidemiological studies was to determine if PAH-DNA adducts and serum antibodies that recognize those adducts can be detected in biological samples from people exposed to high levels of PAHs, for example coke-oven workers, roofers and foundry workers.

Immunoassays for the detection of PAH-DNA adducts

Rabbits were immunized with DNA that had been modified with benzo[*a*]pyrene-7,8-diol 9,10-epoxide (BPDE) (Poirier *et al.*, 1980; Poirier, 1981). The antisera that were produced showed reactivity in antigen competition ELISA towards the immunogen and also towards a group of DNA samples that had been modified with the diol-epoxide derivatives of PAH that are similar to BPDE (Fig. 1) (Weston *et al.*, 1988), but not towards unmodified DNA or aflatoxin B_1-modified (Harris *et al.*, 1985) or 2-acetylaminofluorene-modified DNA (Poirier *et al.*, 1980). Further characterization of the cross-reactive nature of the anti-BPDE-DNA antibody is currently in progress,

Fig. 1. USERIA inhibition curves: inhibition of anti-BPDE-DNA rabbit polyclonal antibody binding to synthetic BPDE-DNA adducts by chrysene diol epoxide-DNA (△), BPDE-DNA (●), and benz[*a*]anthracene diol epoxide-DNA (○)

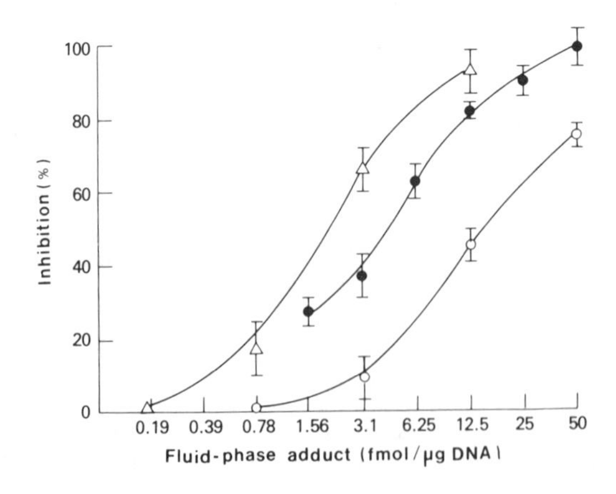

Table 1. Detection of PAH-DNA antigenicity in peripheral white blood cells

Donors	Number	% Positive	Mean adduct level[a]		
			Total cases	Positive cases	Range
Non-industrially exposed volunteers					
Nonsmokers	49	4	0.01	0.3	<0.1–0.4
Smokers	32	9	0.06	0.6	<0.1–0.9
Coke-oven workers					
Los Angeles, USA	27	66	4.72	7.1	<0.1–34.3
Norway	41	34	0.55	1.6	<0.1–13.7
Finland	20	35	0.07	0.2	<0.1–0.4
Roofers	28	25	0.51	2.0	<0.1–2.5

[a] >0.1 fmol BPDE-DNA antigenic equivalents per µg DNA

and comparatively weak activity against benzo[*k*]fluoranthene-diol-epoxide-DNA adducts and dibenz[*a,c*]anthracene-diol-epoxide-DNA adducts has already been observed (about one-tenth that of BPDE-DNA itself). It appears that a positive reaction suggests recognition of DNA modifications that are structurally and dimensionally similar to BPDE-DNA and may demonstrate the presence of covalently bound PAH residues as a class of chemical, but should not be regarded as specific for a single compound. When human peripheral blood lymphocyte DNA samples were tested for the presence of immunological epitopes that were recognized by the rabbit anti-BPDE-DNA antisera, it was clear that individuals exposed industrially to high levels of complex mixtures of PAH more often registered higher competition values in the immunoassays than nonindustrially exposed individuals (Table 1) (Harris *et al.*, 1985; Shamsuddin *et al.*, 1985; Haugen *et al.*, 1986).

In a pilot study, Perera *et al.* (1982) used ELISA and found similar levels of adducts in DNA samples from a series of lung cancer patients; however, there was no correlation between positivity and smoking history. In addition, when Everson *et al.* (1986) measured BPDE-DNA antigenicity in placental DNA, approximately 90% of mothers had adducts, irrespective of smoking status, and although the levels of adducts found were slightly higher in smokers than in nonsmokers this did not prove to be statistically significant.

Immunoassays have been developed for the measurement of alkylated DNA bases (Umbenhauer *et al.*, 1985) and other types of DNA adducts (for example, aflatoxin-DNA adducts, Groopman *et al.*, 1985; and 2-acetylaminofluorene, Hsu *et al.*, 1980; Beland *et al.*, 1987). Furthermore, when antibodies have been raised to an appropriately modified base rather than to DNA, specificity has been obtained by HPLC separation of enzyme hydrolysates of DNA samples of interest (Wild *et al.*, 1986; Beland *et al.*, 1987).

^{32}P-Postlabelling assay for the detection of PAH-DNA adducts

In most ^{32}P-postlabelling studies, known PAH-DNA adducts have not been detected in biological samples (Randerath *et al.*, 1981; Everson *et al.*, 1986; Randerath *et al.*, 1986); the reasons for this are, as yet, unclear. In placental DNA samples which had largely been shown to be antigenically similar to BPDE-DNA using ELISA, Everson *et al.* (1986) found no evidence in the ^{32}P-postlabelling assay of the presence of PAH-DNA adducts arising from seven common parent PAHs (benzo[*a*]pyrene, benzo[*ghi*]perylene, pyrene, chrysene, benz[*a*]anthracene, dibenz[*a,h*]anthracene and fluoranthene), although putative smoking-related adducts were observed. By using modifications of the ^{32}P-postlabelling assays developed by Gupta (1982) and Reddy *et al.* (1984), studies in our laboratory provide evidence for the presence of PAH-DNA-like adducts that cochromatograph with synthetic BPDE-DNA adducts in DNA both from human placental syncytiocytotrophoblasts and from salmon sperm. It has been shown by Dunn and Stich (1986) that micrococcal nuclease digestion of herring sperm DNA gives rise to material that is chromatographically indistinguishable by HPLC and thin-layer chromatography from BPDE-modified 3′-monophosphates of guanine. Taken together, the lack of correlation between ^{32}P-postlabelling and immunoassays that was demonstrated by Everson *et al.* (1986), and these more recent developments, suggest that considerable work must be done before ^{32}P-postlabelling can be used at its full potential in biochemical and molecular epidemiological studies.

Fluorescence assays for the detection of PAH-DNA and PAH-haemoglobin adducts

Fluorescence spectroscopy is a highly sensitive method for the detection of aromatic fluorophores, and the use of synchronous excitation and emission scanning in conjunction with HPLC has enhanced the specificity of this technique (Weston *et al.*, 1988). Briefly, DNA samples (100 μg) from 41 US coke-oven workers were hydrolysed with acid (0.1 N HCl, at 90°C for 3 h) to release PAH-tetrahydrotetrols and to abolish fluorescence quenching by the DNA helix. Spectral analysis of these samples showed that the majority (75%) had synchronous fluorescence emissions in the region expected for pyrene; however, the signals were generally of a broad peak variety that did not permit accurate quantification (Harris *et al.*, 1985). Subsequent analysis of 14 of these samples by SFS following HPLC has confirmed the presence in four of the samples of BPDE-DNA adducts at levels in the order of one adduct in one to two million nucleotides by comparison with a radioactive batch standard. For the ten cases with broad peaks in which the presence of BPDE-DNA adducts was not confirmed, different types of adducts may account for the original observations, or the levels of BPDE-DNA adducts may fall below the detection limit of the HPLC-SFS system, since HPLC analysis reduces sample yield by 30–40%. In another study, DNA samples from tissues of five lung cancer patients were analysed by this method; none of the five tumour samples contained BPDE-DNA adducts, but one of four uninvolved lung specimens from the same patients did contain detectable levels of BPDE-DNA adducts.

Haemoglobin has also been used as a source of chemical-macromolecular adducts, and assays are now available for measuring exposure to ethylene oxide (Farmer *et al.*, 1986; Törnqvist *et al.*, 1986) and to 4-aminobiphenyl (Tannenbaum *et al.*, 1986). Although carcinogen-haemoglobin and other carcinogen-protein adducts are not considered to be pathobiological lesions, these complexes do provide a useful resource in molecular dosimetry of exposure to chemical carcinogens, because erythrocytes have a life span of about 120 days and the haemoglobin adducts are not subject to repair, so that a cumulative exposure is measurable over a period of three to four months. In a smoking cessation study that is currently in progress in our laboratory, evidence for BPDE-haemoglobin adducts has been found in only three of 18 individuals so far studied; the estimated level of adducts in these cases was 1 fmol BPDE/μg haemoglobin when compared to a batch standard.

Immunoassays for the detection of anti-PAH-DNA antibodies in human blood-derived serum

The presence of antibodies directed against synthetic PAH-DNA adducts was assayed in human blood-derived serum and used as a measure of molecular exposure to the metabolically activated chemicals. These experiments showed that different antibodies are produced against a range of PAH-DNA adducts in humans. In each of three separate studies, which included 41 US coke-oven workers (Harris *et al.*, 1985), 38 Norwegian coke-oven workers (Haugen *et al.*, 1986) and 99 nonindustrially exposed individuals, approximately 30% of subjects were PAH-DNA antibody positive. Simple antibody binding patterns obtained with sera with antibody reactivity against a single adduct suggest that these adducts each possess at least one immunologically unique epitope. However, the complex binding patterns that were obtained when reactivity against more than one adduct was observed in the sera may be the result of the presencc in the sera of a range of specific antibodies, each of which recognizes a different adduct, or the presence of an antibody or antibodies that recognize a common or cross-reactive epitope present on all three adducts. In fact, the results of antigen competitive assays indicate the presence in complex sera of both adduct-specific and cross-reactive antibodies. Since the types of adducts studied appear to be closely related structurally, computer simulations (Brookes *et al.*, 1983) of these adducts are being compared in an attempt to understand the results of the ELISA more fully.

The presence or absence of antibodies PAH-DNA adducts was not correlated with gender or current smoking status of the individual serum donors. However, almost all of the nonsmoking contributors of sera to these experiments were reformed smokers; therefore, the possibility that the primary antibody response was produced against adducts formed from PAHs contained in tobacco smoke cannot be precluded. High antibody titres found in coke-oven workers indicate that current exposure is important for the immunological response, and the lack of correlation in the case of smokers and nonsmokers suggests that an analysis of dietary habits and other factors is appropriate for these groups.

In another series of experiments, BPDE-globin adducts were used as the solid-phase antigen, but no reactivity was observed with any of thc sera tested (unpublished). It is known from the work of Tannenbaum *et al.* (1986) that the binding site for the 4-aminobiphenyl adduct of haemoglobin is buried in a hydrophobic pocket within the molecule. If a similar binding mechanism were involved for the BPDE-globin adduct, the proffered antigen may have been inappropriate, since the immunogen would probably contain only a globin fragment or be a more simple adduct such as BPDE-valine.

Discussion

Classical epidemiological studies have made it possible to identify several chemically complex environmental mixtures, including cigarette smoke and coal smoke, as etiological agents (Doll, 1985; Mumford *et al.*, 1987). The presence of mutagenic and carcinogenic PAH in food, especially charbroiled meat and fish (IARC, 1973), has also been recognized. Measurement of chemical carcinogens in the environment and surveys of smoking, drinking and dietary habits are a first step in the determination of human exposure to genotoxic agents. Evidence for bodily exposure to environmental carcinogens has been obtained from analyses of body fluids, such as breast milk, seminal fluid, saliva, cervical mucus, urine and blood-derived serum. Indirect evidence of human exposure to putative environmental carcinogens has been gained by testing these body fluids for mutagenic activity (Ames, 1979; De Meo *et al.*, 1987), whereas direct evidence of human exposure to chemical carcinogens has been obtained from specific chemical analyses of these body fluids for the presence of the substance of interest or of one or more of its metabolites (Campbell *et al.*, 1970; Sasson *et al.*, 1985; LaVoie *et al.*, 1987). These types of epidemiological studies provide a tangible basis for human risk assessment, but they do not reveal the extent to which biologically effective macromolecular damage occurs in an individual, because the biological response of that person to a drug, chemical carcinogen or other xenobiotic agent characterizes the individual. Therefore, genetically determined host factors that control interindividual variations in metabolic activation, detoxication and DNA repair have important roles in the formation of putative pathobiological lesions. Consequently, the development of techniques that can be used to measure the extent to which macromolecular damage, and in particular DNA damage, occurs is a primary approach in accounting for biological variables that are inherent in the study of human populations.

Each of the assay systems used in the studies described above to evaluate human exposure to PAH clearly demonstrates the formation of PAH-DNA or PAH-haemoglobin adducts in some individuals. Quantitative variations were observed, however, in the levels of anti-PAH-DNA antibodies and PAH-DNA adducts detected in samples obtained from individuals with ostensibly similar exposure profiles. These data corroborate previous in-vitro studies using normal human tissues and cells, which also showed wide interindividual differences in the activation of PAH and other chemical carcinogens (Autrup & Harris, 1983), and studies of the metabolism *in vivo* of drugs like debrisoquine, sparteine, phenformin, antipyrine and dapsone (Idle & Ritchie, 1983; Ayesh & Idle, 1985). These differences in metabolic capability are now thought to be due in part to genetic polymorphisms within the population (Distlerath & Guengerich, 1984), and the cloning of appropriate molecular probes will probably lead to the development of techniques for human genotyping for drug metabolism. In addition to genetically determined differences, some individuals may be exposed independently to agents that induce drug metabolizing enzymes and thereby modify metabolic activation of PAH and other carcinogens.

Using polyclonal antibodies raised against BPDE-DNA in competitive immunoassays, it has been shown that a range of PAH–DNA adducts is recognized to varying degrees. Cross-reactivity with PAH-DNA adducts, but not unrelated chemical-DNA adducts (for example, aflatoxin-DNA adducts), actually enhances the sensitivity of the assay and does not limit detection of adducts to a single compound. Efforts are currently in progress to exploit this cross-reactivity phenomenon to concentrate adducted materials when larger amounts of DNA are available. The use of immunocolumns has been attempted previously (Tierney *et al.*, 1986), but with limited success. In our experiments, it is intended to use physicochemical techniques (HPLC, SFS, laser fluorimetry, ^{32}P-labelling thin layer chromatography and gas chromatography-mass spectrometry) to analyse column eluates. The use of a biophysical assay in these studies complements the immunoassays, and each of the assays appears to demonstrate the formation of a range of PAH-DNA adducts in many of the individuals studied. The utility of ^{32}P-postlabelling for detecting PAH-DNA adducts in simple animal models has been demonstrated (Randerath *et al.*, 1981, 1986); however, it appears that application of the ^{32}P-postlabelling assay to highly complex mixtures of low levels of DNA-adducted materials requires further development. It is suggested that with secondary and tertiary preparative techniques (for example, concentration using immunocolumns and separation using HPLC), this methodology could prove to be a powerful tool for the detection of PAH-DNA adducts.

There is at present a fundamental gap in our knowledge between measurement of human exposure to carcinogens and determination of the biologically effective dose levels of those carcinogens. This problem requires an understanding of the relationships between carcinogen dose at the tissue site, carcinogen-DNA adduct load and carcinogen effect. In order to address these questions and to help to bridge this gap it will be necessary to develop reliable methods, of the type described here for PAH, for the detection and characterization of a range of carcinogen-macromolecular adducts in human tissues.

References

Ames, B.N. (1979) Identifying environmental chemicals causing mutations and cancer. *Science, 204,* 587–593

Autrup, H. & Harris, C.C. (1983) *Metabolism of chemical carcinogens by cultured human tissues.* In: Harris, C.C. & Autrup, H., eds, *Human Carcinogenesis,* New York, Academic Press, pp. 169–194

Ayesh, R. & Idle, J.R. (1985) *Evaluation of drug oxidation phenotypes in the biochemical epidemiology of lung cancer risk.* In: Boobis, A.R., Caldwell, J., DeMatteis F. & Elcombe, C.R., eds, *Microsomes and Drug Oxidations,* Philadelphia, Taylor & Francis, pp. 340–346

Balmain, A., Ramsden, M., Bowden, G.T. & Smith, J. (1984) Activation of the mouse cellular Harvey ras gene in chemically induced benign skin papillomas. *Nature, 307,* 658–660

Beland, F.A., Huitfeldt, H.S. & Poirier, M.C., (1987) DNA-adduct formation and removal during chronic administration of carcinogenic aromatic amine. *Prog. exp. Tumor Res., 31,* 33–41

Brookes, B.R., Bruccoleri, R.E., Olafson, B.D., States, D.J., Swaminathan, S., & Karplus, M. (1983) CHARRM: a program for molecular energy, minimization, and dynamics calculations. *J. Computer Chem., 4,* 187–217

Campbell, T.C., Cacedo, J.P., Jr, Salamat, L.A. & Engel, R.W. (1970) Aflatoxin M1 in human urine. *Nature, 227,* 403–404

De Meo, M.P., Dumenil, G., Botta, A.H., Laget, M., Zabaloueff, V. & Mathias, A. (1987) Urine mutagenicity of steel workers exposed to coke oven emissions. *Carcinogenesis, 8,* 363–367

Distlerath, L.M. & Guengerich, F.P. (1984) Characterization of a human liver cytochrome P-450 involved in the oxidation of debrisoquine and other drugs by using antibodies raised to the analogous rat enzyme. *Proc. natl Acad. Sci. USA, 81,* 7348–7352

Doll, R. (1985) *Cancer: a world-wide perspective.* In: Harris, C.C., ed., *Biochemical and Molecular Epidemiology of Cancer,* New York, Alan R. Liss, pp. 111–125

Dubroff, L.M. & Reid, R.J. (1980) Hydralazine-pyrimidine interactions may explain hydralazine-induced lupus erythematosis. *Science, 208,* 404–406

Dunn, B.P. & Stich, H.F. (1986) ^{32}P-Postlabelling analysis of aromatic DNA adducts in human oral mucosal cells. *Carcinogenesis, 7,* 1115–1120

Everson, R.B., Randerath, E., Santella, R.M., Cefalo, R.C., Avitts, T.A. & Randerath, K. (1986) Detection of smoking related covalent DNA adducts in human placenta. *Science, 231,* 54–57

Farmer, P., Bailey, E., Gorf, S.M., Törnqvist, M., Osterman-Golkar, S., Kautiainen, A. & Lewis-Enright, D.P. (1986). Monitoring human exposure to ethylene-oxide by the determination of hemoglobin adducts using gas chromatography-mass spectrometry. *Carcinogenesis, 7,* 637–640

Garner, R.C. (1985) Assessment of carcinogen exposure in man. *Carcinogenesis, 6,* 1071–1078

Groopman, J.D., Donahue, P.R., Zhu, J., Chen, J. & Wogan, G.N. (1985) Aflatoxin metabolism in humans: detection of metabolites and nucleic acid adducts in urine by affinity chromatography. *Proc. natl Acad. Sci. USA, 82,* 6492–6496

Gupta, R.C. (1982) ^{32}P-Postlabelling analysis of non-radioactive aromatic carcinogen-DNA adducts. *Carcinogenesis, 3,* 1081–1092

Harris, C.C., Vähäkangas, K., Newman, M.J., Trivers, G.E., Shamsuddin, A., Sinopoli, N., Mann, D.L. & Wright, W.E. (1985) Detection of benzo[*a*]pyrene diol epoxide-DNA adducts in peripheral blood lymphocytes and antibodies to the adducts in serum from coke oven workers. *Proc. natl Acad. Sci. USA, 82,* 6672–6676

Harris, C.C. (1987) Human tissues and cells in carcinogenesis research. *Cancer Res., 47,* 1–10

Harris, C.C., Weston, A., Willey, J.C., Trivers, G.E. & Mann, D.L. (1987) Biochemical and molecular epidemiology of human cancer: indicators of carcinogen exposure, DNA damage, and genetic predisposition. *Environ. Health Perspectives, 75,* 109–119

Haugen, A., Becher, G., Benstad, C., Vähäkangas, K., Trivers, G.E., Newman, M.J. & Harris C.C. (1986) Determination of polycyclic aromatic hydrocarbons in the urine, benzo(*a*)pyrene diol epoxide-DNA adducts in lymphocyte DNA, and antibodies to the adducts in sera from coke-oven workers exposed to measured amounts of polycyclic aromatic hydrocarbons in the work atmosphere. *Cancer Res., 46,* 4178–4183

Hsu, I.-C., Poirier, M.C., Yuspa, S., Yolken, R. & Harris, C.C. (1980) Ultrasensitive enzymatic radioimmunoassay (USERIA) detects femtomoles of 2-acetyl-amino-fluorene-DNA adducts. *Carcinogenesis, 1,* 455–480

IARC (1973) *IARC Monographs on the Evaluation of Carcinogenic Risk of Chemicals to Man,* Vol. 3, *Certain Polycyclic Aromatic Hydrocarbons and Heterocyclic Compounds,* Lyon

Idle, J.R. & Ritchie, J.C. (1983) *Probing genetically variable carcinogen metabolism using drugs.* In: Harris, C.C. & Autrup, H., eds, *Human Carcinogenesis,* New York, Academic Press, pp. 857–881

Knudsen, A.G. (1985) Hereditary cancer, oncogenes, and antioncogenes. *Cancer Res., 45,* 1437–1443

LaVoie, E.J., Stern, S., Choi, C.-I., Reinhardt, J. & Adams, J.D. (1987) Transfer of the tobacco-specific carcinogens *N'*-nitrosonornicotine and 4-(methylnitrosamino)-(3-pyridyl)-1-butanone and benzo(*a*)pyrene into the milk of lactating rats. *Carcinogenesis, 8,* 433–437

Leitman, S.F., Boltansky, H., Alter, H.J., Pearson, F.C. & Kaliner, M.A. (1986) Allergic reactions in healthy plateletpheresis donors caused by sensitization to ethylene oxide gas. *New Engl. J. Med., 315,* 1192–1196

Marshall, C.J., Vousden, K.H. & Phillips, D.H. (1984) Activation of c-Ha-ras-1 proto-oncogene by *in vitro* modification with a chemical carcinogen, benzo[*a*]pyrene diol-epoxide. *Nature, 310,* 586–589

Mumford, J.L., He, X.Z., Chapman, R.S., Cao, S.R., Harris, D.B., Li, X.M., Xian, Y.L., Jiang, W.Z., Xu, C.W., Chuang, J.C., Wilson, W.E. & Cooke, M. (1987) Lung cancer and indoor air pollution in Xuan Wei, China. *Science, 235,* 217–220

Perera, F.P., Poirier, M.C., Yuspa, S.H., Nakayama, J., Jaretzki, A., Curnen, M.M., Knowles, D.M. & Weinstein, I.B. (1982) A pilot project in molecular cancer epidemiology: determination of benzo(*a*)pyrene-DNA adducts in animal and human tissues by immunoassays. *Carcinogenesis, 3,* 1405–1410

Poirier, M.C., Santella, R., Weinstein, I.B., Grunberger, D. & Yuspa, S.H. (1980) Quantitation of benzo(*a*)pyrene-deoxyguanosine adducts by radioimmunoassay. *Cancer Res., 40,* 412–416

Poirier, M.C. (1981) Antibodies to carcinogen-DNA adducts. *J. natl Cancer Inst., 67,* 515–519

Randerath, K., Reddy, V.J. & Gupta, R.C. (1981) ^{32}P-Labeling test for DNA damage. *Proc. natl Acad. Sci. USA, 78,* 6126–6129

Randerath, E., Avitts, T.A., Reddy, M.V., Miller, R.H., Everson, R.B. & Randerath, K. (1986) Comparative ^{32}P-analysis of cigarette smoke induced DNA damage in human tissue and mouse skin. *Cancer Res., 46,* 5869–5877

Reddy, V.J., Gupta, R.C., Randerath, E. & Randerath, K. (1984) ^{32}P-Postlabelling test for covalent DNA binding of chemicals *in vivo*: application to a variety of aromatic adducts and methylating agents. *Carcinogenesis, 5,* 231–243

Reidenberg, M.M. & Drayer, D.E. (1978) Aromatic amines and hydrazines, drug acetylation and lupus erythematosis. *Human Genet., 1,* 57–63

Sasson, I.M., Haley, N.J., Hoffmann, D., Wynder, E.L., Hellberg, D. & Nilsson, S. (1985) Cigarette smoking and neoplasia of the uterine cervix: smoke constituents in cervical mucus. *New Engl. J. Med., 312,* 315–316

Shamsuddin, A.K.M., Sinopoli, N.T., Hemminki, K., Boesch, R.R. & Harris, C.C. (1985)

Detection of benzo(*a*)pyrene-DNA adducts in human white blood cells. *Cancer Res., 45,* 66–69

Sukumar, S., Notario, V., Martin-Zanca, D. & Barbacid, M. (1983) Induction of mammary carcinomas in rats by nitroso-methylurea involves malignant activation of Ha-ras-1 locus by single point mutation. *Nature, 306,* 658–661

Tannenbaum, S.T., Bryant, M.S., Skipper, P.L. & Maclure, M. (1986) *Hemoglobin adducts of tobacco-related aromatic amines.* In: Hoffmann, D. & Harris, C.C., eds, *Mechanisms in Tobacco Carcinogenesis* (*Banbury Report No. 23*), Cold Spring Harbor, NY, CSH Press, pp. 63–76

Tennant, R.W., Margolin, B.H., Shelby, M.D., Zeiger, E., Haseman, J.K., Spalding, J., Caspary, W., Resnik, M., Stasiewicz, S., Anderson, B. & Minor, R. (1987) Prediction of chemical carcinogenicity in rodents from *in vitro* genetic toxicity assays. *Science, 236,* 933–941

Theall-Arce, G., Allen, J.W., Doerr, C.L., Elmore, E., Hatch, G.G., Moore, M.M., Sharief, Y., Grunberger, D. & Nesnow, S. (1987) Relationships between benzo[*a*]pyrene-DNA adduct levels and genotoxic effects in mammalian cells. *Cancer Res., 47,* 3388–3395

Tierney, B., Benson, A. & Garner, R.C. (1986) Immunoaffinity chromatography of carcinogen DNA adducts with polyclonal antibodies directed against BPDE-DNA. *J. natl Cancer Inst., 77,* 261–267

Törnqvist, M., Osterman-Golkar, S., Kautiainen, A., Jensen, S., Farmer, P.B., & Ehrenberg, L. (1986) Tissue doses of ethylene-oxide in cigarette smokers determined from levels in hemoglobin. *Carcinogenesis, 7,* 1519–1521

Umbenhauer, D., Wild, C.P., Montesano, R., Saffhill, R., Boyle, J.M., Huh, N., Kirstein, U., Thomale, J., Rajewsky, M.F. & Lu, S.H. (1985) *O*-6-Methyldeoxyguanosine in esophageal DNA among individuals at high risk of esophageal cancer. *Int. J. Cancer, 36,* 661–665

Utrecht, J.P., Freeman, R.W. & Woosley, R. (1981) The implications of procainamide metabolism to its induction of lupus. *Arthritis Rheum., 24,* 994–1001

Vousden, K.H., Bos, J.L., Marshall, C.J. & Phillips, D.H. (1986) Mutations activating human c-Ha-ras-1 proto-oncogene induced by chemical carcinogens and depurination. *Proc. natl Acad. Sci. USA, 83,* 1222–1226

Weston, A., Plummer, S.M., Grafstrom, R.C., Trump, B.F. & Harris, C.C. (1986) Genotoxicity of chemical and physical agents in cultured human tissues and cells. *Food chem. Toxicol., 24,* 675–679

Weston, A., Willey, J.C., Newman, M.J., Trivers, G.E., Haugen, A., Manchester, D.K., Choi, J.S., Krontiris, T., Light, B., Mann, D.L. & Harris, C.C. (1988) *Application of biochemical and molecular techniques to the epidemiology of human lung cancer.* In: Miners, J.O., Birkett, D.J., Drew, R., May, B. & McManus, M., eds, *Microsomes and Drug Oxidations,* London, Taylor & Francis (in press)

Wild, C.P., Umbenhauer, D., Chapot, B. & Montesano, R. (1986) *Monitoring of individual human exposure to aflatoxins* (*AF*) *and* N-*nitrosamines* (*NNO*) *by immunoassays.* In: Harris, C.C., ed., *Biochemical and Molecular Epidemiology of Cancer,* New York, A.R. Liss, pp. 43–51

AROMATIC DNA ADDUCTS IN WHITE BLOOD CELLS OF FOUNDRY WORKERS

K. Hemminki,[1] F.P. Perera,[2] D.H. Phillips,[3] K. Randerath,[4] M.V. Reddy[4] & R.M. Santella[2]

[1]*Institute of Occupational Health, Helsinki, Finland;* [2]*Division of Environmental Sciences, School of Public Health, Columbia University, New York, NY, USA;* [3]*Chester Beatty Laboratories, Institute of Cancer Research, London, UK; and* [4]*Department of Pharmacology, Baylor College of Medicine, Houston, TX, USA*

Blood samples were obtained from volunteers working in a Finnish iron foundry who were occupationally exposed to polycyclic aromatic hydrocarbons (PAH) and from control subjects not known to be occupationally exposed to this class of chemical carcinogens. Foundry workers were classified as belonging to high, medium or low exposure groups according to their exposure to airborne benzo[*a*]pyrene: high, >0.2: medium, 0.05–0.2: low, <0.05 µg benzo[*a*]pyrene/m^3 air). Aromatic adducts were found to be present in white blood cell DNA from most of the exposed workers using the enzyme-linked immunosorbent assay (ELISA) to detect aromatic DNA adducts and the ^{32}P-postlabelling technique. There was a dose-response relationship between the estimated exposure and adduct levels by both methods, and a reasonable correlation between the results of the immunoassay and postlabelling carried out in two laboratories. The levels of adducts found in the samples from the high and medium exposure groups by ELISA ranged up to five adducts in 10^7 nucleotides: the aromatic adducts detected by the postlabelling assay were at a level of two adducts/10^8 nucleotides in the high and medium exposure categories. No effect due to age, sex or the smoking habits of the subjects was observed. The results indicate that DNA extracted from white blood cells of highly exposed workers is more likely to contain aromatic DNA adducts than that from workers without occupational exposure to PAH, but large interindividual variations were evident. This study suggests that the antibody and ^{32}P-postlabelling assays may be useful in monitoring human exposure to known and previously unidentified environmental genotoxic agents.

Methods for determining DNA adducts in humans can be validated by comparing populations exposed to different levels of carcinogens. The exposures studied include tobacco smoke (Perera *et al.*, 1982; Everson *et al.*, 1986), PAH in the occupational environment (Harris *et al.*, 1985; Shamsuddin *et al.*, 1985; Vähäkangas *et al.*, 1985; Haugen *et al.*, 1986), and cisplatin in cancer chemotherapy (Poirier *et al.*, 1985; Fichtinger-Schepman *et al.*, 1987). Dose-response data have been obtained only in the case of exposure to cisplatin.

We have undertaken a study among foundry workers to establish dose-response relationships, using the ^{32}P-postlabelling technique and benzo[*a*]pyrene-DNA antibodies. We summarize here the first results of these studies and compare the data obtained by the two methods. Individual studies are reported in detail elsewhere (Phillips *et al.*, 1987; Perera *et al.*, 1988).

Material and methods

Blood samples were obtained from healthy volunteers at a Finnish iron foundry. Industrial hygiene measurements for PAH were carried out in the foundry in the years 1978–1980; as the work processes have remained practically unchanged since then, these measurements were used by two industrial hygienists familiar with the foundry to grade the volunteers for daily exposure by job description. Benzo[*a*]pyrene levels in the workplace atmosphere were used as guidelines to assign the exposure to PAH as high (>0.2 $\mu g/m^3$), medium (0.05–0.2 $\mu g/m^3$) and low (<0.05 $\mu g/m^3$). Control samples were obtained from individuals coming from different parts of Finland to the Institute of Occupational Health for examination, whose job titles did not indicate occupational exposure to PAH. Information on current smoking (number/day) was obtained for all subjects.

Blood (20–50 ml) was withdrawn into heparinized tubes and transported on ice to Helsinki. Cells were collected by centrifugation, and red blood cells were lysed by washing twice with 0.15 M ammonium chloride, followed each time by centrifugation at 1000 *g* for 5 min. DNA was isolated from nuclei by treatment with pancreatic RNase and proteinase K, followed by extraction with phenol and chloroform:isoamyl alcohol (24:1). DNA was precipitated from the aqueous phase with cold ethanol.

Coded samples were assayed by competitive ELISA, essentially as described previously, with fluorescence detection (Perera *et al.*, 1982). Briefly, 96-microwell black plates (MicroFLUOR 'B', Dynatech Laboratories, Alexandria, VA, USA) were coated with 0.5 ng BPDE-I-DNA (0.5% modified). A previously characterized rabbit polyclonal antibody (Poirier *et al.*, 1980) was used at a 1:800 000 dilution. A standard curve was constructed by mixing BPDE-I-DNA modified *in vitro* with carrier unmodified calf thymus DNA, such that 50 μl contained 0.25–25 fmol BPDE-I-deoxyguanine adduct in 50 μg/well after sonication and denaturation by boiling for 3 min and cooling on ice. A conjugate of goat anti-rabbit immunoglobulin G-alkaline phosphatase (Sigma, St Louis, MO, USA) was used at 1:400 dilution. The substrate, 4-methylumbelliferyl phosphate (100 μl, 50 μg/ml 0.1 M diethanolamine, pH 9.6) becomes fluorescent after removal of phosphate. Fluorescence was read on a Microfluor reader (Dynatech Laboratories, Alexandria, VA, USA); samples with greater than 20% inhibition were considered to have detectable levels of adducts. Results are the mean of a single assay with triplicate wells, since it has been established that measured antigenicity may result from multiple diol epoxide adducts (Perera *et al.*, 1987). Modification level is expressed in terms of fmol BPDE-I-deoxyguanine adduct that would cause similar inhibition per μg DNA.

For postlabelling, coded DNA samples (4 μg) were dissolved in 0.1 mM EDTA and digested with micrococcal nuclease (0.14 U, Sigma Chemical Co., Poole, Dorset, UK) and spleen phosphodiesterase (0.6 mU, Boehringer Mannheim, Lewes, East Sussex, UK) in 17 mM sodium succinate, 8 mM $CaCl_2$, pH 6.0 (total volume, 4.8 μl) at 37°C for 20 h. The DNA digests were then further treated with nuclease P_1(1.0 μl, 1 U, Sigma), 0.2 M sodium acetate, pH 5.0 (2.4 μl) and 0.3 mM $ZnCl_2$ (1.4 μl) (Reddy & Randerath, 1986). After incubation at 37°C for 1 h, 0.5 M Tris-base (1.9 μl) was added. The DNA digest was then ^{32}P-labelled as described previously (Gupta *et al.*, 1982). Carrier-free ^{32}P-orthophosphate obtained from Amersham International (Amersham, UK) was used to synthesize [γ-^{32}P]ATP. The reaction was terminated by the addition of potato apyrase (40 mU, Sigma). Resolution of the ^{32}P-labelled adducts was carried out on polyethyleneimine-cellulose thin-layer chromatography sheets (Macherey-Nagel, supplied by Camlab, Cambridge, UK), using solvent systems described previously (Phillips *et al.*, 1986). Adduct spots were detected by autoradiography with

Fig. 1. Benzo[*a*]pyrene antigenicity of white blood cell DNA from foundry workers

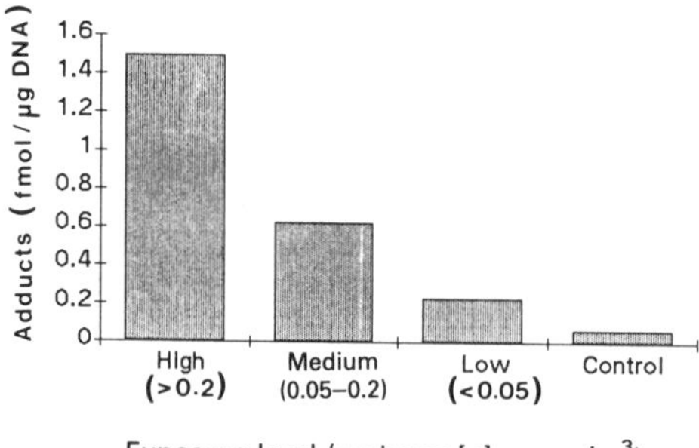

intensifying screens and quantified by Cerenkov counting of the excised areas of the chromatograms (Gupta *et al.*, 1982).

Results

Benzo[*a*]pyrene-DNA antigenicity of DNA samples from foundry workers correlated with the estimated exposure levels, as shown in Figure 1. When adjustment was made for cigarette smoking and time since holiday, benzo[*a*]pyrene exposure was significantly related to adduct levels ($p = 0.0001$; see Perera *et al.*, 1988). Each of the three exposed groups had significantly elevated adduct levels when compared with controls; and the group with low exposure differed significantly from those with high and medium exposure. Smoking, age and sex had no significant effect on aromatic DNA levels.

The samples from foundry workers were assayed by the ^{32}P-postlabelling technique in two laboratories. Laboratory 1 gave the results as the estimated number of adducts/10^8 nucleotides, while laboratory 2 gave an adduct score (0–3). In laboratory 1, the DNA samples from workers in the high and medium exposure categories had an average of 1.8 polycyclic adducts/10^8 nucleotides (Table 1); DNA from the low

Table 1. ^{32}P-Postlabelling of white blood cell DNA

Exposure to benzo[*a*]pyrene	Laboratory 1		Laboratory 2	
	Mean no. adducts/ 10^8 nucleotides	No.	Score[a]	No.
High (>0.2 µg/m³)			2.0	7
Medium	1.8	8		
(0.05–0.2 µg/m³)			1.8	24
Low (<0.05 µg/m³)	0.06	17	0.8	31
Control	0.2	9	0.2	18

[a]Score, 0 = <0.5 adducts; 1 = 0.5–1 adducts; 2 = 1–2 adducts; 3 = >2 adducts/10^8 nucleotides

Table 2. ^{32}P-Postlabelling by age and sex

Age (years)	Sex	No.	Mean score[a]
20–29	M	4	1.0
	F	3	2.7
30–39	M	16	0.9
	F	7	0.6
40–49	M	25	1.3
	F	10	1.4
50–59	M	20	1.2
	F	2	1.0

[a]See footnote to Table 1

exposure category and from the controls had mean adduct levels of 0.06 and 0.2 per 10^8 nucleotides, respectively, which was the approximate limit of sensitivity of the assay. A large number of samples was run by laboratory 2 (Table 1). In the high exposure group, the adduct score was 2.0, that in the medium exposure category, 1.8, that in the low exposure group, 0.8, and that in controls, 0.2, showing a clear correlation with estimated exposure levels at the work place.

Age and sex had no effect on the postlabelling score in the DNA from foundry workers (Table 2).

The correlation coefficients of the results are shown in Table 3. The numerical results of the antibody assay and of the postlabelling assay by laboratory 1 were transformed into logarithmic results; the postlabelling score of laboratory 2 was used as such. The antibody assay and the postlabelling score had a relatively high correlation (0.70), while the correlation with the postlabelling results of laboratory 1 was lower but still reasonable (0.46). The correlation between the postlabelling results of laboratory 1 and 2 was intermediate (0.51).

Discussion

These results are the first that show a correlation between estimated exposure levels to PAH and the amounts of DNA adducts in white blood cells (see Farmer *et al.*, 1987). The two assays differed in terms of estimated exposure levels: in the high exposure category, the adduct levels were five adducts/10^7 nucleotides in the immunoassay and two adducts/10^8 nucleotides in the postlabelling assay. This is interesting, as, on the one hand, the antibodies are raised against benzo[*a*]pyrene diol epoxide-modified DNA, but are known to cross-react with other PAH-modified DNA. Modification levels were determined from a standard curve for BPDE-I-DNA and may be overestimates of the actual levels. The ^{32}P-postlabelling assay on the other hand

Table 3. Correlation coefficients between antibody assay and the two ^{32}P-postlabelling assays (laboratories 1 and 2)

	Antibody (log transformed)	Laboratory 2 (score)
Antibody (log transformed)	—	0.70 (63)
Laboratory 1 (log transformed)	0.46 (23)	0.51 (26)

Number of observations in parentheses

picks up spots of radioactivity on the basis of their migration on thin-layer chromatograms. These results may apply to many types of polycyclic adducts (Reddy & Randerath, 1986).

The correlations between the results of antibody and postlabelling assays were relatively high — 0.70 and 0.46 (with results of laboratories 2 and 1, respectively). In fact, the correlation between the antibody assay and the two postlabelling assays was similar to that between the two postlabelling assays. There was generally a good agreement with the control values, which were at or below the level of detection in the two assays. There was, however, variation in the actual positive values recorded.

The results as reported here suggest that the antibody and postlabelling assays can be used as dosimeters of human exposure to carcinogenic compounds.

Acknowledgements

This study was supported by grants to the Institute of Cancer Research by the British Medical Research Council and the Cancer Research Campaign, the Finnish Work Environment Fund, and by USPHS grant CA 43263 to KR from the National Cancer Institute.

References

Everson, R.B., Randerath, E., Santella, R.M., Cefalo, R.C., Avitts, T.A. & Randerath, K. (1986) Detection of smoking-related covalent DNA adducts in human placenta. *Science, 231,* 54–57

Farmer, P.H., Neumann, H.-G. & Henschler, D. (1987) Estimation of exposure of man to substances reacting covalently with macromolecules. *Arch. Toxicol., 60,* 251–260

Fichtinger-Schepman, A.-M. J., van Oosterom, A.T., Lohman, P.H.M. & Berends, F. (1987) cis-Diamminedichloroplatinum(II)-induced DNA adducts in peripheral leukocytes from seven cancer patients: quantitative immunochemical detection of the adduct induction and removal after a single dose of cis-diamminedichloroplatinum(II). *Cancer Res., 47,* 3000–3004

Gupta, R.C., Reddy, M.V. & Randerath, K. (1982) ^{32}P-Postlabelling analysis of non-radioactive aromatic carcinogen-DNA-adducts. *Carcinogenesis, 3,* 1081–1092

Harris, C.C., Vähäkangas, K., Newman, M.J., Trivers, G.E., Shamsuddin, A., Sinopoli, N., Mann, D.L. & Wright, W.E. (1985) Detection of benzo[*a*]pyrene diol epoxide-DNA adducts in peripheral blood lymphocytes and antibodies to the adducts in serum from coke oven workers. *Proc. natl Acad. Sci. USA, 82,* 6672–6676

Haugen, A., Becher, G., Benestad, C., Vähäkangas, K., Trivers, G.E., Newman, M.J. & Harris, C.C. (1986) Determination of polycyclic aromatic hydrocarbons in the urine, benzo[*a*]pyrene diol epoxide DNA adducts in lymphocyte DNA, and antibodies to the adducts in sera from coke oven workers exposed to measured amounts of polycyclic aromatic hydrocarbons in the work atmosphere. *Cancer Res., 46,* 4178–4183

Perera, F.P., Poirier, M.C., Yuspa, S.H., Nakayama, J., Jaretzki, A., Curnen, M.M., Knowles, D.M. & Weinstein, I.B. (1982) A pilot project in molecular cancer epidemiology: determination of benzo[*a*]pyrene-DNA adducts in animal and human tissues by immunoassays. *Carcinogenesis, 3,* 1405–1410

Perera, F.P., Santella, R.M., Brenner, D., Poirier, M.C., Munshi, A.A., Fischman, H.K. & van Ryzin, J. (1987) DNA adducts, protein adducts and sister chromatid exchange in cigarette smokers and nonsmokers. *J. natl Cancer Inst., 79,* 449–456

Perera, F.P., Hemminki, K., Young, T.L., Santella, R.M., Brenner, D. & Kelly, G. (1988) Detection of polycyclic aromatic hydrocarbon-DNA adducts in white blood cells of foundry workers. *Cancer Res.* (in press)

Phillips, D.H., Hewer, A. & Grover, P.L. (1986) Aromatic DNA adducts in human bone marrow and peripheral blood leukocytes. *Carcinogenesis, 7,* 2071–2075

Phillips, D.H., Hemminki, K., Alhonen, A., Hewer, A. & Grover, P.L. (1987) Monitoring occupational exposure to carcinogens: detection by ^{32}P-postlabelling of aromatic DNA adducts in white blood cells from iron foundry workers. *Mutat. Res.* (in press)

Poirier, M.C., Santella, R., Weinstein, I.B., Grunberger, D. & Yuspa, S.H. (1980) Quantitation of benzo[*a*]pyrene-deoxyguanosine adducts by radioimmunoassay. *Cancer Res., 40,* 412–416

Poirier, M.C., Reed, E., Zwelling, L.A., Ozols, R.F., Litterst, C.L. & Yuspa, S.H. (1985) Polyclonal antibodies to quantitate cis-diamminedichloroplatinum (II)-DNA adducts in cancer patients and animal models. *Environ. Health Perspect., 62,* 89–94

Reddy, M.V. & Randerath, K. (1986) Nuclease P1-mediated enhancement of sensitivity of ^{32}P-postlabelling test for structually diverse DNA adducts. *Carcinogenesis, 7,* 1543–1551

Shamsuddin, A.K.M., Sinopoli, N.T., Hemminki, K., Boesch, R.R. & Harris, C.C. (1985) Detection of benzo[*a*]pyrene:DNA adducts in human white blood cells. *Cancer Res., 45,* 66–68

Vähäkangas, K., Haugen, A. & Harris, C.C. (1985) An applied synchronous spectrophotometric assay to study benzo[*a*]pyrene-diolepoxide-DNA adducts. *Carcinogenesis, 6,* 1109–1116

A COMPARISON OF ^{32}P-POSTLABELLING AND IMMUNOLOGICAL METHODS TO EXAMINE HUMAN LUNG DNA FOR BENZO[*a*]PYRENE ADDUCTS

R.C. Garner[1], B. Tierney[1,3] & D.H. Phillips[2]

[1]Cancer Research Unit, University of York, Heslington, York, UK; and [2]Institute of Cancer Research, Chester Beatty Laboratories, London, UK

Human lung DNA isolated from surgical specimens has been examined for the presence of polycyclic aromatic hydrocarbon-DNA adducts using both ^{32}P-postlabelling and immunological methods. Of 12 samples examined to date, five had detectable amounts of benzo[*a*]pyrene diol epoxide-DNA adducts (BPDE-DNA) as determined by the enzyme-linked immunosorbent assay (ELISA), after immunoaffinity concentration. Values ranged from 3.5 to 11.5 fmol/mg DNA. When the same group of samples was analysed using the ^{32}P-postlabelling technique, adducts could be detected in all the samples examined. There was generally not a good correspondence between the two methods. The number of adducts measured by ^{32}P-postlabelling ranged from 1–100 per 10^8 nucleotides, which is some two orders of magnitude higher than with the immunological method, indicating that the BPDE-DNA adduct is probably not the major adduct present in these samples.

Materials and methods

Chemicals

DNA (calf thymus), sodium dodecylsulphate, Tween 20, DNase (type III), alkaline phosphatase (type I), proteinase K and nuclease P_1 were obtained from Sigma Chemical Co. (Poole, Dorset, UK). Snake venom phosphodiesterase was purchased from Boehringer Corporation (London) Ltd (Lewes, Sussex, UK). Goat anti-rabbit alkaline phosphate conjugate immunoglobulin G was purchased from Miles Laboratories (Stoke Poges, Slough, UK). BPDE-DNA was prepared as described previously (Tierney *et al.*, 1986).

Human samples

Samples were kindly obtained during thoracic surgery at the Bradford Royal Infirmary, Bradford, Yorkshire, UK, by consultant surgeons Mr Mearn and Mr Saunders, and stored at −20°C in sealed containers until transported to York. On arrival at York, samples were stored at −90°C until analysed. Smoking histories of the patients were taken when possible.

Isolation and purification of DNA

Human lung samples were thawed and homogenized in 1% sodium dodecylsulphate 1 mM EDTA with an Ultra-Turrax blender for 15 sec at full speed and then processed

[3] Present address: Cambridge Life Sciences, Cambridge Science Park, Cambridge CB4 4GN, UK

Fig. 1. Protocol for recovering BPDE-DNA adducts

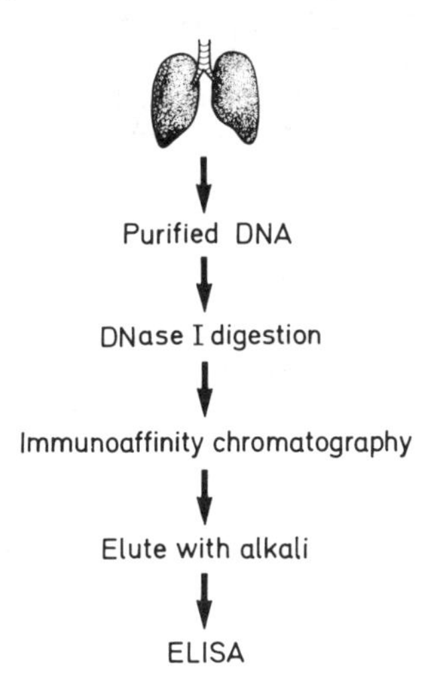

as described by Martin and Garner (1986). The DNA content was determined by measurement of the absorbance at 260 nm ($1A_{260} = 50\ \mu g/ml$ DNA). One aliquot of this DNA was used for postlabelling studies and another for immunoaffinity concentration and ELISA.

Preparation and use of BPDE-DNA antibody immunoaffinity column and ELISA of retained material

These steps were performed as described by Tierney *et al.* (1986) and are outlined in Figure 1.

^{32}P-Postlabelling studies on purified human lung DNA samples

^{32}P-Postlabelling was carried out on 4-μg samples of human lung DNA using the nuclease P_1 digestion method of sensitivity enhancement (Reddy & Randerath, 1986), with minor modifications described elsewhere (Phillips *et al.*, 1986). The limit of detection was calculated to be 0.2 adducts/10^8 nucleotides (6 fmol/mg DNA).

Results

Immunoaffinity concentration and ELISA of purified human lung DNA samples

In order to increase the sensitivity of immunoanalysis of DNA samples, we demonstrated previously that immunoaffinity concentration can be used (Tierney *et al.*, 1986). These studies showed that BPDE-DNA adducts could be concentrated from DNA samples using polyclonal antibodies against BPDE-DNA. A preliminary digestion with DNase I was necessary before applying the sample to the antibody column. Elution of the bound adducts was achieved with dilute alkali (Fig. 1).

Taking this approach, we have examined DNA from 12 lung samples taken from individuals undergoing thoracic surgery. Table 1 gives ELISA results for the material bound to the immunoaffinity column, i.e., that which can be eluted with alkali. Five samples contained anti-BPDE-DNA antibody inhibitory material; four of these

Table 1. Immunoaffinity concentration and ELISA of human lung DNA samples and levels of 'adducts' measured by the ^{32}P-postlabelling method

Sample no.	Sex	Age (years)	Smoking history[a]	BPDE per mg DNA[b] (fmol) (ELISA)	Total packs smoked[c]	No. of adducts/10^8 nucleotides[d] (^{32}P-postlabelling)	
						Expt 1	Expt 2
1	M	51	S	ND	5 642	9	2
2	F	61	S	ND	14 924	15	19
3	F	64	S	ND	?	6	7
4	F	53	S	ND	24 024	8	10
5	M	49	S	3.5	10 920	76	104
6	F	42	S	12.0	2 184	9	18
7	F	63	S	ND	15 652	7	4
8	M	48	FS	11.0	20 384	5	3
10	M	43	NS	ND	—	—[e]	4
11	M	61	NS	ND	—	3	1
12	F	78	?	10.5	?	2	1
13	F	59	FS	7.0	10 647	6	1

[a] S, smoker; FS, former smoker; NS, nonsmoker; ?, unknown
[b] ND, not detected
[c] 1 pack, 20 cigarettes
[d] One adduct/10^8 nucleotides = 30 fmol/mg DNA
[e] Sample lost

samples were from smokers, while the fifth was of unknown smoking history. Taking calf thymus DNA through the procedure revealed no antibody inhibitory material.

^{32}P-Postlabelling studies

It was our original intention to examine the immunoaffinity column-bound material using the ^{32}P-postlabelling technique. It was hoped that this concentration procedure would enable us to detect BPDE-DNA adducts more easily, since other adducts should have been lost during the immuno-concentration process. Unfortunately, it was found that chromatographic spots obtained from blank antibody columns masked any bound material. We therefore had to examine the purified DNA samples without preliminary immunoaffinity chromatography. The results of these analyses are set out in Table 1. Whereas samples 5, 6, 8, 12 and 13 had detectable anti-BPDE-DNA antibody inhibitory activity, samples 2, 4, 5 and 6 contained the most adducts measured by ^{32}P-postlabelling.

Discussion

There have been a number of published reports of use of antibodies to detect BPDE-DNA adducts in human cells (Perera *et al.*, 1982; Harris *et al.*, 1985; Shamsuddin *et al.*, 1985; Haugen *et al.*, 1986). Another procedure using ^{32}P-postlabelling has indicated the presence of polycyclic aromatic hydrocarbon adducts in human placenta, bronchus and trachea (Everson *et al.*, 1986; Randerath *et al.*, 1986) and in leucocytes (Phillips *et al.*, 1986). In one published study (Everson *et al.*, 1986), these two procedures were used on the same human samples. We report here that there is some correspondence between the two methods but that there appears to be a large difference in adduct levels estimated by the two procedures.

With the antibody studies, some positive samples were obtained. Although our polyclonal antibody was raised against BPDE-DNA, we have some evidence that it will cross-react with other PAH-DNA adducts, such as those obtained with 7,12-

dimethylbenz[*a*]anthracene. Thus, the calculated adduct concentration may not be due solely to the presence of BPDE adducts. The adduct values we obtained differ significantly from those reported in other studies, being lower by at least one order of magnitude. This may be a reflection of our immunoaffinity procedure, with which we can concentrate adducts and hence increase the sensitivity of ELISA.

The finding that adduct levels in the ^{32}P-postlabelling studies were at least one order of magnitude higher is not surprising since benzo[*a*]pyrene may not constitute the major DNA-reactive component in cigarette smoke. The ^{32}P-postlabelling thin-layer chromatography maps showed a complex pattern of adduct spots, the majority of which have chromatographic mobilities distinct from those of BPDE-DNA adducts. In Table 1, we have set out an approximation of cigarette exposure on a per pack basis: in this limited study, no correlation could be seen between adduct level and packs smoked. Adduct levels at a particular point in time must be related not only to total exposure but also to DNA repair and cell turnover. It is not surprising, therefore, that no correlation is seen.

In conclusion, further work is required to examine in more detail the correlation between ^{32}P-postlabelling and antibody methods. These preliminary results suggest that the techniques could be used for measuring exposure to carcinogens (Garner, 1985) but that efforts need to be concentrated on the use of ^{32}P-postlabelling on immunoaffinity-concentrated material.

Acknowledgements

This work was partly supported by a grant from the Health and Safety Executive.

References

Everson, R.B., Randerath, E., Santella, R.M., Cefalo, R.C., Avitts, T.A. & Randerath, K. (1986) Detection of smoking-related covalent DNA adducts in human placenta. *Science, 231,* 54–57

Garner, R.C. (1985) Assessment of carcinogen exposure in man. *Carcinogenesis, 6,* 1071–1078

Harris, C.C., Vähäkangas, K., Newman, M.J., Trivers, G.E., Shamsuddin, A., Sinopoli, N., Mann, D.L. & Wright, W.E. (1985) Detection of benzo[*a*]pyrene diol epoxide-DNA adducts in peripheral blood lymphocytes and antibodies to the adducts in serum from coke oven workers. *Proc. natl Acad. Sci. USA, 82,* 6672–6676

Haugen, A., Becher, G., Benestad, C., Vähäkangas, K., Trivers, G.E., Newman, M.J. & Harris, C.C. (1986) Determination of polycyclic aromatic hydrocarbons in the urine, benzo[*a*]pyrene diol epoxide-DNA adducts in lymphocyte DNA and antibodies to the adducts in sera from coke oven workers exposed to measured amounts of polycyclic aromatic hydrocarbons in the work atmosphere. *Cancer Res., 46,* 4178–4183

Martin, C.N. & Garner, R.C. (1986) *The identification and assessment of covalent binding in vitro and in vivo.* In: Snell, K. & Mulloch, B., eds, *Biochemical Toxicology, A Practical Approach,* Oxford, IRL Press, pp. 109–126

Perera, F.P., Poirier, M.C., Yuspa, S.H., Nakayama, J., Jaretzki, A., Curnen, M.M., Knowles, D.M. & Weinstein, I.B. (1982) A pilot project in molecular cancer epidemiology: determination of benzo[*a*]pyrene-DNA adducts in animal and human tissues by immunoassays. *Carcinogenesis, 3,* 1405–1410

Phillips, D.H., Hewer, A. & Grover, P.L. (1986) Aromatic DNA adducts in human bone marrow and peripheral blood leukocytes. *Carcinogenesis, 7,* 2071–2075

Randerath, E., Avitts, T.A., Reddy, M.V., Miller, R.H., Everson, R.B. & Randerath, K. (1986) Comparative ^{32}P-analysis of cigarette smoke-induced DNA damage in human tissues and mouse skin. *Cancer Res., 46,* 5869–5877

Reddy, M.V. & Randerath, K. (1986) Nuclease P_1-mediated enhancement of sensitivity of ^{32}P-postlabelling test for structurally diverse DNA adducts. *Carcinogenesis, 7,* 1543–1551

Shamsuddin, A.K.M., Sinopoli, N.T., Hemminki, K., Boesch, R.R. & Harris, C.C. (1985) Detection of benzo[*a*]pyrene:DNA adducts in human white blood cells. *Cancer Res.*, *45*, 66–68

Tierney, B., Benson, A. & Garner, R.C. (1986) Immunoaffinity chromatography of carcinogen DNA adducts with polyclonal antibodies directed against benzo[*a*]pyrene diol-epoxide-DNA. *J. natl Cancer Inst.*, *77*, 261–267

BINDING EFFICIENCY OF ANTIBODIES TO DNA MODIFIED WITH AROMATIC AMINES AND POLYCYCLIC AROMATIC HYDROCARBONS: IMPLICATIONS FOR QUANTIFICATION OF CARCINOGEN-DNA ADDUCTS *IN VIVO*

E. Kriek, F.J. Van Schooten, M.J.X. Hillebrand & M.C. Welling

Division of Chemical Carcinogenesis, The Netherlands Cancer Institute, Amsterdam, The Netherlands

Antibodies raised against the bovine serum albumin conjugates of *N*-(guanosin-8-yl)-*N*-2-acetylaminofluorene (Guo-8-AAF), the imidazole ring-opened form of *N*-(guanosin-8-yl)-2-aminofluorene (roGuo-8-AF) and the methylated bovine serum albumin complex of DNA modified with (±)*trans*-7,8-dihydroxy-*anti*-9,10-epoxy-7,8,9,10-tetrahydrobenzo[*a*]pyrene (BPDE) have been employed in a highly sensitive enzyme-linked immunosorbent assay (ELISA) to determine their affinity for DNA modified with the corresponding carcinogens at various levels of modification. All antibodies recognized highly modified DNA more efficiently than DNA of low modification. This property, which may be common to all antibodies raised against carcinogen-DNA adducts, has to be taken into account when these antibodies are used to quantify carcinogen-DNA adducts in biological samples. Appropriate DNA preparations of low modification have to be used as reference compounds in immunoassays to allow reliable quantification of adduct levels in DNA from animals and human cells.

A large number of polyclonal and monoclonal antibodies have now been developed to specific carcinogen-DNA adducts (Strickland & Boyle, 1984). These antibodies have been used in highly sensitive enzyme immunoassays to quantify adduct levels in DNA from animal and human tissues. With these techniques, it has been possible to determine adduct levels in the 0.1–1.0 fmol range with antibodies of high specific affinity. This high degree of sensitivity has led a number of laboratories to investigate the feasibility of using these assays to determine DNA adducts in human populations exposed to carcinogens (Poirier & Beland, 1987). We now report on (i) the binding efficiency of some specific antibodies to DNA with different levels of modification, (ii) the instability of specific adducts under conditions of heat denaturation of DNA, (iii) validation of the ELISA for BPDE-DNA with DNA samples from mouse liver following treatment with [^{3}H]BP, and (iv) determination of polycyclic aromatic hydrocarbon (PAH)-DNA adducts in peripheral blood lymphocytes from coke-oven workers.

Antibody affinity

AAF and derivatives

In previous studies, we showed that our polyclonal anti-Guo-8-AAF antibody has highest affinity for the mononucleoside dGuo-8-AAF (Van der Laken *et al.*, 1982;

Table 1. Competitive inhibition of polyclonal antibody binding to AAF-DNA and roAF-DNA[a]

Competitor	Modification (pmol/mg)	Amount of competitor (fmol) causing 50% inhibition of antibody binding[b]	
		Anti-Guo-8-AAF	Anti-roGuo-8-AF
ss AAF-DNA	5300	3.3	500
ds AAF-DNA	5300	8.8	4000
AAF-DNA[c]	5300	4.6	—
ss AAF-DNA	200	12.0	—
ds AAF-DNA	200	13.7	—
AAF-DNA[c]	200	2.2	—
dGuo-8-AAF		2.4	—
roAF-DNA	33 600	—	6.7
roAF-DNA	5000	—	7.8
roAF-DNA	600	9200	17.0
roAF-DNA	43	—	74
roAF-DNA	5	—	78
ss AF-DNA	43	5500	845

[a]ds, double-stranded; ss, single-stranded; ro, ring-opened
[b]Enzyme immunoassays were performed as described previously (Kriek *et al.*, 1984), except that fluorescence values were read in a Microfluor Reader (Dynatech Laboratories Inc., Alexandria, VA, USA); 10 fmol adduct/0.5 ng DNA were coated per microtitre well; antibody dilution was $1:10^6$
[c]Enzymatically hydrolysed preparation (Kriek *et al.*, 1982)

Kriek *et al.*, 1984). The antibody affinity for AAF-DNA was further studied with preparations of different modification levels. The results, presented in Table 1, clearly show that highly modified AAF-DNA is recognized more efficiently than AAF-DNA of low modification. The lower detection limit for dGuo-8-AAF is now 0.02 fmol/μg DNA (i.e., seven adducts per 10^9 nucleotides) and 0.1 fmol/μg for AAF-DNA (35 adducts per 10^9 nucleotides). Anti-Guo-8-AAF showed high cross-reactivity with the *N*-deacetylated adduct, dGuo-8-AF, but the affinity for DNA carrying this adduct appeared to be much lower (Table 1). For this reason, anti-Guo-8-AAF cannot be used to determine dGuo-8-AF in DNA with the same sensitivity as dGuo-8-AAF.

It is now well documented that guanine-8-arylamine adducts are the major reaction products of aromatic amines *in vivo* (Kadlubar & Beland, 1985). Moreover, it was shown that an important group of environmental contaminants, the nitro PAH, show the same target specificity as their amino analogues (Beland *et al.*, 1985) and also lead to the formation of guanine-8-arylamine adducts in DNA (Stanton *et al.*, 1985). Therefore, we wanted to develop a sensitive immunoassay for guanine-8-arylamine adducts in DNA without the requirement of prior enzymatic degradation. In previous studies, we demonstrated that dGuo-8-AF is easily converted into its guanine imidazole ring-opened form, rodGuo-8-AF (Kriek & Westra, 1980; Kriek *et al.*, 1984). This appears to be a general reaction for guanine-8-arylamines, in which the nature of the aromatic hydrocarbon moiety is an important determinant (J.G. Westra *et al.*, unpublished observations). Our antibody, raised against roGuo-8-AF (Kriek *et al.*, 1984), was also tested for its affinity to roAF-DNA at different levels of modification in the competitive ELISA. The results, presented in Table 1, show that the 50% inhibition values are strongly dependent on the level of modification. The lower detection limit for roAF-DNA (1–50 fmol/μg) is 0.3 fmol/μg DNA, equivalent to one rodGuo-8-AF adduct per 10^7 nucleotides. We found that AF-DNA is also converted to the guanine imidazole ring-opened form roAF-DNA upon heat denaturing at 100°C, whereas AAF-DNA is unaffected by this treatment. Interestingly, this reaction is

Fig. 1. Competitive inhibition of anti-roGuo-8-AF antibody binding with AAF-DNA, AF-DNA and roAF-DNA

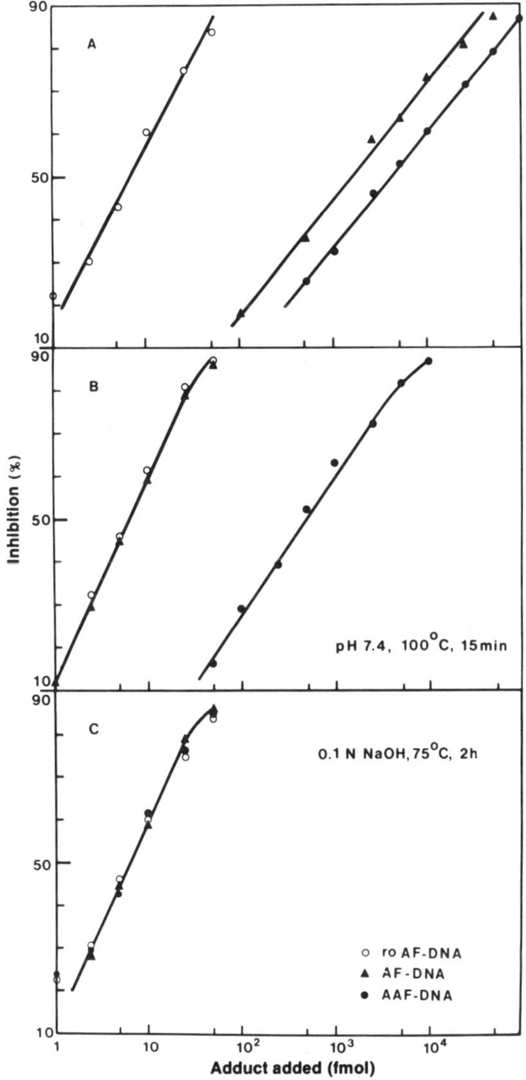

Rabbit-produced antiserum against the bovine serum albumin conjugate of roGuo-8-AF. The microtitre wells were coated with roAF-DNA (10 fmol/0.5 ng); antiserum dilution was $1:10^6$. A, untreated, native AAF- and AF-DNA; B, AAF- and AF-DNA were heated in 1 M NaCl + 0.01 M sodium phosphate pH 7.4 at 100°C for 15 min; C, AAF- and AF-DNA were heated in 0.1 N NaOH at 75°C for 2 h.

dependent on the NaCl concentration and proceeds much faster in 1 M NaCl than in 0.14 M NaCl (phosphate-buffered saline). The formation of roAF-DNA at neutral or alkaline pH can be followed by ELISA using the anti-roGuo-8-AF antibody (Fig. 1). Unfortunately, the cross-reactivity of this antibody does not allow an accurate determination of small amounts of roAF-DNA (10% or less) in the presence of AAF-DNA when (biological) samples of low modification (1–10 fmol/μg) have to be analysed.

Table 2. Competitive inhibition of antibody binding to BPDE-DNA by various competitors[a]

Competitor	Modification (pmol/mg)	Amount of competitor (fmol) causing 50% inhibition			
		Polyclonal antibodies		Monoclonal antibodies	
		F29	F30	41D3	II.E4[b]
ss BPDE-DNA	42 000	4	9	5	45
ds BPDE-DNA	42 000	105	335	17	—
ss BPDE-DNA	0.1–4	17	150	43	100
BPDE-dG	—	370	540	2000	90
BPDE tetrols	—	$>10^5$	$>10^5$	$>10^4$	—
ss CDE-DNA	30 000	10	17	6	—
ss AAF-DNA	18 000	3000	$>10^5$	100	—
ss AF-DNA	33 000	14	—	23	—

[a]BPDE-tetrols, (±)7β,8α,9α,10(α or β)-tetrahydroxy-7,8,9,10-tetrahydrobenzo[*a*]pyrene; CDE-DNA, DNA modified with (±)*trans*-1,2-dihydroxy-*anti*-3,4-epoxy-1,2,3,4-tetrahydrochrysene such that the major adduct is CDE-dG (Hodgson *et al.*, 1983); ss, single-stranded; ds, double-stranded.

[b]Monoclonal anti BPDE-dGMP was a gift from Dr R. A. Baan (Medical Biological Laboratory TNO, Rijswijk, The Netherlands) (Baan *et al.*, 1987). The microtitre wells were coated with denatured BPDE-DNA (15 fmol/25 ng); antibody dilutions were, respectively, 1:200 000 (F29 and F30), 1:100 000 (41D3) and 1:10 000 (II.E4). The 50% inhibition values were taken partly from Van Schooten *et al.* (1987).

Benzo[a]pyrene

A number of polyclonal and monoclonal antibodies raised against BPDE-modified DNA (1–2%) were also tested in our ELISA to quantify BPDE-dG bound to DNA (Van Schooten *et al.*, 1987). All antibodies showed a high affinity for single-stranded BPDE-DNA but had lower affinity for double-stranded BPDE-DNA. There was no affinity for unmodified DNA. Using four different antibodies, we found that the antibody affinity for BPDE-DNA was also dependent on the level of modification: highly modified BPDE-DNA was recognized more efficiently than BPDE-DNA of low modification (Table 2). Even the monoclonal antibody raised against the mononucleotide BPDE-dGMP had the highest affinity for BPDE-DNA of high modification. The affinity for the free mononucleoside BPDE-dG was much lower than that for BPDE-DNA, and no affinity was detected for the BPDE-tetrols. Surprisingly, AAF-DNA, and particularly AF-DNA, cross-react with the anti-BPDE-DNA antibodies, even though the majority of these adducts are at the C-8 position of guanine. We are currently investigating this phenomenon.

DNA adducts in biological samples

Kulkarni and Anderson (1984) demonstrated that in mice treated with [^{3}H]BP, the major adduct in DNA from various tissues was [^{3}H]BPDE-dG. Thus, the mouse would appear to be a suitable experimental animal for validating our ELISA for BPDE-DNA using [^{3}H]BPDE-DNA of low modification as reference compound. When samples of [^{3}H]BP-DNA isolated from the livers of mice treated with different doses of [^{3}H]BP were examined by competitive ELISA, the binding values calculated from the enzyme immunoassay were in good agreement with those from radioactivity measurements. A good coincidence was obtained between the inhibition curves for [^{3}H]BPDE-modified DNA and [^{3}H]BP-DNA from mouse liver (Fig. 2).

Using this inhibition curve as standard, we also examined a number of DNA preparations extracted from the peripheral lymphocytes of coke-oven workers and

Fig. 2. Competitive inhibition of anti-BPDE-DNA antibody binding with [^{3}H]BPDE-DNA (●) and [^{3}H]BP-DNA from the livers of mice treated with [^{3}H]BP (▲)

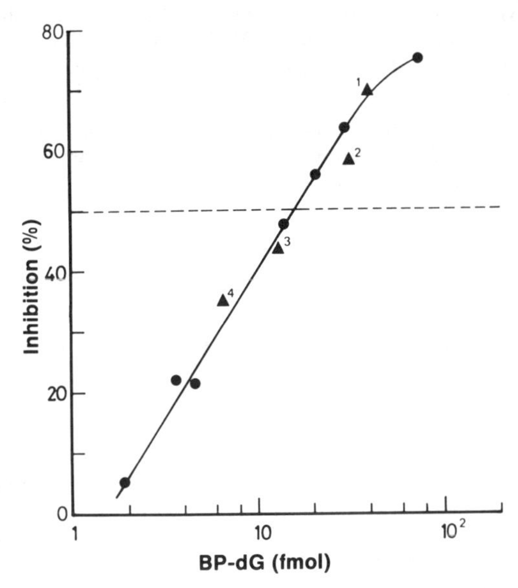

Rabbit-produced antiserum against the methylated bovine serum albumin complex of BPDE-DNA. The microtitre wells were coated with denatured BPDE-DNA (15 fmol/25 ng); antiserum (F29) dilution was 1:200 000. Individually modified [^{3}H]BPDE-DNA preparations (range, 0.1–4 fmol/μg) were used as competitor (25 μg/well). Female BALB/c mice were injected intraperitoneally with a solution of [^{3}H]BP in olive oil at dose levels of 4.0 (1), 2.0 (2), 0.8 (3) and 0.4 (4) μmol per mouse.

unexposed controls. Among a group of 13 coke-oven workers, ten (77%) were found to be carrying significant levels of PAH-DNA adducts (range corresponding to 0.2–1.2 fmol BPDE-deoxyguanine/μg DNA), whereas no such adduct could be detected in the unexposed control group. Our adduct values are in good agreement with those reported for a group of foundry workers (Perera *et al.*, 1987). When urine samples were collected from the same groups of workers and analysed for the presence of 1-hydroxypyrene (Jongeneelen *et al.*, 1987; Bos & Jongeneelen, this volume), significant amounts were found in the urine samples from coke-oven workers but not in the urine of unexposed controls.

Conclusions

Antibodies raised against carcinogen-DNA adducts, or highly modified carcinogen-DNA immunogens, recognize highly modified DNA more efficiently than carcinogen-DNA of low modification (biological samples). This property has now been demonstrated for three different antibodies, and may be common to all antibodies raised against carcinogen-DNA adducts.

DNA preparations of low modification, obtained either *in vitro* by treatment of DNA with ultimate carcinogenic reactants, or from animal tissues following exposure to specific carcinogens, have to be used as reference compounds in immunoassays designed to quantify carcinogen-DNA adducts in human tissues and cells.

Certain carcinogen-DNA adducts, e.g., guanine-8-arylamines, are unstable to heat denaturing of DNA at neutral pH, forming the guanine imidazole ring-opened derivatives as stable end products. In cases of instability of guanine-8-arylamines, it may be advantageous to develop immunoassays for the guanine imidazole ring-opened forms.

A polyclonal anti-BPDE-DNA antibody has been employed successfully in a highly sensitive enzyme immunoassay to quantify BPDE-DNA and other PAH-DNA adducts of related structure in human cells. Although only small groups of individuals have as yet been studied, the data clearly show that levels of BPDE-DNA, and possibly other PAH-DNA adducts, are significantly higher in groups of persons with a high exposure history. Enhanced PAH-DNA adduct levels seem to correlate with an enhanced excretion of 1-hydroxypyrene in the urine.

Acknowledgements

We thank Dr R.A. Baan, Medical-Biological Laboratory TNO, Rijswijk, The Netherlands, for the gift of his monoclonal anti-BPDE-dGMP antibody, and Dr P.L. Grover, Chester Beatty Research Institute, London, UK, for a specimen of DNA modified with chrysene diol epoxide.

References

Baan, R.A., Van Den Berg, P.T.M., Watson, W.P. & Smith, R.J. (1988) In situ detection of DNA adducts formed in cultured cells by benzo[*a*]pyrene diol epoxide (BPDE) with monoclonal antibodies specific for the BP-deoxyguanosine adduct. *Toxicol. environ. Chem., 16,* 325–339

Beland, F.A., Heflich, R.H., Howard, P.C. & Fu, P.P. (1985) *The in vitro metabolic activation of nitro polycyclic aromatic hydrocarbons.* In: Harvey, R.G., ed., *Polycyclic Hydrocarbons and Carcinogenesis (American Chemical Society Symposium Series 283),* Washington DC, American Chemical Society, pp. 371–396

Hodgson, R.M., Cary, P.D., Grover, P.L. & Sims, P. (1983) Metabolic activation of chrysene by hamster embryo cells: evidence for the formation of a 'bay region' diol epoxide-N^2-guanine adduct in RNA. *Carcinogenesis, 4,* 1153–1158

Jongeneelen, F.J., Anzion, R.B.M. & Henderson, P.T. (1987) Determination of hydroxylated metabolites of polycyclic aromatic hydrocarbons in urine. *J. Chromatogr., 413,* 227–232

Kadlubar, F.F. & Beland, F.A. (1985) *Chemical properties of ultimate carcinogenic metabolites of arylamines and arylamides.* In: Harvey, R.G., ed., *Polycyclic Hydrocarbons and Carcinogenesis* (*American Chemical Society Symposium Series 283*), Washington DC, American Chemical Society, pp. 341–370

Kriek, E. & Westra, J.G. (1980) Structural identification of the pyrimidine derivatives formed from *N*-(deoxyguanosin-8-yl)-2-aminofluorene in aqueous solution at alkaline pH. *Carcinogenesis, 1,* 459–468

Kriek, E., Van Der Laken, C.J., Welling, M.C. & Nagel, J. (1982) *Immunological detection and quantification of the reaction products of 2-acetylaminofluorene with guanine in DNA.* In: Bartsch, H. & Armstrong, B., eds, *Host Factors in Human Carcinogenesis* (*IARC Scientific Publications No. 39*), Lyon, International Agency for Research on Cancer, pp. 541–549

Kriek, E., Welling, M.C. & Van Der Laken, C.J. (1984) *Quantification of carcinogen-DNA adducts by a standardized high-sensitive enzyme immunoassay.* In: Berlin, A., Draper, M., Hemminki, K. & Vainio, H., eds, *Monitoring Human Exposure to Carcinogenic and Mutagenic Agents* (*IARC Scientific Publications No. 59*), Lyon, International Agency for Research on Cancer, pp. 297–305

Kulkarni, M.S. & Anderson, M.W. (1984) Persistence of benzo[*a*]pyrene metabolite-DNA adducts in lung and liver of mice. *Cancer Res., 44,* 97–104

Perera, F.P., Hemminki, K.H., Santella, R.M., Brenner, D. & Kelly, G. (1987) DNA adducts in white blood cells of foundry workers. *Proc. Am. Assoc. Cancer Res., 28,* 373

Poirier, M.C. & Beland, F.A., eds (1987) *Carcinogenesis and Adducts in Animals and Humans* (*Progress in Experimental Tumor Research, Vol. 31*), Basel, Karger

Stanton, C.A., Chow, F.L., Phillips, D.H., Grover, P.L., Garner, R.C. & Martin, C.N. (1985)

Evidence for *N*-(deoxyguanosin-8-yl)-1-aminopyrene as a major DNA adduct in female rats treated with 1-nitropyrene. *Carcinogenesis, 6,* 535–538

Strickland, P.T. & Boyle, J.M. (1984) Immunoassay of carcinogen-modified DNA. *Progr. Nucleic Acid Res. Mol. Biol., 31,* 1–58

Van Der Laken, C.J., Hagenaars, A.M., Hermsen, G., Kriek, E., Kuipers, A.C., Nagel, J., Scherer, E. & Welling, M.C. (1982) Measurement of O^6-ethyl-deoxyguanosine and *N*-(deoxyguanosin-8-yl)-*N*-acetyl-2-aminofluorene in DNA by high-sensitive enzyme immunoassays. *Carcinogenesis, 3,* 569–572

Van Schooten, F.J., Kriek, E., Steenwinkel, M.J.S.T., Noteborn, H.P.J.M., Hillebrand, M.J.X. & Van Leeuwen, F.E. (1987) The binding efficiency of polyclonal and monoclonal antibodies to DNA modified with benzo[*a*]pyrene diol epoxide is dependent on the level of modification. Implications for quantification of benzo[*a*]pyrene-DNA adducts *in vivo*. *Carcinogenesis, 8,* 1263–1269

SYNCHRONOUS FLUORESCENCE SPECTROPHOTOMETRY OF BENZO[*a*]PYRENE DIOL EPOXIDE-DNA ADDUCTS: A TOOL FOR DETECTION OF IN-VITRO AND IN-VIVO DNA DAMAGE BY EXPOSURE TO BENZO[*a*]PYRENE

K. Vähäkangas[1], O. Pelkonen[1] & C.C. Harris[2]

[1] *Department of Pharmacology, University of Oulu, Oulu, Finland; and* [2] *Laboratory of Human Carcinogenesis, National Institutes of Health, National Cancer Institute, Bethesda, MD, USA*

We have applied synchronous fluorescence spectrophotometry (SFS) to study benzo[*a*]pyrene diol epoxide (BPDE)-DNA adducts in biological samples. Adducts are measured as benzo[*a*]pyrene (BP) tetrols after acid hydrolysis, and give a peak at 374 nm of emission with a 34-nm wavelength difference. *In vitro* and in animal studies, there is a positive correlation between the amount of adducts and the BP dose. In cell culture studies, the amount of adducts is increased by increasing both the dose of BP and the time of culture with BP. In preliminary human studies, BPDE-DNA has been found by SFS in placental DNA from some but not all smoking mothers and in blood cell DNA from some individuals in occupationally exposed groups.

SFS was originally developed as a tool in analytical chemistry and was used to analyse, e.g., polycyclic aromatic hydrocarbons (PAH) in complex mixtures (Lloyd, 1971; Vo-Dinh, 1978; Vo-Dinh & Martinez, 1981; Vo-Dinh, 1982). Later, it was applied to detection of BPDE-DNA adducts (Vähäkangas *et al.*, 1985a), aflatoxin metabolites and DNA adducts (Autrup *et al.*, 1983; Harris *et al.*, 1986) and 7-methylbenz[*c*]acridine metabolites (Gill & Holder, 1986).

Currently, we use SFS to detect BP-induced DNA damage both *in vivo* and *in vitro*, the main aim being to assess the method for biochemical epidemiology in human populations as a measure of (i) PAH exposure in smokers and in subjects exposed occupationally to PAHs and (ii) individual susceptibility to PAH-induced cancer.

SFS for BPDE-DNA adducts

In conventional fluorescence measurements, the spectra of PAH are usually complex, with many peaks. In synchronous spectra, where emission and excitation are scanned at the same time with a constant wavelength difference, only one peak emerges, if the O-O band difference (so-called Stokes shift) is used as the wavelength difference (see Vo-Dinh, 1982).

Because DNA quenches the fluorescence from BPDE-DNA, we hydrolyse isolated and purified DNA samples with 0.1 M HCl for 3 h at 90°C, after which the samples can be analysed without further purification (Vähäkangas *et al.*, 1985a). This way, DNA-bound BP moieties detach as tetrols, and the fluorescence yield of the sample

Fig. 1. BPDE adducts measured by SFS in cultured human peripheral blood lymphocytes

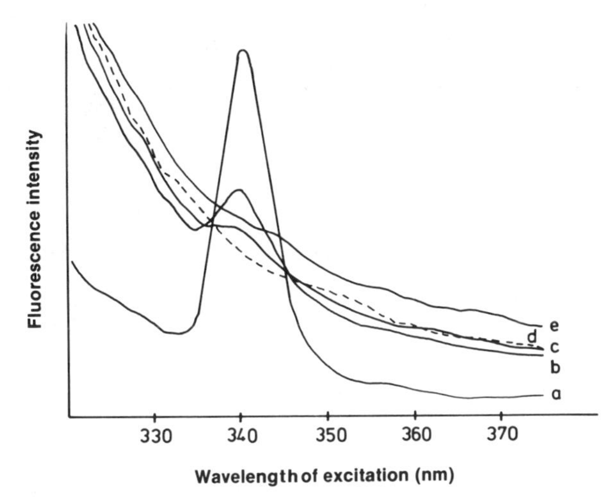

Lymphocytes were separated by gradient centrifugation and cultured with phytohaemagglutinin for two days, after which BP was added to a concentration of 1 μM. DNA was isolated from the cells 7 (e), 24 (c) and 48 (b) h after addition of BP. Control cells (d, dotted line) were cultured for four days without BP. For comparison, a peak from DNA modified *in vitro* with BPDE is shown (a; lower background is due to a different scale).

increases significantly. With 34 nm, delta lambda (the wavelength difference between excitation and emission) BP tetrols give one peak at 374 nm of emission (340 nm of excitation, Fig. 1). The fluorescence yield is in linear positive correlation with the amount of BP moieties in the sample, and the standard curves for in-vitro modified BPDE-DNA and BP tetrol are similar (Vähäkangas *et al.*, 1985b).

Human studies *in vivo*

In preliminary human studies, positive cases have been found by SFS in peripheral blood DNA from occupationally exposed people and in placental DNA from smoking mothers. However, none of the six smokers studied had BPDE adducts in their blood cell DNA (see Table 1). It is notable that only a minor proportion of persons in any group gave positive results. This finding is in contrast with immunochemical measurements, which give a much larger proportion of positive samples (Harris *et al.*, 1985; Haugen *et al.*, 1986).

Wide variability in the amount of adducts within the studied groups may reflect interindividual variation in BP-activating or DNA repair enzyme activities, as well as differences in exposure (see Vähäkangas & Pelkonen, 1987).

Experimental studies

Experimental systems used so far are presented in Table 2. In all of the systems, at least some samples gave positive results after BP treatment, SFS giving the characteristic fluorescence peak at 374 nm.

In-vitro incubation with microsomes: The incubation conditions in these studies were those used by Pelkonen *et al.* (1978), and the medium consisted of the necessary

Table 1. Human studies on BPDE-DNA adducts by SFS after environmental exposure

Probable source of exposure	Tissue studied[a]	No. of positives/ no. studied	Reference
Aluminium plant, smoking	PBL	1/30	Vähäkangas *et al.* (1985a)
Coke oven, smoking	PBL	10/41	Harris *et al.* (1985)
Coke oven, smoking	PBL	4/38	Haugen *et al.* (1986)
Smoking	Placenta	5/27[b]	Vähäkangas *et al.* (unpublished)
Smoking	PBL	0/6	Vähäkangas *et al.* (unpublished)

[a]PBL, peripheral blood lymphocytes
[b]0/31 nonsmokers were positive; smoking status was confirmed by plasma cotinine measurement

cofactors, microsomes, 2 mg calf thymus DNA and 12 nmol BP. The amount of microsomal fraction used varied from 0.1–4 mg microsomal protein in 1 ml incubation medium. The minimal amount of protein needed to detect BPDE-DNA adducts was less than 0.1 mg 3-methylcholanthrene-induced rat liver microsomes, 2 mg human placental microsomes and 2 mg human liver microsomes.

Microsomes from some human placentas and livers were able to activate BP to BPDE-DNA, as measured by SFS. It is interesting that the microsomes able to activate BP *in vitro* were from tissues other than those containing measurable amounts of DNA adducts due to in-vivo exposure. Only placental microsomes with high aryl hydrocarbon hydroxylase (AHH) activity activated BP to BPDE-DNA adducts *in vitro*; two of the placentas with adducts formed *in vivo* had low AHH activity (K. Vähäkangas, M. Pasanen & O. Pelkonen, unpublished). It is important to keep in mind that AHH levels provide only a general indication of the ability of a tissue to metabolize BP, and in-vitro formation of measurable BPDE-DNA adducts is a more specific measure of BP activation. Whether BPDE-DNA adducts are found after in-vivo exposure depends on many additional factors, like DNA repair and the rate of cell proliferation.

Cell culture studies: Both JEG cells (human trophoblastic cancer cell line) and cultured human peripheral blood lymphocytes (for culture conditions, see Kärki *et al.*,

Table 2. Experimental systems in which measurable amounts of BPDE-DNA adducts are formed from added BP

In-vitro incubation with microsomes from
- 3-methylcholanthrene-induced rat liver
- human placenta[a]
- human liver[a]

JEG cells (human trophoblastic cancer cell line)
Cultured human lymphocytes
Mice treated *in vivo* with BP; BPDE-DNA adducts found at least
- in epidermal DNA after skin painting
- in lung and liver DNA after intraperitoneal injection of BP

[a]Only some of the samples could activate BP.

1982) activated BP to BPDE-DNA adducts measurable by SFS when BP was added to the culture medium for 24 h. The response of JEG cells was dose-related, with a linear increase in the amount of adducts up to 1 μM BP in the culture medium (K. Vähäkangas, M. Pasanen & O. Pelkonen, unpublished). In preliminary lymphocyte cultures, there was an increase in the amount of adducts with time (Fig. 1). No adduct was found in human lymphocyte DNA when cells were incubated with BP and cofactors for 2 h, as for measurement of AHH (K. Vähäkangas, N.T. Kärki & O. Pelkonen, unpublished).

In-vivo studies in mice: Measurable amounts of BPDE-DNA adducts were formed in mouse epidermis after local treatment and in lung and liver DNA after intraperitoneal injections. There was a clear dose-response in the formation of the adducts. Interestingly, more BPDE-DNA adducts were formed in mouse lung than in liver, although the basal AHH activity was much higher in the liver (N. Bjelogrlic, O. Pelkonen & K. Vähäkangas, unpublished); however, the inducibility of AHH in the lung was higher than in the liver.

Conclusions

SFS for BPDE-DNA adducts is a simple and sensitive assay for BP-induced DNA damage. Theoretically, it is a more relevant measure of BP activation than AHH activity, which is a measure of hydroxylated metabolites of BP (Nebert & Gelboin, 1968; see also Vähäkangas & Pelkonen, 1987). With in-vitro modified BPDE-DNA, there is a linear, positive correlation between the amount of BP moiety in the sample and the fluorescence intensity (Vähäkangas *et al.*, 1985a). In-vivo mouse studies and cell culture studies show that the amount of BPDE-DNA is dependent on dose of BP; quantification by SFS should also be possible after exposure of humans *in vivo*.

Acknowledgements

Financial support from the Finnish Cancer Society and the Academy of Finland (KV & OP) is acknowledged.

References

Autrup, H., Bradley, K.A., Shamsuddin, A.K.M., Wakhisi, J. & Wasunna, A. (1983) Detection of putative adduct with fluorescence characteristics identical to 2,3-dihydro-2-(7-*N*-guanyl)-3-hydroxyaflatoxin B_1 in human urine collected in Murang'a district, Kenya. *Carcinogenesis, 9,* 1193–1195

Gill, J.H. & Holder, G.M. (1986) Application of synchronous luminescence to the separate determination of cochromatographing metabolites of the carcinogen 7-methylbenz-[*c*]acridine. *J. pharm. biomed. Anal., 4,* 31–36

Harris, C.C., Vähäkangas, K., Newman, M.J., Trivers, G.E., Shamsuddin, A., Sinopoli, N., Mann, D.L. & Wright, W.E. (1985) Detection of benzo[*a*]pyrene diol epoxide-DNA adducts in peripheral blood lymphocytes and antibodies to the adducts in serum from coke oven workers. *Proc. natl Acad. Sci. USA, 82,* 6672–6676

Harris, C.C., LaVeck, G., Groopman, J., Wilson, V.L. & Mann, D. (1986) Measurement of aflatoxin B_1, its metabolites and DNA adducts by synchronous fluorescence spectrophotometry. *Cancer Res., 46,* 3249–3253

Haugen, Å., Becher, G., Benestad, C., Vähäkangas, K., Trivers, G.E., Newman, M.J. & Harris, C.C. (1986) Determination of polycyclic aromatic hydrocarbons in the urine, benzo[*a*]pyrene diol epoxide-DNA adducts in lymphocyte DNA and antibodies to the adducts in sera from coke oven workers exposed to measured amounts of polycyclic aromatic hydrocarbons in the work atmosphere. *Cancer Res., 46,* 4178–4183

Kärki, N.T., Karvonen, J. & Pelkonen, O. (1982) Aryl hydrocarbon hydroxylase activity in cultured lymphocytes of psoriatic patients. *Arch. dermatol. Res., 273,* 103–109

Lloyd, J.B. (1971) The nature and evidential value of the luminescence of automobile engine oils and related materials. I. Synchronous excitation of fluorescence emission. *J. Forensic Sci. Soc., 11,* 83–94

Nebert, D.W. & Gelboin H.V. (1968) Substrate inducible microsomal aryl hydrocarbon hydroxylase in mammalian cell culture. II. Cellular responses during enzyme induction. *J. biol. Chem., 243,* 6250–6261

Pelkonen, O., Boobis, A.R., Yagi, H., Jerina, D.M. & Nebert, D.W. (1978) Tentative identification of benzo[*a*]pyrene metabolite-nucleoside complexes produced *in vitro* by mouse liver microsomes. *Mol. Pharmacol., 14,* 306–322

Vähäkangas, K. & Pelkonen, O. (1987) *Host variations in carcinogen metabolism and DNA-repair.* In: Lynch, H.T. & Hirayama, T., eds, *Familial Cancer* (*CRC crit. Rev. Toxicol.*), Boca Raton, FL, CRC Press (in press)

Vähäkangas, K., Haugen, Å. & Harris, C.C. (1985a) An applied synchronous fluorescence spectrophotometric assay to study benzo(*a*)pyrene-diolepoxide-DNA adducts. *Carcinogenesis, 6,* 1109–1116

Vähäkangas, K., Trivers, G.E., Rowe, M. & Harris, C.C. (1985b) Benzo[*a*]pyrene diolepoxide-DNA adducts detected by synchronous fluorescence spectrophotometry. *Environ. Health Perspectives, 62,* 101–104

Vo-Dinh, T. (1978) Multicomponent analysis by synchronous luminescence spectrometry. *Anal. Chem., 50,* 396–401

Vo-Dinh, T. (1982) Synchronous luminescence spectroscopy: methodology and applicability. *Appl. Spectrosc., 36,* 576–581

Vo-Dinh, T. & Martinez, P.R. (1981) Direct determination of selected polynuclear aromatic hydrocarbons in a coal liquefaction product by synchronous luminescence techniques. *Anal. chim. Acta, 125,* 13–19

MICRONUCLEI IN CYTOKINESIS-BLOCKED LYMPHOCYTES AS AN INDEX OF OCCUPATIONAL EXPOSURE TO ALKYLATING CYTOSTATIC DRUGS

J.W. Yager,[1,2] M. Sorsa[1] & S. Selvin[2]

[1] *Institute of Occupational Health, Helsinki, Finland; and* [2] *Department of Biomedical and Environmental Health Sciences, University of California, Berkeley, CA, USA*

Micronucleated peripheral blood lymphocytes were analysed in cytochalasin-B-treated binucleated lymphocytes using cytocentrifuged preparations. Increased numbers of micronuclei were observed in lymphocytes of groups of workers from industry and hospitals potentially exposed to cyclophosphamide. The finding was independent of the age of the subjects, which was also correlated with micronuclei formation.

Cyclophosphamide is a known alkylating mutagen which requires metabolic activation (IARC, 1987a); it is considered to be an animal and human carcinogen (IARC, 1987b). The purpose of this study was to evaluate cytokinesis-block modification of the lymphocyte micronucleus assay (Fenech & Morley, 1985) as a biological index of worker exposure during industrial formulating and processing, drug manufacture and packaging, and administration to hospital patients of cyclophosphamide.

Initially, in-vitro treatment of cultured lymphocytes with the directly-acting intermediates produced during cyclophosphamide manufacture, CP I (nornitrogen mustard) and CP II (phosphoroxidichloride mustard), was found to produce significant increases in micronucleus induction at treatment concentrations as low as 5 μM ($p = 0.001$) and 2.5 μM ($p = 0.00001$), respectively (Yager & Sorsa, 1987). Parallel cultures without cytokinesis block exhibited considerably less micronucleus induction: treatment with 5 μM CP I and 2.5 μM CP II produced no detectable increase. Thus, the cytokinesis-block modification exhibits considerable advantages over the standard method with regard to increased statistical power and ease of analysis.

Blood samples for analysis were collected from controls and workers at the three work sites described above. Epidemiological information and health and work histories were obtained from each participant, and environmental air sampling for cyclophosphamide was performed (Sorsa *et al.*, 1987). Replicative indices as a measure of cytotoxicity were determined for each sample (Yager & Sorsa, 1987).

Mean micronucleus counts for the four groups — control and three worker groups — are significantly different from one another (Table 1, analysis of variance, $p = 0.05$), the control group having the lowest mean value. Smoking does not appear to have an effect on micronucleus induction, but age does, such that older people have higher lymphocyte micronucleus scores (Fenech & Morley, 1986). Figure 1 shows the mean micronucleus scores and the mean ages of the control group and the three worker groups. The groups differ in mean age, which would interfere with direct comparison of micronucleus means. Regression of $\log_{10}$ micronucleus on age (Fig. 2) for the entire data set showed that age has a significant effect on micronucleus outcome

Table 1. Lymphocyte micronuclei among workers in cyclophosphamide production and manufacture, oncology nurses and controls

Subjects	Sex	Age (years)	Smoking status[a]	Employment time (years)	Micronuclei[b]	Replication index[c]
Controls (n = 9)						
	F	48	+++	—	10	1.80
	F	40	–	—	10	1.67
	F	39	–	—	11	1.88
	F	46	–	—	12	1.54
	F	50	+++	—	10	1.86
	M	27	–	—	7	1.77
	M	31	+++	—	6	1.77
	F	28	–	—	12	1.95
	F	28	–	—	9	1.85
Mean ± SE		37.4 ± 3.1			9.8 ± 0.7	
Positive control[d]						
	M	75	++	—	~77	1.57
Chemical process workers (n = 7)						
	M	38	–	15	7	1.67
	M	35	–	16	8	1.87
	M	25	++	2	6	1.85
	M	22	–	2	4	1.80
	F	41	–	11	17	1.70
	F	49	–	10	43	1.56
	M	31	–	11	6	1.81
Mean ± SE		34.4 ± 3.5			13.0 ± 5.2	
Manufacturing workers (n = 12)						
	F	43	–	2	16	1.74
	F	43	–	3	24	1.95
	F	61	–	2	25	1.56
	F	55	–	2	36	1.82
	F	52	–	4	6	1.89
	F	61	–	4	31	1.78
	F	41	–	3	7	2.00
	F	39	+++	2	12	1.86
	F	47	–	3	9	1.68
	F	44	–	4	9	1.64
	F	46	+++	4	15	1.42
	F	53	–	2	15	1.81
Mean ± SE		48.8 ± 2.2			17.0 ± 2.8	
Oncology nurses (n = 8)						
	F	47	–	20	25	1.79
	F	41	+	18	40	1.75
	F	46	–	5	27	1.61
	F	45	–	5	18	1.60
	F	30	(+)	3	9	1.89
	F	59	–	15	20	1.64
	F	33	–	25	22	1.78
	F	45	–	20	19	1.79
Mean ± SE		43.3 ± 3.2			22.5 ± 3.1	

[a] –, never smoked; (+), occasional (1–2 cigarettes per day); +, smoke ≤10 cigarettes per day; ++, smoke 11–19 cigarettes per day; +++, smoke ≥20 cigarettes per day
[b] Scored in 1000 binucleated lymphocytes (500 per duplicate culture) per sample
[c] [(1 × % mononucleated cells) + (2 × % binucleated cells) + (3 × % cells > binucleated)]/100. Obtained by scoring 200 cells (100 cells per duplicate culture) per sample
[d] Cancer patient treated with cyclophosphamide (2500 mg)

Fig. 1. Bar chart of groups analysed in this study of worker exposure to cyclophosphamide: control, drug manufacturing workers, chemical processing workers and oncology nurses

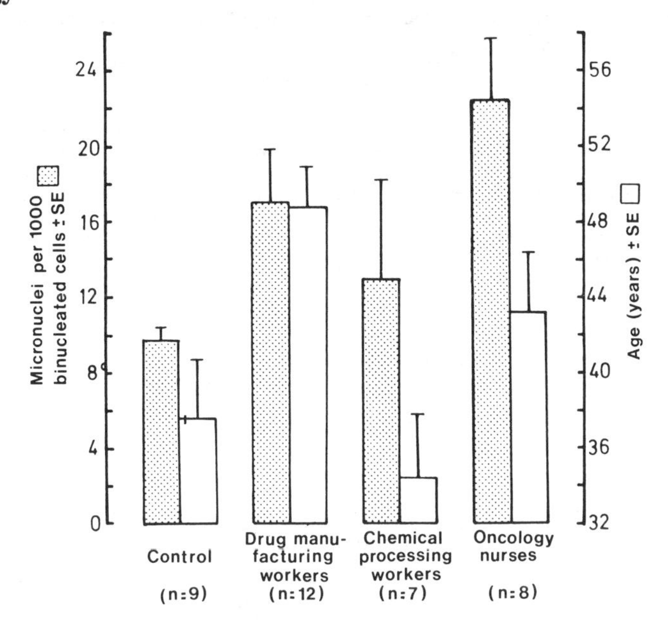

($\hat{y}_i = 0.452 + 0.0158\,x_i$, $p = 0.0001$; correlation = 0.599). The data were then analysed using statistical analysis of covariance techniques (Snedecor & Cochran, 1974), so that the mean number of micronuclei for the control and worker groups could be compared in the absence of the influence of age (Table 2). The untransformed mean of the control group was 9.8 ± 0.7 micronuclei per 1000 binucleated cells, and that of the worker group was 17.3 ± 2.2 (Table 1). Log transforming the data then gave unadjusted means of 0.9791 for the control group and 1.1772 for the worker group

Fig. 2. Regression of age in years on $\log_{10}$ micronuclei for all cases in the data set; $\hat{y}_i = 0.452 + 0.0158\,x_i$, $r = 0.599$

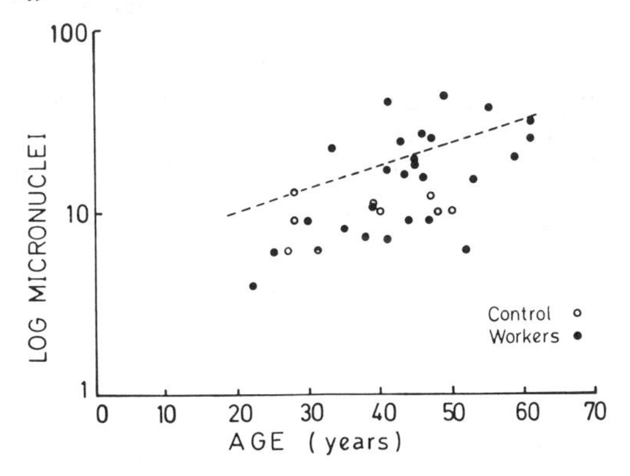

This regression is significant at $p = 0.0001$.

Table 2. Group mean log micronucleus values adjusted to remove the effects of age

Group	No.	Mean log micronuclei ($\bar{y}_i$)[a]	Adjusted mean log micronuclei ($\bar{y}_i$)[a]
Control	9	0.9791	0.9995
Worker	27	1.1772	1.1518[b]

[a]Regressions: control group: $y_{1i} = 0.8045 + 0.00465\ x_i$; worker group: $y_{2i} = 0.4230 + 0.0173\ x_i$; then $\bar{y}_i' = \bar{y}_i + \hat{b}_i(\bar{x}_i - \bar{x})$ and variance $(\bar{y}_i') = \text{var}(\bar{y}_i) + (\bar{x}_i - \bar{x})^2\ \text{var}(\hat{b}_i)$, where $\bar{y}_i'$ is the adjusted mean log micronuclei for the *ith* group, $\bar{y}_i$ is the mean log micronuclei of the *i*th group, $\bar{x}_i$ is the mean age of the *i*th group, and $\bar{x}$ is the overall mean age. Underlying analysis of covariance are certain assumptions that appear to be fulfilled by the log micronuclei data.

[b]Significantly different from control, $t = 2.3685$, df $= 32$, $p = 0.01$

(Table 2). After adjusting the means to remove the influence of age, the adjusted control group mean was 0.9995 and the worker group mean was 1.1518, and the difference between the two was significant ($p = 0.01$).

Conclusion

Enumeration of micronuclei in binucleated lymphocytes on cytocentrifuged preparations appears to provide increased statistical power and ease of analysis over the standard method. This study indicates that the increased numbers of micronuclei present in lymphocytes of groups of workers potentially exposed to cyclophosphamide in the three settings described are separate from the influence of age. To our knowledge, this is the first application of the modified micronucleus method to human monitoring; further studies encompassing larger numbers of observations and exposure data are needed to confirm this finding.

Acknowledgements

The authors are indebted to Ms H. Järver*aus for excellent technical assistance, to colleagues at the Institute of Occupational Health for helpful advice and discussions and to the Academy of Finland (through the Fogarty International Center, National Institutes of Health, USA) for the award of a Visiting Research Fellowship to Dr J.W. Yager.

References

Fenech, M. & Morley, A.A. (1985) Measurement of micronuclei in lymphocytes. *Mutat. Res.*, *147*, 29–36

Fenech, M. & Morley, A.A. (1986) Cytokinesis-block micronucleus method in human lymphocytes: effect of *in vivo* ageing and low dose X-irradiation. *Mutat. Res.*, *161*, 193–198

IARC (1987a) *IARC Monographs on the Evaluation of Carcinogenic Risks to Humans*, Suppl. 6, *Genetic and Related Effects: An Updating of Selected* IARC Monographs *from Volumes 1 to 42*, Lyon

IARC (1987b) *IARC Monographs on the Evaluation of Carcinogenic Risks to Humans*, Suppl. 7, *Overall Evaluations of Carcinogenicity: An Updating of* IARC Monographs *Volumes 1 to 42*, Lyon

Snedecor, G.W. & Cochran, W. (1974) *Statistical Methods*, 7th Ed., Ames, IA, Iowa State Press

Sorsa, M., Pyy, L., Salomaa, S., Nylund, L. & Yager, J.W. (1988) Biological and environmental monitoring of occupational exposure to cyclophosphamide in industry and hospitals. *Mutat. Res.* (in press)

Yager, J.W. & Sorsa, M. (1987) Evaluation of the cytokinesis block modification of the peripheral lymphocyte micronucleus method. *Environ. Mutagenesis*, *9* (*Suppl.* 8), 116

DETECTION OF STYRENE OXIDE-DNA ADDUCTS IN LYMPHOCYTES OF A WORKER EXPOSED TO STYRENE

S.F. Liu,[1] S.M. Rappaport,[1,3] K. Pongracz[2] & W.J. Bodell[2]

[1]*School of Public Health, University of California, Berkeley, CA; and* [2]*Brain Tumour Research Center, University of California, San Francisco, CA, USA*

The ^{32}P-postlabelling procedure has been used to detect styrene oxide (SO)-DNA adducts. Reactions of SO with DNA and dGMP *in vitro* produced adducts that were similar, indicating that dGMP was the primary base for modification in DNA. Two SO adducts were also detected in DNA isolated from lymphocytes of a styrene-exposed worker but not in DNA from an unexposed worker. These results indicate that ^{32}P-postlabelling can be used for quantification of DNA adducts in workers exposed to styrene.

Styrene is a commercially important chemical essential to thc production of reinforced plastics. Occupational exposure to styrene occurs primarily through inhalation of the airborne vapour. Following inhalation, styrene is absorbed into the blood and metabolized to SO, primarily *via* P450 oxidation in the liver (World Health Organization, 1983).

Styrene and SO have been shown to be carcinogenic to rodents (reviewed by Huff, 1984) and to induce sister chromatid exchanges and chromosomal aberrations in human lymphocytes (Meretoja *et al.*, 1977; Fleig & Thiess, 1978; Norppa *et al.*, 1980, 1981). It is presumed for both chemicals that the genotoxic and ultimately carcinogenic species is SO, which has been shown to be mutagenic in prokaryotic test systems (Sugiura & Goto, 1981). SO has been shown to arylalkylate the 7, 2, and O^6 positions of guanosine reacted *in vitro* (Hemminki & Hesso, 1984); 2- and 7-guanine adducts have been detected after reactions of DNA with SO (Savela *et al.*, 1986); and 7-guanine adducts have been identified in the tissues of rats following administration of styrene (Byfält Nordqvist *et al.*, 1985).

These results suggest that occupational exposures to styrene may produce DNA adducts in lymphocytes and tissues of workers. Analytical methods have been developed that have the required sensitivity to measure DNA alkylation products in exposed workers; these are immunoassays (Harris *et al.*, 1985; Shamsuddin *et al.*, 1985) and the ^{32}P-postlabelling method (Everson *et al.*, 1986; Randerath *et al.*, 1986). However, they have not yet been applied to the measurement of SO-DNA adducts. The purpose of this investigation was to apply the ^{32}P-postlabelling procedure to detection of DNA adducts in lymphocytes of workers exposed to styrene.

Reactions of SO with dGMP and DNA

SO was initially reacted with dGMP. ^{32}P-Postlabelling of the reaction mixture resulted in the detection of five SO-dGMP adducts (Fig. 1A) after two-dimensional

[3]To whom correspondence should be sent

Fig. 1. Reactions of SO with dGMP (A) and calf thymus DNA (B) *in vitro*

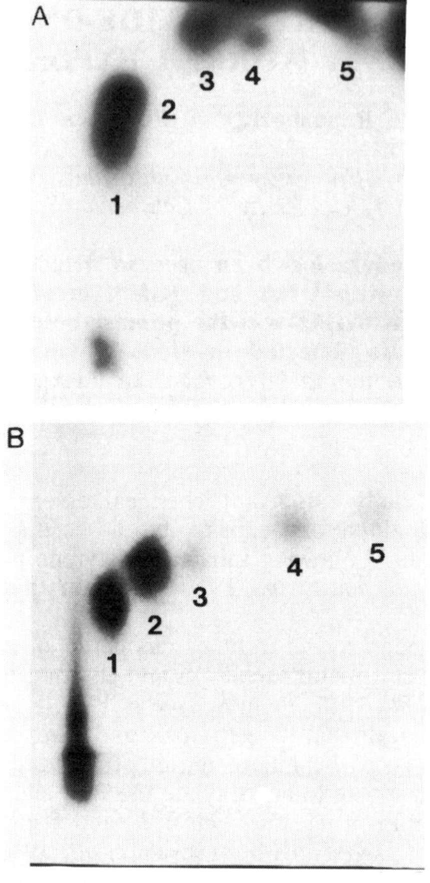

0.5 mg dGMP and 0.5 mg calf thymus DNA, dissolved in 1 ml 10 mM Tris-HCl pH 7.5 and 5 μl SO (97%, Aldrich), were reacted at 37°C for 15 h and then extracted twice with diethyl ether to remove unreacted SO. The concentration of DNA in the aqueous phase was measured by absorption at 260 nm. For the postlabelling procedure, 1 μg DNA was digested enzymatically to 3′-mononucleotides, as described previously (Bodell & Rasmussen, 1984); 0.17 μg digested DNA or dGMP was ^{32}P-postlabelled to form the 5′^{32}P-labelled 3′,5′-nucleotide diphosphate. Initial purification of the ^{32}P-labelled adducts was by C-18-thin-layer chromatography, as described by Randerath *et al.* (1984), using 0.4 M ammonium formate pH 6.0. Contact transfer of ^{32}P radioactivity remaining at the origin of the C-18 plate to the polyethyleneimine (PEI) plate was as described by Randerath *et al.* (1984). Two-dimensional separation of the adducts was performed on the PEI-cellulose plates. The solvent for the first dimension was 1.8 M Li formate, 4.25 M urea pH 3.5, the solvent for the second dimension was 0.36 M LiCl, 0.22 M Tris, 3.8 M urea pH 8.0, followed by a second chromatography with 1.7 M Na_2HPO_4 pH 6.0. The adducts were located by autoradiography at −70°C with XAR-5 film (Kodak). The individual adducts were scraped into scintillation vials containing 5 ml Safety-Solve, and radioactivity was determined by liquid scintillation counting. To determine radioactivity in the normal nucleotides, 1 μl of each ^{32}P-labelled sample was applied to PEI plates. The chromatograms were developed in 1.5 M ammonium formate pH 3.5 (D-1), and 0.3 M $(NH_4)_2SO_4$ pH 6.0 (D-2). The activities of the individual spots were determined by autoradiography, as described above, and collected for liquid scintillation counting.

Table 1. Reactions of styrene oxide[a] *in vitro*

Adduct	dGMP[b]	DNA[c]
1	0.458	0.461 ± 0.074
2	0.451	0.414 ± 0.035
3	0.021	0.077 ± 0.035
4	0.023	0.037 ± 0.027
5	0.025	0.031 ± 0.022

[a]Values expressed as fraction of total
[b]Mean of two determinations
[c]Mean ± SD of 11 determinations

separation of the adducts on polyethyleneimine-cellulose plates. Adducts 1 and 2 were formed in approximately equal amounts, and together accounted for 91% of the total modification (Table 1). Adducts 3–5 were relatively minor products, each representing less than 10% of the total modification. Similar treatment of SO with dAMP revealed two adducts formed at levels that were very low compared to those produced by reaction with dGMP (data not shown). Reaction of SO with dCMP and dTMP did not produce adducts detected by the postlabelling procedure.

Reaction of calf thymus DNA with SO also produced five adducts, as detected by the ^{32}P-postlabelling procedure (Fig. 1B). Again, two major products (adducts 1 and 2) accounted for 88% of the total modification (Table 1). The level of DNA modification was proportional to the SO concentration (data not shown).

The chromatographic behaviour of adducts and levels of modification shown in Figure 1 and Table 1 indicate that dGMP is the primary site of DNA modification. This is consistent with investigations of SO-guanosine adducts (Hemminki & Hesso, 1984) and of SO-DNA adducts (Savela *et al.*, 1986). The principal DNA adduct noted by Savela *et al.* (1986), at the 7 position of guanine, accounted for 80% of the products formed. This is similar to the sum of adducts 1 and 2 detected by the postlabelling assay, which accounted for 90% of the products. Since SO is expected to react by an S_N2 mechanism, which gives predominantly 7-guanine alkylation products, adducts 1 and 2 may be isomers of 7-guanine-SO alkylation products formed by opening of the epoxide ring at the α and β positions. Further work is in progress to assess the structures of these adducts.

We have tentatively identified adduct 3 as O^6-(2-hydroxy-2-phenylethyl)-deoxyguanosine-3′,5′-bisphosphate. This identification was based upon comparison of the chromatographic behaviour of adduct 3 with that of O^6-(2-hydroxyphenyl-ethyl)-2′-deoxyguanosine-3′-monophosphate, which we synthesized. Only one O^6 adduct was detected by ^{32}P-postlabelling instead of the expected two (Hemminki & Hesso, 1984). This may be due to conversion of the O^6-(2-hydroxy-1-phenylethyl)-deoxyguanosine to O^6-(2-hydroxy-2-phenylethyl)deoxyguanosine under the alkaline conditions used for the postlabelling procedure (Moschel *et al.*, 1986). Adducts 4 and 5 may represent SO adducts at the *N*2 position of guanine, but further work is required for a positive structural identification.

Lymphocytes of a worker exposed to styrene

Having applied the postlabelling procedure to SO adducts produced *in vitro*, we were interested in detecting SO adducts in workers occupationally exposed to styrene.

Fig. 2. ^{32}P-Postlabelling analysis of DNA from A, a styrene-exposed worker and B, an unexposed worker

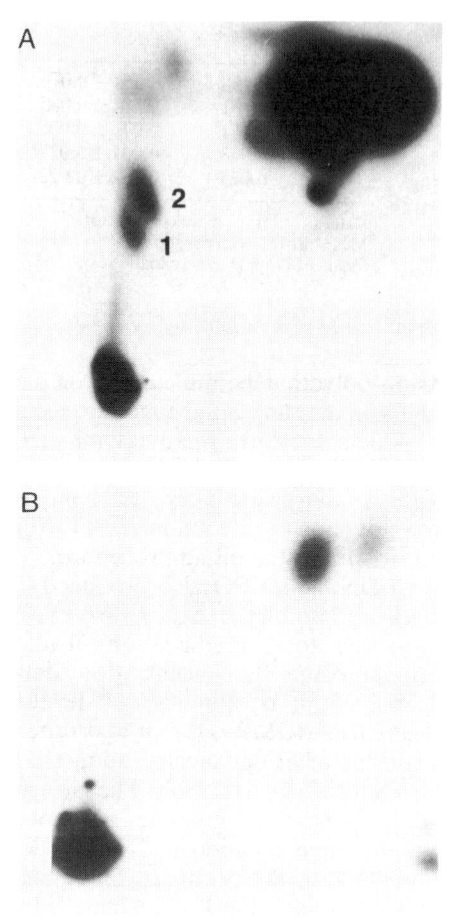

Lymphocytes were isolated from 10 ml of venous blood by centrifugation on a Ficol-1-P gradient (Pharmacia Laboratory Separation). DNA was isolated by a previously described procedure (Bodell & Banerjee, 1976). The ^{32}P-postlabelling and separation procedures were performed as described in the legend to Figure 1, except that 300 μCi ATP were added to the postlabelling mixture.

The results of a single determination are reported here for a styrene-exposed worker and an unexposed worker. Both individuals worked in a facility where reinforced plastics were manufactured; the exposed worker had been employed for eight years in the production department, where styrene-containing resins were sprayed into open moulds; the unexposed worker had been employed for two years in the shipping department, which was outdoors and away from the manufacturing area. The eight-hour, average air concentration measured for the exposed worker on the day of blood collection was 96 ppm; the exposure of the unexposed worker, although not measured, was presumed to be low.

Following isolation of DNA from the two workers' lymphocytes, the ^{32}P-postlabelling procedure was applied. DNA from the exposed worker contained products (Fig. 2A) corresponding to adducts 1 and 2 from the in-vitro modifications described previously for dGMP and DNA. The extent of formation of adducts 1 and 2 was 22×10^{-7}. No SO-DNA adduct was detected in the DNA from the unexposed worker.

This application of the ^{32}P-postlabelling procedure is apparently the first observation of DNA adducts in a worker exposed to styrene. The sensitivity of the method is currently being enhanced, and lymphocytes from additional workers are being analysed to determine the levels of SO-DNA adducts at various intensities of styrene exposure. Such SO-DNA adducts could be responsible for the observed increase in chromosomal aberration frequency in workers occupationally exposed to styrene (Meretoja *et al.*, 1977; Fleig & Thiess, 1978).

Acknowledgements

This project was supported by grant 1R01OH02221 from the US National Institute for Occupational Safety and Health of the Centers for Disease Control and by the Health Effects Component of the University of California's Toxic Substances Research and Teaching Program.

References

Bodell, W.J. & Banerjee, M.R. (1976) Reduced DNA repair in mouse satellite DNA after treatment with methyl methanesulfonate and *N*-methyl-*N*-nitrosourea. *Nucleic Acids Res., 3,* 1689–1701

Bodell, W.J. & Rasmussen, J. (1984) A^{32}P post-labelling assay for determining the incorporation of bromodeoxyuridine into cellular DNA. *Anal. Biochem., 142,* 525–528

Byfält-Nordqvist, M., Lof, A., Osterman-Golkar, S. & Walles, S.A.S. (1985) Covalent binding of styrene and styrene-7,8-oxide to plasma proteins, hemoglobin, and DNA in the mouse. *Chem.-biol. Interactions, 55,* 63–73

Everson, R.B., Randerath, E., Santella, R.M., Cefalo, R.C., Avitts, T.A. & Randerath, K. (1986) Detection of smoking-related covalent DNA adducts in human placenta. *Science, 231,* 54–57

Fleig, I. & Thiess, A.M. (1978) Mutagenicity study of workers employed in the styrene and polystyrene processing and manufacturing industry. *Scand. J. Work environ. Health, 4* (*Suppl.* 2), 254–258

Harris, C.C., Vähäkangas, K., Newman, M.J., Trivers, G.E., Shamsuddin, A., Sinopoli, N., Mann, D.L. & Wright, W.E. (1985) Detection of benzo(*a*)pyrene diol epoxide DNA adducts in peripheral blood lymphocytes and antibodies to the adducts in serum from coke-oven workers. *Proc. natl Acad. Sci. USA, 82,* 6672–6676

Hemminki, K. & Hesso, A. (1984) Reaction products of styrene oxide with guanosine in aqueous media. *Carcinogenesis, 5,* 601–607

Huff, J.E. (1984) Styrene, styrene oxide, polystyrene, and B-nitrostyrene/styrene carcinogenicity in rodents. *Prog. clin. biol. Res., 141,* 227–238

Meretoja, T., Vainio, H., Sorsa, M. & Harkonen, H. (1977) Occupational styrene exposure and chromosomal aberrations. *Mutat. Res., 56,* 193–197

Moschel, R.C., Hemminki, K. & Dipple, A. (1986) Hydrolysis and rearrangement of O^6-substituted guanosine products resulting from reaction of guanosine with styrene oxide. *J. org. Chem., 51,* 2952–2955

Norppa, H., Sorsa, M., Pfaffli, P. & Vainio, H. (1980) Styrene and styrene oxide induce SCEs and are metabolized in human lymphocyte cultures. *Carcinogenesis, 1,* 357–361

Norppa, H., Hemminki, K., Sorsa, M. & Vainio, H. (1981) Effect of monosubstituted epoxides on chromosome aberrations and SCEs in cultured human lymphocytes. *Mutat. Res., 91,* 243–250

Randerath, K., Haglund, R.E., Phillips, D.H. & Reddy, M.V. (1984) ^{32}P Post-labelling analysis of DNA adducts formed in the livers of animals treated with safrole, estragole, and other naturally occurring alkenylbenzenes. I. Adult female CD-1 mice. *Carcinogenesis, 5,* 1613–1622

Randerath, E., Avitts, T.A., Reddy, M.V., Miller, R.H., Everson, R.B. & Randerath, K.

(1986) Comparative ^{32}P analysis of cigarette smoke-induced DNA damage in human tissues and mouse skin. *Cancer Res.*, *46*, 5869–5877

Savela, K., Hesso, A. & Hemminki, K. (1986) Characterization of reaction products between styrene oxide and deoxyribonucleosides and DNA. *Chem.-biol. Interactions*, *60*, 235–246

Shamsuddin, A.K.M., Sinopoli, N.T., Hemminki, K., Boesch, R.R. & Harris, C.C. (1985) Detection of benzo(*a*)pyrene DNA adducts in human white blood cells. *Cancer Res.*, *45*, 66–68

Sugiura, K. & Goto, M. (1981) Mutagenicities of styrene oxide derivatives on bacterial test systems: relationship between mutagenic potencies and chemical reactivity. *Chem.-biol. Interactions*, *35*, 71–91

World Health Organization (1983) *Styrene* (*Environmental Health Criteria 26*), Geneva

SINGLE-STRAND BREAKS IN DNA OF PERIPHERAL LYMPHOCYTES OF STYRENE-EXPOSED WORKERS

S.A.S. Walles,[1] H. Norppa,[2] S. Osterman-Golkar[3] & J. Mäki-Paakkanen[2]

[1] *Unit of Occupational Toxicology, National Institute of Occupational Health, Solna, Sweden;* [2] *Institute of Occupational Health, Helsinki, Finland; and* [3] *Department of Radiobiology, University of Stockholm, Stockholm, Sweden*

Single-strand breaks (SSB) were determined by means of the DNA unwinding technique in peripheral lymphocytes of styrene-exposed workers and referents. A slight increase in SSB was observed among exposed subjects in comparison with a control group of office workers. A correlation was found between SSB, the excretion of mandelic acid and the concentration of styrene glycol in blood. The present work suggests that the DNA unwinding technique can be used for screening workers exposed to genotoxic compounds.

Styrene occurs at high concentrations in certain work places, and there is evidence that it is carcinogenic in animals (Huff, 1984). For the main reactive metabolite of styrene, styrene-7,8-oxide, there is sufficient evidence of carcinogenicity to animals (IARC, 1985), and it has been detected in the blood of animals exposed to styrene experimentally. Low levels of this epoxide have also been demonstrated in experimentally exposed volunteers and in occupationally exposed workers (Wigaeus *et al.*, 1983). The evaluation of the risk of late effects associated with exposure to styrene has therefore become an urgent problem.

Several studies have shown that workers exposed to high concentrations of styrene in the reinforced plastics industry have an increased frequency of structural chromosomal aberrations in their peripheral lymphocytes (World Health Organization, 1983). Some investigations have also indicated that sister chromatid exchanges (SCE) and micronucleus levels are elevated after exposure to styrene (Camurri *et al.*, 1982; World Health Organization, 1983).

Exposure of mice to styrene causes a dose-dependent increase in the level of SSB in the DNA of the kidney (Walles & Orsén, 1983); in other organs, SSB levels correlated with the binding of styrene-7,8-oxide to DNA (Byfält-Nordqvist *et al.*, 1985).

The purpose of this investigation was to study SSB in DNA from the lymphocytes of styrene-exposed workers and to compare this approach with other methods for biological monitoring of genotoxic compounds. The complete study includes analysis of the following endpoints: chromosomal aberrations, SCE, micronuclei, covalent binding to macromolecules, styrene oxide and styrene glycol levels in blood, and mandelic acid levels in urine. In this paper, only the results on SSB, styrene glycol and mandelic acid are reported.

Blood and urine samples were collected at the end of a working shift from workers exposed to styrene in a plant where large containers were manufactured from unsaturated polyester resins. All of the exposed workers were laminators. Office workers outside of the plant were chosen as referents.

Table 1. Levels of SSB ($-\log F_{DS}$) and concentrations of mandelic acid and styrene glycol, duration of employment and smoking habits of nine styrene-exposed workers and eight referents

$-\log F_{DS}$	Mandelic acid (mM)[a]	Styrene glycol (μM)[b]	Duration of employment (years)	Cigarettes per day	Duration of smoking (years)
0.07	<1	<0.6	5	25	15
0.13	8.7	1.6	3	15	16
0.15	21.5	4.5	12	18	30
0.14	9.5	3.2	5	15	20
0.11	14.4	1.9	3	20	15
0.09	9.1	4.0	3	15	12
0.19	6.3	2.2	0.5	20	12
0.18	16.5	3.2	15	20	20
0.11	<1	0.9	8	0	0
0.11	<1	<0.6	0	40	13
0.10	<1	<0.6	0	0	0
0.075	<1	<0.6	0	2	11
0.12	<1	<0.6	0	10	25
0.07	<1	<0.6	0	10	6
0.09	<1	<0.6	0	0	0
0.075	<1	<0.6	0	20	17
0.11	<1	<0.6	0	13	24

[a]Detection limit, 1.0 mM
[b]Detection limit, 0.6 μM

Levels of SSB (in nine workers and eight referents) were determined by the DNA unwinding technique (Ahnström & Erixon, 1973), as modified by Walles and Erixon (1984), using lymphocytes isolated by Ficoll-Paque (Pharmacia Fine Chemicals, Uppsala, Sweden). Styrene glycol was analysed by gas chromatography, essentially as described by Duverger-van Bogaert *et al.* (1978). Mandelic acid was determined according to the method of Engström and Rantanen (1974). The results of measurements of SSB, styrene glycol and mandelic acid are presented in Table 1, together with data on duration of employment and smoking habits.

The mean concentrations of styrene glycol in blood (2.4 ± 0.5 μmol/l) and of mandelic acid in urine (9.6 ± 2.3 mmol/l) indicate that the exposure levels were of the order of 50–100 ppm (see Wigaeus *et al.*, 1983). The two parameters of exposure were strongly correlated, as might be expected.

The mean levels of SSB, expressed as the negative logarithm of the fraction of double-stranded DNA ($-\log F_{DS}$), in exposed workers and referents were 0.13 and 0.09, respectively. A Wilcoxon two-sample (two-sided) test gave $p = 0.06$ (one-sided test gave $p < 0.05$) for a difference between the groups. A multiple regression analysis demonstrates that SSB levels are correlated to mandelic acid concentrations in the urine and to styrene glycol levels in the blood (Fig. 1). Duration of employment and of smoking (cigarettes/day × duration of smoking), tested as covariates in the analysis, did not explain the variation in SSB levels.

In an earlier study, the SSB levels in lymphocyte DNA from exposed workers were not significantly increased over the background level observed in referents. In that case, the air concentration of styrene was about 20 ppm, and the average urinary excretion of mandelic acid by exposed workers was 1.8 mM (unpublished).

Fig. 1. Level of SSB expressed as $-\log F_{DS}$, where F_{DS} = the ratio of the amount of double-stranded DNA and total amount of DNA, as a function of concentration of mandelic acid in urine for nine exposed workers and eight referents. The regression line is: $y = 0.096 + 0.003x$; $r = 0.63$. B, Level of SSB as a function of concentration of styrene glycol in blood. The regression line is: $y = 0.095 + 0.014x$, $r = 0.61$.

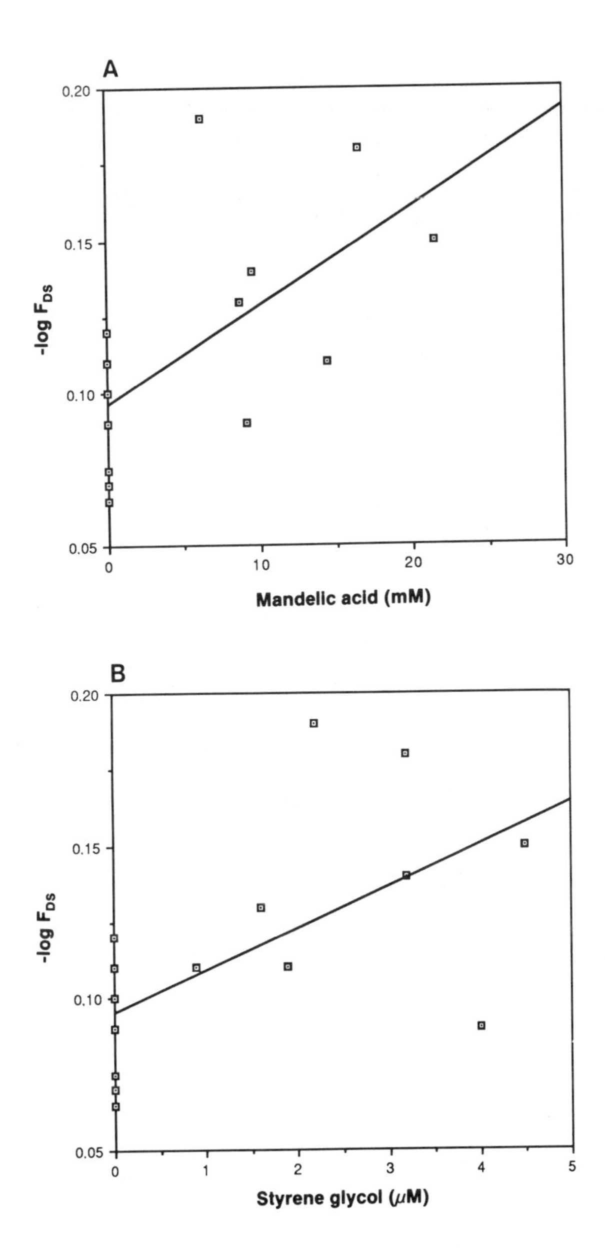

Each point represents one person, except for the following points, which consist of values for two persons: 0.07, 0; 0.11, 0; 0.075, 0.

The present study has shown that the DNA unwinding technique can be applied to lymphocytes of occupationally exposed workers. The method is simple, fast and sensitive and could provide a convenient alternative to the commonly used approaches based mainly on time-consuming chromosome analysis. Analysis of chromosomal aberrations, SCE and micronuclei from the the same blood samples are in progress and will provide comparative data.

Acknowledgements

We are grateful to Ms K. Lindblom, Mrs E. Tardelius-Bengtsson and Mrs H. Järventaus for skilful technical assistance. Ms E. Bergmark is thanked for the gas chromatography-mass spectrometry verification of references and standards for the styrene glycol determinations. Dr A. Palotie and Ms H. Kivistö are acknowledged for mandelic acid analyses. Thanks are due to Mr F. Granath for the performance of the statistical analysis. This work was supported by the Swedish Work Environment Fund, Nos. 82-0104, 84-1269, and 86-0408.

References

Ahnström, G. & Erixon, K. (1973) Radiation induced strand breakage in DNA from mammalian cells. Strand separation in alkaline solution. *Int. J. Radiat. Biol., 23,* 285–289

Byfält-Nordqvist, M., Löf, A., Osterman-Golkar, S. & Walles, S.A.S. (1985) Covalent binding of styrene and styrene-7,8-oxide to plasma proteins, hemoglobin and DNA in the mouse. *Chem.-biol. Interact., 55,* 63–73

Camurri, L., Codeluppi, S., Pedroni, C. & Scardvelli L. (1983) Chromosomal aberrations and sister-chromatid exchanges in workers exposed to styrene. *Mutat. Res., 119,* 361–369

Duverger-van Bogaert, M., Noël, G., Rollman, B., Cumps, J., Roberfroid, M. & Mercier, M. (1978) Determination of oxide synthetase and hydratase activities by a new highly sensitive gas chromatographic method using styrene and styrene oxide as substrates. *Biochim. biophys. Acta, 526,* 77–84

Engström, K. & Rantanen, J. (1974) A new gas chromatographic method for determination of mandelic acid in urine. *Int. Arch. Arbeitsmed., 33,* 163–167

Huff, J.E. (1984) Styrene, styrene oxide, polystyrene, and β-nitrostyrene/styrene carcinogenicity in rodents. *Prog. clin. biol. Res., 141,* 227–238

IARC (1985) *IARC Monographs on the Evaluation of the Carcinogenic Risk of Chemicals to Humans,* Vol. 36, *Allyl Compounds, Aldehydes, Epoxides and Peroxides,* Lyon, pp. 245–263

Walles, S.A.S. & Erixon, K. (1984) Single-strand breaks in DNA of various organs of mice induced by methyl methanesulfonate and dimethylsulfoxide determined by the alkaline unwinding technique. *Carcinogenesis, 5,* 319–323

Walles, S.A.S. & Orsén, I. (1983) Single-strand breaks in DNA of various organs of mice induced by styrene and styrene oxide. *Cancer Lett., 21,* 9–15

Wigaeus, E., Löf, A., Bjurström, R. & Byfält-Nordqvist, M. (1983) Exposure to styrene. Uptake, distribution, metabolism and elimination in man. *Scand. J. Work Environ. Health, 9,* 479–488

World Health Organization (1983) *Styrene* (*Environmental Health Criteria 26*), Geneva

INDUCTION OF SINGLE-STRAND BREAKS IN LIVER DNA OF MICE AFTER INHALATION OF VINYL CHLORIDE

S.A.S. Walles,[1] B. Holmberg,[1] K. Svensson,[2] S. Osterman-Golkar,[2] K. Sigvardsson[1] & K. Lindblom[1]

[1]*National Institute of Occupational Health, Solna; and* [2]*Department of Radiobiology, University of Stockholm, Stockholm, Sweden*

NMRI female mice were exposed to 100, 250 and 500 ppm vinyl chloride (VC). Cell nuclei were prepared from the liver, and single-strand breaks (SSB) were determined by the DNA unwinding technique. Haemoglobin (Hb) was isolated from the blood, and the degree of alkylation was determined as a measure of in-vivo dose by means of a gas chromatography-mass spectrometry (GC-MS) technique. A maximum level of SSB in liver DNA and of adduct levels of Hb was reached at 500 ppm, indicating that saturation of metabolic activation of VC had been achieved. The results demonstrate that VC induces SSB in liver DNA of mice in a dose-dependent manner and that about 80% of the damage is repaired within 20 h.

VC is mutagenic and carcinogenic in various species and has been shown to be an occupational hazard to workers in the production of polyvinyl chloride (IARC, 1979). VC is metabolized to chloroethylene oxide and chloroacetaldehyde, both of which introduce the 2-oxoethyl adduct onto nucleophilic sites in macromolecules. Chloroethylene oxide is the main alkylating intermediate *in vivo* (Osterman-Golkar *et al.*, 1977), and it has been suggested that it is responsible for the carcinogenic potency of VC (Bolt, 1986).

The rates of adduct formation in blood proteins such as Hb and in DNA (in liver or in other organs) are proportional (Ehrenberg & Osterman-Golkar, 1980). Measurements of adduct levels in Hb therefore provide a measure of adduct levels in DNA (Osterman-Golkar *et al.*, 1976, 1977).

When a number of alkylation products undergo repair, transient SSB occur, which can be quantified with high sensitivity. SSB are thus not a primary event but can be used as an indicator of remaining DNA damage in repair studies. It has been shown that VC induces SSB in DNA in different organs of mice (Walles & Holmberg, 1984). The purpose of this investigation was to study the induction and persistence of DNA damage (SSB) in liver DNA of mice after exposure to VC. In parallel, alkylation products of Hb were determined.

NMRI female mice inhaled VC at 100, 250 and 500 ppm for 24 h. Groups of mice were exposed at 250 ppm for 2, 4, 6 and 8 h. The performance of the experiments is described by Walles and Holmberg (1984). In all cases, SSB were determined 1 h after termination of the exposure. In a separate experiment, the repair of DNA damage was studied by determining SSB levels 1 h and 20 h after termination of the exposure at 250 and 500 ppm.

Cell nuclei were prepared from liver, and SSB were determined by the DNA

Fig. 1. Levels of SSB[a], expressed as $-\log F_{DS}$[b], as a function of VC concentration

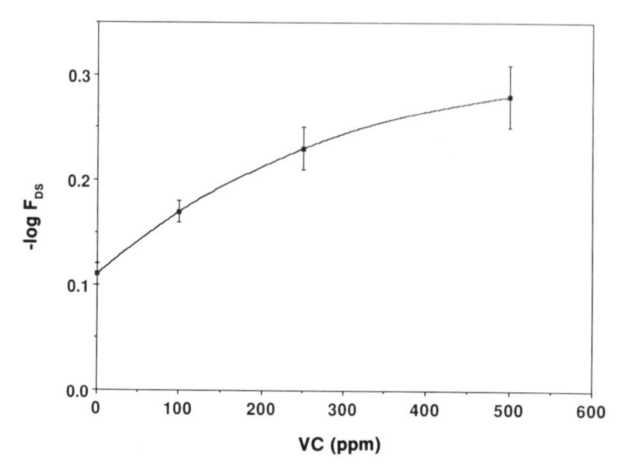

Length of exposure was 6 h/day for 4 days; mice were killed 1 h thereafter. Mean values ± SE are shown.
[a]In the experiments presented in Figure 1, Figure 2 and Table 1, the amount of DNA after hydroxylapatite chromatography was determined by means of the fluorochrome DAPI. In the experiment presented in Table 2, Hoechst 33258 was used as fluorochrome, which gives somewhat lower values of SSB.
[b]F_{DS} = the ratio of the amount of double-stranded (DS) DNA and the total amount of DNA

unwinding technique (Ahnström & Erixon, 1973), as modified by Walles and Erixon (1984). Cell nuclei were incubated in a buffer (0.1 M NaCl/1 mM EDTA/2 mM EGTA/0.02 M Tris-HCl, pH 7.6) (Cohen & Prabhakar, 1983) for 2 h at 37°C in order to optimize the yield of SSB.

Hb was isolated and reduced with sodium borohydride to transfer the 2-oxoethyl adducts to 2-hydroxyethyl adducts, according to the method of Svensson and Osterman-Golkar (1987). The samples were then subjected to total hydrolysis, followed by isolation of alkylated products of histidine, N^{τ}-(2-hydroxyethyl)histidine. Derivatives for GC-MS were prepared and analysed as described by Calleman *et al.* (1978).

The level of SSB increased with VC concentration (Fig. 1). The value at 500 ppm is slightly higher than, or equal to, that at 250 ppm, indicating that saturation of metabolic activation of VC is reached at 500 ppm. The adduct levels in Hb at 250 and 500 ppm (Table 1) are nearly the same.

The SSB levels increased with time for 2–8 h at 250 ppm (Fig. 2); a plateau was reached between 4–8 h of exposure. In another experiment, the levels of SSB at 20 h were compared with those at 1 h after exposure (Table 2). The levels of SSB after 20 h were greatly diminished, indicating that about 80% of the DNA damage had been repaired. Thus, the plateau value obtained in the former experiment may be interpreted as a steady state between induction and repair of DNA damage.

Discussion

The ratio between liver DNA alkylation (7-guanine) and Hb alkylation (N^{τ}-histidine) has been established in experimental animals (CBA mice) after inhalation of radiolabelled VC (Osterman-Golkar *et al.*, 1977). On the basis of this ratio, alkylation of liver DNA has been calculated from the present data, and the

Table 1. Comparison of induced SSB in liver DNA and degree of alkylation of Hb and liver DNA after exposure of mice to VC

VC level[a] (ppm)	$-\Delta \log F_{DS}$[b,c]	Binding to Hb (nmol N^{τ}-OEtHis/ g Hb)[d]	Binding to DNA (nmol 7-OEtGua/ g DNA)[e]	$\frac{-\Delta \log F_{DS}}{\text{DNA binding}} \times 10^4$
100	0.06	5.4	5	120
250	0.12	40	36	33
500	0.17	36	33	52

[a]Length of exposure was 6 h/day for four days; animals were killed 1 h after termination of exposure.
[b]$-\Delta \log F_{DS} = -\log F_{DS(VC)} - (-\log F_{DS(control)})$
[c]See legend to Figure 1, footnote *a*
[d]N^{τ}-OEtHis, N^{τ}-histidine
[e]DNA alkylation was calculated using the ratio N^{τ}-OEtHis (Hb) to 7-OEtGua (liver DNA) from animal (CBA mice) experiments in which radiolabelled VC was used (Osterman-Golkar *et al.*, 1977), assuming that the SSB level was the result of the exposure during the last day, with a possible contribution (~25%) from previous days.

SSB-inducing efficiency has been expressed as the ratio $-\Delta \log F_{DS}$/nmol 7-(2-oxoethyl)guanine per g DNA (see Table 1). The corresponding ratio between SSB induction and 7-guanine alkylation has been determined for other agents: for methyl methanesulphonate, styrene and urethane, the quotients were 19×10^{-4}, 23×10^{-4} and 21×10^{-4}, respectively (S. Walles *et al.*, unpublished data).

With respect to induction of mutations (in *Escherichia coli* Sd-4), compounds that introduce a 2-oxoethyl adduct (chloroethylene oxide and chloroacetaldehyde) are considerably more effective than simple monofunctional alkylating agents such as methyl methanesulphonate (Hussain & Osterman-Golkar, 1976), when compared per alkylation of guanine at O^6 (assumed to be a critical lesion) or, as above, per alkylation of guanine at 7. Also, in carcinogenicity tests of chloroethylene oxide (Zajdela *et al.*, 1980; see also Barbin & Bartsch, 1986), VC (carcinogenicity data from Maltoni *et al.*, 1984), and possibly also urethane (carcinogenicity data from Schmähl *et al.*, 1977), all of which introduce the 2-oxoethyl adduct, seem to be more potent than, e.g., methyl

Fig. 2. Levels of SSB[a] ($-\log F_{DS}$)[b] as a function of length of exposure (h) to 250 ppm

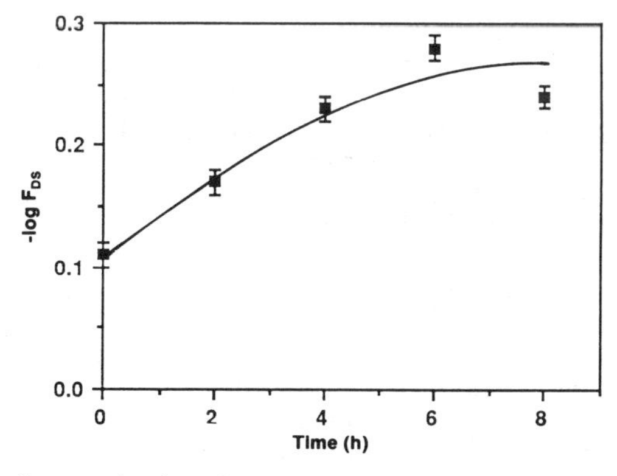

Animals were killed 1 h after termination of exposure. Mean values ± SE are shown.
[a]See Figure 1, footnote *a*
[b]See Figure 1, footnote *b*

Table 2. Levels of SSB 1 and 20 h after exposure to VC (mean values ± SE)

VC level (ppm)	Length of exposure (h)	Time to recovery (h)	$-\log F_{DS}$[a]
250	6	1	0.19 ± 0.01
250	6	20	0.10 ± 0.003
500	6	1	0.19 ± 0.02
500	6	20	0.09 ± 0.002
Control			0.07 ± 0.006

[a]See Figure 1, footnotes *a* and *b*

methanesulphonate (see Barbin & Bartsch, 1986). The high mutagenic efficiency of 2-oxoethylating agents may be related to the secondary reactivity of the 2-oxoethyl adduct and the possibility of forming etheno rings in DNA bases (Singer *et al.*, 1987) or other types of cross-links (Scherer *et al.*, 1986; Svensson & Osterman-Golkar, 1986).

The present study indicates that SSB determination may be a sensitive method for evaluating potential carcinogenic risks for a particular organ. Even after 2 h exposure to 250 ppm VC (500 ppm-h), the SSB level is significantly higher than the normal value. It is also interesting that the SSB level is restored to almost the normal level after 20 h, although SSB are still present. The significance of SSB levels in terms of tumour induction must be elucidated in further studies.

Acknowledgements

We are indebted to Mrs E. Tardelius-Bengtsson for skilful technical assistance. Financial support was received from the Swedish Work Environment Fund (Nos 84-1269 and 86-0408).

References

Ahnström, G. & Erixon, K. (1973) Radiation induced strand breakage in DNA from mammalian cells, strand reparation in alkaline solution. *Int. J. Radiat. Biol.*, *23*, 285–289

Barbin, A. & Bartsch, H. (1986) *Mutagenic and promutagenic properties of DNA adducts formed by vinyl chloride metabolites.* In: Singer, B. & Bartsch, H., eds, *The Role of Cyclic Nucleic Acid Adducts in Carcinogenesis and Mutagenesis* (*IARC Scientific Publications No. 70*), Lyon, International Agency for Research on Cancer, pp. 345–358

Bolt, H.M. (1986) *Metabolic activation of vinyl chloride, formation of nucleic acid adducts and relevance to carcinogenesis.* In: Singer, B. & Bartsch, H., eds, *The Role of Cyclic Nucleic Acid Adducts in Carcinogenesis and Mutagenesis* (*IARC Scientific Publications No. 70*), Lyon, International Agency for Research on Cancer, pp. 261–268

Calleman, C.J., Ehrenberg, L., Jansson, B., Osterman-Golkar, S., Segerbäck, D., Svensson, K. & Wachtmeister, C.A. (1978) Monitoring and risk assessment by means of alkyl groups in hemoglobin in persons occupationally exposed to ethylene oxide. *J. environ. Pathol. Toxicol.*, *2*, 427

Cohen, A.M. & Prabhakar, H. (1983) Carcinogen induced DNA damage in isolated rat liver nuclei. *Cancer Lett.*, *18*, 163–167

Ehrenberg, L. & Osterman-Golkar, S. (1980) Alkylation of macromolecules for detecting mutagenic agents. *Teratog. Carcinog. Mutag.*, *1*, 105–127

Hussain, S. & Osterman-Golkar, S. (1976) Comment on the mutagenic effectiveness of vinyl chloride metabolites. *Chem.-biol. Interactions*, *12*, 265–267

IARC (1979) *IARC Monographs on the Evaluation of the Carcinogenic Risk of Chemicals to Humans*, Vol. 19, *Some Monomers, Plastics and Synthetic Elastomers, and Acrolein*, Lyon, pp. 377–438

Maltoni, C., Lefemine, G., Ciliberti, A. & Carretti, D., eds (1984) *Experimental Research on Vinyl Chloride Carcinogenesis*, Princeton, NJ, Princeton Scientific Publications, pp. 64–65

Osterman-Golkar, S., Ehrenberg, L., Segerbäck, D. & Hällström, I. (1976) Evaluation of genetic risks of alkylating agents. Hemoglobin as a dose monitor. *Mutat. Res.*, *34*, 1–10

Osterman-Golkar, S., Hultmark, D., Segerbäck, D., Calleman, C.J., Göthe, R., Ehrenberg, L. & Wachtmeister, C.A. (1977) Alkylation of DNA and proteins in mice exposed to vinyl chloride. *Biochem. biophys. Res. Commun.*, *76*, 259–266

Scherer, E., Winterwerp, H. & Emmelot, P. (1986) *Modification of DNA and metabolism of ethyl carbamate in vivo: formation of 7-(2-oxoethyl)guanine and its sensitive determination by reductive tritiation using ^{3}H-sodium borohydride*. In: Singer, B. & Bartsch, H., eds, *The Role of Cyclic Nucleic Acid Adducts in Carcinogenesis and Mutagenesis* (*IARC Scientific Publications No. 70*), Lyon, International Agency for Research on Cancer, pp. 109–125

Schmähl, D., Port, R. & Wahrendorf, J. (1977) A dose-response study on urethane carcinogenesis in rats and mice. *Int. J. Cancer*, *19*, 77–80

Singer, B., Spengler, S.J., Chavez, F. & Kusmierek, J.T. (1987) The vinyl chloride-derived nucleoside, N^2,3-ethenoguanosine, is a highly efficient mutagen in transcription. *Carcinogenesis*, *8*, 745–747

Svensson, K. & Osterman-Golkar, S. (1986) *Covalent binding of reactive intermediates to hemoglobin in the mouse as an approach to studying the metabolic pathways of* 1,2-*dichloroethane*. In: Singer, B. & Bartsch, H., eds, *The Role of Cyclic Nucleic Acid Adducts in Carcinogenesis and Mutagenesis* (*IARC Scientific Publications No. 70*), Lyon, International Agency for Research on Cancer, pp. 269–279

Svensson, K. & Osterman-Golkar, S. (1987) *In vivo 2-oxoethyl adducts in hemoglobin and their possible origin*. In: Waters, M.D., Sandhu, S.S. & Claxton, L., eds, *Application of Short-term Bioassays in the Analysis of Complex Environmental Mixtures*, New York, Plenum Press, pp. 49–66

Walles, S.A.S. & Erixon, K. (1984) Single-strand breaks in DNA of various organs of mice induced by methyl methanesulfonate and dimethylsulfoxide determined by the alkaline unwinding technique. *Carcinogenesis*, *5*, 319–323

Walles, S.A.S. & Holmberg, B. (1984) Induction of single-strand breaks in DNA of mice after inhalation of vinyl chloride. *Cancer Lett.*, *25*, 13–18

Zajdela, F., Croisy, A., Barbin, A., Malaveille, C., Tomatis, L. & Bartsch, H. (1980) Carcinogenicity of chloroethylene oxide, an ultimate reactive metabolite of vinyl chloride, and bis(chloromethyl)ether after subcutaneous administration and in initiation-promotion experiments in mice. *Cancer Res.*, *40*, 352–356

DETERMINATION OF SPECIFIC MERCAPTURIC ACIDS IN HUMAN URINE AFTER EXPERIMENTAL EXPOSURE TO TOLUENE OR *o*-XYLENE

Å. Norström,[1,4] B. Andersson,[1] L. Aringer,[2] J.-O. Levin,[1] A. Löf,[3] P. Näslund[3] & M. Wallèn[3]

[1] *National Board of Occupational Safety and Health, Research Department, Chemical Unit in Umeå, Umeå;* [2] *National Board of Occupational Safety and Health, Research Department, Medical Unit; and* [3] *Work Physiology Unit, Solna, Sweden*

Volunteers were exposed to toluene and *o*-xylene for 4 h at 300 mg/m³ and 350 mg/m³, respectively, in an exposure chamber. Urine voided during and after the exposure was analysed for *S*-benzyl-*N*-acetylcysteine (toluene mercapturic acid) and *S*-(*o*-methylbenzyl)-*N*-acetylcysteine (*o*-xylene mercapturic acid). The amount of *S*-(*o*-methylbenzyl)-*N*-acetylcysteine found in urine after exposure corresponded to a metabolic yield of less than about 0.01% of the total uptake. No *S*-benzyl-*N*-acetylcysteine was found. Thus, determination of specific mercapturic acids after exposure to toluene or *o*-xylene can hardly be used for genotoxic monitoring.

Spectrophotometric determination of thioethers (mercapturic acids or *S*-substituted *N*-acetylcysteines) has been used as an indicator of risk for genotoxic exposure in a number of studies, such as that of Malonova and Bardodej (1983). Van Doorn *et al.* (1980) reported data from animal studies which indicated a relatively high yield of *S*-benzyl-*N*-acetylcysteine and *S*-(*o*-methylbenzyl)-*N*-acetylcysteine after exposure to toluene and *o*-xylene, respectively.

We now report on the determination of these two mercapturic acids after experimental human exposure to toluene or *o*-xylene.

Analytical methods

S-Benzyl-*N*-acetylcysteine and *S*-(*o*-methylbenzyl)-*N*-acetylcysteine (Fig. 1) were synthesized according to methods reported by van Bladeren *et al.* (1980) and Van Doorn *et al.* (1980). The analysis was performed by running acidified 10-ml portions of urine samples through Waters C18Sep-Pak. The cartridge was eluted with 0.01 M HCl, benzene and chloroform : acetic acid (99 : 1). The chloroform fraction was evaporated to dryness, redissolved in methanol and analysed by high-performance liquid chromatography (HPLC). The identities of the compounds were confirmed by gas chromatography-mass spectrometry (GC-MS), as reported by Norström *et al.* (1986).

Determination of toluene and *o*-xylene in exhaled air and blood and urinary determination of hippuric acid and *o*-methylhippuric acid were performed as reported by Wallèn (1986) and Norström *et al.* (1988).

[4] To whom correspondence should be sent

Fig. 1. Structures of the two mercapturic acids, *S*-benzyl-*N*-acetylcysteine and *S*-(*o*-methylbenzyl)-*N*-acetylcysteine

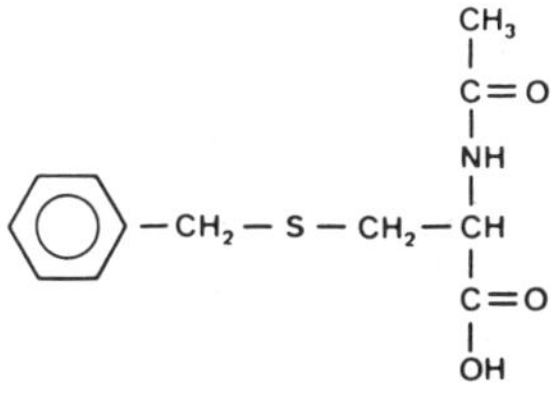

S-Benzyl-*N*-acetylcysteine
(Toluene mercapturic acid)

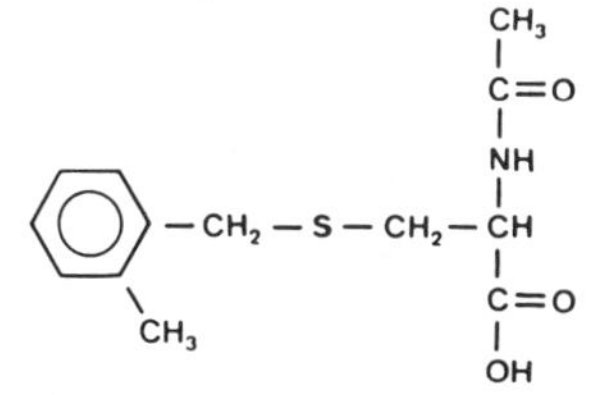

S-(*o*-Methylbenzyl)-*N*-acetylcysteine
(*o*-Xylene mercapturic acid)

Exposure studies

Two experiments with different volunteers were performed in a thermostatted exposure chamber with continuous monitoring of solvent air concentration. (For further details, see Norström *et al.*, 1988.) Four persons, two smokers and two nonsmokers, were chosen from among 26 spray-painters who had taken part in a previous study on toluene. In the study on *o*-xylene, six healthy, nonsmoking men with no previous history of occupational exposure to organic solvents, took part.

Determination of mercapturic acids

The sensitivity of the method used in this investigation was 2 ng/μl and 1 ng/μl for *S*-benzyl-*N*-acetylcysteine and *S*-(*o*-methylbenzyl)-*N*-acetylcysteine, respectively, when using the HPLC method. The corresponding sensitivities for the GC-MS method were 0.5 and 0.3 ng/μl urine, respectively, corresponding to 1 μmol thioether/mmol creatinine (HPLC) and 0.25 μmol thioether/mmol creatinine (GC-MS) for *S*-benzyl-*N*-acetylcysteine. These are well below the detection limit for the spectrophotometric method, with which Aringer *et al.* (1984) found typical background values of about 3 μmol/mmol creatinine for nonsmokers on a standardized diet.

Urine samples from toluene-exposed persons were analysed, but no *S*-benzyl-*N*-acetylcysteine was found either in samples voided immediately after exposure or in samples voided 3 h later. Three hours after exposure to *o*-xylene, trace amounts of *S*-(*o*-methylbenzyl)-*N*-acetylcysteine were found in urine samples. The levels found were very low (0.2 μmol/mmol creatinine), but analysis by GC-MS-multiple-ion detection revealed that the three typical fragments at $m/z = 105$ (100%), 176 (14%), 222 (12%), and the molecular ion at $m/z = 281$ (1%) were present in all samples (Fig. 2). These amounts are just above the detection limit and correspond to a metabolic yield of about 0.01% of the total uptake of *o*-xylene.

Conclusions

In contrast to the results reported from animal studies by Van Doorn *et al.* (1980), humans exposed to the industrial solvents toluene and *o*-xylene excreted no or very little mercapturic acid. Excretion of thioethers or mercapturic acids is considered to be an indicator of exposure to genotoxic compounds and, thereby, an estimate of risk for occupational cancer. At least for these two solvents, it has not been possible to demonstrate the usefulness of mercapturic acid determination for this purpose, but further studies will be carried out on other compounds.

Fig. 2. GC-MS-multiple-ion detection of a urine sample from an *o*-xylene-exposed person

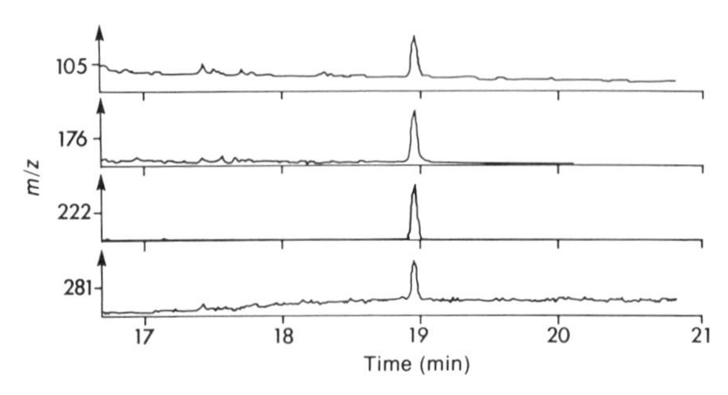

Acknowledgements

The authors are indebted to Professor K. Andersson for critical reading of the manuscript. The skilful technical assistance of Ms C. Bång and Ms E. Lundgren is gratefully acknowledged.

References

Aringer, L., Hogstedt, C., Svensson, E., Lidums, V. & Wrange, R. (1984) *Evaluation of some factors affecting determination of thioethers in urine.* In: Berlin, A., Draper, M., Hemminki, K. & Vainio, H., eds, *Monitoring Human Exposure to Carcinogenic and Mutagenic Agents* (*IARC Scientific Publications No. 59*), Lyon, International Agency for Research on Cancer, p. 443

van Bladeren, P.J., Buys, W., Breimer, D.D. & van der Gen, A. (1980) The synthesis of mercapturic acids and their esters. *Eur. J. Med. Chem. — Chim. therapeut., 15,* 495–497

Malonova, H. & Bardodej, Z. (1983) Urinary excretion of mercapturates as a biological indicator of exposure to electrophilic agents. *J. Hyg. Epidemiol. Microbiol. Immunol., 27,* 319–328

Norström, Å., Andersson, B. & Levin, J.-O. (1986) A procedure for the analysis of *S*-benzyl-*N*-acetylcysteine and *S*-(*o*-methylbenzyl)-*N*-acetylcysteine in human urine. *Xenobiotica, 16,* 525–529

Norström, Å., Andersson, B., Levin, J.-O., Näslund, P.H., Wallèn, M. & Löf, A. (1988) Biological monitoring of *o*-xylene after experimental exposure in man: determination of urinary excretion products. *Int. Arch. occup. environ. Health* (submitted for publication)

Van Doorn, R., Bos, R.P., Brouns, R.M.E., Leijdekkers, C.-M. & Henderson, P.T. (1980) Effect of toluene and xylenes on liver glutathione and their urinary excretion as mercapturic acids in the rat. *Arch. Toxicol., 43,* 293–304

Wallèn, M. (1986) Toxicokinetics of toluene in occupationally exposed volunteers. *Scand. J. Work. Environ. Health, 12,* 588–593

ALKYLATING EXPOSURES

PROSPECTIVE DETECTION AND ASSESSMENT OF GENOTOXIC HAZARDS: A CRITICAL APPRECIATION OF THE CONTRIBUTION OF L. EHRENBERG

A.S. Wright, T.K. Bradshaw & W.P. Watson

Shell Research Ltd, Sittingbourne Research Centre, Sittingbourne, Kent, UK

Advances in our understanding of the mechanisms of chemical carcinogenesis are now being applied to improve the quality of prospective risk assessment. The contribution of Ehrenberg and his colleagues (at the University of Stockholm) probably represents the most comprehensive application of mechanistic knowledge to this field during the past 20 years. The strategic approach developed by the Swedish group was based on the identification of differences between man and experimental risk models in factors that determine the relationships between exposure and biological response and the development of methods to compensate for these differences. Many of the critical stages in chemical carcinogenesis and the cellular determinants of these stages have now been identified. As a first step in seeking to improve risk assessment, Ehrenberg introduced the target dose concept, in which the doses of carcinogens penetrating to the cellular target (DNA) are determined. This approach provides an improved basis for determining exposures to carcinogenic agents and also for compensating for species differences in factors such as metabolism that determine the relationships between exposure dose and the dose at the critical target. The target dose concept is now widely accepted and has led to the development of new biomedical monitoring techniques, based, for example, on the measurement of haemoglobin adducts, which are now being applied to detect and identify genotoxic hazards. The introduction of the target dose concept has led to significant improvements in the quality of prospective risk assessment. Further improvements necessitate procedures to compensate for differences between man and prospective risk models in factors that determine subsequent stages of the carcinogenic process. Ehrenberg has proposed that the rad-equivalence approach may be of value in this respect. Its application has accurately predicted the incidence of leukaemias in occupational cohorts which had exposures to ethylene oxide in common. The possible general applicability of this approach is discussed.

During the last 20 years, the emphasis in cancer research has increasingly been placed on the study of mechanisms, with the aim of developing rational scientific methods for treatment and prevention. Epidemiological evidence indicates that a high proportion of human cancers is caused by environmental factors, including chemicals of industrial and natural origin (Higginson & Muir, 1979). Effective methods to identify and evaluate such agents are essential for the prevention of cancer. So far as is possible, these methods should be prospective in order to minimize human contact and avoid introducing new carcinogens into the environment and the work place.

Until the mid 1970s, the qualitative identification of human carcinogens and estimations of carcinogenic potency were based almost exclusively on the results of long-term carcinogenicity studies in laboratory animals. However, marked quantitative and apparent qualitative species differences in response to chemical carcinogens cautioned against direct extrapolation to man (Wright, 1981). In extreme cases, such species differences in response could reduce even the qualitative prediction of human carcinogens to the level of a lottery.

Fortunately, research during the past 20 years into the mechanisms of chemical carcinogenesis has provided a sound basis for developing improved procedures to detect human carcinogens and evaluate the risks to man. Several important, broad tenets have been established over this period, due principally to the pioneering work of Brookes and Lawley (1964), J.A. and E.C. Miller (1969) and Ames *et al.* (1973). In particular, mutagenic action, i.e., the capacity to induce transmissible alterations in DNA structure, is now viewed as fundamental to the operation of most chemical carcinogens and has led to the classification of such agents as 'genotoxic' (Ehrenberg *et al.*, 1973). DNA is regarded as the key (primary and critical) target of genotoxic carcinogens,[1] although many chemicals classified as genotoxic are in fact precursor agents requiring metabolic conversion into reactive forms (electrophiles) in order to undergo chemical reaction with DNA. Primary products of these reactions are generally promutagenic or lethal. Among these products, DNA base adducts have been singled out for special attention, not only because of their mutagenic propensity, which can vary markedly according to the structure of the adduct, its location and persistence in DNA, but also because of the sensitivity and specificity of the techniques available for their analysis (Randerath *et al.*, 1981; Hemminki & Randerath, 1987; reviewed by Watson, 1987).

The acceptance by the scientific community of a sequential mechanism linking electrophilic reactivity of chemicals or their metabolites with the induction of primary chemical damage to DNA, leading to mutation and cancer, fuelled the development of a diverse series of short-term genotoxicity assays — mainly in-vitro assays — for the prediction of carcinogenic activity. The relative simplicity and low cost of these in-vitro assays has permitted widespread application in screening for environmental and occupational mutagens and carcinogens and in the prescreening of new products, e.g., drugs and agrochemicals.

The most successful of the early in-vitro genotoxicity assays were those designed to detect mutation at specific loci in bacterial cells and mammalian somatic cells or chromosomal damage in cultured somatic cells. These tests are considered to be only qualitative in terms of their value in predicting carcinogenic activity. Cursory analysis might suggest that a purely qualitative determination of carcinogenicity is all that is required. Thus, it is generally accepted that human contact with chemical carcinogens should be minimized (irrespective of potency). Unfortunately, however, these qualitative tests are fallible. Major discrepancies have been reported in the qualitative correlation between the results of in-vitro genotoxicity studies and conventional carcinogenicity studies, and these discrepancies have led to concerns about the value of the in-vitro tests as qualitative indicators of carcinogenic hazards (Tennant *et al.*, 1987). Nevertheless, the evidence that genotoxic action is critical to the initiation of chemical carcinogenesis is very strong. It is, therefore, extremely important to understand the limitations of in-vitro genotoxicity assays and the reasons for the poor correlation with carcinogenicity data. These reasons fall into three distinct categories, which are not

[1] DNA adducts and generalized DNA damage may also be induced by indirect mechanisms, e.g., *via* the generation of reactive oxygen species (Ames, 1983; Kasai & Nishimura, 1986). Indirect genotoxic action may also occur as a consequence of effects on DNA replication or repair functions.

necessarily mutually exclusive:

(1) Differences between the in-vitro model and intact animals, e.g., man, in factors that determine the relationships between exposure dose of a genotoxic chemical and a specific genotoxic effect, e.g., point mutation, can have profound effects on the expression of genotoxicity in the two systems and are likely to give rise to both 'false'-positive and 'false'-negative results *in vitro* (ICPEMC, 1982). These factors include enzyme-mediated bioactivation and detoxification pathways, the fidelity of DNA repair and DNA replication and the rate of DNA repair relative to cell (DNA) replication (Wright, 1983).

(2) The failure of in-vitro genotoxicity assays to respond to nongenotoxic carcinogens or to endogenous or exogenous factors, e.g., promoting agents, that influence the progression of initiated cells into malignant tumours gives rise to 'false'-negative results *in vitro* and is, undoubtedly, a major reason for discrepancies in the qualitative correlation between the results of in-vitro genotoxicity assays and in-vivo carcinogenicity studies.

(3) It is important to recognize that the resolving power of experimental carcinogenicity studies (and epidemiological studies) is very poor and does not permit measurement of carcinogenic effects at a level that may be considered acceptable from the standpoint of human risk, i.e., about 10^{-6}/year (Silbergeld *et al.*, 1987). Indeed, the difference between the limits of detection and the required sensitivity usually exceeds a factor of 10^4. The higher resolving power of in-vitro genotoxicity assays often permits the detection of genotoxic action at doses/exposures below the limits of detection of long-term carcinogenicity studies and may therefore give rise to apparent 'false' positives. Consideration of the higher sensitivity of the in-vitro assays in the light of the inadequate sensitivity of carcinogenicity bioassays and evidence that genotoxic action precedes and is predictive of carcinogenicity suggests that such 'false' positives should not be dismissed lightly. The introduction of new, highly sensitive genotoxicity assays, e.g., postlabelling assays (Randerath *et al.*, 1981), will undoubtedly lead to increases in the occurrence of so-called 'false'-positive results in the future.

Strategies to improve hazard detection and risk assessment

By the mid 1970s, in-vitro genotoxicity assays were becoming widely used as qualitative tests for the prediction of carcinogenic activity. The results of in-vivo carcinogenicity studies in higher animals were employed as reference standards for judging the performance of the genotoxicity assays. In-vivo carcinogenicity studies were also preferred for the quantitative evaluation of carcinogenic risks. However, as discussed above, interpretation of both classes of assay is subject to uncertainties and errors.

Several groups, including Ehrenberg and his colleagues at the University of Stockholm, recognized these deficiencies in the available risk models and sought to develop approaches to solve the problems of interpretation. Some of the groups adopted aspects of Ehrenberg's approach or developed these independently; however, none has analysed the problems in greater depth or developed such a comprehensive approach to deal with them as the Ehrenberg school (Osterman-Golkar *et al.*, 1970; Ehrenberg, 1974; Ehrenberg *et al.*, 1974; Ehrenberg, 1976; Osterman-Golkar *et al.*, 1976; Calleman *et al.*, 1978; Ehrenberg, 1978; Segerbäck *et al.*, 1978; Ehrenberg, 1979; Ehrenberg *et al.*, 1983; Osterman-Golkar, 1983; Osterman-Golkar *et al.*, 1983; Törnqvist *et al.*, 1986a,b).

Ehrenberg focused his attention on genotoxic hazards and determined that the principal requirements were: (i) to develop sensitive and reliable analytical procedures to identify specific genotoxic hazards to man in the environment; and (ii) to develop methods for the quantitative assessment of the genetic and carcinogenic risks posed to man by exposures to genotoxic chemicals. Initially, the Swedish group placed emphasis on the second objective, being guided by the notion that, in order to provide effective protection to man, the methods that were needed to detect and monitor human hazards (objective 1) must ideally be sensitive enough to permit detection at an acceptably low level of risk. Such procedures could then be used to investigate and place in order of priority the hazards, as outlined recently by Garner (1985). Quantification of risk to man is essential in order to achieve these goals.

Quantitative risk assessment

Prospective assessment of carcinogenic risks to man must necessarily be based on quantitative dose-response relationships (risk data) obtained in experimental species/models. As we have seen, this approach necessitates extrapolations (to low doses and from experimental species to man) that are subject to major quantitative and even apparent qualitative errors (Wright, 1981). In order to improve the quality of prospective risk assessment, it is necessary to compensate for differences between the model and man in factors that determine the quantitative relationships between exposure or received dose and the biological effect or response (Wright, 1981, 1983). The effective implementation of such a strategy places a heavy requirement on mechanistic knowledge, and the approach developed by Ehrenberg and his colleagues is based on current perceptions of the nature of chemical carcinogens and the mechanism(s) of chemical carcinogenesis.

A schematic representation of the principal events or stages in chemical carcinogenesis is given in Figure 1. Each of these stages is strongly influenced and often determined by host-dependent factors, which may vary according to cell type, tissue,

Fig. 1. A current view of the principal stages and determinants of chemical carcinogenesis

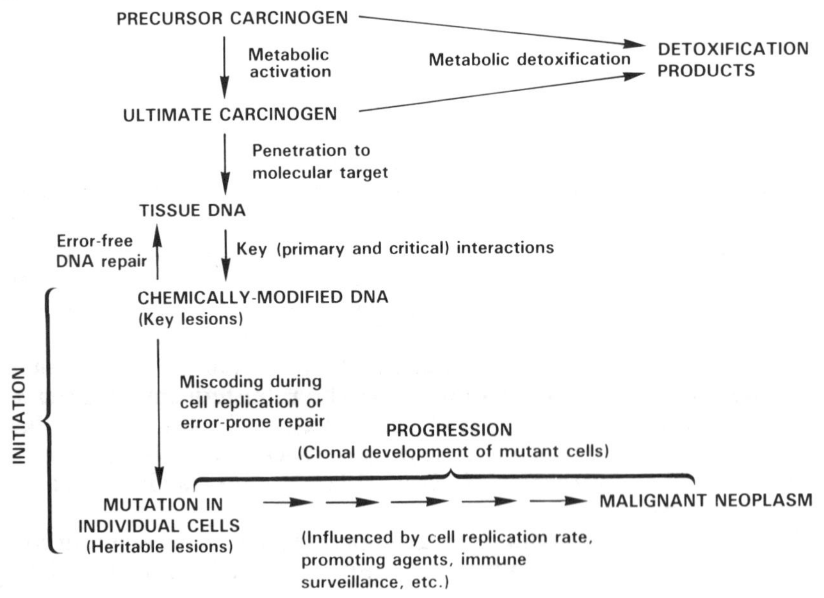

individual, strain and species. In seeking to reduce the errors in extrapolating experimental estimates of risk to man, it is important to correct for species differences among these systemic factors (and general extrinsic factors) — particularly those factors which exert a major influence on the carcinogenic process.

The target dose concept

As an initial step in seeking to improve the quality of prospective risk assessment, Ehrenberg *et al.* (1974) were the first to propose that a new dose concept should be developed to provide a measure of the dose of ultimate genotoxic agent that penetrates to the DNA of tissues or cells. They further suggested that such tissue or target (DNA) dose (Lee, 1976) could be determined by measuring either the primary products of reactions between the genotoxic agent and DNA or an alternative dose monitor, e.g., blood proteins such as haemoglobin (Osterman-Golkar *et al.*, 1976; Calleman *et al.*, 1978). Determination of target dose in the experimental risk model and man would compensate for differences in metabolism and related toxicokinetic and toxicodynamic factors that determine the quantitative relationships between exposure dose and dose of the ultimate toxicant delivered to the target (molecule) (Fig. 1). Thus, the concept may usefully be applied to improve the extrapolation of experimentally determined risk data, i.e., dose-response data, and may also lead to an improved definition of individual risk. The measurement of target dose in man may, therefore, be viewed as an approach towards risk monitoring rather than as an alternative to exposure monitoring.

As previously noted, measurements of the amounts of DNA adducts in tissues or cells provide a basis for determining the doses of ultimate reactive forms of genotoxic chemicals delivered to their key target. The rate constants of the reactions that lead to the formation of the specific adducts and the rates of repair (removal) of these adducts must also be determined in order to transform values of the amounts of adducts into estimates of target dose (Wright, 1981).

Assessment of target (DNA) dose in man

In general, radiolabelled genotoxic agents may be used for determining the amounts of specific DNA adducts in the tissue or cells of laboratory organisms. Such direct methods cannot be applied in man. Ehrenberg and his coworkers (1974, 1983) proposed that indirect dose monitors be developed to determine the doses of genotoxic chemicals delivered to DNA in inaccessible tissues, e.g., human tissues. There are two main possibilities: (i) DNA adducts can be determined in accessible tissues or body fluids, e.g., white blood cells, sperm, placenta or skin; and (ii) the corresponding protein adducts, e.g., with plasma albumin or haemoglobin adducts, can be measured. The problem with such indirect approaches is one of validation. Thus, although the quantitative relationship between dose delivered to DNA in the target tissue and to the dose monitor can be determined in laboratory organisms, these relationships cannot normally be determined in man. Consequently, in considering man, reliance must be placed on the use of experimentally determined coefficients relating doses of genotoxic agent delivered to the target and to the monitor. In order to justify this approach, it is necessary to demonstrate that such coefficients are not subject to significant species differences.

Ehrenberg and his colleagues have investigated a variety of dose monitors, including plasma proteins. The latter may be especially useful with certain genotoxic agents, e.g., aflatoxin B_1. Overall, the Swedish group favoured haemoglobin as a tissue dose monitor, largely because of its availability and relative abundance and also because of the stability of haemoglobin adducts (Osterman-Golkar *et al.*, 1976; Segerbäck *et al.*, 1978) and the longevity of red blood cells, which permit integrated dose measurements (Mollinson, 1983).

Haemoglobin has been thoroughly validated for use as an indirect tissue DNA dose monitor in the case of ethylene oxide (Ehrenberg *et al.*, 1974; Segerbäck *et al.*, 1978; Wright, 1983) and methyl methanesulphonate (Frei & Lawley, 1976). Somewhat more limited validation data have been obtained for vinyl chloride, certain aromatic amines and polycyclic hydrocarbons (reviewed by Ehrenberg *et al.*, 1983 and Neumann, 1984).

Analytical methods

Haemoglobin contains a number of nucleophilic centres which undergo reaction with electrophiles. Usually, the principal targets are the sulphydryl group of cysteine residues, the N_1 and N_3 atoms of histidine residues and the amino groups of the *N*-terminal valine residues of the α- and β-chains. Ehrenberg and coworkers have developed specific and generic gas chromatography-mass spectrometry (GC-MS) techniques to detect, identify and quantify adducts at all of these centres. These methods have been applied in experimental carcinogenicity studies and in human epidemiological studies, e.g., to investigate the hazards posed by exposures to ethylene oxide. The initial approaches were based on the analysis of histidine adducts (Osterman-Golkar *et al.*, 1983) and revealed, for the first time, the presence of 'background' alkylations (methyl, ethyl, 2-hydroxypropyl and 2-hydroxyethyl groups) in the haemoglobin of rodents and humans who had not knowingly been exposed to genotoxic chemicals or their precursors. The initial GC-MS methods were laborious. However, the Swedish group has now developed an improved and simplified procedure, based on the analysis of *N*-terminal valine adducts, which is suitable for routine application (Törnqvist *et al.*, 1986a,b). In this new assay, the adducted *N*-terminal valine residues are cleaved by application of a modified Edman degradation and analysed by GC-MS.

The initial findings of alkyl residues in the haemoglobin of 'unexposed' humans and rodents have been confirmed and extended both in Ehrenberg's and other laboratories (Osterman-Golkar, 1983; Wright, 1983; Van Sittert *et al.*, 1985; Wraith *et al.*, this volume). These alkyl groups are undoubtedly introduced by exposure to alkylating chemicals of endogenous or environmental origin. Methods with a sensitivity sufficient to permit the detection of the corresponding adducts at the level of DNA are being developed. ^{32}P-Postlabelling techniques show great promise in this respect and have already been successfully applied to detect 'background' alkylations in the DNA of rodents and man (Watson, 1987). The ^{32}P-postlabelling techniques are particularly suited to the determination of higher-molecular-weight adducts in DNA. Some modifications are required to obtain similar sensitivity in the assay of low-molecular-weight adducts. Thus, at this stage of their development, the techniques for the determination of protein adducts and DNA adducts are largely complementary (Silbergeld *et al.*, 1987).

Applications to identify human carcinogens

On the premise that causal links have been established between primary chemical damage in DNA and the induction of mutation and cancer, the detection of covalent DNA damage in man provides qualitative evidence of exposure to a carcinogenic hazard. The detection of covalent protein adducts may also signify exposure to chemical mutagens or carcinogens (Ehrenberg, 1980; Hemminki & Randerath, 1987). Biological tests, e.g., the determination of chromosomal aberrations or micronuclei, may also furnish evidence of human exposure to a carcinogenic hazard. However, the detection limits of the latter tests do not permit the determination of levels of risk that may be acceptable from a sociopolitical standpoint, e.g., 10^{-6}/year (Silbergeld *et al.*, 1987), and do not therefore provide a totally acceptable prospective risk indicator in man. Furthermore, these biological procedures *per se* cannot be used to identify specific environmental hazards.

The new analytical procedures for the determination of DNA and protein adducts, however, provide selective approaches to the detection and quantification of specific primary reaction products. These methods are, therefore, potentially of great value in assisting in the identification of agents responsible for the initiation of human cancers and, undoubtedly, will become widely used for this purpose. Furthermore, it is now possible to develop and validate human tissue preparations *in vitro* for application in screening new chemical products for genotoxic activity towards man (Huckle *et al.*, 1986; Watson *et al.*, this volume).

The very high resolving power of these methods suggests that they may have prospective value as qualitative indicators of human carcinogenic hazards *in vivo* and *in vitro*. However, qualitative applications may not be practicable. Thus, it seems likely that the resolving power of these analytical chemistry techniques will permit the determination of adducts with a sensitivity exceeding that required for the determination of acceptable risks (10^{-6}/year). Indeed, the detection limits often exceed one adduct per cell. It seems unlikely, therefore, that such procedures can be applied in a purely qualitative mode — at least in a regulatory sense. Quantification of the risks is essential.

Limitations of the target dose approach

The great sensitivity of the new procedures for determining DNA and protein adducts, i.e., GC-MS, postradiolabelling and immunochemical assays (for a review, see Hemminki & Randerath, 1987), permits applications in assessing tissue DNA doses of genotoxic agents at low environmental or occupational levels in both man and laboratory animals. The target dose approach is designed to compensate for cellular, tissue, individual, strain and species variations in metabolism and related factors that determine the rates of formation of key lesions (Fig. 1). By compensating for such differences between test models and man, the determination of target dose improves the translation of experimental risk data to man and provides a better definition of individual risks. However, in order to improve further the quality of prospective risk assessment, it is necessary to develop procedures to correct for differences between experimental models and man with respect to systemic and general extrinsic factors that determine the progression, firstly, of key lesions into (cancer) mutations and, secondly, of the initiated cells into malignant neoplasms (Fig. 1). Ehrenberg and his colleagues have suggested that the determination of rad-equivalent values[2] for the induction of forward mutation by genotoxic chemicals may be of value in achieving these objectives.

The determination of rad-equivalence

Several authors, including Ehrenberg, have suggested that penetrating ionizing radiation may be used as a reference standard to estimate genetic risks posed by

[2] Ehrenberg (cf. 1980) defines a rad-equivalent as the (target) dose, expressed in concentration × time (e.g., M × h), of an alkylating agent producing the same response as 1 rad of low linear-energy transfer radiation in the low-dose region of the dose-response curve in forward mutation systems.

exposures to genotoxic chemicals. However, the Swedish group was alone in developing effective practical approaches needed to overcome major theoretical and practical problems and test the validity of the approach. The Swedish proposal was regarded as highly controversial and led to much discussion and criticism (reviewed by Golberg, 1986 and ICPEMC, 1983). Most of this criticism is unfounded and is due to superficial analysis that fails to take due account of two essential points. The first point is that the assessment of tissue (or cellular) DNA dose is an essential requirement of the approach (see footnote). The second point is that the proposal is addressed specifically to the determination of low levels of risk, including the determination of risk at a level which may be considered acceptable. Indeed, for reasons discussed below, it is probable that the approach would fail when applied to determining risks (effects) equivalent to those posed by the relatively high doses employed in experimental carcinogenicity studies.

The prospective risk model proposed by Ehrenberg and his colleagues is based on the quantitative determination of the respective capacities of low doses (defined as target DNA dose) of the test chemical and acute γ-radiation to induce the same genetic effect (in both qualitative and quantitative terms) in the same experimental species or system. In-vitro bioassays to detect, e.g., gene mutation, are preferred because of their high resolving power and practicability. In this way, the capacity of a defined target dose of a genotoxic chemical to induce a defined level of genetic damage in a particular organism can be expressed in terms of rads.

Rad-equivalence — mutational risks

The value of experimentally determined rad-equivalance values for mutation hinges on whether or not they can be usefully applied to correct for species differences in the concerted operation of all of the factors, e.g., DNA repair and replication, which determine the quantitative relationship between key lesions, e.g., specific DNA adducts, and mutations. In order to be effective in this respect, the relative mutagenic potencies of unit target doses of radiation and the test chemical must be constant over the dose range(s) of interest. As pointed out by Hathway (1982) in his generally favourable critique, the concept of rad-equivalence stems from the linear (target) dose-mutation response, initiated by both radiation and chemical substances. The resolving power of in-vitro bioassays permits the determination of biological responses in the relatively low target dose range where there is a reasonable expectation for linear target dose-response relationships. However, a second and crucial requirement is that the experimentally-determined rad-equivalence values for mutation must have extrapolative value, i.e., the rad-equivalence value determined for a particular genotoxic chemical in a given experimental species must have a similar numerical value in other species, including man.

Rad-equivalence values for the induction of mutations have been determined for a number of intrinsically reactive, monofunctional alkylating agents in a variety of biological systems, including bacteria, plants and experimental mammals *in vivo* and *in vitro* (Ehrenberg *et al.*, 1974; Ehrenberg, 1976, 1978; Calleman, 1984). The best studied example is ethylene oxide. The rad-equivalence value obtained for a given alkylating agent was approximately the same (within a factor of two) in each of the test systems. On the basis of this evidence, Calleman *et al.* (1978) concluded that there was no reason to presume that a value for rad-equivalence established in these very different test systems would be different in man.

The results obtained in these validation studies are entirely consistent with the theoretical expectation that the key damage inflicted at the target (DNA) is a linear function of the target dose of both radiation and the alkylating agents used in these studies. Since DNA repair and replication are also important quantitative (and qualitative) determinants of mutation, it may be inferred that the influence of these processes in determining the progression of the radiation- and chemical-induced key lesions (to mutation) is proportionately constant in the experimental species employed in these studies and, presumably, in man. There is no obvious reason why such findings and conclusions should not apply to other classes of genotoxic chemicals; however, as implied earlier, the operation of a low capacity, adduct-specific repair function or the induction of a chemical- or radiation-specific repair function at low doses would give rise to errors.

The fact that risk coefficients have not been established for radiation-induced mutation in man precludes the application of rad-equivalence values (mutation) to estimate heritable risks to man. However, risk coefficients have been developed for radiation-induced cancer in man.

Rad-equivalence — carcinogenic risks

In applying the rad-equivalence approach to assessing cancer risks, it may seem important to use experimental rad-equivalence values determined for the induction of cancer rather than of mutation. The aim would be to correct for the influence of all factors that determine the progression of key lesions to (cancer) mutations and of the initiated cells to malignant neoplasms (Fig. 1). However, this approach would be very costly, and it may also be confounded by any promoting action of the test chemical and/or radiation at the relatively high doses needed to produce measurable carcinogenic effects. It would be difficult, if not impossible to correct for variable (nonlinear) influences of this type.

The need to employ a rad-equivalence value determined for cancer rather than for mutation is dependent upon whether or not mutagenic activity is the prime determinant of the carcinogenicity of the test chemical. It is to be expected that any intrinsic promoting activity of a genotoxic agent such as ethylene oxide would be very low at low doses, i.e., initiating activity would be expected to predominate. Contingent, therefore, on the essentially random nature of genotoxic processes induced by chemicals and radiation and on a common basis for initiation, i.e., mutation in a specific set or family of genes or their regulators (oncogenes), it would seem reasonable to suggest that, at low doses, cells that had been initiated by exposure to either radiation or genotoxic chemicals would be subject to the same general promoting and modulating influences that act within an individual. On this basis, rad-equivalence values for mutation may be used in conjunction with an estimate of tissue DNA dose of the alkylating chemical in individuals at risk to estimate human cancer risks by reference to human radiation risk coefficients.

Experimental evidence to support the view that the rad-equivalence approach may compensate for differences between man and laboratory organisms with respect to general promoting pressure and other factors that determine the progression phase of carcinogenesis (Fig. 1) is currently lacking. Experimental approaches are complicated by a number of factors, including the difficulty in ensuring that promoting agents are devoid of genotoxic activity. Nevertheless, experiments are in progress to determine whether the rad-equivalence value for cell transformation by ethylene oxide remains constant when the promotional pressure on C3H 10T1/2 cells is increased by the addition of a phorbol ester (A. Kolman, personal communication).

Application of the rad-equivalence risk model

The risk model developed by the Ehrenberg school has been applied to predict human cancer risks posed by exposures to ethylene oxide (Ehrenberg *et al.*, 1974; Calleman *et al.*, 1978). The quantitative risk estimates were based on experimental rad-equivalence values for genotoxic potency determined in a wide range of experimental models. The results of this prospective analysis were consistent with the incidence of leukaemias observed in occupational cohorts which had exposure to ethylene oxide in common (Hogstedt *et al.*, 1979a,b, 1984; Hogstedt, this volume).

Conclusions

As noted previously, the risk analysis developed by the Ehrenberg school is based on current perceptions of the mechanism(s) of chemical carcinogenesis (Fig. 1). Research is providing increasingly refined and penetrating insights into this process, and the risk model is sufficiently adaptable to accommodate these advances. For example, the development of techniques to determine mutation spectra (Thilly, 1985) and mutation at the same loci in experimental species and man should provide a basis for improving current understanding of the relationships between target dose, primary DNA damage and mutagenic action and on the relationships between mutation and cancer. Such methods may provide a more direct approach to compensate for differences in factors that determine species susceptibility to genotoxic chemicals than is offered by the rad-equivalence approach. At present, however, the determination of rad-equivalence values for genotoxic action appears to provide a valuable approach for evaluating heritable and cancer risks posed by exposures to genotoxic chemicals. Furthermore, the attendant biomedical monitoring techniques, i.e., GC-MS determination of protein adducts, developed by the Swedish group provide both specific and generic methods that may be applied in epidemiological studies in combination with postradiolabelling techniques to detect and identify chemical initiators of human cancer.

References

Ames, B.N. (1983) Dietary carcinogens and anticarcinogens. Oxygen radicals and degenerative diseases. *Science, 221,* 1256–1264

Ames, B.N., Durston, W.E., Yamasaki, E. & Lee, F.E. (1973) Carcinogens are mutagens: a simple test system combining liver homogenates for activation and bacteria for detection. *Proc. natl Acad. Sci. USA, 70,* 2281–2285

Brookes, P. & Lawley, P.D. (1964) Evidence for the binding of polynuclear aromatic hydrocarbons to the nucleic acid of mouse skin: relation between carcinogenic power of hydrocarbons and their binding to DNA. *Nature, 202,* 781–784

Calleman, C.J. (1984) *Haemoglobin as a Dose Monitor and its Application to the Risk Estimation of Ethylene Oxide,* PhD Thesis, University of Stockholm, Sweden, p. 25

Calleman, C.J., Ehrenberg, L., Jansson, B., Osterman-Golkar, S., Segerbäck, D., Svensson, D. & Wachtmeister, C.A. (1978) Monitoring and risk assessment by means of alkyl groups in haemoglobin in persons occupationally exposed to ethylene oxide. *J. environ. Pathol. Toxicol., 2,* 427–442

Ehrenberg, L. (1974) Genotoxicity of environmental chemicals. *Acta biol. yugosl., Ser. F Genetika, 6,* 367–398

Ehrenberg, L. (1976) *Methods of Comparing Effects of Radiation and Chemicals,* Brighton IAEA Consultant Meeting

Ehrenberg, L. (1978) *Purposes and Methods of Comparing Effects of Radiation and Chemicals,* Vienna, International Atomic Energy Agency

Ehrenberg, L. (1979) *Risk assessment of ethylene oxide and other compounds.* In: McElheny,

V.K. & Abrahamson, S., eds, *Assessing Chemical Mutagens: The Risk to Humans* (*Banbury Report 1*), Cold Spring Harbour, NY, CSH Press, pp. 157–190

Ehrenberg, L. (1980) *Methods of comparing risks of radiation and chemicals.* In: *Radiobiological Equivalents of Chemical Pollutants,* Vienna, International Atomic Energy Agency, pp. 11–21

Ehrenberg, L., Brookes, P., Druckrey, H., Lagerlorf, B., Litwin, J. & Williams, G. (1973) *The relation of cancer induction and genetic damage.* In: Ramel, C., ed., *Evaluation of Genetic Risks of Environmental Chemicals* (*Ambio Special Report No. 3*), Stockholm, Royal Swedish Academy of Sciences/Universitetforlarget, pp. 15–16

Ehrenberg, L., Hiesche, K.D., Osterman-Golkar, S. & Wennberg, I. (1974) Evaluation of genetic risks of alkylation agents: tissue doses in the mouse from air contaminated with ethylene oxide. *Mutat. Res., 24,* 83–103

Ehrenberg, L., Moustacchi, E., Osterman-Golkar, S. & Ekman, G. (1983) Dosimetry of genotoxic agents and dose-response relationships of their effects. *Mutat. Res., 123,* 121–182

Frei, J.V. & Lawley, P.D. (1976) Tissue distributions and mode of DNA methylation in mice by methyl methanesulphonate and *N*-methyl-*N'*-nitro-*N*-nitrosoguanidine: lack of thymic lymphoma induction and low extent of methylation of target tissue DNA at O^6 of guanine. *Chem.-biol. Interactions, 13,* 215–222

Garner, R.C. (1985) Assessment of carcinogen exposure in man. *Carcinogenesis, 6,* 1071–1078

Golberg, L. (1986) *Hazard Assessment of Ethylene Oxide,* Boca Raton, FL, CRC Press, pp. 86–89

Hathway, D.E. (1982) *Molecular Aspects of Toxicology,* London, The Royal Society of Chemistry, pp. 38–52

Hemminki, K. & Randerath, K. (1987) *Detection of genetic interaction of chemicals by biochemical methods: determination of DNA and protein adducts.* In: Fowler, B.A., ed., *Mechanisms of Cell Injury: Implications for Human Health,* New York, John Wiley & Sons, pp. 209–227

Higginson, J. & Muir, C.S. (1979) Environmental carcinogenesis: misconceptions and limitations to cancer control. *J. natl Cancer Inst., 63,* 1291–1298

Hogstedt, C., Malmqvist, N. & Wadman, N. (1979a) Leukaemia in workers exposed to ethylene oxide. *J. Am. med. Assoc., 241,* 1132–1133

Hogstedt, C., Rohlen, O., Berndtsson, B.S., Axelson, O. & Ehrenberg, L. (1979b) A cohort study of mortality and cancer incidence in ethylene oxide production workers. *Br. J. ind. Med., 36,* 276–280

Hogstedt, C., Aringer, L. & Gustavsson, A. (1984) Ethylene oxide and cancer — a review and follow up of two epidemiological studies. *Work Health, 49,* 1–37

Huckle, K.R., Smith, R.J., Watson, W.P. & Wright, A.S. (1986) Comparison of hydrocarbon-DNA adducts formed in mouse skin *in vivo* and in organ culture *in vitro* following treatment with benzo[a]pyrene. *Carcinogenesis, 7,* 965–970

ICPEMC (1982) Committee 2 report. Mutagenesis testing as an approach to carcinogenesis. *Mutat. Res., 99,* 73–91

ICPEMC (1983) Committee 4 Report. Estimation of genetic risks and increased incidence of genetic disease due to environmental mutagens. *Mutat. Res., 115,* 255–291

Kasai, H. & Nishimura, S. (1986) Hydroxylation of guanine in nucleosides and DNA at the C-8 position by heated glucose and oxygen radical-forming agents. *Environ. Health Perspectives, 67,* 111–116

Lee, W.R. (1976) Molecular dosimetry of chemical mutagens. Determination of molecular dose to the germ line. *Mutat. Res., 38,* 311–316

Miller, J.A. & Miller, E.C. (1969) In: Bergmann, E.D. & Pullman, B., eds, *The Jerusalem Symposia on Quantum Chemistry and Biochemistry,* Vol. 1, *Physicochemical Mechanisms of Carcinogenesis,* Jerusalem, Israel Academy of Science and Humanities, pp. 237–261

Mollinson, P.L. (1983) *Blood Transfusion in Clinical Medicine,* Oxford, Blackwell Scientific Publications, p. 108

Neumann, H.-G. (1984) Analysis of haemoglobin as a dose monitor for alkylating and arylating agents. *Arch. Toxicol., 56,* 1–6

Osterman-Golkar, S. (1983) *Tissue doses in man: implications in risk assessment.* In: Hayes,

A.W., Schnell, R.C. & Miya, T.S., eds, *Developments in the Science and Practice of Toxicology,* Amsterdam, Elsevier, pp. 289–298

Osterman-Golkar, S., Ehrenberg, L. & Wachtmeister, C.A. (1970) Reaction kinetics and biological action in barley of mono-functional methanesulphonic esters. *Radiat. Bot., 10,* 303–327

Osterman-Golkar, S., Ehrenberg, L., Segerbäck, D. & Hallstrom, I. (1976) Evaluation of genetic risks of alkylating agents. II. Haemoglobin as a dose monitor. *Mutat. Res., 34,* 1–10

Osterman–Golkar, S., Farmer, P.B., Segerbäck, D., Bailey, E., Calleman, C.J., Svensson, K. & Ehrenberg, L. (1983) Dosimetry of ethylene oxide in the rat by quantitation of alkylated histidine in haemoglobin. *Teratog. Carcinog. Mutag., 3,* 395–405

Randerath, K., Reddy, M.V. & Gupta, R.C. (1981) ^{32}P-Labeling test for DNA damage. *Proc. natl Acad. Sci. USA, 78,* 6126–6129

Segerbäck, D., Calleman, C.J., Ehrenberg, L., Lofröth, G. & Osterman-Golkar, S. (1978) Evaluation of genetic risks of alkylating agents. IV. Quantitative determination of alkylated amino acids in haemoglobin as a measure of the dose after treatment of mice with methyl methanesulfonate. *Mutat. Res., 49,* 71

Silbergeld, E.K., Ehrenberg, L.G., Hemminki, K., Hutton, M., Laib, R.J., Lauwerys, R.R., Neumann, H.-G., Nordberg, G.F., Piotrowski, J., Thilly, W.G. & Wright, A.S. (1987) *Exposures: uptake, tissue and target dose.* In: Fowler, B.A., ed., *Mechanisms of Cell Injury: Implications for Human Health,* New York, John Wiley & Sons, pp. 405–429

Tennant, R.W., Margolin, B.H., Shelby, M.D., Zeiger, E., Hazeman, J.K., Spalding, J., Caspary, W., Resnick, M., Stasiewitz, S., Anderson, B. & Minor, R. (1987) Predictions of chemical carcinogenicity in rodents from *in vitro* genetic toxicity assays. *Science, 236,* 933–941

Thilly, W.G. (1985) *The potential use of gradient denaturing gel electrophoresis to obtain mutation spectra in human cells.* In: Huberman, E., ed., *Carcinogenesis,* Vol. 10, *The Role of Chemicals and Radiation in the Etiology of Cancer,* New York, Raven Press, pp. 511–528

Törnqvist, M., Mowrer, J., Jensen, S. & Ehrenberg, L. (1986a) Monitoring of environmental cancer initiators through hemoglobin adducts by a modified Edman degradation method. *Anal. Biochem., 154,* 255–266

Törnqvist, M., Osterman-Golkar, S., Kautianinen, A., Jensen, S., Farmer, P.B. & Ehrenberg, L. (1986b) Tissue doses of ethylene oxide in cigarette smokers determined from adduct levels in haemoglobin. *Carcinogenesis, 7,* 1519–1521

Van Sittert, N.J., De Jong, G., Clare, M.G., Davies, R., Dean, B.J., Wren, L.J. & Wright, A.S. (1985) Cytogenetic, immunological and haematological effects in workers in an ethylene oxide manufacturing plant. *Br. J. ind. Med., 42,* 19–26

Watson, W.P. (1987) Post-radiolabelling for detecting DNA damage. *Mutagenesis, 2,* 339–331

Wright, A.S. (1981) *New strategies in biochemical studies for pesticide toxicity.* In: Bandall, S.K., Marco, G.J., Goldberg, L. & Leng, M.L., eds, *The Pesticide Chemist and Modern Toxicology* (*ACS Symposium Series 160*), Washington DC, American Chemical Society, pp. 285–304

Wright, A.S. (1983) *Molecular dosimetry techniques in human risk assessment: an industrial perspective.* In: Hayes, A.W., Schnell, R.C. & Miya, T.S., eds, *Developments in the Science and Practice of Toxicology,* Amsterdam, Elsevier, pp. 311–318

DOSIMETRY OF ETHYLENE OXIDE

S. Osterman-Golkar

Department of Radiobiology, University of Stockholm, Stockholm, Sweden

Studies by the group of L. Ehrenberg on dosimetry and dose-response relationships of genotoxic compounds have been centred around ethylene oxide. This paper gives a review of these studies — from the early work on mutation induction in barley to the development of procedures for dosimetry and risk estimation for humans.

The key position of adequate concepts and measurements of dose in studies of genotoxic effects of chemicals was recognized early on by L. Ehrenberg at the Department of Radiobiology, Stockholm University (Osterman-Golkar *et al.*, 1970; Ehrenberg *et al.*, 1974) and has since been a governing idea in research at this department. Our studies have concerned low-molecular-weight alkylating agents, and particularly ethylene oxide (EO). EO is of interest as a representative of an important class of genotoxic compounds (epoxides), and in its own right, since it is formed *in vivo* through metabolic activation of ethene, a general contaminant in the environment. Further, EO is used in industry and occurs as the sole genotoxic agent in certain work environments. Studies by the Ehrenberg group, aimed at estimating the risk of late effects associated with exposure to this compound, are presented in Table 1; a brief description of the methods used in these studies to measure (or estimate) chemical doses is given. The expression of chemical dose (in target tissues) in terms of radiation dose equivalents (for the purpose of risk estimation) is discussed in other papers in this volume (A. Kolman *et al.*; A. Wright).

Definition of dose

Dose has by definition the dimension intensity of exposure × time of exposure; e.g., for radiation, $\text{rad h}^{-1} \times \text{h} = \text{rad}$. In the case of chemicals, concentration (C) is the intensity parameter, and dose (D) has thus the dimension concentration × time:

$$D = \int_t C \, dt. \tag{1}$$

The cellular target for a genotoxic agent is DNA (see Brookes, 1966). The dose of concern is accordingly the dose in the environment of DNA, usually denoted as the target dose (see Ehrenberg *et al.*, 1983).

The level of critical lesions in DNA is proportional to the target dose (see below). Theoretically, a dose based on levels of critical lesions and their persistence can be defined; however, the target dose is an endpoint that may be estimated or measured by chemical methods. It is also the dose concept closest to a radiological dose.

Table 1. Review of studies essential for the assessment of risks for genotoxic damage in man of EO (and ethene); nonexhaustive and with emphasis on work by L. Ehrenberg and coworkers on dosimetry and dose-response relationships

1948	Original finding of genotoxic (mutagenic) activity of EO in recessive lethal mutation assay in *Drosophila melanogaster*	Rapoport (1948)
1956	Dose-response curves for chlorophyll mutation, toxicity and sterility established in greenhouse and field tests with barley	Ehrenberg *et al.* (1956, 1959)
1959	Early work of L. Ehrenberg and Å. Gustafsson on use of radiation and chemical mutagens in plant breeding focused attention on the risks to human health. An overview of chemicals in the environment that could carry a genetic risk was submitted to the National Board of Health in Sweden. This overview had some impact on the awareness of risks associated with exposure to chemicals.	Ehrenberg & Gustafsson (1959)
1959–1961	The managers of a factory in Sweden became concerned about the potential 'radiomimetic' effects of EO, and a haematological investigation was initiated. A significant lymphocytosis (apart from one case of leukaemia) was observed among EO exposed personnel. Similar effects had been observed during the 1920s and 1930s in radiological workers frequently exposed to radiation. Increased frequencies of chromosomal aberrations were found in a group of subjects exposed in an accident in 1961.	Ehrenberg & Hällström (1967)
1960	Concepts of mutagenic effectiveness (mutagenic response/unit dose) and mutagenic efficiency (maximum mutation yield attainable; limited, e.g., by cytotoxicity) introduced	Ehrenberg (1960)
1965–1970	Basic work on reaction kinetics of alkylating agents (alkyl methanesulphonates) and their biological action in barley; agents characterized with respect to their absolute reactivity and their *s*-value (selectivity in reactions with nucleophilic atoms of different strength; generally O < N < S). The usefulness of the *s*-value as a descriptor of an alkylating agent was put forward; low *s* generally being linked with high mutagenic efficiency. The mutagenic effectiveness at low doses of the compounds was found to be proportional to their rate of reaction with certain sites of DNA with a low reactivity.	Osterman-Golkar *et al.* (1970). See also Turtóczky & Ehrenberg (1969) Swain & Scott (1953)
1968	EO proved to be a powerful chromosome breaking agent in barley; compared with alkylating agents of several other classes, epoxides are effective inducers of chromosomal aberrations. This may be a general feature of 2-hydroxyalkylating agents, as indicated by chemical and biological comparisons with corresponding 2-methoxyalkylating agents.	Moutschen-Dahmen *et al.* (1968) Walles & Ehrenberg (1968); Lindgren *et al.* (1976)
	Experiments are currently being pursued to study the mechanism on a molecular basis.	Näslund, current work Ahnström, current work
1974	Amount of covalent binding to proteins was used to determine tissue doses (defined according to Eq. 1) in mice exposed to radiolabelled EO. The study showed that EO is rapidly distributed in the body, giving approximately the same dose in different organs. The in-vivo half-life was estimated as 9 min (i.e., the rate constant of elimination, λ, is $4.6\ h^{-1}$).	Ehrenberg *et al.* (1974)

Table 1—(*Continued*)

	Binding to DNA was demonstrated in liver and spleen	
	Allowing for the difference in alveolar ventilation between mouse and man and assuming that a tissue dose of 1 mM × h EO in man is equivalent to 80 rad low linear energy transfer (LET) radiation (provisional estimate from experiments in barley), the genetic risk associated with work at 5 ppm EO for 40 h wk^{-1} was estimated as 4 rad-equivalents wk^{-1} (i.e., 0.13 ppm during working hours would correspond to 100 mrad wk^{-1})	
	In radiological protection, the recommended limit of occupational risk corresponds to whole-body exposure to 5 rad y^{-1} (100 mrad wk^{-1}) of low LET radiation.	International Commission on Radiological Protection (1977)
	The occupational standard for EO in Sweden is 20 ppm.	Swedish Board of Occupational Safety & Health (1974)
1975	Haemoglobin was suggested as a monitor molecule for dosimetry of electrophilic reagents in animals and man.	Osterman-Golkar *et al.* (1976); see also Osterman-Golkar (1975)
	Experiments with a variety of genotoxic compounds have demonstrated that the extent of haemoglobin binding is quantitatively related to binding to DNA in different tissues. Available data for EO show a dose-independent, constant ratio between DNA and haemoglobin alkylation.	See Osterman-Golkar *et al.* (1983) and Segerbäck (1983)
1978	Monitoring and risk assessment of occupational EO exposure. Adducts to histidine in haemoglobin were determined by means of amino acid analysis and gas chromatography-mass spectrometry. The data were considered to be in agreement with the fast elimination of EO from tissues ($\lambda = 4.6\ h^{-1}$) found in mice. Exposure at 1 ppm for 1 h (1 ppm × h) would, based on tissue doses and available values for the rad-equivalence of EO, correspond to 10 mrad γ-radiation. An adjustment of the threshold limit value of EO to 0.25 ppm would thus be implied.	Calleman *et al.* (1978); see also review on epoxides by Ehrenberg & Hussain (1981)
1978, 1981	The occupational standard in Sweden was adjusted to 10 ppm and subsequently to 5 ppm. On new premises, air concentrations below 1 ppm should be aimed at.	Swedish Board of Occupational Safety & Health (1978, 1981)
1977–1987	The formation *in vivo* of EO from ethene was demonstrated in mice after exposure of the animals to ^{14}C-labelled ethene and determination, by radioactivity measurements, of hydroxyethylated amino acid residues in haemoglobin.	Ehrenberg *et al.* (1977)
	In comparative experiments with EO and ethene, including measurements of binding to haemoglobin and to DNA in various organs of mice, activation of ethene to EO was confirmed. At low air levels of ethene, about 8% of the amount contained in the alveolar air was absorbed and metabolized to EO.	Segerbäck (1983)
1984	A method to quantify alkylated *N*-terminal valine residues in haemoglobin by a modified Edman degradation technique using pentafluorophenyl isothiocyanate was developed. In this procedure, alkylated valines are cleaved off in the coupling medium, allowing the modified amino acids to be separated from the rest of the protein.	Törnqvist *et al.* (1986a); Jensen *et al.* (1984)

Table 1—(*Continued*)

	The present detection limit, about 10 pmol hydroxyethylvaline/g haemoglobin is sufficiently high to allow measurements of exposure to EO, even at levels that may be considered acceptable.	
	Haemoglobin from a group of workers in an EO plant was analysed for both histidine and valine adduct levels. A good correlation was observed between the two determinations.	Farmer *et al.* (1986)
	A raised level of hydroxyethylation of *N*-terminal valine of haemoglobin from smokers was demonstrated. The effect was tentatively attributed to ethene in the smoke and was quantitatively compatible with expectation based on the amount of ethene inhaled and on the kinetics of metabolic activation of ethene to ethene oxide in experimental animals. The magnitude of the tissue dose of EO suggests that this agent is a major contributor to smoking-associated cancer risk.	Törnqvist *et al.* (1986b)
1987	Studies on workers occupationally exposed to ethene suggest that the efficiency of the conversion of ethene to EO in man is lower than in experimental animals. This favours alternative explanations (see above) of the raised *N*-(2-hydroxyethyl)valine levels in smokers, one possibility being increased production and/or oxidation of ethene through oxidative processes caused by smoking.	Current work; see also Törnqvist (this volume)
	The genotoxic effectiveness of EO compared with that of γ-radiation was measured by induced transformation of C3H 10T1/2 mouse embryo fibroblasts. The resultant value of the quality factor, $Q \approx 80$ rad/mM × h, is compatible with earlier data for mutation and other endpoints in various test systems. The result justifies this value of the quality factor, to be used in risk estimation of exposure to EO or its precursor, ethene, on the basis of human doses monitored by haemoglobin adducts and cancer risk coefficients established for γ-radiation.	Hussain (1981); D. Segerbäck *et al.* (unpublished); Kolman *et al.* (this volume)
1978–1987	Estimates of cancer and leukaemia risk, based on the rad-equivalence approach, are compatible with epidemiological data.	Calleman *et al.* (1978); Hogstedt *et al.* (1979a,b); Hogstedt (this volume)

Dose determinations in experimental systems

In-vitro systems

In bacteria and suspended cells, near equilibrium is rapidly attained with (uncharged) low-molecular-weight compounds. In such cases, the dose may be approximated by the time integral of the concentration in the medium during the time of treatment. With certain restrictions, evaluated in work with radiolabelled compounds (Walles, 1968), the same reasoning may be applied to plant seeds.

The dose of reactive compound, R_iX, in a medium, assuming first-order kinetics, may be computed according to the equation

$$D = \frac{[R_iX]^0}{k'}(1 - e^{-k't}), \tag{2}$$

which is a solution of Eq. (1), where $[R_iX]^0$ is the administered initial concentration in a single application (or the total applied concentration in repeated treatment), and k' is the first-order rate constant for elimination of the compound. When the time of treatment is short compared to $1/k'$ ($1/k'$ is the mean life span of the molecules in the reaction mixture), Eq. (2) is simplified to

$$D = [R_iX]^0 \times t, \tag{3}$$

and, in the case of a total decay of the compound, to

$$D = [R_iX]^0/k'. \tag{4}$$

For several of the standard compounds used in mutation experiments (including EO), sufficient information on the reaction kinetics is available to allow the calculation (or estimation) of k'. Dosimetry of electrophilic compounds in in-vitro assays may also be based on measurements of product yields (see below).

In-vivo systems

The dose of R_iX in a tissue is determined by the rates of uptake from the environment (formation in tissues), metabolic and chemical reactions, transport and excretion (see Ehrenberg *et al.*, 1983). Variations in concentration of R_iX can only exceptionally be followed.

The rates of formation of reaction products R_iY_j with tissue nucleophiles Y_j are determined by the respective rate constants k_{ij} for the reaction $R_iX + Y_j \rightarrow R_iY_j + X$ and the concentration of R_iX ($[R_iX]$),

$$d[R_iY_j]/dt = k_{ij}[R_iX] \times [Y_j]; \qquad [R_iY_j] \ll [Y_j]. \tag{5}$$

Integration gives

$$[R_iY_j]/[Y_j] = k_{ij} \int [R_iX]\, dt = k_{ij} \times D; \tag{6}$$

i.e., the degree of modification (alkylation, etc.) of Y_j is proportional to the dose. A measurement of stable reaction products in suitable tissue macromolecules (see Table 1; Ehrenberg *et al.*, 1974) thus gives a direct measure of dose. In laboratory experiments in which the high resolving power of radioactive labelling can be used to advantage, it is most straightforward to determine a chemical change in DNA.

Dose determination in man

Dosimetry of electrophilic agents in man by determination of reaction products in (adducts to) macromolecules is based on the general principles discussed above, although restricted to specimens that can be obtained without inconvenience to investigated persons, such as proteins or DNA in blood. Haemoglobin was introduced as a monitor molecule for several reasons, particularly its availability in large quantities and its long life span, which permit the determination of doses accumulated over months (Osterman-Golkar *et al.*, 1976). The methods currently available for adduct measurement are reviewed elsewhere with respect to their applicability and sensitivity (Farmer *et al.*, 1984; Neumann, 1984; Ehrenberg *et al.*, 1986; Farmer *et al.*, 1987).

Doses in target tissues must be estimated on the basis of dose in blood and dose distribution in experimental animals (Ehrenberg & Osterman-Golkar, 1980). An important issue in this context is the establishment of steady-state adduct levels during long-term exposure, and the relation of these adduct levels to dose.

Accumulated adduct levels as a measure of dose

With chronic or intermittent exposure, a steady-state adduct level is attained when the rate of disappearance of adducts is equal to the rate of their formation (see Eq. 6):

$$k_{-ij}[R_iY_j]_{ss} = k_{ij}[R_iX][Y_j]. \quad (7)$$

The steady-state adduct level may then be written

$$\frac{[R_iY_j]_{ss}}{[Y_j]} = \frac{k_{ij}[R_iX]}{k_{-ij}} = \frac{a}{k_{-ij}}, \quad (8)$$

where $[R_iX]$ is the average level of the ultimate carcinogen and a is the increment of the adduct level per unit time. The relation between a and dose per unit time is given by Eq. 6 (see Ehrenberg *et al.*, 1986).

As a function of the constant life span of erythrocytes (t_{er}; ~40 days in mice, ~126 days in man) and of stable haemoglobin adducts, the building up of a steady-state adduct level follows a particular kind of kinetics (see Osterman-Golkar *et al.*, 1976; Segerbäck *et al.*, 1978). The steady-state level attained at $t = t_{er}$ is

$$\frac{[R_iY_j]_{ss}}{[Y_j]} = \frac{a \times t_{er}}{2}. \quad (9)$$

Ethylene oxide

EO is an alkylating agent with a relatively low reactivity in neutral solution (see Ehrenberg & Hussain, 1981). The first-order rate constant, k'_{H_2O}, for hydrolysis at 37°C is $9.1 \times 10^{-3}\,h^{-1}$, i.e., the half-life in pure water is 76 h ($t_{1/2} = 0.693/k'_{H_2O}$). The rates of reaction with buffer components and other nucleophilic groups, Y_j, may be estimated roughly using the Swain–Scott linear free-energy relationship (Swain & Scott, 1953):

$$\log(k_Y/k_{H_2O}) = s \times n, \quad (10)$$

where k_Y and k_{H_2O} ($= k'_{H_2O}/[H_2O]$) are second-order rate constants for reaction with Y and water, respectively, s is 0.96 for EO (see Osterman-Golkar, 1975), and n is a measure of the reactivity of the nucleophile. The rate constant for disappearance of EO from, e.g., a 0.1 M phosphate buffer pH 7 may accordingly be estimated as (using $n_{H_2PO_4^-} = 2.2$ and $n_{HPO_4^{2-}} = 3.52$ in Eq. 10)

$$\begin{aligned} k' &= k'_{H_2O} + k_{H_2PO_4^-}[H_2PO_4^-] + k_{HPO_4^{2-}}[HPO_4^{2-}] \\ &= 9.1 \times 10^{-3} + 0.05 \times 0.02 + 0.05 \times 0.4 = 0.03\,h^{-1}. \end{aligned} \quad (11)$$

In in-vitro assays with treatments of short duration ($\ll 1/k'$), the concentration of EO may be considered constant (losses due to evaporation have to be avoided), and Eq. 3 applied for a calculation of dose.

EO is soluble in both water and organic solvents. It is readily absorbed from alveolar air (see Calleman *et al.*, 1978) and rapidly distributed in the body, giving approximately the same dose in different organs (as demonstrated in experimental animals through determination of binding to DNA; see Segerbäck, 1983, 1985). The animal (human) body may be treated as one compartment and the rate of elimination from the tissues estimated from

$$D = [EO]^0/\lambda, \quad (4a)$$

where λ is the first-order rate constant for the elimination (sum of first-order rate constants for all processes leading to the disappearance of EO from the tissues) and

$[EO]^0$ is the accumulated level of EO absorbed or generated from a precursor (e.g., ethene), respectively.

Values of λ estimated from absorbed/injected amounts of EO and adduct levels in DNA or haemoglobin (Eqs. 7 and 4a) in experimental animals [$\lambda = 4.6$ and $2.5\,h^{-1}$ in male and female CBA mice, respectively, ~4.2 in Fischer rats and 1.3 in dogs (direct determination; Martis *et al.*, 1982); Osterman-Golkar *et al.*, 1976, 1983] indicate moderate differences between sexes and species. The λ-value for man is difficult to assess due to uncertainties in the estimates of uptake of EO ($[EO]^0$). λ-Values in the range $1\text{–}50\,h^{-1}$ have been recorded in different studies (S. Osterman-Golkar & E. Bergmark, unpublished). Since, within each study, there seems to be a reasonable agreement between estimated exposure level and observed haemoglobin adduct levels, probably only part of this variation can be ascribed to a true variation between individuals with respect to clearance rates of EO.

Using a probable value of λ ($4.6\,h^{-1}$), a ventilation rate of $0.2\,l\,kg^{-1}\,min^{-1}$ and the rate constant for reaction of EO with *N*-terminal valine in Hb [$k_{ij} = 0.45 \times 10^{-4}\,l\,(g\,Hb)^{-1}\,h^{-1}$; Segerbäck, 1985], the adduct level in *N* terminal valine, accumulated during long-term exposure to, e.g., 1 ppm EO during working hours, may be estimated as $1.6\,nmol\,(g\,Hb)^{-1}$ ($[R_iY_j]_{ss} \times /[Y_j] = (k_{ij} \times [R_iX] \times t_{er})/(\lambda \times 2)$; see Eqs. 9, 6 and 4a). This value exceeds the present detection limit by two orders of magnitude (see Table 1; Törnqvist *et al.*, 1986a). The practical possibilities for estimating occupational exposure are, however, limited by the background levels found in unexposed subjects [$0.01\text{–}0.1\,nmol\,(g\,Hb)^{-1}$ in nonsmokers and 0.2 $0.7\,nmol\,(g\,Hb)^{-1}$ in heavy smokers; see Table 1; Törnqvist *et al.*, 1986b], probably originating mainly from ethene.

The life span of DNA adducts in white blood cells is not known. A rough estimate (Eq. 7) of the concurrent adduct level in DNA may be based on average repair rates in animal tissues ($k_{-ij} \approx 0.06$ and $0.03\,h^{-1}$, respectively, for elimination of 7-hydroxyethylguanine from DNA in liver and kidney and 0.03 for O^6-hydroxyethylguanine in liver in the mouse; Segerbäck, 1983 and current work) and the rate constant for reaction between EO and guanine-*N*-7 in DNA [$k_{ij} = 1 \times 10^{-4}\,(g\,DNA)^{-1}\,h^{-1}$; Ehrenberg *et al.*, 1974]. The resulting adduct level, $10^{-8}\,mol\,(mol\,DNA\text{-}P)^{-1}$, is outside the scope of presently available methods for analysis of adducts of low molecular weight.

Comments

Experience on human health effects associated with exposure to EO is accumulating. A method to monitor tissue doses of this compound correctly offers the possibility of linking the risk of late effects of a chemical agent to dose. Such a dose-risk relationship would be useful not only for risk assessment of exposure to EO and its precursor ethene but might also serve as a standard for risk assessment of other alkylating agents/metabolites. Better data on exposure dose are needed to establish a more reliable relationship between exposure dose and target dose.

Acknowledgement

Financial support from the Swedish Work Environment Fund is gratefully acknowledged.

References

Brookes, P. (1966) Quantitative aspects of the reaction of some carcinogens with nucleic acids and the possible significance of such reactions in the process of carcinogenesis. *Cancer Res.*, *26*, 1994–2003

Calleman, C.J., Ehrenberg, L., Jansson, B., Osterman-Golkar, S., Segerbäck, D., Svensson, K. & Wachtmeister, C.A. (1978) Monitoring and risk assessment by means of alkyl groups in hemoglobin in persons occupationally exposed to ethylene oxide. *J. environ. Pathol. Toxicol., 2,* 427–442

Ehrenberg, L. (1960) Induced mutation in plants: mechanisms and principles. *Genet. Agr., 12,* 364–389

Ehrenberg, L. & Gustafsson, Å. (1959) *Chemical Mutagens: Their Uses and Hazards in Medicine and Technology. A Report of February* 1959 *to The National Board of Health* (English translation published by Bloms Boktryckeri, Lund, 1970)

Ehrenberg, L. & Hussain, S. (1981) Genetic toxicity of some important epoxides. *Mutat. Res., 86,* 1–113

Ehrenberg, L. & Hällström, T. (1967) *Haematologic studies on persons occupationally exposed to ethylene oxide.* In: *Radiosterilization of Medical Products* (*Report SM 92/96*), Vienna, International Atomic Energy Agency, pp. 327–334

Ehrenberg, L. & Osterman-Golkar, S. (1980) Alkylation of macromolecules for detecting mutagenic agents. *Teratog. Carcinog. Mutagenesis, 1,* 105–127

Ehrenberg, L., Gustafsson, Å. & Lundqvist, U. (1956) Chemically induced mutation and sterility in barley. *Acta chem. scand., 10,* 492–494

Ehrenberg, L., Gustafsson, Å. & Lundqvist, U. (1959) The mutagenic effects of ionizing radiations and reactive ethylene derivatives in barley. *Hereditas, 45,* 351–368

Ehrenberg, L., Hiesche, K.D., Osterman-Golkar, S. & Wennberg, I. (1974) Evaluation of genetic risks of alkylating agents: tissue doses in the mouse from air contaminated with ethylene oxide. *Mutat. Res., 24,* 83–103

Ehrenberg, L., Osterman-Golkar, S., Segerbäck, D., Svensson, K. & Calleman, C.J. (1977) Evaluation of genetic risks of alkylating agents. III. Alkylation of haemoglobin after metabolic conversion of ethene to ethene oxide *in vivo. Mutat. Res., 45,* 175–184

Ehrenberg, L., Moustacchi, E. & Osterman-Golkar, S. (1983) Dosimetry of genotoxic agents and dose-response relationships of their effects. *Mutat. Res., 123,* 121–182

Ehrenberg, L., Osterman-Golkar, S., Segerbäck, D. & Törnqvist, M. (1986) *Power of methods for monitoring exposure to genotoxic chemicals by covalently bound adducts to macromolecules.* In: Notani, N.K. & Chauhan, P.S., eds, *Environmental Mutagenesis and Carcinogenesis,* Bombay, Bhabha Atomic Research Centre, pp. 155–166

Farmer, P.B., Bailey, E. & Campbell, J.B. (1984) *Use of alkylated proteins in the monitoring of exposure to alkylating agents.* In: Berlin, A., Draper, M., Hemminki, K. & Vainio, H., eds., *Monitoring Human Exposure to Carcinogenic and Mutagenic Agents* (*IARC Scientific Publications No. 59*), Lyon, International Agency for Research on Cancer, pp. 189–198

Farmer, P., Bailey, E., Gorf, S.M., Törnqvist, M., Osterman-Golkar, S., Kautiainen, A. & Lewis-Enright, D.P. (1986) Monitoring human exposure to ethylene oxide by the determination of haemoglobin adducts using gas chromatography-mass spectrometry. *Carcinogenesis, 7,* 637–640

Farmer, P.B., Neumann, H.-G. & Henschler, D. (1987) Estimation of exposure of man to substances reacting covalently with macromolecules. *Arch. Toxicol., 60,* 251–260

Hogstedt, C., Rohlen, O., Berndtsson, B.S., Axelson, O. & Ehrenberg, L. (1979a) A cohort study of mortality and cancer incidence in ethylene oxide production workers. *Br. J. ind. Med., 36,* 276–280

Hogstedt, C., Malmqvist, N. & Wadman, B. (1979b) Leukemia in workers exposed to ethylene oxide. *J. Am. med. Assoc., 241,* 1132–1133

Hussain, S.S. (1981) *Mutagenic action of radiation and chemicals: parameters affecting the response of test systems,* Doctoral thesis, Stockholm University, Stockholm

International Commission on Radiological Protection (1977) *Recommendations* [*ICRP Publ. 26* (*Ann. ICRP 1, No. 3*)], Oxford, Pergamon Press

Jensen, S., Törnqvist, M. & Ehrenberg, L. (1984) *Hemoglobin as a dose monitor of alkylating agents. Determination of alkylation products of N-terminal valine.* In: de Serres, F. & Pero, R.W., eds, *Environmental Science Research 30,* New York, Plenum Press, pp. 315–320

Lindgren, K., Ehrenberg, L. & Natarajan, A.T. (1976) The action on barley chromosomes of isopropyl, 2-hydroxyisopropyl and 2-methoxyisopropyl methanesulfonates. *Environ. exp. Bot., 16,* 155–164

Martis, L., Kroes, R., Darby, T.D. & Woods, E.F. (1982) Disposition kinetics of ethylene oxide, ethylene glycol, and 2-chloroethanol in the dog. *J. Toxicol. environ. Health, 10,* 847–856

Moutschen-Dahmen, J., Moutschen-Dahmen, M. & Ehrenberg, L. (1968) Note on the chromosome breaking activity of ethylene oxide and ethylene imine. *Hereditas, 60,* 267–269

Neumann, H.-G. (1984) Analysis of haemoglobin as a dose monitor for alkylating and arylating agents. *Arch. Toxicol., 56,* 1–6

Osterman-Golkar, S. (1975) *Studies on the reaction kinetics of biologically active electrophilic reagents as a basis for risk estimates,* Doctoral thesis, Stockholm University, Stockholm

Osterman-Golkar, S., Ehrenberg, L. & Wachtmeister, C. (1970) Reaction kinetics and biological action in barley of mono-functional methanesulfonic esters. *Radiat. Bot., 10,* 303–327

Osterman-Golkar, S., Ehrenberg, L., Segerbäck, D. & Hällström, I. (1976) Evaluation of genetic risks of alkylating agents. II. Haemoglobin as a dose monitor. *Mutat. Res., 34,* 1–10

Osterman-Golkar, S., Farmer, P.B., Segerbäck, D., Bailey, E., Calleman, C.J., Svensson, K. & Ehrenberg, L. (1983) Dosimetry of ethylene oxide in the rat by quantitation of alkylated histidine in haemoglobin. *Teratog. Carcinog. Mutag. 3,* 395–405

Rapoport, I.A. (1948) Action of ethylene oxide, glycidol and glycols on gene mutation (in Russian). *Dokl. Akad. Nauk SSSR, 60,* 469–472

Segerbäck, D. (1983) Alkylation of DNA and haemoglobin in the mouse following exposure to ethene and ethene oxide. *Chem.-biol. Interact., 45,* 139–151

Segerbäck, D. (1985) *In vivo dosimetry of some alkylating agents as a basis for risk estimation,* Doctoral thesis, Stockholm University, Stockholm

Segerbäck, D., Calleman, C.J., Ehrenberg, L., Löfroth, G. & Osterman-Golkar, S. (1978) Evaluation of genetic risks of alkylating agents. IV. Quantitative determination of alkylated amino acids in haemoglobin as a measure of the dose after treatment of mice with methyl methanesulfonate. *Mutat. Res., 49,* 71–82

Swain, C.G. & Scott, C.B. (1953) Quantitative correlation of relative rates. Comparison of hydroxide ion with other nucleophilic reagents toward alkyl halides, esters, epoxides and acyl halides. *J. Am. chem. Soc., 75,* 141–147

Swedish Board of Occupational Safety and Health (1974) *Arbetarskyddsstyrelsens anvisningar, no. 100,* Stockholm

Swedish Board of Occupational Safety and Health (1978) *Arbetarskyddsstyrelsens anvisningar, no. 100,* revised version, Stockholm

Swedish Board of Occupational Safety and Health (1981) *Arbetarskyddsstyrelsens författningssamling 1981:8,* Stockholm

Turtóczky, I. & Ehrenberg, L. (1969) Reaction rates and biological action of alkylating agents. Preliminary report on bactericidal and mutagenic action in *E. coli. Mutat. Res., 8,* 229–238

Törnqvist, M., Mowrer, J., Jensen, S. & Ehrenberg, L. (1986a) Monitoring of environmental cancer initiators through haemoglobin adducts by a modified Edman degradation method. *Anal. Biochem., 154,* 255–266

Törnqvist, M., Osterman-Golkar, S., Kautiainen, A., Jensen, S., Farmer, P.B. & Ehrenberg, L. (1986b) Tissue doses of ethylene oxide in cigarette smokers determined from adduct levels in haemoglobin. *Carcinogenesis, 7,* 1519–1521

Walles, S. (1968) Studies on the uptake of ethyl methanesulfonate into embryos of barley. *Hereditas, 58,* 95–101

Walles, S. & Ehrenberg, L. (1968) Effects of β-hydroxyethylation and β-methoxyethylation in DNA *in vitro. Acta chem. scand., 22,* 2727–2729

ESTIMATION OF THE CANCER RISK OF GENOTOXIC CHEMICALS BY THE RAD-EQUIVALENCE APPROACH

A. Kolman, D. Segerbäck & S. Osterman-Golkar

Department of Radiobiology, Stockholm University, Stockholm, Sweden

The genotoxic effectiveness of ethylene oxide is compared with that of γ-radiation, as measured by induced transformation of C3H 10T1/2 mouse embryo fibroblasts. The resultant value of the quality factor, $Q \approx 80$ rad/mM × h, is compatible with earlier data on mutation and other endpoints in various test systems. This quality factor can thus be used to estimate the risk of exposures to ethylene oxide or its precursor, ethene, on the basis of human doses monitored by haemoglobin adducts.

Target dose of ultimate mutagens/cancer initiators as a basis for risk estimation

The probability (P) that a genotoxic effect (mutation, cancer initiation) will occur depends primarily on the cumulative frequency of critical changes ('molecular dose', D_{molec}) in the DNA of target cells. The critical changes have not yet been identified for many genotoxic agents, and it is therefore preferable to relate the risk of such chemicals, R_iX, to the target dose, D_{targ}, i.e., the time integral

$$D_{targ} = \int_t [R_iX](t)\,dt \tag{1}$$

of the concentration ($[R_iX]$) of ultimate mutagens/initiators (R_iX) in the vicinity of the nuclear DNA of the target cells (Ehrenberg *et al.*, 1983). D_{targ} is proportional to D_{molec}:

$$D_{molec} = \left(\frac{[R_i - DNA_{crit}]}{[DNA_{crit}]}\right)_{accum} = k_{i,crit} \times D_{targ}, \tag{2}$$

where DNA_{crit} denotes sites in DNA, changes in which lead to an increased probability of effect, and $k_{i,crit}$ is the second-order rate constant for reaction of R_iX with these sites. The dimension of D_{targ} is concentration × time, e.g., mM × h, but this dose may also be expressed in other units, e.g., the (cumulative) frequency of some specific DNA adduct that is suitable for measurement.

Most target organs are not accessible to direct monitoring of D_{targ}. Measurement of the 'tissue dose' or blood dose, D_{bl}, by measuring levels of adducts to blood proteins [haemoglobin (Hb), plasma proteins] or to DNA of leucocytes may give an indirect determination of D_{targ}. For several agents, the levels of DNA adducts have been found to be proportional to the levels of the corresponding Hb adducts, i.e., $D_{targ} \propto D_{bl}$ (Neumann, 1983; Murthy *et al.*, 1984). If a one-compartment model applies for the distribution of R_iX, D_{targ} will be approximately equal to D_{bl}, as indicated for ethylene oxide in experiments with mice (Segerbäck, 1983). When homogeneous distribution cannot be assumed, the ratios $D_{targ} : D_{bl}$ in humans must be estimated from experiments with model animals.

Risk model

Dose monitoring in humans, with derivation of the target dose as a measure proportional to the frequency of critical changes in DNA, eliminates, or decreases, uncertainties about interspecies differences in rates and patterns of metabolism, especially the bioactivation and detoxification of chemical carcinogens. The probability that a critical DNA change will lead to a tumour is determined by a number of factors that cannot at present be measured in human individuals or populations. According to the multistage model of carcinogenesis, initiation and promotion, and possibly further mutational steps during progression (Iversen & Iversen, 1982; Hennings *et al.*, 1983), are assumed to be essential events. In addition, a number of hereditary and acquired factors and conditions exert modifying influences on the probability of disease development. These include the efficiency of DNA repair, which shows great variations between individuals, between organs and tissues and between genes within the genome, and is a primary determinant of the probability that critical DNA damage will lead to initiation.

At the high doses used in experimental carcinogenesis, and especially in carcinogenicity tests in laboratory animals, an initiatior (R_iX, or its precursor, A_i) often exerts not only initiating and other mutagenic events but also promoting and modifying effects. However, at low doses, a genotoxic agent usually acts as an initiator (mutagen), interacting with promoting and modifying conditions that occur for reasons other than exposure to the initiator (von Bahr *et al.*, 1984). To the extent that tumour development requires two (or more) mutations or related events — as in Knudson's (1971) two-mutation hypothesis — it seems unlikely that, at low doses, a genotoxic agent causes more than one of these events; the others are probably components of the load of factors of different origin, collectively called here 'promoting and modifying factors'. Denoting these factors P^0_{pro}, cancer risk is determined by

$$P_{can}(D_{targ}) = P_{ini}(D_{targ}) \cdot P^0_{pro}, \tag{3}$$

where D_{targ} is the target dose. There is at present no method for determining P^0_{pro} in human populations. The rad-equivalence approach (Ehrenberg, 1979, 1980), however, can be used to obtain an implicit estimation of P^0_{pro} by expressing the genotoxic effectiveness of an agent relative to that of γ-radiation, the environmental factor for which cancer risk coefficients, k_γ, are best known. Assuming linear dose-response relationships for initiation by γ-radiation and by chemicals i (see Ehrenberg *et al.*, 1983), we obtain

$$P_{can}(D_\gamma) = P_{ini}(D_\gamma) \cdot P^0_{pro} = k_\gamma \times D_\gamma \tag{4}$$

and

$$P_{can}(D_i) = k_\gamma(\text{rad}^{-1}) \times Q_i\left(\frac{\text{rad}}{\text{mM h}}\right) \times D_i(\text{mM h}), \tag{5}$$

where Q_i is the relative genotoxic effectiveness of i.

A number of observations support the idea that dose-response curves for mutation and related phenomena are dominated, at low doses or dose rates, by a linear component with no zero-effect threshold (Ehrenberg *et al.*, 1983). [In principle, the model does not require that initiation depend linearly on target dose (of chemical or γ-radiation), only that the respective dose-response curves, $P_{ini}(D_{targ})$ (in Eq. 3) and $P_{ini}(D_\gamma)$ (in Eq. 4), have the same shape.] Even if, in some cases, some other dose function applies in the low-dose range, addition to a background dose of the same or related agents would probably lead to linearity of the response to dose increments from a genotoxic agent i (see Crump *et al.*, 1976).

Linearity of dose-response curves has the important consequence that the collective risk from exposure to i becomes a linear function of the mean target dose, $\hat{D}_i$, in a population of size N (see Ehrenberg *et al.*, 1983):

$$\mathrm{Risk}_{coll}(D_{targ}) = k_\gamma \times Q_i \times \hat{D}_i \times N \text{ cases.} \tag{6}$$

Determination of the quality factor, Q

The genotoxic effectiveness, b_i, of an agent i was defined as the response increment per dose unit (Ehrenberg, 1960) in the range of low doses where dose-response curves approach linearity:

$$P(D_i) - a = b_i D_i + \ldots \tag{7a}$$

With a similar expression for the response to γ-radiation in the same biological system:

$$P(D_\gamma) - a = b_\gamma D + \ldots, \tag{7b}$$

the relative genotoxic effectiveness, or quality factor (Q_i), for i, with acute γ-radiation as the standard agent, is defined as the ratio of (linear low-dose components of) slopes

$$Q_i = b_i / b_\gamma. \tag{8}$$

Q_i expresses the γ-radiation dose which, at low doses, gives the same response as the chemical target dose 1 mM h. Q_i thus has the dimension (radiation dose unit)/(chemical dose unit), e.g., rad/mM h.

A main reason for considering Q_i useful for estimating human cancer risk according to equations (5, 6) is the finding, in studies of certain model compounds (see Ehrenberg, 1979), that Q_i for forward mutation and related effects induced by a given alkylating agent i is approximately the same in various test systems, extending from bacteria and plants to mammalian cells and intact mammals, indicating that Q_i would apply also in man (see below).

The approximate constancy of Q_i over phyla of organisms is illustrated in Table 1

Table 1. The quality factor, Q (rad/mM × h) for ethylene oxide in different biological test systems[a]

Test system	Q	Reference
Chinese hamster V79 cells, *hprt* mutation	40	Segerbäck *et al.* (1987)
Escherichia coli Sd-4, forward mutation	120	Hussain (1981)
Salmonella typhimurium SV-3, forward mutation	120	Hussain (1981)
E. coli K-12, prophage induction	80	Hussain & Ehrenberg (1975)
Barley seeds, forward (chlorophyll) mutation	125 (90–160)	Ehrenberg *et al.* (1974)
Mouse fibroblasts, cell transformation of C3H 10T1/2 cells	80	This study
Mouse		
dominant lethal mutations	120 (60–200)	Appelgren *et al.* (1977)
sister chromatid exchanges in spleen lymphocytes	60	Segerbäck *et al.* (1987)
Rat		
dominant lethal mutation	140 (70–200)	Embree (1975)
micronuclei	80 (20–160)	Embree (1975)

[a]From D. Segerbäck *et al.* (unpublished)

Fig. 1. Dose-response curves for transformation by γ-radiation and ethylene oxide in C3H 10T1/2 cells. The quality factor, *Q*, is calculated from the slopes of the curves (b_i, b_γ); eq. (8)

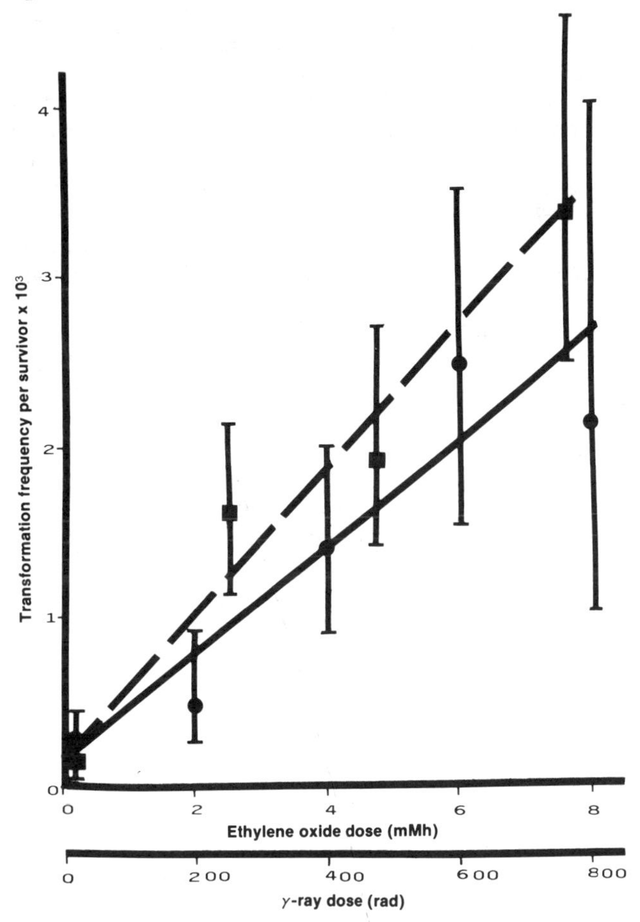

by data for ethylene oxide (from Segerbäck *et al.*, 1987). In the in-vitro tests, the target doses were considered to be equal to the doses in the medium, as corroborated by measurement of the rate of formation of 7-(2-hydroxyethyl)guanine in DNA. In the in-vivo test for sister chromatid exchanges in spleen cells, the doses were monitored by measuring Hb and DNA adducts. The doses in the experiments of Appelgren *et al.* (1977) and Embree (1975) with mice and rats, respectively, were assumed to be equal to doses measured in separate experiments with these species. In the study of barley seeds, treatment was at 21°C, and the value of *Q* for ethylene oxide was therefore recalculated to reaction rates at 37°C.

To the list of test systems can now be added C3H 10T1/2 mouse embryo fibroblasts, with an observed endpoint, cell transformation, that is assumed to be relevant to carcinogenesis. This endpoint gives a value of *Q* for ethylene oxide, ~80 rad/mM h, which is compatible with other test data. Figure 1 illustrates how $Q = b_i/b_\gamma$ was determined in one experiment. Data from repeated experiments will be published elsewhere.

Scientific and technical aspects of the rad-equivalence procedure for risk estimation

Objections to the rad-equivalence approach for risk estimation have included the fact that γ-radiation and the chemicals with which it has been compared act by different mechanisms and the DNA lesions they produce are repaired by different repair functions. However, except for dose-response functions, application of the rad-equivalent does not require equality in mechanisms of cancer initiation or mutation (see Bridges, 1973). With respect to DNA repair, it is not required that the lesions produced by γ-radiation and a chemical be repaired by the same enzymes, only that (efficiency of repair of DNA lesions induced by R_iX)/(efficiency of repair of DNA lesions induced by γ-radiation) $\approx$ constant in systems used for determination of Q_i and in human target cells (Ehrenberg *et al.*, 1983). It is possible that polymorphisms in DNA repair capacity within human populations might lead to variation in individual risks around a mean value that is hopefully reflected by the model as given in equations (5, 6).

These requirements are similar to those that apply when quality factors are used in the risk management of ionizing radiation with biological effectiveness deviating from that of γ-radiation (see International Commission on Radiological Protection, 1977). It should further be stressed that the rad-equivalence model is designed for the estimation of risks at doses or dose rates (i.e., in-vivo concentrations) so low that the (mostly nonspecific) promoting and modifying influences (e.g., through induction or saturation of repair or through immunosuppression) are negligible.

Under these conditions, it seems permissible to assume that given promoting and modifying factors affect an initiated (mutated) cell to the same extent, irrespective of whether initiation (mutation) was induced by γ-radiation or by a chemical. In terms of Eq. (7), Eq. (4) may be written

$$P_{\text{can}}(D_\gamma) - a = b_\gamma \cdot P^0_{\text{pro}} \cdot D_\gamma. \qquad (4a)$$

P^0_{pro} in Eq. (5) is thus implicitly estimated through the equality

$$b_\gamma \cdot P^0_{\text{pro}} = k_\gamma. \qquad (4b)$$

Application and epidemiological confirmation

Occupational exposure to ethylene oxide at an average level of 1 ppm during 40-h/week leads to steady-state levels of approximately 1 nmol *N*-(2-hydroxyethyl)-valine or N^τ-hydroxyethylhistidine per g Hb, corresponding to an annual dose of 0.12 mM h or 10 rad-equivalents. Since ethylene oxide is the sole or predominant genotoxic agent in many work premises the — considerable — cancer risk estimated to be associated with exposure to ethylene oxide could be confirmed, at least to the order of magnitude, in epidemiological studies of the incidence of leukaemia (Calleman *et al.*, 1978; Ehrenberg & Hussain, 1981). The finding by Hogstedt *et al.* (1986) of eight cases of leukaemia among 733 exposed persons, corresponding to a ten-fold increase over expectation, is in good agreement with the prediction from Eq. (6) (data to be published). It should be stressed that exposures have mostly been too recent to permit epidemiological confirmation of the risks of solid tumours, also predicted to be initiated by ethylene oxide.

Unexposed persons exhibit background levels of hydroxyethylations in Hb (van Sittert *et al.*, 1985; Törnqvist *et al.*, 1986a) associated with exposures to tobacco smoking (Törnqvist *et al.*, 1968b), urban air pollution, passive smoking and endogenous production (Törnqvist, this volume). According to current work, the average level in the adult population is of the order of 0.1 nmol *N*-(2-hydroxyethyl)valine/g Hb.

To the extent that this level originates from ethylene oxide — with ethene as precursor (Ehrenberg *et al.*, 1977; Törnqvist *et al.*, 1986b) — the associated cancer risk might amount to some 5–10% of the present incidence in Sweden. This estimate still contains large uncertainties of various kinds. It shows, however, the importance of assessing cancer risks from hydroxyethylating compounds and other low-molecular-weight cancer initiators.

Acknowledgements

We thank collaborators within the department for valuable assistance and viewpoints. Financial support from Shell Internationale Research Maatschappij B.V., the Swedish Work Environment Fund and the National Swedish Environment Protection Board is acknowledged.

References

Appelgren, L.E., Eneroth, G. & Grant, C. (1977) Studies on ethylene oxide; whole-body autoradiography and dominant lethal test in mice. *Proc. Eur. Soc. Toxicol., 18,* 315–317

von Bahr, B., Ehrenberg, L., Scalia-Tomba, G.-P. & Säfwenberg, J.-O. (1984) *Study of Different Models for Dose-response Relationships* (Swed.) (Appendix 9 to the Report of the Governmental Committee on Cancer Prevention, DsS 1984:5) (unpublished)

Bridges, B.A. (1973) Some general principles of mutagenicity screening and a possible framework for testing procedures. *Environ. Health Perspect., 6,* 221–227

Calleman, C.J., Ehrenberg, L., Jansson, B., Osterman-Golkar, S., Segerbäck, D., Svensson, K. & Wachtmeister, C.A. (1978) Monitoring and risk assessment by means of alkyl groups in haemoglobin in persons occupationally exposed to ethylene oxide. *J. environ. Pathol. Toxicol., 2,* 427–442

Crump, K.S., Hoel, D.G., Langley, C.H. & Peto, R. (1976) Fundamental carcinogenic processes and their implications for low dose risk assessment. *Cancer Res., 36,* 2973–2979

Ehrenberg, L. (1960) Induced mutation in plants: mechanisms and principles. *Genet. Agr., 12,* 364–389

Ehrenberg, L. (1979) *Risk assessment of ethylene oxide and other compounds.* In: McElheny, V.K. & Abrahamson, S., eds, *Assessing Chemical Mutagens: The Risk to Humans* (*Banbury Report 1*), Cold Spring Harbor, NY, CSH Press, pp. 157–190

Ehrenberg, L. (1980) *Purposes and methods of comparing effects of radiation and chemicals.* In: *Radiobiological Equivalents of Chemical Pollutants,* Vienna, International Atomic Energy Agency, pp. 23–36

Ehrenberg, L. & Hussain, S. (1981) Genetic toxicity of some important epoxides. *Mutat. Res., 86,* 1–113

Ehrenberg, L., Hiesche, K. D., Osterman-Golkar, S. & Wennberg, I. (1974) Evaluation of genetic risks of alkylating agents: tissue doses in the mouse from air contaminated with ethylene oxide. *Mutat. Res., 24,* 83–103

Ehrenberg, L., Osterman-Golkar, S., Segerbäck, D., Svensson, K. & Calleman, C. J. (1977) Evaluation of genetic risks of alkylating agents. III. Alkylation of haemoglobin after metabolic conversion of ethene to ethene oxide in vivo. *Mutat. Res., 45,* 175–184

Ehrenberg, L., Moustacchi, E. & Osterman-Golkar, S. (1983) Dosimetry of genotoxic agents and dose-response relationships of their effects. *Mutat. Res., 123,* 121–182

Embree, J.W. (1975) *Mutagenicity of ethylene oxide and associated health hazard,* Ph. D. Thesis, University of California, San Francisco, CA, USA

Hennings, H., Shores, R., Wenk, M.L., Spangler, E.F., Tarone, R. & Yuspa, S.H. (1983) Malignant conversion of mouse skin tumours is increased by tumour initiators and unaffected by tumour promoters. *Nature, 304,* 67–69

Hogstedt, C., Aringer, L. & Gustavsson, A. (1986) Epidemiologic support for ethylene oxide as a cancer-causing agent. *J. Am. med. Ass., 255,* 1575–1578

Hussain, S. (1981) *Mutagenic action of radiation and chemicals: parameters affecting the response of test systems,* Ph. D. Thesis, University of Stockholm, Stockholm

Hussain, S. & Ehrenberg, L. (1975) Prophage inductive efficiency of alkylating agents and radiations. *Int. J. radiat. Biol., 27,* 355–362

International Commission on Radiological Protection (1977) *Recommendations* (*ICRP Publication 26*) (Ann. ICRP 1:3), Oxford, Pergamon Press

Iversen, O.H. & Iversen, U.M. (1982) Must initiators come first? Tumorigenic and carcinogenic effects on skin of 3-methylcholanthrene and TPA in various sequences. *Br. J. Cancer, 45,* 912–920

Knudson, A.G. (1971) Mutation and cancer: statistical study of retinoblastoma. *Proc. natl Acad. Sci. USA, 68,* 820–823

Murthy, M.S.S., Calleman, C.J., Osterman-Golkar, S., Segerbäck, D. & Svensson, K. (1984) Relationships between ethylation of hemoglobin, ethylation of DNA and administered amount of ethyl methanesulfonate in the mouse. *Mutat. Res., 127,* 1–8

Neumann, H.-G. (1983) *The dose dependence of DNA interactions of aminostilbene derivatives and other chemical carcinogens.* In: Hayes, A.W., Schnell, R.C. & Miya, T.S., eds, *Developments in the Science and Practice of Toxicology,* Amsterdam, Elsevier, pp. 135–144

Segerbäck, D. (1983) Alkylation of DNA and hemoglobin in the mouse following exposure to ethene and ethene oxide. *Chem.-biol. Interact., 45,* 139–151

van Sittert, N.J., de Jong, G., Clare, M.C., Davies, R., Dean, B.J., Wren, L.J. & Wright, A.S. (1985) Cytogenetic, immunological, and haematological effects in workers in an ethylene oxide manufacturing plant. *Br. J. ind. Med., 42,* 19–26

Törnqvist, M., Mowrer, J., Jensen, S. & Ehrenberg, L. (1986a) Monitoring of environmental cancer initiators through hemoglobin adducts by a modified Edman degradation method. *Anal. Biochem., 154,* 255–266

Törnqvist, M., Osterman-Golkar, S., Kautiainen, A., Jensen, S., Farmer, P.B. & Ehrenberg, L. (1986b) Tissue doses of ethylene oxide in cigarette smokers determined from adduct levels in hemoglobin. *Carcinogenesis, 9,* 1519–1521

EPIDEMIOLOGICAL STUDIES ON ETHYLENE OXIDE AND CANCER: AN UPDATING

L.C. Hogstedt

Departments of Occupational Medicine, National Institute of Occupational Health, Solna and Karolinska Hospital, Stockholm, Sweden

In 1959, Dr Lars Ehrenberg and a coworker warned the Swedish authorities that ethylene oxide, a common chemical, constituted a potential cancer hazard. Twenty years later, the first epidemiological study and case reports were published indicating an increased cancer risk after occupational exposure to ethylene oxide. An updating of three small Swedish cohorts comprising 709 employees revealed 33 deaths from cancer whereas 20 were expected from national average rates. The excess was due mainly to an increased risk of stomach cancer in one production plant and an excess of blood and lymphatic malignancies in all three cohorts. The results are in accordance with the results of clastogenic, animal and short-term tests and support Professor Ehrenberg's hypothesis, formulated 28 years ago.

Ethylene oxide has been produced commercially since the First World War and is an important intermediate product in the chemical industry. It is also used for sterilizing medical products, hospital equipment and food. In 1981, about four million tonnes of ethylene oxide were used in the USA, western Europe and Japan. In 1983, the US Occupational Safety and Health Administration estimated that 80 000 US workers were directly exposed to ethylene oxide and that another 144 000 were incidentally exposed (IARC, 1985).

In 1959, Dr L. Ehrenberg and Professor Å. Gustafsson submitted a report, on their own initiative, to the Swedish National Board of Health entitled 'Chemical mutagens: their uses and hazards in medicine and technology' with warnings of the potential cancer hazards of ethylene oxide, a directly acting epoxide and an alkylating agent, as well as of certain other chemicals. They stated that 'Especially well clarified with regard to their mutagenic properties are the substances that contain epoxy and epimino groups. The simplest chemically are the ethylene derivatives (ethylene oxide and ethyleneimine). The mutagenic effects of the substances were demonstrated originally by the Russian investigator Rapoport in the fruit fly and have since been confirmed in a number of plant and animal species. . . . considerable medical risks exist in the manufacture of these epoxides and epimino compounds, as well as in working with them. Protection to the worker must therefore be given special attention.'

Ethylene oxide was then widely tested and gave positive results in most test systems (Ehrenberg & Hussain, 1981), but, in spite of all these results, no case report or epidemiological or animal carcinogenicity study was published before 1979, when we reported two cases of leukaemia and one case of morbus Waldenström (a malignant lymphoma) among a small group of workers exposed to ethylene oxide (Hogstedt *et al.*, 1979a).

The importance of this small cluster would not have been obvious if the strong mutagenic and clastogenic potential of ethylene oxide in various organisms and species had not been known. Thus, experimental results indicating that ethylene oxide was a carcinogen preceded the first case reports of cancer by several decades.

In cooperation with Professor Ehrenberg, we performed a retrospective cohort study of production and maintenance workers in the chemical factory originally investigated in 1959–1961 by cytogenetic analyses (Ehrenberg & Hällström, 1967). The follow-up of 89 ethylene oxide operators revealed a significant excess mortality compared with national statistics, due mainly to increased mortality from tumours, especially three cases of stomach cancer and two of leukaemia, as well as from diseases of the circulatory system (Hogstedt *et al.*, 1979b). The workers had been exposed to other chemicals in the production of ethylene oxide, but the results gave further support to the hypothesis that ethylene oxide is a carcinogen.

The results of an updating through 1982 of the two Swedish cohorts and an additional group of ethylene oxide-exposed production workers in southern Sweden have since been published (Hogstedt *et al.*, 1986). Altogether, eight cases of leukaemia, compared with 0.8 expected, and six cases of stomach cancer, compared with 0.65 cases expected, had occurred.

This paper presents results from a further extended follow-up of the three small Swedish cohorts.

Study groups and methods

For this updating, we have excluded persons employed for less than one year; the total study group thus comprises 539 men and 170 women. Of these, 167 operators and repairmen, all men, had been employed in the old production plant in Örnsköldsvik in northern Sweden, where ethylene oxide was synthesized by the chlorohydrin method. Another 322 men and 19 women had been employed in the new production plant in Stenungsund in southern Sweden, where ethylene oxide is produced by the direct oxidation process in which ethylene is oxidized to ethylene oxide using oxygen. A further 151 female and 50 male nonchemical factory workers have been employed in a plant in Hallsberg where hospital equipment was sterilized. These workers were exposed to ethylene oxide vapours leaking from large sterilized boxes stored for weekly periods in the factory hall.

Everyone ever employed in the two plants in Stenungsund and Hallsberg with exposure to ethylene oxide for more than one year was included. However, the cohort from Örnsköldsvik consisted only of those who participated in the cross-sectional study performed by Ehrenberg and coworkers in 1959–1961. Many more had been exposed in this plant, but the earlier personnel records were not complete. As all the subjects were actively working and were interviewed in about 1960, the work histories are of high quality.

The same methods of follow-up and calculation of expected numbers from national rates by the person-year method were applied. Details of methods, work and exposure assessments are given in earlier publications (Hogstedt *et al.*, 1979a,b, 1986).

The study period for mortality now includes 1985 and that for cancer incidence 1983 (later records are not available from the Cancer Registry). The follow-up has therefore been extended by three and two years, respectively.

Results

A total of 85 men and women had died, compared with 79 expected from the national rates; 33 had died from cancer, with 20 expected, and there was an excess of eight stomach cancers and six leukaemias (Table 1).

In men in all three industries, there was a distinct increase in the risk for stomach

Table 1. Mortality among 709 ethylene oxide-exposed men and women in three Swedish industries

Cause of death	Length of employment								
	1–9 years			⩾10 years			All		
	Obs	Exp	SMR	Obs	Exp	SMR	Obs	Exp	SMR
All causes	39	37.4	104	46	41.1	112	85	78.5	108
Neoplasms	17	9.5	179	16	10.2	157	33	19.8	167
Stomach	4	0.8	482	6	1.0	577	10	1.8	546
Blood and lymphatic	5	1.0	500	4	1.0	417	9	2.0	459
Leukaemia	3	0.4	769	4	0.3	1250	7	0.8	921

SMR, standardized mortality ratio

cancer (10 observed, 1.7 expected) and leukaemia (4 observed, 0.6 expected). Furthermore, a doubled risk was noted for cerebrovascular diseases (9 observed, 4.9 expected) but not for ischaemic heart disease (Table 2). There is no obvious relation between length of employment and risk for stomach cancer, but most of the leukaemias occurred in the subgroup exposed for more than ten years.

Most of the excess mortality derives from the plant in Örnsköldsvik (Table 3). The continuously exposed operators had a large excess of cancers, but the drastic excess of stomach cancer is now also evident among maintenance workers with intermittent exposure to ethylene oxide, as well as to many other chemicals. During the last three years, three fatal cases of stomach cancer have occurred in this small group (0.1 expected). Only one case of stomach cancer has been observed outside the Örnsköldsvik plant.

Another fatal case of lymphatic tumour (reticular-cell sarcoma) has occurred in the Stenungsund plant, and an incident case of polycythaemia vera has been found in the Hallsberg plant (Table 4).

Table 2. Mortality among 539 ethylene oxide-exposed men in three industries

Cause of death	Length of employment									
	1–9 years			⩾10 years			All			
	Obs	Exp	SMR	Obs	Exp	SMR	Obs	Exp	SMR	95% confidence interval
All causes	33	31.3	105	44	39.1	112	77	70.5	109	0.9–1.4
Neoplasms	12	7.0	171	14	9.4	148	26	16.5	158	1.0–2.3
Stomach	4	0.7	597	6	1.0	608	10	1.7	602	2.9–11.1
Lung	1	1.3	77	0	1.8	0	1	3.1	32	0.0–1.8
Blood and lymphatic	3	0.8	380	3	0.9	330	6	1.7	354	1.3–7.7
Leukaemia	1	0.3	322	3	0.3	880	4	0.6	611	1.7–15.7
Circulatory system	13	13.5	96	24	20.0	120	37	33.5	110	0.8–1.5
Ischaemic heart diseases	8	9.8	82	16	14.4	111	24	24.2	99	0.6–1.5
Cerebrovascular diseases	4	2.0	205	5	2.9	172	9	4.9	185	0.8–3.5
Respiratory system	2	1.3	159	1	1.9	52	3	3.2	95	0.2–2.8
Violent deaths	5	5.4	92	2	3.3	61	7	8.7	80	0.3–1.6

Table 3. Mortality among ethylene oxide-exposed male operators and repairmen and unexposed chemical industry workers in Örnsköldsvik, 1962–1985

Cause of death	Operators (89)		Repairmen (78)		Both (167)	Unexposed (66)		
	Obs	Exp	Obs	Exp	SMR	Obs	Exp	SMR
All causes	34	25.0	23	24.2	115	16	15.5	103
Neoplasms	14	6.1	6	5.8	168	3	3.6	83
Oesophageal	1	0.1	1	0.1	755	0	0	0
Stomach	5	0.6	4	0.6	707	1	0.4	263
Leukaemia	2	0.2	1	0.2	703	0	0.1	0
Cerebrovascular diseases	5	1.8	2	1.9	189	2	1.1	179

Since the initial cluster at Hallsberg, 37 incident cancer cases have been reported in the three cohorts, with 27 expected. Seven more lymphatic and blood malignancies have been observed (2.2 expected), five of them leukaemias (0.8 expected).

Comments and evaluation

The prolonged follow-up of Swedish ethylene oxide-exposed cohorts, comprising most of the ethylene oxide-exposed people in Sweden, supports the conclusion of an increased cancer risk associated with exposure to ethylene oxide. In these cohorts, the risk for leukaemia is increased five to ten times, while the risk for tumours of the lymphatic system appears to be more moderately increased. The large number of cytogenetic studies that have demonstrated increased numbers of chromosomal aberrations and sister chromatid exchanges at low-level exposure to ethylene oxide (IARC, 1985) indicate that the lymphatic and haematopoietic system is particularly sensitive to the genotoxic effects of ethylene oxide.

The dramatic excess of stomach cancer seen in the Örnsköldsvik plant seems to be of the same magnitude among full-time exposed operators as among intermittently

Table 4. Cases of blood and lymphatic malignancies in workers in three industries

Tumour (ICD 8)	Year born	Year died	Year diagnosed	Ethylene oxide exposure[a]	Sex
200.0 Reticular-cell sarcoma	1927	1983	1983	Production (S) 1974–1982	M
202.2 Primary tumour in lymphatic tissue (macroglobulinaemia Waldenström)	1918	1976	1974	Sterilizer (H) 1968–1976	M
204.1 Chronic lymphocytic leukaemia	1918	1965	1959	Operator (Ö) 1941–1965	M
204.1 Chronic lymphocytic leukaemia	1916	1973	1972	Repairman (Ö) 1942–1973	M
205.0 Acute myelogenous leukaemia	1940	1979	1977	Sterilizer (H) 1968–1976	F
205.0 Acute myelogenous leukaemia	1934	1982	1976	Repairman (S) 1963–1982	M
205.1 Chronic myelogenous leukaemia	1921	1977	1972	Sterilizer (H) 1968–1976	F
205.1 Chronic myelogenous leukaemia	1923	—	1979	Operator (Ö) 1943–1959	M
207.0 Acute leukaemia (not otherwise specified)	1913	1971	1971	Operator (Ö) 1942–1944	M
207.0 Acute leukaemia (not otherwise specified)	1923	1979	1979	Sterilizer (H) 1969–1972	F
208.0 Polycythaemia vera	1916	—	1982	Sterilizer (H) 1968–1971	F

[a]S, Stenungsund; H, Hallsberg; Ö, Örnsköldsvik

exposed repairmen. This fact, and the absence of stomach cancer excess in the other plants, indicates that other chemical exposures play an important role in combination with ethylene oxide in the excess risk for stomach cancer.

In our earlier publications on the Örnsköldsvik cohort, we began calculating expected numbers of deaths '10 years after the first day of exposure to take into account the induction latency time' (Hogstedt *et al.*, 1986), which caused some concern. The rationale for this procedure was to cover the fact that only those who had survived the 1940s and 1950s could still be employed in 1959 and therefore included in the exposed cohorts. Several persons might already have been dead if there had been a serious cancer risk operating with a short induction time. No induction latency requirements have been applied in this updating, as the procedure had very little influence on the results and might have confused some readers.

Morgan *et al.* (1981) reported a retrospective cohort study on 767 men employed between 1955 and 1977 for at least five years and 'potentially exposed to ethylene oxide' in an outdoor reaction system at a chemical plant in Texas, USA. In the total cohort, 46 deaths occurred compared to 80 expected on the basis of US vital statistics; 11 malignant neoplasms were observed, with 15.2 expected. Excesses were found for pancreatic cancer (3 observed, 0.8 expected) and Hodgkin's disease (2 observed, 0.4 expected), but no death from leukaemia was found. According to the authors, only a ten-fold or greater increase in the risk of leukaemia deaths is likely to have been detected in this study, and the exposure was low; the study was thus inconclusive.

Thiess *et al.* (1981) reported a mortality study of 602 male active and former employees who had worked for six months or more in an area of alkylene oxide production in the Federal Republic of Germany, and had been exposed to ethylene oxide and propylene oxide, as well as other chemicals. Four stomach cancers (2.7 expected), one case of myeloid leukaemia (0.15 expected) and one case of lymphatic sarcoma were observed.

An increased frequency of spontaneous abortions was correlated with exposure to ethylene oxide among Finnish hospital staff engaged in sterilizing instruments (Hemminki *et al.*, 1982).

Both the epidemiological and the toxicological data (IARC, 1985) strongly indicate a causal relation between ethylene oxide and the induction of malignancies, even at low-level and intermittent exposures. This should be seriously considered by industry and regulating authorities, as Professor Ehrenberg suggested 28 years ago.

Acknowledgement

This updating could not have been done without the skilful assistance of A. Gustavsson.

References

Ehrenberg, L. & Gustafsson, Å. (1959) *Chemical Mutagens: Their Uses and Hazards in Medicine and Technology. A report of February 1959 to the National Board of Health* (English translation published by Bloms Boletryckeri, Lund, 1970)

Ehrenberg, L. & Hällström, T. (1967) *Haematologic studies on persons occupationally exposed to ethylene oxide.* In: *Radiosterilization of Medical Products* (*Report SM 92/96*), Vienna, International Atomic Energy Agency, pp. 327–334

Ehrenberg, L. & Hussain, S. (1981) Genetic toxicity of some important epoxides. *Mutat. Res.*, *86*, 1–113

Hemminki, K., Mutanen, P., Saloniemi, I., Niemi, M.-L. & Vainio, H. (1982) Spontaneous abortions in hospital staff engaged in sterilising instruments with chemical agents. *Br. med. J.*, *285*, 1461–1463

Hogstedt, C., Malmqvist, N. & Wadman, B. (1979a) Leukemia in workers exposed to ethylene oxide. *J. Am. med. Assoc., 241,* 1132–1133

Hogstedt, C., Rohlén, O., Berndtsson, B.S., Axelson, O. & Ehrenberg, L. (1979b) A cohort study of mortality and cancer incidence in ethylene oxide production workers. *Br. J. ind. Med., 36,* 276–280

Hogstedt, C., Aringer, L. & Gustavsson, A. (1986) Epidemiologic support of ethylene oxide as a cancer-causing agent. *J. Am. med. Assoc., 255,* 1575–1578

IARC (1985) *IARC Monographs on the Evaluation of the Carcinogenic Risk of Chemicals to Humans,* Vol. 36, *Allyl Compounds, Aldehydes, Epoxides and Peroxides,* Lyon, pp. 189–226

Morgan, R.W., Claxton, K.W., Divine, B.J., Kaplan, S.D. & Harris, V.B. (1981) Mortality among ethylene oxide workers. *J. occup. Med., 23,* 767–770

Thiess, A.M., Frentzel-Beyme, R., Link, R. & Stocker, W.G. (1981) *Mortality study on employees exposed to alkylene oxides (ethylene oxide/propylene oxide) and their derivatives.* In: *Prevention of Occupational Cancer (Occupational Safety and Health Series No. 46),* Geneva, International Labour Office, pp. 249–259

AN IMMUNOASSAY FOR MONITORING HUMAN EXPOSURE TO ETHYLENE OXIDE

M.J. Wraith,[1] W.P. Watson,[1] C.V. Eadsforth,[2] N.J. van Sittert[2], M. Törnqvist[3] & A.S. Wright[1]

[1]Shell Research Ltd, Sittingbourne Research Centre, Sittingbourne, Kent, UK; [2]Health, Safety and Environment Division, Shell International Petroleum Maatschappij, The Hague, The Netherlands; and [3]Department of Radiobiology, University of Stockholm, Stockholm, Sweden

A novel immunochemical approach has been developed to monitor human exposures to ethylene oxide (EO). The method exploits the interaction of EO with the amino function of the *N*-terminal valine residue of the α-chain of human haemoglobin (Hb). Antibodies were raised against the adducted valine in the form of the *N*-terminal tryptic heptapeptide and have been used to develop a sensitive radioimmunoassay (RIA) for the adducted heptapeptide. This method has been fully validated against a gas chromatography-mass spectrometry (GC-MS) method and has been applied to the monitoring of EO exposure in a group of sterilization workers.

Molecular dosimetry techniques are now widely employed to provide improved estimates of exposure to genotoxic agents. In experimental species, the main focus has been on the application of such techniques to determine tissue DNA doses. However, human DNA is generally inaccessible for experimental purposes, and therefore indirect assessments of human DNA doses of genotoxicants are valuable.

Human exposure monitoring

The benefits of measuring Hb adducts to estimate DNA doses were first suggested by Osterman-Golkar *et al.* (1976). The validation of such an approach must be based, of necessity, on correlations determined in experimental species. Inhalation studies in rats exposed to ^{14}C-EO indicate a rapid absorption and equilibration of EO throughout the tissues (Wright, 1983). The results of these studies are consistent with the view that Hb would be an effective tissue DNA dose monitor for EO exposure.

Peptide approach

An immunochemical procedure was developed to provide a specific, sensitive and rapid assay for use in the biomedical monitoring of EO exposure. The principal adducted amino acids formed in human Hb by reaction with EO are *N*-(2-hydroxyethyl)valine (α- and β-chain), N_1-(2-hydroxyethyl)histidine, N_3-(2-hydroxyethyl)histidine and *S*-(2-hydroxyethyl)cysteine. The possibility of developing antibodies against the adducted amino acids was judged to be low as their small molecular size offers only limited antigenicity. An alternative approach was to raise antibodies against a peptide from human Hb which contained an EO-adducted amino acid. The peptide selected was the *N*-terminal heptapeptide released from the α-chains of human Hb by the action of trypsin (Fig. 1).

Fig. 1. Hydroxyethylated Val-Leu-Ser-Pro-Ala-Asp-Lys

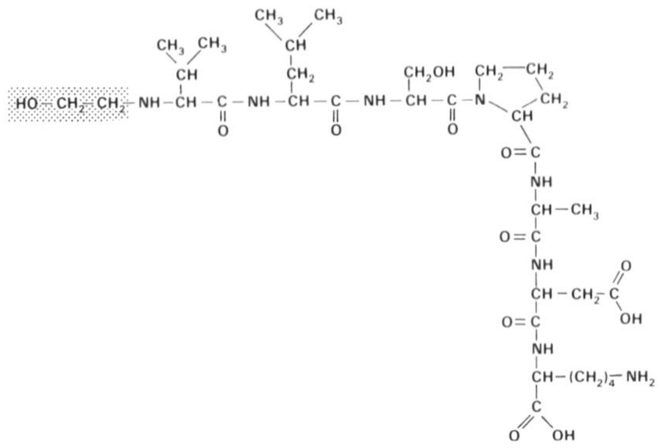

The *N*-terminal heptapeptide released from the α-chain of ethylene oxide-treated haemoglobin by trypsin hydrolysis

The adducted heptapeptide and the unmodified analogue were synthesized chemically. Both peptides were analysed by high-performance liquid chromatography and were homogeneous on two different stationary phases. In addition, fast atom bombardment-mass spectrometry of both peptides gave $(M + H)^+$ ions consistent with the required molecular weights.

Radioimmunoassay development

The hydroxyethylated (HOEt) peptide was radioiodinated at the amino group of the C-terminal lysine using the conjugation procedure developed by Bolton and Hunter (1973). This radioactive tracer was used in the optimized RIA and to monitor incorporation of the HOEt peptide during preparation of the immunogen. The HOEt peptide was coupled to horse albumin using 1-ethyl-3-(3-dimethylamino-propyl)carbodiimide, resulting in approximately 16 mol adducted peptide per mol immunogen. Four rabbits were immunized, and antisera from one animal demonstrated sufficiently low cross-reactivity when tested against the non-HOEt peptide, native human Hb and the peptides from trypsin-hydrolysed Hb, to be useful for the development of an RIA. The cross-reactivity results indicated that it was possible to quantify the HOEt peptide in the presence of a 10^6-fold excess of the non-HOEt peptide. During development of the assay, attempts were made to analyse native Hb treated with EO; however, very low recoveries indicated that the antibody was capable of binding the HOEt peptide only after its release by trypsin hydrolysis. Additional assessments of specificity indicated that the antibody bound equally well to the equivalent HOEt peptide from rat Hb. This was probably due to the sequence homology of the first three amino acids of the α-chains of human and rat Hb (also rabbit, mouse and chimpanzee Hb). In contrast, the antibody did not bind the analogous propylene oxide adducted peptide, indicating high specificity for the HOEt modification.

The lower limit of detection of the optimized RIA was 25 fmol/50 μl sample, which in conjunction with the cross-reactivity data gave an overall sensitivity of 0.14 pmol HOEt peptide/g globin.

Biomonitoring study

The RIA was validated in a study of hospital workers potentially exposed to EO. Samples of blood were obtained from a group of operatives employed in EO

Fig. 2. Levels of α-chain *N*-(2-hydroxyethyl)valine

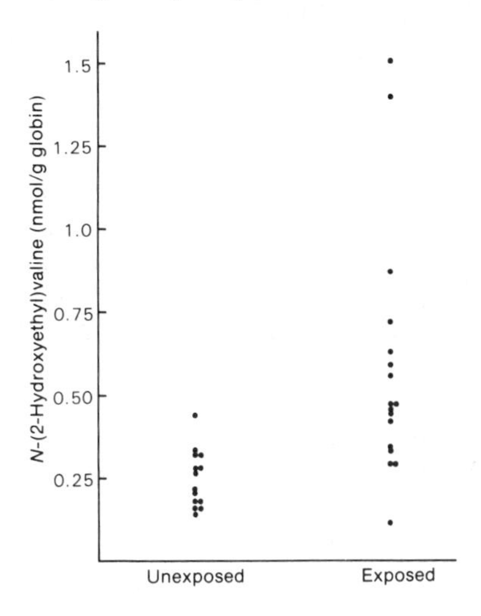

Background levels of α-chain *N*-(2-hydroxyethyl)valine ranged from 0.14–0.44 nmol/g globin in samples from the unexposed group, the mean level being 0.25 nmol/g globin (SD, 0.09; $n = 14$). Samples from potentially exposed operatives gave corresponding values ranging from 0.11–1.51 nmol/g globin, the mean level being 0.58 nmol/g globin (SD, 0.37; $n = 17$).

sterilization of medical equipment and supplies. Blood samples were also obtained from a group not involved in sterilization work. Test and control samples were analysed using the RIA procedure and were also analysed independently using a GC-MS method for *N*-(2-hydroxyethyl)valine reported by Törnqvist *et al.* (1986). Significant differences were found between potentially exposed workers and the

Fig. 3. Comparison of RIA and GC-MS results

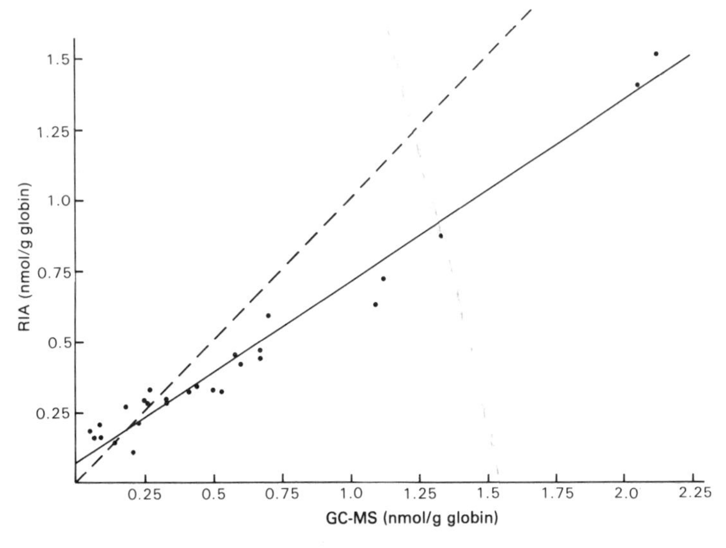

The linear regression correlation coefficient was 0.97, with a slope of 0.64 and a *y* intercept of 0.07; dashed line, theoretical line of agreement.

unexposed control group (Fig. 2). Background levels of hydroxyethylation were also found in the unexposed group, in agreement with earlier findings (Calleman, 1986). The independent GC-MS analysis gave results that were in very good agreement with the RIA data (Fig. 3).

Conclusions

A novel, sensitive, specific immunochemical biomonitoring method for EO exposure has been developed. The RIA has been validated against an existing GC-MS method, and the two widely differing analytical methods showed very good agreement. At the outset of this study, it was anticipated that the *N*-terminal valine residues of the α- and β-chains of Hb would display similar reactivities towards EO, but the correlation shown in Figure 3 indicates that this may not be the case. Therefore, in order to calculate the EO dose reaching Hb, it will be necessary to establish the kinetics of interaction at the α- and β-centres.

The observation of background levels of hydroxyethylation of Hb suggests the possible occurrence of the corresponding adducts as a 'background' in DNA. The origins and significance of background alkylations are under investigation.

References

Bolton, A.E. & Hunter, W.M. (1973) The labelling of proteins to high specific radioactivities by conjugation to a ^{125}I-containing acylating agent. *Biochem. J., 133,* 529–538

Calleman, C.J. (1986) Monitoring of background levels of hydroxyethyl adducts in human hemoglobin. *Progr. clin. biol. Res., 109B,* 261–270

Osterman-Golkar, S., Ehrenberg, L., Segerbäck, O. & Hällström, I. (1976) Evaluation of genetic risks of alkylating agents. II. Haemoglobin as a dose monitor. *Mutat. Res., 34,* 1–10

Törnqvist, M., Mowrer, S., Jensen, S. & Ehrenberg, L. (1986) Monitoring of environmental cancer initiators through hemoglobin adducts by a modified Edman degradation method. *Anal. Biochem., 154,* 255–266

Wright, A.S. (1983) *Molecular dosimetry techniques in human risk assessment.* In: Hayes, A.W., Schnell, R.C. & Miya, T.S., eds, *Developments in the Science and Practice of Toxicology,* Amsterdam, Elsevier, pp. 311–318

DETERMINATION OF SPECIFIC URINARY THIOETHERS DERIVED FROM ACRYLONITRILE AND ETHYLENE OXIDE

M. Gérin, R. Tardif & J. Brodeur

Department of Occupational and Environmental Health, Université de Montréal, Montréal, Québec, Canada

A simple analytical method has been developed for the determination of specific urinary thioethers. Mercapturic acids present in urine are first de-acetylated enzymatically or by acid hydrolysis. Cysteine conjugates thus formed are reacted with *o*-phthalaldehyde and mercaptoethanol, yielding fluorescent derivatives, which are further separated by high-performance liquid chromatography (HPLC) in a reverse-phase system. The method is applicable to the determination of the following thioethers: 2-hydroxyethylmercapturic acid (*N*-acetyl-*S*-2-hydroxyethyl-L-cysteine; HMA), carboxymethylmercapturic acid (*N*-acetyl-*S*-carboxymethyl-L-cysteine; CAMA) and 2-cyanoethylmercapturic acid (*N*-acetyl-*S*-2-cyanoethyl-L-cysteine; CYMA), and the corresponding cysteine conjugates, *S*-2-hydroxyethyl-L-cysteine (HCYS), *S*-2-carboxymethyl-L-cysteine (CACYS) and *S*-2-cyanoethyl-L-cysteine (CYCYS). The excretion pattern of these thioethers in rats exposed to acrylonitrile and various rodents exposed to ethylene oxide is reported. Thioether excretion can be dependent on species and route of administration. The analytical procedure may prove to be applicable to the biological monitoring in humans of exposure to acrylonitrile, ethylene oxide, vinyl chloride, 1,2-dibromoethane and 1,2-dichloroethane.

A common route of elimination of electrophilic compounds is *via* their conjugation with glutathione and the resulting excretion of urinary thioethers in the form of mercapturic acids or cysteine conjugates (Chasseaud, 1979). Complementary to the traditional nonspecific urinary thioether assay (van Doorn *et al.*, 1981), methods to determine specific thioethers have been judged to be of considerable interest as potential tools for monitoring human exposure to specific carcinogenic and mutagenic agents (Berlin *et al.*, 1984). We report here the use of a novel analytical method for the determination of specific urinary thioethers derived from acrylonitrile and ethylene oxide in rodents.

Analytical method

The method is based on the HPLC separation and fluorescence detection of thioethers after chemical derivatization. Mercapturic acids present in urine are first de-acetylated enzymatically or by acid hydrolysis. The resulting cysteine conjugates are reacted with *o*-phthalaldehyde and mercaptoethanol, yielding fluorescent derivatives, which are further separated by HPLC in a reverse-phase system. Cysteine conjugates present in urine can be analysed without the de-acetylation step.

Details of the determination of HMA, HCYS, CAMA and CACYS have been

Fig. 1. Representative HPLC chromatogram of rat urine

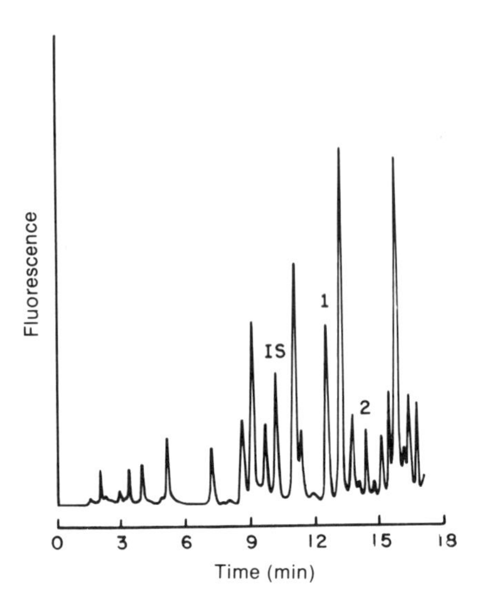

Rat exposed to acrylonitrile (6-h inhalation at 100 ppm, 24-h urine). Urine sample treated as described in Methods section. Column: Supelcosil LC-18, 150 × 4.6 mm; flow rate, 1 ml/min; injection volume, 10 μl. Fluorescence detector set with a 338-nm band pass filter for excitation and a 425-nm long-pass filter for emission. IS, fluorescent derivative of internal standard; 1 and 2, fluorescent derivatives of *S*-2-hydroxyethyl-L-cysteine and *S*-2-cyanoethyl-L-cysteine, respectively

reported by Gérin and Tardif (1986) and by Tardif *et al.* (1987). Details are given here only for the analysis of CYMA.

CYMA in urine is de-acetylated by incubation (30°C, 90 min) of 1 ml urine with 1 ml (1800 units) of an acylase 1 (Grade 1, Sigma) solution (pH 7.0, phosphate buffer). Proteins are precipitated by addition of 400 μl of a 0.2% (vol/vol) mercaptoethanol solution in methanol to 100 μl incubation mixture. After centrifugation, 20 μl of the supernatant solution are mixed with 100 μl *o*-phthalaldehyde-mercaptoethanol reagent (Turnell & Cooper, 1982). During the subsequent HPLC analysis, the composition of the mobile phase changes linearly from 0 to 10 min from solvent A (20% methanol, 78% 0.2 M sodium acetate in water, 2% tetrahydrofuran) to 56% solvent A and 44% solvent B (80% methanol, 19.5% 0.2 M sodium acetate in water, 0.5% tetrahydrofuran), then to 100% solvent B at 12.5 min, to return to 100% solvent A at 14.5 min to be held constant until 17 min from injection (see footnote to Fig. 1 for other details). A calibration curve is prepared with synthesized CYMA obtained by mixing equimolar amounts of acrylonitrile and *N*-acetyl-L-cysteine in alkaline water (m.p., 117°C). L-Homoserine is used as an internal standard and is added to urine samples as 250 μl of a 2 mM solution in water. The standard curve for CYMA is linear (0.1 to 3 mM). Within-day coefficients of variation ($n = 5$) were 2 and 4% at 1.6 and 0.2 mM; while day-to-day coefficients of variation (five days) were 2.7 and 7.3%. The detection limit, defined as three standard deviations of blank, was 3 μM. The method allows the simultaneous determination of CYMA and CYCYS.

Specific thioether determination in rodents exposed to acrylonitrile or ethylene oxide

Excretion of specific thioethers following exposure of various rodents to ethylene oxide has already been reported by Gérin and Tardif (1986) and Tardif *et al.* (1987). We report here on excretion in rats exposed to acrylonitrile by various routes of administration.

Groups of five adult male Sprague-Dawley rats were administered intraperitoneally or intravenously through the caudal vein 0.1 ml/100 g body weight of a solution of acrylonitrile dissolved in saline, for a total dose of 15.0 mg/kg. Urine was collected over 24 h from animals placed in metabolism cages.

In inhalation experiments, a group of five rats was exposed to acrylonitrile at a target concentration of 100 ppm for 6 h and urine collected over 24 h following removal from the exposure chamber. The experimental design was as described by Rouisse *et al.* (1986). All exposed animals excreted both CYMA and HMA; the amounts (μmol/kg $\pm$ SD) excreted in 24 h were: 113 $\pm$ 12 (i.p.); 97 $\pm$ 12 (i.v.); 12.5 $\pm$ 2.7 (inh.) for CYMA and 11.1 $\pm$ 0.9 (i.p.); 15.0 $\pm$ 2.7 (i.v.); 24.2 $\pm$ 3.7 (inh.) for HMA. Even though no absolute comparison is possible between the inhalation and injection techniques, the results indicate that the amounts of the two thioethers excreted can vary greatly with the route of administration.

Determinations performed without the deacetylation step indicated no detectable amount of the corresponding cysteine conjugates. None of the thioethers could be detected in control animals. These results confirm those of van Bladeren *et al.* (1981) on the simultaneous excretion of the two mercapturic acids by acrylonitrile-treated rats.

Experiments with ethylene oxide-treated rodents (Tardif *et al.*, 1987) indicated previously the influence of species on specific thioether excretion: three of the thioethers investigated were excreted by mice, one by rats and none by rabbits. (See Table 1 for combined results on acrylonitrile and ethylene oxide.)

Table 1. Specific 24-h urinary thioether excretion (% administered dose) of rodents treated intravenously with ethylene oxide or acrylonitrile

Thioether	Ethylene oxide			Acrylonitrile
	Rats[a]	Mice[b]	Rabbits[c]	Rats[d]
HMA	30.8[e]	8.3	ND[f]	5.3
HCYS	ND	5.8	ND	ND
CACYS[g]	ND	1.9	ND	—[h]
CYMA[i]	—	—	—	34.4

[a] 20 mg/kg ($n = 5$)
[b] 20 mg/kg ($n = 10$)
[c] 20 mg/kg ($n = 3$)
[d] 15 mg/kg ($n = 5$)
[e] Average % of the administered dose excreted in 24 h
[f] ND, not detected
[g] No carboxymethylmercapturic acid detected
[h] —, not investigated
[i] No *S*-2-cyanoethyl-L-cysteine detected in acrylonitrile-treated rats

Conclusion

The analytical method described allows the determination of a number of specific thioethers potentially related to exposure to acrylonitrile and ethylene oxide in man. Some of these thioethers have been measured as important metabolites in the urine of exposed rodents. The specific thioether excretion pattern has been shown, however, to be dependent on species and route of administration. The analytical procedure presented is simple, specific and has a limit of detection in the range of 10^{-5} M. It may also prove to be applicable to the biological monitoring of human exposure to acrylonitrile, ethylene oxide, vinyl chloride, 1,2-dibromoethane and 1,2-dichloroethane.

Acknowledgement

This research was supported by grants RS-83-27 and RS-85-39 from the Institut de recherche en santé et sécurité du travail du Québec.

References

Berlin, A., Draper, M., Hemminki, D. & Vainio, H. (1984) International seminar on methods of monitoring human exposure to carcinogenic and mutagenic agents. *Int. Arch. occup. environ. Health, 54,* 369–375

van Bladeren, P.J., Delbressine, L.P.C., Hoogeterp, J.J., Beaumont, A.H.G.M., Breimer, D.D., Seutter-Berlage, F. & van der Gen, A. (1981) Formation of mercapturic acids from acrylonitrile, crotononitrile and cinnamonitrile by direct conjugation and *via* an intermediate oxidation process. *Drug Metab. Dispos., 54,* 369–375

Chasseaud, L.F. (1979) The role of glutathione and glutathione *S*-transferases in the metabolism of chemical carcinogens and other electrophilic agents. *Adv. Cancer Res., 29,* 175–274

van Doorn, R., Leijdekkers, C.M., Bos, R.P., Brouns, R.M.E. & Henderson, P.T. (1981) Detection of human exposure to electrophilic compounds by assay of thioether detoxication products in urine. *Ann. occup. Hyg., 24,* 77–92

Gérin, M. & Tardif, R. (1986) Urinary *N*-acetyl-*S*-2-hydroxyethyl-L-cysteine in rats as biological indicator of ethylene oxide exposure. *Fundam. appl. Toxicol., 7,* 419–423

Rouisse, L., Chakrabarti, S. & Tuchweher, B. (1986) Acute nephrotoxic potential of acrylonitrile in Fischer-344 rats. *Res. Commun. Chem. Pathol. Pharmacol., 53,* 347–360

Tardif, R., Goyal, R., Brodeur, J. & Gérin, M. (1987) Species differences in the urinary disposition of some metabolites of ethylene oxide. *Fundam. appl. Toxicol., 9;* 448–453

Turnell, D.C. & Cooper, J.D.H. (1982) Rapid assay for amino acids in serum or urine by pre-column derivatization and reverse phase liquid chromatography. *Clin. Chem., 28,* 527–531

2-HYDROXYETHYLATION OF HAEMOGLOBIN IN MAN

B.J. Passingham,[1] P.B. Farmer,[1] E. Bailey,[1] A.G.F. Brooks[1] & D.W. Yates[2]

[1]*MRC Toxicology Unit, Medical Research Council Laboratories, Carshalton, Surrey; and* [2]*Hope Hospital, Salford, UK*

Exposure of humans to 2-hydroxyethylating agents, such as ethylene oxide, results in the formation of *N*-(2-hydroxyethyl)valine (HOEtVal) at the *N*-terminal amino acid of haemoglobin. A novel method using gas chromatography-mass spectrometry (GC-MS) has been used to monitor the presence of this adduct in smokers and control subjects, and dose-response relationships were investigated between HOEtVal in haemoglobin, number of cigarettes smoked per day and plasma cotinine levels.

2-Hydroxyethylation of histidine

In the pioneering work of Osterman-Golkar *et al.* (1976), the concept of using alkylated haemoglobin as a dosimeter for exposure to alkylating agents was first evaluated using mice treated with two simple alkylating agents — *N*-nitroso-dimethylamine and ethylene oxide. For the latter compound, the adducts 1- and 3-(2-hydroxyethyl)histidine (HOEtHis) were identified in the protein hydrolysates. The same adducts, and also *S*-(2-hydroxyethyl)cysteine (HOEtCys), were found in the haemoglobin of mice exposed to ethylene, indicating that ethylene oxide is produced metabolically from this olefin (Ehrenberg *et al.*, 1977). More detailed, later studies by Segerbäck (1983) characterized the spectrum of hydroxyethylated adducts in the haemoglobin of mice treated with ethylene and ethylene oxide, and showed that the two compounds give similar proportions of the major alkylated amino acids, confirming the intermediacy of ethylene oxide in the metabolic pathway of ethylene. Thus, determination of hydroxyethylated amino acids in haemoglobin may be used to monitor exposure to either of these compounds, and there has been sustained interest in the use of this technique for human monitoring in view of the known mutagenicity and possible human carcinogenicity of ethylene oxide.

The validity of the technique was first established with experiments on rats exposed to controlled concentrations of ethylene oxide, using determinations of 3-HOEtHis as the dose monitor (Osterman-Golkar *et al.*, 1983). An almost linear dose-response relationship was observed. The alkylated amino acid was determined in acid-hydrolysed globin after purification by ion-exchange chromatography and derivatization to the *N,O*-bisheptafluorobutyryl methyl ester. Quantification was done by GC-MS selected-ion monitoring, using the tetradeuterated analogue of 3-HOEtHis as internal standard. The analytical procedure was satisfactory for analyses down to 0.5 nmol/g globin, although the method is rather tedious and time-consuming. Despite this, it has been used in three studies to monitor humans exposed occupationally to ethylene oxide (Calleman *et al.*, 1978; Van Sittert *et al.*, 1985; Farmer *et al.*, 1986). In the investigation of Van Sittert *et al.* (1985), on 32 plant workers and 31 controls, no difference was observed in the HOEtHis levels in globin; but in the more limited study of Farmer *et al.* (1986), a partial correlation of HOEtHis levels with ethylene oxide exposure was observed. Of particular note is the presence of low (about 1 nmol/g globin) levels of 3-HOEtHis in the globin of control personnel. This background is of unknown origin and limits the sensitivity of the method for the determination of low levels of exposure.

2-Hydroxyethylation of valine

A major breakthrough in the analytical procedure came with the work of Törnqvist *et al.* (1986a), who developed a rapid method for isolating a derivative of the hydroxyethylated *N*-terminal amino acid of haemoglobin (HOEtVal) by a modified Edman degradation. In addition to HOEtHis and HOEtCys, HOEtVal is a major alkylation product in hydrolysates of globin from humans exposed to ethylene oxide (Farmer *et al.*, 1986), and its background levels are lower (normally less than 100 pmol/g globin) than those of 3-HOEtHis. Furthermore, the modified Edman degradation procedure allows the handling of larger amounts of protein (i.e., greater sensitivity of HOEtVal determination), without introducing excessive contamination into the analysis. Results obtained from HOEtVal determinations were consistent with HOEtHis levels in ethylene oxide-exposed individuals (Farmer *et al.*, 1986). In a separate study, HOEtVal levels were shown to be greater in the haemoglobin of smokers (389 ± 138 pmol/g, mean ± SD) than in nonsmokers (58 ± 25 pmol/g, mean ± SD) (Törnqvist *et al.*, 1986b).

The product from the modified Edman degradation used for the HOEtVal determination described above was the pentafluorophenylthiohydantoin derivative of HOEtVal (HOEtVal-PFPTH), which was quantified by Törnqvist and co-workers using GC with negative chemical ionization MS. We have now further modified the procedure, improving the GC properties of HOEtVal-PFPTH by eliminating its possible chromatographic adsorption at low concentrations during GC-MS analysis. This was achieved by reacting the product with *N,O*-bis(trimethylsilyl)-trifluoroacetamide to yield the trimethylsilyl ether derivative of HOEtVal-PFPTH. The structure of this compound and its analogue labelled with four deuterium atoms is shown in Figure 1. Quantification was done by electron impact selective ion recording of its molecular ion (*m/z* 440), using the deuterated analogue as internal standard (*m/z* 444). The electron impact mass spectrum of the trimethylsilyl ether derivative of HOEtVal-PFPTH is shown in Figure 2.

A further alteration to the method of Törnqvist *et al.* (1986a) involved the use of a falling-needle type solid injection device to introduce the analytical sample into the GC capillary column. This has the advantage over on-column liquid injection of protecting the capillary column from contamination by nonvolatile by-products, which can cause deterioration in GC column performance. In addition, it allows a much larger fraction of the sample extract to be analysed by GC-MS, improving the sensitivity of the assay.

Fig. 1. Structure of the trimethylsilyl ether derivative of HOEtVal-PFPTH (A) and of its d_4-labelled analogue (B)

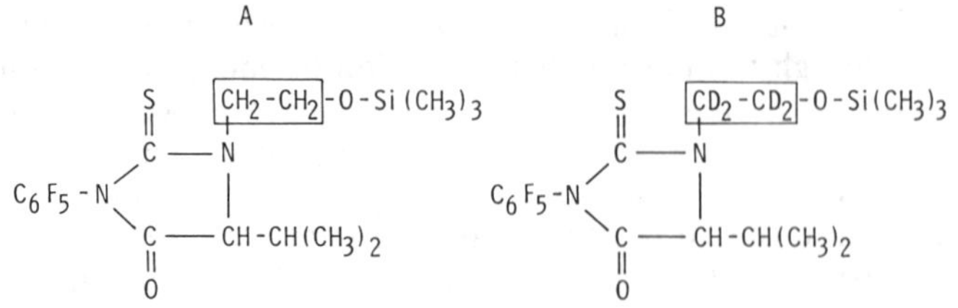

A

S, $CH_2-CH_2-O-Si(CH_3)_3$, C_6F_5-N, C, N, C, $CH-CH(CH_3)_2$, O

B

S, $CD_2-CD_2-O-Si(CH_3)_3$, C_6F_5-N, C, N, C, $CH-CH(CH_3)_2$, O

Fig. 2. Electron impact mass spectrum of the trimethylsilyl ether derivative of HOEtVal-PFPTH (HOEtVal-PFPTH-TMSi)

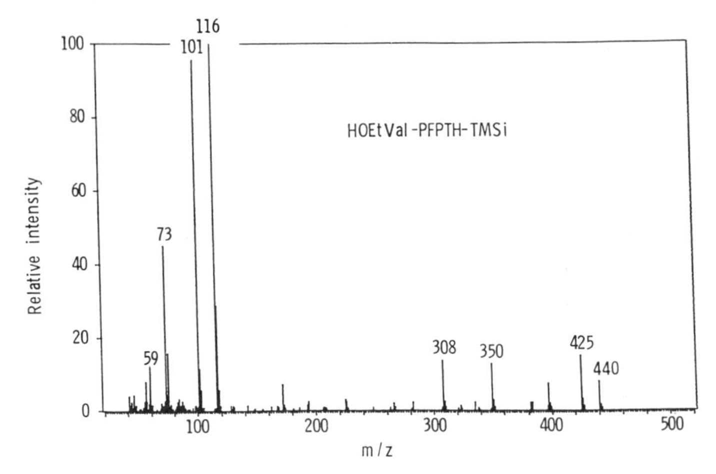

Linear calibration curves were obtained (see Fig. 3) for the analysis of mixtures of varying amounts of hydroxyethylated globin with a fixed amount of internal standard (d_4-hydroxyethylated globin containing 72.3 pmol d_4-HOEtVal).

Blood samples were drawn into heparinized tubes from a group of cigarette smokers ($n = 17$) and nonsmokers ($n = 13$). The erythrocytes were centrifuged, washed three times with isotonic saline, lysed and the supernatant from the lysate treated with 1% HCl in acetone to precipitate the globin. Aliquots of 50 mg globin were taken for analysis of HOEtVal. Typical selective ion monitor traces from the analysis of

Fig. 3. Calibration curve for the analysis of HOEtVal in globin

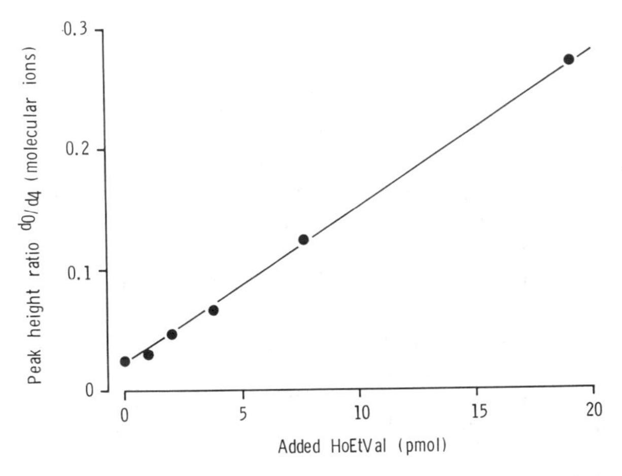

d_4-Labelled globin internal standard, containing 72.3 pmol d_4-HOEtVal, was added to aliquots (50 mg) of human globin containing 0–19 pmol added unlabelled HOEtVal. The endogenous content of the globin was 1.75 pmol. The ratio of the peak heights for the molecular ions of the d_0 and d_4-trimethylsilyl ether derivative of HOEtVal-PFPTH is linearly related to the amount of unlabelled adduct in the globin.

Fig. 4. Selective ion monitor trace (*m/z* 440, 444) for the analysis of HOEtVal in the globin from a smoker (30 cigarettes/day) (A) and from a nonsmoker (B)

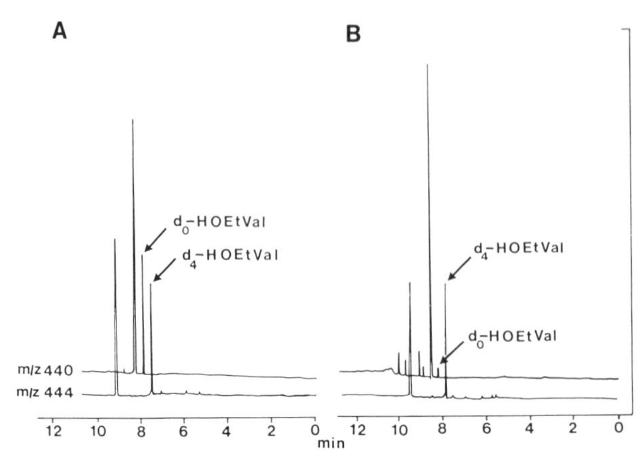

The smoker's globin contained 501 pmol/g HOEtVal. GC-MS analysis: Carlo Erba HRGC 5160 GC linked to a VG Analytical 70–70 F double focusing MS; 25 m × 0.32 mm SE-S2 fused silica capillary; column temperature: 100°C/min, then programmed at 30°C/min to 280°C; MS source temperature: 225°C; electron energy, 70 eV

haemoglobin samples from a smoker and from a control subject are illustrated in Figure 4.

Widely used alternative indicators of tobacco smoke intake are levels in plasma, saliva or urine of cotinine (a major metabolite of nicotine) (IARC, 1986). In contrast to data obtained by monitoring HOEtVal, which give an estimate of smoke exposure over the lifetime of the globin molecule (120 days in man), cotinine levels give an indication of the most recent daily exposure levels (plasma half-life, 20–30 h). We have developed a method for estimating cotinine in plasma based on capillary GC with nitrogen-phosphorus detection. Plasma was chosen as the most appropriate biological fluid for this estimation in view of its ready availability as a by-product from the work-up procedure for globin preparation. The plasma was extracted with methylene chloride after the addition of alkali and the internal standard 2-methyl-4-nitroaniline. Samples were analysed on an OV-1-coated fused-silica capillary column using splitless liquid injection. Quantification down to levels of 5 ng/ml plasma was possible using this assay procedure. The plasma cotinine and haemoglobin HOEtVal levels determined in blood samples taken from 17 subjects who smoked two to 30 cigarettes per day are presented in Table 1, and a comparison of the HOEtVal levels and smoking intensity is presented in Figure 5. These data clearly demonstrate that there is a relationship between the extent of smoking and production of HOEtVal in globin ($r^2 = 58.7\%$; $p < 0.01$). However, comparison of HOEtVal and cotinine values shows a less significant level of correlation ($r^2 = 34.4\%$; $p < 0.05$). It should be noted also that one sample that contained an unexpectedly low level of HOEtVal on the basis of a linear relationship between cigarettes smoked and globin alkylation (subject no. 8, Table 1) also had a depressed plasma cotinine concentration. This discrepancy could be due to interindividual variations in smoke inhalation characteristics.

Conclusion

The source of the 2-hydroxyethyl group in HOEtVal is unknown, although both ethylene oxide and its metabolic precursor ethylene are present in cigarette smoke

Table 1. Content of HOEtVal in haemoglobin (Hb) and of cotinine in plasma from cigarette smokers

Subject no.	Average no. cigarettes smoked/day	HOEtVal (pmol/g Hb)	Cotinine (ng/ml plasma)
1	2	44	<5
2	3	38	16
3	5	175	153
4	6	118	152
5	10	258	101
6	10	248	171
7	10	129	364
8	10	42	19
9	10	297	472
10	10	212	248
11	10	304	429
12	20	159	174
13	20	225	114
14	20	329	196
15	25	347	192
16	30	377	352
17	30	501	279

(about 5 μg and 250 μg per cigarette, respectively; Binder & Lindner, 1972; Elmenhorst & Schultz, 1968). Cigarette smoke also contains *N*-nitrosodiethanolamine at amounts up to 36 ng in nonfilter cigarettes and 24 ng in filter cigarettes (Hoffmann *et al.*, 1984). *N*-Nitrosodiethanolamine is known to 2-hydroxyethylate nucleic acids (Farrelly *et al.*, 1987) and could therefore be a further precursor for the hydroxyethylation of valine residues in haemoglobin. *N*-Nitrosodiethylamine is another constituent of cigarette smoke, with levels up to 2.8 ng in nonfiltered cigarettes and 7.6 ng in filtered cigarettes (Hoffmann *et al.*, 1984). We have recently demonstrated (I. White, P.B. Farmer, E. Bailey, B.J. Passingham, unpublished) that this nitrosamine also yields HOEtVal at

Fig. 5. Content of HOEtVal (mean values) in the haemoglobin of smokers

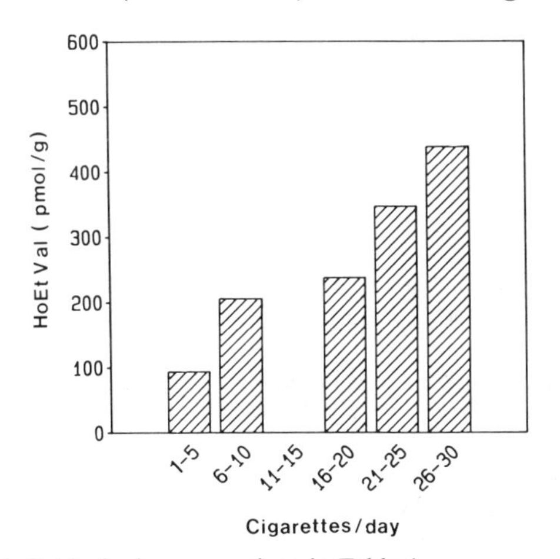

The smoking habits of the individuals shown are given in Table 1.

the *N*-terminal position of haemoglobin following its administration to rats. However, it clearly cannot make a major contribution to the HOEtVal levels in cigarette smoke.

The smoking-related increase in globin HOEtVal levels determined in this study is 114 pmol/g globin per ten cigarettes/day, which compares well with the earlier reported value of 110 (Törnqvist *et al.*, 1986b). Of particular interest in this study, however, is the demonstration of measurable levels (mean, 57.4; range, 23.8–105.7 pmol/g globin) in nonsmoking control subjects ($n = 13$). The precision of the method was examined by analysing globin from three nonsmoking individuals: the mean background levels of HOEtVal (±SD) were 42.5 ± 7.3, 47.4 ± 12.8 and 59.2 ± 5.3 pmol/g globin ($n = 6$). It would be of great interest to study the contribution to these background levels of other sources of hydroxyethylating agents in the diet and the environment (e.g., urban pollution, sidestream tobacco smoke). Such investigations may require further improvements in the precision and accuracy of determinations at low levels.

Acknowledgement

The authors wish to thank J.H. Lamb for assistance in the GC-MS analyses.

References

Binder, H. & Lindner, W. (1972) Bestimmung von Äthylenoxid im Rauch garantiert unbegaster Zigarettes. *Fachliche Mitt. Austria Tabakwerke AG, 13,* 215–220

Calleman, C.J., Ehrenberg, L., Jansson, B., Osterman-Golkar, S., Segerbäck, D., Svensson, K. & Wachtmeister, C.A. (1978) Monitoring and risk assessment by means of alkyl groups in hemoglobin in persons occupationally exposed to ethylene oxide. *J. environ. Pathol. Toxicol., 2,* 427–442

Ehrenberg, L., Osterman-Golkar, S., Segerbäck, D., Svensson, K. & Calleman, C.J. (1977) Evaluation of genetic risks of alkylating agents. III. Alkylation of haemoglobin after metabolic conversion of ethene to ethene oxide *in vivo. Mutat. Res., 45,* 175–184

Elmenhorst, H. & Schultz, C. (1968) Flüchtige Inhaltsstoffe des Tabakrauches. *Beitr. Tabakforsch., 4,* 90–122

Farmer, P.B., Bailey, E., Gorf, S.M., Törnqvist, M., Osterman-Golkar, S., Kautiainen, A. & Lewis-Enright, D.P. (1986) Monitoring human exposure to ethylene oxide by the determination of haemoglobin adducts using gas chromatography-mass spectrometry. *Carcinogenesis, 7,* 637–640

Farrelly, J.G., Thomas, B.J. & Lijinsky, W. (1987) *Metabolism and cellular interactions of* N-*nitrosodiethanolamine.* In: Bartsch, H., O'Neill, I.K. & Schulte-Hermann, R., eds, *The Relevance of* N-*Nitroso Compounds to Human Cancer: Exposures and Mechanisms* (*IARC Scientific Publications No. 84*), Lyon, International Agency for Research on Cancer, pp. 87–90

Hoffmann, D., Brunnemann, K.D., Adams, J.D. & Hecht, S.S. (1984) *Formation and analysis of* N-*nitrosamines in tobacco products and their endogenous formation in tobacco consumers.* In: O'Neill, I.K., von Borstel, R.C., Miller, C.T., Long, J. & Bartsch, H., eds, N-*Nitroso Compounds: Occurrence, Biological Effects and Relevance to Human Cancer* (*IARC Scientific Publications No. 57*), Lyon, International Agency for Research on Cancer, pp. 743–762

IARC (1986) *IARC Monographs on the Evaluation of the Carcinogenic Risk of Chemicals to Humans,* Vol. 38, *Tobacco Smoking,* Lyon, pp. 166–167

Osterman-Golkar, S., Ehrenberg, L., Segerbäck, D. & Hällström, I. (1976) Evaluation of genetic risks of alkylating agents. II. Haemoglobin as a dose monitor. *Mutat. Res., 34,* 1–10

Osterman-Golkar, S., Farmer, P.B., Segerbäck, D., Bailey, E., Calleman, C.J., Svensson, K. & Ehrenberg, L. (1983) Dosimetry of ethylene oxide in the rat by quantitation of alkylated histidine in hemoglobin. *Teratog. Carcinog. Mutag., 3,* 395–405

Segerbäck, D. (1983) Alkylation of DNA and hemoglobin in the mouse following exposure to ethene and ethene oxide. *Chem.-biol. Interactions, 45,* 139–151

Törnqvist, M., Mowrer, J., Jensen, S. & Ehrenberg, L. (1986a) Monitoring of environmental cancer initiators through hemoglobin adducts by a modified Edman degradation method. *Anal. Biochem., 154,* 255–266

Törnqvist, M., Osterman-Golkar, S., Kautiainen, A., Jensen, S., Farmer, P.B. & Ehrenberg, L. (1986b) Tissue doses of ethylene oxide in cigarette smokers determined from adduct levels in hemoglobin. *Carcinogenesis, 7,* 1519–1521

Van Sittert, N.J., De Jong, G., Clam, M.G., Davies, R., Dean, B.J., Wren, C.J. & Wright, A.S. (1985) Cytogenetic, immunological and haematological effects in workers in an ethylene oxide manufacturing plant. *Br. J. ind. Med., 42,* 19–26

IMMUNOCYTOCHEMICAL ANALYSIS OF DNA ADDUCTS IN SINGLE CELLS: A NEW TOOL FOR EXPERIMENTAL CARCINOGENESIS, CHEMOTHERAPY AND MOLECULAR EPIDEMIOLOGY

E. Scherer, J. Van Benthem, P.M.A.B. Terheggen, E. Vermeulen, H.H.K. Winterwerp & L. Den Engelse

Division of Chemical Carcinogenesis, The Netherlands Cancer Institute, Amsterdam, The Netherlands

The immunocytochemical staining of carcinogen-DNA adducts by a double peroxidase-anti-peroxidase (PAP) method is critically described. It is a powerful new tool for the investigation of the initial processes of chemical carcinogenesis — such as metabolic activation of carcinogens, and modification/repair of DNA — at the level of individual, putative target cell types. It is the method of choice if the cell populations are too small for determination of adducts in isolated DNA, or if information on the tissue distribution of DNA damage is needed. Advantages of the peroxidase staining endpoint over immunofluorescence are its stability on storage of slides and the possibility of evaluation by a conventional microscope. First attempts to quantify staining intensity by microdensitometric equipment are described.

Since our first publications on the double PAP immunocytochemical method for the visualization of carcinogen-DNA adducts in tissue sections (Heyting *et al.*, 1983; Menkveld *et al.*, 1985), several steps have been changed in order to improve the sensitivity and tissue preservation. In the present communication, we describe and comment on the currently used protocol and report on attempts to quantify the immunocytochemical staining intensity.

Mounting of tissue specimens for immunocytochemistry

Tissues to be compared are mounted on the same block and frozen quickly by placing the aluminium tissue holder on dry ice. Blocks are packed in aluminium sheet and stored at −80°C until cut. Although freeze substitution techniques can be used (Menkveld *et al.*, 1985), we have concentrated on cryostat sections for practical reasons.

Coating of microscope slides

Precleaned slides (ColorFrost, Menzel, Federal Republic of Germany) are used without further purification. Slides are coated with 0.5% egg albumin (in demineralized water, 60 μl for 2.6 × 6 cm), quickly dried on a hot metal plate (60°C), baked overnight at 80°C, and stored at room temperature. The relatively thick ovalbumin layer is necessary to prevent loss of sections during treatment with sodium hydroxide.

Cutting of cryostat sections

Sections of 5–10 μm are thawed on ovalbumin-coated slides and quickly dried in a stream of air. The dry sections (packed back-to-back in aluminium sheet) are stored at −80°C. When sections are removed from the freezer, condensation of water should be avoided, and the sections must be processed immediately.

Buffers and media used

Phosphate-buffered saline (PBS): 140 mM NaCl in 10 mM phosphate buffer pH 7.4; RNase buffer: 10 mM Tris, 10 mM EDTA, pH 8.0; incubation medium: 10% normal goat serum (inactivated at 56°C, 20 min) in PBS containing 0.04% Triton X-100; wash buffer: 50 mM Tris, 150 mM NaCl, 0.25% gelatine, 5 mM EDTA, 0.05% Triton X-100, pH 7.4; 3,3-diaminobenzidine (DAB) buffer: 50 mM Tris-HCl, 10 mM imidazole, pH 7.5; DAB reagent: DAB buffer containing 0.05% H_2O_2, 0.5 mg/ml DAB · 4 HCl (Sigma, St Louis, MO, USA)

Pretreatment of cryostat sections

Dry sections are placed for 45 min in dry methanol containing 0.3% H_2O_2, rehydrated in decreasing concentrations of ethanol in demineralized water (90%, 60%, 30% H_2O), placed in RNase buffer for 5 min and incubated for 1 h at 37°C in RNase buffer containing 200 μg/ml RNase A (bovine pancreas Type 1-A; Sigma) and 50 U/ml RNase T1 (from *Aspergillus oryzae*; Boehringer, Mannheim, Federal Republic of Germany). RNase A is predigested at 70°C for 20 min at a stock concentration of 40 mg/ml in PBS. Sections are then washed in demineralized water and placed for 1 min in 40% ethanol, followed by 10 min in 50 mM NaOH in 40% ethanol. Subsequently the sections are rinsed briefly in 40% ethanol and in 5% acetic acid in 40% ethanol (essential to neutralize remaining NaOH), carefully washed with demineralized water, placed for 5 min in wash buffer, and rinsed in PBS.

The methanol/H_2O_2 step serves to inactivate endogenous tissue peroxidase and fixes the section. Attempts to use aldehyde fixation were unsuccessful, as modified nuclei could no longer be stained immunocytochemically. The RNase step might be omitted in view of the high sensitivity of RNA to alkali and the alcoholic NaOH step which follows. Until now, the alkali step has proven to be essential for good immunocytochemical results. This may be related to observations that all the antibodies we have used show a much higher affinity for single-stranded than for double-stranded DNA. Another explanation might be that the DNA is made more accessible by removal of part of the nuclear proteins. It is a critical step, since sections may be lost if the slides were not really clean on coating or the layer of ovalbumin was too thin. In the original protocol, this step was carried out for only 30 sec in aqueous 70 mM NaOH; however, Feulgen staining indicated a considerable loss (about 70%) of DNA. An additional argument was that penetration of the section would be far from complete in 30 sec. At lower aqueous alkali concentrations and longer incubation times, the loss of DNA was diminished, but the immunostaining intensity also decreased dramatically. We therefore tried alcoholic NaOH solutions; the ethanol concentration was tested between 20 and 70%, at incubation times of up to 30 min. According to Feulgen staining, DNA loss was acceptable (0–20%) at ethanol concentrations ⩾30%; immunostaining intensity decreased at ethanol concentrations ⩾60%. We chose 40% ethanol since both small and large nuclei in the rat liver proved to be optimally accessible to immunostaining at that concentration. For quantitative immunocytochemistry, it remains to be examined whether the optimal NaOH conditions depend on the concentration of DNA in nuclei, on the cell type, or on a cut nuclear membrane.

When formamide (50 to 100%, 50°–80°C, 10 min) was used instead of NaOH to denature DNA, no immunostaining was obtained. Because of the acid sensitivity of many DNA modifications, we did not incorporate a HCl step (Wynford-Thomas & Williams, 1986; Yang *et al.*, 1987) in our staining procedure.

Incubation with immunoreagents

Incubations are performed in drops of immunoreagent on horizontally placed slides; spreading of the drop over the slide is prevented by drawing a rim of vacuum grease around the section; drying out of the section is avoided by incubation in moisture chambers. This type of incubation gave good results in semiquantitative staining. For quantitative purposes, it is probably less suitable, since rate-limiting concentrations of immunoreagents are used. This implies that the concentration of immunoreagent during incubation is not identical for all sections processed but depends on the amount of modification present. We therefore developed incubation chambers in which sections on two opposing slides are reacted simultaneously with the same slowly mixing immunoreagent. This type of incubation is currently being used for reaction of all slides in a run (up to 30) with the same solutions. The concentrations are kept as low as possible in order that nuclear background staining be very low to invisible. Typical dilutions for good rabbit antisera raised against DNA modifications are 1:10 000 to 1:50 000. The second antibody and the PAP complex are also highly diluted [goat-anti-rabbit, Campro Benelux G-102, at 1:640 and PAP (rabbit) complex, American Qualex, at 1:3200]. Of the many batches tested (second antibody or PAP), only a few were suitable (high titre, low background) for our purposes.

Antiserum specific for carcinogen-DNA adducts

The sections are preincubated (to block nonspecific binding) for 1 h at 37°C in 150 μl incubation medium placed on each slide; the incubation medium is removed (no washing step); 150 μl of the first antiserum (suitably diluted in incubation medium) are added; and incubation is performed overnight at 4°C. The sections are washed gently—twice up and down for each step—in PBS, in wash buffer and in PBS. Diluted rabbit preimmunserum is used as a control. Overnight incubation at 4°C led to better results than incubation for 2 h at 37°C.

Peroxidase staining

The sections are incubated for 45 min at 37°C with 150 μl goat-anti-rabbit serum diluted in incubation medium, gently washed in PBS, wash buffer and PBS (as above), incubated for 45 min at 37°C with 150 μl PAP complex diluted in incubation medium, and gently washed in PBS, wash buffer, PBS (as above). These steps are repeated once (double PAP). The colour reaction is performed in the conventional way: the sections are placed for 5 min in DAB buffer, for 5 min in DAB buffer containing 0.5 mg/ml DAB, incubated for 15 min in the dark with 150 μl of 0.5 mg/ml DAB, 0.05% H_2O_2 in DAB buffer, washed five times in demineralized water, dehydrated through increasing concentrations of ethanol, placed for 5 min in two changes of xylene, and mounted in DePeX (BDH Chemicals Ltd, Poole, UK). If incubation chambers are used, an equivalent volume per slide is added to the chamber at each incubation step. Slow rotation of the partially filled chamber ensures equal distribution of the immunoreagent all over the sections.

It cannot be excluded that the immunostaining obtained under these conditions is not equal for modified DNA in cut and uncut nuclei or in the upper and lower parts of the cryostat section due to collapse of the sections during air drying, resulting in a much denser layer of proteins and nucleic acids, eventually hampering penetration of the immunoreagents and the substrate for the peroxidase reaction. This complication should be absent with paraffin- or plastic-embedded sections.

Quantification of immunostaining intensity

In collaboration with Dr J.S. Ploem (State University, Leiden, The Netherlands), the quantitative aspects of immunocytochemical staining for carcinogen-DNA adducts in tissue sections are currently under investigation. Computer-assisted microdensitometric scanning equipment is used, based on the HIDACSYS system developed at Leiden University (Van der Ploeg *et al.*, 1977). A modern version of this program for Atari ST microcomputers (Atari, Sunnyvale, USA) has been purchased by Microscan (Leiden, The Netherlands). Its graphic user interphase (Fig. 1) greatly facilitates the operation and the evaluation of the scanned pictures.

Semiquantitative data have recently been collected using this equipment on *N*-nitrosomethylbenzylamine-modified oesophageal epithelium (compare Van Benthem *et al.*, this volume, and Fig. 1), on human buccal cells treated either *in vitro* (Fig. 2) or *in vivo* with cisplatin, and on rat liver modified *in vivo* with aflatoxin B_1 (C.P. Wild, unpublished). These results are promising, since clear dose-staining intensity relationships have been observed; in the case of aflatoxin B_1, the immunocytochemical data fitted well with data obtained for isolated DNA after modification with radiolabelled aflatoxin.

The immunocytochemical analysis of carcinogen-DNA adducts makes it possible to investigate the initial processes of chemical carcinogenesis — such as metabolic activation of carcinogens, as well as modification and repair of DNA — at the level of

Fig. 1. Scanning of a cryostat section from rat oesophagus modified *in vivo* with *N*-nitrosomethylbenzylamine (4 mg/kg) and stained immunocytochemically for the presence of O^6-methylguanine in DNA

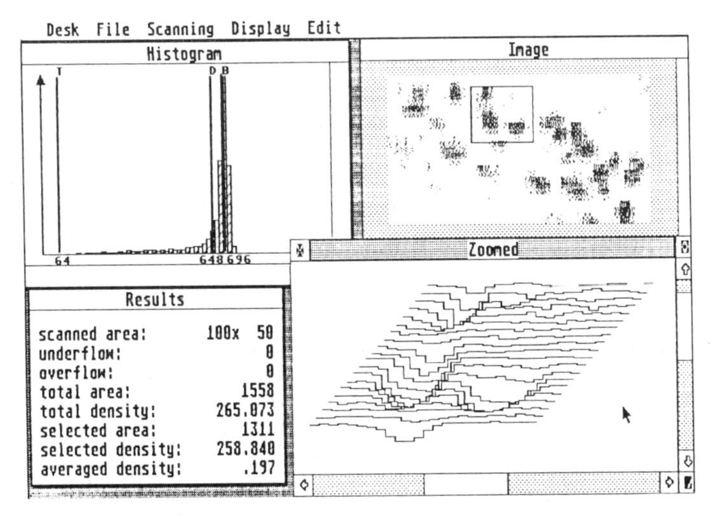

The scanned area, showing the mucosal epithelium, is displayed as pixels in the image window and as a 3-D plot in the zoomed window. Results are also displayed as a histogram and as calculated densities.

Fig. 2. Quantification of staining intensity of human buccal cells treated *in vitro* (A) with various concentrations of cisplatin for 1 h, or (B) for 1–6 h with 2 µg cisplatin/ml

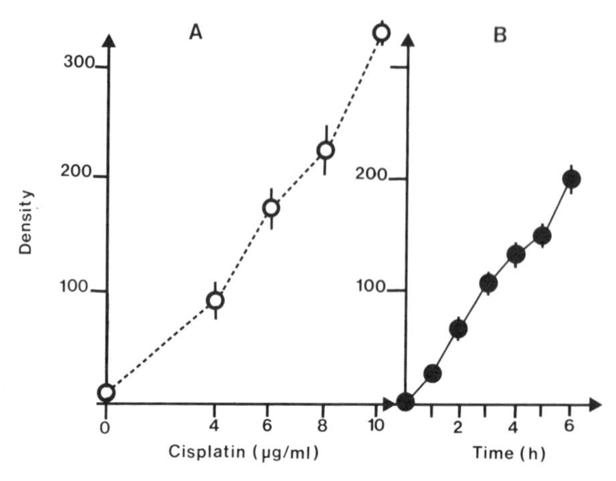

The cells were sedimented onto coated slides and processed immunocytochemically using a rabbit antiserum raised against cisplatin-modified DNA.

the putative target cell type, even when target cell populations are small (e.g., a few hundred buccal cells or tracheal glands) or confined to a particular differentiation state (centrilobular and periportal hepatocytes, Menkveld *et al.*, 1985). In comparison with other cytological methods, such as autoradiography after application of radioactively labelled carcinogens (Johansson & Tjälve, 1978; Kraft & Tannenbaum, 1980; P. Kleihues, personal communication), immunocytochemistry offers not only superior sensitivity and specificity in visualizing well-defined DNA modifications (instead of reaction mostly with proteins), but also better cytological resolution. It has been possible to visualize in rodent tissue sections all the DNA adducts for which a potent antiserum is available: guanine modified at the 7- or O^6-position by a methyl or ethyl group, guanine modified by 2-acetylaminofluorene, aflatoxin B_1 or benzo[*a*]pyrene, or cisplatin-modified DNA (Terheggen *et al.*, 1987). The sensitivity obtained depends on the DNA adduct and on the quality of the antiserum and is in the range of one modified base per 10^5 to 10^7 unmodified DNA bases. The sensitivity is sufficient to allow the experimental study of body distribution, metabolic activation and DNA modification and repair at relevant doses of carcinogens and other DNA-modifying agents.

References

Heyting, C., Van Der Laken, C.J., Van Raamsdonk, W. & Pool, C. W. (1983) Immunohistochemical detection of O^6-ethyldeoxyguanosine in the rat brain after *in vivo* applications of *N*-ethyl-*N*-nitrosourea. *Cancer Res.*, *43*, 2935–2941

Johansson, E.B. & Tjälve, H. (1978) The distribution of [^{14}C]dimethylnitrosamine in mice. Autoradiographic studies in mice with inhibited and non-inhibited dimethylnitrosamine metabolism and a comparison with the distribution of [^{14}C]formaldehyde. *Toxicol. appl. Pharmacol.*, *45*, 565–575

Kraft, P.L. & Tannenbaum, S.R. (1980) The distribution of *N*-nitrosomethylbenzylamine evaluated by whole-body radioautography and densitometry. *Cancer Res.*, *40*, 1921–1927

Menkveld, G.J., Van Der Laken, C.J., Hermsen, T., Kriek, E., Scherer, E. & Den Engelse, L. (1985) Immunohistochemical localization of O^6-ethyldeoxyguanosine and deoxyguanosin-8-yl-(acetyl)aminofluorene in liver sections of rats treated with diethylnitrosamine, ethylnitrosourea or *N*-acetylaminofluorene, *Carcinogenesis, 6,* 263–270

Terheggen, P.M.A.B., Floot, B.G.J., Scherer, E., Begg, A.C., Fichtinger-Schepman, A.M.J. & Den Engelse, L. (1987) Immunocytochemical detection of interaction products of cis-diamminedichloroplatinum(II) and cis-diammine(1,1-cyclobutanedicarboxylato)-platinum(II) with DNA in rodent tissue sections. *Cancer Res., 47,* 6719–6725

Van Der Ploeg, M., Van den Broek, K., Smulders, A.W.M., Vossepoel, A.M. & Van Duin, P. (1977) HIDACSYS: computer programs for interactive scanning cytophotometry. *Histochemistry, 54,* 273–288

Wynford-Thomas, D. & Williams, E.D. (1986) Use of bromodeoxyuridine for cell kinetic studies in intact animals. *Cell Tissue Kinet., 19,* 179–182

Yang, X.Y., DeLeo, V. & Santella, R.M. (1987) Immunological detection and visualization of 8-methoxypsoralen-DNA photoadducts. *Cancer Res., 47,* 2451–2455

DETECTION OF O^4-ETHYLTHYMINE IN HUMAN LIVER DNA

N. Huh,[1] M.S. Satoh,[1] J. Shiga[2] & T. Kuroki[1]

[1]*Department of Cancer Cell Research, Institute of Medical Science, University of Tokyo, Tokyo; and* [2]*Department of Pathology, Faculty of Medicine, University of Tokyo, Tokyo, Japan*

O^4-Ethyl-2′-deoxythymidine (O^4-EtdThy) in human liver DNA was quantified, in order to monitor possible human exposure to ethylating agents, using a highly sensitive immunological detection method. In 30 of 33 cases analysed, O^4-EtdThy was detected at above the detection limit (i.e., 3×10^{-8} O^4-EtdThy/2′-deoxythymidine[dThy]), indicating actual chronic exposure to ethylating agents. The mean content of O^4-EtdThy in 19 cases of malignant tumours was significantly higher than that in 11 nonmalignant cases ($p < 0.05$).

Detection of DNA adducts in human materials is one of the most reliable parameters of exposure to environmental carcinogens (Perera, 1987). Preferable conditions for such studies include chemically and biologically stable DNA adducts (i.e., not repaired rapidly), DNA adducts relevant to neoplastic transformation, a highly sensitive method for detecting the adducts, absent or low cell proliferation in the tissues from which DNA is prepared (thus minimizing dilution by DNA replication), and tissues capable of metabolizing chemical carcinogens. In view of the high incidence of liver cancer in Japan, we examined the amount of O^4-EtdThy in human liver DNA obtained from 33 autopsy specimens from persons with malignant and nonmalignant diseases.

DNA extraction and quantification

DNA was extracted by a modification of the conventional phenol-chloroform method: 30–50 g liver tissue (nonmalignant portion in cases of liver cancer) were cut into thin slices and digested with 100 μg/ml proteinase K and 1% sodium dodecyl sulphate in six volumes of TE5 buffer (10 mM Tris/HCl, 5 mM EDTA, pH 7.5) at 37°C for at least 24 h. Undigested tissue fragments and high-molecular-weight substances were precipitated with 2.5 volumes of ethanol and 1/10 volume of 3 M NaCl, and digested again in 50 ml of the same solution for another 24 h. The resulting digests were extracted twice with phenol:chloroform:isoamyl alcohol (25:24:1) and twice with chloroform:isoamyl alcohol (24:1), and the DNA was precipitated with ethanol:NaCl. The precipitates were dissolved in 20 ml TE5, treated with a mixture of heat-treated RNase A (100 μg/ml) and RNase Tl (100 U/ml) at 37°C overnight, and again precipitated with ethanol:NaCl. The DNA was digested enzymatically to 2′-deoxynucleosides by successive treatments with DNase I (100 μg/ml), phosphodiesterase (100 μg/ml) and alkaline phosphatase (100 μg/ml), as described previously (Huh & Rajewsky, 1986). After hydrolysis, the enzyme protein was precipitated with 2.5 volumes of ethanol, and the supernatant was dried *in vacuo* and redissolved in TBS (10 mM Tris/HCl, 140 mM NaCl, 3 mM NaN_3, pH 7.5) for further analysis.

DNA hydrolysates were quantified by high-performance liquid chromatography (HPLC) using a 4 × 150-mm reverse-phase column (Lichrosorb RP18, 5 μm; Merck, FRG). The elution profile was a linearly increasing concentration of methanol from 5% to 15% in 0.1 M ammonium formate buffer, pH 5.0 for the first 15 min, and then from 15% to 40% in the buffer for 5 min. The content of dThy in the samples was determined by comparing the peak area with that of the authentic nucleoside, and the amount of DNA was calculated by assuming that 1 mg human DNA contains 0.971 μmol dThy. The yield of DNA from human liver tissues varied from 0.5–1.0 mg per g wet weight, partly depending on the extent of fibrosis and autolysis.

Fractionation and radioimmunoassay of O^4-EtdThy

Since the O^4-EtdThy content of the samples was very low, the DNA hydrolysates were fractionated by HPLC before radioimmunoassay (RIA). The DNA hydrolysates were injected into an HPLC system equipped with 2-ml sample loop and two 4 × 250-mm reverse-phase columns (Lichrosorb RP18, 10 μm, Merck, FRG) in series. The columns were eluted with linearly increasing concentrations of methanol from 25% to 40% in 0.1N ammonium acetate, pH 5.0, for 35 min and 100% methanol for 10 min at 40°C. The flow rate was 1 ml/min, and the fraction size was 0.5 min/fraction. The retention times and peak profiles of authentic ^{3}H-O^4-EtdThy and of ^{3}H-O^6-ethyl-2′-deoxyguanosine (O^6-EtdGuo) were not changed even when 20-50 mg of DNA hydrolysates were injected simultaneously. Fractions (0.5 min/fraction) corresponding to O^4-EtdThy were pooled and dried *in vacuo*. O^4-EtdThy was quantified by a competitive RIA using anti-O^4-EtdThy monoclonal antibody, ER-O-1 (affinity constant for O^4-EtdThy, 1.3×10^9 l/mol; Adamkiewicz *et al.*, 1982), as described by Swenberg *et al.* (1984). The content of O^4-EtdThy in the sample DNA was expressed as the molar ratio of O^4-EtdThy:dThy. The detection limit of O^4-EtdThy in our standard RIA was 6×10^{-13} mol by reading 25% inhibition of antibody tracer binding on the standard curve, corresponding 3×10^{-8} O^4-EtdThy:dThy ratio when 20 mg DNA were available. The antibody, ER-O-1, authentic O^4-EtdThy and O^4-ethyl[6-^{3}H]-deoxythymidine (17 Ci/mmol) were provided by M.F. Rajewsky (University of Essen, FRG) in the framework of mutual collaboration between the two laboratories.

O^4-EtdThy in human liver DNA

We analysed liver DNA from 13 liver cancer cases, eight cases of cancer other than the liver, and 12 nonmalignant cases (Table 1). O^4-EtdThy was detected in all cases but one in each group. All of the nonmalignant cases had a O^4-EtdThy:dThy ratio below 3×10^{-7}, with a mean value of 1.17×10^{-7} (among 11 cases). However, five of the 13 liver cancer cases and three of the eight cases of cancer other than liver had O^4-EtdThy levels significantly higher than 3×10^{-7}. The highest ratio, 20.6×10^{-7}, was found in a case of renal cancer. A comparison of the O^4-EtdThy levels in 12 liver cancer cases, or in 19 total malignant cases, with those in 11 nonmalignant cases showed a statistically significant difference ($p < 0.05$; Welch's t-test). The difference between cancer cases other than the liver and nonmalignant cases was not significant, possibly because only a small number of cases was analysed.

Implications

O^4-EtdThy was removed hardly at all and therefore accumulated to a level of 10^{-5} O^4-EtdThy:dThy in rat liver DNA during continuous feeding of rats with 2.5–3.5 mg/kg per day N-nitrosodiethylamine, while the O^6-EtdGuo content remained at a

Table 1. O^4-EtdThy in human liver DNA

Cases	No. of cases analysed	O^4-EtdThy:dThy ($\times 10^7$)		
		No. of cases above detection limit[a]	Mean ± SD[b]	Range
I Liver cancer	13	12	3.99 ± 4.02	0.36–11.3
II Cancer in other organs	8	7	5.43 ± 7.40	0.34–20.6
III Nonmalignancy	12	11	1.17 ± 0.65	0.34–2.58

[a] Detection limit for O^4-EtdThy:dThy, 0.3×10^{-7}
[b] Difference between I and III: $p < 0.05$; difference between I + II and III: $p < 0.05$

level nearly two orders of magnitude lower than that of O^4-EtdThy (Swenberg *et al.*, 1984). Thus, O^4-EtdThy, and not O^6-EtdGuo, is considered to be relevant to induction of liver cancer under the experimental conditions, in which a small dose of the carcinogen was applied chronically. The lower O^6-EtdGuo content is due to active removal from rat liver, because *N*-nitrosodiethylamine is known to form amounts of O^6-EtdGuo nearly one order of magnitude higher than O^4-EtdThy. Umbenhauer *et al.* (1985) detected O^6-methyl-2′-deoxyguanosine, but not O^6-EtdGuo, in human oesophageal tissues, with a detection limit of 2×10^{-8} O^6-EtdGuo:dGuo. These and our present results indicate that humans are exposed chronically to small amounts of ethylating agents, eventually giving rise to significant accumulation of O^4-EtdThy, a premutagenic and pretransformational DNA adduct.

Although a statistically significant difference in O^4-EtdThy levels was found between the malignant and nonmalignant groups in this study, the limited number of cases analysed and the long latent period of human cancer preclude any definite assessment of a causal link between higher O^4-EtdThy level and induction of cancer. However, the biological behaviour of O^4-EtdThy and the availability of a highly sensitive detection method certainly provide a promising tool for molecular epidemiological analysis in the near future.

Acknowledgements

This work was performed in collaboration with Dr M.F. Rajewsky, University of Essen, Federal Republic of Germany, and supported in part by a Grant for Cancer Research from the Ministry of Education, Science, and Culture of Japan.

References

Adamkiewicz, J., Drosdziok, W., Eberhardt, W., Langenberg, U. & Rajewsky, M.F. (1982) *High-affinity monoclonal antibodies specific for DNA components structurally modified by alkylating agents.* In: Bridges, B.A., Butterworth, B.E. & Weinstein, I.B., eds, *Indicators of Genotoxic Exposure* (*Banbury Report 13*), Cold Spring Harbor, NY, CSH Press, pp. 265–267

Huh, N. & Rajewsky, M.F. (1986) Enzymatic elimination of O^6-ethylguanine and stability of O^4-ethylthymine in the DNA of malignant neural cell lines exposed to *N*-ethyl-*N*-nitrosourea in culture. *Carcinogenesis, 7,* 435–439

Perera, F.P. (1987) Molecular cancer epidemiology: a new tool in cancer prevention. *J. natl Cancer Inst., 78,* 887–898

Swenberg, J.A., Dyroff, M.C., Bedell, M.A., Popp, J.A., Huh, N., Kirstein, U. & Rajewsky, M.F. (1984) O^4-Ethyldeoxythymidine, but not O^6-ethyldeoxyguanosine, accumulates in hepatocyte DNA of rats exposed continuously to diethylnitrosamine. *Proc. natl Acad. Sci. USA, 81,* 1692–1695

Umbenhauer, D., Wild, C.P., Montesano, R., Saffhill, R., Boyle, J.M., Huh, N., Kirstein, U., Thomale, J., Rajewsky, M.F. & Lu, S.H. (1985) O^6-Methyldeoxyguanosine in oesophageal DNA among individuals at high risk of oesophageal cancer. *Int. J. Cancer, 36,* 661–665

DETERMINATION OF 7-METHYLGUANINE BY IMMUNOASSAY

D.E.G. Shuker

International Agency for Research on Cancer, Lyon, France

Antisera to 7-methylguanine (7-meGua) were obtained using a novel analogue of 7-meGua which was bound covalently to methylated bovine serum albumin (mBSA) and was used as an antigen. Rabbit antisera at high dilution (1 in 10^5) recognized 7-meGua-ovalbumin, and the free base itself was detected at low levels (1 pmol/well) in a competitive enzyme-linked immunosorbent assay (ELISA). A number of purines have been screened for cross-reactivity, and only those of closely related structure (e.g., 7-methylxanthine) cross-reacted to any appreciable extent (50% inhibition at 250 pmol/well). The ELISA procedure was used to quantify 7-meGua in DNA following neutral thermal hydrolysis. Currently, the sensitivity of this assay is of the order of 10–100 µmol 7-meGua/mol Gua.

Antibodies have been raised against many carcinogen-DNA adducts (Perera & Weinstein, 1982). The main methods available involve use of antigens based on modified DNA (e.g., melphalan-DNA) or nucleoside-based antigens derived from stable adducts (e.g., O^6-methylguanosine). For bulky adducts, antibodies that recognize the carcinogen cross-react sufficiently well with the adduct to provide useful immunoassays (e.g., aflatoxins, benzo[*a*]pyrene). In the case of low-molecular-weight adducts (e.g., methyl), the major products in DNA are unstable lesions which are rapidly repaired to give the corresponding alkylated base (e.g., 7-methyldeoxyguanosine→7-meGua and 3-methyldeoxyadenosine→3-methyladenine [3-meAde]). It is, therefore, of some interest to raise antibodies that are capable of recognizing such small purine-based adducts.

A strategy has been devised in which 7-alkylguanine and 3-alkyladenine analogues are used to prepare purine-protein antigens. This approach has been used, so far, for 3-meAde (Shuker & Farmer, this volume) and 7-meGua. The preliminary results for 7-meGua are described in this paper.

Preparation of antisera to 7-meGua

In order to optimize the possibility of obtaining specific antisera for such a small hapten as 7-meGua, the antigen (i.e., hapten + protein) was prepared using an analogue of the purine, namely, *N*2-carboxymethyl-7-methylguanine (CM-7meGua; D. Shuker, manuscript in preparation). The structures of 7-meGua and CM-7-meGua are shown in Figure 1. The synthesis of the antigen is summarized in Figure 2.

Two rabbits were immunized with 7-meGua-mBSA [0.5 mg adsorbed onto aluminium hydroxide in complete Freund's adjuvant (1 ml), prepared according to Muller and Rajewsky, 1980] by subcutaneous injection at multiple sites on the shaved back. After eight weeks, a booster dose of the same amount of antigen in adjuvant, without aluminium hydroxide, was given by the same route. At 16 weeks, each rabbit received

Fig. 1. Structures of 7-meGua and CM-7-meGua

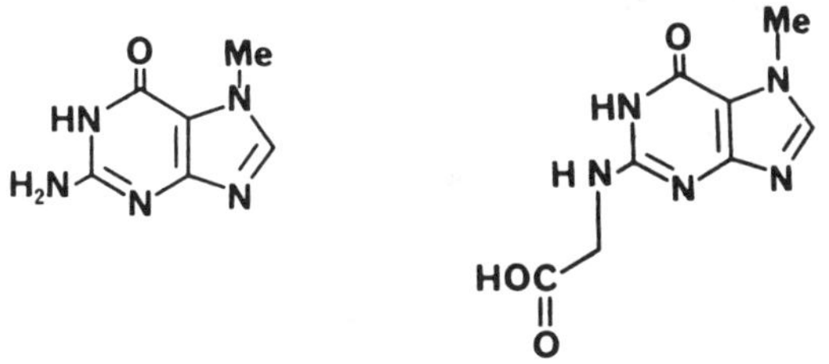

the same amount of antigen in incomplete Freund's adjuvant injected into the hindquarters. One week after the booster injection, both rabbits were bled from the lateral ear vein (local anaesthesia) and serum was prepared. Sera were assayed for anti-7-meGua activity by ELISA, using a checkerboard procedure: 96-well plates were coated with 7-meGua-ovalbumin over a range from 1 ng to 10 μg per well, and sera were diluted with phosphate-buffered saline from 1 in 10 to 1 in 10^6. Binding of rabbit immunoglobulin G was detected using horseradish peroxidase-linked goat anti-rabbit immunoglobulin G with 3,3′,5,5′-tetramethylbenzidine as substrate. Both rabbits produced active antisera of approximately equal titre.

ELISA for 7-meGua

The checkerboard procedure established conditions for optimal colour development (with a limiting dilution of antiserum) of 4 ng coating antigen per well and a working serum dilution of 1 in 10^5. Unmodified ovalbumin did not show appreciable binding.

The ability of the antisera to detect 7-meGua was determined using a competitive ELISA procedure, in which solutions of 7-meGua were added to antisera just prior to plating out. A standard curve for antiserum 1 is shown in Figure 3 (antiserum 2 gave an almost identical curve). 7-MeGua was reliably detected (>20% inhibition) at 1 pmol per well with a 50% inhibition at 10 pmol per well. A number of structurally related purines have been examined for cross-rcactivity (Table 1), and only those of closely related structure cross-react to any appreciable extent. Interestingly, the modified analogue, CM-7-meGua was detected at much lower levels than 7-meGua. Also, 7-methylxanthine was detected quite readily, whereas 7-methyluric acid was barely detected, indicating that the intact 7-methylimidazole moiety is important in the recognition.

Fig. 2. Synthesis of 7-meGua-protein conjugates using the 7-meGua analogue, CM-7-meGua

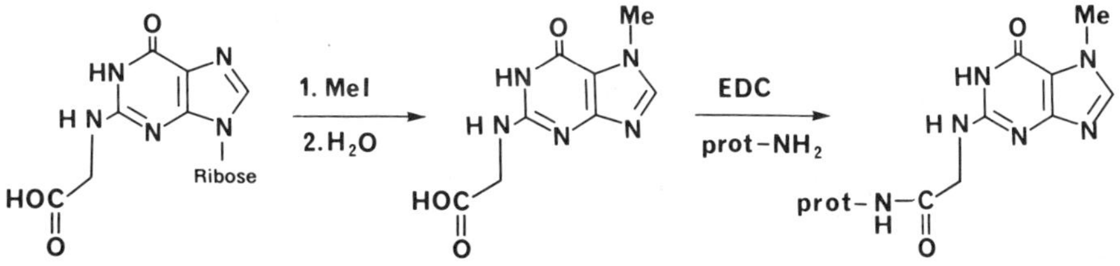

1. MeI
2. H_2O

EDC
prot-NH_2

Protein was either mBSA (for immunogen) or ovalbumin (for coating antigen).

Fig. 3. Inhibition curve for 7-meGua using antiserum 1 (1 in 10^5 dilution) and 7-meGua-ovalbumin (4 ng/well)

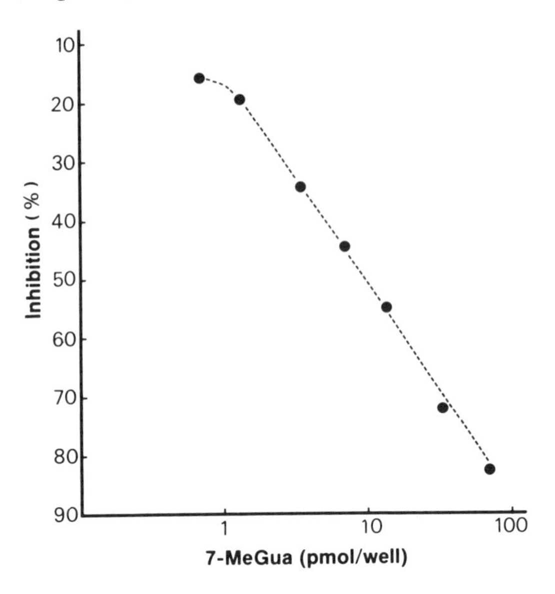

Determination of 7-meGua in DNA using ELISA

7-MeGua is readily removed from DNA by heating at neutral pH. A procedure has been developed in which 7-meGua in the supernatant is determined by ELISA (Fig. 4). Standard amounts of 7-meGua (0–50 pmol) were added to aliquots (125 μl) of DNA solution (2.4 mg/ml). The standard curves obtained in this way did not differ from those obtained with untreated 7-meGua standards, indicating that the presence of DNA does not interfere with the assay. At the level of detection of 1 pmol per well, the sensitivity of the assay corresponds to a level of modification of one 7-meGua residue per 10 000 guanine residues. This is clearly not sensitive enough for detection of methylation in human samples, where a sensitivity sufficient to detect one 7-meGua per 10^7 or 10^8 guanine residues is required. However, having shown that antibodies can

Table 1. Cross-reactivity of anti-7-meGua serum 1

Purine	Concentration for 50% inhibition (pmol/well)	Concentration for 20% inhibition (pmol/well)
CM-7-meGua	0.04	0.006
7-meGua	10	1.6
7-meXanthine	250	28
7-meUric acid	>10 000	>10 000
7-meAdenine	>10 000	2 500
Guanine	>10 000	3 100
Adenine	>10 000	6 300

Fig. 4. Flow chart for the determination of 7-meGua in DNA by ELISA

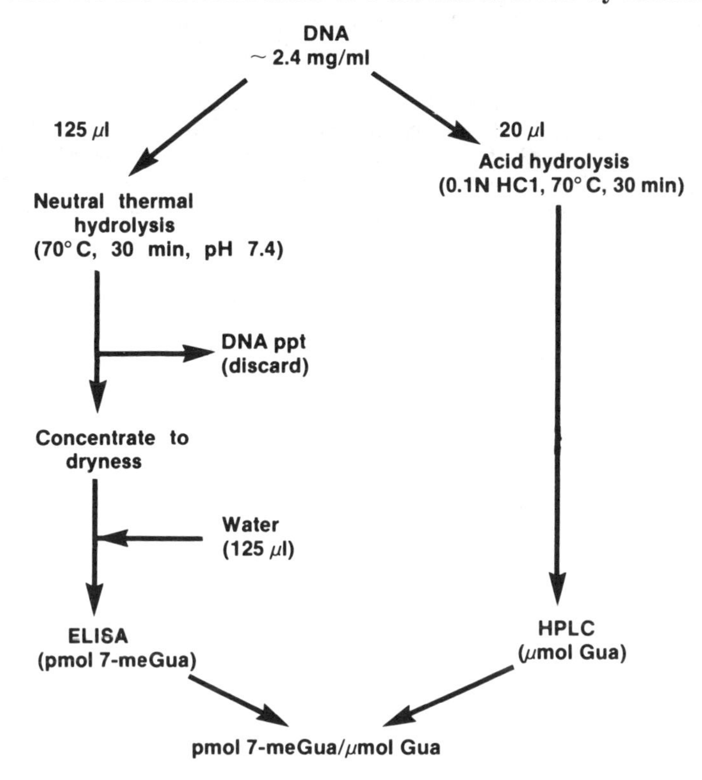

be raised against 7-meGua, a good, high affinity, monoclonal antibody might be obtained which would make possible much more sensitive assays.

Conclusions

Antisera have been raised against 7-meGua-protein conjugates which also recognize the free base. An ELISA procedure has been developed which can be used to detect reliably 1 pmol 7-meGua per well. This is a level of sensitivity approximately equal to that of a recently developed gas chromatography-mass spectrometry technique (Shuker *et al.*, 1984). The ELISA procedure was also used to determine excised 7-meGua in the presence of DNA.

The method for raising 7-meGua antisera is applicable to a range of 7-alkylguanines, and current work is directed towards producing a battery of specific antisera.

Acknowledgements

The author gratefully acknowledges the award of a Royal Society European Science Exchange Programme Fellowship and the continuing support and encouragement of Dr H. Bartsch, Chief, Unit of Environmental Carcinogens and Host Factors, IARC. It is also a pleasure to acknowledge the assistance of Mlle B. Chapot in the preparation of antisera and Dr C.P. Wild for stimulating discussions.

References

Müller, R. & Rajewsky, M.F. (1980) Immunological quantification by high affinity antibodies of O^6-ethyldeoxyguanosine in DNA exposed to *N*-ethyl-*N*-nitrosourea. *Cancer Res., 40,* 887–896

Perera, F.P. & Weinstein, I.B. (1982) Molecular epidemiology and carcinogen-DNA adducts: new approaches to studies of human cancer causation. *J. chron. Dis., 35,* 581–600

Shuker, D.E.G., Bailey, E., Gorf, S.M., Lamb, J. & Farmer, P.B. (1984) Determination of *N*-7-[$^2H_3$]methylguanine in rat urine by gas chromatography-mass spectrometry following administration of trideuteromethylating agents or precursors. *Anal. Biochem., 140,* 270–275

DETECTION OF O^6-METHYLGUANINE IN HUMAN DNA

R. Saffhill,[1] A.F. Badawi[1] & C.N. Hall[2]

[1]*Paterson Institute for Cancer Research, Christie Hospital and Holt Radium Institute; and* [2]*Manchester Royal Infirmary, Manchester, UK*

Using very specific antibodies in sensitive radioimmunoassays for O^6-methyldeoxyguanosine (O^6-medGuo), we have been able to detect the presence of this modification in human DNA. Since O^6-medGuo is not likely to be a normal component of DNA, its presence must be due to exposure to environmental alkylating agents. We have studied two groups of samples, one of which appears to have received exposure to an alkylating agent (so that most members of the group show a detectable level of O^6-medGuo). Although some individuals in the other group showed detectable O^6-medGuo (sometimes at levels exceeding those observed in the first group), the majority showed undetectable levels. This may indicate that they had much less exposure to environmental alkylating agents than the first group and that some individuals may have received additional exposures due to other factors such as life style or drugs that they may have been given.

With the development of immunological methods for the detection of DNA-carcinogen adducts, it has been possible to detect them at extremely low levels and to contemplate experimentation that would not be feasible by other methods. We have introduced very sensitive, specific radioimmunoassays (RIA) (Saffhill *et al.*, 1982; Wild *et al.*, 1983) for several alkylated DNA bases that are potential promutagenic lesions (Saffhill *et al.*, 1985). These involve mouse monoclonal antibodies with a high affinity for the specific adduct to be measured and show a very high level of selectivity in the presence of normal components of DNA and with other alkyl adducts that may be produced during reaction with an alkylating carcinogen. Using these techniques, we have studied the formation and repair of O^6-butylguanine and O^6-methylguanine (O^6-meGua) in the small amounts of DNA that can be extracted from cell cultures containing about 1×10^6 cells (Boyle *et al.*, 1986a,b, 1987) or from mitochondria (Myers *et al.*, 1986).

The monoclonal antibody specific for O^6-medGuo has an affinity constant of 3×10^9 l/mol. The most sensitive RIAs in which this antibody is used have a 50% inhibition value of 95 fmol and a 10% inhibition value of only 10 fmol (which is the practical limit of the assay). Other alkyl adducts have to be present at 10^5–10^6 fold, and the normal components of DNA by as much as 10^6, before they interfere with the assay (Wild *et al.*, 1983).

In addition to facilitating the analysis of small amounts of DNA, the high sensitivity of the RIA can be exploited to detect very low levels of reaction by analysing milligram quantities of DNA. Such low levels of reaction may be expected to arise from exposure of human populations to environmental alkylating agents. Using these methods, the presence of O^6-meGua has already been demonstrated in human DNA prepared from oesophagus and stomach tissues in an area of China (Linxian county) with a high incidence of oesophageal cancer (Yang, 1980; Umbenhauer *et al.*, 1985). However, in that study, it was difficult to obtain adequate controls from low-incidence areas, and the results from Linxian county were compared only with tissue samples from Europe.

With the possible exception of a case of suspected nitrosamine poisoning (Shank & Herron, 1982), there is little information available concerning the levels of O^6-meGua to be found in DNA prepared from human tissues. We now report the levels of O^6-meGua found in human DNA obtained from patients undergoing surgery in a region of south-east Asia, mostly for oesophageal and stomach cancer, or within the Manchester, UK, area, primarily for cancers of the stomach and bowel. A few post-mortem tissues were also included in the last group to investigate the possibility of analysing tissues from this source.

Experimental procedures

All but six tissue samples were removed during surgery; six samples were taken *post mortem* within 24 h of death. The tissues studied were oesophagus and stomach (from south-east Asia) and stomach and bowel (from the Manchester area). When possible, a region of uninvolved tissue was removed at the same time as the tumour for separate analysis. Tissues were frozen on dry ice and stored at −80°C. All experimental procedures have been described elsewhere and are only outlined here (Wild *et al.*, 1983; Umbenhauer *et al.*, 1985; Boyle *et al.*, 1986a,b; Myers *et al.*, 1986; R. Saffhill, S. Fida, M. Bromley & P.J. O'Connor, in preparation). DNA was prepared by phenol extraction, according to the method of Kirby and Cook (1957). The DNA was digested to nucleosides with DNase I, phosphodiesterase and alkaline phosphatase in the presence of 2′-deoxycoformycin. The latter is necessary to inhibit adenosine deaminase (Fox & Kelly, 1978), which demethylates O^6-medGuo (O'Connor & Saffhill, 1979) and can contaminate some DNA preparations. The hydrolysate is chromatographed on Aminex A-7 (25 × 1 cm) at 50°C, eluting with ammonium carbonate. On this column, at least 10 mg DNA may be chromatographed to separate the four deoxynucleosides of DNA, thus allowing base analysis and quantification of the DNA by spectrophotometric procedures, and separation of the O^6-medGuo. Since the four ribonucleotides from any contaminating RNA also separate from the deoxynucleosides, and O^6-merGuo separates from O^6-medGuo in this system, the effects of any contaminating RNA are also eliminated. The region of the chromatograph corresponding to O^6-medGuo is lyophilized and redissolved for determination by RIA. (The use of ammonium carbonate, which is volatile during the lyophilization, ensures that there is no concentration of buffer that may interfere with the RIA.) Results are expressed as the ratio of O^6-medGuo, determined by RIA, to deoxyadenosine, determined by ultraviolet spectrophotometry. In practice, we have found the determination of deoxyadenosine to be more reproducible and reliable than that of the three other deoxynucleosides.

Levels of O^6-methylguanine observed

The results of the analyses of the DNA samples from south-east Asia are summarized in Table 1. The levels of alkylation observed were of similar magnitude to those observed in samples from an area of high oesophageal cancer incidence within China (Umbenhauer *et al.*, 1985), O^6-meGua being detectable in most samples. No difference in the level of O^6-meGua was found between oesophagus and stomach, for malignant or benign tumour, or uninvolved tissue.

The results of the analyses of the DNA samples from the Manchester area are summarized in Table 2. As in tissue samples from Europe analysed previously

Table 1. Presence of O^6-meGua in tissue DNA from south-east Asia

Amount of O^6-meGua $\times 10^6$	No. of samples
None detected	5
0.016–0.020	6
0.021–0.040	9
0.041–0.060	16
0.061–0.080	9
Total samples	45

Mean value of detectable O^6-meGua/dA = $0.041 \pm 0.018 \times 10^{-6}$

(Umbenhauer *et al.*, 1985), the levels of O^6-meGua were much lower than those from Asia and were undetectable in most samples. However, the promutagenic lesion was detected in some samples and in a few cases at very high levels (up to a ratio of O^6-meGua:deoxyadenosine of $>0.2 \times 10^{-6}$). Of the six tissue samples taken *post mortem*, two did not yield DNA, which compares favourably with the five surgical samples (out of 59) which yielded no DNA. This source of samples is thus interesting for further investigations.

Discussion

These results, together with those reported previously (Umbenhauer *et al.*, 1985), show that O^6-meGua may be detected in human DNA. In general, the results obtained from the two sources reported here agree with those reported from south-east Asia, in which O^6-meGua occurred at consistently detectable levels. In many of the tissues from the Manchester area there were undetectable levels of O^6-meGua (limit, O^6-meGua:deoxyadenosine, $<0.01 \times 10^{-6}$). However, a number of samples had detectable levels of O^6-meGua, which, in a few instances, exceeded those in the tissues from south-east Asia.

Table 2. Presence of O^6-meGua in tissue DNA from Manchester

O^6-meGua:deoxyadenosine $\times 10^6$	No. of samples
None detected	28
0.011–0.020	4[a]
0.021–0.040	4
0.041–0.060	2
0.061–0.080	4
0.081–0.100	3
0.101–0.200	5
>0.200	3
DNA not precipitated	7
Too little DNA to analyse[b]	4

[a]Includes the pooled sample, see footnote b
[b]<1 mg DNA obtained; samples were pooled and analysed as one

Since O^6-meGua is not known to be a normal component of DNA, we presume that it has arisen in these samples as the result of exposure to an alkylating agent. (It is reasonable to expect that O^6-meGua is not a normal component of DNA due to its promutagenic nature (Saffhill *et al.*, 1985) and the effects of DNA precursor pools in determining the mutagenic efficiency (Saffhill, 1986).) The most likely environmental alkylating agents are nitrosamines (including the simplest, *N*-nitrosodimethylamine), which occur widely in the environment. The exact amount to which an individual is exposed depends upon many factors, which include life style (Doll & Peto, 1981), e.g., smoking (IARC, 1986) and the eating of particular foods that may contain preformed nitrosamines or their precursors (amines and nitrite), which can form nitrosamines in the acidic conditions of the stomach (Preussmann *et al.*, 1979; Swann, 1982). Since it is unlikely that exposure to alkylating agents from such sources can be totally eliminated, O^6-meGua is probably present in essentially all human tissues to some extent (even at undetectable levels). Presumably, it is to cope with such exposures that efficient repair mechanisms have evolved (Saffhill *et al.*, 1985).

Other sources of alkylating agents, particularly for persons receiving medication, may be drugs. Thus, many drugs contain amine groups that can be nitrosated within the body to yield simple nitrosamines (Lijinsky & Epstein, 1970). Also, some drugs can undergo a series of complicated metabolic and chemical changes to generate alkylating species (e.g. isoniazid; Saffhill *et al.*, 1988), which can yield unexpectedly high levels of alkylation when administered to experimental animals. One possible explanation of the differences between the two sets of results described here is that, in one set, O^6-meGua is almost always detectable, indicating widespread exposure to an (environmental) alkylating agent. In the other set, a large number of samples had an undetectable level of O^6-meGua while others had detectable levels, due perhaps to differences in life style, including use of medications. However, the drug histories of the patients are yet to be determined.

O^6-MeGua in DNA is efficiently repaired by the O^6-meGua methyltransferase protein (Saffhill *et al.*, 1985) which is present at high levels in human cells (Umbenhauer *et al.*, 1985). However, recent experiments have indicated that, following very low doses of *N*-nitrosodimethylamine, O^6-meGua is repaired less efficiently than may otherwise be expected (A.F. Badawi & R. Saffhill, unpublished), indicating the possible existence of an alkylation threshold before efficient repair can commence. Repair still occurs, but at a slower rate. If the O^6-meGua determined in these samples is uniformly distributed throughout the tissue, it must be the result of a recent exposure. Alternatively, it could have arisen over a period of time in a small population of repair-deficient cells. With the advent of immunological staining procedures for O^6-meGua (Saffhill *et al.*, 1988; P.J. O'Connor, S. Fida & R. Saffhill, in preparation) this may be resolved.

Since human tissues normally contain high levels of alkyltransferase, the enzyme responsible for the repair of O^6-meGua, the amounts of O^6-meGua determined in these studies are unlikely to be related to accumulated dose. Elucidation of this aspect of environmental alkylation of human DNA must await the development of sensitive assays to measure the presence of phosphotriesters in DNA, which appear to be repaired only slowly, if at all, in mammalian cells (Saffhill *et al.*, 1985).

In view of the strong mutagenic potential of O^6-meGua and the wide body of evidence implicating the involvement of this modified base in the carcinogenic process (Saffhill *et al.*, 1985), studies of this kind not only demonstrate the possibility that human DNA can be alkylated (almost certainly as the result of exposure to an external agent) but are expected to strengthen the growing link between exposure to alkylating agents and human cancer.

Acknowledgements

We thank Mrs J. Smith for skilled technical assistance and the British Council for sponsoring one of the authors (A.F.B.). This work is supported by grants from the Cancer Research Campaign.

References

Boyle, J.M., Saffhill, R., Margison, G.P. & Fox, M. (1986a) A comparison of cell survival, mutation and persistence of putative promutagenic lesions in Chinese hamster cells exposed to BNU or MUN. *Carcinogenesis, 7,* 1981–1985

Boyle, J.M., Margison, G.P. & Saffhill, R. (1986b) Evidence for the excision repair of O^6-n-butyldeoxyguanosine in human cells. *Carcinogenesis, 7,* 1987–1990

Boyle, J.M., Durrant, L.G., Wild, C.P., Saffhill, R. & Margison, G.P. (1987) Genetic evidence of nucleotide excision repair of O^6-alkylguanine in mammalian cells. *J. Cell Sci., Suppl. 6,* 147–160

Doll, R. & Peto, R. (1981) The causes of cancer: quantitative estimates of avoidable risk of cancer in the United States today. *J. natl Cancer Inst., 66,* 1192–1308

Fox, H. & Kelly, W.N. (1978) The role of adenosine and 2′-deoxyadenosine in mammalian cells. *Ann. Rev. Biochem.,* 655–686

IARC (1986) *IARC Monographs on the Evaluation of the Carcinogenic Risk of Chemicals to Humans,* Vol. 38, *Tobacco Smoking,* Lyon

Kirby, K.S. & Cook, E.A. (1957) A new method for the isolation of deoxynucleic acid. *Biochem. J., 66,* 459–504

Lijinsky, W. & Epstein, S.S. (1970) Nitrosamines as environmental carcinogens. *Nature, 225,* 21–23

Myers, K.A., Saffhill, R. & O'Connor, P.J. (1986) Evidence for the repair of the promutagenic lesion O^6-methylguanine from mitochondrial DNA. *Proc. biochem. Soc., 14,* 266–267

O'Connor, P.J. & Saffhill, R. (1979) The action of rat cytosol enzymes on some methylated nucleic acid components produced by the carcinogenic *N*-nitroso compounds. *Chem.-biol. Interactions, 26,* 91–102

Preussmann, R., Eisenbrand, G. & Spiegelhalder, B. (1979) *Occurrence and formation of* N-*nitroso compounds in the environment and* in vivo. In: Emmelot, P. & Kriek, E., eds, *Environmental Carcinogenesis: Occurrence, Risk Evaluation and Mechanisms,* Amsterdam, Elsevier, pp. 51–71

Saffhill, R. (1986) The competitive miscoding of O^6-methylguanine and O^6-ethylguanine and the possible importance of cellular deoxynucleoside 5′-triphosphate pool sizes in mutagenesis and carcinogenesis. *Biochim. biophys. Acta, 866,* 53–60

Saffhill, R., Strickland, P.T. & Boyle, J.M. (1982) Sensitive radioimmunoassays for O^6-n-butyldeoxyguanosine, O^2-butylthymidine and O^4-butylthymidine. *Carcinogenesis, 3,* 547–552

Saffhill, R., Margison, G.P. & O'Connor, P.J. (1985) Mechanisms of carcinogenesis induced by alkylating agents. *Biochim. biophys. Acta, 823,* 111–145

Saffhill, R., Fida, S., Bromley, M. & O'Connor, P.J. (1988) Promutagenic alkyl lesions are induced in the tissue DNA of animals treated with isoniazid. *Human Toxicol.* (in press)

Shank, R.C. & Herron, D.C. (1982) *Methylation of human liver DNA after probable dimethylnitrosamine poisoning.* In: Magee, P.N., ed., N-*Nitroso Compounds* (*Banbury Report 12*), Cold Spring Harbor, NY, CSH Press, pp. 153–162

Swann, P.F. (1982) *Metabolism of nitrosamines: observations on the effect of alcohol on nitrosamine metabolism and on human cancer.* In: Magee, P.N., ed., N-*Nitroso Compounds* (*Banbury Report 12*), Cold Spring Harbor, NY, CSH Press, pp. 53–68

Umbenhauer, D., Wild, C.P., Montesano, R., Saffhill, R., Boyle, J.M., Huh, N., Kirstein, U., Thomale, J., Rajewsky, M.F. & Lu, S.H. (1985) O^6-Methyldeoxyguanosine in oesophageal DNA among individuals at high risk of oesophageal cancer. *Int. J. Cancer, 36,* 53–57

Wild, C.P., Smart, G., Saffhill, R. & Boyle, J.M. (1983) Radioimmunoassay of O^6-methyldeoxyguanosine in DNA of cells alkylated *in vitro* and *in vivo. Carcinogenesis, 4,* 1605–1609

Yang, C.S. (1980) Research in esophageal cancer in China: a review. *Cancer Res., 40,* 2633–2644

DETECTION OF DNA ADDUCTS BY POSTLABELLING WITH ^{3}H-ACETIC ANHYDRIDE

K. Hemminki, K. Savela, E. Linkola & A. Hesso

Institute of Occupational Health, Helsinki, Finland

A method has been developed for detecting 7-alkylguanines which is based on derivatization of hydroxyl and primary amino groups with ^{3}H-acetic anhydride. The derivatization procedure is devised for 7-methylguanine, with examination of reaction kinetics. The method is applied to the detection of the isomeric styrene oxide-guanine adducts; styrene-7,8-oxide is the active metabolite of styrene. The procedure can be used as a postlabelling method for the simultaneous detection of several adducts after selective depurination of 7-alkylguanines from DNA.

7-Methylguanine

7-Methylguanine was incubated with ^{3}H-acetic anhydride in pyridine-toluene solution (Blau & King, 1977) in closed glass capillaries at 100°C overnight. After the capillaries had been broken, the reaction mixture was evaporated to dryness, and its formate buffer solution was purified on a strong-cation exchanger. The product was separated by high-performance liquid chromatography (HPLC) on a Spherisorb ODS 2 column and a 0–60% methanol gradient in 0.1 M ammonium formate pH 5. Fractions were collected for liquid scintillation counting, and the N^2-[^{3}H]acetyl-7-methylguanine peak was localized by ultraviolet detection of nonradioactive N^2-acetyl-7-methylguanine, used as internal standard.

The pK values of N^2-acetyl-7-methylguanine were 2.0 (basic) and 9.0 (acidic), as determined by a phase-separation method (Moore & Koreeda, 1976). The acetylation of 7-methylguanine followed second-order kinetics, reacting through a bimolecular substitution mechanism.

Adducts between styrene-7,8-oxide and guanosine

The active metabolite of styrene is styrene-7,8-oxide (SO), which binds covalently to nucleic acids. It reacts with guanine mainly in the *N*7 position, and two isomeric products can be formed with the α- and β-carbons of SO (Savela *et al.*, 1986).

7-SO-guanosine [7-(2-hydroxyphenylethyl)guanosine] adducts were prepared by incubation in acetic acid (Savela & Hemminki, 1986), and LH-20 Sephadex chromatography was used to prefractionate them. The α- and β-isomers were isolated by HPLC. The sugar residue was hydrolysed at 100°C for 2 h, and the purified and characterized (by ultraviolet, mass spectrometry and mass spectrometry/mass spectrometry methods, Hesso *et al.*, 1986) α-isomer was used for acetylation experiments.

The nonradioactive acetylation of 7-SO-guanine [7-(2-hydroxyphenylethyl)guanine] was carried out by incubation with acetic anhydride at 100°C, and the acetyl derivatives were isolated by HPLC and used later as internal standards.

The acetylation kinetics of 7-SO-guanine were studied at 37°C in acetic anhydride (Fig. 1). The hydroxyl group of the SO residue was acetylated faster, and the amount

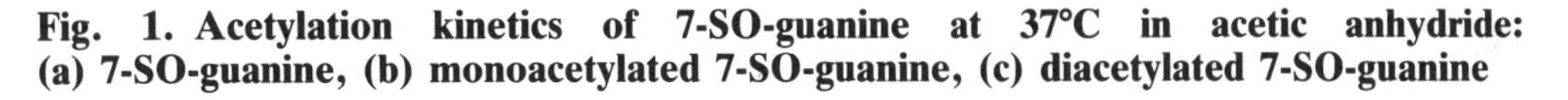

Fig. 1. Acetylation kinetics of 7-SO-guanine at 37°C in acetic anhydride: (a) 7-SO-guanine, (b) monoacetylated 7-SO-guanine, (c) diacetylated 7-SO-guanine

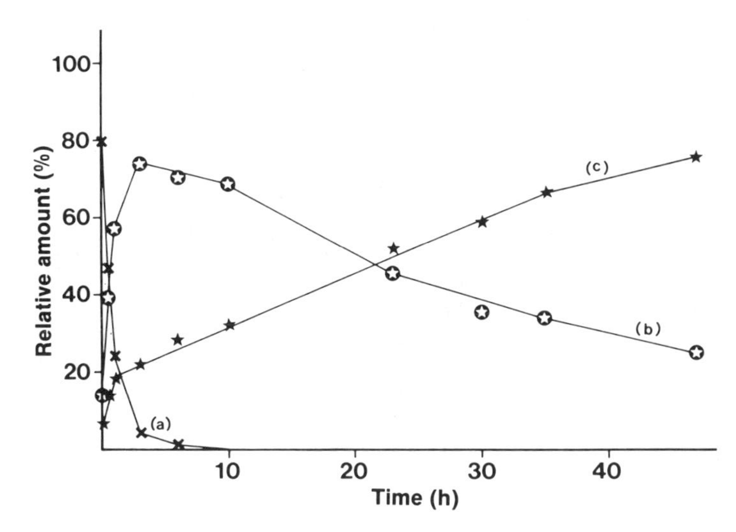

of monoacetyl product increased rapidly up to 6 h. A diacetyl product was formed simultaneously but more slowly, and after 20 h it was the main product. Mass spectral analysis of the diacetyl product (Fig. 2) established the acetylation in the *N*2 position of guanine and in the hydroxyl group of the SO residue. The ultraviolet maximum of the diacetyl product at pH 6.5 is 265 nm (Fig. 3).

The radioactive acetyl derivative of SO-guanine was prepared by incubation of SO-guanine with ^{3}H-acetic anhydride and nonradioactive acetic anhydride in toluene.

Fig. 2. Daughter ion mass spectrum of protonated α-isomer of diacetylated 7-SO-guanine

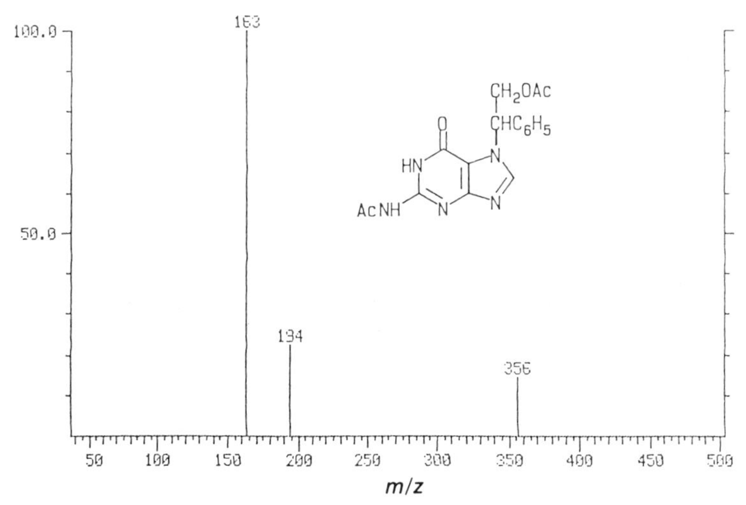

Precursor ion is *m/z* 356 (MH^{+}); *m/z* 163 indicates the acetylated SO residue, *m/z* 194 the N^{2}-acetylated guanine residue.

Fig. 3. Ultraviolet spectra of α-isomer of diacetylated 7-SO-guanine, in neutral (——), acid (- - - -) and alkaline (. . .) solution

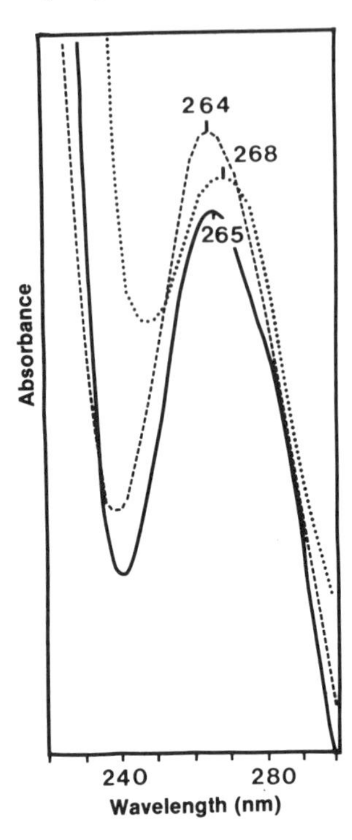

Reaction was carried out in microvials at 100°C for 16 h, and the acetylation product was purified on a strong cation exchanger. After HPLC separation, fractions were collected for liquid scintillation counting, and the product peaks were localized by ultraviolet detection of nonradioactive acetyl-SO-guanines added as internal standard. Under these conditions, radioactive mono- and diacetyl derivatives were detected in almost equal amounts.

Conclusions

Acetylation reactions have commonly been used to protect the primary *N*2-amino group of guanine in nucleoside synthesis (Lohrman & Khorana, 1964; Bridson *et al.*, 1977). ^{14}C-Acetic anhydride has been used to quantify compounds in problematic sample matrices (Benson & Turner, 1960; Lawrence & Frei, 1976). In this study, we used acetylation with high specific activity ^{3}H-acetic anhydride to prepare tritiated acetyl derivatives of guanine adducts for detection by scintillation counting after HPLC separation. In these preliminary experiments, we were able to detect nanogram amounts of 7-methylguanine and 7-SO-guanines.

Acknowledgements

The study was supported by the Work Environment Funds of Finland and Sweden.

References

Benson, R.H. & Turner, R.B. (1960) Acetyl group determination using (C^{14}) acetic anhydride. *Anal. Chem.*, *32*, 1464–1465

Blau, K. & King, G.S. (1977) *Acetylation methods.* In: Blau, K. & King, G.S., eds, *Handbook of Derivatives for Chromatography,* London, Heyden & Son, pp. 111–115

Bridson, P.K., Markiewicz, W.T. & Reese, C.B. (1977) Acylation of 2′,3′,5′-tri-*O*-acetylguanosine. *J. chem. Soc. chem. Commun.*, 791–792

Hesso, A., Kostiainen, R., Kotiaho, T. & Hemminki, K. (1986) *Characterization of styrene oxide-adducts of guanosine by DEP/MS/MS.* In: Todd, J.F.J., ed., *Advances in Mass Spectrometry 1985,* New York, John Wiley, pp. 1465–1466

Lawrence, J.F. & Frei, R.W. (1976) *Acetylation with (^{14}C) acetic anhydride.* In: Lawrence, J.F. & Frei, R.W., eds, *Chemical Derivatization in Liquid Chromatography* (*Journal of Chromatography Library, 7*), Amsterdam, Elsevier, p. 204

Lohrman, R. & Khorana, H.G. (1964) Studies on polynucleotides. XXXIV. The specific synthesis of $C^{3\prime}C^{5\prime}$-linked ribooligonucleotides. New protected derivatives of ribonucleotides and ribonucleoside 3′-phosphates. Further syntheses of diribonucleoside phosphates. *J. Am. chem. Soc.*, *86,* 4188–4194

Moore, P.D. & Koreeda, M. (1976) Application of the change in partition coefficient with pH to the structure determination of alkyl substituted guanosines. *Biochem. biophys. Res. Commun.*, *73,* 459–464

Savela, K. & Hemminki, K. (1986) Reaction products of styrene oxide with deoxynucleosides and DNA *in vitro. Arch. Toxicol., Suppl. 9,* 281–285

Savela, K., Hesso, A. & Hemminki, K. (1986) Characterization of reaction products between styrene oxide and deoxynucleosides and DNA. *Chem.-biol. Interactions, 60,* 235–246

MEDICINAL EXPOSURES

DNA ADDUCTS OF CISPLATIN AND CARBOPLATIN IN TISSUES OF CANCER PATIENTS

M.C. Poirier,[1] M.J. Egorin,[2] A.M.J. Fichtinger-Schepman,[3] S.H. Yuspa[1] & E. Reed[1]

[1]National Cancer Institute, National Institutes of Health, Bethesda, MD; [2]University of Maryland Cancer Center, Baltimore, MD, USA; and [3]TNO Medical Biological Laboratory, Rijswijk, The Netherlands

An enzyme-linked immunosorbent assay (ELISA) has been developed with an antiserum elicited against cisplatin-modified DNA and used to quantify the intra-strand bidentate d(GpG)- and d(ApG)-diammineplatinum adducts in DNA samples prepared from nucleated blood cells and tissues of cancer patients receiving cisplatin or carboplatin chemotherapy. In nucleated blood cell DNA, adducts accumulated with increasing dose administered over a period of months, and a correlation was observed between the ability of a patient to form high levels of adduct and the frequency of tumour remission. Thus, many patients who did not form adducts also did not respond to therapy. Adduct distribution was shown to be widespread in many human tissues, and similar quantities of adducts were formed in peripheral blood cell DNA and tumour tissue. In addition, evidence suggests that residues of persistent adducts remain in many tissues weeks and even months after treatment. All of the above observations were obtained with the cisplatin-DNA ELISA; however, in comparison with other published data, the adduct levels reported are low. It now appears certain that the cisplatin-DNA ELISA results in an underestimation of adduct values in biological samples, since some human samples have been assayed by both this and two other procedures — the G-Pt-GMP ELISA and atomic absorbance spectroscopy. Values obtained with the two other procedures compare well with each other, but those obtained with the cisplatin-DNA ELISA for three human samples are 10–300-fold lower. The factors that result in this discrepancy are still under investigation.

The possibility of detecting exposure of humans to chemicals by monitoring the formation of macromolecular adducts became a reality in the last decade through the development of highly sensitive methods for the determination of DNA, RNA and protein adducts (Neumann, 1984; Poirier, 1984; Randerath *et al.*, 1985). At present, human exposure to several different classes of chemical carcinogens has been documented in cohorts exposed through occupation, environment, life style and medical treatment (Poirier & Beland, 1987). Progress to date represents extensive efforts on the part of many investigators to validate highly sensitive, and often highly variable, new procedures in the face of formidable problems. Not the least of these, encountered by many investigators, includes difficulty in procuring tissues from unexposed controls, lack of a quantifiable dose-response relationship and inability to confirm the initial data using an alternative method. Thus, validation of existing assays has proved to be more complex than previously anticipated. Nevertheless, great progress has been achieved since the first IARC sponsored workshop on this topic was held in Essen, Federal Republic of Germany, in 1981. At that time, human biomonitoring was a possibility but not a reality.

Among the first methods applied to this problem were quantitative immunoassays using antisera specific for carcinogen-DNA adducts and carcinogen-modified DNA (Poirier, 1984). In particular, a pilot study performed by Perera *et al.* (1982) showed that DNA from some human lung tumours and surrounding tissue reacted with an antiserum now known to be specific for a broad spectrum of polycyclic aromatic hydrocarbon-DNA adducts (Weston *et al.*, this volume). While performing these studies, the investigators became acutely aware of the necessity to validate the assay system with dose-response data, and chose to study adduct formation in cancer patients receiving cisplatin chemotherapy. An antiserum was elicited against DNA modified with cisplatin, a drug used extensively in the treatment of testicular and ovarian malignancies (Einhorn *et al.*, 1985; Calvert, 1986) but known to be carcinogenic to rodents (Leopold *et al.*, 1979; Barnhart & Bowden, 1985). The data presented here were obtained with an ELISA utilizing an antiserum specific for the intrastrand N^7-d(GpG)- and N^7-d(ApG)-diammineplatinum adducts (Poirier *et al.*, 1982), which are the major DNA adducts formed *in vivo* (Plooy *et al.*, 1985). The ELISA measures dose-related increases in cisplatin-DNA intrastrand adducts in tissues of human cancer patients and cisplatin-treated animals. However, the antiserum was elicited against a highly modified DNA immunogen, and its use significantly underestimates the much lower modification in a biological sample. The material presented here focuses on results obtained with human tissues, and we discuss various aspects of assay validation.

Monitoring of platinum-DNA adducts in tissues of cancer patients

The earliest studies on platinum-DNA adducts in patients involved procuring blood samples from patients receiving five-day infusions of cisplatin as a single agent or in combination therapy (Reed *et al.*, 1986). Blood samples were drawn with heparin 24 h after the last drug infusion, and nucleated blood cells (WBC) were obtained by aspiration of the layer between serum and red blood cells (the buffy coat) after slow-speed centrifugation. Since the inception of this work, a total of 231 blood samples from 88 patients receiving platinum drug therapy have been assayed by ELISA, and about one-half of these had measurable adducts (Poirier *et al.*, 1987). Of 23 samples obtained from normal volunteers, patients on nonplatinum chemotherapy or patients before treatment, none contained adducts. Adduct values in positive samples ranged between 25 and 450 amol/μg DNA. A dose-response relationship was evident in a subset of 97 samples from previously untreated ovarian and testicular cancer patients, although only about one-half of these samples contained adducts measurable by the ELISA (Fig. 1). Since the total dose shown was administered over a period of four to five months, adducts accumulated with time in the WBC DNA of some patients. Thus, removal of cisplatin-DNA adducts may be relatively slow, or adduct formation may be particularly efficient in those patients who show accumulation of adducts with dose. Conversely, there may be heterogeneity with respect to either metabolism for adduct formation or the kinetics of adduct removal in those patients who do not form measurable adducts.

In order to correlate adduct formation as an indication of biologically-effective dose with chemotherapeutic efficacy, disease-response data were extracted from the medical records of 55 patients with ovarian cancer and of 17 testicular cancer patients with a poor prognosis. Data for clinical response are expressed as complete response (absence of visible tumour), partial response (a greater than 50% diminution in tumour size), and no response (less than 50% reduction in tumour size). The 55 ovarian cancer

Fig. 1. Cisplatin-DNA adduct values in ELISA for WBC DNA prepared on CsCl buoyant density gradients

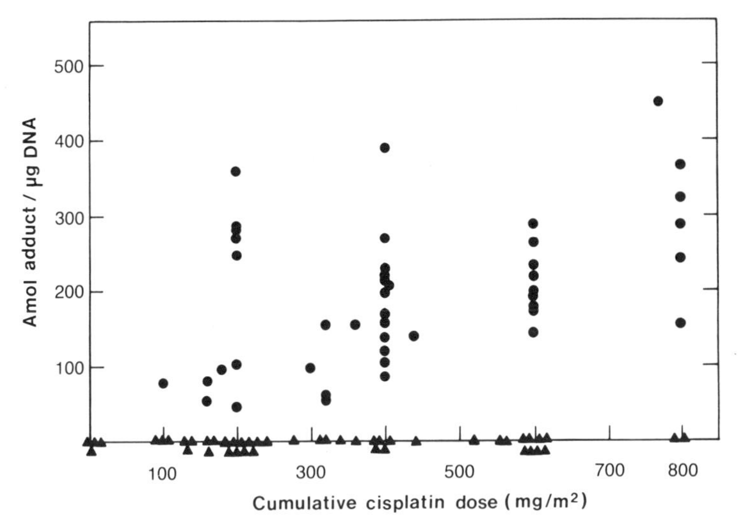

Blood samples (35–50 ml) were obtained 24 h after the last of five drug infusions from previously untreated ovarian or testicular cancer patients given chemotherapy on 21- or 28-day cycles at the NIH Clinical Center. Adduct levels (ordinate) are plotted as a function of total cumulative cisplatin dose. Positive (●) and negative (▲) samples are from a total of 77 patients.

patients received cisplatin or carboplatin, either as single-agent therapy (Fig. 2A, HD-DDP and CBDCA) or as part of a combination protocol (Fig. 2A, CTX-DDP and CHIPS) (Reed *et al.*, 1987). In this study, 101 WBC samples were obtained and the highest or peak adduct value from each patient was chosen for statistical analysis. Values for median adduct levels (Fig. 2A, heavy lines) in patients grouped by complete, partial and no response were 212, 193 and 62 amol/μg DNA, and this trend was shown to be statistically significant by a two-sided Jonckherre test ($p = 0.03$) (Armitage, 1971). Of particular interest is the fact that several individuals who did not form adducts also did not respond to therapy.

The second group of patients for whom disease response was correlated with the extent of adduct formation are nonseminomatous testicular cancer patients. These individuals are rare, usually have bulky disease and metastases, and lack the high favourable response rates seen for testicular cancer patients in general. The 17 patients included in this study had not received previous chemotherapy and were given cisplatin at two different doses in combination with other drugs. The patients receiving the PVB protocol were given 20 μg/m^2 cisplatin per day for five days for each cycle, and the PVeBV patients received twice that quantity of cisplatin per cycle (Reed *et al.*, 1986). The data are shown in Figure 2B. The median adduct level in patients with a complete response was 170 amol/μg DNA, and that for those with a partial response was 78 amol/μg DNA. Although these differences are not statistically significant, it is interesting to note that all of the complete responses occurred among individuals receiving the higher dose and with the highest median adduct levels. Thus, for cohorts of both ovarian and testicular cancer patients receiving platinum drug therapy, a high rate of complete response is associated with high levels of WBC cisplatin-DNA adducts, and lack of response is associated with inability to form measurable levels of adducts.

Fig. 2. Comparison of WBC DNA adduct levels with disease response of ovarian (A) and poor-prognosis testicular (B) cancer patients

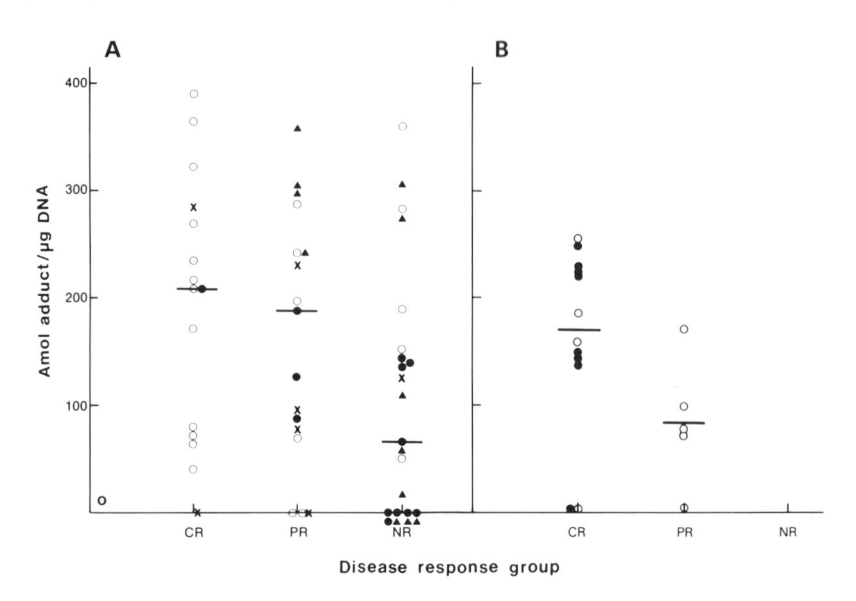

All samples were assayed by cisplatin-DNA ELISA. CR, complete response; PR, partial response; NR, no response. A, the highest, or 'peak', adduct value obtained from each of 55 ovarian cancer patients is plotted as a function of disease response. The HD-DDP (●) and CBDCA (▲) groups received single agent therapy and CTX-DDP (○) and CHIPS (×) groups received combination therapy. B, peak adduct value from each of 17 poor-prognosis testicular cancer patients plotted as a function of disease response. Treatment groups are PVB (○) and PVeBV (●). For both A and B, the median adduct level is shown by a heavy line.

Although these studies show correlations between platinum-DNA adducts in WBC and both drug dose and disease response, they do not show that adducts form in the target tumour tissue. Recently, it has been possible to obtain samples of cervical tumours and blood before and 24 h after administration of a single dose of carboplatin to previously untreated patients (Egorin *et al.*, 1985). In some cases, a second blood sample was drawn eight days after the original treatment. The results are shown in Table 1. Pretreatment samples uniformly showed no adduct formation in either tissue, and adduct levels 24 h after the single dose ranged between 83 and 250 amol/μg DNA for the two patients who received the full dose. Thus, in these patients, adduct levels in WBC were similar to those in the tumour 24 h after treatment. In addition, WBC adduct levels did not change much during the week after treatment, suggesting persistence of adducts for at least that period of time. One patient did not tolerate the drug and was given approximately one-half the usual dose; this individual had measurable adducts in WBC only at 24 h. These data, and other investigations of adduct distribution in tissues obtained at autopsy (Poirier *et al.*, 1987; Reed *et al.*, 1987), suggest that intrastrand cisplatin-DNA adducts are widely distributed in human tissues as a result of platinum drug chemotherapy, and that these adducts are highly persistent for days, and even weeks, after the last treatment. Adduct persistence may, in fact, be the cause of the chemotherapeutic efficacy, the toxic side effects and the malignant potential (Leopold *et al.*, 1979) of these platinum drugs.

Table 1. Platinum-DNA adducts (amol/μg DNA) in cervical tumour tissue and WBC obtained before and after treatment with carboplatin

Patient	Dose (mg/m^2)	Tumour of the cervix		WBC		
		Before treatment	After treatment (24 h)	Before treatment	After treatment	
					24 h	8 days
A	399	0	97	0	160	137
B	393	0	250	0	83	127
C	226	0	ND	0	51	ND

ND, not detectable

Validation of the cisplatin-DNA ELISA

The cisplatin-DNA ELISA used for these studies (Poirier *et al.*, 1982) was developed with a rabbit polyclonal antiserum raised against a highly-modified cisplatin DNA (4.4 adducts in 100 bases). This high level of platination is necessary to obtain an immunogenic response to the adduct without eliciting concomitant anti-DNA activity. Recent studies have shown that antisera raised against a highly-modified immunogen DNA may have lower affinity for a biological sample modified *in vivo* than for the immunogen DNA (Van Schooten *et al.*, 1987). This may reflect different physical conformations of highly modified and less modified DNA, and can result in significant underestimation of the number of adducts in a biological sample when assayed against a standard curve obtained with highly modified DNA. Thus, quantification by ELISA of an antiserum elicited against highly modified DNA should include competition with DNA at various levels of modification and, if possible, the use of another technique to confirm the modification level determined by ELISA in a biological sample. These controls are essential because some modified DNA antisera recognize all of the adducts present in a biological sample (Santella *et al.*, this volume), while others do not.

When the anti-cisplatin-DNA antiserum was first elicited against calf thymus DNA modified *in vitro*, there was no other equally sensitive method available with which to confirm the results with biological samples. Recently, however, two methods have been developed with sufficient sensitivity to measure cisplatin-DNA adducts at modification levels in the range of one adduct in 10^6–10^8 bases. A battery of immunoassays employs antisera raised against G-Pt-GMP and other specific adducts, and utilizes column fractions after enzymatic digestion of the DNA and subsequent chromatography (Fichtinger-Schepman *et al.*, 1987). This procedure allows for determination of multiple adducts and yields results which compare well with values for total DNA-bound platinum measured by atomic absorbance spectroscopy (AAS). Table 2 shows the results of a comparison in which a highly-modified DNA sample was assayed

Table 2. Assay of cisplatin-DNA adducts in DNA modified *in vitro*

DNA	Modification level (adducts/nucleotides)	Cisplatin-DNA ELISA	G-Pt-GMP ELISA	AAS
Highly modified	1/100	159 pmol/μg DNA	169 pmol/μg DNA	157 pmol/μg DNA
Little modified	$1/1.7 \times 10^6$	Not detectable	2.5 fmol/μg DNA	Not assayed

Table 3. Platinum-DNA adducts in human tissues 24 h after drug infusion

Tissue	Assay	Range of adduct values (fmol/μg DNA)	Reference
Ovarian ascites cells	AAS	10.0–11.0	Roberts *et al.* (1986)
Blood cells	G-Pt-GMP ELISA	0–2.5	Fichtinger-Schepman *et al.* (1987)
Blood cells	Cisplatin-DNA ELISA	0–0.45	Reed *et al.* (1986); Poirier *et al.* (1987)

by the cisplatin-DNA ELISA, the G-Pt-GMP ELISA and AAS. The last method has also been adapted for greater sensitivity and can now be used in conjunction with the immunoassays to measure biological samples (Roberts *et al.,* 1986). Table 3 summarizes results obtained when using these three methods to assay for adducts in tissues of cancer patients 24 h after platinum-based therapy. Table 3 suggests that the cisplatin-DNA ELISA may underestimate adducts in human samples, since the highest values were five and 20 times lower than those obtained with the G-Pt-GMP ELISA and AAS, respectively. When samples of human brain, kidney and lung were analysed by both immunoassays, the G-Pt-GMP ELISA detected 13-, 100- and 324-fold more adducts than the cisplatin-DNA ELISA, confirming the underestimation by the latter assay. The extent of underestimation does not appear to be consistent for these few human samples; however, experiments with rodent tissues, in which adduct values are higher, have demonstrated a 500–1000-fold discrepancy between values determined by the cisplatin-DNA ELISA and the other procedures. For example, rat kidney DNA contained 175 ± 9 amol/μg DNA by the cisplatin-DNA ELISA, and 137 ± 45 and 99 ± 45 fmol/μg DNA by the G-Pt-GMP ELISA and AAS, respectively. Thus, for biological samples, the AAS agrees well with the G-Pt-GMP ELISA, and significantly fewer adducts are detected by the cisplatin-DNA ELISA.

We are currently unable to determine whether the cisplatin-DNA ELISA recognizes primarily clusters of intrastrand adducts within specific regions of the DNA, all of the relatively isolated, biologically formed adducts with lower affinity or just a certain conformation of the DNA. However, it is clear that the adducts measurable by this ELISA correlate with administered dose, are removed with time after treatment and predict biological response in both human cancer patients and animal models. Thus, the results appear to be internally consistent, even though the adducts are not assayed quantitatively. We are presently developing radioimmunoassays for the d(ApG)-and d(GpG)-diammineplatinum adducts, which will be used in conjunction with AAS to determine these adducts quantitatively.

The ability to monitor macromolecules from human tissues for evidence of exposure to hazardous chemicals has advanced significantly in the last decade. In some systems, including that reported here, it is possible to demonstrate a dose-response relationship for DNA adduct formation, as well as adduct persistence and removal. However, the technologies used are still relatively new. In the future, optimal results may be obtained using combinations of the techniques currently in use, and it should be possible to confirm most data with more than one type of assay. For the present, many of us are still committed to the development of more accurate and sensitive assays, and to the interlaboratory collaboration necessary for their standardization.

References

Armitage, P. (1971) *Statistical Methods in Medical Research,* Oxford, Blackwell Scientific Publications

Barnhart, K.M. & Bowden, G.T. (1985) Cisplatin as an initiating agent in two-stage mouse skin carcinogenesis. *Cancer Lett., 29,* 101–105

Calvert, A.H. (1986) *Clinical application of platinum metal complexes.* In: McBrien, D.C.H. & Slater, T.F., eds, *Biochemical Mechanisms of Platinum Antitumor Drugs,* Oxford, IRL Press, pp. 307–315

Egorin, M.J., Van Echo, D.A., Olman, E.A., Whitacre, M.Y., Forrest, A. & Aisner, J. (1985) Prospective validation of a pharmacologically based dosing scheme for the *cis*-diamminedichloroplatinum(II) analogue diamminecyclobutanedicarboxylatoplatinum. *Cancer Res., 45,* 6502–6506

Einhorn, L.H., Donohue, J.P., Peckham, M.J., Williams, S.D. & Loehrer, P.J. (1985) *Cancer of the testes.* In: DeVita, V.T., Hellman, S. & Rosenberg, S.A., eds, *Cancer — Principles and Practice of Oncology,* Philadelphia, Lippincott, pp. 979–1011

Fichtinger-Schepman, A.M.J., Van Oosterom, A.T., Lohman, P.H.M. & Berends, F. (1987) Cisplatin-induced DNA adducts in peripheral leukocytes from seven cancer patients: quantitative immunochemical detection of the adduct induction and removal after a single dose of cisplatin. *Cancer Res., 47,* 3000–3004

Leopold, W.R., Miller, E.C. & Miller, J.A. (1979) Carcinogenicity of antitumor *cis*-platinum(II) coordination complexes in the mouse and rat. *Cancer Res., 39,* 913–918

Neumann, H.-G. (1984) Analysis of hemoglobin as a dose monitor for alkylating and arylating agents. *Arch. Toxicol., 56,* 1–6

Perera, F.P., Poirier, M.C., Yuspa, S.H., Nakayama, J., Jaretzski, A., Curnen, M.M., Knowles, D.M. & Weinstein, I.B. (1982) A pilot project in molecular cancer epidemiology: determination of benzo(*a*)pyrene-DNA adducts in animal and human tissues by immunoassays. *Carcinogenesis, 3,* 1405–1410

Plooy, A.C.M., Fichtinger-Schepman, A.M.J., Schutte, H.H., van Dijk, M. & Lohman, P.H.M. (1985) The quantitative detection of various Pt-DNA adducts in Chinese hamster ovary cells treated with cisplatin: application of immunochemical techniques. *Carcinogenesis, 6,* 561–566

Poirier, M.C. (1984) The use of carcinogen-DNA adduct antisera for quantitation and localization of genomic damage in animal models and the human population. *Environ. Mutagenesis, 6,* 879–887

Poirier, M.C. & Beland, F.A. (1987) Determination of carcinogen-induced macromolecular adducts in animals and humans. *Prog. exp. Tumor Res., 31,* 1–10

Poirier, M.C., Lippard, S.J., Zwelling, L.A., Ushay, H.M., Kerrigan, D., Thill, C.C., Santella, R.M., Grunberger, D. & Yuspa, S.H. (1982) Antibodies elicited against *cis*-diamminedichloroplatinum(II)-modified DNA are specific for *cis*-diamminedichloroplatinum(II)-DNA adducts formed *in vivo* and *in vitro. Proc. natl Acad. Sci. USA, 79,* 6443–6447

Poirier, M.C., Reed, E., Ozols, R.F., Fasy, T. & Yuspa, S.H. (1987) DNA adducts of cisplatin in nucleated peripheral blood cells and tissues of cancer patients. *Prog. exp. Tumor Res., 31,* 104–113

Randerath, K., Randerath, E., Agrawal, H.P., Gupta, R.C., Schurdak, M.E. & Reddy, M.V. (1985) Postlabeling methods for carcinogen-DNA adduct analysis. *Environ. Health Perspect., 62,* 57–65

Reed, E., Yuspa, S.H., Zwelling, L.A., Ozols, R.F. & Poirier, M.C. (1986) Quantitation of *cis*-diamminedichloroplatinum(II) (Cisplatin)-DNA intrastrand adducts in testicular and ovarian cancer patients receiving cisplatin chemotherapy. *J. clin. Invest., 77,* 545–550

Reed, E., Ozols, R.F, Tarone, R., Yuspa, S.H. & Poirier, M.C. (1987) Platinum-DNA adducts in leukocyte DNA correlate with disease response in ovarian cancer patients receiving platinum-based chemotherapy. *Proc. natl Acad. Sci. USA, 84,* 5024–5028

Roberts, J.J., Knox, R.J., Friedlos, F. & Lydall, D.A. (1986) *DNA as the target for the cytotoxic and antitumor action of platinum coordination complexes: comparative in vitro and in vivo studies of cisplatin and carboplatin*. In: McBrien, D.C.H. & Slater, T.F., eds, *Biochemical Mechanisms of Platinum Antitumour Drugs,* Oxford, IRL Press, pp. 29–64

Van Schooten, F.J., Kriek, E., Steenwinkel, M.J.S.T., Noteborn, H.P.J.M., Hillebrand, M.J.X. & Van Leeuwen, F.E. (1987) The binding efficiency of polyclonal and monoclonal antibodies to DNA modified with benzo(*a*)pyrene diol epoxide is dependent on the level of modification. Implications for quantitation of benzo(*a*)pyrene-DNA adducts *in vivo*. *Carcinogenesis, 8,* 1263–1269

INDUCTION AND REMOVAL OF CISPLATIN-DNA ADDUCTS IN HUMAN CELLS *IN VIVO* AND *IN VITRO* AS MEASURED BY IMMUNOCHEMICAL TECHNIQUES

A.M.J. Fichtinger-Schepman,[1] F.J. Dijt,[2] P. Bedford,[3] A.T. van Oosterom,[4] B.T. Hill[3] & F. Berends[1]

[1]TNO Medical Biological Laboratory, Rijswijk; [2]Department of Chemistry, Gorlaeus Laboratories, State University of Leiden, Leiden, The Netherlands; [3]Laboratory of Cellular Chemotherapy, Imperial Cancer Research Fund Laboratories, Lincoln's Inn Fields, London, UK; and [4]Department of Oncology, Antwerpen University Hospital, Edegem, Belgium

The same spectrum of cisplatin adducts was detected in DNA isolated from white blood cells of a cisplatin-treated cancer patient as had been found in cisplatin-treated DNA *in vitro*. The adducts were quantified in femtomole amounts by competitive enzyme-linked immunosorbent assay (ELISA) with three antisera raised against synthetic cisplatin-containing (oligo)nucleotides. For this assay, DNA samples digested with nucleases were fractionated by ion-exchange chromatography; the fractions were used as inhibitors of antibody binding. Determinations of the main adduct formed, cis-Pt$(NH_3)_2$d(pGpG), in patients immediately after a first treatment with equal doses of cisplatin showed interindividual differences in the platination levels of the white blood cells. These differences were found to correlate with those found after in-vitro exposure to cisplatin of blood samples taken from patients before treatment. *In vivo,* about 75% of the adducts formed after the first treatment were removed within 24 h. During a five-day course, the amounts of the main adduct increased after the first three administrations; no increase was seen on day 4 or 5. By day 6, considerable removal of adducts had occurred. Analysis of the formation and repair of the cis-Pt$(NH_3)_2$d(pGpG) adducts in cultured cells, i.e., human fibroblasts with different DNA repair capacities and one bladder and two testicular human cancer cell lines, indicated that both the amounts of adducts formed and the ability of the cells to repair the adducts can differ. These differences appear to determine the susceptibility of the cells for the cytotoxic action of cisplatin.

cis-Diamminedichloroplatinum(II) (cisplatin) is an antineoplastic agent used in (combined) chemotherapy of cancer patients. It is especially successful in the treatment of testicular and ovarian tumours. Because its activity is generally ascribed to interaction with the DNA in tumour cells, knowledge about the resulting DNA modifications (adducts) and their persistence will help to elucidate the working mechanism of this agent. Upon interaction with DNA, cisplatin forms various adducts (Fig. 1) which can be detected and quantified after chromatography of enzymatically digested DNA samples (Fichtinger-Schepman *et al.*, 1985a,b, 1987a). Here, we report results of cisplatin-adduct measurements in DNA isolated from human cells after cisplatin treatment *in vivo* or *in vitro*.

Fig. 1. Known cisplatin-DNA adducts

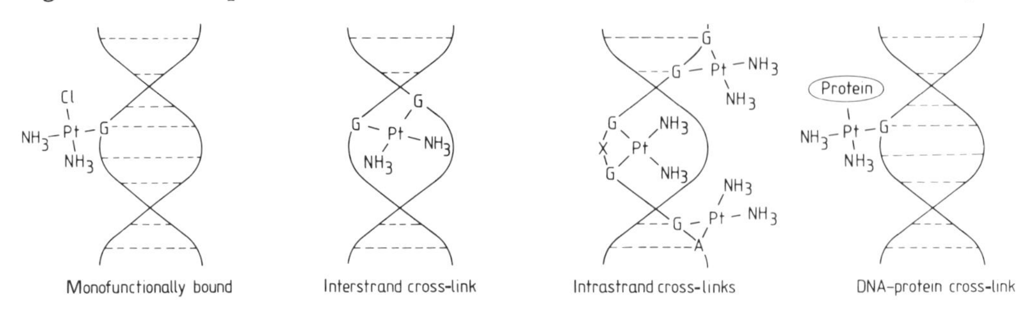

Immunochemical quantification of cisplatin-DNA adducts

For studies on the induction and removal of cisplatin-adducts, DNA samples were enzymatically degraded to the unmodified mononucleotides dCMP, dAMP, dTMP and dGMP and four Pt-containing products (Fig. 1):

1. $Pt(NH_3)_3dGMP$ [Pt-G], derived from DNA adducts with the cisplatin moiety monofunctionally bound to guanine and the remaining reactive coordination site of cisplatin blocked by subsequent exposure of the DNA to NH_4HCO_3 (Fichtinger-Schepman *et al.*, 1984, 1988);
2. *cis*-$Pt(NH_3)_2d(pApG)$ [Pt-AG], derived from cisplatin bound bifunctionally to the nucleobases adenine and guanine in the sequence pApG;
3. *cis*-$Pt(NH_3)_2d(pGpG)$ [Pt-GG], from intrastrand cross-links on two vicinal guanines in the sequence pGpG; and
4. *cis*-$Pt(NH_3)_2d(GMP)_2$ [G-Pt-G], from intrastrand cross-links on base sequences $pG(pX)_npG$ and from interstrand cross-links on two guanines.

After anion-exchange chromatography of the mixture of these cisplatin-DNA digestion products, the Pt products can be determined in the appropriate fractions of the column eluate. Quantification can be performed with atomic absorption spectroscopy when at least picomole amounts of Pt are present. In studies on human cells, as described in this paper, in which the cells have been treated with biologically relevant dosages of cisplatin, the Pt adducts are present in only femtomole amounts per assayable quantity of DNA. For assays at these low levels, immunochemical methods have been developed, with polyclonal antibodies raised against cisplatin-containing nucleotides coupled to immunogenic carrier proteins. With three rabbit sera, we are now able to quantify at the femtomole level the four Pt-DNA digestion products in a competitive ELISA (Fichtinger-Schepman *et al.*, 1985b, 1987a).

Figure 2 shows the results of the application of these methods to a digested DNA sample isolated from circulating white blood cells from a cancer patient treated with cisplatin for the first time. They clearly demonstrate the presence of the same four cisplatin-DNA digestion products as obtained with cisplatin-treated DNA *in vitro* (Fichtinger-Schepman *et al.*, 1985a). The Pt-GG adduct is by far the predominant adduct, a phenomenon observed in all DNAs tested so far.

Cisplatin-DNA adducts in human blood cells

We have started an investigation on the induction and removal of cisplatin-DNA adducts in white blood cells of cancer patients, in the hope that the data obtained will

Fig. 2. Immunochemical quantification of digestion products derived from cisplatin-DNA adducts in eluate fractions obtained after anion-exchange column chromatography of the digest

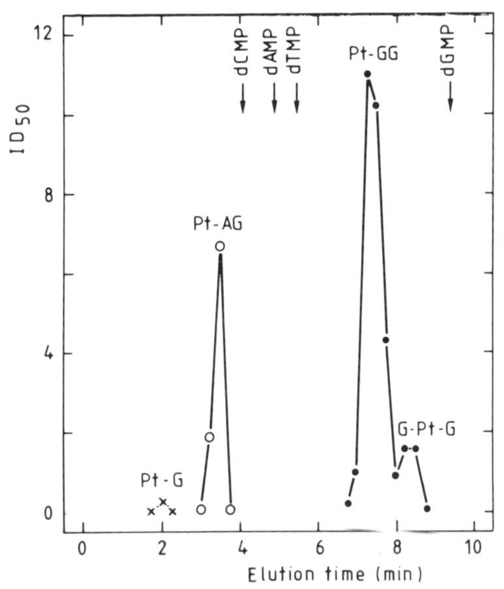

The DNA was isolated from white blood cells in a blood sample from a cisplatin-treated patient. Dilutions of the fractions that give 50% inhibition of the relevant antiserum in the competitive ELISA (ID50): ×, serum 3/43 elicited against Pt-G; ○, serum 3/65 elicited against Pt-AG; and ●, serum W101 raised against cisPt$(NH_3)_2$GuoGMP. Arrows indicate the elution positions of the unmodified nucleotides, which were detected by their ultraviolet absorbance.

reflect the interactions of cisplatin with DNA inside the tumour cells (for underlying rationale, see Fichtinger-Schepman *et al.*, 1986, 1987a).

Figure 3 shows the amounts per microgram of DNA of the main adduct, Pt-GG, assayed in the DNA of blood samples obtained from six male patients treated with cisplatin for the first time. The data show clear differences in the levels of adducts between patients, initially and after 24 h, which might be correlated with variations in the response of the tumour to the cisplatin treatment (Poirier *et al.*, 1985). From these results it is also evident that most of the Pt-GG adducts are removed within a day after treatment. As has been reported elsewhere (Fichtinger-Schepman *et al.*, 1987b), the amounts of Pt-GG adducts found in the blood of patients immediately after a first infusion of cisplatin correlate very well with the levels measured in blood samples taken from the same patients before treatment and exposed to cisplatin *in vitro* (correlation coefficient, 0.91). This result bolsters our hope of developing a predictive assay for the efficacy of cisplatin as an antineoplastic agent in individual patients.

Because cisplatin is often applied in five-day courses, we also studied the levels of Pt-GG adducts on consecutive days. The results for an individual patient are given in Figure 4, showing that over the first three days the number of adducts increased; no further increase was measured, however, after the infusions on days 4 and 5. To study whether Pt-GG adducts are still removed after repeated administrations, we also analysed blood samples taken the day after the final infusion: substantial removal had occurred (Fig. 4). We found higher amounts of adducts at day 5 than after the first treatment in blood samples from all patients tested, but the increase was rather

Fig. 3. Induction and removal of the main cisplatin-DNA adduct (Pt-GG) in white blood cells from six cancer patients suffering from different types of tumours after their first treatment with cisplatin at dosages of 20, 30 or 100 mg/m² body area

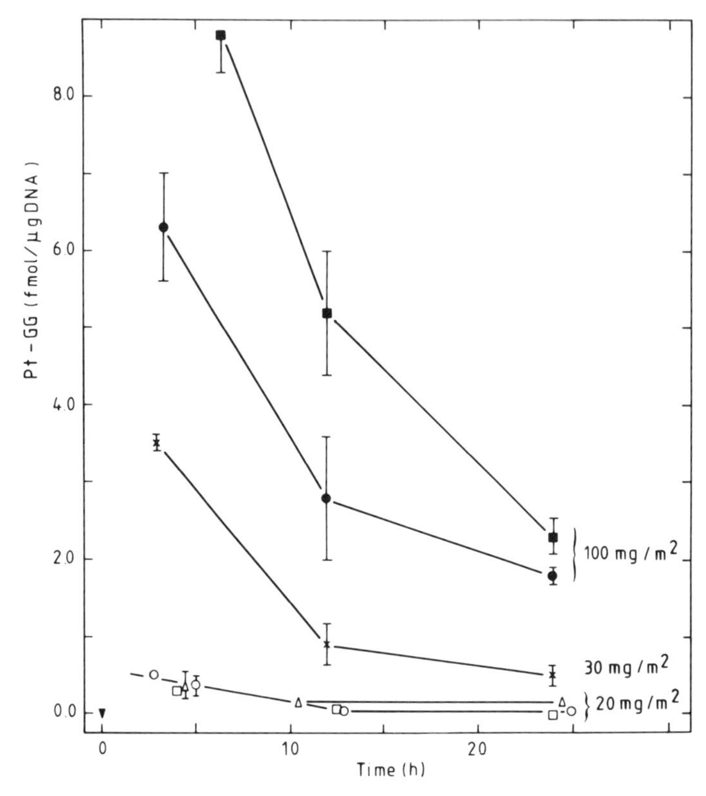

The error bars represent the range of the data. At the start ($t = 0$) of the 3-h infusion, no adduct could be detected in the samples (▼).

irregular. Variation between the patients was also observed with regard to the day at which the highest adduct level was reached.

Cisplatin-DNA adducts in cultured human fibroblasts

Because the variation in response of patients to cisplatin chemotherapy may be due in part to interindividual differences in the ability to repair damaged DNA, we studied the removal of cisplatin-DNA adducts in human fibroblasts. Cell lines from a 'normal' person were compared with those from a Fanconi's anaemia (FA) patient and a patient with xeroderma pigmentosum (XP, complementation group A); such cells are very sensitive to cisplatin. The sensitivity of FA cells is assumed to result from insufficient capacity to repair interstrand DNA cross-links, and that of XP cells from deficient DNA excision repair.

Identical cisplatin treatments resulted in approximately 50% more platination in FA fibroblasts than in normal and XP cells. In normal and FA cells, a substantial proportion of the adducts was removed during a 4.5-h post-treatment incubation, in contrast to the XP cells which showed hardly any removal. Figure 5 shows the relative

Fig. 4. Level of Pt-GG adducts (mean ± range) in white blood cells from a cancer patient on five successive days of cisplatin treatment at a dose of 20 mg/m² per day

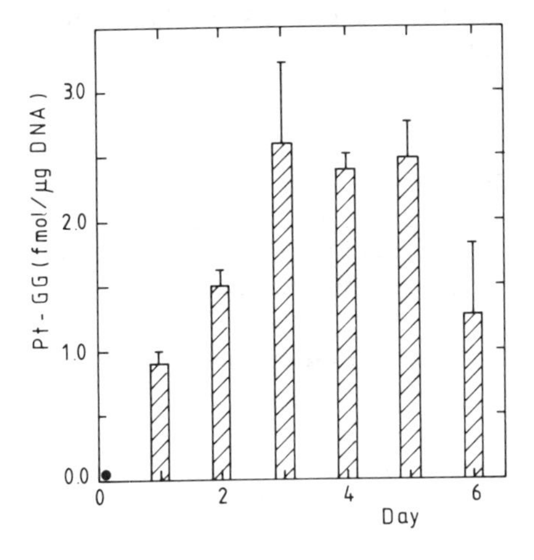

Blood samples were taken immediately after the end of each administration and about 24 h after the start of the final (5th) 3-h infusion period (day 6). A control blood sample was taken before treatment (●, day 0)

amounts of the main adduct, Pt-GG, directly after treatment (100%) and after 4.5-h and 22-h repair periods. Apparently, rapid removal in normal and FA cells is restricted to the initial period; the limited decrease seen after this period may be ascribed to dilution by DNA synthesis during the postincubation period (Bedford *et al.*, 1988). The same process might be responsible in part for the decrease seen in XP cells. The repair deficiency of XP cells is reflected in the persistence of the DNA adducts. Evidently, excision repair is the major pathway for the removal of Pt-GG adducts. It appears highly plausible that the persistence of the DNA adducts can be correlated with the extreme sensitivity of XP to cell killing by cisplatin.

Fig. 5. Percentages of Pt-GG adducts originally induced ($t = 0$; 100%) remaining in DNA of cultured human fibroblasts after repair incubation periods following cisplatin treatment at 19 µg/ml for 70 min

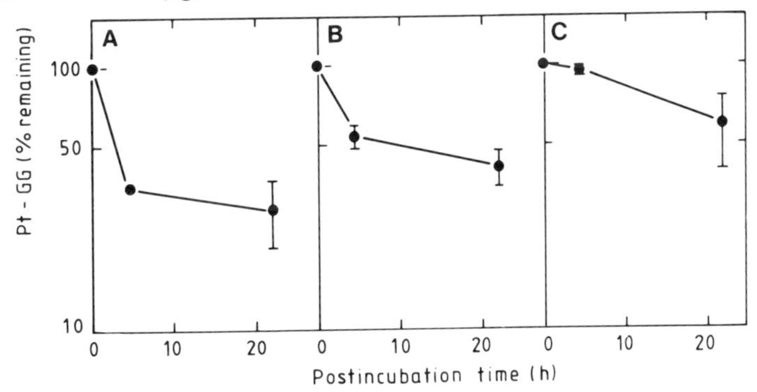

A, normal cells; B, FA cells; C, XP cells. Initial adduct levels: normal cells, 20 fmol/µg DNA; FA cells, 30 fmol/µg DNA; XP cells, 22 fmol/µg DNA

For FA cells, the relation is less clear. During the initial period of fast repair, the amount of adducts removed is similar to that in wild-type fibroblasts, but because of the higher levels originally induced, by 22 h as many adducts remain as in XP cells, and this might be the cause of the high cytotoxicity of cisplatin for FA cells. The persistence of interstrand cross-links formed by cisplatin (Plooy *et al.*, 1985) may also play a crucial role in the high sensitivity of FA cells.

The results indicate that the cytotoxicity of cisplatin can be influenced to a large extent by the repair capacity of the cells under consideration. As the persistence of DNA adducts can give a fair reflection of this capacity, it appears worthwhile to study not only the levels of adduct induction in biological materials from cisplatin-treated patients, but also the rapidity of adduct removal.

Cisplatin-DNA adducts in cultured tumour cells

In cancer chemotherapy, a wide variation has been observed among tumour types in response to cisplatin treatment. In order to study this phenomenon, a human bladder carcinoma cell line (RT112) and two cell lines derived from human germ-cell tumours of the testis (SUSA and 833K) have been investigated for the induction and repair of cisplatin-DNA adducts (Bedford *et al.*, 1988). The RT112 cells are substantially less sensitive to cisplatin than either testicular cell line; the difference in cytotoxicity is about three to five fold, the SUSA proving the more sensitive.

After cisplatin treatment of these three cell lines, all four cisplatin products were found in the DNA digests, Pt-GG being the major adduct. Figure 6 gives the amounts of the Pt-GG adducts in the cells immediately after exposure to cisplatin and after an 18-h post-treatment incubation. The latter data have been corrected for dilution by DNA synthesis during this period (dilution factor, 0.7–0.8).

After identical cisplatin treatment, different Pt-GG levels were reached in the three cell lines which did not correlate with differences in cytotoxicity; these levels ranked SUSA < RT112 < 833K. After the post-incubation period, the ratio had changed to RT112 < SUSA = 833K. In particular, the difference between the two sensitive testicular tumour cell lines is striking: in SUSA, the level of induction is relatively low,

Fig. 6. Pt-GG adducts (mean ± range) in cultured human bladder carcinoma (RT112) and testicular tumour (SUSA and 833K) cells after a 1-h exposure to 5 μg cisplatin/ml (hatched bars)

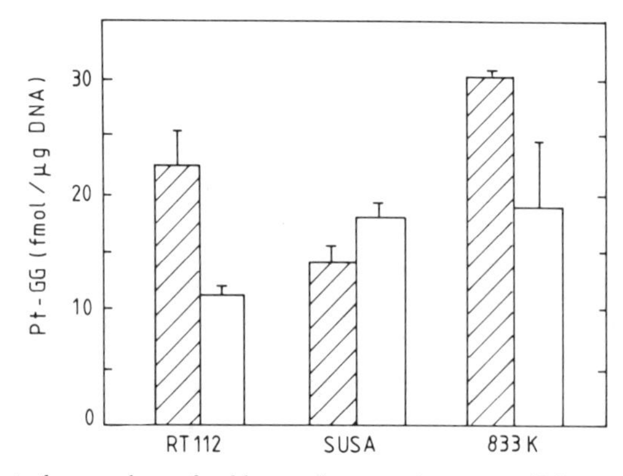

The open bars represent the number of adducts after a subsequent 18-h post-incubation period. Data adapted from Bedford *et al.* (1987)

but these cells appear to be devoid of any repair activity; in 833K the induction is twice as high, but, owing to repair proficiency, after 18 h the level has been reduced to the same value as in SUSA. The much less cisplatin-susceptible RT112 cells are at an intermediate level with regard to adduct induction, but after the 18-h repair period fewer adducts remain than in the testis-derived cells.

More extensive data will be needed on the persistence of the DNA lesions before a valid conclusion with regard to the relation with sensitivity to cisplatin can be reached. On the basis of the present information it appears likely that a difference in the capacity to process the original DNA adducts has an important influence on the susceptibility of cells to the cytostatic action of cisplatin, a notion consistent with the results obtained with repair-deficient XP and FA cells.

Conclusions

With the currently available immunochemical techniques and antibodies, it is possible to analyse the induction and repair of cisplatin-DNA adducts in a range of cell lines treated *in vitro* or *in vivo* with biologically relevant dosages of cisplatin. This opens new possibilities for gathering information about the mode of action of cisplatin on DNA. Knowledge about the formation of adducts and their repair may lead to the identification of the 'antitumour lesion', and give insight into differences in cisplatin susceptibility between individuals, between tumour types, and between cisplatin-sensitive tumour cells and their resistant daughter cells. Data on the behaviour of adducts in various tissues and organs might help to explain why certain sites are more susceptible to the undesirable side-effects of cisplatin than others.

These research aims may be realized within the next few years if attempts in progress (Fichtinger-Schepman *et al.*, 1986) to obtain (monoclonal) antibodies that recognize selectively and sensitively adducts in DNA in single cells are successful, allowing analysis of large numbers of very small samples and discrimination of cell types in tissue slices.

Acknowledgements

The authors thank their colleagues and collaborators, whose names are mentioned as coauthors in the list of references. In addition, W.C.M. van Dijk-Knijnenburg, S.D. Visser and G.E. v.d. Hout are acknowledged for their technical assistance. These investigations were supported by the Netherlands Cancer Foundation (Koningin Wilhelmina Fonds), grants KWF-MBL 79–1, 83–1 and 87–1 and in part by the Imperial Cancer Research Fund (London).

References

Bedford, P., Fichtinger-Schepman, A.M.J., Walker, M.C., Masters, J.R.W. & Hill, B.T. (1988) Differential repair of platinum-DNA adducts in human bladder and testicular tumor continuous cell lines. *Cancer Res.* (in press)

Fichtinger-Schepman, A.M.J., Van der Veer, J.L., Lohman, P.H.M. & Reedijk, J. (1984) A simple method for the inactivation of monofunctionally DNA-bound *cis*-diamminedichloroplatinum(II). *J. inorg. Biochem., 21,* 103–112

Fichtinger-Schepman, A.M.J., Van der Veer, J.L., Den Hartog, J.H.J., Lohman, P.H.M. & Reedijk, J. (1985a) Adducts of the antitumor drug *cis*-diamminedichloroplatinum(II) with DNA: formation, identification, and quantitation. *Biochemistry, 24,* 707–713

Fichtinger-Schepman, A.M.J., Baan, R.A., Luiten-Schuite, A., Van Dijk, M. & Lohman, P.H.M. (1985b) Immunochemical quantitation of adducts induced in DNA by *cis*-diamminedichloroplatinum(II) and analysis of adduct-related DNA-unwinding. *Chem.-biol. Interactions, 55,* 275–288

Fichtinger-Schepman, A.M.J., Lohman, P.H.M., Berends, F., Reedijk, J. & Van Oosterom,

A.T. (1986) *Interactions of the antitumour drug cisplatin with DNA* in vitro *and* in vivo. In: Schmähl, D. & Kaldor, J.M., eds, *Carcinogenicity of Alkylating Cytostatic Drugs (IARC Scientific Publications No. 78),* Lyon, International Agency for Research on Cancer, pp. 83–99

Fichtinger-Schepman, A.M.J., Van Oosterom, A.T., Lohman, P.H.M. & Berends, F. (1987a) Cisplatin-induced DNA adducts in peripheral leucocytes from seven cancer patients: quantitative immunochemical detection of the adduct induction and removal after a single dose of cisplatin. *Cancer Res., 47,* 3000–3004

Fichtinger-Schepman, A.M.J., Van Oosterom, A.T., Lohman, P.H.M. & Berends, F. (1987b) Interindividual human variation in cisplatinum sensitivity, predictable in an in vitro assay? *Mutat. Res., 190,* 59–62

Fichtinger-Schepman, A.M.J., Dijt, F.J., De Jong, W.H., Van Oosterom, A.T. & Berends, F. (1988) *In vivo cis-diamminedichloroplatinum(II)-DNA adduct formation and removal as measured with immunochemical techniques.* In: Nicolini, M., ed., *Platinum and Other Metal Coordination Compounds in Cancer Chemotherapy,* Boston, Martinus Nijhoff, pp. 32–46

Plooy, A.C.M., Van Dijk, M., Berends, F. & Lohman, P.H.M. (1985) Formation and repair of DNA interstrand cross-links in relation to cytotoxicity and unscheduled DNA synthesis induced in control and mutant human cells treated with *cis*-diamminedichloroplatinum(II). *Cancer Res., 45,* 4178–4184

Poirier, M.C., Reed, E., Zwelling, L.A., Ozols, R.F., Litterst, C.L. & Yuspa, S.H. (1985) Polyclonal antibodies to quantitate *cis*-diamminedichloroplatinum(II)-DNA adducts in cancer patients and animal models. *Environ. Health Perspect., 62,* 89–94

DETERMINATION OF CISPLATIN IN BLOOD COMPARTMENTS OF CANCER PATIENTS

R. Mustonen,[1] K. Hemminki,[1] A. Alhonen,[1] P. Hietanen[2] & M. Kiilunen[1]

[1]*Institute of Occupational Health, Helsinki; and* [2]*Department of Radiotherapy and Oncology, University Central Hospital, Helsinki, Finland*

A new approach has been developed to determine the levels of cisplatin in different blood compartments of treated cancer patients. The cisplatin content of plasma, plasma proteins, red blood cells and white blood cell DNA can be measured by atomic absorption spectroscopy (AAS). The approximate levels of cisplatin were 10, 500 and 100 ng Pt/ml blood in plasma, plasma proteins and haemoglobin, respectively; in white blood cell DNA, the level of cisplatin was about 1 pg/µg DNA. Preliminary data indicate that cancer patients have measurable amounts of cisplatin in their blood compartments. Furthermore, antibodies have been raised against cisplatin-DNA, with which 50% inhibition occurs at 50–100 fmol cisplatin. The detection limit is about 1–10 fmol cisplatin/µg DNA. Enzyme immunoassay techniques will be used to detect cisplatin-DNA adducts in white blood cell DNA of cancer patients.

Cisplatin is used in the treatment especially of ovarian and testicular cancers (Einhorn & Donohue, 1977; Young *et al.*, 1979). The mechanism by which it exerts its antineoplastic activity is associated with damage to DNA: it binds to DNA and produces both monofunctional and bifunctional adducts. A new approach has been developed to determine the levels of cisplatin in different blood compartments of cancer patients. The cisplatin content of plasma, plasma proteins, red blood cells and white blood cell DNA can be measured by AAS. Furthermore, antibodies have been raised against cisplatin-DNA.

Determination of cisplatin by AAS in different blood compartments of cancer patients

Blood samples from individuals receiving cancer chemotherapy were fractionated according to the scheme shown in Figure 1. Plasma and blood cells were separated from heparinized blood. The cisplatin concentration in red blood cells was measured in cells lysed with NH_4Cl; DNA was purified from white blood cells by RNase and proteinase treatment, phenol and chloroform/isoamyl alcohol extraction and ethanol precipitation. Unbound cisplatin in plasma was measured after ethanol precipitation of plasma proteins; protein-bound cisplatin was measured in the dialysate.

Our preliminary results are based on 20 blood samples from patients with different types of cancer. The total dose of cisplatin varied with the treatment protocol but was between 60 and 180 mg. Blood samples were taken one to two days after the latest dose of cisplatin. Approximate levels found were 10, 500 and 100 ng cisplatin/ml blood in plasma, plasma protein and red blood cell fractions, respectively. White blood cell

Fig. 1. Determination of cisplatin in blood compartments

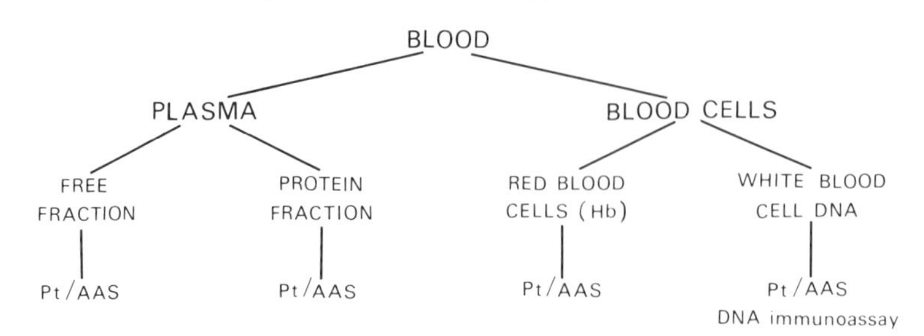

DNA contained a level of about 1 pg/μg DNA. The levels correspond well to values obtained previously with AAS (Knox *et al.*, 1986) and by immunoassay (Fichtinger-Schepman *et al.*, 1987), although the values reported by Reed *et al.* (1986) were lower. In the present study, several-fold increases in cisplatin content were detected in plasma protein and red blood cell fractions in treated cancer patients as compared to levels before treatment (data not shown).

These preliminary results indicate that cancer patients receiving cisplatin chemotherapy have measurable amounts of cisplatin in their blood compartments. This information could be used in monitoring cisplatin levels during chemotherapy.

Development of enzyme immunoassay for detecting cisplatin-DNA adducts

A sensitive enzyme-linked immunosorbent assay (ELISA) for detecting cisplatin-DNA adducts has been developed, and polyclonal and monoclonal antibodies have been raised against cisplatin-DNA. Noncompetitive and competitive ELISA have been used to determine the specificity and sensitivity of these antibodies.

Noncompetitive ELISA showed that antibodies against cisplatin-DNA react with native as well as heat-denatured, cisplatin-modified DNA (Fig. 2): the antibodies do not bind to control calf thymus DNA. The lowest amount that can be differentiated from control DNA is 100 pg cisplatin-DNA. Competitive ELISA was used to

Fig. 2. Antigen titration using noncompetitive ELISA

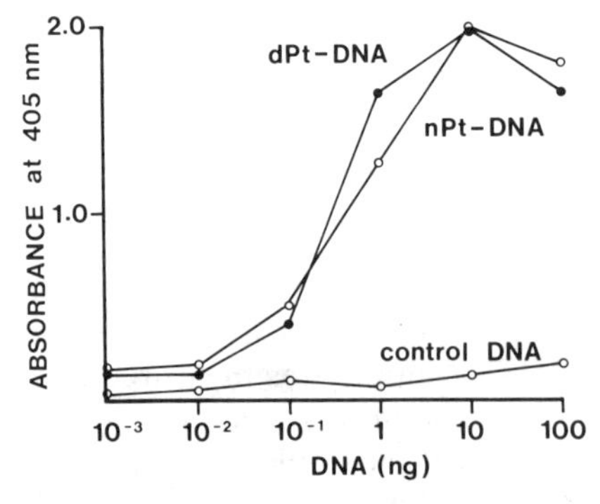

Plates were coated with increasing amounts of control DNA or native (○) or heat-denatured (●) cisplatin-modified DNA (the level of modification was 5.5%). The polyclonal antiserum against cisplatin-DNA was added at a dilution of 1:80 000. Antigen-antibody reactions were determined by spectrometric measurement of the enzymatic product, *p*-nitrophenol.

Fig. 3. Inhibition of antibody binding to plate *versus* molar concentration of cisplatin as determined by competitive ELISA

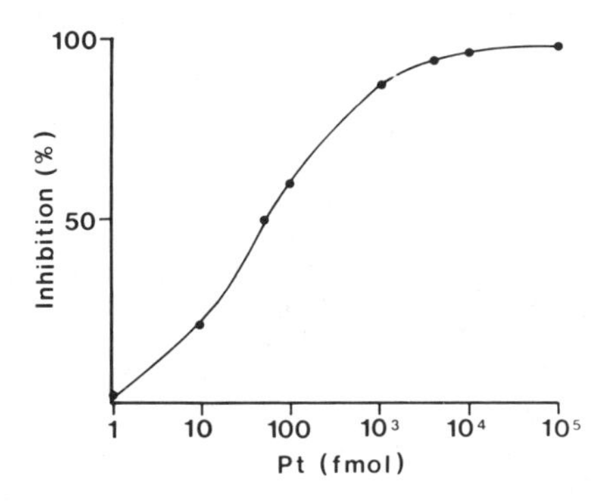

Plates were coated with 10 ng cisplatin-modified DNA (the level of modification was 5.5%). In separate tubes, increasing amounts of cisplatin were mixed with anti-cisplatin-DNA antibodies (dilution, 1:80 000). The total amount of DNA was 10 μg/well. The mixtures were added to the plates and percent inhibition was calculated. Antibody sensitivity is expressed as 50% inhibition.

determine the sensitivity of the antibodies; 50% inhibition occurs at 50–100 fmol cisplatin (Fig. 3). The detection limit is about 1–10 fmol of cisplatin/μg DNA.

Enzymatic degradation (nuclease P1 and alkaline phosphatase; anion-exchange chromatography) of cisplatin did not improve the sensitivity of the antibodies directed toward cisplatin-DNA (Fig. 4)

Our antibodies raised against cisplatin-modified DNA recognize the native conformation of cisplatin-DNA better than enzymatically digested shorter segments of cisplatin-DNA. These data can be applied in raising antibodies against cisplatin-DNA adducts. Furthermore, this method can be used in detecting cisplatin-induced modifications in DNA of cancer patients receiving cisplatin chemotherapy.

Fig. 4. Effects of enzymatic digestion of DNA on sensitivity of antibodies

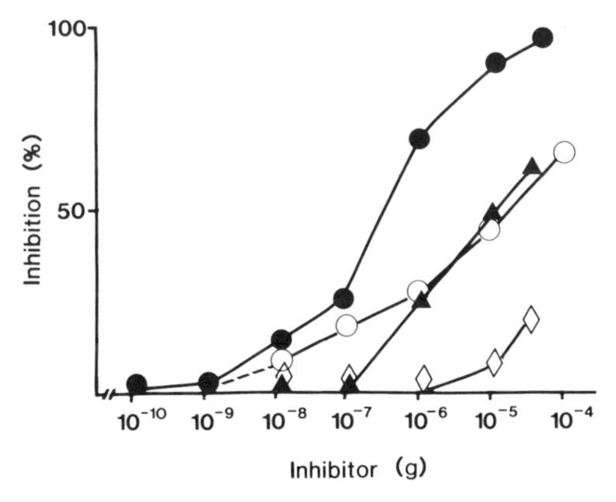

Plates were coated with 10 ng cisplatin-DNA per well. Native (●) and enzymatically digested cisplatin-DNA (▲) as well as chromatographically purified cisplatin-DNA adduct fraction (○) were used as inhibitors. The modification level of cisplatin-DNA was 1%. Enzymatically digested control DNA (◇) was also included. The dilution of anti-cisplatin-DNA was 1:80 000.

References

Einhorn, L.H. & Donohue, J. (1977) *cis*-Diamminedichloroplatinum, vinblastine, and bleomycin combination chemotherapy in disseminated testicular cancer. *Ann. intern. Med., 87,* 293–298

Fichtinger-Schepman, A.M.J., van Oosterom, A.T., Lohman, P.H.M. & Berends, F. (1987) *cis*-Diamminedichloroplatinum(II)-induced DNA adducts in peripheral leukocytes from seven cancer patients: quantitative immunochemical detection of the adduct induction and removal after a single dose of *cis*-diamminedichloroplatinum. *Cancer Res., 47,* 3000–3004

Knox, R.J., Friedlos, F., Lydall, D.A. & Roberts, J.J. (1986) Mechanisms of cytotoxicity of anticancer platinum drugs: evidence that *cis*-diamminedichloroplatinum(II) and *cis*-diammine(1,1-cyclobutanedicarboxylato)platinum(II) differ only in the kinetics of their interaction with DNA. *Cancer Res., 46,* 1972–1979

Reed, E., Yuspa, S.H., Zwelling, L.A., Ozols, R.F. & Poirier, M. (1986) Quantitation of *cis*-diamminedichloroplatinum(II) (cisplatin)-DNA intrastrand adducts in testicular and ovarian cancer patients receiving cisplatin chemotherapy. *J. clin. Invest., 77,* 545–550

Young, R.C., von Hoff, D.D., Gormley, R., Makuch, J., Cassidy, J., Howser, D. & Bull, J.M. (1979) *cis*-Dichlorodiammineplatinum(II) for treatment of advanced ovarian cancer. *Cancer Treat. Rev., 63,* 1539–1544

DETECTION AND QUANTIFICATION OF 8-METHOXYPSORALEN-DNA ADDUCTS

R.M. Santella,[1] X.Y. Yang,[1] V.A. DeLeo[2] & F.P. Gasparro[3]

[1] Comprehensive Cancer Center and Division of Environmental Sciences, School of Public Health; and [2] Department of Dermatology, Columbia University, New York, NY; and [3] Department of Dermatology, Yale University, New Haven, CT, USA

8-Methoxypsoralen (8-MOP) is a photoactivated drug used clinically in the treatment of psoriasis and cutaneous T-cell lymphoma (CTCL). We have developed monoclonal antibodies which specifically recognize 8-MOP-modified DNA and do not cross-react with unmodified DNA or free 8-MOP. Highly sensitive, competitive enzyme-linked immunosorbent assays (ELISA) have been developed with both colour and fluorescence endpoint detection for quantification of DNA adducts in biological samples. In addition, immunofluorescence and flow cytometric techniques have been developed to visualize adducts in tissues and cells. These techniques have been validated in keratinocytes treated in culture and in animals treated *in vivo* with 8-MOP and ultraviolet A (UVA) light. Adduct levels have also been monitored in skin biopsies and lymphocytes of patients with psoriasis and lymphoma.

The combination of 8-MOP and UVA (320–400 nm), termed PUVA, is used clinically in the treatment of psoriasis, a hyperproliferative disease of the epidermis (Parrish *et al.*, 1974). More recently, it has been used extracorporeally as a cytoreductive treatment in the leukaemic phase of CTCL (Edelson *et al.*, 1987). PUVA patients have provided an ideal population for the development and validation of techniques to monitor human exposure to carcinogenic and mutagenic chemicals by measurement of DNA adducts: high, well-defined doses are administered and unexposed controls are readily obtained.

Psoralen photoreacts primarily with the thymines in DNA, forming both monoadducts and interstrand cross-linked adducts (Song & Tapley, 1979; Ben-hur & Song, 1984). We have developed a panel of monoclonal antibodies that specifically recognize 8-MOP-modified DNA but not free psoralen or unmodified DNA (Santella *et al.*, 1985). Characterization of the most sensitive antibody, 8Gl, indicated preferential reaction with the 4′,5′ monoadduct. A sensitive, competitive ELISA was developed to quantify levels of the adduct in biological samples. Details of the assay have been published (Santella *et al.*, 1985; Yang *et al.*, 1987): 96-microwell plates were coated with 10 ng of DNA modified *in vitro* with 8-MOP; antibody 8G1 was diluted $1:3 \times 10^4$ and mixed with serial dilutions of standard in-vitro modified DNA or unknown before addition to the wells. Specific antibody binding to the well was quantified with a goat anti-mouse immunoglobin G-alkaline phosphatase conjugate. For the colour ELISA, the enzyme substrate was *p*-nitrophenyl phosphate, and 50% inhibition occurred at 17 fmol 8-MOP-DNA adduct. Alternatively, a fluorescent ELISA can be used in which the plates are coated with lower amounts (0.5 ng) of 8-MOP-DNA and the antibody diluted $1:6 \times 10^5$. The substrate for alkaline phosphatase is 4-methylumbelliferyl

Fig. 1. Level of 8-MOP-DNA adducts in lymphocytes after treatment *in vitro* with 400 ng/ml ^{3}H and increasing doses of UVA

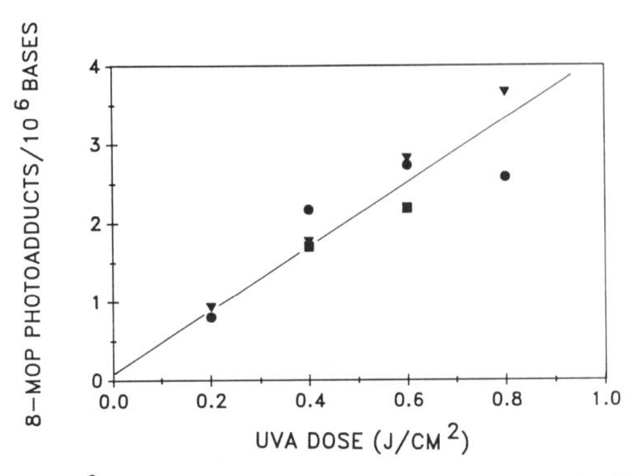

Adduct levels were determined by ^{3}H counting (▼) or competitive ELISA (■,●) with antibody 8G1.

phosphate, and this assay has 50% inhibition at 4 fmol adduct. Since 50 μg of DNA can be assayed per microwell, one adduct in 10^{8} normal nucleotides can be detected.

To validate the ELISA, lymphocytes were treated with 400 ng/ml ^{3}H-8-MOP and increasing doses of UVA. DNA was isolated from the cells and assayed for adducts by competitive ELISA and scintillation counting. The two techniques provided similar data and indicate a UVA dose-dependent increase in adduct level (Fig. 1).

Immunofluorescence detection of 8-MOP-DNA adducts

We have also developed techniques for visualizing 8-MOP-DNA adducts in tissue samples by immunofluorescence techniques (Yang *et al.*, 1987). Antibody 9D8 (Santella *et al.*, 1985) was found to be the most sensitive for these studies. Initially, human keratinocytes were treated with 0.25–10 μg/ml 8-MOP and 12 J/cm^{2} UVA. Ethanol-fixed cells were treated with RNase A to eliminate potential cross-reactivity with RNA adducts, and with proteinase K to release proteins from the DNA to enhance antibody binding. This was followed by treatment with 4 N HCl to denature the DNA *in situ* and increase sensitivity. Specific nuclear staining could be detected at all concentrations of 8-MOP tested. With treatment above 2.5 μg/ml 8-MOP, nuclear staining was homogeneous (Fig. 2A). Below 0.5 μg/ml 8-MOP treatment, staining was granular and weaker. No staining of the cytoplasm was detected. Controls, including cells treated with dimethyl sulphoxide and 8-MOP-treated cells stained with nonspecific antibody or treated with DNase before staining, were all unstained. DNA was isolated from cells in a number of samples and the level of adducts measured by ^{3}H and/or competitive ELISA. Table 1 indicates that the lowest level of 8-MOP treatment that results in positive immunofluorescence staining is 0.25 μg/ml. This treatment corresponds to 9 fmol adduct/μg DNA or 2.9 adducts/10^{6} nucleotides. More recently, we have used photon counting immunofluorescence to increase the level of sensitivity over that of conventional immunofluorescence. Keratinocytes were treated with lower doses of 8-MOP and fluorescence measured after staining. Figure 3 gives relative fluorescence intensity (the average of measurements on five single cells) as a function of 8-MOP treatment. With this technique, it is possible to detect adducts after treatment with 10–100 ng/ml 8-MOP.

Animals were also treated *in vivo* with 8-MOP by either intraperitoneal or

Fig. 2. Immunofluorescence staining of human keratinocytes treated with 10 μg/ml 8-MOP and 12 J/cm² UVA (A) and skin from mice treated intraperitoneally with 30 μg/g 8-MOP(B).

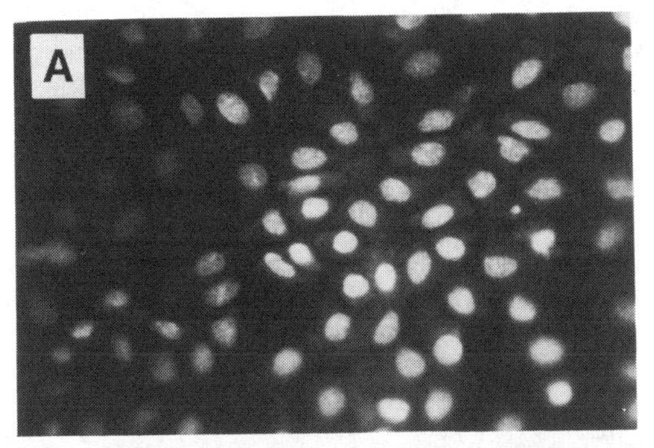

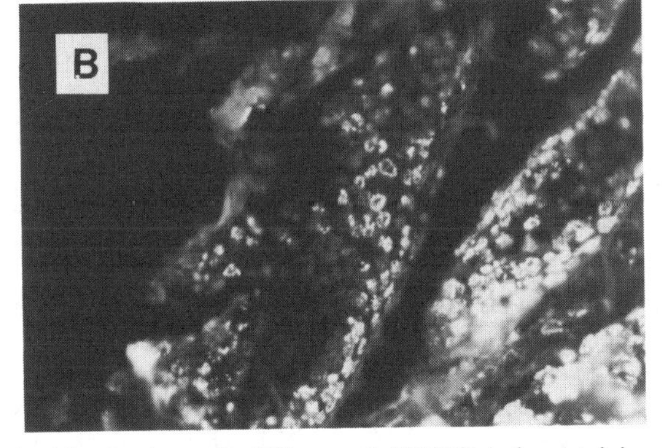

Fixed slides were treated with proteinase K, RNase and 4 N HCl before staining with anti-8-MOP-DNA antibody (9D8, 1:100 dilution). Goat anti-mouse immunoglobulin G-fluorescein-conjugated antibody was used at 1:40 dilution.

Table 1. Immunofluorescence staining and 8-MOP-DNA adduct levels

Sample	Dose		Relative immuno-fluorescence staining	8-MOP-adduct level (fmol adduct/μg DNA)	
	8-MOP	UVA (J/cm^2)		^3H	ELISA
In vitro					
human keratinocyte	2.5 μg/ml	12	+ + +	103 ± 1	104 ± 31
human keratinocyte	0.5 μg/ml	12	+ +	23 ± 1	39 ± 2
human keratinocyte	0.25 μg/ml	12	+		9 ± 6
human keratinocyte	0	12	−		
In vivo					
mouse skin intradermally	100 μg/cm²	12	+ +		59 ± 9
mouse skin intraperitoneally	30 μg/g	12	+ +		32 ± 7
human skin topically	75 μg/cm²	22	+ + +		
human skin orally	0.6 mg/kg	9	+		

Fig. 3. Relative fluorescence of keratinocytes treated with increasing doses of 8-MOP and 12 J/cm² UVA

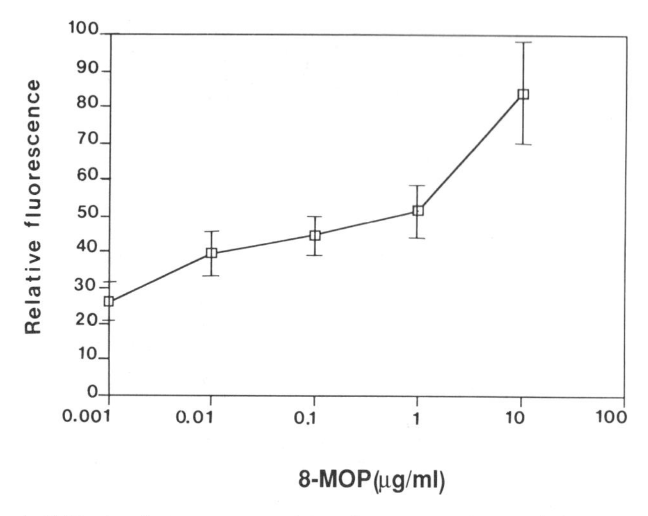

Fluorescence in five individual cells was measured by photon counting, and data are expressed as mean ±SD.

intradermal injection, followed by skin irradiation (Table 1). Immunohistochemical staining of cryostat sections of skin from a mouse treated intraperitoneally with 30 μg 8-MOP/g body weight indicated that the adducts were localized mainly in the keratinocytes of the epidermal layer (Fig. 2B). DNA isolated from the skin sections contained adducts at a level of 32 fmol/μg DNA when measured by ELISA (Table 1).

In order to develop the immunofluorescence technique for use on human samples, a volunteer received 8-MOP (30 μl 0.1% solution/4 cm²) topically, and a 4-mm punch biopsy was removed after 30 min. This biopsy was then irradiated *in vitro* with 2 or 22 J/cm². Fixed 4-μm sections were stained with antibody 9D8 without acid denaturing or enzyme treatment and the slides incubated with goat anti-mouse immunoglobulin G-fluorescein-conjugated antibody. Specific nuclear staining was visible in the stratified squamous epithelium of the epidermis (Fig. 4A). No staining was visible in the cytoplasm or the dermis. A control, treated with nonspecific antiserum and fluorescein-conjugated second antibody, was not stained (Fig. 4B).

Flow cytometric methods for 8-MOP-DNA adduct detection

Flow cytometry is a convenient method for the detection and quantification of cellular antigens. We have adapted these techniques for use with 8-MOP-DNA antibodies. Simultaneous measurement of adducts with fluorescein-labelled antibodies and of cellular DNA content with propidium iodide was possible. Initial studies were done with human keratinocytes treated in culture with 8-MOP and UVA. Cells were fixed in 70% ethanol then treated with RNase A, proteinase K and 1.5 N HCl before staining with specific antibody 9D8 and fluorescein-conjugated goat anti-mouse immunoglobulin G. Finally, cells were incubated with propidium iodide (Dolbeare *et al.*, 1983) to measure DNA content and analysed with a Becton Dickinson FACS IV flow cytometer. With this approach, cell size can be determined from the scattered light (Fig. 5A and D), the level of 8-MOP-DNA antibody binding from the fluorescein

Fig. 4. Immunofluorescence staining of a human skin biopsy from a volunteer treated topically with 8-MOP and biopsied 30 min later

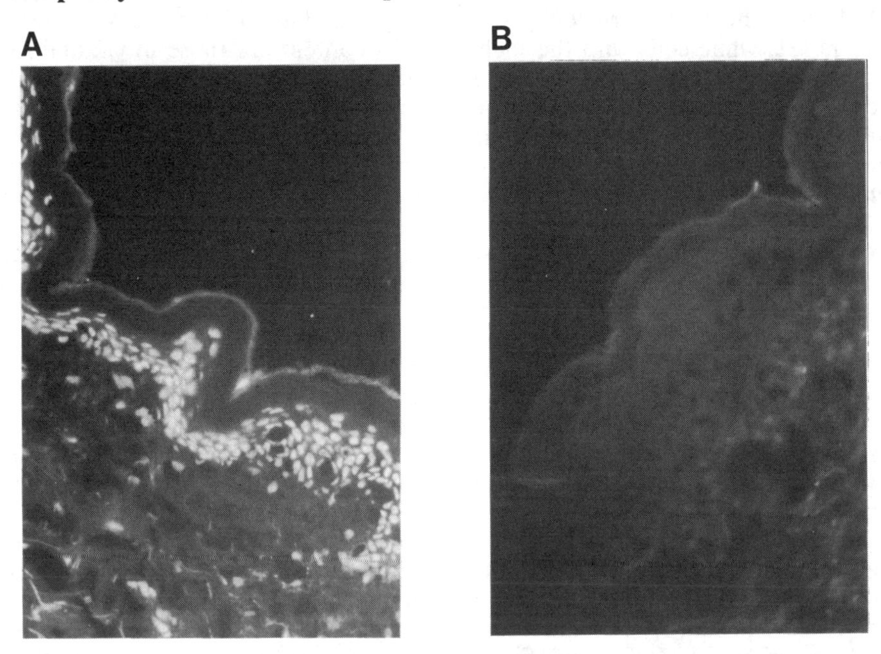

The biopsy was irradiated *in vitro* with 22 J/cm^2. Fixed 4-μm sections were stained directly with antibody 9D8 (1:10) and goat anti-mouse IgG-fluorescein antibody (1:10). (A) Biopsy stained with specific antiserum; (B) biopsy stained with nonspecific antiserum

Fig. 5. Flow cytometric analysis of human keratinocytes treated with 1 µg/ml 8-MOP and 12 J/cm² UVA

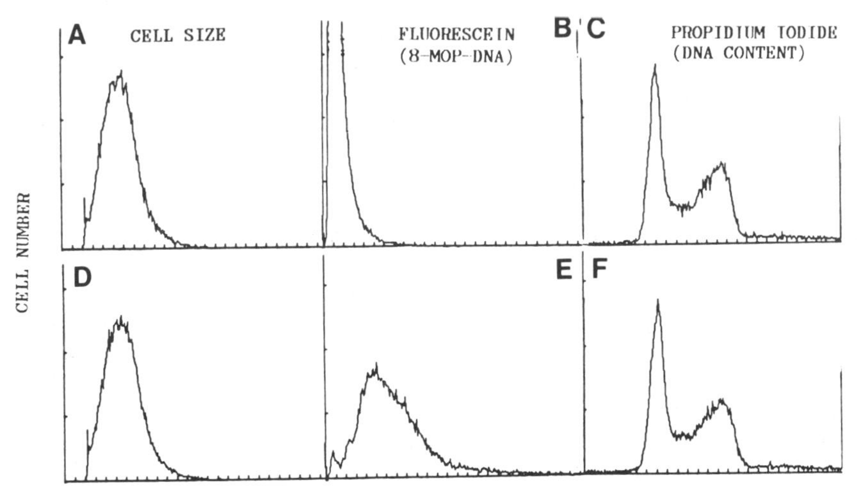

Fixed cells were treated with RNase, proteinase K and 1.5 N HCl before staining with antibody 9D8 (A–C) or nonspecific serum (D–F). In panels A and D, cell size is monitored by relative light scattering, B and E give fluorescein content, a measure of 8-MOP-DNA levels, and C and F, propidium iodide content, a measure of DNA content.

levels (5B and E) and relative DNA content from the propidium iodide content (5C and F). Cell size and DNA content do not change after treatment with 1 μg/ml 8-MOP and 12 J/cm^2. The characteristic DNA content profile has an initial peak corresponding to cells in G1, while cells with the highest DNA content are those in G2 and M. Cells in S appear between these two peaks. In contrast to the results on cell size and DNA content, cells treated with 8-MOP have much higher mean fluorescein levels (relative fluorescence, 71, Fig. 5B) than control cells (relative fluorescence, 17, Fig. 5E). Currently, the lowest level of 8-MOP treatment that can be detected by flow cytometry is 1 μg/ml and 12 J/cm^2, which generally produces about one adduct/10^5 nucleotides. It should be possible to improve these sensitivities with biotin-streptavidin-labelled fluorescent probes. While improved sensitivities would be necessary for monitoring adducts in human lymphocytes, current levels may be applicable to animal studies.

Detection of 8-MOP-DNA adducts in psoriasis patients

PUVA treatment of psoriasis patients is carried out by oral administration of 8-MOP followed 2 h later by UVA irradiation of the skin. We have obtained skin biopsies and blood samples from five patients within 20 min of UVA treatment. Treatment history varied from 27 to 186 previous treatments. Three out of five skin biopsies were found to show weak positive staining when tested with 8-MOP-DNA adduct specific antibody. On the basis of the limits of sensitivity of the immunofluorescence technique, determined from the keratinocyte studies, these positive samples probably contain around one adduct/10^6 nucleotides.

Differences in skin DNA adduct levels between patients could be due to differences in the levels of free 8-MOP circulating in the blood at the time of exposure to UVA. It is known that after oral administration of 8-MOP, plasma levels can vary depending on individual rates of absorption and excretion (Herfst & DeWolff, 1983). Plasma levels of free 8-MOP were, therefore, measured by solid-phase absorption of the drug, followed by quantification by high-performance liquid chromatography, and found to range from 17 to 104 g/ml. Drug plasma levels were not, however, related to positive immunofluorescence, since the patient with the highest plasma level of free 8-MOP did not show detectable immunofluorescence. Perhaps other factors, such as variation in DNA repair or in skin pigmentation, which influence UVA absorption, affect the extent of DNA adduct formation.

We were also interested in determining whether 8-MOP-DNA adducts could be detected in circulating lymphocytes of psoriasis patients. Previous studies indicated that patients undergoing PUVA therapy for one to six years have a small but statistically significant increase in levels of sister chromatid exchange in their peripheral lymphocytes (Bredberg *et al.*, 1983). Another study demonstrated an increased level of mutant lymphocytes in PUVA patients (Strauss *et al.*, 1979). These studies suggest that a significant fraction of administered UVA radiation can reach the peripheral lymphocytes. To determine whether 8-MOP-DNA adducts could be measured in the lymphocytes of psoriasis patients given PUVA therapy, DNA was isolated from the white blood cells of the five patients and assayed by competitive ELISA with fluorescence detection (Santella *et al.*, 1986). All samples gave negative results, indicating that the level must be below one adduct in 10^8 nucleotides, the limit of sensitivity of the assay. Lymphocytes are exposed to approximately 1–5% of the skin surface dose of UVA (Kraemer & Weinstein, 1977); prolonged treatment results in skin pigmentation, which may decrease the effective UVA dose. Since we estimate that the level of DNA adducts in the skin of treated patients is about one adduct/10^6 nucleotides, adduct levels in lymphocytes would be expected to be less than one/10^8.

Detection of 8-MOP-DNA adducts in CTCL patients

Recently, the use of PUVA has been extended to the treatment of CTCL patients; during the leukaemic phase of CTCL, PUVA is used as a cytoreductive treatment. Patients take 8-MOP orally and 2 h later are attached to a leucophoresis apparatus for separation of lymphocytes. These cells, about 20% of the total lymphocytes, are mixed with the 8-MOP-containing plasma of the patients and irradiated *in vitro*; the cells are then reinfused into the patient. This treatment has given a positive response in 27 of 37 patients (Edelson *et al.*, 1987). To monitor 8-MOP-DNA adducts in CTCL patients, lymphocytes were obtained for DNA isolation from the photopheresis apparatus after irradiation and before reinfusion. Adduct levels in ten patients were easily quantified by competitive colour ELISA and found to range from 0.1 to 15 adducts/10^6 nucleotides. We have also followed the levels of adducts in patients during repeated treatments. These studies are complicated, however, by the variable number of previous treatments and the fact that only 20% of the lymphocytes are irradiated during any one treatment cycle. We are currently enrolling new patients and will obtain blood samples from the patients several hours after reinfusion of irradiated blood instead of from the photopheresis apparatus.

Conclusion

These studies demonstrate that monoclonal antibodies that recognize 8-MOP-DNA adducts can be used in highly sensitive immunoassays to quantify levels of adducts in animals and humans. In addition, immunofluorescence techniques allow visualization of adducts in specific cell types and tissues.

Psoriasis patients treated with PUVA have a ten-fold elevated risk for cutaneous squamous-cell carcinoma (Stern *et al.*, 1984). Our studies demonstrate that PUVA treatment results in detectable levels of 8-MOP-DNA adducts in the skin of psoriasis patients in the range of one/10^6. These adducts may be related to the elevated risk of skin cancer in these patients. No adduct could be detected in the circulating lymphocytes of these patients. In contrast, high levels of adducts were detected in the lymphocytes of CTCL patients undergoing extracorporeal photopheresis. This is not surprising, since lymphocytes are irradiated *in vitro* during this treatment.

DNA adduct detection may be of use to clinicians in determining what level of adduct formation is required for therapeutic effect and for adjusting the dose of 8-MOP and UVA to maximize clinical response and minimize potential risk. In addition, biological monitoring of PUVA patients may help in our understanding of the relationship between adduct levels and tumour induction relevant to other carcinogens.

Acknowledgements

This investigation was supported by grants from NIEHS ES03881, NCI CA21111 (RMS), NIH AR 37629 (FPG) and NIH AM34365 (VAD). The authors thank T. Delohery for assistance with the flow cytometry studies and R. Yang for secretarial assistance.

References

Ben-hur, E. & Song, P.S. (1984) The photochemistry and photobiology of furocoumarines (psoralens). *Adv. Radiat. Biol.*, *11*, 131–171

Bredberg, A., Lambert, B., Lindblad, A., Swanbeck, G. & Wennersten, G. (1983) Studies of DNA and chromosome damage in skin fibroblasts and blood lymphocytes from psoriasis patients treated with 8-methoxypsoralen and UVA irradiation. *J. invest. Dermatol.*, *81*, 93–97

Dolbeare, F., Gratzner, H., Pallavicini, M.G. & Gray, J.W. (1983) Flow cytometric measurement of total DNA content and incorporated bromodeoxyuridine. *Proc. natl Acad. Sci. USA, 80,* 5573–5577

Edelson, R., Berger, C., Gasparro, F., Jegasothy, B., Heald, P., Wintroub, B., Vonderheid, E. & Knobler, R. (1987) Treatment of cutaneous T-cell lymphoma by extracorporeal photochemotherapy. *New Engl. J. Med., 316,* 297–303

Herfst, M.J. & DeWolff, F.A. (1983) Intraindividual and interindividual variability in 8-methoxypsoralen kinetics and effect in psoriatic patients. *Clin. Pharm. Ther., 34,* 117

Kraemer, K.H. & Weinstein, G.D. (1977) Decreased thymidine incorporation in circulating leucocytes after treatment of psoriasis with psoralen and long-wave ultraviolet light. *J. invest. Dermatol., 69,* 211–214

Parrish, J.A., Fitzpatrick, T.B., Pathak, M.A. & Tannenbaum, L. (1974) Photochemotherapy of psoriasis with oral methoxysalen and long wave ultraviolet light. *New Engl. J. Med., 291,* 1207–1211

Santella, R.M., Dharmaraja, N., Gasparro, F.P. & Edelson, R.L. (1985) Monoclonal antibodies to DNA modified by 8-methoxypsoralen and ultraviolet A light. *Nucleic Acids Res., 13,* 2533–2544

Santella, R.M., Gasparro, F.P. & Edelson, R.L. (1986) *Quantification of methoxsalen-DNA adducts with specific antibodies.* In: Schmähl, D. & Kaldor, J.M., eds, *Carcinogenicity of Alkylating Cytostatic Drugs (IARC Scientific Publications No. 83),* Lyon, International Agency for Research on Cancer, pp. 127–139

Song, P.S. & Tapley, J.K. (1979) Photochemistry and photobiology of psoralens. *Photobiology, 29,* 1177–1197

Stern, R.S., Laird, N., Melski, J., Parrish, J.A., Fitzpatrick, T.B. & Blech, H.I. (1984) Cutaneous squamous-cell carcinoma in patients treated with PUVA. *New Engl. J. Med., 310,* 1156–1161

Strauss, G.H., Albertini, R.J., Krusinski, P.A. & Baughman, R.D. (1979) 6-Thioguanine resistant peripheral blood lymphocytes in humans following psoralen, long wave ultraviolet light (PUVA) therapy. *J. invest. Dermatol., 73,* 211–216

Yang, X.Y., DeLeo, V. & Santella, R.M. (1987) Immunological detection and visualization of 8-methoxypsoralen-DNA photoadducts. *Cancer Res., 47,* 2451–2455

IMMUNOASSAY OF DITHYMIDINE CYCLOBUTANE DIMERS IN NANOGRAM QUANTITIES OF DNA

P.T. Strickland & J.S. Creasey

Division of Occupational Medicine, Department of Environmental Health Sciences, The Johns Hopkins University School of Hygiene and Public Health, Baltimore, MD, USA

Immunological assays for DNA photoproducts, which have been in use for a number of years, provide a rapid means for quantifying photodamage in DNA. Radioimmunoassays (RIA) are capable of measuring photoproducts in microgram quantities (1–10 μg) of DNA extracted from biological samples, e.g., animal skin or cultured cells. We have refined an indirect enzyme-linked immunosorbent assay (ELISA) for the analysis of cyclobutadithymidine (T-T) photoproducts in nanogram quantities (3–100 ng) of DNA. T-T photoproducts were detectable in 30 ng DNA irradiated with a minimum of 10 J/m^2 of 254 nm ultraviolet light (UV). The efficiency of T-T induction in DNA was approximately 4000-fold lower per unit dose using a solar-simulating UV radiation source (290–400 nm) than with 254 nm radiation. The assay was used to measure T-T photoproducts induced by solar-simulated UV radiation in hamster skin and human skin and should be useful for the routine analysis of photoproducts in small amounts of DNA (500–1000 ng) available from human skin biopsies.

Immunological assays for DNA photoproducts provide a rapid, simple approach to assessing DNA damage caused by UV radiation. Antibodies specific for DNA photoproducts have been used in RIA, direct ELISA and immunofluorescent assay (IFA) to quantify these photoproducts in cultured cells or animal skin (Strickland, 1985). In the present study we have determined optimal conditions for a competitive (indirect) ELISA to measure photoproducts in nanogram quantities (3–100 ng) of DNA.

Monoclonal antibody

Characterization of monoclonal antibody UVssDNA-1 (Strickland & Boyle, 1981) by RIA (Fig. 1 and unpublished) indicates that it is specific for cyclobuta-T-T and does not cross-react with cytosine containing cyclobutadipyrimidines (C-T or C-C). The antibody binds to T-T within an oligonucleotide sequence at least four nucleotides long. This is consistent with the size of the binding site for other antibodies to DNA moieties and is slightly smaller than the approximate size (3 nm) of the immunoglobulin combining site (Kabat, 1976).

ELISA for T-T photoproducts

Costar 96-well culture clusters were prepared by drying 100 μl phosphate-buffered saline (PBS) containing 3 ng UV-irradiated (1000 J/m^2, 254 nm), heat-denatured calf thymus DNA in each well (16 h/37°C). Prior to use, plates were incubated (1 h/37°C)

Fig. 1. Antibody specificity

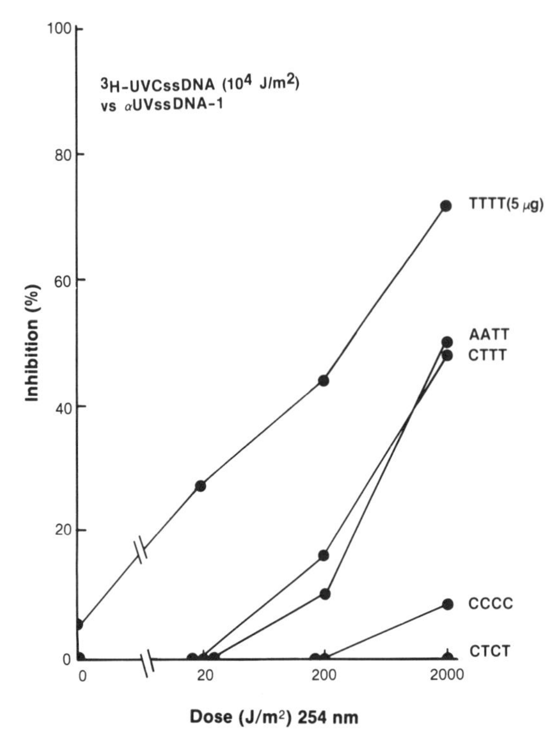

Inhibition of antibody binding to ^{3}H-labelled UV-irradiated DNA (100 ng) by defined sequence tetranucleotides (5 μg) irradiated with indicated UV doses. RIA performed as described by Strickland and Boyle (1981). The antibody binds specifically with photoproducts formed by adjacent thymidines and does not cross-react with cytosine-containing photoproducts.

with 200 μl/well of 30 mM PBS containing 1% bovine serum albumin (BSA) and 15 ng/ml of unirradiated calf thymus DNA to block nonspecific antibody binding. Wells were emptied and washed four times with 30 mM PBS. Experimental DNA samples (3–100 ng/50 μl) or DNA standards containing known amounts of photoproducts were premixed with monoclonal antibody (1 ng/50 μl) specific for T-T and immediately added (100 μl/well) to microtitre plates (5 wells per sample) and incubated (30 min/37°C). After washing three times with 30 mM PBS containing 1% BSA and 0.05% Tween 20, anti-mouse immunoglobin G-peroxidase conjugate (Biorad, 100 μl of 1:500 dilution), was added and incubated (1 h/37°C). After washing wells four times with PBS/BSA/Tween, 100 μl enzyme substrate (0.2% *o*-phenylenediamine and 0.002% H_2O_2) were added and incubated (1 h/37°C). The reaction was stopped by adding 3 M H_2SO_4, and the absorption at 495 nm measured with a Cambridge Technology plate reader. Percent inhibition was calculated from quintuplicate wells as follows:

$$\text{Percent inhibition} = 100 \times (E_2 - E_1)/(E_2 - E_3),$$

where E_1 is the average OD_{495} of the wells containing inhibitor; E_2 is the average OD_{495} of the wells without inhibitor (0% inhibition); and E_3 is the average OD_{495} of the wells without antigen (100% inhibition).

Fig. 2. ELISA standard curve

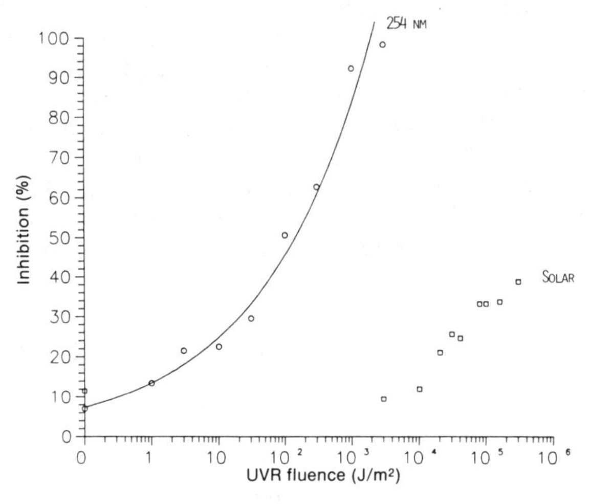

Inhibition of antibody binding to immobilized UV-irradiated DNA (3 ng) by DNA (30 ng) irradiated with indicated doses of 254 nm or solar-simulated UV radiation. ELISA performed as described in text

Photoproduct measurements

The competitive ELISA was optimized for analysis of T-T in small amounts of DNA, and a standard curve was generated using DNA irradiated with 254 nm UV radiation (Fig. 2). Using 30 ng inhibitor DNA, T-T induced by 3–10 J/m^2 of 254 nm UV were detectable. This corresponds to 0.03–0.1 T-T per kilobase (kb) as determined by paper chromatography of hydrolysed radiolabelled DNA or high-performance liquid chromatography of unlabelled DNA. DNA irradiated with a solar-simulating light source (290–400 nm) was also assayed by competitive ELISA. This light source produced T-T photoproducts approximately 4000 times less efficiently than 254 nm UV (Fig. 2).

Fig. 3. T-T induction in hamster skin irradiated with solar UV

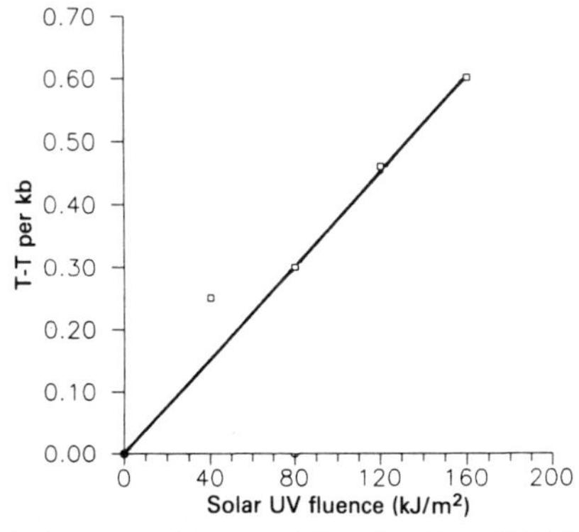

Cyclobuta T-T per kb of DNA in hamster skin was determined by ELISA. Hamster skin was irradiated *in vivo* with indicated doses of solar UV and DNA extracted and analysed as described in text

Syrian hamsters were irradiated with the solar simulating light source (0, 40, 80, 120, 160 kJ/m^2) following removal of dorsal hair with electric clippers. Dorsal skin samples were collected at 0 h and 24 h after irradiation and the epidermis removed. High-molecular-weight DNA was extracted from isolated epidermis with phenol and chloroform/isoamyl alcohol (Maniatis *et al.*, 1982) and analysed by ELISA. T-T were induced at a rate of 3.8×10^{-3} T-T per kb per kJ/m^2 by solar radiation (Fig. 3). At 24 h after irradiation, 0–35% (average, 20%) of T-T remained in extractable DNA (data not shown). Preliminary measurement of T-T in human skin samples irradiated with solar radiation (18 kJ/m^2) and cultured *in vitro* indicates that 0.013 T-T per kb were induced per kJ/m^2 and complete loss of photoproducts within 24 h.

Future studies

Future studies will focus on induction and persistence of T-T and other photoproducts in human skin exposed to simulated or natural solar radiation. Photoproducts will be examined in (i) surgical skin samples exposed *in vitro*, (ii) skin biopsies obtained from individuals exposed experimentally to solar radiation, and (iii) skin biopsies from individuals exposed occupationally to natural sunlight. This approach will allow us to obtain information on individual repair efficiences *in vivo* and *in vitro*.

Acknowledgements

Supported in part by the Mellon Foundation and DHHS grants ES03841 and RR05445.

References

Kabat, E. (1976) *Structural Concepts in Immunology and Immunochemistry*, New York, Holt, p. 127

Maniatis, T., Fritsch, E.F. & Sambrook, J. (1982) *Molecular Cloning*, Cold Spring Harbor, NY, CSH Press, p. 458

Strickland, P.T. (1985) Immunoassay of DNA modified by ultraviolet radiation. *Environ. Mutag.*, *7*, 599–607

Strickland, P.T. & Boyle, J.M. (1981) Characterization of two monoclonal antibodies specific for dimerised and non-dimerised adjacent thymidines in single stranded DNA. *Photochem. Photobiol.*, *34*, 595–601

UNIDENTIFIED DNA DAMAGING AGENTS

NOVEL USES OF MASS SPECTROMETRY IN STUDIES OF ADDUCTS OF ALKYLATING AGENTS WITH NUCLEIC ACIDS AND PROTEINS

P.B. Farmer,[1] J. Lamb[1] & P.D. Lawley[2]

[1]*MRC Toxicology Unit, MRC Laboratories, Carshalton, Surrey, UK; and* [2]*Chester Beatty Laboratories, London, UK*

Several recent and major advances in the technology of mass spectrometry (MS) have greatly promoted use of this technique for the study of the interaction of alkylating agents with biomolecules. MS, in combination with gas chromatography (GC), may now be used to quantify adducts of carcinogens with proteins at levels down to 20 pmol/g protein. Soft ionization techniques have proved invaluable in determining the structure of carcinogen adducts with both DNA and proteins, and the newly developed tandem MS promises to be of considerable use in the characterization of complex carcinogen adduct mixtures.

In recent years, there has been a dramatic upsurge in the use of MS as a biological tool, owing to several major technological advances which have transformed the scope of the technique for biomolecular study. These include (i) fast atom bombardment (FAB) (an ionization technique for polar molecules); (ii) high-performance liquid chromatography (HPLC)-MS interfaces, which allow the direct transfer of materials from an HPLC separation into the MS; (iii) supercritical fluid chromatography (SFC)-MS, which provides a complementary technique to GC-MS and HPLC-MS for analysis of moderately polar compounds which are thermally labile or of relatively high molecular weight; and (iv) tandem mass spectrometry (MS-MS), in which ions separated by an MS are then fragmented for structural study in a second, linked MS. In parallel with these developments there have been great improvements in magnet technology and MS resolution allowing the transmission, resolution and mass measurement of ions of mass as high as 15 000 at full accelerating voltage.

In this paper, the application of FAB, high-resolution selective ion monitoring and MS-MS in the identification and quantification of carcinogen adducts in biomolecules is discussed.

Structure determination of adducts of carcinogens with nucleic acids and proteins using FAB

A powerful technique was added to the methods available for structure determination with the introduction of FAB (Barber *et al.*, 1981; Surman & Vickerman, 1981). In this mode of ionization, the sample is dissolved in a relatively nonvolatile solvent (e.g., glycerol, thioglycerol) and is bombarded by a beam of atoms, normally of xenon or argon. Similar liquid secondary ion MS has been carried out using a beam of caesium ions. Sample ions are expelled into the MS source from the surface of the solution. There is no requirement for the sample itself to be volatile, and excellent spectra, often containing intense molecular ions (MH^+ in the positive ion mode or

Fig. 1. Examples of adducts of alkylating agents with proteins or nucleic acids where FAB MS has been used in structure elucidation

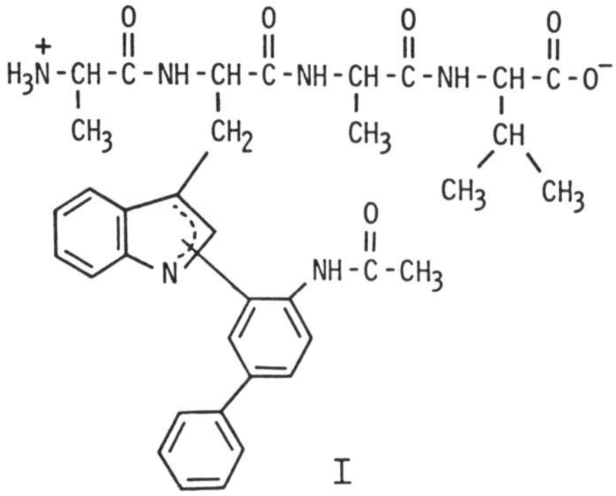

I, Peptide from 4-aminobiphenyl-rat serum albumin adduct (Skipper *et al.*, 1985); II, aflatoxin B_1 adduct with lysine in rat serum albumin (Sabbioni *et al.*, 1987); III, cross-linked adduct of reductively activated mitomycin C with DNA (Tomasz *et al.*, 1987); IV, adduct of dehydroretronecine with deoxyguanosine 5′-monophosphate (Wickramanayake *et al.*, 1985); d-Rib, 2′-deoxyribose

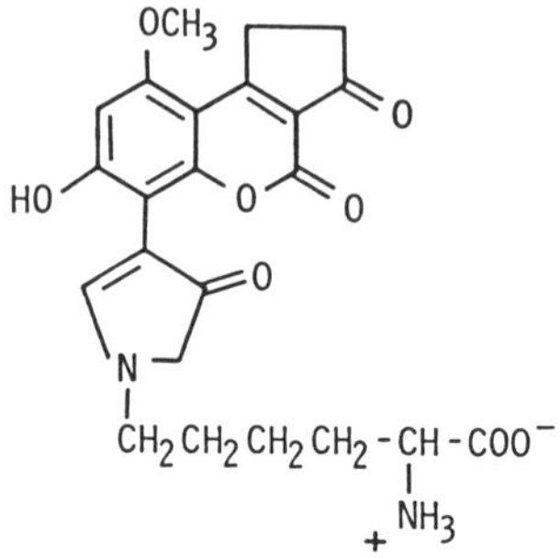

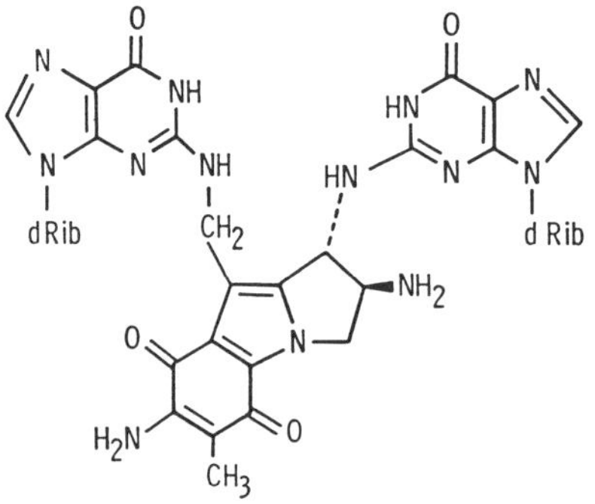

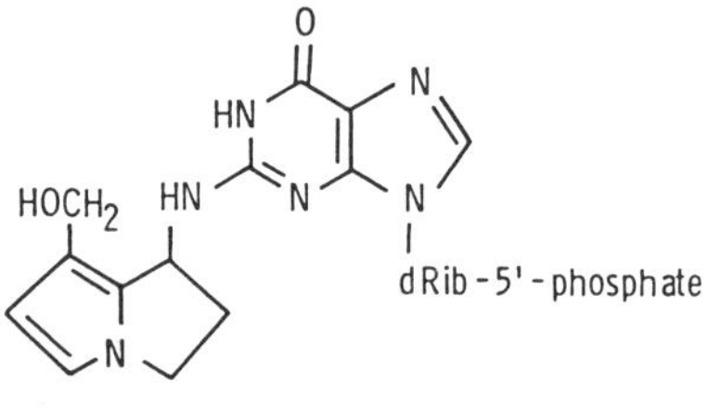

$M - H^-$ in the negative ion mode), may be obtained from many polar molecules, e.g., peptides and nucleotides.

The interaction of the carcinogen 4-aminobiphenyl with rat serum albumin was investigated by Skipper *et al.* (1985) using positive-ion FAB of a peptide after pronase digestion of the carcinogen-modified protein. The protonated molecular ion (m/z 655) that was observed indicated that the amine was acetylated and covalently bound to the tryptophan residue of a tetrapeptide, Ala Trp Ala Val (structure 1, Fig. 1). More recently, the adduct of aflatoxin B_1 with rat serum albumin, isolated after pronase digestion of the protein, was shown by positive FAB to have a molecular weight corresponding to a conjugate with lysine together with a loss of the elements of water from the molecule (Sabbioni *et al.*, 1987). The structure (II), deduced from MS, ultraviolet and nuclear magnetic resonance data, is shown in Figure 1.

Nucleic acid adducts may similarly be identified by FAB MS. Two recent examples are the cross-linked adduct with DNA produced by mitomycin C after reductive

Fig. 2. FAB MS (xenon) of the major cross-linked DNA product formed by melphalan (glycerol matrix)

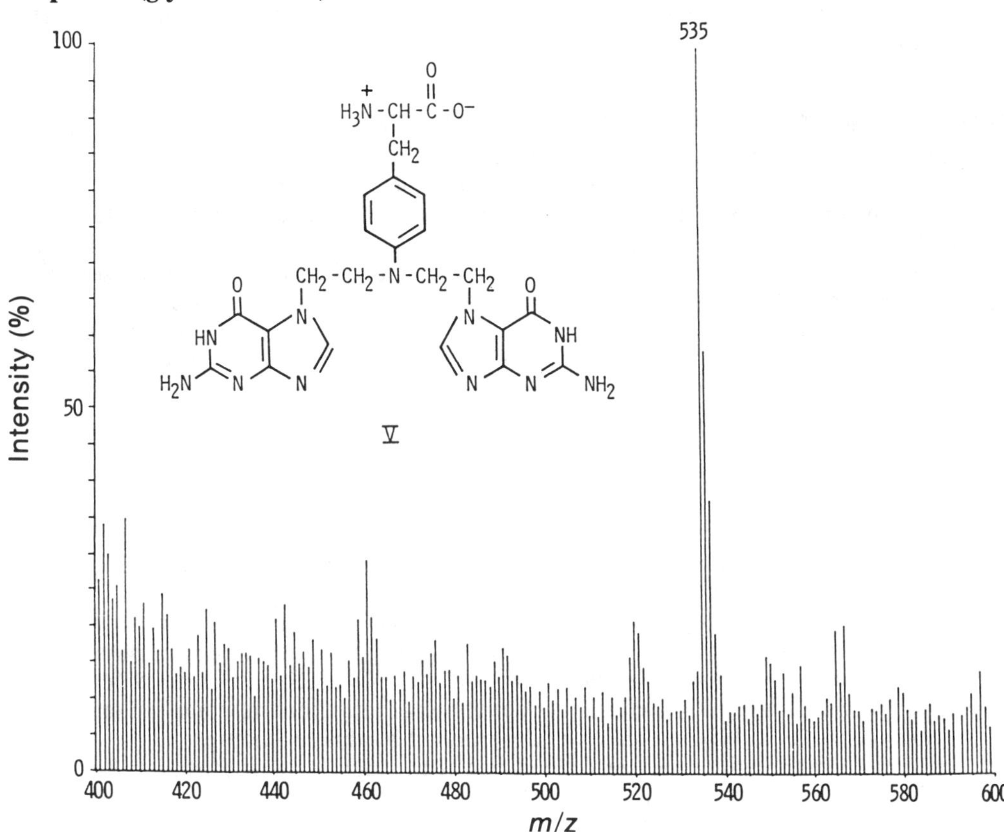

Instrument conditions: VG70-SEQ; accelerating voltage, 8 kV; resolution, 1000

activation (III, Fig. 1) (Tomasz *et al.*, 1987) using FAB (xenon), and the nucleoside and nucleotide adducts of dehydroretronecine (an alkylating agent thought to be associated with the toxic action of some naturally occurring pyrrolizidine alkaloids), where Cs^+ ion bombardment was used (e.g., IV, Fig. 1) (Wickramanayake *et al.*, 1985).

We have recently investigated the nature of the DNA cross-linked product produced by the bifunctional alkylating agent melphalan (P.D. Lawley, M. Tilby and P.B. Farmer, to be published). The molecular ion region of the purified major cross-linked adduct is shown in the positive FAB spectrum in Figure 2, confirming the conclusion (also based on ultraviolet spectral data) that the compound is a bis(7-guaninyl) adduct (V). We obtained similar results for the analogous product from nitrogen mustard(HN2)-treated DNA.

Thus, FAB is of great value in identifying the exact nature of carcinogen adducts within macromolecules, although to date these adducts have not been monitored in exposed humans, owing partly to the fact that FAB MS requires a relatively large amount (often in the microgram range) of pure analyte. However, the analysis of mixtures may now be considered either with HPLC-MS interfaces (which often give

FAB-like mass spectra) or with collision-induced decomposition techniques (see below). The use of selective ion monitoring with either of these techniques may provide a suitable approach for identifying and quantifying human carcinogen exposures.

Quantitative procedures using selective ion monitoring

Quantitative procedures for measuring carcinogen adducts with MS have invariably been carried out by selective ion monitoring techniques on compounds eluted from GC columns. In these procedures, an ion specific for the compound of interest is monitored, and quantification is carried out by comparing the intensity of this ion signal with that for an analogous ion from an internal standard. This is the most sensitive form of MS and allows detection, per GC injection, of adduct amounts as low as 1 pg. Two recently developed GC-MS methods exemplify the use of selective ion monitoring for detecting carcinogen exposure. The adducts of metabolically activated aromatic amines with cysteine in haemoglobin may be identified following basic hydrolysis of the protein, which yields the free amines. These amines may be analysed, after derivatization, by negative-ion chemical ionization GC-MS. Using this technique, Stillwell *et al.* (1987) showed that adduct levels of 4-aminobiphenyl, 2-naphthylamine and *o*- and *p*-toluidine were higher in smokers than in nonsmokers. A second method, which may be advantageous for determining the spectrum of alkylations at the *N*-terminal position of haemoglobin, is the modified Edman degradation procedure designed by Törnqvist *et al.* (1986). In this procedure, alkylated adducts at the *N*-terminal valine of haemoglobin are cleaved from the protein chain by reaction with pentafluorophenylisothiocyanate, and the resulting pentafluorophenylthiohydantoin can be analysed by negative-ion chemical ionization GC-MS. Exposure to ethylene oxide, propylene oxide, styrene oxide, methylating and other low-molecular-weight alkylating agents could be identified and determined by this technique.

The main concern with analyses carried out at the high level of sensitivity required for adduct detection relates to their selectivity, i.e., are the ions that are being monitored really derived from the adduct to be determined? Firstly, it is essential to use a chromatographic system of the highest available resolution, and only capillary GC columns should be considered. Secondly, for the mass spectral determination, it is generally true that the higher the mass monitored the less likelihood there will be of contamination from other components. Thus, the use of soft ionization techniques, which promote the intensity of molecular ion species in the mass spectra, is also to be recommended for increasing the selectivity of the analysis. Furthermore, any remaining contamination to the ion being monitored for the carcinogen adduct may effectively be eliminated by increasing the mass resolution of the MS, as the ions due to any impurities will generally have different formulae (and hence different accurate masses) from the adduct ion. Consequently, less purification of the adduct may be required prior to its analysis.

An example of this approach for the analysis of exposure to acrylamide is shown in Figure 3. The monitored adduct, derived from an acid hydrolysate of acrylamide-exposed haemoglobin, is 2-carboxyethylcysteine, which is converted to its bismethyl ester heptafluorobutyryl derivative for GC-MS analysis, and quantified using a trideuterated internal standard (Bailey *et al.*, 1986). Electron impact selective ion monitoring yields GC-MS traces of extreme complexity, prompting the use of chemical ionization (isobutane) for the analysis. However, as can be seen in Figure 3A, the analysis is not selective at low mass resolution even in chemical ionization, as other components contribute to the m/z 386 peak being monitored for 2-carboxy-ethylcysteine. Increasing the mass resolution to 10 000 (Fig. 3B), however, results in a

Fig. 3. A, Low mass resolution (1000) selective ion monitoring (m/z 386) of the bismethyl ester heptafluorobutyryl derivative of 2-carboxyethylcysteine from haemoglobin exposed to acrylamide. B, Equivalent high mass resolution (10 000) selective ion monitoring trace

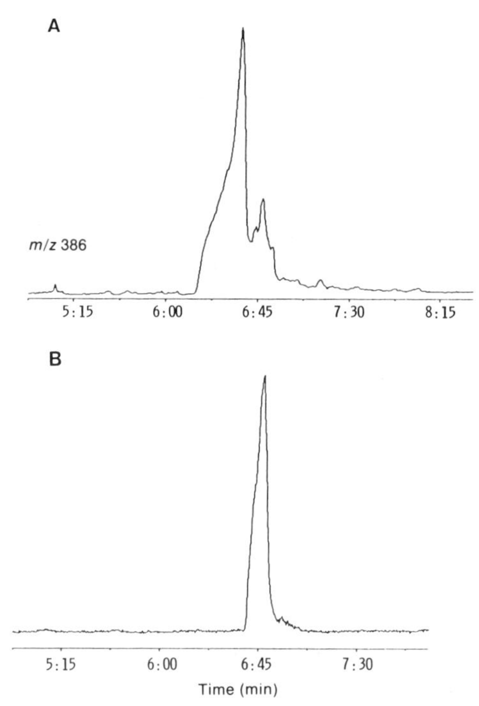

Instrument conditions: VG70-SEQ; chemical ionization (isobutane) accelerating voltage 8 kV; 25 m × 0.32 mm GC column coated with OV1; on-column injector, 70°C/min, 30°C/min to 290°C

specific analysis for the adduct. An analysis at this resolution of 2-carboxyethylcysteine from the globin of a worker exposed to acrylamide is shown in Figure 4.

Identification of alkylating agents after unknown exposures using MS-MS

The methods described above depend for their success on a prior knowledge of the nature of the alkylating species involved in the exposure. In practice, this is frequently not the situation, and techniques are therefore required for identifying 'unknown' genotoxic compounds, in particular following exposures to complex mixtures of substances such as cigarette smoke or urban pollution.

A technique of exceptional promise for analysis of complex mixtures of carcinogen adducts is MS-MS, in which an ion from the sample mixture is selected and focused through an MS into a collision cell filled with a gas at low pressure (e.g., argon or air). Fragments resulting from this collision are passed into, and scanned by a second MS ('collision spectra'). High mass resolution may also be used in the first MS to increase the selectivity of the ion focusing and eliminate contaminants. Thus, components in complex mixtures may be analysed directly without any, or with only minimal, purification by prior chromatographic separation. The technique may also be used in reverse, i.e., a single fragment ion may be focused in the second MS, and the first MS

Fig. 4. High mass resolution (10 000) selective ion monitoring GC-MS trace of 2-carboxyethylcysteine from the haemoglobin of a worker exposed to acrylamide

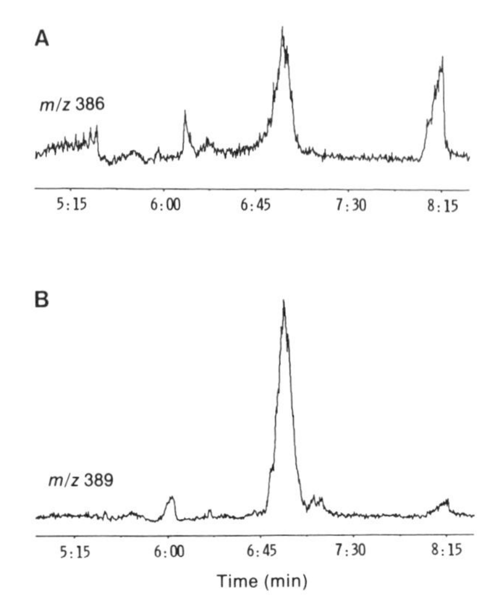

A, m/z 386 for the bismethyl ester heptafluorobutyryl derivative of the adduct; B, m/z 389 for the d_3-labelled internal standard. Instrument conditions as in legend to Figure 3

may be scanned to determine what ions have fragmented in the collision cell to yield the chosen fragment ('parent spectra').

We are now exploring the potential of MS-MS using collision-activated fragmentation for the identification of 7-alkylguanines in human urine to monitor unknown or complex exposure situations, e.g., to carcinogens from dietary sources or from smoke. 7-Alkylguanines formed during exposure to carcinogens are liberated from nucleic acids, and their urinary excretion has been used to monitor exposure to methylating agents (Farmer *et al.*, 1986) and to aflatoxin B_1 (Autrup *et al.*, 1985; Groopman *et al.*, 1985).

Our analyses are carried out using a VG70 SEQ MS with electron impact or FAB for sample ionization and air as the gas in the collision cell. 7-Methylguanine, which is a natural constituent of urine arising from t-RNA turnover, may be identified directly in a sample of urine by the collision spectrum of its molecular ion (m/z 165) (Fig. 5A); identical spectra are obtained to those of an authentic sample (Fig. 5B). For the higher alkylated guanines, 7-ethylguanine and 7-(2-hydroxyethyl)guanine, the collision spectra from the molecular ions (179 and 195, respectively) contain the common fragment m/z 151 (guanine$^+$), and we are now investigating the possibility that this ion might be a useful marker for these and other alkylated guanines. Parent ion spectra for the ion at m/z 151 from a purine fraction isolated from a 25-ml aliquot of human urine by a simple silver nitrate precipitation procedure, revealed several ions, including predominantly m/z 179, 180 and 195 (Fig. 6). The identity of the 179 and 195 peaks is under study in an attempt to confirm their structures as 7-ethyl and 7-hydroxyethylguanines. The peak at m/z 180 appears to be due to theophylline or its isomers theobromine and paraxanthine. The identity of the remaining peaks which *may* be derived from 7-alkylguanines is also being investigated.

Fig. 5. A, Collision activated decomposition spectrum of the molecular ion (*m*/*z* 165) of 7-methylguanine in human urine; B, collision activated decomposition spectrum of the molecular ion (*m*/*z* 165) of authentic 7-methylguanine

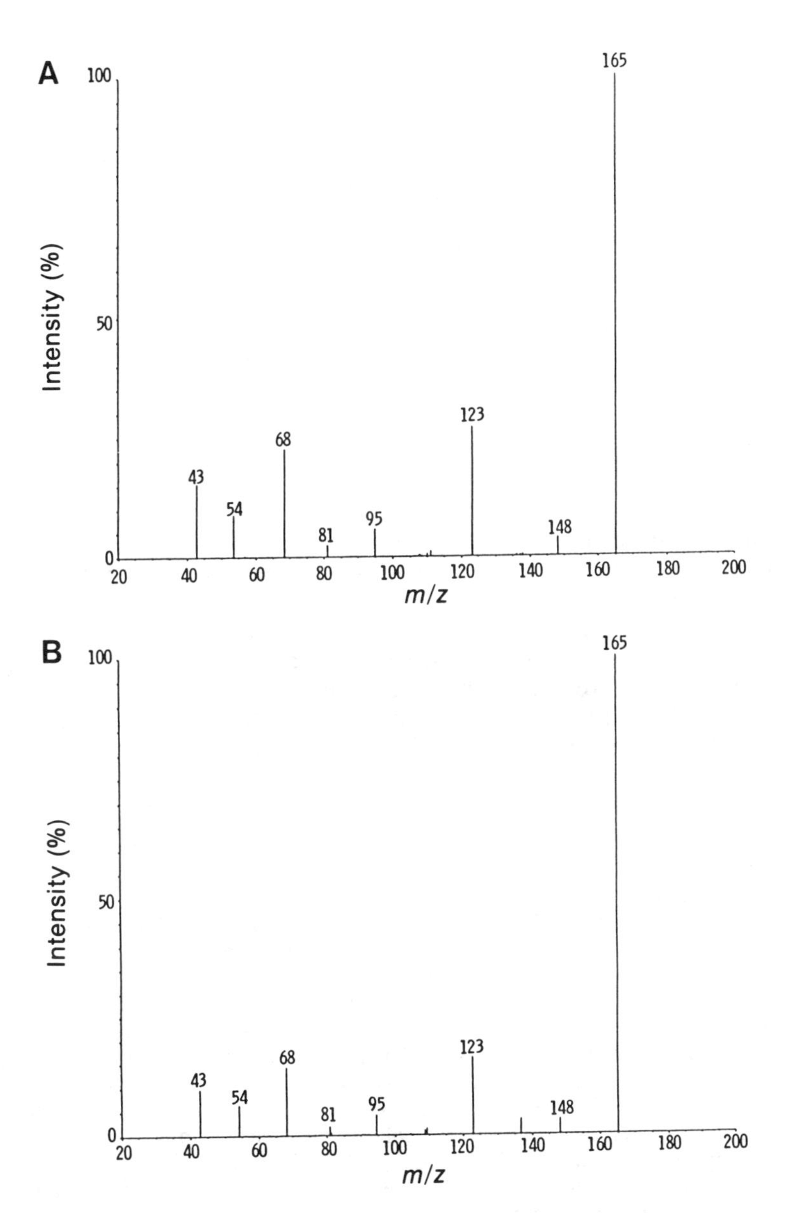

Instrument conditions: VG70-SEQ; electron impact; resolution 1000; accelerating voltage, 8 kV; collision gas, air; collision energy, 120 eV

Fig. 6. Parent ion scan of m/z 151 for a purine fraction isolated from human urine

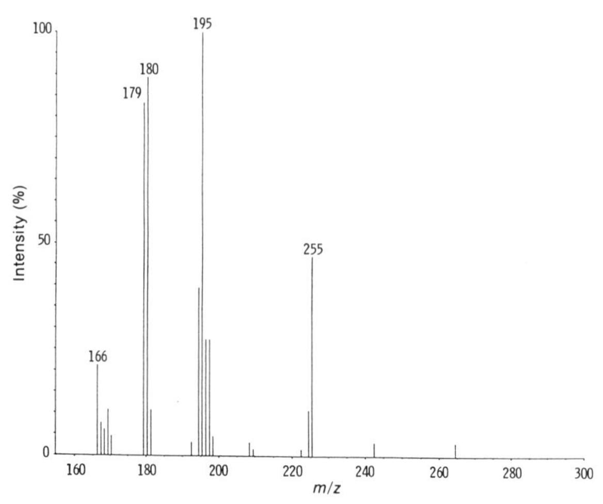

Instrument conditions as in legend to Figure 5

Conclusion

In addition to MS techniques, procedures used to monitor exposure to carcinogens by examining adduct formation have included immunoassay, DNA postlabelling and HPLC coupled with ultraviolet or fluorescence measurements. Although MS has the advantage over these techniques that it identifies unequivocally the chemical nature of the adduct, it has the disadvantage of great expense and slow sample throughput. For monitoring purposes, therefore, perhaps less complex techniques (e.g., immunoassay) should be used, whereas MS should be considered as the reference analytical method for confirming the specificity and selectivity of these more routine techniques.

References

Autrup, H., Wakhisi, J., Vähäkangas, K., Wasunna, A. & Harris, C.C. (1985) Detection of 8,9-dihydro-(7′-guanyl)-9-hydroxyaflatoxin B_1 in human urine. *Environ. Health Perspectives, 62,* 105–108

Bailey, E., Farmer, P.B., Bird, I., Lamb, J.H. & Peal, J.A. (1986) Monitoring exposure to acrylamide by the determination of S-(2-carboxyethyl)cysteine in hydrolyzed haemoglobin by gas chromatography-mass spectrometry. *Anal. Biochem., 157,* 241–248

Barber, M., Bordoli, R.S., Sedgwick, R.D., Tyler, A.N. & Green, B.N. (1981) A new ion source for mass spectrometry. *J. chem. Soc. chem. Commun.,* 325–327

Farmer, P.B., Shuker, D.E.G. & Bird, I. (1986) DNA and protein adducts as indicators of *in vivo* methylation by nitrosatable drugs. *Carcinogenesis, 7,* 49–52

Groopman, J.D., Donahue, P.R., Zhu, J., Chen, J. & Wogan, G.N. (1985) Aflatoxin metabolism in humans: detection of metabolites and nucleic acid adducts in urine by affinity chromatography. *Proc. natl Acad. Sci. USA, 82,* 6492–6496

Sabbioni, G., Skipper, P.L., Buchi, G. & Tannenbaum, S.R. (1987) Isolation and characterisation of the major serum albumin adduct formed by aflatoxin B_1 *in vivo* in rats. *Carcinogenesis, 8,* 819–824

Skipper, P.L., Obiedzinski, M.W., Tannenbaum, S.R., Miller, D.W., Mitchum, R.K. & Kadlubar, F.F. (1985) Identification of the major serum albumin adduct formed by 4-aminobiphenyl *in vivo* in rats. *Cancer Res., 45,* 5122–5127

Stillwell, W.G., Bryant, M.S. & Wishnok, J.S. (1987) GC/MS analyses of biologically important aromatic amines. Application to human dosimetry. *Biomed. environ. Mass Spectrom.*, *14*, 221–223

Surman, D.J. & Vickerman, J.C. (1981) Fast atom bombardment quadrupole mass spectrometry. *J. chem. Soc. chem. Commun.*, 324–325

Tomasz, M., Lipman, R., Chowdary, D., Powlak, J., Verdine, G.L. & Nakanishi, K. (1987) Isolation and structure of a covalent cross-link adduct between mitomycin C and DNA. *Science*, *235*, 1204–1208

Törnqvist, M., Mowrer, J., Jensen, S. & Ehrenberg, L. (1986) Monitoring of environmental cancer initiators through haemoglobin adducts by a modified Edman degradation method. *Anal. Biochem.*, *154*, 255–266

Wickramanayake, P.P., Arbogast, B.L., Buhler, D.R., Deinzer, M.L. & Burlingame, A.L. (1985) Alkylation of nucleosides and nucleotides by dehydroretronecine: characterization of covalent adducts by liquid secondary ion mass spectrometry. *J. Am. chem. Soc.*, *107*, 2485–2488

PENTAFLUOROBENZYLATION OF ALKYL AND RELATED DNA BASE ADDUCTS FACILITATES THEIR DETERMINATION BY ELECTROPHORE DETECTION

M. Saha, O. Minnetian, D. Fisher, E. Rogers, R. Annan, G. Kresbach, P. Vouros & R. Giese

Department of Medicinal Chemistry in the College of Pharmacy and Allied Health Professions, Chemistry Department and Barnett Institute of Chemical Analysis and Materials Science, Northeastern University, Boston, MA, USA

The DNA adduct *O*⁴-ethylthymine can be alkylated under mild conditions with pentafluorobenzyl bromide. The product has good gas chromatographic characteristics and also forms a structurally characteristic anion in high yield when subjected to electron capture mass spectrometry. Related adducts for other DNA bases behave similarly. These properties stimulated us to develop a general analytical method based on the derivatization reaction. Important for this method is an oxidation-elimination reaction that mildly releases a base from a nucleoside. Thus, a general analytical method based on pentafluorobenzylation is now available for determining many alkyl and related DNA adducts.

Electrophore postlabelling is an emerging technique for quantifying DNA adducts. Typically in this technique, the DNA adduct is first isolated as a modified base or nucleoside and then derivatized with an electrophore. After interferences are removed, e.g., chromatographically, the derivatized adduct is quantified by electrophore detection. The latter step may comprise electron capture detection (ECD) or electron capture mass spectrometry (ECMS).

The purpose of this paper is to point out the usefulness of a particular form of this technique in which the DNA adduct is isolated as a modified base by acid hydrolysis or oxidation elimination and then derivatized with pentafluorobenzyl bromide for subsequent chromatography and electrophore detection.

O^4-Ethyl-*N*1-pentafluorobenzylthymine

A mixture containing 0.028 g (0.09 mmol) O^4-ethylthymine, 0.12 g K_2CO_3 and 0.1 ml pentafluorobenzyl bromide was heated at 70°C for 12 h. The cooled reaction mixture was filtered and the residue was purified by preparative thin-layer chromatography. The product (0.035 g, 57% yield) was obtained as a white solid after recrystallization from ethyl acetate/hexane. It was identified by its spectral properties.

Electrophore detection

An electrophore is a substance which avidly combines with a low energy electron in the gas phase, forming an anion radical. The latter may persist or undergo fragmentation, forming anionic, radical or neutral degradation products. In ECD, the loss of a thermal electron (because of capture by the analyte) is detected, whereas by ECMS one sorts and detects the anions, whether parent or daughter, that arise in this

Fig. 1. Pentafluorobenzylation of O^4-ethylthymine

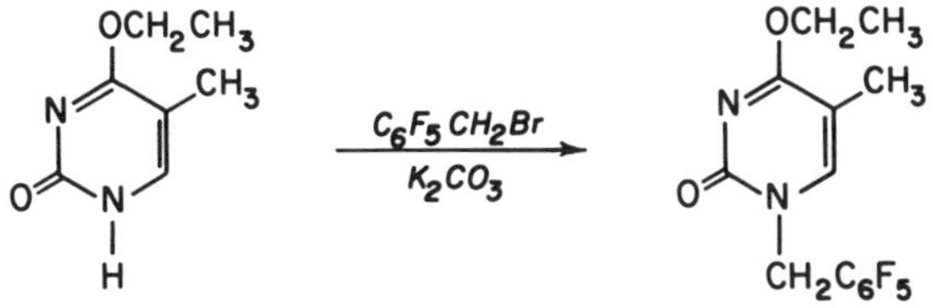

process. The availability of both ECD and ECMS detectors is advantageous because the former is low in cost and simple to use, whereas ECMS lacks these advantages but overcomes residual sample interferences and can provide structural information about the analyte. Both forms of electrophore detection can be highly sensitive, reaching detection limits of 10^{-18} mol for standards (Corkill *et al.*, 1982; Mohamed *et al.*, 1984).

Because of these useful features electrophore postlabelling is being pursued for the quantification of DNA adducts. The best methods mildly form a stable derivative of the analyte in a good yield, and this derivative efficiently forms a structurally characteristic anion when subjected to ECMS.

We have found that pentafluorobenzylation of alkyl and related DNA adducts in their base form satisfies these criteria for successful electrophore postlabelling. This is illustrated for O^4-ethylthymine in Figure 1.

Although the pentafluorobenzyl group is not inherently a strong electrophore, a corresponding DNA base alkylated on one or more of its ring NH sites with a pentafluorobenzyl group has excellent ECD and ECMS properties because dissociative electron capture takes place, leading to a base anion. For example, the gas chromatography (GC)-ECD chromatogram shown in Figure 2 indicates the high response from O^4-ethyl-*N*1-pentafluorobenzylthymine. The electron capture mass spectrum of this compound (not shown) is dominated by an intense peak corresponding to loss of a pentafluorobenzyl radical. Other modified DNA bases that have similar characteristics when alkylated with pentafluorobenzyl bromide are O^2-ethylthymine, N^4-pivalylcytosine, O^6-methylguanine, O^6-hydroxyethylguanine and N^6-methyladenine (unpublished results).

Some nucleosides can also be derivatized with pentafluorobenzyl bromide, if stronger conditions such as alkaline phase transfer are used, yielding products that are highly sensitive to GC-ECD (Adams *et al.*, 1986). However, these products give little analyte-characteristic anion by ECMS (unpublished results).

Base release by oxidation-elimination

Since pentafluorobenzylation on a ring NH site of a DNA base is attractive for the determination of DNA adduct by electrophore postlabelling, it is desirable to establish a mild procedure for obtaining a DNA adduct in this form, starting from DNA. This problem translates into a method for mildly releasing a base from a nucleoside, since the latter is available by enzymatic hydrolysis of DNA.

Acid hydrolysis of DNA or of its nucleosides only partly solves this problem. Mild acid hydrolysis releases only certain bases, such as O^6-methylguanine (Singer & Grunberger, 1983). Strong acid, which can release all of the bases, destroys some adducts, such as O^4-ethylthymine (which is hydrolysed under these conditions to thymine).

We modified a reaction reported by Albright and Goldman (1965) to establish a general procedure for mildly releasing a DNA base from its nucleoside. In this reaction

Fig. 2. Gas chromatograph-ECD chromatogram of 62.5 pg O^4-ethyl-*N*1-pentafluorobenzylthymine (3), with 16.5 pg each of lindane (1) and aldrin (2) as reference compounds

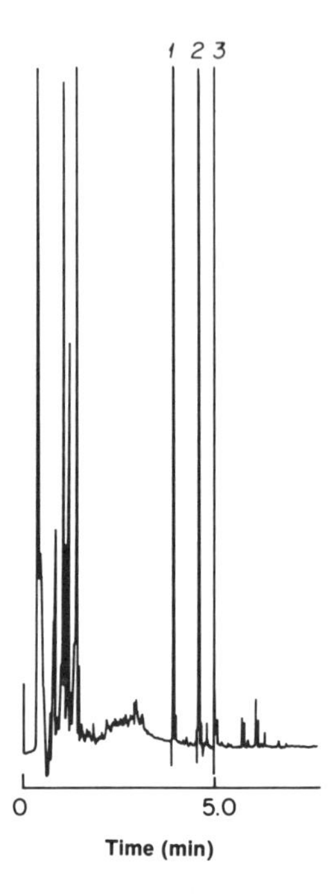

Column: HP-Ultra capillary column (Hewlett-Packard), 5% phenylmethylsilicone cross-linked, 25 m × 32 mm, 0.25 μm film, 110–280°C at a setting of 80°C/min, then a 2-min hold

sequence, illustrated in Figure 3, a nucleoside is first oxidized with dimethylsulphoxide-acetic anhydride at 40°C for 3 h. Exposure to 1 N ammonium hydroxide for 1 h at 80°C then releases the base from the oxidized nucleoside by an elimination reaction. For example, O^4-ethylthymine is formed in a 96% yield from O^4-ethylthymidine, and several other bases, including the four bases of DNA, are obtained from their corresponding deoxynucleosides in a 57–97% yield (unpublished results).

Applications

To date, we have determined two DNA adducts (or adduct models) starting from DNA by electrophore postlabelling: hydroxymethyluracil (HMU) and 5-methylcytosine (5-MC). The adduct HMU is a consequence of damage to thymine sites on DNA by

Fig. 3. Scheme for release of a base from a nucleoside by an oxidation-elimination reaction sequence

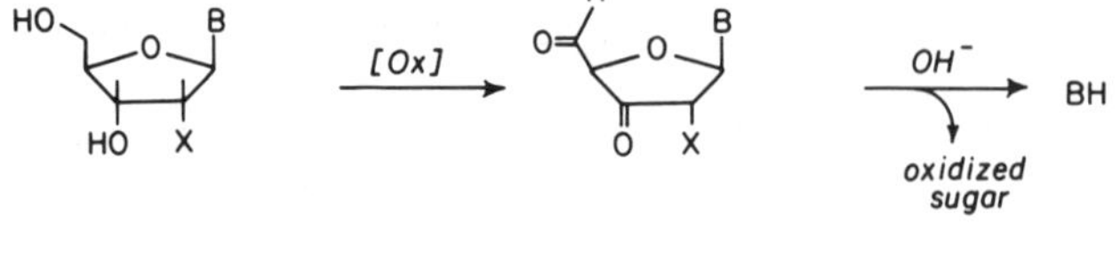

HO B O HO X —[Ox]→ H O= B O O X —OH^-→ BH (oxidized sugar)

Fig. 4. GC-ECD chromatogram of N^4-pivalyl-*N*1-pentafluorobenzyl-5-methylcytosine (peak 1) derived from calf thymus DNA; peak 2 is N^4-pivalyl-*N*1-(2,3,5,6-tetrafluorobenzyl)-5-methylcytosine, the internal standard.

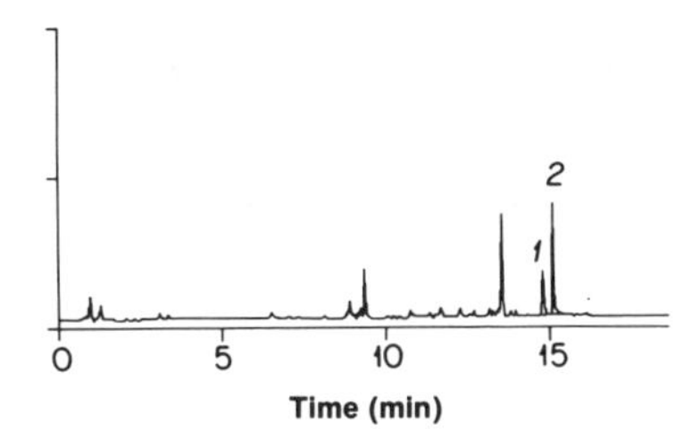

ionizing radiation, mediated apparently by hydroxy radicals. 5-MC, an adduct model, is a natural minor base, comprising, for example, about 1% of the bases in human DNA.

For the determination of HMU, we spiked 78 ng to 7.8 mg of calf thymus DNA with an equivalent of 125 pg to 1.5 ng of HMU. The DNA was acid hydrolysed (prior to our developing the oxidation-elimination reaction described above), and the released HMU was purified on C18-silica, derivatized with pentafluorobenzyl bromide, and purified further by high-performance liquid chromatography (HPLC) prior to GC with electrophore detection. By GC-ECD, the sensitivity was equivalent to about one HMU adduct in 10^7 bases; by GC-ECMS it was one adduct in 10^{10} bases. We also demonstrated the usefulness of HPLC directly coupled to ECMS on a Finnigan 4000 instrument fitted with a moving belt interface and a home-built liquid transfer device. The minimal amount of derivatized HMU detected corresponds to 185 fg of the base (unpublished results).

A similar electrophore postlabelling procedure was used to determine 5-MC in calf thymus and human leucocyte DNA. For example, the mol% of 5-MC in 360 ng calf thymus DNA was determined to be $1.2 \pm 0.1\%$ by GC-ECD (unpublished results). A representative chromatogram is shown in Figure 4.

Conclusion

A general procedure is now available for quantifying an alkyl or related DNA adduct by electrophore postlabelling. In this procedure, the DNA adduct is obtained as a base and alkylated with pentafluorobenzyl bromide. It is then separated chromatographically (GC or HPLC) and quantified by electrophore detection.

Acknowledgements

This work was supported by National Cancer Institute Grants CA35843 and CA43012, and Grant CR812740 from the Reproductive Effects Assessment Group of the US Environmental Protection Agency. Publication No. 321 from the Barnett Institute.

References

Adams, J., David, M. & Giese, R.W. (1986) Pentafluorobenzylation of O^4-ethylthymidine and analogues by phase-transfer catalysis for determination by gas chromatography with electron capture detection. *Anal. Chem., 58,* 345–348

Albright, J.D. & Goldman, L. (1965) Dimethyl sulfoxide-acid anhydride mixtures. New reagents for oxidation of alcohols. *J. Am. chem. Soc., 87,* 4214–4216

Corkill, J.A., Joppich, M., Kuttab, S.H. & Giese, R.W. (1982) Attogram-level detection and relative response of strong electrophores by gas chromatography with electron capture detection. *Anal. Chem.*, *54*, 481–485

Mohamed, G.B., Nazareth, A., Hayes, M.J., Giese, R.W. & Vouros, P. (1984) Gas chromatography-mass spectrometry characteristics of methylated perfluoroacyl derivatives of cytosine and 5-methylcytosine. *J. Chromatogr.*, *314*, 211–217

Singer, B. & Grunberger, D. (1983) *Molecular Biology of Mutagens and Carcinogens*, New York, Plenum Press, p. 19

MONITORING HUMAN EXPOSURE TO CARCINOGENS BY ULTRASENSITIVE POSTLABELLING ASSAYS: APPLICATION TO UNIDENTIFIED GENOTOXICANTS

K. Randerath,[1,3] R.H. Miller,[2], D. Mittal[1] & E. Randerath[1]

[1]*Department of Pharmacology and* [2]*Department of Otorhinolaryngology, Baylor College of Medicine, Houston, TX, USA*

The ^{32}P-postlabelling assay is a recently developed analytical tool for the detection and measurement of nucleic acid (DNA and RNA) adducts formed by covalent binding of identified or unidentified electrophiles. The detection limit of the assay for many adducts is as low as 0.3 amol adduct/µg DNA (=one adduct/10^{10} DNA nucleotides, or one adduct per mammalian genome). As presented here, the method can be applied to DNA alterations elicited by (i) complex mixtures of genotoxicants (e.g., cigarette smoke, occupational exposures), (ii) oestrogens (i.e., hormones that cause DNA damage *via* the formation of unidentified electrophiles), and (iii) DNA-reactive chemicals that may be formed metabolically in animal tissues without known exposure and give rise to adduct-like DNA alterations (termed I compounds).

Many lines of evidence have implicated DNA adducts as a key element in the initiation of chemical carcinogenesis (Brookes & Lawley, 1964; Miller & Miller, 1981; Hemminki, 1983; Kriek *et al.*, 1984; Wogan & Gorelick, 1985). The ^{32}P-postlabelling assay developed in our laboratory (Randerath *et al.*, 1981; Gupta *et al.*, 1982; Reddy *et al.*, 1984) allows one to detect extremely low levels of DNA adducts formed by the reaction of chemical carcinogens with DNA in both experimental animals (Randerath *et al.*, 1981; Reddy *et al.*, 1981; Gupta *et al.*, 1982; Gupta, 1984; Gupta & Dighe, 1984; Phillips *et al.*, 1984; Randerath *et al.*, 1984; Reddy *et al.*, 1984; Liehr *et al.*, 1985; Randerath *et al.*, 1985a; Reddy *et al.*, 1985; Schurdak & Randerath, 1985; Liehr *et al.*, 1986; Lu *et al.*, 1986; Randerath *et al.*, 1986a,b,c; Reddy & Randerath, 1986; Schoepe *et al.*, 1986; Liehr *et al.*, 1987; Randerath *et al.*, 1988; *Schurdak et al.*, 1987) and humans (Everson *et al.*, 1986; Parks *et al.*, 1986; Randerath *et al.*, 1986a,b,c; Everson *et al.*, 1987; Randerath *et al.*, 1987; Reddy *et al.*, 1987). It is the purpose of this article to review recent applications of the assay to adducts formed by the reaction of DNA *in vivo* with incompletely identified or unidentified carcinogens. The results demonstrate the general potential of the assay for the detection of unusual components in DNA samples from animal or human tissues.

General method

The basic method (Randerath *et al.*, 1981) entails nucleolytic digestion of DNA to 3′-mononucleotides, subsequent enzymatic conversion of the digestion products to 5′-^{32}P-labelled deoxyribonucleoside 3′,5′-bisphosphates with [γ-^{32}P]ATP as the donor

[3] To whom correspondence should be addressed

of the label and T4 polynucleotide kinase as the catalyst, separation of the labelled nucleotides by polyethyleneimine (PEI)-cellulose thin-layer chromatography (TLC), autoradiography to detect adducts and normal nucleotides, and scintillation counting of excised TLC fractions for quantification of adduct levels (Randerath *et al.*, 1981; Gupta *et al.*, 1982; Randerath *et al.*, 1985b). Several modifications of the basic assay have been developed to enhance its sensitivity, involving adduct intensification (Randerath *et al.*, 1985a), butanol enrichment (Gupta, 1985) and nuclease P_1 enhancement (Reddy & Randerath, 1986).

Cigarette smoke-induced DNA lesions

Cigarette smoke-induced DNA adducts were detected initially in both experimental animals (Randerath *et al.*, 1986a,c) and humans (Everson *et al.*, 1986; Randerath *et al.*, 1986c; Everson *et al.*, 1987) by the adduct intensification version of the postlabelling assay. The recently developed, highly sensitive nuclease P_1-enhanced version (Reddy & Randerath, 1986) of the method has allowed us to detect much lower levels of smoking-related DNA adducts than heretofore possible. In these experiments, the DNA preparation to be tested for the presence of such lesions was digested to 3′-mononucleotides, normal nucleotides but not adducts were specifically removed by enzymatic 3′-dephosphorylation with nuclease P_1 (Reddy & Randerath, 1986), the remaining adducts were ^{32}P-labelled, separated by TLC, autoradiographed, and quantified by scintillation counting. Results obtained by this procedure in both humans and mice (Fig. 1) provide evidence for the presence of DNA adducts. The cigarette smoke-related ^{32}P-labelled material occupied two extensive zones (DRZ 1 and DRZ 2, Fig. 1) on the autoradiograms. We have postulated (Randerath *et al.*, 1986a) that these results are indicative of the presence in the DNA samples of numerous, only partially resolved, presumably aromatic DNA adducts of widely varying polarity. The similar patterns of ^{32}P-labelled adducts in human and mouse tissues (Fig. 1) suggest common characteristics of tobacco-associated carcinogen activation and adduct formation in different mammalian species. The mouse thus provides a suitable model for studying DNA lesions in smoking-associated carcinogenesis (Randerath *et al.*, 1988). High adduct levels in lung (Fig. 1, panel E) and heart (panel G) of mice treated topically with cigarette smoke condensate were also observed in smokers (Randerath *et al.*, 1987; K. Randerath, H. Dunsford, D. Mittal and E. Randerath, unpublished). Adduct levels in other tissues, such as liver, kidney and spleen, were considerably lower. The mechanism(s) underlying this peculiar tissue distribution has not yet been elucidated, but may in part entail free radical reactions (Church & Pryor, 1985). Tobacco-associated DNA adducts in lung are likely to play a key role in pulmonary carcinogenesis. In view of the possible contribution of DNA adducts to degenerative tissue changes (Yamashita *et al.*, 1986), the DNA alterations observed in our experiments may also be associated with lung and heart pathology that is not directly related to cancer, such as chronic obstructive lung disease (US Department of Health and Human Services, 1984) and cardiomyopathy (Hartz *et al.*, 1984).

The use of the ^{32}P-postlabelling assay for the analysis of DNA adducts in white blood cells of foundry workers is discussed in another article in this volume (Hemminki *et al.*, this volume).

Oestrogen-induced DNA lesions

A further example of the application of the ^{32}P-postlabelling assay to chemically unidentified adducts is provided by the analysis of kidney DNA from oestrogen-treated male Syrian hamsters (Liehr *et al.*, 1985, 1986, 1987). These animals develop a high

Fig. 1. Autoradiograms of PEI-cellulose TLC separations of adducts in lung DNA of human smokers and lung and heart DNA of mice treated dermally with cigarette smoke condensate

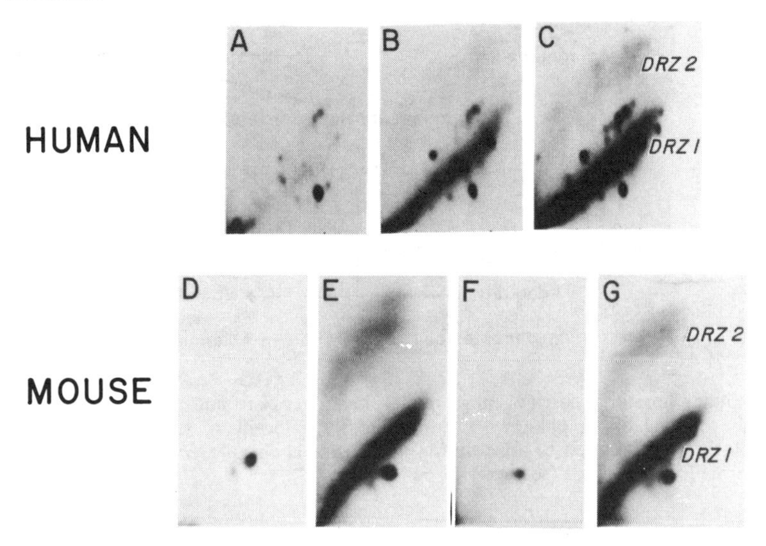

A, nonsmoker; B, smoker (50 pack-years); C, smoker (75 pack-years); D, mouse lung control; E, lung after exposure to six daily doses of cigarette smoke condensate corresponding to 0.75 cigarette each; F, heart control; G, heart after treatment as in E. Human and mouse DNA adducts were mapped similarly, as described previously (Randerath *et al.*, 1986a), except that in the first dimension (bottom to top), development was performed in three steps: water to 2 cm above the bottom edge, 2.5 M sodium formate, 6.25 M urea, pH 3.8, to 6 cm, and 3.0 M sodium formate, 7.5 M urea, pH 3.8, to 13 cm (upper row) or 18.5 cm (lower row). Adducts were purified by nuclease P_1 treatment (Reddy & Randerath, 1986). Autoradiography was at −80°C for 8 h, employing Kodak XAR-5 film and Du Pont Lightning Plus intensifying screens. DRZ, diagonal radioactive zone (see text). Total levels of (presumably aromatic) DNA modification in the exposed samples were estimated to be approximately 1 adduct in 10^7 DNA nucleotides.

incidence of renal carcinoma after six to eight months of exposure to natural or synthetic oestrogens (Kirkman, 1959). Our initial finding (Liehr *et al.*, 1985) of DNA adducts in kidneys (but not liver) of diethylstilboestrol-treated hamsters was interpreted as indicating the binding of reactive diethylstilboestrol metabolites to DNA. Subsequent work, however, showed that stilbene and steroid oestrogens of widely divergent structures gave rise to a set of chromatographically identical DNA adducts (Liehr *et al.*, 1986). Thus, the adducts did not contain bound oestrogen moieties but rather were formed by the binding of an oestrogen-induced, unknown electrophilic metabolite (or metabolites) to kidney DNA; such adducts were not observed in other tissues of the oestrogenized hamsters (Liehr *et al.*, 1986). We have postulated that such unidentified metabolites are involved in the initiation of renal cancer in the hamster kidney (Fig. 2). The results presented in the following section suggest that additional mechanisms may exist by which 'endogenous' electrophiles lead to DNA adducts in animal and human tissues (see also Fig. 2).

I compounds

Tissue DNA of ageing animals contains modifications, termed I compounds (Randerath *et al.*, 1986b), which were detected by ^{32}P-postlabelling assay. The

Fig. 2. Formation of DNA adducts *via* 'endogenous' electrophiles

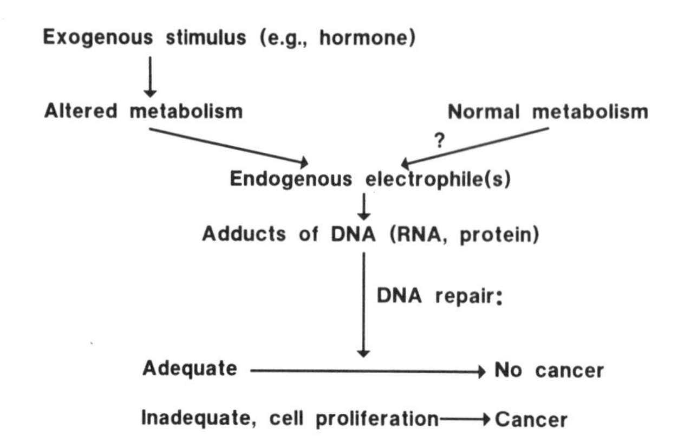

In the hamster kidney hormonal carcinogenesis model, oestrogens are not part of the adduct structure but rather induce the binding of an unknown electrophile(s) specifically to kidney DNA. Likewise, DNA-reactive metabolites are postulated to arise during normal metabolism, giving rise to age-dependent DNA modifications, termed I compounds (see text).

amounts and patterns of I compounds were found to be dependent not only on age but also on species, strain or stock, tissue type and sex of the animals (Randerath *et al.*, 1986b; and unpublished). I compound-like DNA modifications have also been observed in human tissues (Reddy *et al.*, 1987). Analogous DNA modifications have not been detected in newborn rat and mouse tissues. While I compounds resemble aromatic DNA adducts in their chromatographic properties, their origin and chemistry have not as yet been elucidated. No exogenous source of I compounds has been identified. We have speculated that the I compounds may play a role in natural ageing and in 'spontaneous' tumour formation (Randerath *et al.*, 1986b).

Conclusions

We have shown that the ^{32}P-postlabelling assay, originally developed as a general method for studying the binding of authentic carcinogens to DNA (Randerath *et al.*, 1981), is well suited for detecting and measuring covalent DNA alterations elicited by unidentified chemicals (oestrogen-induced adducts and I compounds) or by only partially characterized complex mixtures (cigarette smoke components). While the capacity of the assay to detect DNA alterations of unknown sources, in addition to its great sensitivity, is one of its important assets, additional methods will be needed for the chemical characterization of such DNA alterations. Current spectroscopic methods for structural identification are far too insensitive for the small amounts of adducts detected in the ^{32}P assay. (Note that in order to prepare 1 μg of an adduct present at a level of 10^{-8} mol/mol DNA-P, ~100 g DNA or ~100 kg tissue would be required.) It appears essential, therefore, to develop methods for the structural characterization of subnanogram amounts of adducts. It is hoped that the combination of the ^{32}P-postlabelling assay with ultrasensitive techniques for structural analysis will be a powerful tool in the characterization of the origin and nature of human adducts from exogenous and endogenous sources.

Acknowledgements

Work in the authors' laboratory has been supported by US Public Health Service Grants (CA 10893 (P6), CA 25590, CA 32157, and CA 43263) and by an Occupational and Environmental Health Grant from E.I. du Pont de Nemours & Co.

References

Brookes, P. & Lawley, P.D. (1964) Evidence for the binding of polynuclear aromatic hydrocarbons to the nucleic acids of mouse skin: relation between carcinogenic power of hydrocarbons and their binding to deoxyribonucleic acid. *Nature, 202,* 781–784

Church, D.F. & Pryor, W.A. (1985) Free-radical chemistry of cigarette smoke and its toxicological implications. *Environ. Health Perspect., 64,* 111–126

Everson, R.B., Randerath, E., Santella, R.M., Cefalo, R.C., Avitts, T.A. & Randerath, K. (1986) Detection of smoking-related covalent DNA adducts in human placenta. *Science, 231,* 54–57

Everson, R.B., Randerath, E., Avitts, T.A., Schut, H.A.J. & Randerath, K. (1987) Preliminary investigations of tissue specificity, species specificity, and strategies for identifying chemicals causing DNA adducts in human placenta. *Progr. exp. Tumour Res., 31,* 86–103

Gupta, R.C. (1984) Nonrandom binding of the carcinogen *N*-hydroxy-2-acetylaminofluorene to repetitive sequences of rat liver DNA in vivo. *Proc. natl Acad. Sci. USA, 81,* 6943–6947

Gupta, R.C. (1985) Enhanced sensitivity of ^{32}P-postlabelling analysis of aromatic carcinogen-DNA adducts. *Cancer Res., 45,* 5656 5662

Gupta, R.C. & Dighe, N.R. (1984) Formation and removal of DNA adducts in rat liver treated with *N*-hydroxy derivatives of 2-acetylaminofluorene, 4-acetylaminobiphenyl and 2-acetylaminophenanthrene. *Carcinogenesis, 5,* 343–349

Gupta, R.C., Reddy, M.V. & Randerath, K. (1982) ^{32}P-Postlabelling analysis of non-radioactive aromatic carcinogen-DNA adducts. *Carcinogenesis, 3,* 1081–1092

Hartz, A.J., Anderson, A.J., Brooks, H.L., Manley, J.C., Parent, G.T. & Barboriak, J.J. (1984) The association of smoking with cardiomyopathy. *New Engl. J. Med., 311,* 1201–1206

Hemminki, K. (1983) Nucleic acid adducts of chemical carcinogens and mutagens. *Arch. Toxicol., 52,* 249–285

Kirkman, H. (1959) Estrogen-induced tumours of the kidney. Growth characteristics in the Syrian hamster. *Natl Cancer Inst. Monogr., 1,* 1–57

Kriek, E., Den Engelse, L., Scherer, E. & Westra, J.G. (1984) Formation of DNA modifications by chemical carcinogens: identification, localization and quantification. *Biochim. biophys. Acta, 738,* 181–201

Liehr, J.G., Randerath, K. & Randerath, E. (1985) Target organ-specific covalent DNA damage preceding diethylstilbestrol-induced carcinogenesis. *Carcinogenesis, 6,* 1067–1069

Liehr, J.G., Avitts, T.A., Randerath, E. & Randerath, K. (1986) Estrogen-induced endogenous DNA adduction: possible mechanism of hormonal cancer. *Proc. natl Acad. Sci. USA, 83,* 5301–5305

Liehr, J.G., Hall, E.R., Avitts, T.A., Randerath, E. & Randerath, K. (1987) Localization of estrogen-induced DNA adducts and cytochrome P-450 activity at the site of renal carcinogenesis in the hamster kidney. *Cancer Res., 47,* 2156–2159

Lu, L.-J.W., Disher, R.M., Reddy, M.V. & Randerath, K. (1986) ^{32}P-Postlabelling assay of transplacental DNA damage induced by the environmental carcinogens safrole, 4-aminobiphenyl, and benzo[*a*]pyrene. *Cancer Res., 46,* 3046–3054

Miller, E.C. & Miller, J.A. (1981) Searches for ultimate chemical carcinogens and their reactions with cellular macromolecules. *Cancer, 47,* 2327–2345

Parks, W.C., Schurdak, M.E., Randerath, K., Maher, V.M. & McCormick, J.J. (1986) Human cell mediated cytotoxicity, mutagenicity, and DNA adduct formation of 7H-dibenzo[c,g]-carbazole and its *N*-methyl derivative in diploid human fibroblasts. *Cancer Res., 46,* 4706–4711

Phillips, D.H., Reddy, M.V. & Randerath, K. (1984) ^{32}P-Postlabelling analysis of DNA adducts formed in the livers of animals treated with safrole, estragole and other naturally occurring alkenylbenzenes. II. Newborn male B6C3F$_1$ mice. *Carcinogenesis, 5,* 1623–1628

Randerath, K., Reddy, M.V. & Gupta, R.C. (1981) ^{32}P-Labeling test for DNA damage. *Proc. natl Acad. Sci. USA, 78,* 6126–6129

Randerath, K., Haglund, R.E., Phillips, D.H. & Reddy, M.V. (1984) ^{32}P-Postlabelling analysis of DNA adducts formed in the livers of animals treated with safrole, estragole and other naturally-occurring alkenylbenzenes. I. Adult female CD-1 mice. *Carcinogenesis, 5,* 1613–1622

Randerath, E., Agrawal, H.P., Weaver, J.A., Bordelon, C.B. & Randerath, K. (1985a) ^{32}P-Postlabelling analysis of DNA adducts persisting for up to 42 weeks in the skin, epidermis and dermis of mice treated topically with 7,12-dimethylbenz[*a*]anthracene. *Carcinogenesis, 6,* 1117–1126

Randerath, K., Randerath, E., Agrawal, H.P., Gupta, R.C., Schurdak, M.E. & Reddy, M.V. (1985b) Postlabeling methods for carcinogen-DNA adduct analysis. *Environ. Health Perspect., 62,* 57–65

Randerath, E., Avitts, T.A., Reddy, M.V., Miller, R.H., Everson, R.B. & Randerath, K. (1986a) Comparative ^{32}P-analysis of cigarette smoke-induced DNA damage in human tissues and mouse skin. *Cancer Res., 46,* 5869–5877

Randerath, K., Reddy, M.V. & Disher, R.M. (1986b) Age- and tissue-related DNA modification in untreated rats: detection by ^{32}P-postlabelling assay and possible significance for spontaneous tumour induction and ageing. *Carcinogenesis, 7,* 1615–1617

Randerath, K., Reddy, M.V., Avitts, T.A., Miller, R.H., Everson, R.B., & Randerath, E. (1986c) *^{32}P-Postlabeling test for smoking-related DNA adducts in animal and human tissues.* In: Hoffmann, D. & Harris, C.C., eds, *Mechanisms in Tobacco Carcinogenesis* (*Banbury Report 23*), Cold Spring Harbor, NY, CSH Press, pp. 85–95

Randerath, E., Mittal, D. & Randerath, K. (1988) Tissue distribution of covalent DNA damage in mice treated dermally with cigarette 'tar': preference for lung and heart DNA. *Carcinogenesis, 9,* 75–80

Randerath, K., Avitts, T.A., Miller, R.H. & Randerath, E. (1987b) ^{32}P-Postlabeling analysis of cigarette smoking-related DNA adducts in target tissues of human carcinogenesis. *Proc. Am. Assoc. Cancer Res., 28,* 98

Reddy, M.V. & Randerath, K. (1986) Nuclease P$_1$-mediated enhancement of sensitivity of ^{32}P-postlabelling test for structurally diverse DNA adducts. *Carcinogenesis, 7,* 1543–1551

Reddy, M.V., Gupta, R.C., & Randerath, K. (1981) ^{32}P-Base analysis of DNA. *Anal. Biochem., 117,* 271–279

Reddy, M.V., Gupta, R.C., Randerath, E. & Randerath, K. (1984) ^{32}P-Postlabelling test for covalent DNA binding of chemicals in vivo: application to a variety of aromatic carcinogens and methylating agents. *Carcinogenesis, 5,* 231–243

Reddy, M.V., Irvin, T.R. & Randerath, K. (1985) Formation and persistence of sterigmatocystin-DNA adducts in rat liver determined via ^{32}P-postlabelling analysis. *Mutat. Res., 152,* 85–96

Reddy, M.V., Kenny, P.C. & Randerath, K. (1987) ^{32}P-Assay of DNA adducts in white blood cells (WBC) and placentas of pregnant women exposed to residential wood combustion (RWC) smoke. *Proc. Am. Assoc. Cancer Res., 28,* 97

Schoepe, K.-B., Friesel, H., Schurdak, M.E., Randerath, K. & Hecker, E. (1986) Comparative DNA binding of 7,12-dimethylbenz[*a*]anthracene and some of its metabolites in mouse epidermis in vivo as revealed by the ^{32}P-postlabelling technique. *Carcinogenesis, 7,* 535–540

Schurdak, M.E. & Randerath, K. (1985) Tissue-specific DNA adduct formation in mice treated with the environmental carcinogen, 7H-dibenzo[*c,g*]carbazole, *Carcinogenesis, 6,* 1271–1274

Schurdak, M.E., Stong, D.B., Warshawsky, D. & Randerath, K. (1987) ^{32}P-Postlabelling analysis of DNA adduction in mice by synthetic metabolites of the environmental carcinogen, 7H-dibenzo[*c,g*]carbazole: chromatographic evidence for 3-hydroxy-7H-dibenzo[*c,g*]carbazole being a proximate genotoxicant in liver but not skin. *Carcinogenesis, 8,* 591–597

US Department of Health and Human Services (1984) *The Health Consequences of Smoking*:

Chronic Obstructive Lung Disease. A Report of the Surgeon General, Rockville, MD, Office on Smoking and Health

Wogan, G.N. & Gorelick, N.J. (1985) Chemical and biochemical dosimetry of exposure to genotoxic chemicals. *Environ. Health Perspect., 62,* 5–18

Yamashita, K., Takayama, S., Nagao, M., Sato, S. & Sugimura, T. (1986) Amino-methyl-α-carboline induced DNA modification in rat salivary glands and pancreas detected by ^{32}P-postlabelling method. *Proc. Jpn Acad., 62* (Ser. B), 45–48

AN AROMATIC DNA ADDUCT IN COLONIC MUCOSA FROM PATIENTS WITH COLORECTAL CANCER

D.H. Phillips,[1] A. Hewer,[1] P.L. Grover[1] & J.R. Jass[2]

[1]*Chester Beatty Laboratories, Institute of Cancer Research, London; and* [2]*Department of Pathology, St Mark's Hospital, London, UK*

The incidence of colorectal cancer in the western hemisphere is thought to be the result, in part, of environmental agents, and many studies strongly implicate diet as a determining factor. It is conceivable that the ingestion of genotoxic chemicals present in food or the endogenous formation of such substances in the gut may initiate colorectal cancer in humans. In the present study, ^{32}P-postlabelling has been used to examine DNA from normal-appearing colonic mucosa obtained from (i) patients undergoing surgery for colorectal cancer and (ii) adult and fetal controls for the presence of aromatic DNA adducts.

^{32}P-Postlabelling (Gupta *et al.*, 1982) is a method for the detection of carcinogen-DNA adducts that requires neither the use of radiolabelled carcinogens nor prior characterization of an adduct's structure in order to detect it. The technique is thus readily suited to the analysis of DNA from human tissues for possible prior exposure to both known and unidentified genotoxic agents.

DNA was isolated from morphologically normal colorectal mucosal specimens by phenol/chloroform extraction and RNase treatment (Gupta, 1984). DNA samples (4 μg) were digested with micrococcal nuclease and spleen phosphodiesterase (Gupta *et al.*, 1982) and then incubated with nuclease P_1 (Phillips *et al.*, 1986; Reddy & Randerath, 1986). The resulting digest, consisting of normal nucleosides and 3′-mononucleotides of any aromatic adducts that might be present and resistant to nuclease P_1 digestion, was incubated with [γ-^{32}P]ATP (150 μCi) and T4 polynucleotide kinase (Gupta *et al.*, 1982; Phillips *et al.*, 1986). ^{32}P-Labelled adducts were then resolved on polyethyleneimine-cellulose thin-layer chromatograms, using modifications (Phillips *et al.*, 1986) of the original multidirectional procedures (Gupta *et al.*, 1982; Everson *et al.*, 1986). ^{32}P-Labelled adduct spots were detected by autoradiography at −70°C with intensifying screens.

DNA samples from 44 patients with colorectal tumours were analysed, and representative examples of the polyethyleneimine-cellulose thin-layer chromatographic maps of adduct spots obtained are shown in Figure 1. All samples displayed the radioactive spot marked X; we have previously reported the presence of this material in all DNA samples, regardless of origin, and its detection when the ^{32}P-postlabelling analysis is carried out with carrier-free [γ-^{32}P]ATP and following nuclease P_1 digestion (Phillips *et al.*, 1986, 1987). A fast-migrating spot (1, Fig. 1A) was reproducibly observed in the chromatograms of 43% (19/44) of the samples. The level at which the adduct was present was estimated to be at up to one adduct in 10^8 nucleotides, the lower limit of detection in these experiments being approximately 0.2 adducts in 10^8 nucleotides.

Figure 1B shows an example of the analysis of a sample of DNA from a cancer

Fig. 1. Polyethyleneimine-cellulose thin-layer chromatograms of ^{32}P-labelled digests of human colonic DNA

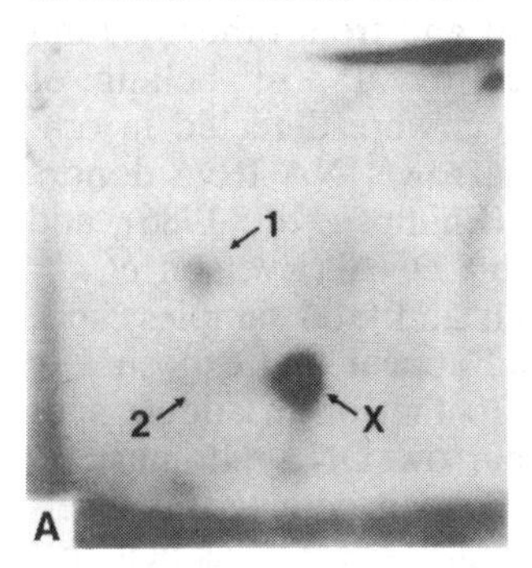

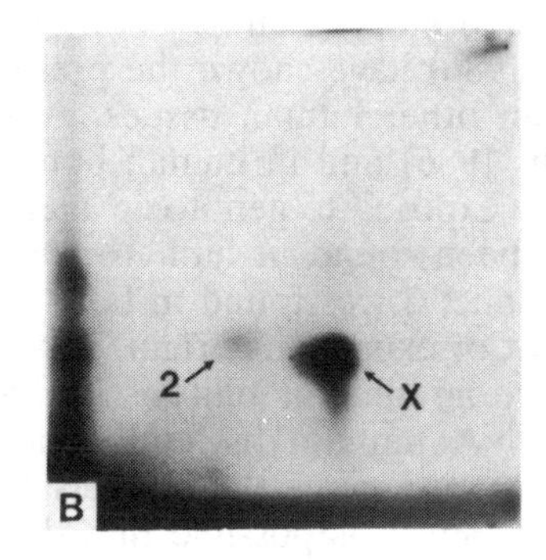

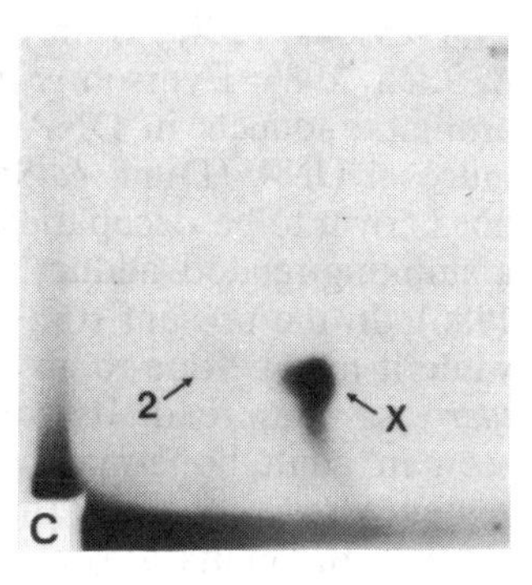

A, adult sample HC1 (56-year-old female with cancer of the sigmoid colon); B, adult sample HC5 (71-year-old male with cancer of the rectum); C, fetal sample FC3. Autoradiography was for 2.5 days at −70°C. The origins, located in the bottom left-hand corners of the chromatographs, were excised prior to autoradiography

patient that did not contain adduct 1. However, all samples contained a slower migrating adduct, designated adduct 2. Analysis of samples of colonic mucosal DNA from four noncancer patients revealed the presence of adduct 2 but not of adduct 1. Further, ten of 11 DNA samples obtained from fetal colon contained adduct 2 and not adduct 1 (see, for example, Fig. 1C). The eleventh sample contained both adducts.

Examination of the cancer patients' case histories indicated that samples from 3/11 patients with poorly-differentiated tumours contained adduct 1, as did 12/26 from patients with moderately-differentiated tumours and 4/7 from individuals with well-differentiated tumours (Table 1). Thus, the patients with poorly-differentiated tumours were less likely to show evidence of adduct 1 than those with moderately- or well-differentiated tumours, but, because of the small size of groups studied to date, the differences are not statistically significant. No correlation of age, sex or tumour site with the presence of adduct 1 was noted. Adduct 1 was identified slightly less frequently in patients with adenomas (5/15) than in patients without adenomas (12/29) but was detected with the same frequency in patients with metaplastic polyps (7/19) as in those without (12/25).

While there is evidence for the role of environmental agents in the etiology of colorectal cancer, no specific agent has so far been identified. Analysis of colonic DNA for prior exposure to potentially carcinogenic compounds that have bound covalently to the DNA is a rational approach to establishing the existence of such an initiating agent or agents. The role of diet as a risk factor for colon carcinogenesis and the

Table 1. Presence of adduct 1 in the DNA of human colonic mucosa

Source	No. with adduct 1	No. without adduct 1	Total
Cancer patients — poorly differentiated tumours	3	8	11
— moderately-differentiated tumours	12	14	26
— well-differentiated tumours	4	3	7
Noncancer patients	0	4	4
Fetal colon	1	10	11

known occurrence of a number of potent carcinogens and mutagens in food prompted us to examine colonic mucosal DNA for the presence of aromatic adducts.

Previous studies from our laboratory (Phillips *et al.*, 1986) and from others (Dunn & Stich, 1986; Everson *et al.*, 1986) have shown the presence, by ^{32}P-postlabelling, of aromatic adducts in DNA from other human tissues. Adducts were detected in oral mucosal DNA (Dunn & Stich, 1986) and in human bone-marrow DNA from donors not known to be occupationally exposed to genotoxic agents (Phillips *et al.*, 1986), and a smoking-related adduct has been found in human placental DNA (Everson *et al.*, 1986). In the present study, adduct 2 was found in both adult and fetal samples, and, while it may be due to an agent of exogenous origin, it would appear that exposure *in utero* is widespread. Its chromatographic mobility is similar to that of an adduct spot seen in adult bone-marrow DNA but not in fetal bone-marrow DNA. It must be remembered, however, that in many cases the adducts have been detected at levels close to the limits of sensitivity, and it is possible that the samples scored as negative contained lower, undetectable levels of a particular adduct. Similarly, the detection of adduct 1 in 19/44 samples of uninvolved colonic mucosa from cancer patients but in only 1/11 fetal samples does not rule out the possibility of quantitative rather than qualitative differences between the two groups. The chromatographic mobility of adduct 1 is greater than that of the previously observed bone-marrow DNA adducts (Phillips *et al.*, 1986), and it has not yet been identified by correspondence with known carcinogen-DNA adducts.

Epidemiological studies have demonstrated an excess of poorly-differentiated large-bowel cancers in low-risk populations (Elmasry & Boulos, 1975), in whom inherited factors may assume greater etiological importance than environmental factors. A confounding factor is that not all poorly-differentiated cancers would be genetically determined. Similarly, the appearance of adenomas in addition to carcinomas may be taken in some cases as an indication of genetically determined disease (Burt *et al.*, 1985); but, again, the correlation between the presence of adduct 1 and the absence of adenomas was not statistically significant in the present study.

This preliminary study has provided evidence of the presence of an aromatic DNA adduct in adult colonic mucosa from 43% of cancer patients that was not seen in any of four adult noncancer patients or in 10/11 samples of fetal colon. Further investigations will be required to determine (i) the significance of this finding and (ii) the origin and possible role of this adduct in the initiation of large-bowel cancer in man.

Acknowledgements

This work was supported by grants to the Institute of Cancer Research from the Medical Research Council and the Cancer Research Campaign.

References

Burt, R.W., Bishop, T., Cannon, L.A., Dowdl, M.A., Lee, R.G. & Skolnick, M.H. (1985) Dominant inheritance of adenomatous colon polyps and colorectal cancer. *New Engl. J. Med., 312,* 1540–1544

Dunn, B.P. & Stich, H.F. (1986) ^{32}P-Postlabelling analysis of aromatic DNA adducts in human and mucosal cells. *Carcinogenesis, 7,* 1115–1120

Elmasry, S.H. & Boulos, P.B. (1975) Carcinoma of the large bowel in the Sudan. *Br. J. Surg., 62,* 284–286

Everson, R.B., Randerath, E., Santella, R.M., Cefalo, R.C., Avitts, T.A. & Randerath, K. (1986) Detection of smoking-related covalent DNA adducts in human placenta. *Science, 231,* 54–57

Gupta, R.C. (1984) Non-random binding of the carcinogen *N*-hydroxy-2-acetylaminofluorene to repetitive sequences of rat liver DNA *in vivo*. *Proc. natl Acad. Sci. USA, 81,* 6943–6947

Gupta, R.C., Reddy, M.V. & Randerath, K. (1982) ^{32}P-Postlabelling analysis of non-radioactive aromatic carcinogen-DNA adducts. *Carcinogenesis, 3,* 1081–1092

Phillips, D.H., Hewer, A. & Grover, P.L. (1986) Aromatic DNA adducts in human bone marrow and peripheral blood leukocytes. *Carcinogenesis, 7,* 2071–2075

Phillips, D.H., Hewer, A. & Grover, P.L. (1987) Formation of DNA adducts in mouse skin treated with metabolites of chrysene. *Cancer Lett., 35,* 207–214

Reddy, M.V. & Randerath, K. (1986) Nuclease P1-mediated enhancement of sensitivity of ^{32}P-postlabelling test for structurally diverse DNA adducts. *Carcinogenesis, 7,* 1543–1551

ENHANCEMENT OF SENSITIVITY OF FLUORESCENCE LINE NARROWING SPECTROMETRY FOR DETECTION OF CARCINOGEN-DNA ADDUCTS

R. Jankowiak,[1] R.S. Cooper,[1] D. Zamzow,[1] G.J. Small,[1] G. Doskocil[2] & A.M. Jeffrey[2]

[1]*Ames Laboratory, Department of Energy and Department of Chemistry, Iowa State University, IA; and* [2]*Institute of Cancer Research and Division of Environmental Sciences, Columbia University, New York, NY, USA*

Fluorescence line narrowing (FLN), is a method by which highly characteristic spectra have been obtained for a large number of polycyclic aromatic hydrocarbon (PAH)-DNA adducts and is well suited for the analysis of exposures to complex mixtures of PAH. The basic method is described and recent improvements discussed which overcome one of the major limitations of the method to its wider application to biological samples, that is its sensitivity.

Over the past several years, there has been considerable interest in the development of new analytical methods for the detection of both DNA- and protein-carcinogen adducts. Techniques that take advantage of the fluorescence of PAH for the detection of PAH adducts resulting from exposure to individual hydrocarbons include both conventional low-temperature fluorescence techniques (Ivanovic *et al.*, 1976) and synchronous fluorescence, in which excitation and emission frequencies are scanned simultaneously maintaining a fixed wavelength between the two (Vähäkangas *et al.*, 1985). While these approaches provide good sensitivity, the line widths of the fluorescence bands are too broad to allow resolution of multiple adducts in the case of exposure to complex mixtures of PAH derivatives.

We have previously investigated laser excitation of benzo[*a*]pyrene diol epoxide (BPDE)-modified DNA samples, which were cooled to 4.2 K in water-glycerol-ethanol glasses such that only a narrow subset, or isochromat, of sites absorbed the narrow bandwidth light (about 1 cm^{-3}) (Heisig *et al.*, 1984). Results showed dramatic FLN of the spectra (Fig. 1). These spectra are highly characteristic and can provide structural information about the DNA adducts present in samples by comparison with metabolites of the hydrocarbons. The latter are more readily available and more easily characterized than the adducts themselves. Careful inspection of the metabolite spectra reveals subtle differences which allow distinction between the DNA adduct and the free compound.

Resolution of mixtures with similar chromophores can be achieved with this method, but not with conventional techniques. Figure 2 (Sanders *et al.*, 1986) illustrates the resolution by FLN of a mixture of modified DNA samples in which three sets of the adducts have a phenanthrene chromophore in common. Thus, the method provides for direct analysis of multiple DNA adducts within a single DNA sample

Fig. 1. FLN spectra of molecules embedded in an ethanol-glycerol-water glass at 4.2 K excited with a nitrogen pumped dye laser at 3748 cm^{-1} for 7,8,9,10-tetrahydrobenzo[*a*]pyrene (THBP) (top), 3771 cm^{-1} for 7,8,9,10-tetra-7,8,9,10-tetrahydroxybenzo[*a*]pyrene (tetrol) (middle), and 3748 cm^{-1} for BPDE-modified DNA (bottom). From Heisig *et al.* (1984)

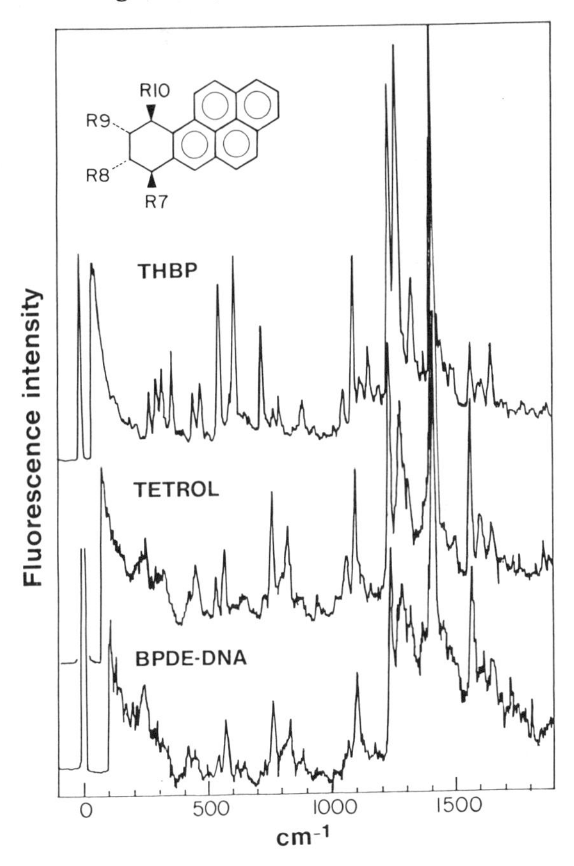

without degradation of the DNA, and offers the possibility of obtaining structural information about the adducts present.

The major current limitation to this method is the sensitivity of detection. Since only an isochromat (Fig. 3, top) is excited, most of the molecules bound to the DNA are not analysed. This earlier placed a limit of detection of about one adduct per 10^6, which is two and four orders of magnitude less sensitive than the ranges needed to measure exposure levels in samples of human DNA and by the ^{32}P-postlabelling technique, respectively. Increasing the intensity of the laser light source to 400 mW/cm^2 resulted in photochemical degradation of the sample (Fig. 3, middle), as observed previously (Boles & Hogan, 1986), although at much higher temperatures. This problem could be eliminated by the use of anaerobic conditions; however, as the laser intensity was increased, nonphotochemical hole burning (Fig. 3, bottom) (Jankowiak & Small, 1987) was observed in the BPDE-DNA samples. Excited molecules relaxed into slightly different microenvironments, so that they no longer absorbed the laser light. This resulted in a decrease in observed fluorescence intensity with respect to time

Fig. 2. FLN spectra of a mixture of the five DNA adducts obtained with (1, 0) type excitation at 343.1 nm, $T = 4.2$ K. Peaks labelled A, B and C are due to the three individual adducts pictured. From Sanders *et al.* (1986)

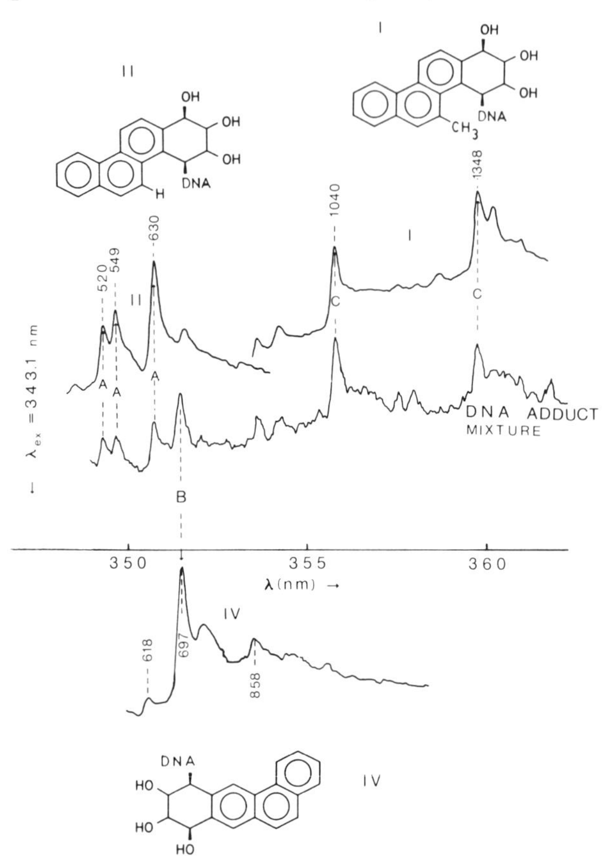

(Fig. 4). Unlike photochemical decomposition, this process can be reversed by warming the sample to allow re-equilibration. If spectra are collected over time and the differences measured, an improvement in selectivity can be obtained by observing the sharp negative signals of the (0–0) sites of the phonon bands that have been burnt out. Using a novel approach (Jankowiak *et al.*, 1987), which effectively minimizes fluorescence intensity loss due to hole burning, we are now able to detect DNA damage at the level of three adducts/10^8 bases, using as little as a femtomole of adduct in DNA samples isolated from C3H 10T1/2 mouse embryo fibroblast cells exposed to 40 ng/ml ^{3}H-benzo[*a*]pyrene for 24 h (Brown *et al.*, 1979).

Conclusions

This method of analysis for PAH-DNA adducts has several advantages over other procedures. The structures of the adducts need not be known, the DNA need not be digested prior to analysis, and the potential for obtaining structural information regarding the PAH moiety is high. The main obstacle, which was the sensitivity of

Fig. 3. (a) Inhomogeneous line broadening; (b) top: photochemical decomposition; (b) bottom: new excitation spectrum resulting from redistribution of configurations of previously excited adducts; PHB, photochemical hole burning; NPHB nonphotochemical hole burning. From Jankowiak & Small (1987)

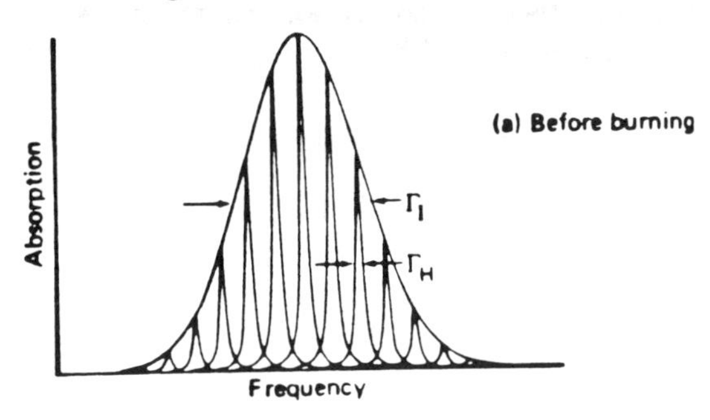

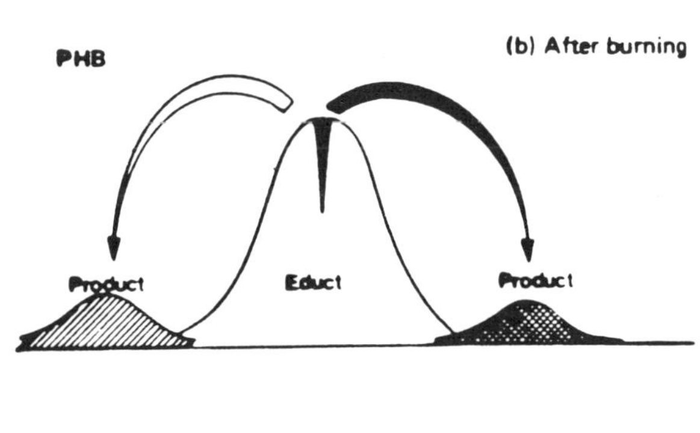

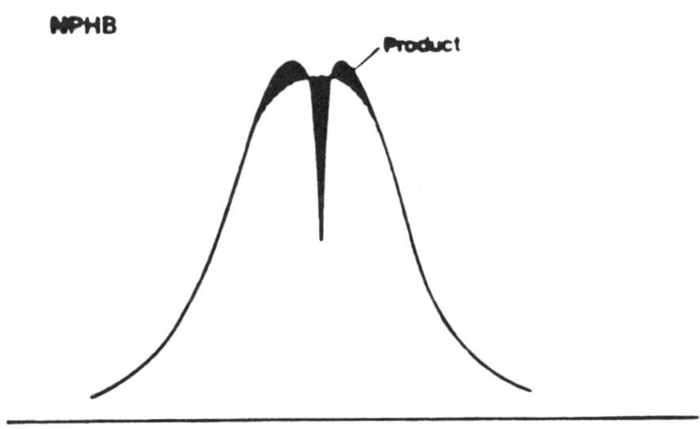

detection, is overcome, so that FLN spectrometry is now comparable in this respect with many of the other techniques currently employed. The prospect for further improvements is present, and these studies are under way.

Acknowledgements

This work was supported by grants from the Office of Health and Environmental Research, Office of Energy Research, and by NCI Grant CA 02111.

Fig. 4. (a) FLN spectrum of BPDE-DNA (degassed); excitation laser intensity $I = 85\ mW/cm^2$, ex = 369.6 nm. The numbers correspond to the excited state vibrations (in cm^{-1}); (b) decay of integrated fluorescence intensity of 579 cm^{-1} peak for a laser intensity of 400 mW/cm^2; (c) resulting FLN spectrum after burning for 30 min, showing complete loss of spectrum. Spectrum is restored if sample is warmed to 60 K before reanalysis at 4 K

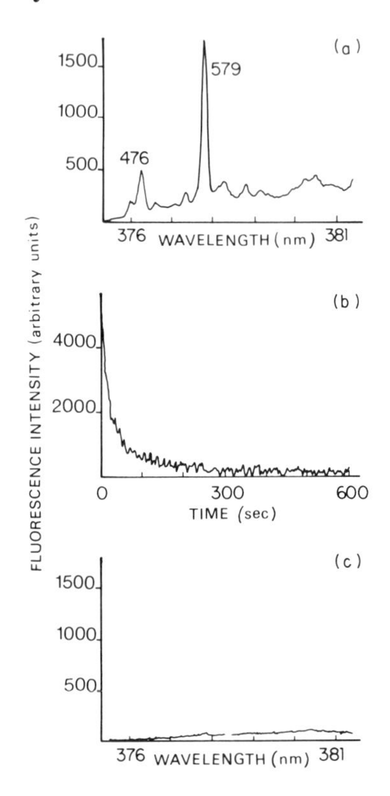

References

Boles, T.C. & Hogan, M.E. (1986) High-resolution mapping of carcinogen binding sites on DNA. *Biochemistry, 25,* 3039–3043

Brown, H.S., Jeffrey, A.M. & Weinstein, I.B. (1979) Formation of DNA adducts in 10T1/2 mouse embryo fibroblasts incubated with benzo[*a*]pyrene or dihydrodiol oxide derivatives. *Cancer Res., 39,* 1673–1677

Heisig, V., Jeffrey, A.M., McGlade, M.J. & Small, G.J. (1984) Fluorescence line narrowed spectra of polycyclic aromatic carcinogen-DNA adducts. *Science, 223,* 289–291

Ivanovic, V., Geacintov, N. & Weinstein, I.B. (1976) Cellular binding of benzo[*a*]pyrene to DNA characterized by low temperature fluorescence. *Biochem. biophys. Res. Commun., 70,* 1172–1179

Jankowiak, R. & Small, G.J. (1987) Hole-burning spectroscopy and relaxation dynamics of amorphous solids at low temperatures. *Science, 237,* 618–625

Jankowiak, R., Cooper, R.S., Zamzow, D., Small, G.J., Doskocil, G. & Jeffrey, A.M. (1988) Fluorescence line narrowing–nonphotochemical hole burning spectrometry: femtomole detection and high selectivity for intact DNA-PAH adducts. *Chem. Res. Toxicol., 1,* 60–68

Sanders, M.J., Cooper, R.S., Jankowiak, R., Small, G.J., Heisig, V. & Jeffrey, A.M. (1986) Identification of polycyclic aromatic hydrocarbon metabolites and DNA adducts in mixtures using fluorescence line narrowing spectrometry. *Anal. Chem.*, *58*, 816–820

Vähäkangas, K., Haugen, A. & Harris, C.C. (1985) An applied synchronous fluorescence spectrophotometric assay to study benzo[*a*]pyrene-diolepoxide-DNA adducts. *Carcinogenesis*, *6*, 1109–1116

SEARCH FOR UNKNOWN ADDUCTS: INCREASE OF SENSITIVITY THROUGH PRESELECTION BY BIOCHEMICAL PARAMETERS

M. Törnqvist

Department of Radiobiology, University of Stockholm, Stockholm, Sweden

Electrophilically reactive compounds or metabolites that occur in humans can be monitored through their reaction products with proteins and DNA. The identification of genotoxic risk factors, e.g., by haemoglobin (Hb) adducts, comprises both the identification of the causative factor and an estimate of the contribution to risk. The problems involved are illustrated by efforts to trace the origin of the observed background hydroxyethylations in Hb from persons without occupational exposures. Earlier work showed influences of smoking and, in animal studies, of dietary fat and intestinal flora on the levels of hydroxyethyl adducts to *N*-terminal valine. Efforts to measure adducts originating from ethene in environmental tobacco smoke and urban air show, however, that the resolving power of the methods used for identifying unknown risk factors must be increased. Studies of groups with excessive living habits and biochemical parameters that favour higher tissue doses of electrophiles would also enhance the possibilities of identifying risk factors.

Although it has been suggested that 70–90% of cancer incidence in western countries is due to environmental factors (in a broad sense) (Higginson & Muir, 1979), our present knowledge is limited, especially as concerns the initiators that are presumed to act in the induction of tumours. This is true, for instance, for the main portion of tumours that are ascribed to dietary and life-style factors. Except for radiation and initiators in tobacco smoke, we can account for 2% at most of current cancer incidence in Sweden, the main contributions coming from work environments and urban air pollution (Cancer Committee, 1984).

Most known chemical initiators possess electrophilic reactivity (Miller & Miller, 1966) and are also mutagens (Miller & Miller, 1971). One means of determining these reactive compounds *in vivo* is to analyse their chemically stable products of reaction with proteins and DNA. Mapping of such adducts to macromolecules in humans, especially in unexposed members of the public, might shed light on the current load of initiators/mutagens.

One such approach is the determination of adducts to Hb (Osterman-Golkar *et al.*, 1976). Chemical analytical methods for Hb adduct determination in humans have so far been successfully applied to low-molecular-weight alkylating agents (and Schiff bases) and aromatic amines (Farmer *et al.*, 1987). These methods, which offer certain advantages with respect to identification of the chemical structure of adducts, are at an initial stage of use for the screening of unknown adducts. They must be developed further to permit the determination of adducts of other classes.

Identification of risk factors

The identification of chemical risk factors, for instance, using adducts to Hb as an endpoint in epidemiological studies, comprises several steps:

(1) Identification of the chemical structure of the observed adduct and quantification of the level of adducts (e.g., expressed in pmol/g Hb); in some cases the observed adducts are the same as those seen to be associated with specific exposures and occur as background levels in persons without such exposures; in other cases the adducts are 'new'.

(2) Identification of the causative electrophile: especially for low-molecular-weight adducts, several options may be available. For example, 2-oxoethyl may originate from metabolites of urethane, vinyl chloride and, *via* Schiff bases, from glycolaldehyde (Svensson & Osterman-Golkar, 1987). To some extent, determination of the spectrum of adducts over nucleophilic sites in the Hb molecule may give guidance (Calleman *et al.*, 1986). For instance, ethylene oxide and 2-hydroxyethylnitrosourea give different patterns of 2-hydroxyethyl adducts, in accordance with reaction kinetic laws (Segerbäck, 1985).

(3) Establishing the source of the electrophile: this aspect of the work involves trial and error. Knowledge of metabolism and organic chemistry has to be combined with careful information from donors of samples participating in epidemiologic studies. A-priori and a-posteriori hypotheses have to be tested in animal experiments.

(4) Estimation of contribution to cancer risk: this must be based on knowledge about current doses and genotoxic potency, estimated, for instance, by application of rad-equivalence (Kolman *et al.*, this volume).

Required sensitivity of analytical methods

It is desirable that the analytical methods used to determine adducts possess such sensitivity that unacceptable risks do not elude detection. There is of course no generalizable limit between acceptable and unacceptable risks, but one could apply the radiation risk philosophy to define a desired detection level for individual compounds. The International Commission on Radiological Protection (1977) considers risks of cancer mortality due to ionizing radiation of 10^{-6}–10^{-5}/year to be acceptable to members of the public. On the basis of rad-equivalence (see Kolman *et al.*, this volume) of chemical dose, this is estimated to correspond to levels of adducts to the *N* terminals of Hb in the range of 1–10 pmol/g Hb. This value would be valid for ethylene oxide and alkylating agents with similar reaction patterns [*s* values in the sense of Swain and Scott (1953) in the range 0.8–1.2]. A method for determining adducts to *N* terminals (valine) in Hb, based on the Edman degradation for protein sequencing, has been worked out to attain this sensitivity (Törnqvist *et al.*, 1986a).

Origin of the background hydroxyethylations: an example of risk identification

Several adducts to Hb, such as those with methyl (Bailey *et al.*, 1981; Törnqvist *et al.*, 1988a), 2-hydroxyethyl (Osterman-Golkar, 1983; van Sittert *et al.*, 1985; Törnqvist *et al.*, 1986a,b), 2-oxoethyl (Svensson & Osterman-Golkar, 1987) 2-hydroxypropyl (unpublished), several aldehydes (unpublished) and aromatic amines (Bryant *et al.*, 1987; Stillwell *et al.*, 1987), have been observed in unexposed persons.

Problems involved in risk identification can be illustrated by efforts to trace the origin of the observed background hydroxyethylations in persons with no occupational exposure (Table 1). Current measurements (using the modified Edman method) indicate that the average level of hydroxyethylvaline in Hb is about 100 pmol/g among

Table 1. Source of hydroxyethylations in Hb as indicated by measurements of hydroxyethylvaline (HOEtVal) in humans and rodents

Species	Exposures	'Dose' (level of chronic exposure)	HOEtVal (pmol/g Hb)	Reference
Man	Tobacco smoke	10 cigarettes/day; about 2.5 mg ethene/day (average Swedish smoker)	80[a]	Törnqvist *et al.* (1986b)
Man	Urban air	10–50 ppb ethene	0 ± 15 (exp, 5–50)	Törnqvist *et al.* (in preparation)
Rodents	Motor exhaust	100–200 ppm-h ethene/week	350–620	Törnqvist *et al.* (1988b)
Man	Passive smoking	>10 ppb ethene	0 ± 15 (exp, >5)	Persson *et al.* (1988); Törnqvist *et al.* (in preparation)
Mouse	Unsaturated fat in diet	Unsaturation of fats, ethene from lipid peroxidation	15 ± 5	Törnqvist *et al.* (1988c)
Mouse	Intestinal flora	Ethene or other factor	11 ± 5	Törnqvist *et al.* (1988c)

[a]Value adjusted to new (linear) calibration curve (Törnqvist *et al.*, 1988b), about 30% lower than given in the reference

Swedish male adults, which, to the extent that the alkylations are caused by ethylene oxide or compounds with similar reaction patterns, seems to correspond to an unacceptable risk (see Törnqvist *et al.*, 1986a). A considerable portion of this level could be ascribed to cigarette smoking, with ethene, a precursor of ethylene oxide (Ehrenberg *et al.*, 1977), as a probable source (Törnqvist *et al.*, 1986b). Other possible exogenous sources are ethene in urban air, mainly from motor exhausts, and in environmental tobacco smoke. In a current epidemiological pilot study, comprising about 80 male nonsmokers, it was, however, impossible to observe significant effects of urban air and passive smoking on the level of this adduct. Exploratory animal experiments indicate that further sources of the residual background level could be ethene formed during lipid peroxidation and some metabolite of intestinal bacteria (see Table 1).

The conclusion that ethene/ethylene oxide is the likely source of excess hydroxyethylation in smokers was based on observations of two kinds (Törnqvist *et al.*, 1986b): (i) the rate of conversion of ethene is quantitatively compatible with that in animal experiments; and (ii) the ratio of histidine to valine alkylation is compatible with that obtained for ethylene oxide. This interpretation meets, however, a basic difficulty: the content of propene in tobacco smoke is nearly as high as that of ethene (Elmenhorst & Schultz, 1968; Vickroy, 1976; Persson *et al.*, 1988). Although propene was found in animal experiments to be metabolized to propylene oxide at a rate comparable to that of ethene (Svensson & Osterman-Golkar, 1984), no significant difference in 2-hydroxypropylvaline levels could be found in comparisons of smokers and nonsmokers (unpublished). This might be taken to favour alternative interpretations of the raised hydroxyethylvaline levels in smokers, one possibility being increased lipid peroxidation (Frank *et al.*, 1980) caused by the smoking. This possibility is also favoured by the observation that the incremental tissue dose of ethylene oxide in fruit-store workers, although significant (which shows that this conversion occurs in humans), is smaller than that expected from the estimated ethene exposure (unpublished). These deliberations illustrate procedures and difficulties in work aimed at identifying risk factors from macromolecular adducts in blood samples. The hydroxyethyl adduct is a simple case, considering that its chemical structure is known and that the adduct levels are relatively high. However, the adduct may be formed by different electrophiles (Törnqvist & Ehrenberg, 1985; Törnqvist *et al.*, 1986b), perhaps in some cases from more than one precursor. In addition, ethene may be derived from several sources.

Methodological difficulties encountered

As exemplified above, adduct levels expected from known sources of alkylation could not be detected in an epidemiological pilot study of unexposed subjects. Apart from the possibility that the true levels are lower than the expected ones, failure to detect adducts may be due to the following:

(1) The low resolving power of the questionnaire used in the study, with respect to grading subjects according to degree of urbanization, degree of passive smoking, relevant dietary and other living habits, is due particularly to subjectivity of answers and to lack of a relevant scale of degree of urbanization.

(2) There are variable contributions from different sources to the level of the same adduct.

(3) When close to the detection level of the gas chromatographic-mass spectrometric method used, the analytical error of small increments will be large (see Table 1).

(4) A specific problem in measuring adducts close to the detection limit is created by contamination of samples due to previous work on the laboratory premises with higher levels of compounds.

(5) Formation of artefacts (Calleman *et al.*, 1979) is decreased by using the modified Edman degradation method for the determination of adducts to the *N*-terminal valine, due to the mild conditions during preparation and derivatization of samples (Törnqvist *et al.*, 1986a), but precautions must be taken during storage and handling of samples, especially in work with low-molecular-weight adducts (for instance, aldehydes).

(6) Laboratory personnel working with blood must be protected against virus infection. If viral activity is eliminated by preheating freshly drawn blood, the possibilities of artefact formation and loss of adducts through hydrolysis of, e.g., esters will increase.

Means of increasing resolving power in risk identification

In order to overcome such difficulties and to potentiate identification and quantification of unknown risk factors, more effective, more sensitive analytical procedures are a primary requirement, e.g., by pre-isolation of adducted subfractions of Hb. Further, objective work on random samples should be supplemented by studies of preselected groups with extreme levels of easily measured parameters judged to be determinants of tissue doses of electrophiles. Preselection of this kind is essential in view of the work and costs involved in measurement of low levels. Besides extreme living habits (e.g., vegetarians, alcoholics, social outcasts) and extreme conditions of residence (highly urbanized and rural areas), a number of biochemical parameters might be of guidance in such work. The activities of enzymes in blood and certain metabolites in urine and expired air can provide information relevant for preselection. These include: bioactivating enzymes (Kellermann *et al.*, 1973; Vähäkangas *et al.*, 1984); detoxicating enzymes, especially glutathione transferase and epoxide hydrolase activities, tested with different substrates (Vähäkangas *et al.*, 1984; Glatt *et al.*, 1985); urine metabolites that indicate what to look for in Hb and, in persons with known exposures, that help to identify persons with deficient detoxification functions; volatile compounds in exhaled air, as measures of metabolic products from intestinal bacteria and of lipid peroxidation (Törnqvist *et al.*, 1988c); cotinine in blood plasma, as a measure of passive smoking (Curvall & Enzell, 1986 and references therein); and newly diagnosed cancer cases, which possibly represent increased sensitivity associated with higher in-vivo doses (Vähäkangas *et al.*, 1984).

Identification and quantification of risk factors in subpopulations with biochemical conditions that result in high tissue doses of electrophiles may be used to estimate risks for individual members of the subpopulation. The general collective risk can then be estimated from the identified factors by adjustment to the population mean values of the respective biochemical conditions.

Acknowledgements

Financial support was obtained from the National Swedish Environment Protection Board and Shell Internationale Research Maatschappij B.V.

References

Bailey, E., Connors, T.A., Farmer, P.B., Gorf, S.M. & Rickard, J. (1981) Methylation of cysteine in haemoglobin following exposure to methylating agents. *Cancer Res., 41,* 2514–2527

Bryant, M.S., Skipper, P.L., Tannenbaum, S.R. & Maclure, M. (1987) Hemoglobin adducts of 4-aminobiphenyl in smokers and nonsmokers. *Cancer Res., 47,* 602–608

Calleman, C.J., Ehrenberg, L., Osterman-Golkar, S. & Segerbäck, D. (1979) Formation of *S*-alkylcysteines as artifacts in acid protein hydrolysis, in the absence and in the presence of 2-mercaptoethanol. *Acta chem. scand., B33,* 488–494

Calleman, C.J., Bergmark, E., Ehrenberg, L., Kautiainen, A., Osterman-Golkar, S., Segerbäck, D., Svensson, K. & Törnqvist, M. (1986) *Monitoring of background levels of hydroxyethyl adducts in human hemoglobin.* In: Ramel, C., Lambert, B. & Magnussson, J., eds, *Genetic Toxicology of Environmental Chemicals,* Part B, *Genetic Effects and Applied Mutagenesis,* New York, Alan R. Liss, pp. 261–270

Cancer Committee (1984) *Cancer: Causes, Prevention, etc.* (*SOU 1984:67*), Stockholm (in Swedish, English version in press, London, Taylor & Francis)

Curvall, M. & Enzell, C. R. (1986) Monitoring absorption by means of determination of nicotine and cotinine. *Arch. Toxicol., Suppl. 9,* 88–102

Ehrenberg, L., Osterman-Golkar, S., Segerbäck, D., Svensson, K. & Calleman, C.J. (1977) Evaluation of genetic risks of alkylating agents. III. Alkylation of haemoglobin after metabolic conversion of ethene to ethene oxide *in vivo. Mutat. Res., 45,* 175–184

Elmenhorst, H. & Schultz, C. (1968) Flüchtige Inhaltsstoffe des Tabakrauches. *Beitr. Tabakforsch., 4,* 90–122

Farmer, P.B., Neumann, H.-G. & Henschler, D. (1987) Estimation of exposure of man to substances reacting covalently with macromolecules. *Arch. Toxicol., 60,* 251–260

Frank, H., Hintze, T., Bimboes, D. & Remmer, H. (1980) Monitoring lipid peroxidation by breath analysis; endogenous hydrocarbons and their metabolic elimination. *Toxicol. appl. Pharmacol., 56,* 377–344

Glatt, H.R., Halfer-Wirkus, H., Herborn, J., Lehrbach, E., Löffler, S., Porn, W., Setiabudi, F., Wölfel, T., Gemperlein-Mertes, I., Doerjer, G. & Oesch, F. (1985) *Interindividual variations in epoxide-detoxifying enzymes.* In: Müller, H. & Weber, W., eds, *Familial Cancer,* Basel, Karger, pp. 242–247

Higginson, J. & Muir, C.S. (1979) Environmental carcinogenesis: misconceptions and limitations to cancer control. *J. natl Cancer Inst., 63,* 1291–1298

International Commission on Radiological Protection (1977) *Recommendations* (*ICRP Publication 26*), Oxford, Pergamon Press

Kellermann, G., Shaw, C.R. & Luyten-Kellerman, M. (1973) Aryl hydrocarbon hydroxylase inducibility and bronchogenic carcinoma. *New Engl. J. Med., 289,* 934–937

Miller, E.C. & Miller, J.A. (1966) Mechanisms of chemical carcinogenesis: nature of proximate carcinogens and interactions with macromolecules. *Pharmacol. Rev., 18,* 805–838

Miller, E.C. & Miller, J.A. (1971) *The mutagenicity of chemical carcinogens: correlations, problems and interpretations.* In: Hollaender, A., ed., *Chemical Mutagens,* Vol. 1, New York, Plenum Press, pp. 83–119

Osterman-Golkar, S. (1983) *Tissue doses in man: implications in risk assessment.* In: Hayes, A.W., Schnell, R.C. & Miya, T.S., eds, *Developments in the Science and Practice of Toxicology,* Amsterdam, Elsevier, pp. 289–298

Osterman-Golkar, S., Ehrenberg, L., Segerbäck, D. & Hällström, I. (1976) Evaluation of genetic risks of alkylating agents. II. Haemoglobin as a dose monitor. *Mutat. Res., 34,* 1–10

Persson, K.-A., Berg, S., Törnqvist, M., Scalia-Tomba, G.-P. & Ehrenberg, L. (1987) Note on low-molecular weight alkenes in environmental tobacco smoke (submitted for publication)

Segerbäck, D. (1985) *In-vivo Dosimetry of Some Alkylating Agents as a Basis for Risk Estimation,* Doctoral Thesis, Stockholm University, Stockholm

van Sittert, N.J., de Jong, G., Clare, M.G., Davies, R., Dean, B.J., Wren, L.J. & Wright, A.S. (1985) Cytogenetic, immunological and haemotological effects in workers in an ethylene oxide manufacturing plant. *Br. J. ind. Med., 42,* 19–26

Stillwell, W.G., Bryant, M.S. & Wishnok, J.S. (1987) Gas chromatographic/mass spectrometric analysis of biologically important aromatic amines. Application in human dosimetry. *Biomed. environ. Mass Spectrom., 14,* 221–227

Svensson, K. & Osterman-Golkar, S. (1984) Kinetics of metabolism of propene and covalent binding to macromolecules in the mouse. *Toxicol. appl. Pharmacol., 73,* 363–372

Svensson, K. & Osterman-Golkar, S. (1987) *In vivo 2-oxoethyl adducts in haemoglobin and their possible origin.* In: Sandhu, S.S., DeMarini, D.M., Mass, M.J., Moore, M.M. & Mumford, J.S., eds, *Application of Short-term Bioassays in the Analysis of Complex Environmental Mixtures V,* New York, Plenum Press, pp. 49–66

Swain, C.G. & Scott, C.B. (1953) Quantitative correlation of relative rates. Comparison of hydroxide ion with other nucleophilic reagents toward alkyl halides, esters, epoxides and acyl halides. *J. Am. chem. Soc., 75,* 141–147

Törnqvist, M. & Ehrenberg, L. (1985) Risk estimation of urban air pollution: information sources and methods. *Environ. Int., 11,* 401–406

Törnqvist, M., Mowrer, J., Jensen, S. & Ehrenberg, L. (1986a) Monitoring of environmental cancer initiators through haemoglobin adducts by a modified Edman degradation method. *Anal. Biochem., 154,* 255–266

Törnqvist, M., Osterman-Golkar, S., Kautiainen, A., Jensen, S., Farmer, P.B. & Ehrenberg, L. (1986b) Tissue doses of ethylene oxide in cigarette smokers determined from adduct levels in haemoglobin. *Carcinogenesis, 7,* 1519–1521

Törnqvist, M., Osterman-Golkar, S., Kautiainen, A., Näslund, M., Calleman, C.J. & Ehrenberg, L. (1988a) Methylations in human haemoglobin. *Mutat. Res., 204,* 521–529

Törnqvist, M., Kautiainen, A., Gatz, R.N. & Ehrenberg, L. (1988b) Haemoglobin adducts in animals exposed to gasoline and diesel exhausts. 1. Alkenes. *J. appl. Toxicol., 8* (in press)

Törnqvist, M., Kautiainen, A., Harms-Ringdahl, M., Gustafsson, B. & Ehrenberg, L. (1988c) Unsaturated lipids and intestinal bacteria as sources of endogenous production of ethene and ethylene oxide (to be published)

Vähäkangas, K., Autrup, H. & Harris, C.C. (1984) *Interindividual variation in carcinogen metabolism, DNA damage and DNA repair.* In: Berlin, A., Draper, M., Hemminki, K. & Vainio, H., eds, *Monitoring Human Exposure to Carcinogenic and Mutagenic Agents* (*IARC Scientific Publications No. 59*), Lyon, International Agency for Research on Cancer, pp. 85–98

Vickroy, D.G. (1976) The characterization of cigarette smoke from Cytrel smoking products and its comparison to smoke from flue-cured tobacco. 1. Vapor phase analysis. *Beitr. Tabakforsch., 8,* 415–421

HUMAN ORGAN CULTURE TECHNIQUES FOR THE DETECTION AND EVALUATION OF GENOTOXIC AGENTS

W.P. Watson, R.J. Smith, K.R. Huckle & A.S. Wright

Shell Research Ltd, Sittingbourne Research Centre, Sittingbourne, Kent, UK

In order to obtain information on the genotoxic metabolism of carcinogens in human skin *in vivo,* model in-vitro systems have been developed to mimic in-vivo metabolism qualitatively. Direct labelling (^{3}H and ^{14}C) and ^{32}P-postlabelling analyses of benzo[*a*]pyrene (BP)-DNA adducts in human skin explants, CD1 mouse skin explants and CD1 mouse skin *in vivo* have thus allowed comparisons of the genotoxic metabolism of BP in mouse and human skin.

Postradiolabelling methods for detecting DNA adducts provide the basis of very sensitive methods for the assessment of primary chemical damage to DNA (Randerath *et al.*, 1981; Watson, 1987). Coupled with tissue explant systems, these new techniques offer possibilities for assessing human risk of genotoxic chemicals. Species differences in the metabolism of carcinogens can, however, confound extrapolations to man, and therefore valid comparisons of metabolism in experimental models and man are particularly important (Wright, 1980, 1983). In order to obtain information on human genotoxic metabolism *in vivo,* model in-vitro systems have been developed to mimic in-vivo metabolism qualitatively (Huckle *et al.*, 1986). However, human in-vitro systems are clearly not amenable to direct validation *in vivo*. Recent studies in this laboratory using direct labelling have validated the CD1 mouse skin model by demonstrating that ^{3}H- and ^{14}C-BP-DNA adducts formed *in vitro* in tissue preparations are qualitatively very similar to those formed *in vivo* (Huckle *et al.*, 1986). This approach has now been used to compare the genotoxic metabolism of BP in the epidermis of CD1 mouse and human skin.

In-vivo studies

After removal of their dorsal hair, female CD1 mice (Charles River, Manston, Kent, UK; age, eight to nine weeks) were treated with a single topical dose (400 nmol) of ^{3}H-BP (500 μCi) or ^{14}C-BP (23 μCi) in acetone, as described previously (Huckle *et al.*, 1986).

In-vitro studies

Discs of excised skin were prepared as described previously (Huckle *et al.*, 1986). Human skin obtained from mastectomy patients was initially maintained at 0°C for approximately 1.5 h after excision, prior to preparation of skin discs. Maintenance of viable skin explants was based on techniques described by Kao *et al.* (1983), but Dulbecco's modified essential medium was employed containing glutamine and 10% v/v fetal calf serum. The epidermal surface of each skin disc was treated with

^{3}H-BP (33 nmol/cm^2) in acetone, and explant cultures were then incubated (37°C, 95% air, 5% CO_2, 100% humidity) for 24 h under aseptic conditions

Analyses of BP-DNA adducts

Epidermis was removed from skin samples and the DNA was isolated by chloroform/phenol extraction then purified by hydroxylapatite column chromatography (Adriaenssens *et al.*, 1982; Huckle *et al.*, 1986).

Direct labelling studies

Samples of BP-modified epidermal DNA from mouse and human skin treated with ^{3}H- or ^{14}C-BP were hydrolysed to nucleosides (bovine DNase I, snake venom phosphodiesterase and bacterial alkaline phosphatase) according to the method of Baird and Brookes (1973). BP-DNA adducts were bulk separated (Sephadex LH20) from unmodified deoxyribonucleosides, then resolved by reverse-phase high-performance liquid chromatography (HPLC) (Huckle *et al.*, 1986).

^{32}P-Postlabelling studies

BP-modified DNA was hydrolysed to nucleoside 3′-monophosphates (micrococcal nuclease and spleen phosphodiesterase II), and adducts were extracted into *n*-butanol in the presence of tetrabutyl ammonium chloride (Gupta, 1985). After evaporation of the butanol extract, the adduct residue from DNA (1 μg) was labelled using carrier-free [γ-^{32}P]ATP (80 μCi; specific activity, 5000 Ci/mmol) in the presence of T4 polynucleotide kinase. The ^{32}P-labelled 3′,5′-bisphosphate adducts were then separated by anion exchange chromatography on polyethyleneimine-cellulose layers and located by autoradiography.

Our previous studies (Huckle *et al.*, 1986) established that CD1 mouse skin explants qualitatively mimic the in-vivo genotoxic metabolism of BP, and that organ skin culture techniques are thus valid models for in-vivo metabolism. The principal adduct was (+)-N^2-(7*R*,8*S*,9*R*-trihydroxy-7,8,9,10-tetrahydrobenzo[*a*]pyrene-10*S*-yl)-2′-deoxyguanosinc [(+)-7*R*-*trans*-(*anti*)-BPDE-dGuo] in both cases. Quantitatively, the amount of BP-DNA adducts formed in mouse epidermis *in vitro* was about 50% of that occurring *in vivo*. There was also a close similarity in the ratios of different adducts formed *in vivo* and *in vitro*, particularly for the N^2-deoxyguanosine adducts derived from *anti*-BPDE and *syn*-BPDE (Peaks A and B, Fig. 1A; only data for mouse skin *in vivo* shown). The HPLC profile of BP-DNA adducts obtained by ^{3}H direct labelling studies in human skin *in vitro* was qualitatively very similar to those obtained in mouse skin *in vivo* and *in vitro* (Fig. 1B). All of the ^{3}H-BP-DNA adducts obtained from human skin cochromatographed on HPLC with analogous adducts from mouse skin *in vivo*. In addition, cochromatography of ^{14}C-BP-DNA adducts obtained from mouse skin *in vivo* and ^{3}H-BP-DNA adducts from human skin gave qualitatively the same profiles of adducts. Retention times of ^{3}H adducts were slightly shorter than those of corresponding ^{14}C adducts due to isotope effects. These direct labelling studies therefore strongly suggest that the genotoxic metabolism of BP in mouse and human skin is qualitatively very similar.

However, recent results of ^{32}P-postlabelling analyses of DNA from mouse and human skin treated with BP have revealed quantitative differences in the formation of DNA adducts in these tissues. Thus, ^{32}P-postlabelling of BP-modified epidermal DNA from mouse skin *in vivo* and *in vitro* followed by polyethyleneimine-cellulose thin-layer chromatography analysis revealed the formation of two major adducts (1 and 2; Fig. 2A) in a ratio of about 3:1. The maps of adducts from in-vivo and in-vitro experiments

Fig. 1. HPLC profiles of ^{3}H-BP-deoxyribonucleoside adducts obtained from epidermal DNA of (A) CD1 mouse skin *in vivo* and (B) human skin explant each treated (24 h) with ^{3}H-BP

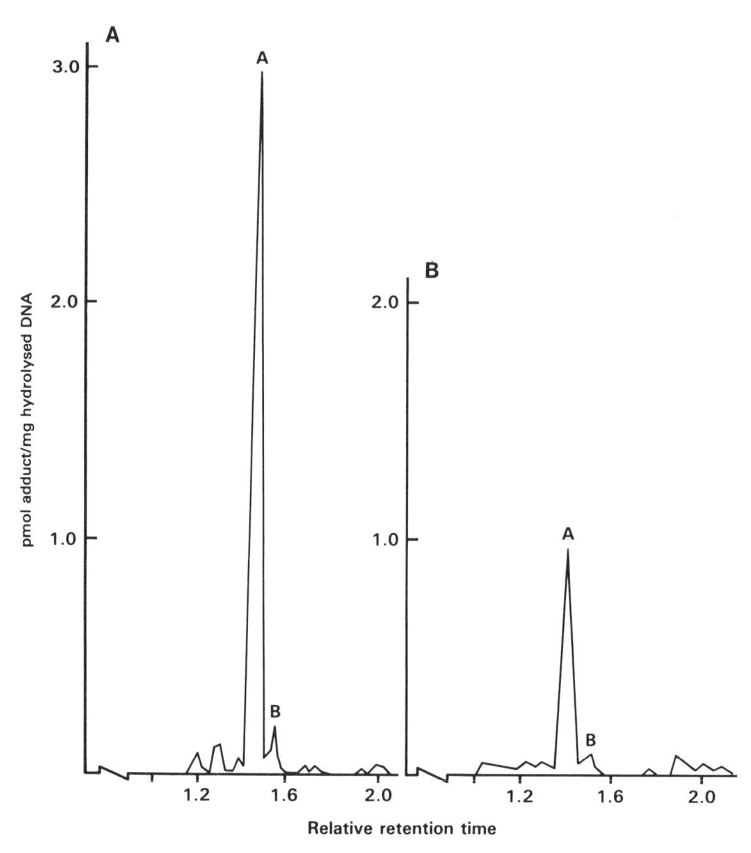

Reverse-phase HPLC carried out on a 5-μm Altex Ultrasphere ODS column (250 × 4.6 mm) eluted with methanol-water gradients (Huckle *et al.*, 1986). Fractions were collected at 1-min intervals and radioactivity determined by liquid scintillation counting. Retention times of eluting radioactivity were related to an internal standard [($\pm$)-7β,8α,9α,10β-tetrahydroxy-7,8,9,10-tetrahydro BP].

were qualitatively the same; only data from in-vivo treatments are shown. The principal adduct (1) was the 3′,5′-bisphosphate of (+)-7*R*-*trans*-(*anti*)-BPDE-dGuo. This assignment was confirmed by ^{32}P-postlabelling of calf thymus DNA modified by reaction with ($\pm$)-*anti*-BPDE and known to contain (+)-7*R*-*trans*-(*anti*)-BPDE-dGuo as the principal adduct. This also established that adduct 2 was not derived from *anti*-BPDE. Interestingly, it was difficult to detect adduct 2 after ^{32}P-postlabelling analysis of BP-modified epidermal DNA from human skin explants. ^{32}P-Postlabelling has thus revealed quantitative differences in the formation of BP-DNA adducts in mouse and human skin not shown by direct prelabelling studies. The identity of adduct 2 is not yet known, but it appears to be an efficiently repaired adduct previously observed by ^{32}P-postlabelling analysis of skin DNA from BALB/c mice treated *in vivo* with BP (Randerath *et al.*, 1983).

There are several possible reasons why a nucleoside adduct corresponding to adduct 2 was not observed in direct prelabelling studies. These analyses were carried out at the

Fig. 2. ^{32}P-Postlabelling maps of BP-modified epidermal DNA from (A) CD1 mouse skin *in vivo* and (B) human skin explant

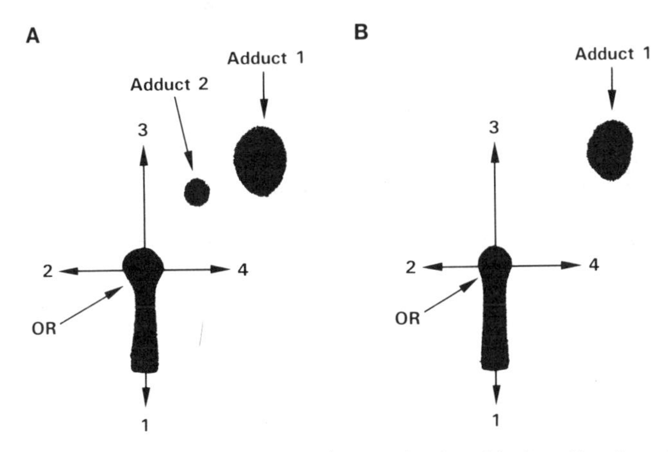

Polyethyleneimine-cellulose layers (Merck, ART 5579) were developed in four directions: D1, 1.1 M lithium chloride, pH 6; D2, 2.5 M ammonium formate, pH 3.5; D3, 3 M lithium formate/8 M urea, pH 3.5; D4, 0.8 M lithium chloride/0.5 M Tris-HCl/8 M urea, pH 8.0. Autoradiography (24 h, RT) was on Kodak XAR-films.

deoxyribonucleoside level, involving lengthy enzymatic and chromatographic techniques, and it is possible that the adduct was not stable under these conditions. The formation of 3′,5′-bisphosphates in the ^{32}P-postlabelling assay might increase the stability of this adduct. Alternatively, ^{32}P-postlabelling catalysed by polynucleotide kinase could preferentially label adduct 2 over adduct 1, thus giving a misleading indication of the relative amounts of the two adducts (Randerath *et al.*, 1985). Appropriate control experiments showed that adduct 2 was not present in DNA from untreated mouse skin and that it was not an artefact of the labelling procedure.

These studies have shown that skin organ culture techniques are valid models for in-vivo metabolism and that the techniques are directly applicable to human skin. Further studies are necessary to identify the nature of adduct 2.

References

Adriaenssens, P.I., Bixler, C.J. & Anderson, M.W. (1982) Isolation and quantitation of DNA-bound benzo(a)pyrene metabolites: comparison of hydroxylapatite and precipitation procedures. *Anal. Biochem.*, *123*, 162–169

Baird, W.M. & Brookes, P. (1973) Isolation of the hydrocarbon-deoxyribonucleoside products from the DNA of mouse embryo cells treated in culture with 7-methylbenzo(*a*)anthracene-^{3}H. *Cancer Res.*, *33*, 2378–2385

Gupta, R.C. (1985) Enhanced sensitivity of ^{32}P-postlabeling analysis of aromatic carcinogen: DNA adducts. *Cancer Res.*, *45*, 5656–5662

Huckle, K.R., Smith, R.J., Watson, W.P. & Wright, A.S. (1986) Comparison of hydrocarbon-DNA adducts formed in mouse skin *in vivo* and in organ culture *in vitro* following treatment with benzo[*a*]pyrene. *Carcinogenesis*, *7*, 965–970

Kao, J., Hall, J. & Holland, J.M. (1983) Quantitation of cutaneous toxicity: an *in-vitro* approach using skin organ culture. *Toxicol. appl. Pharmacol.*, *68*, 206–217

Randerath, K., Reddy, M.V. & Gupta, R.C. (1981) ^{32}P-Labeling test for DNA damage. *Proc. natl Acad. Sci. USA*, *78*, 6126–6129

Randerath, E., Agrawal, H.P., Reddy, M.V. & Randerath, K. (1983) Highly persistent polycyclic hydrocarbon-DNA adducts in mouse skin: detection by ^{32}P-postlabeling analysis. *Cancer Lett.*, *20,* 109–114

Randerath, E., Agrawal, H.P., Weaver, J.A., Bordelon, B.C. & Randerath, K. (1985). ^{32}P-Postlabeling analysis of DNA adducts persisting for up to 42 weeks in the skin, epidermis and dermis of mice treated topically with 7,12-dimethylbenz(a)anthracene. *Carcinogenesis, 6,* 1117–1126

Watson, W.P. (1987) Post-radiolabelling for detecting DNA damage. *Mutagenesis, 2,* 319–331

Wright, A.S. (1980) The role of metabolism in chemical mutagenesis and chemical carcinogenesis. *Mutat. Res.*, *75*, 215–241

Wright, A.S. (1983) *Molecular dosimetry techniques in human risk assessment: an industrial perspective.* In: Hayes, A.W., Schnell, R.C. & Miya, T.S., eds, *Developments in the Science and Practice of Toxicology,* Amsterdam, Elsevier, pp. 311–318

NONSELECTIVE AND SELECTIVE METHODS FOR BIOLOGICAL MONITORING OF EXPOSURE TO COAL-TAR PRODUCTS

R.P. Bos & F.J. Jongeneelen

Department of Toxicology, Faculty of Medicine, University of Nijmegen, Nijmegen, The Netherlands

Biological monitoring of exposure to coal-tar products has been carried out using the nonselective urinary mutagenicity and thioether assays and a selective method for the detection of 1-hydroxypyrene in urine. The sensitivities of the three methods have been compared. In several work environments, no increase that could be related to exposure to coal-tar products was found with the nonselective methods, whereas the levels of 1-hydroxypyrene in urine were enhanced. The applicability of the method for the detection of 1-hydroxypyrene in urine is further demonstrated.

A frequent exposure to a mixture of genotoxic chemical compounds is that of workers to coal-tar and coal-tar-derived products, which contain relatively high concentrations of polycyclic aromatic hydrocarbons (PAH). This family of closely related compounds contains many mutagens and carcinogens. Few methods have been available to date for biological monitoring of exposure to PAH-containing products in urine: the urinary mutagenicity assay (Bos, 1984), the thioether assay (Henderson *et al.*, 1984) and a method for the detection of 1-hydroxypyrene in urine (Jongeneelen, 1987). Methods for monitoring biological effects such as formation of DNA adducts are being developed by several other research groups. In this paper, we present results obtained after application of the three tests in situations in which patients or workers were exposed to PAH-containing products.

Urinary mutagenicity testing

Mutagenicity was determined in urine samples as described by Bos (1984). Urine samples were concentrated on Amberlite XAD-2, eluted with acetone, the acetone evaporated and the residue taken up in dimethyl sulphoxide (DMSO). These extracts were tested for mutagenicity towards *Salmonella typhimurium* TA98 in the presence of a 9000 *g* supernatant of the liver of Aroclor 1254-induced rats and of β-glucuronidase (150 U per plate).

Thioether assay

This assay has been described previously (Henderson *et al.*, 1984). In short, urine samples are acidified (pH 1.5–2.0) and extracted with ethyl acetate; after evaporation of the ethyl acetate, the residue is taken up in water. After alkaline hydrolysis and neutralization, the SH concentration is determined according to Ellman (1959). The mean concentration in the urine of about 200 control persons was 3.8 mmol SH per mol creatinine; the upper limit of the 95 percentile was 5.9 mmol SH per mol creatinine (Henderson *et al.*, 1984).

1-Hydroxypyrene in urine

The high-performance liquid chromatography method for the detection of 1-hydroxypyrene in urine after enzymatic hydrolysis and solid-phase extraction has been described by Jongeneelen (1987).

Reference population: Urine samples were taken from 90 male referents—38 smokers and 52 nonsmokers—divided into four smoking categories: 0, 0–10, 11–20 and >20 cigarettes/day, and analysed for the presence of 1-hydroxypyrene (Fig. 1). The level of 1-hydroxypyrene was not significantly greater in the urine of smokers than in that of nonsmokers (Jongeneelen, 1987).

Dermatological patients treated with a coal-tar ointment

Five female patients (A and B were nonsmokers, C, D and E were smokers) suffering from eczematous dermatitis on the arms and legs were treated for several days with an ointment containing 10% pix lithanthracis dermata (coal-tar), representing 16.7 mg/g pyrene and 7.0 mg/g benzo[*a*]pyrene. During the treatment, the ointment was removed daily with arachis oil, and a fresh dose of approximately 40 g was rubbed in. The patients volunteered to collect urine samples—one before application of the ointment and two during the day (morning and evening)—for the first three days of treatment.

Because of the high concentrations of toxic compounds in the urine samples, we were not able to measure any mutagenicity, even after diluting by several factors. One of the two nonsmoking patients excreted a considerably larger amount of

Fig. 1. Levels of 1-hydroxypyrene in urine of control subjects

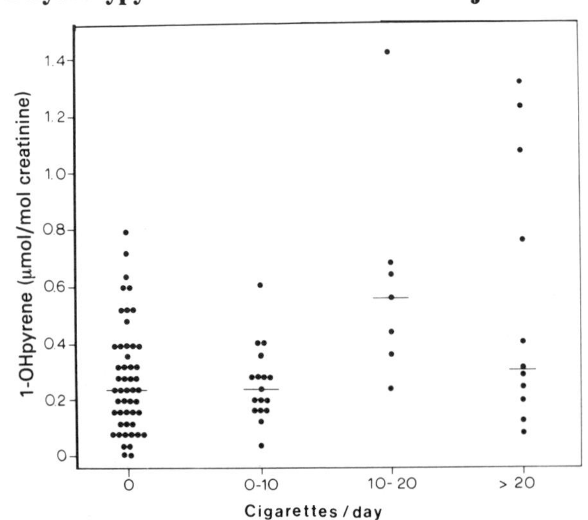

The control population is subdivided according to smoking habits. The median value for each smoking category is indicated.

Fig. 2. Urinary excretion of 1-hydroxypyrene by five patients undergoing topical coal-tar treatment

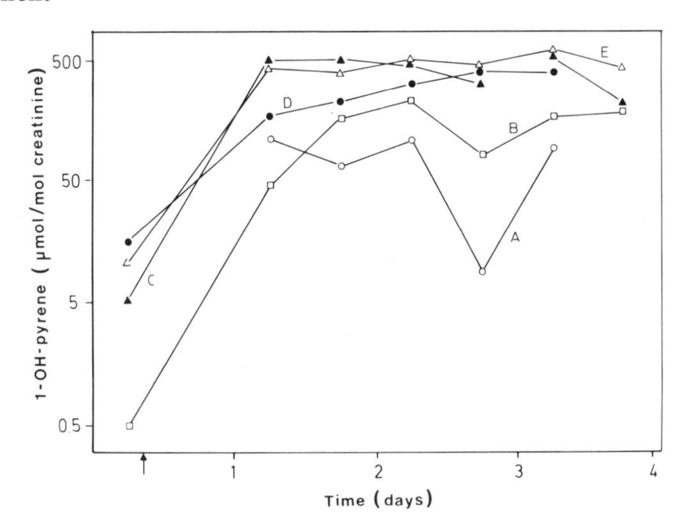

The arrow indicates the start of application.

thioether products during the beginning of the treatment: 19.6 and 24.5 mmol SH/mol creatinine were measured in the evening urine sample of the first day of treatment and in the morning urine sample of the second day of treatment, respectively. In the evening of the second day and on the morning of the third day, the concentrations were 5.4 and 3.3 mmol SH/mol creatinine, which are within the control range. No similar increase in thioether excretion during coal-tar treatment was found for the other patients, whose values were within or slightly above the normal range.

The excretion pattern of 1-hydroxypyrene in the urine of these patients is shown in Figure 2. After the beginning of treatment, the concentration of 1-hydroxypyrene rose rapidly to over 100 times the control value. The levels in the 'pretreatment' urine samples of patients C, D and E are somewhat higher than those for control smokers (Fig. 1).

Workers in a creosote impregnating plant

We studied three workers involved in creosoting wood in a wood-preserving industry. One (A) operated an open-air creosote vacuum-pressure cylinder; another (B) moved wood in and out of the cylinder; and worker C was the chief operator, with most probably the lowest exposure. Workers A and B were especially exposed to creosote vapours when the cylinder door was opened after treatment of the wood. Worker A was a smoker, B and C nonsmokers. Creosote is a coal-tar distillation product that contains many PAH, and pyrene, benzo[*a*]pyrene and benz[*a*]anthracene were measured at concentrations of 20 mg/g, 1.8 mg/g and 11 mg/g, respectively. Urine samples were collected from the workers over ten consecutive days, including two free weekends, in two portions: a morning urine sample and a sample collected between 10:00 and 16:00 h.

No increase in urinary mutagenicity or thioether concentration was detected that could be related to work exposure. Mutagenicity and thioether concentration were higher in the urine of the one smoker than in that of the two nonsmokers.

Fig. 3. Urinary excretion of 1-hydroxypyrene by three workers creosoting wood

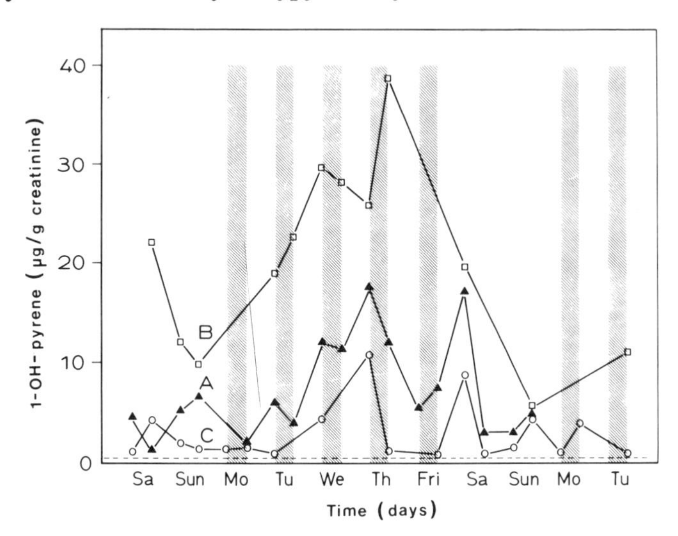

The dashed line is the mean control level of 1-hydroxypyrene. The shaded areas are the working periods. A, cylinder operator; B, wood handler; C, chief operator

The 1-hydroxypyrene levels in the urine of the three workers are shown in Figure 3. Worker B had the highest 1-hydroxypyrene levels, and the chief operator C the lowest. For workers A and B, an increased excretion was seen during the week and a decrease over the weekend.

Workers in a coal-tar distillation plant

In a study of 14 workers at a coal-tar plant, no increase in mutagens or thioethers in urine could be determined, either in the course of the shift or in the course of the seven-day working period. Figure 4 shows, as an example, the mutagenicity values in pre- and post-work urine samples. The only factor that significantly influenced the urinary mutagenicity was smoking. Urine samples were available from four workers for determination of 1-hydroxypyrene, and Table 1 shows the enhanced levels found.

Road asphalt-paving workers

Pre-work and post-work urine samples were collected from 31 road asphalt-paving workers active at ten different paving sites and analysed for 1-hydroxypyrene. Table 2 shows an increase in 1-hydroxypyrene in urine over the day; 55% of the pre-work samples and 70% of the post-work samples showed 1-hydroxypyrene levels over the upper 95 percentile of those in urine of a smoking control group (1.3 µmol/mol creatinine). At paving site H, petroleum-based asphalt was handled; at the nine other sites, blends of bitumen with refined coal-tar were applied.

Coke-oven workers

Urine samples from coke-oven workers were also analysed for the presence of 1-hydroxypyrene. After a work-free period of 56 h, urine samples were collected during three consecutive 8-h working days from 25 coke-oven workers at the beginning and at the end of the shift. Control samples were taken from 43 workers active in the

Fig. 4. Urinary mutagenicity in workers at a coal-tar distillation plant. Pre- and post-work values are depicted separately for nonsmoking and smoking workers.

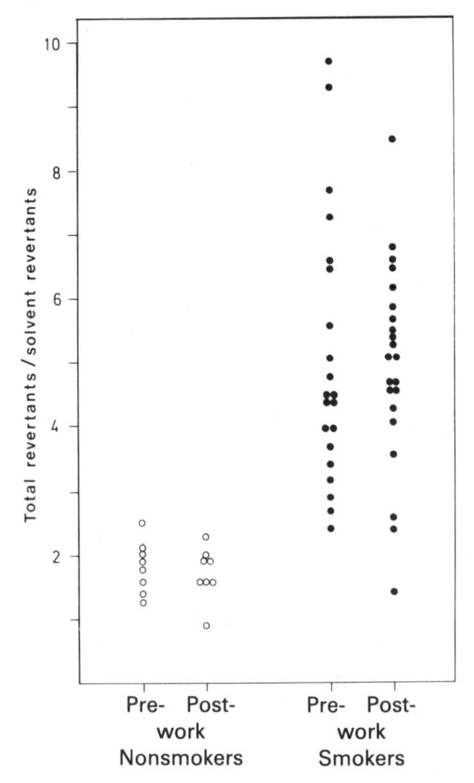

steel-rolling department. Thc first, preliminary results of this study are presented in Figure 5. Urinary levels of 1-hydroxypyrene in workers were enhanced compared to controls and were higher at the end of work than before work. The results also suggest an accumulation of pyrene during the exposure period.

Conclusions

The applicability of the nonselective urinary mutagenicity assay and the thioether assay for the detection of exposure to PAH appears to be limited. In a previous paper,

Table 1. 1-Hydroxypyrene in the urine of four workers at a coal-tar distillation plant

Function	Smoking status	Number of samples	1-Hydroxypyrene level (mean μmol/mol creatinine)
Operator, pitch unit	Nonsmoker	2	3.7
Operator, batch distillation	Smoker	2	11.8
Operator, pumping station	Nonsmoker	4	4.0
Cleaner	Smoker	4	4.6

Table 2. 1-Hydroxypyrene in urine of asphalt-paving workers

Paving site	No. of measuring days	No. of workers per site	Pre-work urine sample (mean μmol/mol creatinine)	Post-work urine sample
A	3	3	2.3	3.1
B	5	6	1.5	1.6
C	1	2	1.5	2.8
D	1	4	1.1	0.9
E	1	2	3.1	3.2
F	3	3	1.4	2.2
G	2	3	1.5	3.1
H	1	3	0.5	0.6
I	1	3	2.1	2.8
K	1	2	0.8	1.2

we presented the first evidence that the urinary mutagenicity assay could not be used to detect exposure to relatively high levels of PAH (Bos *et al.*, 1984). PAH have a rather complex metabolism (Phillips & Grover, 1984), and after exposure to (mutagenic, potentially electrophilic) PAH, mutagens and thioether compounds could appear in urine (Fig. 6). However, the excretion of mutagens in urine (arrow 1) depends on several toxicokinetic factors: (i) whether chemical compounds after passage through the organism still possess (potentially) mutagenic properties and (ii) the extent to which the (biotransformation) products are excreted *via* the urine. The excretion of thioether products in the urine after exposure to PAH (arrow 2) depends on: (i) bioactivation of PAH to electrophilic intermediates; (ii) subsequent detoxification *via* glutathione or other SH molecules; and (iii) excretion of the thioether detoxification products *via* urine. Because of relatively high background levels, neither the urinary mutagenicity assay nor the thioether assay is sensitive enough for the detection of small increases in urinary levels of mutagens or thioethers. Besides, smoking is a major determinant in both assays. The method for the detection of 1-hydroxypyrene in urine, however, is sensitive, and smoking habits are only a minor determinant. The presence of 1-hydroxypyrene in urine in fact reflects exposure to pyrene. Since PAH always occur in the environment as a family, 1-hydroxypyrene in urine can be considered a

Fig. 5. Levels of 1-hydroxypyrene in urine of coke-oven workers

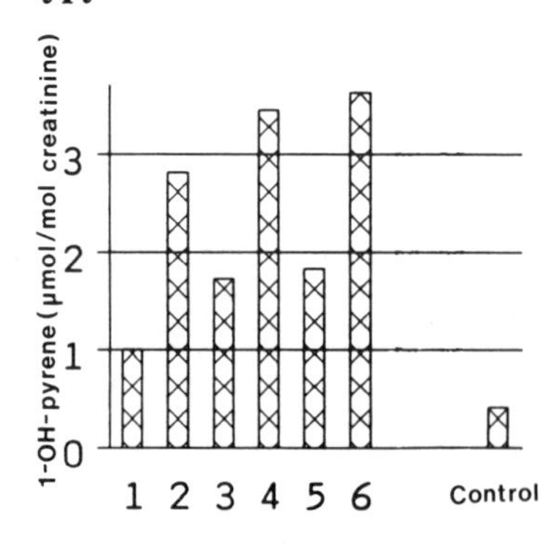

1, 3 and 5, levels in pre-work urine; 2, 4 and 6, corresponding end of work values.

Fig. 6. Products in urine after exposure to mutagenic chemical compounds

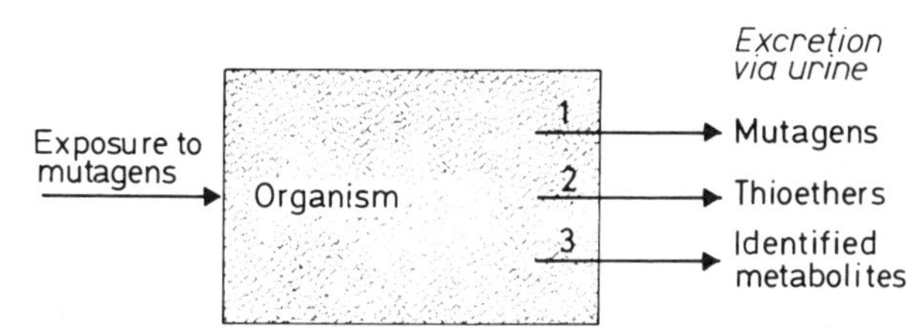

marker for exposure to PAH. It is important that, for a more detailed evaluation of exposure to genotoxic PAH, the profile of hydrocarbons in the environment be considered.

Acknowledgement

Financial support was obtained from the General Directorate of Labour, Dutch Ministry of Social Affairs and Employment.

References

Bos, R.P. (1984) *Application of Bacterial Mutagenicity Assays in Genotoxicity Studies,* Thesis, University of Nijmegen

Bos, R.P., Jongeneelen, F.J., Theuws, J.L.G. & Henderson, P.T. (1984) *Exposure to mutagenic aromatic hydrocarbons of workers creosoting wood.* In: Berlin, A., Draper, M., Hemminki, K. & Vainio, H., eds, *Monitoring Human Exposure to Carcinogenic and Mutagenic Agents (IARC Scientific Publications No. 59),* Lyon, International Agency for Research on Cancer, pp. 279–288

Ellman, G.L. (1959) Tissue sulfhydryl groups. *Arch. Biochem. Biophys., 82,* 70–77

Henderson, P.T., van Doorn, R., Leijdekkers, C.-M. & Bos, R.P. (1984) *Excretion of thioethers in urine after exposure to electrophilic chemicals.* In: Berlin, A., Draper, M., Hemminki, K. & Vainio, H., eds, *Monitoring Human Exposure to Carcinogenic and Mutagenic Agents (IARC Scientific Publications No. 59),* Lyon, International Agency for Research on Cancer, pp. 173–187

Jongeneelen, F.J. (1987) *Biological Monitoring of Occupational Exposure to Polycyclic Aromatic Hydrocarbons,* Thesis, University of Nijmegen

Phillips, D.H. & Grover, P.L. (1984) *Biologically-active and chemically-reactive polycyclic aromatic hydrocarbon metabolites.* In: Berlin, A., Draper, M., Hemminki, K. & Vainio, H., eds, *Monitoring Human Exposure to Carcinogenic and Mutagenic Agents (IARC Scientific Publications No. 59),* Lyon, International Agency for Research on Cancer, pp. 47–61

AN IMPROVED STANDARDIZED PROCEDURE FOR URINE MUTAGENICITY TESTING

A. Rannug[1], M. Olsson[2], L. Aringer[2] & G. Brunius[1]

[1]*Department of Toxicology; and* [2]*Department of Occupational Medicine, National Institute of Occupational Health, Solna, Sweden*

The aim of the present study was to optimize the procedures for urinary mutagenicity testing in order to lower the baseline variation in mutagenic activity found in urine from unexposed subjects and to increase the sensitivity of the method. This was accomplished by using urine from nonsmokers and smokers as well as chemically spiked nonsmokers' urine. Diet was standardized. The number of mutants per ml of urine calculated from the linear portion of the dose-response curve was used as a measure of mutagenicity. The parameters investigated were (i) the total volume of urine per resin volume, (ii) the flow rate, (iii) the pH, (iv) the ionic strength of the urine, and (v) elimination of histidine. XAD-2 and C18 Sep-Pak resins recovered mutagens in smokers' urine and in chemically spiked urine with the same efficiency when an optimized procedure was adopted. The optimized procedure using a maximum volume of 50 ml acidified urine per Sep-Pak cartridge, or equal amount of XAD-2 resin, gave well over ten times greater recovery of mutagens from smokers' urine than in earlier reports. Histidine was effectively eliminated, and the background variation was also lowered.

Testing human urine for mutagenic activity towards bacteria has proven to be a useful means for identifying genotoxic exposures. Earlier studies, mainly of occupational exposures, have frequently given inconsistent results (Everson, 1986, and references therein), owing to several technical difficulties, such as overloading of the resins used to concentrate the urine samples. The influence of known confounding factors like certain food components and smoking has not always been accounted for.

Mutagenicity testing

Mutagenicity tests were conducted according to the standard plate assay (Ames *et al.*, 1975). *Salmonella typhimurium* TA98 was used throughout the study, except for tests on the mutagens benzo[*a*]pyrene, styrene oxide and sodium azide, where TA100 was used. A 9000 *g* supernatant from Aroclor 1254-treated rats (S9) was used for activation. In tests of human urine, 20 μl S9 were added per plate. Results of one representative experiment are presented.

Comparison of XAD-2 and C18 Sep-Pak resins

Glass columns (0.7 × 10 cm) packed with XAD-2 (Serva, Heidelberg, FRG), 1.5 cm^3 bed volume, and C18 Sep-Pak (Waters Assoc.) columns were compared with

Table 1. Sorption of ^{14}C-labelled histidine on XAD-2 and C18 Sep-Pak columns[a]

Aliquot	Percentage of totally recovered[b] histidine			
	Urine pH 6–7		Urine pH 2	
	XAD-2	Sep-Pak	XAD-2	Sep-Pak
Urine after passing column	16.2	72.5	20.7	67.0
Rinse 1[c]	74.1	26.3	68.4	28.9
Rinse 2[c]	7.9	0.7	8.4	3.4
Acetone eluate	1.2	0.4	2.4	0.7

[a] ^{14}C-Histidine was added to each urine sample that was passed through XAD-2 or C18 Sep-Pak columns. Residual radioactivity was measured in the eluates.
[b] Total recovery, approximately 80%
[c] XAD-2 was rinsed with 5 ml water; Sep-Pak was rinsed with 2 ml water.

regard to histidine retention (Table 1) and ability to sorb chemical mutagens from urine (Table 2).

Optimizing concentration procedure

Using Sep-Pak columns, the influence of parameters related to the chromatographic procedure was examined, with the recovery of mutagens from smokers' urine as a measure of efficiency (Table 3). Furthermore, the effect of diet on the mutagenicity of nonsmokers' urine was studied (Table 4). The subjects were instructed to eat only white bread, to drink milk with meals and to avoid coffee. A commercially available deep-frozen chicken meal was taken, unless otherwise stated. Using the optimal procedure, the mutagenicity of urine samples from five smokers and 13 nonsmokers was estimated, and the results are presented in Figure 1, where a comparison is made with data from earlier reports on the mutagenicity of smokers' urine in TA98 in the presence of S9.

Table 2. Recovery of mutagens from smokers' urine and from chemically spiked nonsmokers' urine[a]

	Revertants per ml smokers' urine	
	XAD-2	Sep-PakC_{18}
Smokers' urine, acidified to pH 2	31.8	35.6
Mutagen added to urine (amount per 100 μl)	Recovery of mutagens (%)	
	XAD-2	Sep-PakC_{18}
Benzo[*a*]pyrene, 5 μg	31.6	49.6
2-Aminoanthracene, 0.2 μg	48.2	39.2
Trp-P-1, 0.5 μg	97.3	100
Trp-P-2, 0.5 μg	95.9	94.3
Styrene oxide, 120 μg	0	0
Sodium azide, 1.5 μg	0	0

[a] Mutagens were added to 50 ml of urine to yield the indicated amounts per 100 μl. Samples (100 μl) were tested from each urine before the chromatographic procedure. The sorbed mutagens were eluted with acetone, diluted to 50 ml with water, and 100 μl were tested for mutagenicity on TA98.

Table 3. Optimizing procedures with regard to total load, flow rate, pH, and ionic strength in tests of smokers' urine

Load	Volume of urine (ml)	Revertants per ml	
	150	15.0	
	100	18.6	
	50	21.3	
Flow rate	(ml/min)	Revertants per ml in two serially adapted columns	
		Column 1	Column 2
	2–3	37.0	6.2
	>10	28.4	9.2
pH	Subject	Revertants per ml	
		pH 2.0	pH 7.0
	Smoker	21.3	16.4
	Smoker	26.7	20.1
	Nonsmoker	2.1	3.4
Ionic strength	Samples	Revertants per ml	
	50 ml urine	26.2	
	50 ml urine + 150 ml H_2O	28.3	

Standardized procedure

Sampling: Urine is sampled from subjects on a defined diet (nonmutagenic); urine samples are immediately frozen and stored at −20°C for a maximum of four weeks.

Concentration of mutagens: Precipitate is removed by centrifugation and the urine acidified to pH 2.0; 50 ml acidified urine are slowly (2–3 ml/min) passed through one (or two serially adapted) Sep-Pak cartridges. After rinsing with 5 ml water (pH 2.0), the columns are flushed with nitrogen to remove excess water and then eluted with 10 ml acetone. The acetone is evaporated under a stream of nitrogen with very gentle heating (30–35°C) and 1.0 ml dimethyl sulphoxide is added to the acetone residue, resulting in a 50× concentrate, in which 20 μl corresponds to 1 ml of the original urine.

Table 4. Influence of diet on urine mutagenicity

Diet[a]	*n*	Revertants per ml
Fish with almonds	1	3.1
Fish with caviar	1	2.2
Fried pork	1	3.8
Bacon with egg	1	12.0
Chicken with mashed potatoes	13	2.3 ± 1.0 (SE)

[a] Standardized for 18 h before sampling

Fig. 1. Mutagenic activity of smokers' and nonsmokers' urine in *S. typhimurium* TA98 in the presence of S9

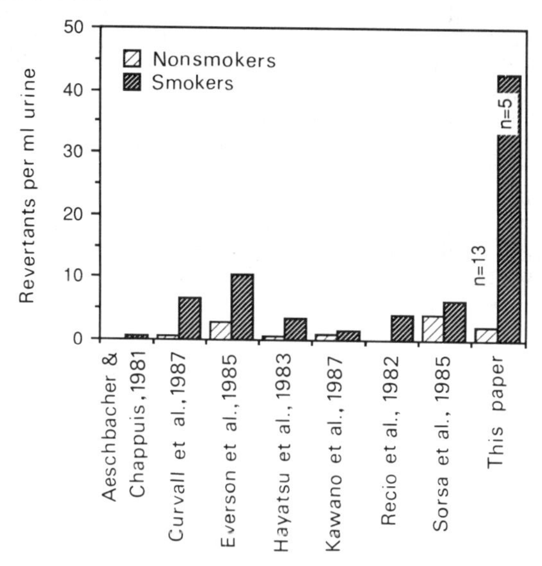

Comparison with data from recent publications (see reference list)

Mutagenicity testing: The appropriate test procedure (plate, microsuspension or fluctuation test) with appropriate tester strains and metabolic activation is chosen, depending on the type of exposure that is being investigated. A dose-response relationship is established for each urine concentrate by testing several doses. A simple linear regression is performed over the linear part of the curve, and the slope (number of mutant colonies per ml urine) is used as an estimate of the mutagenic potency of each sample.

Conclusions

Histidine was efficiently removed from both XAD-2 and Sep-Pak columns, residual histidine being lowest on the Sep-Pak columns. Recovery of six model mutagens from nonsmokers' urine and of mutagens from smokers' urine did not differ substantially between the two resins. By acidifying the urine, not overloading the columns and using a slow flow rate, the recovery of mutagens from smokers' urine was optimal. The influence of diet on the results was diminished by standardizing it.

The standardized procedure for urine mutagenicity testing presented here, involving a 50× concentration of the urine sample, is simple to perform, minimizes background variation, produces data on mutagenic potency that are based on tests of several doses of urine concentrate, and is highly sensitive towards mutagens in smokers' urine.

References

Aeschbacher, H.U. & Chappuis, C. (1981) Non-mutagenicity of urine from coffee drinkers compared with that from cigarette smokers. *Mutat. Res., 89,* 161–177

Ames, B.N., McCann, J. & Yamasaki, E. (1975) Methods for detecting carcinogens and mutagens with the *Salmonella*/mammalian-microsome mutagenicity test. *Mutat. Res., 31,* 347–363

Curvall, M., Romert, L., Norle'n, E. & Enzell, C. R. (1987) Mutagen levels in urine from snuff users, cigarette smokers and non-tobacco users—a comparison. *Mutat. Res., 188,* 105–110

Everson, R.B., Ratcliffe, J.M., Flack, P.M., Hoffman, D.M. & Watanabe, A.S. (1985) Detection of low levels of urinary mutagen excretion by chemotherapy workers which was not related to occupational drug exposures. *Cancer Res., 45,* 6487–6497

Everson, R.B. (1986) Detection of occupational and environmental exposures by bacterial mutagenesis assays of human body fluids. *J. occup. Med., 28,* 647–655

Hayatsu, H., Oka, T., Wakata, A., Ohara, Y., Hayatsu, T., Kobayashi, H. & Arimoto, S. (1983) Adsorption of mutagens to cotton bearing covalently bound trisulfo-copper-phthalocyanine. *Mutat. Res., 119,* 233–238

Kawano, H., Inamasu, T., Ishizawa, M., Ishinishi, N. & Kumazawa, J. (1987) Mutagenicity of urine from young male smokers and nonsmokers. *Int. Arch. occup. environ. Health, 59,* 1–9

Recio, L., Enoch, H. & Hannan, M.A. (1982) Parameters affecting the mutagenic activity of cigarette smokers urine. *J. appl. Toxicol., 2,* 241–246

Sorsa, M., Einistö, P., Husgafvel-Pursiainen, K., Järventaus, H., Kivistö, H., Peltonen, Y., Tuomi, T., Valkonen, S. & Pelkonen, O. (1985) Passive and active exposure to cigarette smoke in a smoking experiment. *J. Toxicol. environ. Health, 16,* 523–534

PROBLEMS IN MONITORING MUTAGENICITY OF HUMAN URINE

H. Hayatsu, T. Hayatsu, Q.L. Zheng, Y. Ohara & S. Arimoto

Faculty of Pharmaceutical Sciences, Okayama University, Tsushima, Okayama, Japan

Blue-cotton (-rayon) adsorbable fractions of human urines were examined for mutagenicity in *Salmonella typhimurium* TA98 with metabolic activation. Ingestion of cooked beef caused significant increases in urinary mutagenicity that were comparable to that caused by cigarette smoking. When a sample obtained after ingestion of cooked beef was passed through a carboxymethyl cellulose column, the mutagenicity of the eluate was found to be almost one order of magnitude greater than that of the original sample, suggesting the presence of antimutagenic factors in the sample. The oleic acid content of the sample was not great enough to account for this phenomenon. Other urine samples subjected to column fractionation were found to contain the putative antimutagenic factors. This finding further confounds the monitoring of urinary mutagenicity.

Urinary mutagenicity is a measure of human exposure to genotoxic substances. The use of blue cotton, a polycyclic-compound specific adsorbent (Hayatsu *et al.*, 1983a), has facilitated the monitoring of polycyclic mutagens in human urine (Kobayashi & Hayatsu, 1984; Hayatsu *et al.*, 1985a; Mohtashamipur *et al.*, 1985). This adsorbent has been used to detect heteropolycyclic amines in foods (Hayatsu *et al.*, 1983b; Takahashi *et al.*, 1985; Kikugawa *et al.*, 1986) and in tobacco smoke (Yamashita *et al.*, 1986). It has also been used in various studies of polycyclic amine mutagens (Jägerstad *et al.*, 1984; Hayatsu *et al.*, 1985b, 1987; Bashir *et al.*, 1987).

In this paper, we discuss the effect of diet, especially the ingestion of cooked meat, on urinary mutagenicity, as compared with the effect of cigarette smoking. Furthermore, we show that blue-cotton extracts of urines often contain antimutagenic factors which can be removed by passing the extract through a carboxymethyl cellulose column.

Ingestion of cooked beef

Heated meat contains mutagenic heteropolycyclic amines, most of which are proven carcinogens in rodents (Sugimura, 1985). Soon after a meal of cooked beef, the urine becomes mutagenic, as assayed in *S. typhimurium* TA98 with metabolic activation; the mutagenicity disappears 12 h after ingestion (Hayatsu *et al.*, 1985a).

We collected 6-h urine samples from volunteers before and after ingestion of cooked ground beef. The two previous meals that the volunteers had eaten did not contain cooked meat. The urines were treated with blue rayon (which is an improved preparation of blue cotton), and the rayon was eluted with methanol-ammonia. The eluate was then subjected to carboxymethyl cellulose column chromatography to

Fig. 1. Urinary mutagenicity due to ingestion of fried ground beef

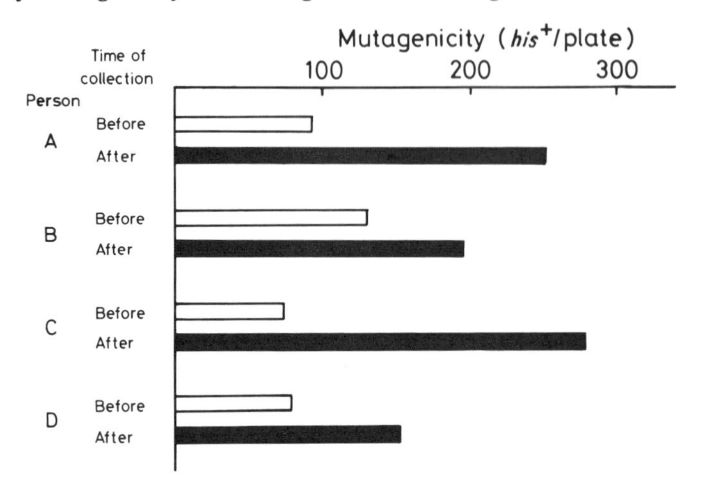

Urines were collected over 6 h before and for 6 h after a meal containing two patties of commercially available fried ground beef, which weighed 68 g. The total urine samples were treated with blue rayon (25 μmol copper phthalocyanine trisulphonate per gram rayon) at 0.3 g/100 ml urine. After 30 min of gentle shaking, the rayon was collected, washed with water, dried and eluted with methanol: ammonia (50:1). The eluate was evaporated to dryness, and the residue was subjected to carboxymethyl cellulose column chromatography (Hayatsu *et al.*, 1983b). The first 10-ml fraction, which was obtained by eluting the column with formic acid at pH 3, was evaporated to dryness and assayed for mutagenicity in *S. typhimurium* TA98 with metabolic activation using the preincubation method (Yahagi *et al.*, 1977).

remove antimutagenic factors, and the mutagenicity was measured in *S. typhimurium* TA98 with S9 mix.

Examples are given in Figure 1. In four people, the urines were more mutagenic after eating the cooked meat than before the meal.

Antimutagenic factors in urine

During the analysis of urinary mutagenicity after the meat meal, we became aware that the samples contained antimutagenic components. Thus, the mutagenicity of a blue-rayon extract was much lower (254 his^+ revertants per plate) than that of a sample obtained by further passing the rayon extract through a carboxymethyl cellulose column (1850 revertants). We concluded that there must be some components in the blue-rayon extract that inhibited mutagenic activity and could be removed by passage through the carboxymethyl cellulose column. Our previous studies on the antimutagenic activity of fatty acids (Hayatsu *et al.*, 1981a,b, 1983c; Negishi & Hayatsu, 1984) suggested to us that the blue-rayon extract might contain fatty acids that are inhibitory for mutagenesis. However, that was not the case. Although the blue-rayon extracts were shown to contain oleic acid, as analysed by gas chromatography (Hayatsu *et al.*, 1981a), the amounts (1.2 μg/100 ml urine equivalent blue-rayon extract) were too small to account for the inhibitory activity. The components have not yet been characterized.

In addition, enhancement of mutagenicity by passing the samples through the carboxymethyl cellulose column was observed in only five of six urine samples collected after meat meals; the sixth showed a decrease, indicating that the putative antimutagenic factors occur variably among samples.

Table 1. Urinary mutagenicity arising from ingestion of cooked beef and from cigarette smoking

Cause	Subject	Urine volume (ml)	Net increase in number of his^+ revertants[a] (TA98, + S9) A	Mutagenicity of ingested beef[b] B	Output [A]/ Input [B] × 100
Fried ground beef	A[c]	530	158	1420	11.1
	B[c]	150	64	1420	4.5
	C[c]	300	207	1420	14.6
	D[c]	880	74	1420	5.2
	A	530	172	2840	6.1
	B	290	64	2840	2.3
	C	340	41	2840	1.4
	D	190	533	2840	18.8
	A	1160	755	8900	8.5
	E	1350	1210	8900	14.8
Cigarettes[d]	F	1220	690	(7)[d]	
	G	—	920	(7)[d]	
	G	800	1320	(9)[d]	

[a] [No. of revertants after meat meal or smoking] − [No. of revertants before meat meal or smoking]
[b] Measured by the method of Hayatsu *et al.* (1983b)
[c] Data obtained from the results in Figure 1
[d] Results obtained by Kobayashi & Hayatsu (1984); numbers in the parentheses, number of cigarettes smoked

Comparison of meat and cigarettes with respect to their effect on urinary mutagenicity

The studies described above indicate that the mutagenicity that arises from ingesting cooked meat is substantial. Table 1 shows that the urinary mutagenicity caused by meat ingestion is similar to that due to cigarette smoking. It is noteworthy that about 10% of ingested mutagenicity in meat was excreted in the urine; however, the chemical structures of the mutagenic components in the meat and in the urine are probably different (Hayatsu *et al.*, 1985a, 1987).

Ingestion of cooked meat is thus a major cause of increased urinary mutagenicity. An important consequence of this finding is that any attempt to correlate urinary mutagenicity with a given causative agent should take into account the effect of diet.

Acknowledgements

This work was supported by a Grant-in-Aid for Scientific Research on Priority Areas (62614525) from the Ministry of Education, Science and Culture of Japan, and by a grant from Nissan Science Foundation.

References

Bashir, M., Kingston, D.G.I., Carman, R.J., van Tassell, R.L. & Wilkins, T.D. (1987) Anaerobic metabolism of 2-amino-3-methyl-3*H*-imidazo[4,5-*f*]quinoline (IQ) by human fecal flora. *Mutat. Res.*, *190*, 187–190

Hayatsu, H., Arimoto, S., Togawa, K. & Makita, M. (1981a) Inhibitory effect of the ether extract of human feces on activities of mutagens: inhibition by oleic and linoleic acids. *Mutat. Res.*, *81*, 287–293

Hayatsu, H., Inoue, K., Ohta, H., Namba, T., Togawa, K., Hayatsu, T., Makita, M. &

Wataya, Y. (1981b) Inhibition of the mutagenicity of cooked-beef basic fraction by its acidic fraction. *Mutat. Res., 91,* 437–442

Hayatsu, H., Oka, T., Wakata, A., Ohara, Y., Hayatsu, T., Kobayashi, H. & Arimoto, S. (1983a) Adsorption of mutagens to cotton bearing covalently bound trisulfo-copper-phthalocyanine. *Mutat. Res., 119,* 233–238

Hayatsu, H., Matsui, Y., Ohara, Y., Oka, T. & Hayatsu, T. (1983b) Characterization of mutagenic fractions in beef extract and in cooked ground beef. Use of blue-cotton for efficient extraction. *Gann, 74,* 472–482

Hayatsu, H., Hamasaki, K., Togawa, K., Arimoto, S. & Negishi, T. (1983c) *Antimutagenic activity in extracts of human feces.* In: Stich, H.F., ed., *Carcinogens and Mutagens in the Environment, Vol.* 2, *Naturally Occurring Compounds,* Boca Raton, FL, CRC Press, pp. 91–99

Hayatsu, H., Hayatsu, Y. & Ohara, Y. (1985a) Mutagenicity of human urine caused by ingestion of fried ground beef. *Jpn. J. Cancer Res., 76,* 445–448

Hayatsu, H., Hayatsu, T., Wataya, Y. & Mower, H.F. (1985b) Fecal mutagenicity arising from ingestion of fried ground beef in the human. *Mutat. Res., 143,* 207–211

Hayatsu, H., Kasai, H., Yokoyama, S., Miyazawa, T., Yamaizumi, Z., Sato, S., Nishimura, S., Arimoto, S., Hayatsu, T. & Ohara, Y. (1987) Mutagenic metabolites in urine and feces of rats fed with 2-amino-3,8-dimethyl-imidazo[4,5-*f*]quinoxaline, a carcinogenic mutagen present in cooked meat. *Cancer Res., 47,* 791–794

Jägerstad, M., Olsson, K., Grivas, S., Negishi, C., Wakabayashi, K., Tsuda, M., Sato, S. & Sugimura, T. (1984) Formation of 2-amino-3,8-dimethyl-imidazo[4,5-*f*]quinoline in a model system by heating creatinine, glycine and glucose. *Mutat Res., 126,* 239–244

Kikugawa, K., Kato, T. & Hayatsu, H. (1986) The presence of 2-amino-3,8-dimethylimidazo-[4,5-*f*]quinoxaline in smoked dry bonito (katsuobushi). *Jpn. J. Cancer Res., 77,* 99–102

Kobayashi, H. & Hayatsu, H. (1984) A time-course study on the mutagenicity of smokers' urine. *Gann, 75,* 489–493

Mohtashamipur, E., Norpoth, K. & Lieder, F. (1985) Isolation of frameshift mutagens from smokers' urine: experiences with three concentration methods. *Carcinogenesis, 6,* 783–788

Negishi, T. & Hayatsu, H. (1984) Inhibitory effect of saturated fatty acids on the mutagenicity of N-nitrosodimethylamine. *Mutat. Res., 135,* 87–96

Sugimura, T. (1985) Carcinogenicity of mutagenic heterocyclic amines formed during the cooking process. *Mutat. Res., 150,* 33–41

Takahashi, M., Wakabayashi, K., Nagao, M., Yamamoto, M., Masui, T., Goto, T., Kinae, N., Tomita, I. & Sugimura, T. (1985) Quantification of 2-amino-3-methylimidazo[4,5-*f*]-quinoline (IQ) and 2-amino-3,8-dimethylimidazo[4,5-*f*]quinoxaline (MeIQx) in beef extracts by liquid chromatography with electrochemical detection (LCEC). *Carcinogenesis, 6,* 1195–1199

Yahagi, T., Nagao, M., Seino, Y., Matsushima, T., Sugimura, T. & Okada, M. (1977) Mutagenicities of N-nitrosamines on Salmonella. *Mutat. Res., 48,* 121–130

Yamashita, M., Wakabayashi, K., Nagao, M., Sato, S., Yamaizumi, Z., Takahashi, M., Kinae, N., Tomita, I. & Sugimura, T. (1986) Detection of 2-amino-3-methylimidazo[4,5-*f*]-quinoline in cigarette smoke condensate. *Jpn. J. Cancer Res., 77,* 419–422

OXIDATIVE DAMAGE

MEASURING OXIDATIVE DAMAGE IN HUMANS: RELATION TO CANCER AND AGEING

B.N. Ames

Department of Biochemistry, University of California, Berkeley, CA, USA

Many uncertainties remain about the free-radical theory of ageing and the role of oxidative damage to DNA in cancer. The chemistry and biochemistry of radical-induced DNA damage are now well characterized *in vitro*, but the complexity of in-vivo systems leaves this area still largely unexplored. Measurement of thymine and thymidine glycols in urine may be a means of assaying background levels of radical-induced DNA damage in live organisms. Similar approaches may prove useful for testing some of the predictions of the free-radical theory of ageing and of the contribution of free radicals to cancer.

Cumulative cancer risk increases with approximately the fourth power of age (Fig. 1), both in short-lived species such as rats and mice (about 30% have cancer by the end of their two- to three-year life span) and in long-lived species such as humans (about 30% have cancer by the end of their 85-year life span). Thus, the marked increase in life span that has occurred in 60 million years of primate evolution has been accompanied by a marked decrease in age-specific cancer rates. One important factor in longevity appears to be basal metabolic rate (Tolmasoff *et al.*, 1980; Cutler, 1984), which is much lower in man than in rodents and could markedly affect the level of endogenous mutagens produced by normal metabolism. Oxidative DNA damage could be one major contributor to cancer and ageing related to metabolic rate (Totter, 1980; Cutler, 1984).

Sources of free radicals *in vivo*

Numerous mechanisms are capable of generating free radicals *in vivo*. The reactive O_2 species $O_2^{\bar{}}$, H_2O_2, and ·OH are generated *in vivo* during normal metabolism and are also the active agents of DNA damage produced by ionizing radiation (Pryor, 1976–1984; Ames, 1983; Nygaard & Simic, 1983). The interaction of certain cellular components with these O_2 species could contribute both to ageing and age-dependent diseases such as cancer (Ames, 1983; Cerutti, 1985; Ames & Saul, 1987). ·OH has been shown to produce both base damage and strand breaks in DNA. Oxidants in the presence of metal ions initiate lipid peroxidation, which produces a variety of mutagens, carcinogens and promoters (Fig. 2) (Bischoff, 1969; Pryor, 1976–1984; Demopoulos *et al.*, 1980; Ferrali *et al.*, 1980; Imai *et al.*, 1980; Simic & Karel, 1980; Petrakis *et al.*, 1981; Shorland *et al.*, 1981; Bird *et al.*, 1982; Levin *et al.*, 1982), such as fatty acid hydroperoxides, cholesterol hydroperoxide, endoperoxides, cholesterol and fatty acid epoxides, enals and other aldehydes, and alkoxy and hydroperoxy radicals, as well as reactive O_2 species. The digestive tract is exposed to a variety of these fat-derived carcinogens. Human breast fluid can contain high levels (up to 780 μmol/l) of cholesterol epoxide (a mutagenic and carcinogenic oxidative product of cholesterol)

Fig. 1. Cumulative net risk of death from cancer for rats and humans

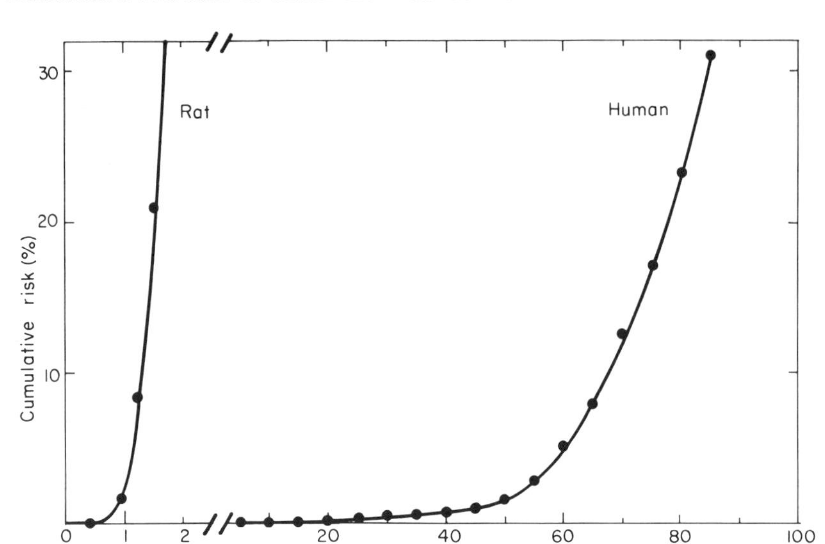

Data on rats are for male Charles River COBS-CD outbred rats (Ross *et al.*, 1982). Human data are from life tables for the year 1964 for men from 11 developed countries (Preston *et al.*, 1972). From Ames *et al.* (1985), with permission from Raven Press

Fig. 2. Lipid peroxidation

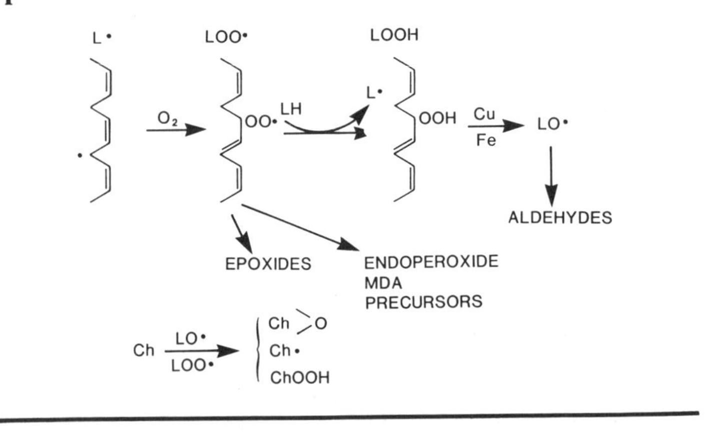

Mutagens/Carcinogens

L• LO• LOO• Ch•

LOOH ChOOH

L>O Ch>O α,β

L• = lipid radical; LO• = alkoxy lipid radical; LOO• = hydroperoxy lipid radical; LOOH = lipid hydroperoxide; Ch = cholesterol; Ch>O = cholesterol epoxide; L>O = lipid epoxide; MDA = malondialdehyde

(Petrakis *et al.*, 1981; Sevanian & Peterson, 1984), which could originate either from ingested oxidized fat or from oxidative processes in body lipids.

Peroxisomes oxidize appreciable amounts of dietary fatty acids, and removal of each two-carbon unit generates one molecule of H_2O_2—itself a mutagen, promoter and carcinogen (Plaine, 1955; Hirota & Yokoyama, 1981; Tsuda, 1981; Ito *et al.*, 1982; Reddy *et al.*, 1982; Speit *et al.*, 1982; Reddy & Lalwani, 1983). Low levels of H_2O_2 escape the catalase in the peroxisome (Chance *et al.*, 1979; Reddy *et al.*, 1982; Reddy & Lalwani, 1983), thus contributing to the pool of O_2-derived species that can damage DNA and can also initiate lipid peroxidation, leading to the production of the mutagens and carcinogens listed previously. Phagocytosis by neutrophils generates $O_2^{\bar{}}$, H_2O_2, HOCl and chloramine (oxidizing and chlorinating agents). Exposure to light can raise the energy of many compounds, leading to excitation of ground state O_2 to highly reactive singlet O_2 (Halliwell & Gutteridge, 1984), a potent mutagen and inducer of lipid and protein oxidation.

Antioxidant defences

Many defence mechanisms within the organism have evolved to limit the levels of reactive O_2 species and the damage they induce. Among the defences are superoxide dismutase, catalase and glutathione peroxidase, as well as the antioxidants β-carotene, tocopherols and vitamin C (Pryor, 1976–1984; Ames, 1983; Porter & Whelan, 1983). We have discussed several previously unappreciated antioxidants that have appeared during evolution. Haem is degraded to biliverdin, which we have shown is a powerful antioxidant (Stocker *et al.*, 1987a,b; Stocker & Ames, 1987). In mammals, biliverdin is converted to bilirubin, which is also a powerful antioxidant. The bilirubin in human blood is bound at a specific site on albumin at a concentration of 20 μM (Stocker *et al.*, 1987a,b; Stocker & Ames, 1987), which is a much higher level than in rat blood. Conjugated bilirubin also appears to be the most important antioxidant in bile, and with the copper ions present in bile forms a powerful redox system for oxidizing xenobiotics and destroying hydroperoxides (Stocker & Ames, 1987).

We have previously discussed uric acid as a powerful antioxidant that appeared during primate evolution concomitant with the development of a long life span and a large, metabolically active brain (Ames *et al.*, 1981). Uric acid is the main antioxidant in saliva, and occurs at 300 μM in human blood. It is present in much lower amounts in animals before the primates. Uric acid levels increased during primate evolution at about the same time as we lost the ability to synthesize ascorbic acid, so that these events may be related.

We are currently working on another antioxidant, carnosine, present in high concentrations in human muscle and brain (R. Kohn, Y. Yamamoto, K. Cundy and B.N. Ames, in preparation).

Because of the finite time between generation of a radical species and its destruction by a defence mechanism, low levels of reactive O_2 species can exist for sufficient time to produce damage to cellular macromolecules (Chance *et al.*, 1979). For nuclear DNA, however, the mammalian cell has three more levels of defence. First, nuclear DNA is compartmentalized away from mitochondria and peroxisomes, where most radicals are probably generated. Second, most nonreplicating nuclear DNA is surrounded by histones and polyamines, which may protect against radicals. Finally, most of the types of DNA damage produced can be repaired by efficient enzyme systems. The net result of this multilevel defence is that nuclear DNA is very well, but not completely, protected from radicals. One view of the somatic damage theory of ageing is that the amount of maintenance and repair of somatic tissues is always less than required for indefinite survival. Thus, some DNA damage induced in somatic cells by radicals will accumulate with time.

Measuring lipid hydroperoxides

We have developed a new, very sensitive method for measuring lipid hydroperoxides in tissues by coupling a chemiluminescent detection system for hydroperoxides with high-performance liquid chromatography (HPLC) separation (Yamamoto *et al.*, 1987). We find hydrogen peroxide in human blood at a level of 5 μM. Preliminary evidence suggests that this level increases three or four fold when normal subjects get a cold, presumably due to the respiratory burst from neutrophils during phagocytosis (Y. Yamamoto and B.N. Ames, unpublished). We find two peaks of hydroperoxide in normal plasma, X-OOH and Y-OOH. We have tentatively identified X-OOH as a long-chain hydroperoxide oxidative breakdown product of fatty acids. Y-OOH is present at 0.3 μM in normal plasma and has been tentatively identified as a cholesterol ester of a fatty acid hydroperoxide (Y. Yamamoto and B.N. Ames, in preparation). Rat tissues contain appreciable levels of lipid hydroperoxides, which appear in the triglyceride fraction (G. Bartoli, Y. Yamamoto and B.N. Ames, in preparation). We believe that these new methods and findings are relevant to heart disease, cancer and ageing.

Detection of oxidative DNA damage

Oxidative damage of cellular DNA has been detected by chemical, physical, enzymatic and immunochemical methods. The chemical methods can be made highly specific but are relatively insensitive. Specific chemical assays have been developed for the free-radical DNA damage products thymine glycol (Frenkel *et al.*, 1981; Cathcart *et al.*, 1984), 5-hydroxymethyluracil (Teebor *et al.*, 1982) and 8-hydroxyguanine (Dizdaroglu, 1985). Some radioactive labelling techniques have been developed that promise to be useful for detecting low levels of many other lesions (Teebor *et al.*, 1982). Physical methods such as alkali elution and agarose gel electrophoresis have been used to detect strand breaks in DNA. These techniques can be used to detect subtle changes in small genomes but are of limited usefulness in mammalian cells. Enyzmatic methods have been used to produce nicks at damage sites so that physical methods can be used to detect the strand breaks (Paterson *et al.*, 1981). Finally, immunochemical methods have been developed for a number of specific lesions, including 8-hydroxyadenine and thymine glycol (West *et al.*, 1982a,b; Leadon & Hanawalt, 1983).

Using these techniques, several products of DNA damage have been detected in DNA from cells exposed to ionizing radiation and other treatments that produce radicals. Mammalian cells exposed to kilorad doses of ionizing radiation were shown to contain increased levels of thymine glycol and 5-hydroxymethyluracil (Fig. 3) (Leadon & Hanawalt, 1983; Frenkel *et al.*, 1985). Recently, Kaneko and Leadon (1986) showed that exposure of human fibroblasts to *N*-hydroxy-2-naphthylamine, a bladder carcinogen that generates free radicals in the cell, leads to a dramatic increase in thymine glycol in the DNA.

The methods cited have been sensitive enough to detect DNA damage induced by severe stresses such as kilorad doses of radiation but have not been useful for examining the background levels of damage products formed from normal aerobic metabolism. In our laboratory, we have overcome this problem by using an approach based on the pathways shown in Figure 4. Specific repair enzymes excise most DNA

Fig. 3. Thymine residues in DNA can be oxidized by a variety of agents to yield thymine glycol or hydroxymethyluracil

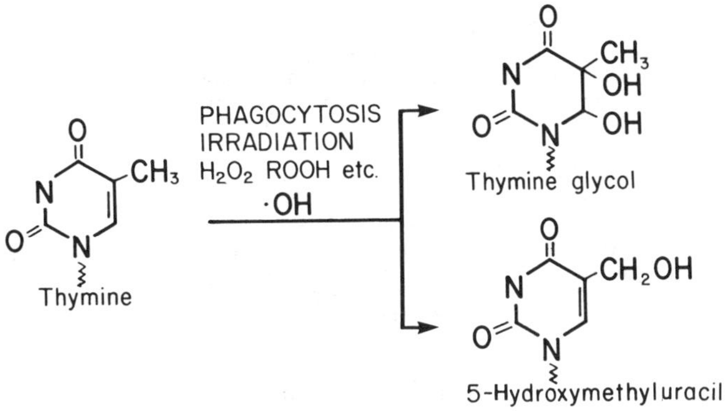

Thymine

PHAGOCYTOSIS
IRRADIATION
H_2O_2 ROOH etc.
·OH

Thymine glycol

5-Hydroxymethyluracil

From Ames *et al.* (1985), with permission from Raven Press

lesions to release a free base or a deoxynucleotide. Deoxynucleotides may lose phosphate to become deoxynucleosides, which are not further metabolized and may be recovered in the urine. Two such products of oxidative damage of DNA are thymine glycol and 5-hydroxymethyluracil. We have recently described a specific DNA repair enzyme, a DNA glycosylase from mouse cells, which repairs 5-hydroxymethyluracil and differs from the specific DNA glycosylase repair enzyme for thymine glycol in mouse cells (Hollstein *et al.*, 1984). The existence of these specific repair enzymes points to the importance of this type of DNA damage *in vivo*.

Fig. 4. Repair of a damaged DNA base by excision repair or a specific glycosylase

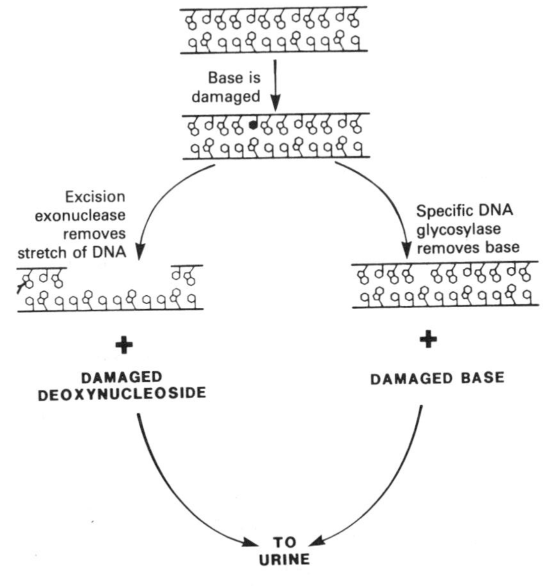

If these products are removed from the cell, are not further metabolized and are fairly hydrophilic, they may be recovered in the urine. From Ames *et al.* (1985), with permission from Raven Press

Table 1. Average background levels of various DNA damage products in human urine[a]

Compound[b]	Level	
	nmol/kg per day	molecules/cell per day
Tg	0.39	270
dTg	0.10	70
HMU	0.9	620
HMdU	trace	trace

[a]Data from Cathcart *et al.* (1984), and unpublished. From Ames and Saul (1987)
[b]Tg, thymine glycol; dTg, thymidine glycol; HMU, 5-hydroxy-methyluracil; HMdU, 5-hydroxymethyl-2′-deoxyuridine

Our method suffers from being an indirect measurement of what was in the DNA and being potentially subject to artefacts. Nevertheless, it has two very powerful advantages. First, it can be made extremely sensitive, in part because DNA lesions from all the cells of the body are concentrated in a relatively small volume of urine. Second, this technique is noninvasive.

In order to quantify the daily removal of these lesions from DNA, we developed an HPLC assay for thymine glycol, thymidine glycol, hydroxymethyluracil and hydroxy-methyldeoxyuridine in urine (Cathcart *et al.*, 1984; unpublished observations). Our results indicate that normal humans excrete a total of about 100 nmol/day of the first three compounds (Table 1). We have considerable evidence that most of this total is derived from repair of oxidized DNA, rather than from alternative sources, such as diet and bacterial flora (Cathcart *et al.*, 1984; Ames & Saul, 1987). This total may therefore represent an average of about 10^3 oxidized thymine residues per day for each of the body's 6×10^{13} cells. Because these products are only three of a considerable number of possible products of oxidative damage of DNA (Scholes, 1983; Cadet & Berger, 1985), the total number of all types of oxidative hits on DNA per cell per day in man may be much more than 10^3.

8-Hydroxy-2′-deoxyguanosine in urine

Our urinary thymine glycol assay is difficult to perform and takes about three weeks. We are thus developing a simpler urinary assay for 8-hydroxy-2′-deoxyguano-sine, which has been measured in DNA by electrochemical detection at about 1000 times the sensitivity of ultraviolet detection (Floyd *et al.*, 1986; Kasai *et al.*, 1986; Kuchino *et al.*, 1987). This assay (Cundy *et al.*, 1987) can be done in an afternoon and will thus enable a much more rapid and simpler assay of urine for oxidized DNA. The levels in human urine appear to be of the same order as those of thymine glycol. We also have preliminary evidence for the presence of 8-hydroxy-2′-deoxyguanosine in urine of humans with chronic granulomatous disease. These people lack the respiratory burst from phagocytic cells, yet they still produce 8-hydroxyguanosine in their urine, indicating that the presence of oxidized DNA bases in urine is not a consequence of normal cell turnover.

Testing the relationships among radical-induced DNA damage, cancer and ageing

Although considerable speculation about the relationships among free radicals, cancer and ageing abounds, the involvement of radicals remains obscure. Perhaps the best circumstantial evidence implicating free radicals in ageing is the impressive inverse

Fig. 5. Average urinary output of thymine glycol (■) and thymidine glycol (○) by three species, expressed as a function of the specific metabolic rate of that species

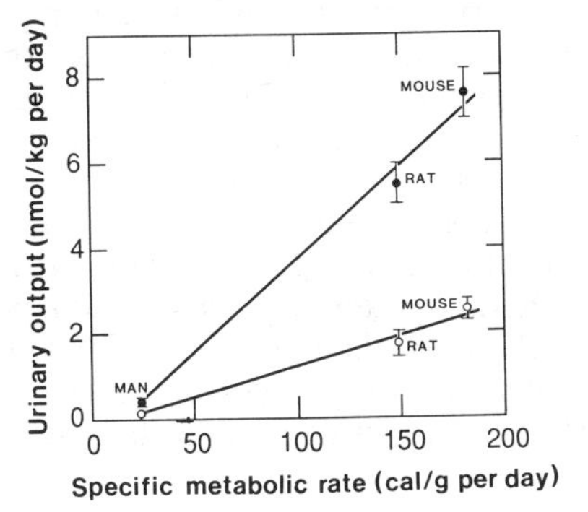

Urine levels are mean values ± SE for untreated humans (ten subjects; see Cathcart *et al.*, 1984), rats (nine animals; see Cathcart *et al.*, 1984), and mice (four animals; unpublished). From Saul *et al.* (1987), with permission from Raven Press

interspecies correlation between specific metabolic rate and rate of ageing, i.e., species with high metabolic rates also have a high age-specific cancer incidence (Fig. 1.) The faster rate of ageing and the faster accumulation of carcinogenic events in mammals with higher specific metabolic rates may be explained by assuming that these species have higher rates of production of free radicals per cell, leading to faster accumulation of somatic damage, carcinogenic events and ageing. We now have data on three species that provide additional circumstantial evidence for this theory and are consistent with the possibility that DNA is a critical target in ageing. Rats, which have a higher specific metabolic rate and a shorter life span than humans, excrete about 15 times more thymine glycol and thymidine glycol per kg of body weight than do humans (Fig. 5) (Cathcart *et al.*, 1984). Mice have an even higher metabolic rate and a shorter life span, and they have higher levels of thymine and thymidine glycols than rats. Data recently obtained with monkeys are consistent with this relationship (R. Adelman, R. Saul and B.N. Ames, in preparation).

We have now tested urines from normal human volunteers aged 22 to 84 years for thymine and thymidine glycols (Saul *et al.*, 1987). We conclude that their urinary outputs are age-independent, which suggests that the rate of oxidative DNA damage in man does not change substantially with age. This is consistent with age-independent somatic damage theories of ageing (Peto *et al.*, 1985).

Mechanisms relating free radicals, DNA damage, cancer and ageing

Several models relate free radicals, DNA damage, cancer and ageing. One possibility is that free radicals react with nuclear DNA to produce somatic mutations. The sources of radicals in this case could include: long-lived reactive species generated outside the nucleus and capable of crossing the nuclear membrane, lipid-soluble radicals generated in the nuclear membrane itself, and radicals generated within the nucleus. The types of DNA lesion that might lead to somatic mutation include: base changes to give point mutations, frameshift mutations and deletions, and strand breaks to give chromosomal rearrangements. Somatic mutation could disrupt the cell by altering gene products or by altering their regulation.

Other types of DNA damage could impair the cell without producing a mutagenic event. Many types of lesion, even in noncoding sequences, can prevent DNA replication and thus prevent cell proliferation. Certain unrepaired lesions in coding sequences might impair transcription and decrease protein synthesis. This type of DNA damage would be particularly important for terminally-differentiated cells that do not normally undergo DNA replication. Free radicals could also cause loss of 5-methylcytosine. Since 5-methylcytosine may be important in turning off genes during differentiation, oxidative DNA damage could prevent this and cause de-differentiation and contribute to cancer and ageing (Doerfler, 1984; Denda *et al.*, 1985). By analogy with 5-hydroxymethyluracil, 5-hydroxymethylcytosine could be formed by oxidative damage. This could lead to lack of methylation after DNA replication if the DNA methylase did not recognize 5-hydroxymethylcytosine as 5-methylcytosine. Similarly, 8-hydroxyguanine might interfere with methylation at its base-paired cytosine.

Acknowledgements

This work was supported by US National Cancer Institute Outstanding Investigator Grant CA39910 to B.N.A. and by US National Institute of Environmental Health Sciences Center Grant ES01896. This paper was adapted in part from Ames *et al.* (1985), and Ames and Saul (1987).

References

Ames, B.N. (1983) Dietary carcinogens and anticarcinogens. Oxygen radicals and degenerative diseases. *Science, 221,* 1256–1264

Ames, B.N. & Saul, R.L. (1987) *Cancer, ageing, and oxidative DNA damage.* In: Iversen, O.H., ed., *Theories of Carcinogenesis,* New York, Hemisphere (in press)

Ames, B.N., Cathcart, R., Schwiers, E. & Hochstein, P. (1981) Uric acid provides an antioxidant defense in humans against oxidant- and radical-caused aging and cancer: a hypothesis. *Proc. natl Acad. Sci. USA, 78,* 6858–6862

Ames, B.N., Saul, R.L., Schwiers, E., Adelman, R. & Cathcart, R. (1985) *Oxidative DNA damage as related to cancer and aging: the assay of thymine glycol, thymidine glycol, and hydroxymethyluracil in human and rat urine.* In: Sohal, R.S., Birnbaum, L.S. & Cutler, R.G., eds, *Molecular Biology of Aging: Gene Stability and Gene Expression,* New York, Raven Press, pp. 137–144

Bird, R.P., Draper, H.H. & Basrur, P.K. (1982) Effect of malonaldehyde and acetaldehyde on cultured mammalian cells. Production of micronuclei and chromosomal aberrations. *Mutat. Res., 101,* 237–246

Bischoff, F. (1969) *Carcinogenic effects of steroids.* In: Paoletti, R. & Dritchevsky, D., eds, *Advances in Lipid Research,* New York, Academic Press, pp. 164–244

Cadet, J. & Berger, M. (1985) Radiation-induced decomposition of the purine bases within DNA and related model compounds. *Int. J. Radiat. Biol., 47,* 127–143

Cathcart, R., Schwiers, E., Saul, R.L. & Ames, B.N. (1984) Thymine glycol and thymidine glycol in human and rat urine: a possible assay for oxidative DNA damage. *Proc. natl Acad. Sci. USA, 81,* 5633–5637

Cerutti, P.A. (1985) Prooxidant states and tumor promotion. *Science, 227,* 375–380

Chance, B., Sies, H. & Boveris, A. (1979) Hydroperoxide metabolism in mammalian organs. *Physiol. Rev., 59,* 527–605

Cundy, K., Kohen, R. & Ames, B.N. (1987) *Determination of 8-hydroxydeoxyguanosine in human urine: a possible assay for in vivo oxidative DNA damage* (Abstract). In: *4th International Congress on Oxygen Radicals, La Jolla, CA, June 27–July 3, 1987*

Cutler, R.G. (1984) *Antioxidants, aging, and longevity.* In: Pryor, W.A., ed., *Free Radicals in Biology,* Vol. 6. New York, Academic Press, pp. 371–428

Demopoulos, H.B., Pietronigro, D.D., Flamm, E.S. & Seligman, M.L. (1980) *The possible role of free radical reactions in carcinogenesis.* In: Demopoulos, H.B. & Melman, M.A., eds, *Cancer and the Environment,* Park Forest South, IL, Pathotox, pp. 273–303

Denda, A., Rao, P.M., Rajalakshmi, S. & Sarma, D.S.R. (1985) 5-Azacytidine potentiates initiation induced by carcinogens in rat liver. *Carcinogenesis, 6,* 145–146

Dizdaroglu, M. (1985) Formation of an 8-hydroxyguanine moiety in deoxyribonucleic acid on γ-irradiation in aqueous solution. *Biochemistry, 24,* 4476–4481

Doerfler, W. (1984) DNA methylation: site-specific methylations cause gene inactivation. *Angew. Chem. int. Ed. Engl., 23,* 919–931

Ferrali, M., Fulceri, R., Benedetti, A. & Comporti, M. (1980) Effects of carbonyl compounds (4-hydroxyalkenals) originating from the peroxidation of liver microsomal lipids on various microsomal enzyme activities of the liver. *Res. Commun. Chem. Pathol. Pharmacol., 30,* 99–112

Floyd, R.A., Watson, J.J. & Wong, P.K. (1986) Hydroxyl free radical adduct of deoxyguanosine: sensitive detection and mechanisms of formation. *Free Radical Res. Commun., 1,* 163–172

Frenkel, K., Goldstein, M.S. & Teebor, G.W. (1981) Identification of the cis-thymine glycol moiety in chemically oxidized and γ-irradiated deoxyribonucleic acid by high-pressure liquid chromatography analysis. *Biochemistry, 20,* 7566–7571.

Frenkel, K., Cummings, A., Solomon, J., Cadet, J., Steinberg, J.J. & Teebor, G.W. (1985) Quantitative determination of the 5-(hydroxymethyl)uracil moiety in the DNA of γ-irradiated cells. *Biochemistry, 24,* 4527–4533

Halliwell, B. & Gutteridge, J.M.C. (1984) *Free Radicals in Biology and Medicine,* Oxford, Clarendon Press, p. 346

Hirota, N. & Yokoyama, T. (1981) Enchancing effect of hydrogen peroxide upon duodenal and upper jejunal carcinogenesis in rats. *Gann, 72,* 811–812

Hollstein, M.C., Brooks, P., Linn, S. & Ames, B.N. (1984) Hydroxymethyluracil DNA glycosylase in mammalian cells. *Proc. natl Acad. Sci. USA, 81,* 4003–4007

Imai, H., Werthessen, N.T., Subramanyam, V., LeQuesne, P.W., Soloway, A.H. & Kanisawa, M. (1980) Angiotoxicity of oxygenated sterols and possible precursors. *Science, 207,* 651–653

Ito, A., Naito, M., Naito, Y. & Watanabe, H. (1982) Induction and characterization of gastro-duodenal lesions in mice given continuous oral administration of hydrogen peroxide. *Gann, 73,* 315–322

Kaneko, M. & Leadon, S.A. (1986) Production of thymine glycols in DNA by N-hydroxy-2-naphthylamine as detected by a monoclonal antibody. *Cancer Res., 46,* 71–75

Kasai, H., Crain, P.F., Kuchino, Y., Nishimura, S., Ootsuyama, A. & Tanooka, H. (1986) Formation of 8-hydroxyguanine moiety in cellular DNA by agents producing oxygen radicals and evidence for its repair. *Carcinogenesis, 7,* 1849–1851

Kuchino, Y., Mori, F., Kasai, H., Inoue, H., Iwai, S., Miura, K., Ohtsuka, E. & Nishimura, S. (1987) Misreading of DNA templates containing 8-hydroxydeoxyguanosine at the modified base and at adjacent residues. *Nature, 327,* 77–79

Leadon, S.A. & Hanawalt, P.C. (1983) Monoclonal antibody to DNA containing thymine glycol. *Mutat Res., 112,* 191–200

Levin, D.E., Hollstein, M., Christman, M.F., Schwiers, E. & Ames, B.N. (1982) A new *Salmonella* tester strain (TA102), with A:T base pairs at the site of mutation, detects oxidative mutagens. *Proc. natl Acad. Sci. USA, 79,* 7445–7449

Nygaard, O.F. & Simic, M.G., eds (1983) *Radioprotectors and Anticarcinogens,* New York, Academic Press

Paterson, M.C., Smith, B.P. & Smith, P.J. (1981) *Measurement of enzyme-sensitive sites in UV- or X-irradiated human cells using* Micrococcus luteus *extracts.* In: Friedberg, E.C. & Hanawalt, P., eds, *DNA Repair, A Laboratory Manual of Research Procedures,* New York, Marcel Dekker, pp. 99–111

Peto, R., Parish, S.E. & Gray, R.G. (1985) *There is no such thing as ageing, and cancer is not related to it.* In: Likhachev, A., Anisimov, V. & Montesano, R., eds, *Age-related Factors in Carcinogenesis* (*IARC Scientific Publications No. 58*), Lyon, International Agency for Research on Cancer, pp. 43–53

Petrakis, N.L., Gruenke, L.D. & Craig, J.C. (1981) Cholesterol and cholesterol epoxides in nipple aspirates of human breast fluid. *Cancer Res., 41,* 2563–2565

Plaine, H.L. (1955) The effect of oxygen and of hydrogen peroxide on the action of a specific gene and on tumor induction in *Drosophila melanogaster*. *Genetics, 40,* 268–280

Porter, R. & Whelan, J. (1983) *Biology of Vitamin E (CIBA Symposium 101),* Bath, Pitman Press

Preston, S.H., Keyfitz, N. & Schoen, R. (1972) *Causes of Death, Life Tables for National Populations,* New York, Seminar Press

Pryor, W.A., ed. (1976–1984) *Free Radicals in Biology,* Vols 1–6, New York, Academic Press

Reddy, J.K. & Lalwani, N.D. (1983) Carcinogenesis by hepatic peroxisome proliferators: evaluation of the risk of hypolipidemic drugs and industrial plasticizers to humans. *CRC crit. Rev. Toxicol., 12,* 1–58

Reddy, J.K., Warren, J.R., Reddy, M.K. & Lalwani, N.D. (1982) Hepatic and renal effects of peroxisome proliferators: biological implications. *Ann. N.Y. Acad. Sci., 386,* 81–110

Ross, M.H., Lustbader, E.D. & Bras, G. (1982) Dietary practices of early life and spontaneous tumors of the rat. *Nutr. Cancer, 3,* 150–167

Saul, R.L., Gee, P. & Ames, B.N. (1987) *Free radicals, DNA damage, and aging*. In: Warner, H.R., *et al.*, eds, *Modern Biological Theories of Aging,* New York, Raven Press (in press)

Scholes, G. (1983) Radiation effects on DNA. *Br. J. Radiol., 56,* 221–231

Sevanian, A. & Peterson, A.R. (1984) Cholesterol epoxide is a direct-acting mutagen. *Proc. natl Acad. Sci. USA, 81,* 4198–4202

Shorland, F.B., Igene, J.O., Pearson, A.M., Thomas, J.W., McGuffey, R.K. & Aldridge, A.E. (1981) Effects of dietary fat and vitamin E on the lipid composition and stability of veal during frozen storage. *J. Agric. Food Chem., 29,* 863–871

Simic, M.G. & Karel, M., eds (1980) *Autoxidation in Food and Biological Systems,* New York, Plenum Press

Speit, G., Vogel, W. & Wolf, M. (1982) Characterization of sister chromatid exchange induction by hydrogen peroxide. *Environ. Mutagenesis, 4,* 135–142

Stocker, R. & Ames, B.N. (1987) Potential role of conjugated bilirubin and copper in the metabolism of lipid peroxides in bile. *Proc. natl Acad. Sci. USA, 84* (in press)

Stocker, R., Yamamoto, Y., McDonagh, A.F., Glazer, A.N. & Ames, B.N. (1987a) Bilirubin is an antioxidant of possible physiological importance. *Science, 235,* 1043–1046

Stocker, R., Glazer, A.N. & Ames, B.N. (1987b) Antioxidant activity of albumin-bound bilirubin. *Proc. natl Acad. Sci. USA, 84* (in press)

Teebor, G.W., Frenkel, K. & Goldstein, M.S (1982) Identification of radiation-induced thymine derivatives in DNA. *Adv. Enzyme Regul., 20,* 39–54

Tolmasoff, J.M., Ono, T. & Cutler, R.G. (1980) Superoxide dismutase: correlation with life-span and specific metabolic rate in primate species. *Proc. natl Acad. Sci. USA, 77,* 2777–2781

Totter, J.R. (1980) Spontaneous cancer and its possible relationship to oxygen metabolism. *Proc. natl Acad. Sci. USA, 77,* 1763–1767

Tsuda, H. (1981) Chromosomal aberrations induced by hydrogen peroxide in cultured mammalian cells. *Jpn. J. Genet., 56,* 1–8

West, G.C., West, I.W.-L & Ward, J.F. (1982a) Radioimmunoassay of a thymine glycol. *Radiat. Res., 90,* 595–608

West, G.C., West, I.W.-L. & Ward, J.F. (1982b) Radioimmunoassay of 7,8-dihydro-8-oxoadenine (8-hydroxyadenine). *Int. J. Radiat. Biol., 42,* 481–490

Yamamoto, Y., Brodsky, M.H., Baker, J.C. & Ames, B.N. (1987) Detection and characterization of lipid hydroperoxides at picomole leves by high performance liquid chromatography. *Anal. Biochem., 160,* 7–13

FORMATION OF REACTIVE OXYGEN SPECIES AND OF 8-HYDROXY-2′-DEOXYGUANOSINE IN DNA *IN VITRO* WITH BETEL-QUID INGREDIENTS

M. Friesen[1], G. Maru[1], V. Bussachini[1], H. Bartsch[1], U. Nair[2], J. Nair[2] & R.A. Floyd[3]

[1]*International Agency for Research on Cancer, Lyon, France;* [2]*Cancer Research Institute, Tata Memorial Centre, Parel, Bombay, India; and* [3]*Oklahoma Medical Research Foundation, Oklahoma City, OK, USA*

Using a chemiluminescence technique, superoxide anion ($O_2^{\bar{}}$) and H_2O_2 were shown to be formed *in vitro,* above pH 9.5, from betel-quid (BQ) ingredients, such as areca-nut extract and catechu. The formation of $O_2^{\bar{}}$ was enhanced by Fe^{2+}, Fe^{3+} and Cu^{2+} and inhibited by Mn^{2+}. Saliva was found to inhibit both $O_2^{\bar{}}$ and H_2O_2 formation from BQ ingredients. Upon incubation of DNA at alkaline pH with areca-nut extract or catechu, in the presence or absence of Fe^{3+}, 8-hydroxy-2′-deoxyguanosine was formed, as quantified by high-performance liquid chromatography. The data suggest a possible role of reactive oxygen species (ROS) in the etiology of oral cancer in betel quid chewers.

Epidemiological studies have associated chewing of BQ with tobacco (BQT) or tobacco alone with an increased risk for cancer of the oral cavity and of the oesophagus in India and other Asian countries (IARC, 1985). Even though several tobacco-specific areca-nut-specific nitrosamines, all shown to exert genetic effects in at least one short-term test, have been identified in BQT and in the saliva of BQT chewers (Nair *et al.*, 1985; Brunnemann *et al.*, 1986), the major causative agents in BQT-associated cancers have not yet been identified. Increasing evidence that oxidative damage to DNA by ROS may play a role in cancer initiation and promotion prompted us to explore whether BQ ingredients can generate ROS capable of reacting with DNA.

Formation of $O_2^{\bar{}}$ and H_2O_2 by BQ ingredients *in vitro*

The formation of ROS *in vitro* from BQ ingredients was detected following their reaction with lucigenin to produce chemiluminescence. Aqueous areca-nut extract (Fig. 1) produced a sharp peak for $O_2^{\bar{}}$ at about 2 sec, which was completely inhibited by superoxide dismutase (Fig. 2), and a later broad peak for H_2O_2 which was completely inhibited by catalase (Fig. 1). An optimal response for $O_2^{\bar{}}$ was found for 50 μg aqueous areca-nut extract, beyond which there was inhibition (Fig. 3). Various fractions obtained from areca nut, such as areca-nut tannins, areca-nut flavonoids, areca-nut catechin and catechu, were also found to produce $O_2^{\bar{}}$ under these conditions (Fig. 4).

The formation of $O_2^{\bar{}}$ was enhanced by Fe^{2+}, Fe^{3+} and Cu^{2+} but inhibited by Mn^{2+}

[1] To whom correspondence should be sent

Fig. 1. Formation of ROS from areca-nut extract

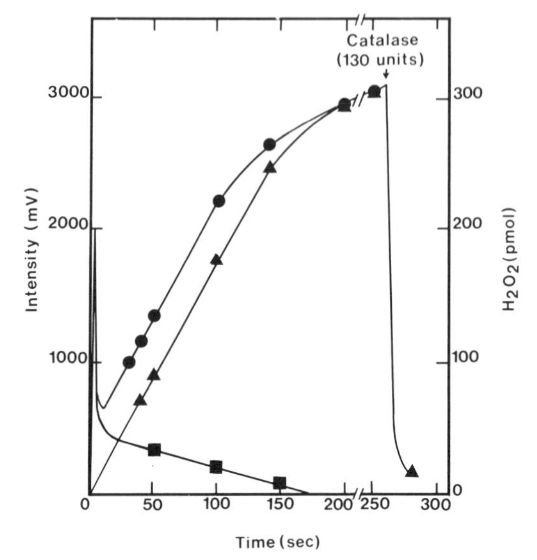

●, 50 μg areca-nut extract; ▲, 50 μg areca-nut extract + 20 units superoxide dismutase; ■, 50 μg areca-nut extract + 130 units catalase. Areca-nut extract was prepared from 10 g dry areca nuts, powdered in a blender and extracted with 250 ml distilled H_2O for 1 h. After filtration through a sintered glass funnel, the filtrate was freeze-dried. Samples containing 25 μl lucigenin (5 mM) and the test substances were diluted to 360 μl with water in a cuvette. After addition of 40 μl 0.02 M Na_2CO_3 to initiate the reaction, response was monitored with a computer-controlled luminometer.

Fig. 2. Concentration-dependent inhibition of $O_2^{\overline{\cdot}}$ formation by superoxide dismutase from 50 μg areca-nut extract

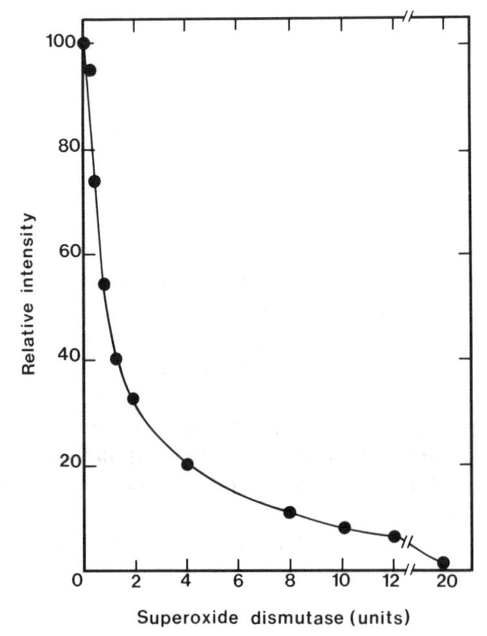

Fig. 3. Formation of $O_2^{\cdot-}$ from areca-nut extract

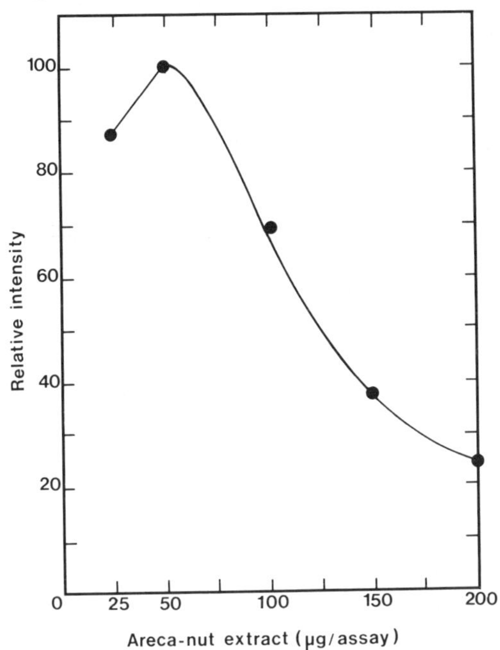

Fig. 4. Formation of $O_2^{\cdot-}$ from areca-nut extracts (ANE), catechu and tannin

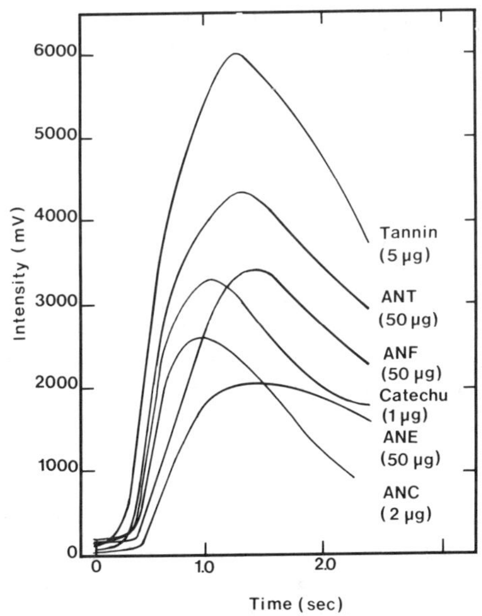

Tannin (ANT), catechin (ANC) and flavonoid (ANF) fractions of areca nut were generously provided by Dr H.F. Stich (University of British Columbia, Vancouver, Canada) and were prepared by the method of Stich *et al.* (1983). Tannin was obtained commercially. Catechu (resin of *Acacia catechu*) and areca nut were obtained from local shops in Bombay, India.

Fig. 5. Effect of metal ions on the formation of $O_2^{\bar{}}$ from 50 µg areca-nut extract

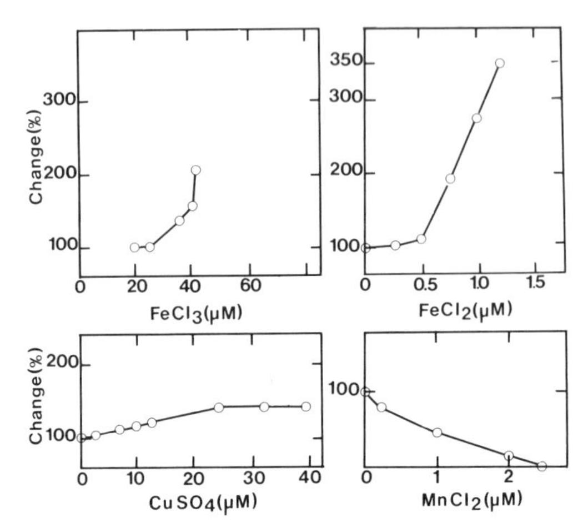

(Fig. 5). None of these metal ions was observed to have an effect on the formation of H_2O_2. Saliva was found to inhibit both $O_2^{\bar{}}$ and H_2O_2 formation from BQ ingredients under these conditions. Tobacco extract produced $O_2^{\bar{}}$ only in the presence of metal ions.

Formation of 8-hydroxy-2′-deoxyguanosine in DNA by BQ ingredients *in vitro*

The formation of 8-hydroxy-2′-deoxyguanosine in calf thymus DNA, following incubation with BQ ingredients at alkaline pH, was quantified by high-performance liquid chromatography with electrochemical detection (0.8 V) after enzymic hydrolysis (Floyd *et al.*, 1986). The method allows measurement of one modified base in about 10^6 molecules deoxyguanosine.

For the assays listed in Table 1, 15 mg calf thymus DNA were dissolved in 10 ml distilled water containing the BQ ingredients as listed. After initiation of the reaction with 3 ml 0.2 M Na_2CO_3 (pH 10.9–11), the mixtures were incubated for 60 min at 37°C. Following extraction with 3 × 10 ml ethyl acetate and 2 × 10 ml isoamyl alcohol:chloroform mixture (3:1), the DNA in the aqueous phase was precipitated with excess ethanol, washed with cold ethanol and dried under vacuum. DNA hydrolysis was carried out by the procedure of Beland *et al.* (1979).

In agreement with enhanced ROS formation, areca-nut extract and catechu, in the presence or absence of Fe^{3+}, led to an increased yield of 8-hydroxy-2′-deoxyguanosine (Table 1). Yields were highest when the phenolics were replaced by H_2O_2, indicating the latter as a possible intermediate.

Conclusions

Our study demonstrates that, under alkaline conditions, ROS are generated from areca nut and catechu, essential ingredients used by BQ chewers. The lime that is taken with BQT, BQ or tobacco can elevate the pH in the microenvironment of the oral cavity, and the likelihood that ROS are generated locally from the BQ ingredients during the chewing process is high. Thus, the ROS concentration may be sufficiently high to temporarily overload the ability of saliva to destroy it. The possibility of oxidative damage is further substantiated by our findings that 8-hydroxy-2′-

Table 1. Formation of 8-hydroxy-2′-deoxyguanosine in DNA *in vitro*

Reaction mixture	8-Hydroxy-2′-deoxyguanosine (μmol/mg DNA)
DNA + Na_2CO_3	50
DNA + Na_2CO_3 + areca-nut extract (15 mg/10 ml)	250
DNA + Na_2CO_3 + catechu (15 mg/10 ml)	400
DNA + Na_2CO_3 + Fe^{3+} (5 μM)	150
DNA + Na_2CO_3 + Fe^{3+} (5 μM) + areca-nut extract (15 mg/10 ml)	350
DNA + Na_2CO_3 + Fe^{3+} (5 μM) + catechu (4.4 mg/10 ml)	350
DNA + Na_2CO_3 + Fe^{3+} (5 μM) + H_2O_2 (50 μl of 30%/10 ml)	1250
DNA + Na_2CO_3 + Fe^{2+} (1 μM)	250
DNA + $NaCO_3$ + Fe^{2+} (1 μM) + areca-nut extract (15 mg/10 ml)	200
DNA + Na_2CO_3 + Fe^{2+} (1 μM) + H_2O_2 (50 μl of 30%/10 ml)	750

deoxyguanosine is formed at alkaline pH in DNA upon incubation of BQ ingredients, such as areca-nut extract or catechu, with or without Fe^{3+}.

If such a situation occurs, the ROS formed *in vivo* could lead to oxidative damage of DNA in the target cells of the oral cavity of BQ and BQT chewers. Attempts are under way to analyse buccal tissue specimens from BQ, BQT and tobacco chewers for the presence of DNA bases modified by ROS and other BQ and BQT constituents.

References

Beland, F.A., Doodley, K.L. & Cascano, D.A. (1979) Rapid isolation of carcinogen-bound DNA and RNA by hydroxyapatite chromatography. *J. Chromatogr.*, *174*, 177–186

Brunnemann, K.D., Prokopczyk, B., Hoffman, D., Nair, J., Ohshima, H. & Bartsch, H. (1986) *Laboratory studies on oral cancer and smokeless tobacco.* In: Hoffmann, D. & Harris, C.C., eds, *New Aspects of Tobacco Carcinogenesis* (*Banbury Report No. 23*), Cold Spring Harbor, NY, CSH Press, pp. 197–213

Floyd, R.A., Watson, J.J., Wong, P.K., Altmiller, D.H. & Rickard, R.C. (1986) Hydroxyl free radical adduct of deoxyguanosine: sensitive detection and mechanisms of formation. *Free Radical Res. Commun.*, *1*, 163–172

IARC (1985) *IARC Monographs on the Evaluation of the Carcinogenic Risk of Chemicals to Humans*, Vol. 37, *Tobacco Habits other than Smoking*; *Betel-quid and Areca-nut Chewing*; *and some Related Nitrosamines*, Lyon

Nair, J., Ohshima, H., Friesen, M., Croisy, A., Bhide, S.V. & Bartsch, H. (1985) Tobacoo specific and betel-nut specific *N*-nitroso compounds: occurrence in saliva and urine of betel quid chewers and formation *in vitro* by nitrosation of betel quid. *Carcinogenesis*, *6*, 295–303

Stich, H.F., Ohshima, H., Pignatelli, B., Michelon, J. & Bartsch, H. (1983) Inhibitory effect of betel nut extracts on endogenous nitrosation in humans. *J. natl Cancer Inst.*, *70*, 1047–1050

FORMATION OF DNA ADDUCT 8-HYDROXY-2′-DEOXYGUANOSINE INDUCED BY MAN-MADE MINERAL FIBRES

P. Leanderson, P. Söderkvist, C. Tagesson & O. Axelson

Department of Occupational Medicine, Faculty of Health Sciences, Linköping, Sweden

Two man-made mineral fibres, rockwool and glasswool, were found to mediate hydroxylation of deoxyguanosine and calf thymus DNA to form the DNA adduct 8-hydroxy-2′-deoxyguanosine. The modification of the nucleoside is probably mediated by hydroxyl radicals and may play a role in fibre-induced carcinogenesis.

Epidemiological studies have suggested that an excess of lung cancer is associated with exposure to man-made mineral fibres (Enterline & Marsh, 1984; Saracci *et al.*, 1984), as observed in experimental animals (IARC, 1987). The underlying mechanism of this carcinogenic activity is not fully understood; however, one possibility is that the fibres induce formation of active oxygen species that interact with DNA. Such oxygen radical-mediated DNA damage has been shown to occur with asbestos fibres (Kasai & Nishimura, 1984). We now demonstrate that not only asbestos but also man-made mineral fibres are able to modify nucleosides.

Experimental

First, a mixture (total volume, 2.0 ml) containing 10.0 mM deoxyguanosine in 100 mM phosphate buffer, pH 7.5, and rockwool or glasswool was incubated in the dark at 37°C in a shaking water bath. After different times of incubation, aliquots were removed and analysed for 8-hydroxy-2′-deoxyguanosine, according to Floyd *et al.* (1986).

Second, calf thymus DNA (2.5 mg/ml) and 20 mg rockwool or glasswool were incubated as described above. After different times of incubation, the samples were centrifuged, the supernatants were collected and DNA was digested with nuclease P1 for 2 h at 37°C, followed by incubation with alkaline phosphatase for 4 h (Kasai & Nishimura, 1984). The resulting deoxynucleoside mixture was injected into a high-performance liquid chromatography apparatus, and deoxyguanosine and 8-hydroxy-2′-deoxyguanosine were identified by ultraviolet and electrochemical detection, respectively (Floyd *et al.*, 1986), the latter at subpicomole levels.

Incubation of deoxyguanosine in the presence of rockwool or glasswool *in vitro* resulted in a time-dependent increase in the level of 8-hydroxy-2′-deoxyguanosine (Figs 1 and 2). No major quantitative difference in 8-hydroxy-2′-deoxyguanosine formation was seen between the two fibre species. After exposure of calf thymus DNA to rockwool or glasswool *in vitro* for different times, an enhanced formation of 8-hydroxy-2′-deoxyguanosine was observed (Fig. 3). To our knowledge this is the first time that man-made mineral fibres have been shown to generate this type of DNA adduct.

Fig. 1. Hydroxylation of deoxyguanosine to 8-hydroxy-2′-deoxyguanosine

Oxygen radical

Deoxyguanosine

8-Hydroxy-2′-deoxyguanosine

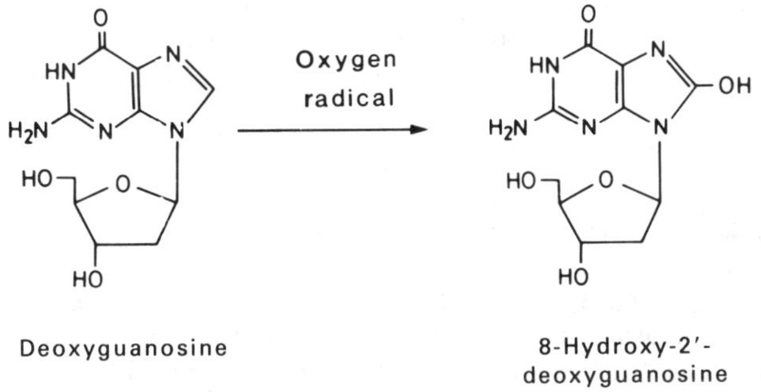

Fig. 2. Formation of 8-hydroxy-2′-deoxyguanosine with time in mixtures containing deoxyguanosine and rockwool (●), glasswool (■) and with no fibre (▼)

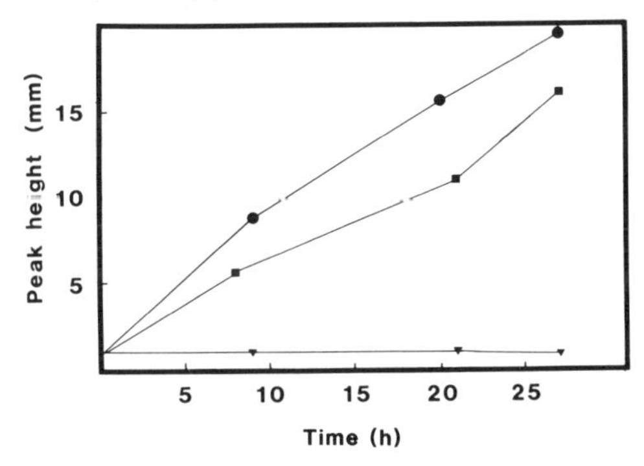

Fig. 3. Formation of 8-hydroxy-2′-deoxyguanosine in calf thymus DNA incubated with rockwool (●), glasswool (■) and with no fibre (▼)

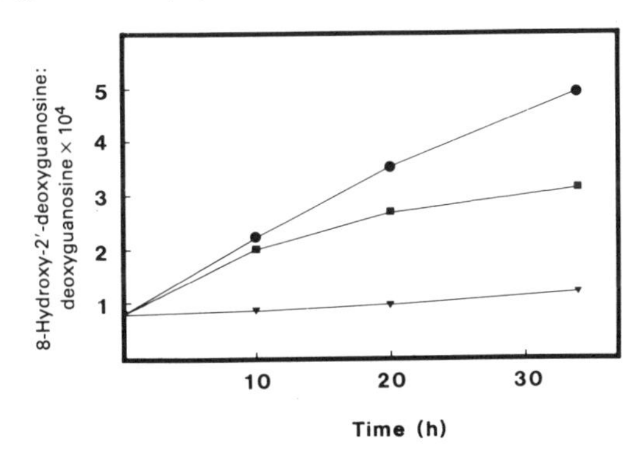

Conclusion

Oxidative stress is believed to be of importance in tumour initiation and promotion (Cerutti, 1985). To the extent that the carcinogenic effect of asbestos is due to production of oxygen radicals, with subsequent damage of DNA, it is tempting to speculate that the carcinogenic activity of man-made mineral fibres seen in animal experiments (Wagner *et al.*, 1984), and as suggested by epidemiological studies, may operate in the same way as for asbestos.

References

Cerutti, P.A. (1985) Prooxidant states and tumor promotion. *Science, 227,* 375–380

Enterline, P.E. & Marsh, G.M. (1984) *The health of workers in the MMMF industry.* In: *Biological Effects of Man-made Mineral Fibres,* Copenhagen, World Health Organization, pp. 311–339

Floyd, R.A., Watson, J.J., Wong, P.K., Altmiller, D.H. & Rickard, R.C (1986) Hydroxyl free radical adduct of deoxyguanosine: sensitive detection and mechanism of formation. *Free Radical Res. Commun., 1,* 163–169

Kasai, H. & Nishimura, S. (1984) DNA damage induced by asbestos in the presence of hydrogen peroxide. *Gann, 75,* 841–844

Kasai, H., Crain, P.F., Kuchino, Y., Nishimura, S., Ootsuyama, A. & Tanooka, H. (1986) Formation of 8-hydroxyguanine moiety in cellular DNA by agents producing oxygen radicals and evidence for its repair. *Carcinogenesis, 7,* 1849–1851

Saracci, R., Simonato, L., Acheson, E.D., Andersen, A., Bertazzi, P.A., Claude, J., Charnay, N., Estève, J., Frentzel-Beyme, R.R., Gardner, M.J., Jensen, O.M., Maasing, R., Olsen, J.H., Teppo, L., Westerholm, P. & Zocchetti, C. (1984) Mortaility and incidence of cancer in man-made vitreous fibre producing industry: an international investigation at 13 European plants. *Br. J. ind. Med., 41,* 425–436

Wagner J.C., Berry, G.B., Hill, R.J., Munday, D.I. & Skidmore, J.W. (1984) *Animal experiments with MM(V)F — effects of inhalation and intrapleural inoculation in rats.* In: *Biological Effects of Man-made Mineral Fibres,* Copenhagen, World Health Organization, pp. 209–233

APPLICATIONS IN MOLECULAR EPIDEMIOLOGY AND CANCER ETIOLOGY

PROSPECTS FOR EPIDEMIOLOGICAL STUDIES ON HEPATOCELLULAR CANCER AS A MODEL FOR ASSESSING VIRAL AND CHEMICAL INTERACTIONS

F.X. Bosch & N. Muñoz

International Agency for Research on Cancer, Lyon, France

Chronic infection with hepatitis B virus (HBV) accounts for 1–10% of hepatocellular carcinoma (HCC) in low-risk countries and for 56–94% in high-risk populations. However, although HBV is perhaps the second most important human carcinogen so far identified, chronic HBV infection is neither a sufficient nor a necessary cause of HCC. Other factors must be causally related to HCC, and some of them have been identified: aflatoxins, tobacco smoking, and use of alcohol and oral contraceptives. The evidence for an association between these factors and HCC is reviewed, as well as their joint effects. Finally, prospects for epidemiological research on HCC, and specifically the assessment of viral and chemical interactions, are discussed.

Worldwide, HCC is the seventh most common form of cancer in males and the ninth in females. It is the most common malignant tumour among males in western, middle and eastern Africa, the second most common in southern Africa and south-east Asia and the third most common among males in China. It is a relatively rare tumour in most parts of America, Europe, northern Africa and middle and eastern Asia (Parkin *et al.*, 1984).

During the last decade, a series of epidemiological and laboratory investigations has established an association between HBV and HCC. The association is restricted to chronically persistent forms of HBV infection and is strong, specific and consistent. Most epidemiologists accept that the association is causal. However, the fact that hepatitis B infection is not a necessary cause is indicated by the existence of varying proportions of HCC cases that do not have the hepatitis B surface antigen (HBsAg). In countries with high HCC incidence, 10–30% of the cases are found to be HBsAg-negative; but in intermediate and low-risk countries, the fraction of cases without HBsAg can be as high as 50–85%. That the viral infection itself is not a sufficient cause is indicated by the fact that only a small fraction of HBsAg carriers eventually develop HCC. Therefore, other factors must be causally related to HCC, and some have already been identified. However, the evidence that supports these associations is still less well defined than the evidence for HBV. Laboratory and epidemiological studies indicate that aflatoxins play an important role in the development of HCC in certain areas of the developing world. Other studies show an increased risk among cigarette smokers, alcohol consumers and long-term oral contraceptive users. The different relative contributions of each of these putative risk factors in high- and low-risk areas for HCC and uncertainties about the magnitude of the risk, make HCC one of the most promising models for epidemiological assessment of the interactions between viral and chemical factors in human cancer.

The evidence for HBV

The association between HCC and HBV has been reviewed extensively (Szmuness, 1978; Beasley, 1982; Trichopoulos *et al.*, 1982; Blumberg & London, 1985; Nishioka, 1985; Muñoz & Bosch, 1987). In summary, correlation studies have demonstrated that there is a strong positive correlation between the incidence of or mortality from HCC and the prevalence of HBsAg carrier state (Szmuness, 1978; Muñoz & Linsell, 1982). There are some as yet unexplained exceptions to this general pattern: for example, a high prevalence of HBsAg carriers and a low incidence rate of HCC have been reported among Greenland Eskimos (Skinhøj *et al.*, 1978; Melbye *et al.*, 1984).

Case-control studies in high- and low-risk populations have shown that the relative risk (RR) associated with the presence of HBsAg in sera ranges from ten to infinity, and in most studies from ten to 20 (Prince *et al.*, 1975; Kew *et al.*, 1979; Yarrish *et al.*, 1980; Lingao *et al.*, 1981; Lam *et al.*, 1982; Yeh *et al.*, 1985a; Austin *et al.*, 1986; Trichopoulos *et al.*, 1987).

In cohort studies, the occurrence of HCC among HBsAg carriers has been compared with that of noncarrier control populations. Estimates of the RR vary from seven to over 100, with 95% confidence limits ranging from 2 to 212 (Muñoz & Bosch, 1987). These prospective cohort studies provide unequivocal proof that HBV infection precedes the development of HCC. That this association is specific as well as strong is suggested by the lack of association of HBV with other cancers (Prince *et al.*, 1975), including metastatic liver cancer (Trichopoulos *et al.*, 1978). In populations at high risk for HCC, the attributable risk for chronic HBV infection ranges from 56 to 94%, and that in low-risk populations from 1 to 10% (Muñoz & Bosch, 1987).

The evidence for aflatoxins

Aflatoxins are mycotoxins elaborated by *Aspergillus flavus* and *Aspergillus parasiticus* fungi. Human exposure can occur following ingestion of contaminated food or of products derived from animals that have consumed aflatoxin-contaminated feeds. The main sources of aflatoxin in most countries are peanuts, peanut derivatives and corn. There are four major aflatoxins, B_1, B_2, G_1 and G_2, for which there is strong evidence of a carcinogenic effect in many animal species, aflatoxin B_1 being the most potent (Busby & Wogan, 1984).

The evidence for their carcinogenic role in humans is less conclusive because of the lack of appropriate methods to assess past exposures at the individual level. The development of sensitive tests for detecting recent exposure by measuring aflatoxin metabolites or DNA adducts in urine, sera and milk could be the first step in this direction (Garner *et al.*, 1985; Groopman *et al.*, 1986).

The results of recent animal experiments suggest that after chronic exposure to dietary aflatoxin B_1 there is cumulative binding to albumin (up to 6–7% of the administered dose) and that the levels correlate with those of aflatoxin-DNA adducts in liver cells. Following acute (single-dose) exposure, 1–3% aflatoxin B_1 was found to be bound to plasma albumin. If these albumin adducts are also stable in humans, such measurements would reflect exposure to aflatoxin during the four to six weeks prior to sampling (that is, the average lifetime for human albumin). Therefore, this assay is presently the most promising means for assessing aflatoxin exposure in epidemiological studies (Wild *et al.*, 1986; Sabbioni *et al.*, 1987). The epidemiological evidence relating aflatoxins to HCC consists of population correlation studies, sometimes referred to as ecological studies (Shank *et al.*, 1972; Peers & Linsell, 1973; Peers *et al.*, 1976; Van Rensburg *et al.*, 1985; Peers *et al.*, 1987), and case-control studies (Bulatao-Jayme *et al.*, 1982; Lam *et al.*, 1982).

Fig. 1. Correlation between incidence of HCC in males over 15 years of age and estimates of aflatoxin B_1 exposure: Kenya, 1973; Swaziland, 1976, 1987; Mozambique and Transkei, 1985; Thailand, 1972

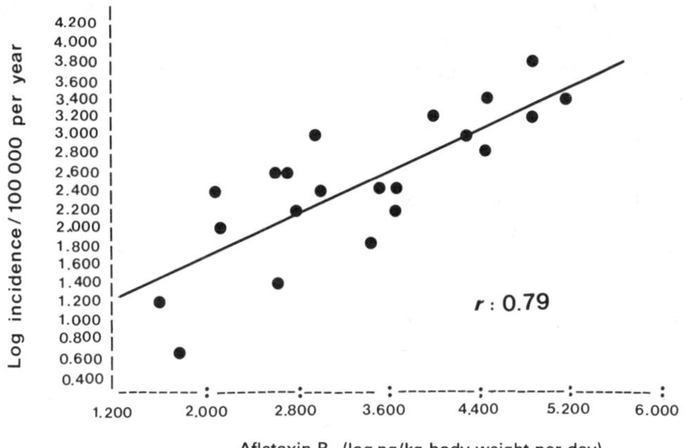

Figure 1 summarizes the correlation between estimates of exposure to aflatoxin B_1 in various high-risk countries for HCC and the corresponding HCC incidence rates in males over 15 years of age. Although the techniques used to estimate aflatoxin exposure and to estimate HCC incidence rates vary from study to study, the overall correlation is highly significant. There are some indications that the strength of the correlation might be different in other contexts. Some surveys in parts of South and Central America suggest that aflatoxin contamination of crops is high (Varsavsky & Sommer, 1977; De Campos & Olszyma-Marzys, 1979; Ochomogo, 1979), and the occurrence of HCC is rare in these populations (Parkin, 1986). In a study in the USA, a weak correlation was found between mortality from HCC in the periods 1968–1971 and 1973–1976 and extrapolated estimates of exposure to aflatoxin in the previous eight to 66 years for white males resident in south-eastern rural areas (estimated exposure, 13–197 ng/kg bw per day) and corresponding males in northern and western areas (estimated exposure, 0.2–0.3 ng/kg bw per day) (Stoloff, 1983). The main difficulties in interpreting these correlation studies in terms of causality are that (i) the assessment of aflatoxin exposure does not take into account individual variation or variation in risk among the population of a geographic area; (ii) there is no temporal relation between current levels of aflatoxin exposure and current incidence rates of HCC; (iii) the possibility of confounding due to some other unmeasured factor cannot be ruled out; and (iv) in most of the countries in which these studies have been performed, the completeness of cancer registration and the diagnostic accuracy are inadequate.

In two case-control studies it was attempted to determine aflatoxin exposure using dietary questionnaires and aflatoxin measurements in local foodstuffs. In the Philippines, RR of 17.0 and 13.9 were reported for individuals classified as receiving an overall mean load of aflatoxin described as 'very heavy' or 'moderately heavy', as compared to those with a light mean load ($p < 0.05$). The effect of aflatoxin was higher among heavy alcohol consumers (RR = 35.5) (Bulatato-Jayme *et al.*, 1982).

A study in Hong Kong (Lam *et al.*, 1982) showed no effect of aflatoxin, as measured by the frequency of consumption of corn and beans, the two major sources of aflatoxin in foodstuffs in Hong Kong (Shank *et al.*, 1972). The two major drawbacks of published case-control studies on aflatoxin and HCC are, firstly, that the methods used for assessing exposure to aflatoxin (dietary questionnaires and surveys of aflatoxin in foodstuffs) are still extremely crude, and, secondly, that concurrent exposure to HBV has to be taken into account. The methodological difficulties are threefold. (i) Dietary questionnaires tend to provide estimates of recent diets, and these are highly influenced by seasonal and secular changes as well as by disease status. (ii) In most case-control studies, measurements of the aflatoxin contents of foodstuffs obtained from market or household surveys have been used to estimate aflatoxin exposure, reflecting current or recent aflatoxin contents of the most common foodstuffs. Even if these surveys sometimes reflect seasonal variation, no reliable information is available on the applicability of current results to the more distant past, which is probably the relevant period in terms of carcinogenesis for current HCC cases. This is particularly important in countries where changes in agricultural practices (improvements in storage, introduction of pest-resistant strains of maize) have been introduced recently. (iii) Aflatoxin occurs in only a fraction of a particular foodstuff (i.e., 4% of maize samples and 20% of peanut samples were contaminated in Swaziland, whereas 84 and 99% of the same food items were contaminated in the Philippines). Therefore, in populations with a monotonous diet (as is the case in African settings), most people would have been exposed to the foodstuff, whereas only a few would have been exposed to aflatoxin.

Finally, a proper assessment of the contribution of aflatoxin to HCC must take into account the strong causal association between HBV and HCC.

Studies on the joint effects of exposure to aflatoxin and HBV

Because of methodological difficulties in assessing chronic exposure to aflatoxin at the individual level, our knowledge of the joint effects of HBV and aflatoxin in the development of HCC are still limited. Only a few studies in China and two in Africa have addressed the issue, but the results are inconclusive.

Studies in China

HCC is the third most important cause of death from cancer in China. High-incidence areas have been identified in the south-eastern coastal areas. The average world standardized mortality rates have been reported to be as high as 111.75 and 20.9/100 000 for males and females, respectively, in some districts of Guanxi province (Yeh *et al.*, 1985b). In this province and in Quidong county, another high-risk area located in the Changjiang (Yangtse) River delta, the epidemiology of HCC has been studied extensively; the risk linked to HBsAg carrier status has been documented repeatedly and is consistent with findings elsewhere. For example, a case-control study in Guangxi province showed a matched odds ratio of 17.0 [95% confidence interval (CI), 4.3–99.4] for HBsAg carriers. The corresponding attributable risk can be estimated as at least 80% (Yeh *et al.*, 1985a). In a cohort study in Quidong province, about 2500 HBsAg carriers and 12 000 non-HBsAg carriers were followed up for 5.5 years; the ratio of incidence rates for HCC was 18 (Sun & Chu, 1984).

In some studies, aflatoxin exposure was also considered. In one study, mortality from HCC was strongly correlated both with the presence and concentration of aflatoxin B_1 in food and with the prevalence of HBsAg in males (Yaobin *et al.*, 1983). However, in some areas, pronounced gradients in HCC mortality rates do not parallel corresponding changes in the prevalence of HBsAg. Because in some of these areas a correlation between dietary aflatoxin and HCC was found, it was suggested that differences in aflatoxin exposure would offer a better explanation for the variations in

HCC than HBsAg prevalence (Armstrong, 1980; Sun & Chu, 1984; Sun *et al.*, 1986). It has also been suggested that mortality rates for HCC among HBsAg carriers resident in areas with heavy aflatoxin B_1 food contamination were higher than corresponding rates for HBsAg carriers residing in areas with light aflatoxin B_1 contamination (Yeh *et al.*, 1985b). In none of these studies was the evaluation of exposure to aflatoxin made at the individual level, and no information on other risk factors was reported. In addition, some of these studies are difficult to evaluate due to insufficient information on the methods used.

Studies in Africa

Two reports from Africa have discussed the effects of joint exposure to HBV and aflatoxin in HCC. In Mozambique (Van Rensburg *et al.*, 1985), the incidence of HCC was estimated from various sources, and a significant correlation was found between aflatoxin contamination of cooked food and HCC risk in six areas of high HCC incidence (six districts of Inhambane province), in an area of intermediate incidence (Manhica-Magude) and in a low-risk area (Transkei). These results are in contrast to the lack of geographical correlation between HCC incidence and HBsAg prevalence rates in various countries and in territories of southern Africa including Mozambique (P. Cook–Mozaffari, in preparation). The authors also report a declining time trend for HCC in various population groups in South Africa, which was attributed to increased awareness of the risk of mouldy foodstuffs and improvements in the food supply chain. Similar observations have been made by other investigators in a South African mining population (Harington *et al.*, 1975; Bradshaw *et al.*, 1982). Van Rensburg *et al.* (1985) concluded that the HBsAg carrier state was an early-stage event that would increase individual susceptibility to HCC and that aflatoxin (particularly transient exposures to very high levels of aflatoxin) acts as a late-stage carcinogen, leading to HCC. In Swaziland (Peers *et al.*, 1987), a study was conducted to assess specifically the relationship between aflatoxin exposure, HBV infection and the incidence of HCC, which is the most commonly occurring malignancy among males. In brief, levels of aflatoxin intake were evaluated in dietary samples from households across the country. The prevalence of hepatitis B markers was estimated in blood donors, and HCC incidence was reported through a national system of hospital-based cancer registration. Across four broad geographical regions, there was a more than five-fold variation in the estimated daily intake of aflatoxin, ranging from 3.1 to 17.5 μg. The proportion of HBV-exposed individuals was very high (86% in men), but varied relatively little by geographical region; the prevalence of HBsAg carriers was 23% in men, and varied from 21 to 28%, which are among the highest rates so far reported. HCC incidence varied over a five-fold range and was strongly associated with estimated levels of aflatoxin in the diet. In an analysis involving ten smaller subregions, aflatoxin exposure emerged as a more important determinant of the variation in HCC incidence than the prevalence of HBsAg. However, in addition to the shortcomings of the correlation studies described above, these results should be interpreted with caution in view of the underreporting of HCC cases to the registry and the crudeness of the estimates of aflatoxin exposure. The strong association found between aflatoxin and HCC does not contradict the strong effect of HBsAg carrier status on HCC; indeed, within the study period, a case-control study was also conducted, and a RR of 19.0 (95% CI, 2.5–149.3) was found for HBsAg carriers. The HBV attributable risk was 80.5% among males. Although virtually all of the HCC cases were exposed to HBV, only half were HBsAg positive, which might suggest that a proportion of the HCC cases, ranging from 20% to perhaps 50%, would not be explained by HBV and aflatoxin could be acting as an independent risk factor for HCC rather than simply as a cofactor with HBV.

Tobacco and alcohol

Moderate excesses of HCC have been observed in the data from some of the major cohort studies on smoking and cancer (Garfinkel, 1980; Hirayama, 1981; Oshima *et al.*, 1984). In four cohort studies of groups identified as having excessive consumption of alcohol in Norway, Finland, Denmark and Japan, increased risks for HCC have been observed (Tuyns, 1980). In a case-control study within a cohort in Japan, a strong positive association with alcohol and a dose-response relationship were demonstrated: for heavy drinkers (more than 80 ml alcohol/day), a RR of 8.0 (95% CI, 1.3–49.5) was found, and this increased risk persisted after adjusting for smoking (Oshima *et al.*, 1984). In some cohort studies among alcoholics, however, no increased risk for HCC has been demonstrated (Rothman, 1980). These results should be interpreted with caution, because most of these studies are based on death certificates, which are known to be unreliable as diagnostic sources of primary liver cancer. In addition, in cohort studies, confounding with other unmeasured risk factors cannot be ruled out. Most of the evidence available on the interaction between HBV and chemicals is therefore based on case-control studies.

Joint effects of HBV, tobacco and alcohol

Four studies have addressed the joint effects of HBV, cigarette smoking and alcohol (Table 1). Lam *et al.* (1982) conducted a case-control study on HCC in Hong Kong in which HBV status and smoking and drinking habits were assessed. Their conclusion was that HBsAg and cigarette smoking are both independent risk factors for HCC. The association between cigarette smoking and HCC was restricted to HBsAg-negative HCC cases and was stronger for cases older than 50 years. A RR of 3.3 (95% CI, 1.0–13.4) was reported for the HBsAg-negative cases of all ages, and the corresponding parameters for cases older than 50 were RR 8.2 (95% CI, 1.5–91.9). However, in this study, only 19 HCC cases were negative for HBsAg; therefore, to

Table 1. Joint effects of HBsAg, alcohol and cigarette smoking on risk for HCC

Reference	Cases/Controls	Factor		RR (95% CI)
Lam *et al.* (1982)	107/107	HBsAg		21.3 (10.1–45.9)
	19/88 (HBsAg−)	Smoking	(>50 years)	8.2 (1.5–91.9)
		Smoking	(All ages)	3.3 (1.0–13.4)
		Alcohol		1.6[a]
Austin *et al.* (1986)	67/63	HBsAg	(in Males)	— (3.8–∞)
		Alcohol		3.3[a]
Trichopoulos *et al.* (1987)	194/456	HBsAg		13.7 (8.0–23.5)
	104/456 (HBsAg−)	Smoking	(20–30 cigarettes/day)	2.4[a]
			(+ 30 cigarettes/day)	7.3[a]
Yu *et al.* (1983)	78/78	Hepatitis		13.0 (2.2–272.4)
		Alcohol		4.2 (1.3–13.8)
		Smoking		2.6 (1.0–6.7)
		Smoking	(Alcohol ≥ 80 g/day)	14.0 (1.7–113.9)

[a] Adjusted by age, sex and either smoking or alcohol

establish comparisons, the group of heavy smokers was defined as current smokers of 20 or more cigarettes per day and the reference group consisted of nonsmokers and smokers of 1–19 cigarettes per day combined. In addition, the number of drinkers in the study group was very low. A more recent evaluation of the HBsAg-negative subjects in the same study showed a RR of 1.6 for a cumulative drinking exposure of over 65 drinks/year *versus* nondrinkers and a RR of 1.9 for smokers of 25+ pack-years *versus* nonsmokers. These estimates were adjusted for sex, age and either tobacco or alcohol use (Yu & Henderson, 1987).

Yu *et al.* (1983) published a case-control study conducted in Los Angeles, USA in which the RR for a history of hepatitis was 13.0 (95% CI, 2.2–272.4) and that for a history of blood transfusion, 7.0 (95% CI, 1.7–43.1). HBV markers were not measured in the study population. The RRs for heavy exposure to alcohol (80+ g/day *versus* 0–9 g/day) and cigarette smoking (current smokers of more than one pack/day *versus* non- and ex-smokers combined) were 4.2 (95% CI, 1.3–13.8) and 2.6 (95% CI, 1.0–6.7). A stratified analysis showed a RR of 14.0 (95% CI, 1.7–113.9) for heavy alcohol consumption (80+ g of ethanol per day) among heavy smokers (>1 pack/day), whereas the RR was only 1.8 (95% CI, 0.7–5.0) for the group of heavy smokers with low alcohol consumption (0–79 g/day). On a multivariate analysis, these significant increases in risk remained after adjusting for alcohol, tobacco and history of hepatitis and/or blood transfusion. In this study, the effect of cigarette smoking was significant for all HCC cases, of which 19% had a history of hepatitis and 29% a history of having received a blood transfusion.

Austin *et al.* (1986) reported a RR of 3.3 and a significant dose-response relationship for drinkers of 65 or more drink-years (a measure of cumulative consumption) compared to nondrinkers. No effect was found for cigarette smoking. For all subjects, the relative rate of HCC among smokers compared with nonsmokers after adjustment for alcohol was 1.0 (95% CI, 0.5–1.8). For HBsAg-negative cases, the relative rate was 1.1 (95% CI, 0.5–2.4). The risk linked to HBsAg had a lower 95% confidence limit of 3.8. HBsAg-positive HCC cases (18%) were younger than HBsAg-negative HCC cases ($p < 0.01$), and all were males ($p = 0.01$). None of the female HCC cases was HBsAg-positive. No significant difference was found between white and nonwhite HCC patients in respect to HBV status, although the proportions of HBsAg positivity were 17% for whites and 47% for nonwhites ($p = 0.08$).

Trichopoulos *et al.* (1987) reported the largest case-control study of HCC in which the independent contributions of HBV, smoking and alcohol have been assessed. This study is an expansion of a previous report in which a possible role of tobacco was first postulated for HBV-negative cases of HCC (Trichopoulos *et al.*, 1980). The RR for HCC among HBsAg carriers was 13.7 (95% CI, 8.0–23.5); and in the presence of cirrhosis it was considerably higher (30.7 *versus* 7.1 for HBsAg carriers without cirrhosis). A significant dose-response effect of cigarette smoking was found among HBsAg-negative HCC cases, which remained after adjusting for alcohol consumption. For HBsAg-positive cases, a nonsignificant increased risk for HCC was found, but there was no clear dose-response relationship. No effect was found for alcohol. Neither of the two factors had a significant effect on risk in the presence of HBsAg. In this hospital-based study, 42% of HCC cases had concurrent cirrhosis — lower than the proportion found in other European series. For example, in the UK, about 70% of HCC patients have concurrent cirrhosis (Melia *et al.*, 1984). Twenty per cent of the cases and 14% of the controls were in the high alcohol consumption group (70+ g of ethanol per day, equivalent to one litre of wine). The prevalence of heavy drinkers (80+ g per day) in Italy, Spain, Switzerland and France is two to three times higher than in Greece (G. Pequignot *et al.*, unpublished results).

Two other studies have investigated the effects of cigarette smoking and/or alcohol intake, but markers of HBV were not determined. One study showed an effect of alcohol consumption on HCC and a small effect of tobacco, which disappeared after adjusting for alcohol (Hardell *et al.*, 1984); in the other report, no significant effect was identified, although a nonsignificant dose-response for alcohol was found (Stemhagen *et al.*, 1983).

The evaluation of the joint effects of tobacco, alcohol and HBV can be affected by study design, particularly by the source of cases and controls, the prevalence of the risk factors and the size of the study. These methodological problems might explain some of the discrepancies seen. The finding of Austin *et al.* (1986) of a significant difference in the prevalence of HBsAg among HCC cases according to sex, age and perhaps race, indicates that these factors should be taken into account in any future study. The evidence for an effect of cigarette smoking on HCC seems to be convincing for HBsAg-negative cases. The results obtained among HBsAg-positive cases are compatible with an increase in risk added by cigarette smoking; however, because the association of HCC with HBsAg is so strong, more powerful studies than the ones so far conducted might be required to detect the effect of any other additional risk factor.

Oral contraceptives

The occurrence of benign liver adenomas among oral contraceptive users has been reported repeatedly as case reports, and an increased risk among users has been found in two case-control studies. Case reports of HCC among oral contraceptive users have also been published (IARC, 1979). More recently, three case-control studies of HCC have been reported; the results are summarized in Table 2. A case-control study based on 11 cases showed a strong association between average duration of oral contraceptive use and HCC (Henderson *et al.*, 1983). Two other case-control studies showed that the risk for HCC among oral contraceptive users was significantly increased, had a dose-response relationship with duration of exposure and was present in women with HCC who were either positive or negative for HBV markers (Forman *et al.*, 1986; Neuberger *et al.*, 1986).

An analysis of the time-trends in mortality from HCC in England and Wales showed a small but consistent increase in young women in the last decade. However, the trend was not found for other countries with similar patterns of oral contraceptive use (Forman *et al.*, 1983). The overall evidence so far available appears to indicate an association between use of oral contraceptives and HCC. However, these data should be interpreted with caution in view of the relatively small size of the studies involved and the resulting impossibility of controlling for possible confounders.

Prospects for epidemiological research on HCC

The role of chronic HBV infection in the causation of HCC has been demonstrated convincingly in different countries and among different populations. HBV infection

Table 2. Oral contraceptives and HCC: case-control studies

Reference	No. of cases	Years of use	RR (95% CI)
Henderson *et al.* (1983)	11	5.4 *vs.* 1.6	13.5 (1.2–152.2)
Neuberger *et al.* (1986)	26	≥8	4.4 (1.5–12.8)
	22[a]	≥8	7.2 (2.0–25.7)
Forman *et al.* (1986)	19	ever	3.8[b] (1.0–14.6)
		≥8	20.1 (2.3–175.7)

[a]Excluding four women with HBV markers
[b]Adjusted for age at diagnosis and year of birth

seems to be responsible for 5–95% of HCC cases; however, only a fraction of the HBsAg carriers in high-risk countries develop HCC (less than 10%) over their average life span. Therefore, although it is perhaps the second most important human carcinogen, viral infection is neither a sufficient nor a necessary cause of HCC.

The role of chemicals in the causation of HCC is a relevant area of research because it can help in understanding the carcinogenic process in the liver and because it might provide opportunities for HCC prevention other than HBV vaccination.

Exposure to oral contraceptives, cigarette smoking and alcohol consumption is reliably assessed in most epidemiological studies by means of questionnaires and/or medical records. In contrast, assessment of aflatoxin exposure based on measures of aflatoxin metabolites or adducts in biological specimens can be done at the individual level only for exposures taking place a few days to a few weeks prior to sampling. Therefore, one has to rely on repeated measurements of short-term aflatoxin markers in individuals followed up prospectively. Pilot studies should be undertaken to help to establish the best methods for monitoring aflatoxin exposure in epidemiological field work. For example, animal experiments indicate that aflatoxin-albumin adducts closely reflect aflatoxin binding to hepatocytes (Wild *et al.*, 1986), but these adducts have not yet been measured in humans exposed acutely to known quantities of aflatoxin. While these methods are being developed, the relevant specimens could be collected and stored at −70°C for a defined cohort of subjects in whom exposure to all known risk factors for HCC is being determined.

Such a study has now been initiated in Thailand, with IARC support. The main objective of the study is to determine the roles of HBV, aflatoxin, tobacco, alcohol, steroid hormones and *N*-nitroso compounds in the development of HCC and other chronic liver diseases, such as chronic hepatitis and cirrhosis. The study design is a prospective five-year follow-up of approximately 3000 male HBsAg carriers aged over 30. The cohort has been identified from among first-time blood donors between 1985–1987 and army recruits. Follow-up will be done yearly. On the first and last visits, a questionnaire, a physical examination and blood and urine samples will be collected; ultrasound scanning of the liver will be performed at each visit. Follow-up visits will include all the above except the questionnaire. At the end of the five-year period, incident HCC cases occurring in this cohort will be compared with those in an appropriate set of controls. The questionnaires will assess exposures to cigarettes and alcohol, as well as crude estimates of intake of aflatoxins and *N*-nitroso compounds. Blood analysis will allow comparisons of HBV markers with aflatoxin adducts to albumin and other aflatoxin markers that might become available by the end of the study period, of haemoglobin adducts to other chemicals, of profiles of steroid hormones and of α-fetoprotein levels. Urinary analysis will allow comparisons of the excretion of certain *N*-nitroso compounds, aflatoxin metabolites and 3-methyladenine adducts. The cohort comprises HBsAg carriers only, so the results will provide information specifically on the role of aflatoxin in promoting the HBsAg chronic carrier state to more advanced liver conditions, including HCC, in the presence or absence of several other chemicals. Prospective monitoring of cohorts of non-HBsAg carriers with different levels of aflatoxin exposure could be envisaged in order to assess the effects of aflatoxin on HCC risk in the absence of HBV. However, this approach would require very large sample sizes in order to yield a significant number of cases, since the proportion of HBsAg-negative HCC cases is sizeable (40–60%) in most countries where both HCC incidence and aflatoxin in the diet are low.

Studies to clarify the role and mechanism of action of smoking, alcohol and oral contraceptives are needed. These studies should provide detailed quantitative information on exposure to cigarette smoking, other forms of tobacco and alcohol, including information on age at first exposure and time since last exposure among individuals who have abandoned these habits. The study design could be case-control studies with emphasis on these aspects or follow-up studies of cohorts of individuals who have given up smoking and/or drinking. Sample size and length of follow-up would have to be estimated according to the magnitude of risk provided by current results. Considering that the number of women using oral contraceptives is enormous in developed countries and is increasing in some developing countries, the assessment of any increases in risk should be viewed as an important public health issue on which further research is needed. In order to evaluate the role of oral contraceptives in countries with low risk for HCC, large collaborative case-control studies could be undertaken. In high-risk countries for HCC, studies of smaller size can be envisaged, such as case-control studies of HCC or follow-up of oral contraceptive users. In such studies, the role of aflatoxin should be assessed simultaneously.

Finally, intervention studies would provide valuable information on the effects of eliminating or reducing exposure to HBV, aflatoxin, smoking and alcohol on the HCC risk.

References

Armstrong, B. (1980) The epidemiology of cancer in the People's Republic of China. *Int. J. Epidemiol., 9,* 305–315

Austin, H., Delzell, E., Grufferman, S., Levine, R., Morrison, A.S., Stolley, P.D. & Cole, P. (1986) A case-control study of hepatocellular carcinoma and the hepatitis B virus, cigarette smoking, and alcohol consumption. *Cancer Res., 46,* 962–966

Beasley, R.P. (1982) Hepatitis B virus as the etiologic agent in hepatocellular carcinoma—epidemiologic considerations. *Hepatology, 2,* 21S–26S

Blumberg, B.S. & London, W.T. (1985) Hepatitis B virus and prevention of primary cancer of the liver. *J. natl Cancer Inst., 74,* 267–273

Bradshaw, E., McGlashand, N.D., Fitzgerald, D. & Harington, J.S. (1982) Analyses of cancer incidence in black gold miners from Southern Africa (1964–79). *Br. J. Cancer, 46,* 737–748

Bulatao-Jayme, J., Almero, E.M., Castro, C.A., Jardeleza, T.R. & Salamat, L.A. (1982) A case-control dietary study of primary liver cancer risk from aflatoxin exposure. *Int. J. Epidemiol., 11,* 112–119

Busby, W.F. & Wogan, G.N. (1984) *Aflatoxins.* In: Searle, C.E., ed., *Chemical Carcinogens,* 2nd ed., Washington DC, American Chemical Society, pp. 945–1136

De Campos, M. & Olszyma-Marzys, A.E. (1979) Aflatoxin contamination in grains and grain products during the dry season in Guatemala. *Bull. environ. Contam. Toxicol., 22,* 350–356

Forman, D., Doll, R. & Peto, R. (1983) Trends in mortality from carcinoma of the liver and the use of oral contraceptives. *Br. J. Cancer, 48,* 349–354

Forman, D., Vincent, T.J. & Doll, R. (1986) Cancer of the liver and the use of oral contraceptives. *Br. med. J., 292,* 1357–1361

Garfinkel, L. (1980) Cancer mortality in non-smokers. Prospective study by the American Cancer Society. *J. natl Cancer Inst., 65,* 1169–1173

Garner, C., Ryder, R. & Montesano, R. (1985) Monitoring of aflatoxins in human body fluids and application to field studies. *Cancer Res., 45,* 922–928

Groopman, J.D., Busby, W.F., Donahue, P.R. & Wogan, G.N. (1986) *Aflatoxins as risk factors for liver cancer: an application of monoclonal antibodies to monitor human exposure.* In: Harris, C.C., ed., *Biochemical and Molecular Epidemiology of Cancer,* New York, Alan R. Liss, pp. 233–256

Hardell, L., Bengtsson, N.O., Jonsson, U., Eriksson, S. & Larsson, L.G. (1984) Aetiological aspects on primary liver cancer with special regard to alcohol, organic solvents and acute intermittent porphyria—an epidemiological investigation. *Br. J. Cancer, 50,* 389–397

Harington, J.S., McGlashan, N.D., Bradshaw, E., Geddes, E.W. & Purves, L.R. (1975) A spatial and temporal analysis of four cancers in African gold miners from Southern Africa (1964–79). *Br. J. Cancer, 31,* 665–678

Henderson, B.E., Preston-Martin, S., Edmondson, H.A., Peters, R.L. & Pike, M.C. (1983) Hepatocellular carcinoma and oral contraceptives. *Br. J. Cancer, 48,* 437–440

Hirayama, T. (1981) *A large-scale cohort study on the relationship between diet and selected cancers of digestive organs.* In: Bruce, W.R., Correa, P., Lipkin, M., Tannenbaum, S. & Wilkins, T., eds, *Gastrointestinal Cancer: Endogenous Factors* (*Banbury Report* 7), Cold Spring Harbor, NY, CSH Press, pp. 409–426

IARC (1979) *IARC Monographs on the Evaluation of the Carcinogenic Risk of Chemicals to Humans,* Vol. 21, *Sex Hormones* (*II*), Lyon, pp. 114–115

Kew, M.C., Desmyter, J., Bradburne, A.F. & Macnab, G.M. (1979) Hepatitis B virus infection in southern African blacks with hepatocellular cancer. *J. natl Cancer Inst., 62,* 517–520

Lam, K.C., Yu, M.C., Leung, J.W.C. & Henderson, B.E. (1982) Hepatitis B virus and cigarette smoking: risk factors for hepatocellular carcinoma in Hong Kong. *Cancer Res., 42,* 5246–5248

Lingao, A.L., Domingo, E.O. & Nishioka, K. (1981) Hepatitis B virus profile of hepatocellular carcinoma in the Philippines. *Cancer, 48,* 1590–1595

Melbye, M., Skinhøj, P., Højgaard Nielsen, N., Vestergaard, B.F., Ebbesen, P., Hart Hansen, J.P. & Biggar, R.J. (1984) Virus-associated cancers in Greenland: frequent hepatitis B virus infection but low primary hepatocellular carcinoma incidence. *J. natl Cancer Inst., 73,* 1267–1272

Melia, W.M., Wilkinson, M.L., Portmann, B.C., Johnson, P.J. & Williams, R. (1984) Hepatocellular carcinoma in the non-cirrhotic liver: a comparison with that complicating cirrhosis. *Q. J. Med., 53,* 391

Muñoz, N. & Bosch, F.X. (1987) *Epidemiology of hepatocellular carcinoma.* In: Okuda, K. & Ishak, K.G., eds, *Neoplasms of the Liver,* Tokyo, Springer, pp. 3–19

Muñoz, N. & Linsell, A. (1982) *Epidemiology of primary liver cancer.* In: Correa, P. & Haenszel, W., eds, *Epidemiology of Cancer of the Digestive Tract,* The Hague, Nijhoff, pp. 161–195

Neuberger, J., Forman, D., Doll, R. & Williams, R. (1986) Oral contraceptives and hepatocellular carcinoma. *Br. med. J., 292,* 1355–1357

Nishioka, K. (1985) Hepatitis B virus and hepatocellular carcinoma: postulates for an etiological relationship. *Adv. viral Oncol., 5,* 173–199

Ochomogo, M.G. (1979) Detection and detoxification of aflatoxin in corn and peanuts. *Diss. Abstr. Int., 39,* 3742

Oshima, A., Tsukuma, H., Hiyama, T., Fujimoto, I., Yamano, H. & Tanaka, M. (1984) Follow-up study of HBsAg positive blood donors with special reference to effect of drinking and smoking on development of liver cancer. *Int. J. Cancer, 34,* 775–779

Parkin, D.M. (1986) *Cancer Occurrence in Developing Countries* (*IARC Scientific Publications No. 75*), Lyon, International Agency for Research on Cancer, pp. 133–189.

Parkin, D.M., Stjernswärd, J. & Muir, C.S. (1984) Estimates of the worldwide frequency of twelve major cancers. *Bull. World Health Organ., 62,* 163–182

Peers, F.G. & Linsell, C.A. (1973) Dietary aflatoxins and liver cancer. A population based study in Kenya. *Br. J. Cancer, 27,* 473–484

Peers, F.G., Gilman, G.A. & Linsell, C.A. (1976) Dietary aflatoxins and human liver cancer. A study in Swaziland. *Int. J. Cancer, 17,* 167–176

Peers, F., Bosch, X., Kaldor, J., Linsell, A. & Pluijmen, M. (1987) Aflatoxin exposure, hepatitis B virus infection and liver cancer in Swaziland. *Int. J. Cancer, 39,* 545–553

Prince, A.M., Szmuness, W., Michon, J., Desmaille, J., Diebolt, G., Linhard, J., Quenum, C. & Sankale, M. (1975) A case-control study of the association between primary liver cancer and hepatitis B infection in Senegal. *Int. J. Cancer, 16,* 376–383

Rothman, K.J. (1980) The proportion of cancer attributable to alcohol consumption. *Prev. Med., 9,* 174–179

Sabbioni, G., Skipper, P.L., Büchi, G. & Tannenbaum, S.R. (1987) Isolation and characterization of the major serum albumin adduct formed by aflatoxin B_1 *in vivo* in rats. *Carcinogenesis, 8,* 819–824

Shank, R.C., Gordon, J.E., Wogan, G.N., Nondasuta, A. & Subhamani, B. (1972) Dietary aflatoxins and human liver cancer: III. Field survey of rural Thai families for ingested aflatoxins. *Food Cosmet. Toxicol., 10,* 71–84

Skinhøj, P., Hart Hansen, J.P., Højgaard Nielsen, N. & Mikkelsen, F. (1978) Occurrence of cirrhosis and primary liver cancer in an Eskimo population hyperendemically infected with hepatitis B virus. *Am. J. Epidemiol., 108,* 121–125

Stemhagen, A., Slade, J., Altman, R. & Bill, J. (1983) Occupational risk factors and liver cancer. A retrospective case-control study of primary liver cancer in New Jersey. *Am. J. Epidemiol., 117,* 443–454

Stoloff, L. (1983) Aflatoxin as a cause of primary liver-cell cancer in the United States: a probability study. *Nutr. Cancer, 5,* 165–186

Sun, T.T. & Chu, Y.Y. (1984) Carcinogenesis and prevention strategy of liver cancer in areas of prevalence. *J. cell. Physiol., Suppl. 3,* 39–44

Sun, T.T., Chu, Y.R., Hsia, C.C., Wei, Y.P. & Wu, S.M. (1986) *Strategies and current trends of etiologic prevention of liver cancer.* In: Harris, C.C., ed., *Biochemical and Molecular Epidemiology of Cancer,* New York, Alan R. Liss, pp. 283–292

Szmuness, W. (1978) Hepatocellular carcinoma and the hepatıtis B virus: evidence for a causal association. *Prog. med. Virol., 24,* 40–69

Trichopoulos, D., Tabor, E., Gerety, R.J., Xirouchaki, E., Sparros, L., Muñoz, N. & Linsell, A. (1978) Hepatitis B and primary hepatocellular carcinoma in a European population. *Lancet, ii,* 1217–1219

Trichopoulos, D., MacMahon, B., Sparros, L. & Merikas, G. (1980) Smoking and hepatitis B-negative primary hepatocellular carcinoma. *J. natl Cancer Inst., 65,* 111–114

Trichopoulos, D., Kremastinou, J. & Tzonou, A. (1982) *Does hepatitis B virus cause hepatocellular carcinoma*? In: Bartsch, H. & Armstrong, B., eds, *Host Factors in Human Carcinogenesis* (*IARC Scientific Publications No. 39*), Lyon, International Agency for Research on Cancer, pp. 317–332

Trichopoulos, D., Day, N.E., Kaklamani, E., Tzonou, A., Muñoz, N., Zavitsanos, X., Koumantaki, Y. & Trichopoulou, A. (1987) Hepatitis B virus, tobacco smoking and ethanol consumption in the etiology of hepatocellular carcinoma. *Int. J. Cancer, 39,* 45–49

Tuyns, A. (1980) *Alcohol.* In: Schottenfeld, D. & Fraumeni, J.F., eds, *Cancer Epidemiology and Prevention,* Philadelphia, Saunders, pp. 293–303

Van Rensburg, S.J., Cook-Mozaffari, P., van Schalkwyk, D.J., van der Watt, J.J., Vincent, T.J. & Purchase, I.F. (1985) Hepatocellular carcinoma and dietary aflatoxin in Mozambique and Transkei. *Br. J. Cancer, 51,* 713–726

Varsavsky, E. & Sommer, S.E. (1977) Determination of aflatoxins in peanuts. *Ann. Nutr. Aliment., 31,* 539–544

Wild, C., Garner, R.C., Montesano, R. & Tursi, F. (1986) Aflatoxin B_1 binding to plasma albumin and liver DNA upon chronic administration ro rats. *Carcinogenesis, 7,* 853–858

Yaobin, W., Lizun, L., Benfa, Y., Yaochu, X., Yunyuan, L. & Wenguang, L. (1983) Relation between geographical distribution of liver cancer and climate — aflatoxin B_1 in China. *Sci. Sin* (*Ser. B*), *26,* 1166–1175

Yarrish, R.L., Werner, B.G. & Blumberg, B.S. (1980) Association of hepatitis B virus infection with hepatocellular carcinoma in American patients. *Int. J. Cancer, 26,* 711–715

Yeh, F.S., Mo, C.C., Luo, S., Henderson, B.E., Tong, M.J. & Yu, M.C. (1985a) A serological case-control study of primary hepatocellular carcinoma in Guangxi, China. *Cancer Res., 45,* 872–873

Yeh, F.S., Mo, C.C. & Yen, R.C. (1985b) Risk factors for hepatocellular carcinoma in Guangxi, People's Republic of China. *Natl Cancer Inst. Monogr., 69,* 47–48

Yu, M.C. & Henderson, B. (1987) Correspondence re: Harland Austin *et al.* A case-control study of hepatocellular carcinoma and the hepatitis B virus, cigarette smoking and alcohol consumption. *Cancer Res., 46:* 962–966, 1986. *Cancer Res., 47,* 645–655

Yu, M.C., Mack, T., Hanisch, R., Peters, R.L., Henderson, B.E. & Pike, M.C. (1983) Hepatitis, alcohol consumption, cigarette smoking and hepatocellular carcinoma in Los Angeles. *Cancer Res., 43,* 6077–6079

DNA RESTRICTION FRAGMENT LENGTH POLYMORPHISM ANALYSIS OF HUMAN BRONCHOGENIC CARCINOMA

J.C. Willey[1], A. Weston[1], A. Haugen[1], T. Krontiris[2], J. Resau[3], E. McDowell[3], B. Trump[3] & C.C. Harris[1]

[1]*Laboratory of Human Carcinogenesis, Division of Cancer Etiology, National Cancer Institute, National Institutes of Health, Bethesda, MD;* [2]*Tufts-New England Medical Center, Boston, MA; and* [3]*Department of Pathology, University of Maryland Medical Center, Baltimore, MD, USA*

DNA restriction fragment length polymorphism (RFLP) analysis has been successfully applied both to evaluate possible human susceptibility factors as well as to identify genes involved in malignant transformation of human cells. In this report we review previous applications of RFLP analysis to evaluation of human malignancies and discuss our RFLP studies of human bronchogenic carcinoma, which are in progress. Preliminary results of our analysis of the RFLP associated with the cHa *ras* variable tandem repeat (VTR) indicate that rare alleles of this VTR are more frequent in patients with bronchogenic carcinoma than in controls.

Bronchogenic carcinoma is the most common cause of cancer mortality in both men and women in the USA (Silverberg & Lubera, 1987). Epidemiological data have made the causative association between cigarette smoking and bronchogenic carcinoma abundantly clear (US Public Health Service, 1982), yet some important questions regarding the etiology of this disease remain (Harris, 1987), including why some heavy cigarette smokers do not develop bronchogenic carcinoma and some nonsmokers do. In addition, while it is clear that many agents in cigarette smoke may damage DNA through a variety of mechanisms, it is not clear which genes must be altered in order for a cell to be malignantly transformed. Because bronchogenic carcinoma generally occurs later in life and usually after many years of exposure to xenobiotics, it has been proposed that a progenitor epithelial cell is transformed only after multiple genes have been altered (Ashley, 1969; Fraumeni, 1975; Doll, 1978), and, further, that perhaps a particular combination of activated proto-oncogenes and inactivated 'tumour suppressor' genes is involved (Klein & Klein, 1986). Therefore, if a number of genes are postulated to control bronchial epithelial cell proliferation, and an individual inherits one or two abnormal ones, the individual may be at increased risk for developing bronchogenic carcinoma without manifesting an easily discernible inherited susceptibility pattern. In agreement with this, epidemiological studies of bronchogenic carcinoma as well as other common adult tumours have found a small but consistent two- to four-fold increase in cancer incidence in the families of cancer patients, when other factors are controlled (Anderson, 1975).

One approach to improving our understanding of mechanisms involved in malignant transformation has been to study hereditary conditions of increased cancer incidence. Several such conditions have been described (Table 1; Ponder, 1984; Schneider *et al.*, 1986). These diseases may be transmitted by single-gene, polygenic or chromosomal mechanisms. Some hereditary conditions predispose only one type of

Table 1. Examples of inherited disorders associated with an increased risk of cancer[a]

Constitutional chromosomal abnormalities
  Down's syndrome
  Klinefelter's syndrome

Mendelian traits
(1) Inherited cancer syndromes
    Retinoblastoma
    Familial polyposis coli
    Multiple endocrine neoplasia syndromes II, III
    Wilms' tumour
(2) Inherited preneoplastic states
    DNA repair defects and chromosomal instability syndromes: xeroderma pigmentosum; Fanconi's anaemia; Bloom's syndrome; ataxia telangiectasia; Werner's syndrome
    Disturbances of tissue organization
      Hamartomatous syndromes: Peutz-Jegher's; Cowden; neurofibromatosis
      Other conditions in which disturbance of tissue proliferation, differentiation or organization is associated with increased cancer risk: disorders of skin keratinization (?); α-1-antitrypsin deficiency
    Immune deficiency syndromes
    Metabolic variation: albinism; aryl hydrocarbon hydroxylase inducibility (?); variations in oestrogen metabolism

Multifactorial predisposition
  Ethnic cancer differences
  Familial cancer aggregations

[a]From Ponder (1984)

tissue to neoplasia (e.g., actinic keratosis predisposes to basal-cell carcinomas of the skin and familial polyposis coli predisposes to tumours of the gastrointestinal tract), while others predispose to tumours in different tissues (e.g., hereditary retinoblastoma, Bloom's syndrome and ataxia telangiectasia).

Description of restriction fragment length polymorphism analysis

New experimental approaches to identifying people genetically predisposed to cancer are being evaluated, e.g., restriction enzyme DNA fragment length analysis of genetic polymorphism (RFLP). Until recently, genetic predisposition could be assessed only by measuring gene products, e.g., histocompatibility antigens on cell surfaces, isoenzymes and pharmacogenetic phenotypes. Advances in molecular biology now make it possible to measure genetic polymorphism at the DNA level, which greatly expands the potential for assays of polymorphisms in biomedical research. Genetic polymorphism can be measured by restriction enzyme digestion and DNA hybridization (Fig. 1; White *et al.*, 1983). RFLPs are of two types — site or insertion/deletion polymorphism. In the case of site polymorphism, the recognition site for a given restriction enzyme in a particular region of DNA either appears or disappears as the result of point mutation. Accordingly (Fig. 1), in the site-absent genotype, one large fragment instead of two smaller fragments is observed. Conversely, in the site-present genotype, three fragments (one of expected size and two smaller) instead of two fragments are seen. In the case of insertion/deletion polymorphism, variation in fragment length is the result of insertion or deletion of DNA sequences and can be

Fig. 1. DNA polymorphism as detected by restriction enzyme digestion, DNA electrophoresis and DNA hybridization

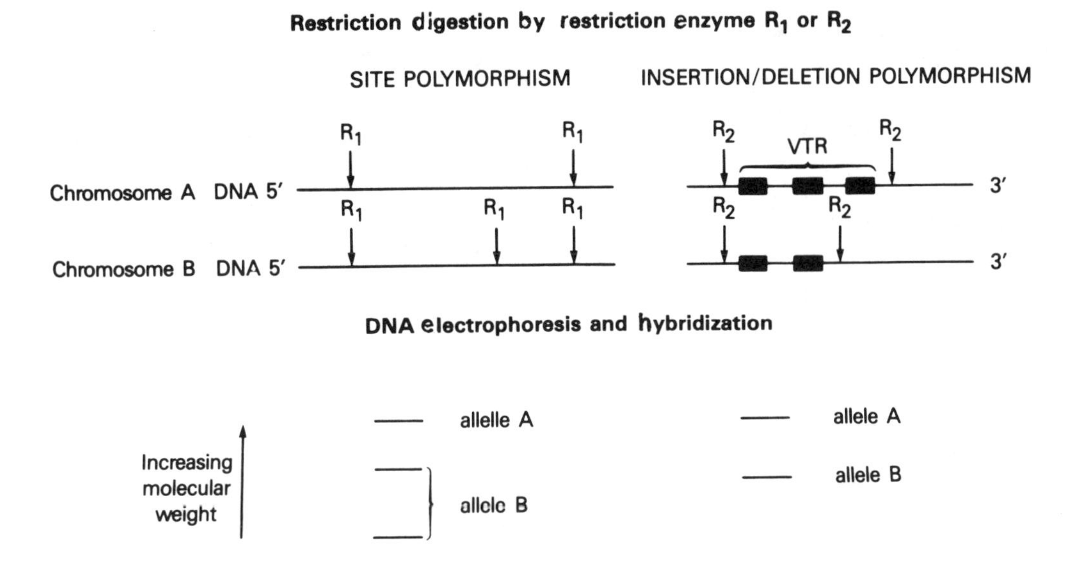

A site polymorphism is determined by the presence or absence of a restriction site and is therefore detected only by restriction enzymes that recognize that site. An insertion/deletion polymorphism is determined by the presence of varying numbers of copies of a repeat sequence (solid boxes) and may be detected by any restriction enzymes that have sites relatively close to either side of the variable tandem repeat (VTR)

detected by any restriction endonuclease that possesses recognition sites that tightly span the region of sequence alteration. Because the insertions or deletions can assume a continuum of lengths, multiple alleles are possible.

RFLP detection of hereditary susceptibility to cancer

This approach has already proven beneficial for identifying individuals with a genetic predisposition to a variety of diseases, including retinoblastoma, noninsulin-dependent diabetes mellitus, Huntington's disease, and haemoglobinopathies (Francomano & Kazazian, 1986). The same molecular approach, using specific DNA probes, e.g., for oncogenes and ectopic hormones, is now being studied to determine its potential for identifying persons with a genetic predisposition to cancer. The human Ha-*ras* proto-oncogene may prove useful for these studies because it contains a hypervariable insertion/deletion polymorphism. The molecular basis for this insertion/deletion polymorphism is a region of 30 to 100 tandem nucleotide repeats with a 28-base pair consensus sequence aligned head-to-tail that is approximately one kilobase downstream from the structural gene of Ha-*ras*. The function of this variable tandem repeat (VTR) is unknown, but it may have gene enhancer activity and appears to be inherited in a Mendelian fashion. It has been determined that there are four common alleles, a_1–a_4, and numerous alleles of intermediate or rare occurrence in the population (Fig. 2). An increased frequency of rare alleles at this locus has been found in certain cancer patients, including those with bladder carcinoma (Krontiris *et al.*, 1986) and breast carcinoma (Lidereau *et al.*, 1986). Preliminary data indicate that rare

Fig. 2. Basis for cHa-*ras* polymorphism

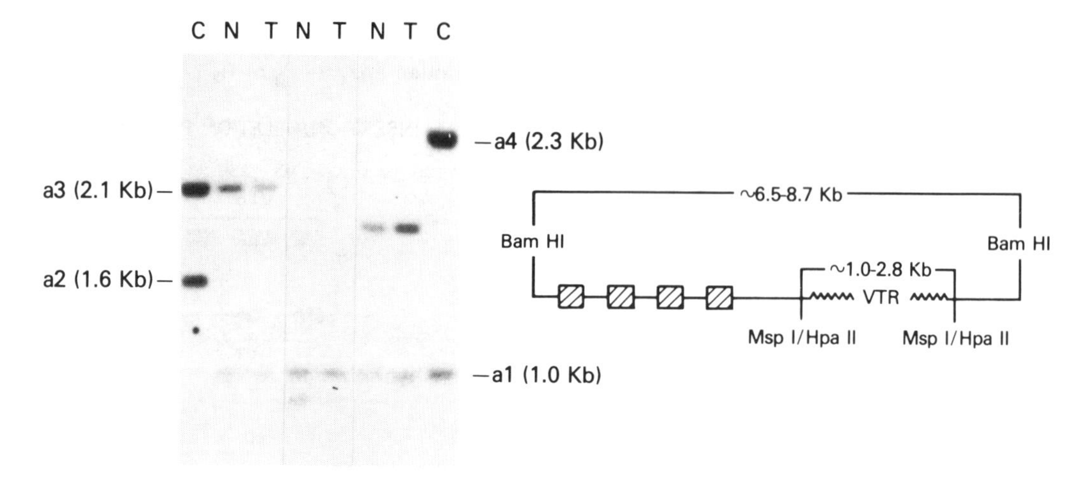

The VTR region associated with cHa-*ras* is located 3′ to the structural gene. There are four common alleles, a_1, a_2, a_3 and a_4. The autoradiogram includes control DNA (lanes C) from patients that had the alleles a_2 and a_3 or the alleles a_1 and a_4, and DNA from three matched pairs of human normal and bronchogenic carcinoma DNA; N, DNA from normal tissue; T, DNA from tumour tissue. In the first case, the patient possessed alleles a_1 and a_3; in the second case, allele a_1 and a smaller rare allele; and in the third case, allele a_1 and a rare allele

alleles are also more frequent in patients with bronchogenic carcinoma (A. Weston *et al.*, unpublished). Evidence for (Krontiris *et al.*, 1985) and against (Gerhard *et al.*, 1987) the involvement of Ha-*ras* in sporadic and familial melanoma has been reported.

Evidence for tumour suppressor genes

Most known hereditary causes of tumours were discovered because they manifested as dramatic clinical syndromes with a Mendelian pattern of inheritance. In some of these diseases, an increased rate of spontaneous mutation has been observed to be associated with chromosomal instability. This instability may be due to defective DNA repair in the case of xeroderma pigmentosum (Cleaver & Bootsma, 1975) or to defective ligase activity in the cases of Fanconi's anaemia and Bloom's syndrome (Hirsch-Kauffmann *et al.*, 1978; Willis & Lindahl, 1987). In contrast, for some diseases that confer hereditary susceptibility to childhood tumours, such as retinoblastoma, analysis of certain epidemiological data has suggested a 'two-hit' mutation model of carcinogenesis (Moolgavkar & Knudson, 1981) (Fig. 3). According to this model, genetic predisposition may be due to inheritance of the first autosomal mutation, so that only one additional genetic event is required. Therefore, these individuals may already have cells predisposed to develop cancer due to a germline genetic lesion. By inference, the gene(s) involved may play a role in cellular homeostasis, including controlling proliferation, terminal differentiation and angiogenesis. In retinoblastoma, there is now substantial cytogenetic and molecular evidence for a putative 'tumour suppressor' gene located on the long arm of chromosome 13 (13q) (Murphree & Benedict, 1985). RFLP techniques have been used both to link the hereditary susceptibility for retinoblastoma to other genes previously mapped to the long arm of

Fig. 3. Two-hit mutation model of carcinogenesis

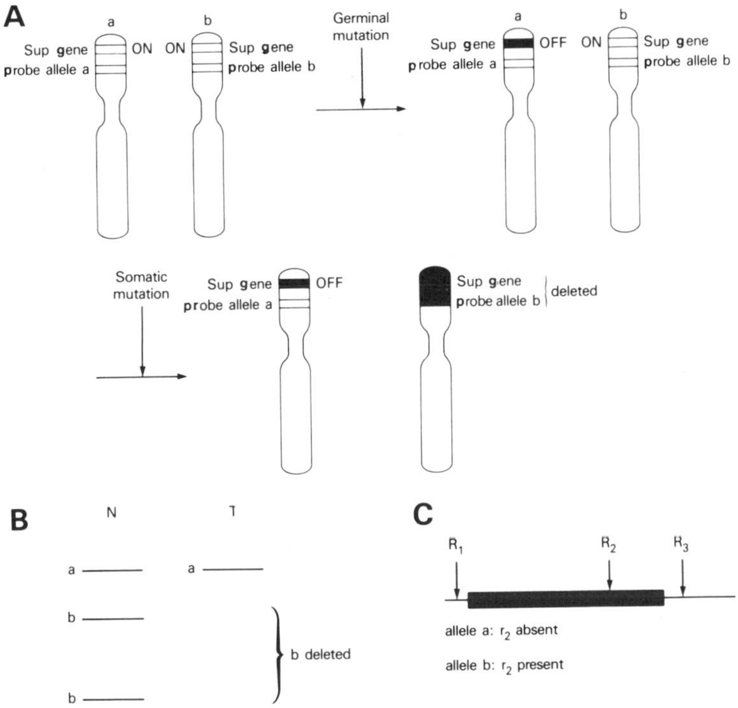

A, An individual inherits the first mutated gene that controls nephroblast cell proliferation, and the second allele is inactivated through a somatic cell event, in this case a deletion of a portion of the short arm of chromosome b. B, If either the germinal or the somatic mutation event involves chromosome nondisjunction or a deletion on the short arm of chromosome 11 (11p), it may be detected by digesting genomic DNA from the affected individual with the restriction enzyme R, and after electrophoresis and Southern blotting, hybridizing with a ^{32}P-labelled DNA probe. If a deletion of the short arm, including the locus homologous to the probe, has occurred on one of the chromosomes, the allele on that chromosome, in this case allele b, will not be represented on the autoradiogram taken of the hybridized Southern blot. C, An R site absent in allele a and present in allele b explains bands observed in B

chromosome 13, and to find somatic mutations on 13q that are specific to retinoblastoma tumour cells by comparing them with normal cells from the same patient. Possible mechanisms to produce somatic mutation at the retinoblastoma locus in tumour cells include (1) chromosomal nondisjunction, (2) nondisjunction and reduplication, (3) mitotic recombination with gene rearrangement, (4) deletion, (5) gene inactivation and (6) mutation (Cavenee *et al.*, 1983; Fig. 4). Mutations caused by mechanisms 1–4 may be detected by RFLP analysis as a reduction from heterozygosity to homozygosity (1 and 2) or hemizygosity (4) or a change in the band pattern due to gene rearrangement (3).

Further indirect evidence for these presumed dominant-acting 'tumour suppressor' genes includes (1) karyotypic and RFLP data for a reduction to homozygosity or hemizygosity on the short arm of chromosome 11 in Wilms' tumour (Fearon *et al.*, 1984; Koufos *et al.*, 1984; Orkin *et al.*, 1984; Reeve *et al.*, 1984) and bladder cancer (Fearon *et al.*, 1985), on chromosome 22 in acoustic neuroma (Seizinger *et al.*, 1986)

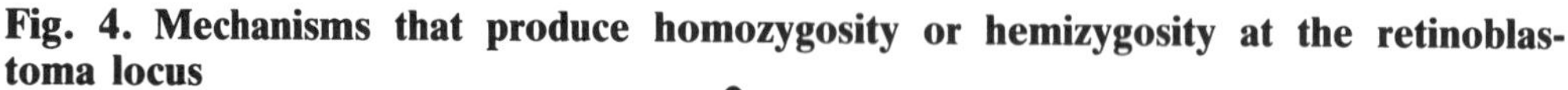

Fig. 4. Mechanisms that produce homozygosity or hemizygosity at the retinoblastoma locus

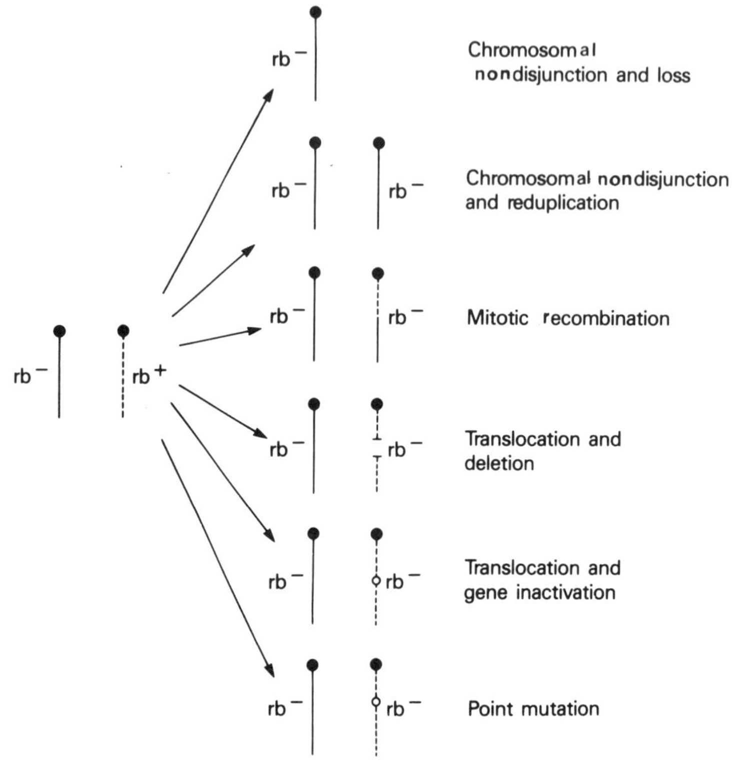

An individual has inherited only one active allele at the retinoblastoma locus. Inactivation at the second allele in a somatic cell, leading to loss of replication control, may occur by any of the illustrated mechanisms.

and on chromosome 3p in renal-cell carcinoma and small-cell carcinoma of the lung (Zbar *et al.*, 1987; Naylor *et al.*, 1988; Table 2) and (2) evidence from studies with human cell hybrids (Stanbridge *et al.*, 1983; Sager, 1985; Kaelbling & Klinger, 1986); e.g., Stanbridge and coworkers (Geiser *et al.*, 1986) have reported suppression of tumorigenicity with continued expression of the cHa-*ras* oncogene in EJ bladder carcinoma-human fibroblast hybrid cells. This finding suggests that, even in the presence of an activated proto-oncogene, 'tumour suppressor' genes are 'dominant-acting'. In addition, the same group has reported that introduction of normal human

Table 2. Chromosomal location of putative suppressor genes for specific human malignancies

Malignancy	Chromosome location	Reference
Retinoblastoma	13q14.1	Friend *et al.* (1986) Lee *et al.* (1987)
Osteosarcoma	13q14.1	Fung *et al.* (1987)
Wilms' tumour	11p13	Riccardi *et al.* (1987)
Renal-cell carcinoma	3p11–21	Zbar *et al.* (1987)
Small-cell carcinoma	3p11–21	Whang-Peng *et al.* (1982)
Neuroblastoma	22	Seizinger *et al.* (1987)
Bladder carcinoma	11p	Fearon *et al.* (1985)

chromosome 11 into Wilms' tumour cells *via* microcell transfer controls tumorigenicity (Weissman *et al.*, 1987). Recently, a DNA sequence with properties of the Rb gene has been isolated, loss of which correlates with the development of retinoblastoma and osteosarcoma (Friend *et al.*, 1986; Fung *et al.*, 1987; Lee *et al.*, 1987). One step towards proving that this isolated gene in fact suppresses tumorigenicity will be to conduct gene transfection studies and to observe resumption of growth control in the transfected cells.

RFLP detection of somatic mutations in bronchogenic carcinoma

RFLP analysis may prove valuable for locating somatic gene mutations involved in bronchial epithelial cell transformation, as it has for retinoblastoma. An important difference in the approach to investigation of hereditary tumours from that for common solid tumours is that, because there is no clear hereditary component in most cases of the latter, it has not been possible to conduct linkage analysis studies in order to localize on the chromosome the gene or genes responsible. Therefore, selection of probes to be used in RFLP analysis of bronchogenic carcinoma is based on other criteria, including proximity to chromosomal areas associated with other types of cancer (Table 1), evidence from somatic cell hybridization and gene transfection studies (Yoakum *et al.*, 1985), and, in some cases, such as small-cell carcinoma of the lung, cytogenetic evidence (Whang-Peng *et al.*, 1982). In addition, it is important to select probes with a high polymorphic information content (PIC). The PIC value is dependent on the number of alleles revealed by the probe and their relative frequency in the gene pool, and it refers to the likelihood that genomic DNA from any individual, digested with a particular restriction enzyme, will be heterozygous at the region probed. The higher the PIC the more likely it is that an individual will be heterozygous. VTR probes invariably have a high PIC because they have many alleles. As our initial probes we chose six DNA fragments localized on chromosome 11 (HRAS, INS, CALC, CAT, INT2, HBB) (Fig. 5) and six fragments localized on six other chromosomes: 3(*raf*), 13(p9A7), 15(DOSLC3), 17(DOSLC4), 18(DOSLC6) and 20(DOSLC2). We are presently investigating matched pairs of normal and tumour tissue DNA from more than 50 non-small-cell lung cancer patients by RFLP analysis. These experiments are designed to test the hypothesis that the loss of as yet undefined chromosomal genetic elements that may be candidates for 'tumour suppressor' genes are associated with human bronchogenic carcinoma. Matched-pair DNA is digested with restriction enzymes and subjected to agarose gel electrophoresis in adjacent lanes. After Southern transfer, blots are hybridized to ^{32}P-labelled probes and subjected to autoradiography. As depicted in Figure 1, it is possible to determine whether a person is heterozygous from the size and number of bands present. Bands that appear on the autoradiograms are analysed densitometrically. If an allele is represented by more than one band, one is selected as that to be analysed; this selection is usually based on sequence homology to the probe (a higher sequence homology produces a more intense band). The ratio of the densitometric values for the bands representing the two alleles in the normal tissue is divided by the same ratio obtained for the tumour tissue. This ratio of a ratio, or odds ratio, should be 1.0 if there is no deletion of either allele in the tumour; it should be greater than or less than 1.0 if one of the alleles is deleted.

An important consideration when doing RFLP analysis of primary solid tumours is that they contain varying amounts of normal cells, such as fibroblasts, lymphocytes, macrophages and cells associated with blood vessels, including endothelial cells and muscle cells. As a result, if there is a deletion of the probed region in the tumour cells, because of the contribution of DNA from normal cells, a band or bands representing

Fig. 5. Relative location of polymorphic loci on chromosome 11

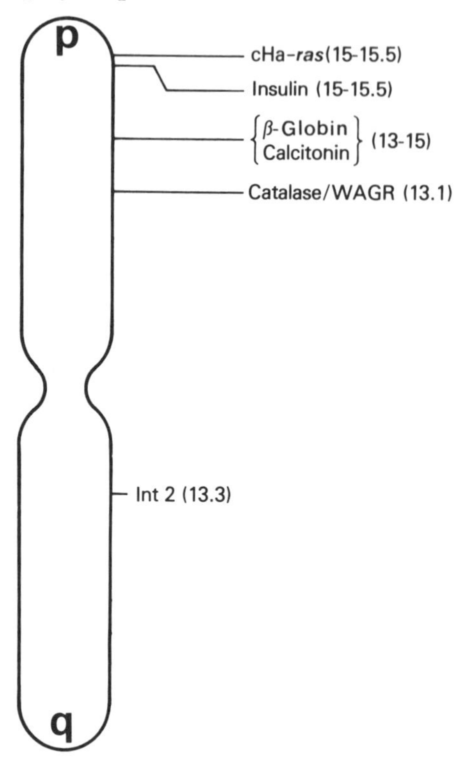

one allele may be diminished but not absent. For this reason, an important control is evaluation of haematoxalin-and-eosin-stained sections from the same block of tissue in order to estimate the percentage of normal cells included in the block. It would be expected that the densitometric value for the band representing the allele deleted in the tumour would be directly related to the percentage of stroma present on the slide, assuming that the section on the slide is representative of the piece of tumour from which the DNA was extracted.

There are at least two other possible reasons that a presumably deleted allele might remain partially represented, both having to do with tumour heterogeneity. Firstly, although the somatic mutation model of carcinogenesis predicts that human tumours are monoclonal in origin (Burnet, 1974), and there is evidence to support this from animal and human studies (Fialkow, 1978; Iannaccone *et al.*, 1978; Cossman *et al.*, 1983; Desforges, 1985), there are few data on the percentage of each human tumour type that is in fact monoclonal, and there are no direct data for bronchogenic carcinoma. If a tumour were polyclonal in origin, then even though the same 'tumour suppressor' gene may have been doubly mutated in all progenitor cells contributing to the tumour, if deletions played a role in each case, the size of the deletion would vary for each progenitor cell. Since the probed region is some distance from the putative suppressor gene, the probed region may be deleted in some progenitor cells but not in others. Secondly, a known cause of tumour-cell heterogeneity is karyotypic instability. Tumours frequently comprise a variety of cell populations with differing karyotypes. However, if it is assumed that a tumour is monoclonal in origin, then even if the karotype is unstable and varies from cell to cell it is unlikely to lead to a mixed population with respect to a genetic region bearing a 'tumour suppressor' gene. If the initial transforming event involved a deletion that included the probed region, then that allele would be irretrievably lost from all daughter cells. Alternatively, if the initial deletion spared the probed region, it would probably lose its linkage to the putative suppressor gene; although, conceivably, chromosomal areas adjacent to a previously deleted area are more likely to be deleted than are other areas. In conclusion, when a signal representing an allele is diminished but not entirely absent, the residual signal is probably due to DNA contributed by nonmalignant cells that were in the tumour specimen.

Future studies

We are also investigating other chromosomes and conducting multivariate analysis of the data in order to discern somatic mutations associated with bronchogenic carcinoma. We will include in future studies probes for chromosomal regions that contain genes associated with other forms of hereditary predisposition to cancer, including acoustic neuroma associated with neurofibromatosis (chromosome 22) and familial ductal carcinoma of the breast (chromosome 13), since some of these may also be involved in bronchogenic carcinoma. Recent developments in the field are likely to make future RFLP analysis more efficient. For example, a method for locating new VTR probes was recently described (Nakamura *et al.*, 1987), and many such probes should become available soon. As discussed above, VTR probes have a higher PIC than non-VTR probes and should provide more information from each case studied. These VTR probes will be widely available through the American Type Culture Collection. In addition, newly acquired gene mapping data and polymorphic probe information are now rapidly filed with the Yale gene bank, access to which is widely available.

After locating chromosomal areas associated with cancer, using the methods described, it will be necessary to use more powerful techniques to isolate the actual genes. Recently described cloning and electrophoresis methods that have proved useful in locating other genes through 'reverse genetics' (Friend *et al.*, 1986; Monaco *et al.*, 1986; Royer-Pokora *et al.*, 1986; Estivill *et al.*, 1987) may prove useful in this project. Although RFLP analysis should help to identify the general location of genes involved in common malignancies, in most cases more powerful tools will be required to identify them precisely. The recent development of pulsed-field electrophoresis combined with innovative cloning techniques (Poustka & Lehrach, 1986; Barlow & Lehrach, 1987; Burke *et al.*, 1987) should prove valuable in gene isolation.

References

Anderson, D.E. (1975) *Familial susceptibility*. In: Fraumeni, J.F., ed., *Persons at High Risk of Cancer, An Approach to Etiology and Control,* New York, Academic Press, pp. 39–54

Ashley, D.J.B. (1969) The two 'hit' and multiple 'hit' theories of carcinogenesis. *Br. J. Cancer, 23,* 313–328

Barlow, D.P. & Lehrach, H. (1987) Genetics by gel electrophoresis: the impact of pulsed field gel electrophoresis on mammalian genetics. *Trends Genet., 3,* 167–171

Burke, D.T., Carle, G.F. & Olson, M.V. (1987) Cloning of large segments of exogenous DNA into yeast by means of artificial chromosome vectors. *Science, 236,* 806–812

Burnet, M. (1974) *The biology of cancer*. In: German, J. ed., *Chromosomes and Cancer,* New York, Wiley, pp. 21–38

Cavenee, W.K., Dryja, T.P., Phillips, R.A., Benedict, W.F., Godbout, R. Gallie, B.C., Murphree, A.L., Strong, L.C. & White, R.C. (1983) Expression of recessive alleles by chromosomal mechanisms in retinoblastoma. *Nature, 305,* 779–784

Cleaver, J.E. & Bootsma, D. (1975) Xeroderma pigmentosum: biochemical and genetic characteristics. *Ann. Rev. Genet., 9,* 19–38

Cossman, A.A., Bakhshi, A., Jaffe, E.S., Waldmann, T.A. & Korsmeyer, S.J. (1983) Immunoglobulin gene rearrangements as unique clonal markers in human lymphoid neoplasms. *New Engl. J. Med., 309,* 1593–1599

Desforges, J.F. (1985) T-cell receptors. *New Engl. J. Med., 313,* 576–577

Doll, R. (1978) An epidemiological perspective of the biology of cancer. *Cancer Res., 38,* 3573–3583

Estivill, X., Farrall, M., Scambler, P.J., Bell, G.M., Hawley, K.M.F., Lench, N.J., Bates, G.P., Kruyer, H.C., Frederick, P.A., Stanier, P., Watson, E.K., Williamson, R. & Wainwright, B.J. (1987) A candidate for the cystic fibrosis locus isolated by selection for methylation-free islands. *Nature, 326,* 840–845

Fearon, E.R., Vogelstein, B. & Feinberg, A.P. (1984) Somatic deletion and duplication of genes on chromosome 11 in Wilms' tumour. *Nature, 309,* 176–178

Fearon, E.R., Feinberg, A.P., Hamilton, S.H. & Vogelstein, B. (1985) Loss of genes on the short arm of chromosome 11 in bladder cancer. *Nature, 318,* 377–380

Fialkow, P.J. (1978) *Clonal origin and stem cell evolution of human tumors.* In: Mulvihill, J.L., Miller, R.W. & Fraumeni, J.F., eds, *Genetics of Human Cancer,* New York, Raven Press, pp. 439–453

Francomano, C. & Kazazian, H.H., Jr (1986) DNA analysis in genetic disorders. *Ann. Rev. Med., 37,* 377–395

Fraumeni, J.F., ed. (1975) *Persons at High Risk of Cancer, An Approach to Etiology and Control,* New York, Academic Press

Friend, S.H., Bernards, R., Rogelj, S., Weinberg, R.A., Rapaport, J.M., Albert, D.M. & Dryja, T.P. (1986) A human DNA segment with properties of the gene that predisposes to retinoblastoma and osteosarcoma. *Nature, 323,* 643–646

Fung, Y.K.T., Murphree, A.L., Tang, A., Qian, J., Hinrichs, S.H. & Benedict, W.F. (1987) Structured evidence for the authenticity of the human retinoblastoma gene. *Science, 236,* 1657–1661

Geiser, A.G., Der, C.J., Marshall, C.J. & Stanbridge, E.J. (1986) Suppression of tumorigenicity with continued expression of the c-ha-*ras* oncogene in EJ bladder carcinoma-human fibroblast hybrid cells. *Proc. natl Acad. Sci. USA, 83,* 5209–5213

Gerhard, D.S., Dracopali, N.C., Bale, J.J., Haughton, A.N., Watkins, P., Payne, C.E., Green, M.H. & Houseman, D.E. (1987) Evidence against Ha-*ras*-1 involvement in sporadic and familial melanoma. *Nature, 325,* 73–75

Harris, C.C. (1987) Tobacco smoke, lung disease: who is susceptible? *Ann. intern. Med., 105,* 607–609

Hirsch-Kauffmann, M., Schweiger, M., Wagner, E.F. & Sperling, K. (1978) Deficiency of DNA ligase activity in Fanconi's anemia. *Human Genet., 233,* 1–8

Iannaccone, P.M., Gardner, R.L. & Harris, H. (1978) The cellular origin of chemically induced tumours. *J. Cell Sci., 29,* 249–269

Kaelbling, M. & Klinger, H.P. (1986) Suppression of tumorigenicity in somatic cell hybrids. III. Cosegregation of human chromosome 11 of a normal cell and suppression of tumorigencity in intraspecies hybrids of normal diploid × malignant cells. *Cytogenet. Cell Genet., 41,* 65–70

Klein, G. & Klein, E. (1986) Evolution of tumours and the impact of molecular oncology. *Nature, 315,* 190–195

Koufos, A., Hansen, M.F., Lampkin, B.C., Workman, M.L., Copeland, N.G., Jenkins, N.A. & Cavenee, W.K. (1984) Loss of alleles at loci on human chromosome 11 during genesis of Wilms' tumour. *Nature, 309,* 170–172

Krontiris, T.G., DiMartino, N.A., Colb, M., Mitcheson, D.H. & Parkinson, D.R. (1986) *Human restriction fragment length polymorphism and cancer risk.* In: Harris, C.C., ed., *Biochemical and Molecular Epidemiology of Cancer,* New York, Alan R. Liss, pp. 99–110

Lee, W.H., Bookstein, R., Hong, F., Young, L.J., Shew, J.Y. & Lee, E.Y.H.P. (1987) Human retinoblastoma susceptibility gene: cloning, identification and sequence. *Science, 235,* 1394–1399

Lidereau, R., Escot, C., Theillet, C., Champeme, M., Brunet, M., Gest, J. & Callahan, R. (1986) High frequency of rare alleles of the human c-Ha-*ras*-1 proto-oncogene in breast cancer patients. *J. natl Cancer Inst., 77,* 697–701

Monaco, A.P., Neve, R.L., Colletti-Feenei, C., Bertelson, C.J., Kurnit, D.M. & Kunkel, L.M. (1986) Isolation of candidate cDNAs for portions of the Duchenne muscular dystrophy gene. *Nature, 323,* 646–650

Moolgavkar, S. & Knudson, A. (1981) Mutation and cancer: a model for human carcinogenesis. *J. natl Cancer Inst., 66,* 1037–1052

Murphree, A. & Benedict, W. (1985) Retinoblastoma: clues to human oncogenesis. *Science, 223,* 1028–1033

Nakamura, Y., Leppert, M., O'Connell, P., Wolff, R., Holm, T., Culver, M., Martin, C., Fujimoto, E., Hoff, M., Kumlin, E. & White, R. (1987) Variable number of tandem repeat (VNTR) markers for human gene mapping. *Science, 235,* 1616–1622

Naylor, S.L., Johnson, B.E., Minna, J.D. & Sakaguchi, A.Y. (1988) Loss of heterozygosity of chromosome 3p markers in small-cell lung cancer. *Nature, 329,* 451–454

Orkin, S.H., Goldman, D.S. & Sallan, S.E. (1984) Development of homozygosity for chromosome 11p markers in Wilms' tumour. *Nature, 309,* 172–174

Ponder, B.A.J. (1984) *Role of genetic and familial factors.* In: Stoll, B.A., ed., *Risk Factors and Multiple Cancer,* London, Wiley, pp. 177–204

Poustka, A. & Lehrach, H. (1986) Jumping libraries and linking libraries: the next generation of molecular tools in mammalian genetics. *Trends Genet., 2,* 174–179

Reeve, A.E., Housiax, P.J., Gardner, R.J.M., Chewings, W.E., Grindley, R.M. & Millow, L.J. (1984) Loss of a Harvey *ras* allele in sporadic Wilms' tumor. *Nature, 309,* 174–176

Riccardi, V.M., Sujansky, E., Smith, A.C. & Francke, Y. (1987) Chromosomal imbalance in the aniridia-Wilms' tumor association: 11P interstitial deletions. *Pediatrics, 61,* 604–610

Royer-Pokora, B., Kunkel, L.M., Monaco, A.P., Goff, S.C., Newburger, P.E., Baehner, R.L., Cole, F.S., Curnutte, J.T. & Orkin, S.H. (1986) Cloning the gene for an inherited human disorder — chronic granulomatous disease — on the basis of its chromosomal location. *Nature, 322,* 32–38

Sager, R. (1985) Genetic suppression of tumor formation. *Adv. Cancer Res., 44,* 43–69

Schneider, N.R., Williams, W.R. & Chaganti, R.S.K. (1986) Genetic epidemiology of familial aggregation of cancer. *Adv. Cancer Res., 47,* 1–36

Seizinger, B.R., Martuza, R.L. & Cousella, J.F. (1987) Loss of genes of chromosome 22 in tumorigenesis of human acoustic neuroma. *Nature, 322,* 644–646

Silverberg, E. & Lubera, J. (1987) Cancer statistics, 1986. *Ca — Cancer J. Clin., 36,* 9–25

Stanbridge, E.J., Der, C.J., Doersen, C., Nishimi, R.Y., Peehl, D.M., Weissman, B.E. & Wilkinson, J.E. (1983) Human cell hybrids: analysis of transformation and tumorigenicity. *Science, 215,* 252–259

US Public Health Service (1982) *Smoking and Health. A Report of the Surgeon General,* Washington DC, US Department of Health and Human Services, Office on Smoking and Health

Weissman, B.E., Saxon, P.J., Pasquale, S.R., Jones, G.R., Geiser, A.G. & Stanbridge, E.J. (1987) Introduction of a normal human chromosome 11 into a Wilms' tumor cell line controls its tumorigenic expression. *Science, 236,* 175–180

Whang-Peng, J., Bunn, P.A., Jr, Kao-Shan, C.S., Lee, E.C., Carney, D.N., Gazdar, A. & Minna, J.D. (1982) A non-random chromosomal abnormality, del 3p(14–23) in human small cell lung cancer (SCLC). *Cancer Genet. Cytogenet., 6,* 119–134

White, R., Barker, D., Holm, T., Berkowitz, J., Leppert, M., Cavenee, W., Leach, R. & Drayna, D. (1983) *Approaches to linkage analysis in the human.* In: Caskey, C.T. & White, R.L., eds, *Recombinant DNA Applications to Human Disease* (*Banbury Report No. 14*), Cold Spring Harbor, NY, CSH Press, pp. 235–250

Willis, A.E. & Lindahl, T. (1987) DNA ligase I deficiency in Bloom's syndrome. *Nature, 325,* 355–358

Yoakum, G.H., Lechner, J.F., Gabrielson, E.W., Korba, B.E., Malan-Shibley, L., Willey, J.C., Valerio, M.G., Shamsuddin, A.M., Trump, B.F. & Harris, C.C. (1985) Transformation of human bronchial epithelial cells transfected by Harvey *ras* oncogene. *Science, 227,* 1174

Zbar, B., Branch, H., Talmadge, C. & Linehan, M. (1987) Loss of alleles of loci on the short arm of chromosome 3 in renal cell carcinoma. *Nature, 327,* 721–724

APPLICATION OF BIOLOGICAL MARKERS TO THE STUDY OF LUNG CANCER CAUSATION AND PREVENTION

F.P. Perera, R.M. Santella, D. Brenner, T.-L. Young & I.B. Weinstein

School of Public Health, Columbia University, New York, NY, USA

Lung cancer is now the major cause of cancer deaths in the USA and is an increasingly significant cancer worldwide. Biological markers could be used to prevent lung cancer by allowing more timely and precise understanding of the role of environmental factors. So far, biological markers that can serve as carcinogen dosimeters have been investigated in only a small number of pilot studies of populations with current or past exposure to lung carcinogens. We describe several of our collaborative studies involving smokers, various worker populations, lung cancer cases and controls in order to illustrate the advantages and the limitations of 'molecular epidemiology'. The enzyme-linked immunosorbent assay (ELISA) with antibodies to polycyclic aromatic hydrocarbon (PAH)-DNA adducts has been used in conjunction with one or more of the following: physicochemical techniques to monitor carcinogen adducts on haemoglobin, cytogenetic methods to quantify sister chromatid exchange (SCE) and chromosomal aberrations, and Southern and western blot analyses of oncogene activation.

Increased levels of markers of biologically effective doses have generally been seen in exposed and high-risk groups when compared to controls. We have also observed significant background levels of such markers and interindividual variation in levels of certain biological markers resulting from exposures to carcinogens. Thus, these methods may be particularly useful in identifying segments of the population that have received a significant effective dose and hence can be considered to be at elevated risk of cancer. Such studies are necessary to validate laboratory methods and lay the groundwork for more definitive molecular epidemiological investigations of lung cancer.

Introduction: biological markers and lung cancer

In keeping with the goals of this volume, we discuss the potential usefulness of biological markers in the study of human lung cancer causation, in risk assessment and in prevention. To illustrate the possibilities and the pitfalls of the approach known as 'molecular' or 'biochemical' epidemiology, we summarize the results of several of our pilot investigations involving populations with documented current or past exposure to a class of airborne carcinogens. Reference is also made to related research by other investigators in the field.

The need for biological markers

Conventional epidemiological methods are inadequate to identify causal factors in common cancers such as that of the lung. The long latency of lung cancer generally precludes reliable estimation of exposure, from which one can identify qualitative risks (hazards) and establish dose-response relationships and quantitative estimates of risk.

Moreover, epidemiological studies afford limited opportunity to evaluate mechanisms by which carcinogens (interacting with other environmental and host factors) exert their effect and to identify groups with increased susceptibility resulting from genetic and environmental factors. Most importantly, conventional epidemiology precludes prevention, as it relies on tumour incidence or mortality as the dependent variable — an approach sometimes referred to as 'body counting'. Although in its infancy, molecular epidemiology could allow more precise, timely identification of risk factors and of the mechanisms by which cancer occurs (for review see Bridges, 1980; Perera & Weinstein, 1982; Wogan & Gorelick, 1985; Harris, 1985; Ehrenberg *et al.*, 1986; Perera, 1987).

Biological markers can achieve these goals in several ways: (i) by indicating host factors (e.g., health status and genetic traits) that predispose individuals to cancer; (ii) by acting as markers of an internal or biologically effective dose of carcinogens; and (iii) by indicating a very early preclinical stage in cancer development, obviating the need to await tumour formation. The focus of this paper is on the second point — the usefulness of biological markers as dosimeters of lung carcinogens — because, in our view, this role is currently the best developed of the three. This is not to minimize progress in the study of potential markers of susceptibility such as arylhydrocarbon hydroxylase activity, shown in some but not all investigations to be significantly higher in lung cancer cases than in controls (Paigen *et al.*, 1977; Kouri *et al.*, 1982).

By acting as dosimeters, biological markers such as DNA and protein adducts, cytogenetic effects, DNA damage and somatic cell mutation can also enhance the power of a study to identify causal relationships by reducing misclassification error. That is, they can differentiate those individuals who have sustained a biologically significant dose of carcinogen from those who, albeit with comparable external exposure, have not. The importance of such differentiation is underscored by studies that demonstrate a high degree of variation in levels of DNA and protein adducts among individuals with similar exposures in the work place environment or clinical setting (Shamsuddin *et al.*, 1985; Bryant *et al.*, 1986; Reed *et al.*, 1986; Perera *et al.*, 1987, 1988).

Markers of early preclinical response would also increase the power of epidemiological studies by providing a more frequent and earlier endpoint than tumours. At present, however, available biological markers can serve as dosimeters but have not yet been definitively correlated with tumour incidence. This will require prospective or nested case-control studies to establish their predictive relationship to cancer risk.

A recognized limitation of available markers is that they cannot provide information on all stages of the carcinogenic process, since they largely reflect exposures to that subset of carcinogens that act *via* genetic toxicity. There is, however, general agreement that a group of individuals shown in properly controlled studies to have experienced a significant biologically effective dose of a carcinogen can be presumed to be at higher risk of cancer and that this information can serve as a valuable early warning system or 'red flag' to those concerned with surveillance and prevention (Vainio *et al.*, 1983).

There is a major need for markers of biologically effective dose in the study of lung cancer. Lung cancer is the leading cause of cancer deaths in the USA (136 000/year) and a common cancer worldwide (Muir & Nectoux, 1982). The US annual incidence rate (150 000) is very similar to the mortality rate, reflecting the low five-year survival rate of lung cancer patients (13% for all stages). During the past three decades (1957 to 1981–83), US males experienced a 184% increase in age-adjusted lung cancer death rates, while in women lung cancer mortality escalated at an even greater rate (360%) (American Cancer Society, 1987). In most countries, there has been a consistent increase in lung cancer over the past 50 years (Muir & Nectoux, 1982).

It is estimated that active cigarette smoking is responsible for the great majority of lung cancer (Surgeon General, 1986) and for about 30% of all cancer (Tomatis, 1985). However, a significant fraction is believed to be attributable to other environmental factors (Doll & Peto, 1981; Toxic Substances Strategy Committee, 1980); with respect to lung cancer, one of these is passive smoking (Surgeon General, 1986; National Research Council, 1986). Other known or suspected environmental risk factors include exposures to synthetic organic chemicals, products of fossil fuel combustion (e.g., PAH), metals and fibres, and radon (Decouflé, 1982). Workers exposed to these agents have significantly increased risks of lung cancer; but risk is not confined to the work place: there is growing evidence that man-made pollutants in the ambient air contribute to lung cancer incidence in the general population. (See Perera & Weinstein, 1982, for review.)

Frequently, experimental data demonstrating the carcinogenicity of a chemical have preceded epidemiological findings. This is true for aflatoxin, 4-aminobiphenyl, bis(chloromethyl)ether, diethylstilboestrol, melphalan, mustard gas and vinyl chloride (Tomatis, 1979; Huff *et al.*, 1986). Unfortunately, in a number of cases, steps to prevent or reduce human exposure were taken only after epidemiological evidence of causality became available.

Lack of biomonitoring data on human lung carcinogens

Could human biomonitoring data have predicted—qualitatively or quantitatively—the increased human cancer risks that have been found to be associated with specific environmental exposures by conventional epidemiological studies? Unfortunately, as shown in Table 1, human biomonitoring data on known or probable human lung carcinogens are still too limited to evaluate this question. In

Table 1. Summary of available data on biological markers in humans exposed to known or suspected chemical lung carcinogens[a]

Chemical	Chromosomal aberrations	SCE	Adducts
Acrylonitrile	−		
Amitrole			
Arsenic	− −	− +	
Asbestos		(+)	
Bis(chloromethyl)ether	(+)		
Beryllium			
Cadmium	(+)(−)(+)(+)	−	
Chromium (VI)	+ − −	− −	
Chloromethylmethylether	(+)		
Cutting oils			
Dimethyl sulphate			
Lead	− + −	(+) − +	
Mustard gas			
Nickel	(+)(+)	− −	
PAH[b]	+[c]	+[c,d]	+
Vinyl chloride	+ + + + +	(+) − −	
	+ + + + +		+

[a]List of carcinogens: Decouflé (1982); data: M. Waters, F. Stack & A. Brady (US Environmental Protection Agency Activity Profile Data Base, personal communication) (except for PAH)
[b]As components of mixtures
[c]Studies of cigarette smokers
[d]Miner *et al.* (1983)
(+), weakly positive or positive in a study of limited quality; (−), weakly negative or negative in a study of limited quality

Table 1 we have listed the human data (*in vivo*) for chemicals for which there was epidemiological evidence of a causal association with lung cancer as of 1982 (Decouflé, 1982). This list is not complete. For example, there is now epidemiological evidence that formaldehyde is a probable human carcinogen (IARC, 1987), with the lung as a target site.

Some representative studies of biological markers

Our research at Columbia University has focused on PAH-DNA adducts, protein adducts, cytogenetic effects and the occurrence of activated oncogenes in cross-sectional and serial studies of exposed populations at risk for lung cancer and, retrospectively, in lung cancer cases — all with appropriate controls.

Standardization studies

Levels of PAH-DNA adducts measured by ELISA, using a polyclonal antibody which recognizes diol epoxide-DNA adducts of benzo[*a*]pyrene (BPDE-I) and related PAH, are higher than reported previously (Perera *et al.*, 1987). These values reflect improved quantification based on studies in which antibody recovery of adducts was compared in low and highly modified standards. Specifically, in collaborative studies with M. Poirier and C.C. Harris and his colleagues at the National Cancer Institute, USA, highly modified (15 pmol/μg) and low-modified (4.4 fmol/μg) samples of BPDE-I-DNA were used to validate adduct recovery by colour and fluorescent ELISA under the conditions of our standard assay (R. Santella *et al.*, in preparation). The antibody detected adducts in the low-modified DNA sample (4.4 fmol/μg) with 2.5-fold (colour) and ten-fold (fluorescent) lower efficiency than in the more highly modified DNA sample (15 pmol/μg). Previously, we routinely used the highly modified DNA sample to generate standard curves; however, since human samples have a modification level close to the 4.4 fmol adduct/μg standard, it is more appropriate to use the low-modified standard in assaying human samples. Therefore, values for adducts in the biological samples have been 'corrected' by multiplying by factors of 2.5 (colour) and 10 (fluorescence). Currently, we are routinely using a low-modified standard in assays of human samples to eliminate the need for this correction. However, these corrected values are based on BPDE-DNA standard curves; the antibodies are known to recognize other PAH-DNA adducts with varying affinities, so the data provide an approximate indication of total PAH-DNA adducts.

Studies of cigarette smokers

We have studied 22 smokers and 24 nonsmokers using a battery of markers (Perera *et al.*, 1987): PAH-DNA adducts in white blood cells using the ELISA described above; 4-aminobiphenyl-haemoglobin (4-ABP-Hb) adducts measured by negative-ion chemical ionization mass spectrometry (collaboratively with M. Bryant and S. Tannenbaum and his colleagues at the Massachusetts Institute of Technology, Boston, MA, USA); SCE in cultured lymphocytes; and cotinine in plasma measured by radioimmunoassay (in collaboration with N. Haley of the Naylor Dana Institute for Disease Prevention, Valhalla, NY, USA). The results for one or more blood samples drawn from the same individual are given in Table 2. Values are given for samples with detectable levels of adducts (i.e., corresponding to 25% or greater inhibition).

Briefly, all four markers were found at higher levels in smokers than in nonsmokers. The difference between the means for detectable levels of PAH-DNA adducts in white blood cells from smokers and from nonsmokers was of only borderline significance (sample 2), possibly because of the large number of 'background'

Table 2. Levels of biological markers in healthy volunteers: mean (SE)[a,b]

Marker	Smokers	Nonsmokers
PAH-DNA (fmol/μg)		
Sample 1	0.33 (0.053)	0.29 (0.035)
	5/22[c]	7/24
Sample 2	0.32 (0.068)†	0.16 (0.033)
	4/21	4/21
4-ABP-Hb (pg/g)		
Sample 1	154.5 (11.3)**	32.2 (2.9)
(n)	(19)	(18)
Sample 2	139.0 (7.9)**	36.2 (5.02)
(n)	(9)	(6)
SCE (average no./metaphase)	10.8 (0.603)*	8.1 (0.47)
(n)	(11)	(10)
Cotinine (ng/ml)	419.2 (47.4)**	0.3 (0.032)
(n)	(10)	(10)

[a]One or two successive blood samples per subject
[b]Smoker/nonsmoker difference: †, borderline significance; * $p < 0.01$; ** $p < 0.001$
[c]Fraction of individuals (samples) with detectable levels of adducts

(noncigarette related) exposures to PAH. In smokers, the levels of PAH-DNA adducts were highly correlated with several indices of cigarette smoking (pack-years, lifetime tar intake) and with dietary intake of charcoal. We also observed an effect of passive smoking on DNA adducts in nonsmokers.

Smokers had significantly higher levels of the other three markers than controls (Table 2). Reduced lapse of time since smoking had a significant effect on both 4-ABP-Hb and cotinine levels. Interestingly, 4-ABP-Hb was significantly correlated with cotinine, SCE and DNA adducts. These results indicate the usefulness of a battery of markers with varying specificities for the exposure of interest.

Studies of foundry and coke-oven workers

Another study (carried out collaboratively with K. Hemminki, Institute of Occupational Health, Finland) has demonstrated a dose-response relationship between exposure of 35 iron-foundry workers and PAH-DNA adducts in white blood cells (Perera *et al.*, 1988; Hemminki *et al*, this volume). (See Table 3.) Workers were classified as having low (<0.05 μg/m^3), medium (0.05–0.2 μg/m^3), or high (>0.2 μg/m^3) exposure to benzo[*a*]pyrene (as an indicator of total PAH).

These results are roughly consistent with the PAH-DNA adduct levels seen in our study of smokers of about one pack per day (mean 0.32 fmol/μg), who are exposed daily to approximately 0.5 μg benzo[*a*]pyrene, since foundry workers with estimated medium exposure (and mean adduct levels of 0.62 fmol/μg) would be expected to inhale about 1 μg benzo[*a*]pyrene daily (approximately a two-fold greater dose than that of the smokers), assuming that workers inhale 10 m^3 air during an 8-h shift with no physiological filtration of benzo[*a*]pyrene.

Shamsuddin *et al.* (1985) reported a range of PAH-DNA adducts (0.06–0.4 fmol/μg) in 7/20 white blood cell samples from foundry workers, using a

Table 3. PAH-DNA adducts in Finnish foundry workers: mean (SE)

Exposure group	Adduct level (fmol/μg)
High	1.50 (0.55) 4/4[a]
Medium	0.62 (0.147) 13/13
Low	0.32 (0.065) 13/18
Control	0.22 (0.077) 2/10

[a]Fraction of samples (individuals) with detectable levels of adducts

competitive ultrasensitive enzymatic radioimmunoassay. However, they did not classify workers as to work place exposure.

We have also carried out a small study of coke-oven workers (in collaboration with M. Bender, R. Leonard and O. White of Brookhaven National Laboratory, USA). The ELISA with colorimetric detection was used to quantify PAH-DNA adducts; SCE and chromosomal aberrations were scored in cultured lymphocytes. Four of 24 exposed workers and none of 11 laboratory worker controls had detectable levels of adducts (mean and SE, 0.45(0.05)). This value is low compared to the range of adduct levels reported by Harris *et al.* (1985) in coke-oven workers (0.4–34.3 fmol/μg). The mean SCE value, adjusted for smoking, was significantly higher ($p = 0.009$) in 27 coke-oven workers (including the subjects studied with respect to PAH-DNA binding) than in controls, who comprised six unexposed steel workers and 50 Brookhaven National Laboratory employees. Chromatid-type, but not chromosome-type aberrations, were also significantly increased in incidence in the exposed group (deletions, $p = 0.0002$; exchanges, $p = 0.05$) (Bender *et al.*, 1988).

Studies of patients with lung cancer

In studies in progress of lung cancer cases and controls, we are investigating the relationship between various molecular markers (adducts, SCE and altered oncogenes) and lung cancer risk, in contrast to previous investigations of the relationship between estimated exposure and these markers. This research follows a pilot study in which detectable levels of PAH-DNA adducts were seen in tumour tissue, tissue adjacent to tumours and white blood cells of 5/15 primary lung cancer patients and in none of eight controls (Perera *et al.*, 1982).

Currently, we are investigating whether the lung cancer cases have higher levels of adducts, SCE and activated oncogenes than subjects without cancer (orthopaedic patients) with similar smoking histories. Such a finding would constitute circumstantial evidence that the propensity to activate and bind carcinogens to DNA and to sustain other types of genetic damage is a risk factor for lung cancer.

There is considerable evidence that activation of proto-oncogenes plays an important role in the neoplastic transformation of cells and may be triggered by a variety of types of genetic damage (point mutations, translocations, deletions and abnormal amplification). Structurally altered or amplified oncogenes of the *myc* and *ras* families have been detected in a number of human tumours, including carcinoma of the lung. (See Bishop, 1987, for review.) *c-myc* amplification has frequently been seen in various epithelial tumours (including lung) in association with advanced or aggressive tumours (Yokota *et al.*, 1986). Of particular interest to us is the possible relationship between environmental exposure to mutagens/carcinogens (e.g., in cigarette smoke) and oncogene activation in the causation of lung cancer.

In collaboration with D. Grunberger and M. Goldfarb at Columbia University, New York, USA, we are investigating the prevalence of genetic alterations in the *c-myc* and *c-ras* proto-oncogenes in lung tumours and in tissue adjacent to tumours from surgical lung cancer cases and in lung tissue from autopsy controls. DNA samples are being analysed by Southern blot hybridization using several *c-myc* probes as well as the standard NIH 3T3 DNA transfection focus assay. We are evaluating the relationship between the occurrence of altered proto-oncogenes, smoking history and clinicopathological characteristics of each subject.

Using a complementary approach to the study of oncogene activation in human cancer, in collaboration with P. Brandt-Rauf at Columbia University, USA, antibodies directed against protein sequences encoded by several oncogenes (including *c-myc* and *c-ras*) have been used to assay sera from 18 of the lung cancer cases enrolled in the DNA adduct studies described above. The rationale is that oncogene activation in target tissues might result in the release of related protein into the peripheral circulation, thus providing a convenient marker. The method, a modified western blot, has been adapted from that previously applied by Niman *et al.* (1985) to urine samples. Thus far, the sera of 18 lung cancer patients, half of whom were also enrolled in the study of oncogenes in lung tissue, have been evaluated, as well as sera from 20 controls (healthy volunteers whose smoking status is unknown). Evidence for the presence of the *c*-Ha-*ras* P21 protein was seen in a large proportion of cancer patients (83% compared to 10% of controls). Thus far, we have not seen evidence of *c-myc* in the sera of cancer patients; further studies are under way. The *c*-Ha-*ras* P21 protein was present in sera from two of the 20 controls. Prior studies have detected *ras* protein products in sera from smokers, suggesting that exposure to carcinogens may activate oncogenes (Niman *et al.*, 1985). Further studies are now being carried out on a more comparable group of controls (orthopaedic patients).

Conclusion

These studies demonstrate a significant variability between individuals in the levels of various biological markers in DNA and in haemoglobin. For example, in smokers of one to two packs per day, the range of PAH-DNA adducts was none detected to 0.5 fmol/μg; while the levels of 4-ABP-Hb adducts ranged from 75–256 pg/g. Other studies have demonstrated that patients treated with the same cumulative dose of the chemotherapeutic drug cisplatin (400 mg/m^2) form varying levels of adducts (none detected to 0.18 fmol/μg; Reed *et al.*, 1986). Thus, as mentioned above, these methods may prove to be valuable in evaluating carcinogenic hazard and risk to segments of the population who, albeit with low exposure to environmental pollutants, sustain a significant, biologically effective dose of carcinogens.

These studies also showed elevations in the levels of various biological markers in exposed populations (smokers and workers) compared to appropriate control groups. In the smokers, these increases were highly significant for 4-ABP-Hb and cotinine and of borderline significance for PAH-DNA adducts, possibly reflecting the substantially greater 'background' exposure to benzo[*a*]pyrene and other PAH compared with 4-ABP and nicotine. The study of foundry workers has demonstrated the clearest dose-response relationship between an environmental exposure and DNA adducts. However, the levels of adducts observed in several groups with comparable estimated exposure to PAH (e.g., heavy smokers and workers with low exposure) are generally consistent.

Acknowledgements

We gratefully acknowledge the valuable contributions of our colleagues Dr A. Jaretzki, Dr D. Carberry, Dr J. Kelsey, Dr G. Kelly, Dr J. Van Ryzin (deceased), Dr M. Grimes, Dr S. Grantham, S. Hearne, J. Orazem, Y. Mucha, Dr K. Fischman, Dr A. Munshi, B. Lee and O. Osafradu. We also warmly thank Dr A. Weston and Dr G. Trivers of the National Cancer Institute, and Dr M. Waters and colleagues at the Environmental Protection Agency for their important contributions to the research cited. We are grateful to R. Ryan for secretarial help with this manuscript. Support has been provided by NIHCA35809, NIHCA39174, NIHCA43013, NIEHSESO3881, and ACS- SIG-13.

References

American Cancer Society (1987) *Cancer Facts and Figures,* New York

Bender, M.A., Leonard, R.C., White, O., Jr, Constantine, J.P. & Redmond, C.K. (1988) Chromosomal aberrations and SCEs in lymphocytes from coke oven workers. (Abstract 27). *Environ. mol. Mutag., 11,* 12

Bishop, J.M. (1987) The molecular genetics of cancer. *Science, 235,* 305–311

Bridges, B.A. (1980) An approach to the assessment of the risk to man from DNA damaging agents. *Arch. Toxicol., Suppl. 3,* 271–281

Bryant, M.S., Skipper, P.L. & Tannenbaum, S.R. (1986) Haemoglobin adducts of 4-aminobiphenyl in smokers and nonsmokers. *Cancer Res., 47,* 602–608

Decouflé, P. (1982) *Occupation.* In: Schottenfeld, D. & Fraumeni, J.F., eds, *Cancer Epidemiology and Prevention,* Philadelphia, W.B. Saunders, pp. 318–335

Doll, R. & Peto, R. (1981) The causes of cancer: quantitative estimates of avoidable risks of cancer in the United States today. *J. natl Cancer Inst., 66,* 1191–1309

Ehrenberg, L., Osterman-Golkar, S. & Törnqvist, M. (1986) *Macromolecule adducts, target dose, and risk assessment.* In: Ramel, C., Lambert, B. & Magnusson, J., eds, *Genetic Toxicology of Environmental Chemicals, Part B, Genetic Effects and Applied Mutagenesis,* New York, Alan R. Liss, pp. 253–260

Harris, C.C. (1985) Future directions in the use of DNA adducts as internal dosimeters for monitoring human exposure to environmental mutagens and carcinogens. *Environ. Health Perspectives, 62,* 185–191

Harris, C.C., Vähäkangas, K., Newman, M.J., Trivers, G.E., Shamsuddin, A., Sinopoli, N., Mann, D.L. & Wright, W.E. (1985) Detection of benzo(*a*)pyrene diol epoxide-DNA adducts in peripheral blood lymphocytes and antibodies to the adducts in serum from coke oven workers. *Proc. natl Acad. Sci. USA, 82,* 6672–6676

Huff, J.E., Haseman, J.K., McConnell, E.E. & Moore, J.A. (1986) *The National Toxicology Program, toxicology data evaluation techniques, and long-term carcinogenesis studies.* In: Lloyd, W.E., ed., *Safety Evaluation of Drugs and Chemicals,* New York, Hemisphere, pp. 411–446

IARC (1987) *IARC Monographs on the Evaluation of Carcinogenic Risks to Humans,* Suppl. 7, *Overall Evaluations of Carcinogenicity. An Updating of IARC Monographs Volumes 1 to 42,* Lyon.

Kouri, R.E., McKinney, C.E., Slomiany, D.J., Snodgrass, D.R., Wray, N.P. & McLemore, T.L. (1982) Positive correlation between high aryl hydrocarbon hydroxylase activity and primary lung cancer as analyzed in cryopreserved lymphocytes. *Cancer Res., 42,* 5030–5037

Miner, J.K., Rom, W.N., Livingston, G.K. & Lyon, J.L. (1983) Lymphocyte sister chromatid exchange (SCE) frequencies in coke oven workers. *J. occup. Med., 25,* 30–33

Muir, C.S. & Nectoux, J. (1982) *International patterns of cancer.* In: Schottenfeld, D. & Fraumeni, J.F., eds, *Cancer Epidemiology and Prevention,* Philadelphia, W.B. Saunders, pp. 119–138

National Research Council (1986) *Environmental Tobacco Smoke: Measuring Exposures and Assessing Health Effects,* Washington DC, National Academy Press

Niman, H.L., Thompson, A.M.H., Yu, A., Markham, M., Willems, J.J., Herwig, K.R., Habib,

N.A., Wood, C.B., Houghten, R.A. & Lerner, R.A. (1985) Anti-peptide antibodies detect oncogene-related proteins in urine. *Proc. natl Acad. Sci. USA, 82,* 7924–7928

Paigen, B., Gurtoo, H.L., Minowada, J., Houten, L., Vincent, R., Paigen, K., Parker, N.B., Ward, E. & Hayner, N.T. (1977) Questionable relation of aryl hydrocarbon hydroxylase to lung cancer risk. *New Engl. J. Med., 297,* 346–350

Perera, F.P. (1987) Molecular cancer epidemiology: a new tool in cancer prevention. *J. natl Cancer Inst., 78,* 887–898

Perera, F.P. & Weinstein, I.B. (1982) Molecular epidemiology and carcinogen-DNA adduct detection: new approaches to studies of human cancer causation. *J. chronic Dis., 3,* 581–600

Perera, F.P., Poirier, M.C., Yuspa, S.H., Nakayama, J., Jaretzki, A., Curnen, M.M., Knowles, D.M. & Weinstein, I.B. (1982) A pilot project in molecular cancer epidemiology: determination of benzo(*a*)pyrene-DNA adducts in animal and human tissues by immunoassays. *Carcinogenesis, 3,* 1405–1410

Perera, F.P., Santella, R.M., Brenner, D., Poirier, M.C., Munshi, A.A., Fischman, H.K. & Van Ryzin, J. (1987) DNA adducts, protein adducts and sister chromatid exchange in cigarette smokers and nonsmokers. *J. natl Cancer Inst., 79,* 449–456

Perera, F.P., Hemminki, K., Young, T.-L., Brenner, D., Kelly, G. & Santella, R.M. (1988) Detection of polycyclic aromatic hydrocarbon-DNA adducts in white blood cells of foundry workers. *Cancer Res.* (in press)

Reed, E., Yuspa, S.H., Zwelling, L.A., Ozols, R.F. & Poirier, M.C. (1986) Quantitation of *cis*-diamminedichloroplatinum II (*cis*platin)-DNA-intrastrand adducts in testicular and ovarian cancer patients receiving cisplatin chemotherapy. *J. clin. Invest., 77,* 545–550

Shamsuddin, A.K., Sinopoli, N.T., Hemminki, K., Boesch, R.R. & Harris, C.C. (1985) Detection of benzo(*a*)pyrene-DNA adducts in human white blood cells. *Cancer Res., 45,* 66–68

Surgeon General (1986) *The Health Consequences of Involuntary Smoking: A Report of the Surgeon General,* Washington DC, US Government Printing Office

Tomatis, L. (1979) The predictive value of rodent carcinogenicity tests in the evaluation of human risks. *Ann. Rev. Pharmacol., 19,* 522–530

Tomatis, L. (1985) The contribution of epidemiological and experimental data to the control of environmental carcinogens. *Cancer Lett., 26,* 5–16

Toxic Substances Strategy Committee (1980) *Toxic Chemicals and Public Protection: A Report to the President by the Toxic Substances Strategy Committee,* Washington DC, US Government Printing Office

Vainio, H., Sorsa, M. & Hemminki, K. (1983) Biological monitoring in surveillance of exposure to genotoxicants. *Am. J. ind. Med., 4,* 87–103

Wogan, G.N. & Gorelick, N.J. (1985) Chemical and biochemical dosimetry of exposure to genotoxic chemicals. *Environ. Health Perspectives, 62,* 5–18

Yokota, J., Tsunetsugu-Yokota, Y., Battifora, H., Le Fevre, C. & Cline, M.J. (1986) Alterations of *myc, myb* and *ras* proto-oncogenes in cancers are frequent and show clinical correlation. *Science, 231,* 261–265

EPIDEMIOLOGICAL STUDIES OF THE RELATIONSHIP BETWEEN CARCINOGENICITY AND DNA DAMAGE

J. Kaldor[1] & N.E. Day[2]

[1]International Agency for Research on Cancer, Lyon, France; and [2]MRC Biostatistics Unit, Cambridge, UK

The new methods for studying DNA damage *in vivo* offer cancer epidemiologists both a new study endpoint and a refinement of exposure measurement. Up to the present, they have not been applied in studies that satisfy standard epidemiological criteria. In particular, more attention must be paid to the selection of study samples and to ensuring adequate numbers of study subjects. There are substantial problems in studying the link between DNA damage and carcinogenesis at the individual level, including the long latency and the multifactorial etiology of most cancers. Acute leukaemia induced by alkylating agent therapy represents one of the few human situations in which such studies are feasible, since patients are exposed at carefully measured and recorded doses, the risk is high and specific, and follow-up is good. A considerable saving in resources could be achieved by storing samples from a large number of treated patients and carrying out analyses of DNA damage only in those from people who develop leukaemia and from suitably chosen matched controls.

Epidemiologists search for etiological agents by studying the association between a disease state and various other characteristics, in a study population large enough to eliminate the effects of chance, and chosen in such a way as to avoid the effects of bias. The new methods for detecting DNA damage *in vivo* offer exciting possibilities to epidemiologists, both as a way of defining a rather specific disease state on the basis of the presence or absence of damage, or as a refinement of exposure measurement using DNA damage as a biological dosimeter. The term 'molecular epidemiology' has been adopted to describe studies in which either of these strategies is used (Perera, 1987).

This field is nonetheless very young and has so far yielded few studies that would stand up to a rigorous examination by epidemiological criteria. It is the purpose of this paper to examine ways in which molecular epidemiology could be strengthened, by indicating how the study designs fail to satisfy these criteria, and suggesting how they should be modified to enable them to do so. We then put forward our proposal for studies of patients treated with cytotoxic drug therapy, which we believe offer the best hope of yielding useful epidemiological results on the relationship between DNA damage (both mutational events and gene rearrangements) and carcinogenicity.

Studies of DNA damage as an endpoint

So far, most epidemiological studies in which DNA damage has been measured have treated it as an endpoint. Groups of exposed and unexposed individuals are selected and then compared with regard to some parameter of DNA damage. Exposures studied have included cigarette smoking (IARC, 1985: Everson *et al.*, 1986), benzo[*a*]pyrene in occupational settings (Harris *et al.*, this volume; Hemminki *et al.*, this volume; Perera *et al.*, this volume) and cytotoxic drugs used in the therapy of malignant and nonmalignant disease (Raposa, 1978). The goal of these studies is to relate the external dose to the level of DNA damage.

In planning these studies, several basic epidemiological principles should be respected. The first is in the choice of subjects. As far as possible, exposed individuals should be comparable to those unexposed with respect to factors that may influence levels of DNA damage, apart from the exposure itself. The most important such factor is probably age, although sex, ethnicity and, in certain cases, smoking status may be important. The requirement of comparability applies equally in situations in which unexposed individuals are unavailable for study and comparisons must be made between groups of individuals exposed to different dose levels. This may be the case if the study subjects are cancer patients, all of whom are receiving the same DNA damaging drug.

Another important design principle is that of an adequate sample size. Too often, studies of DNA damage are based on a handful of individuals, and, even if significant differences are seen between exposed and unexposed, confidence intervals on the differences are still rather wide.

Finally, the statistical treatment of these studies has been rather simplistic. Typical analyses do not go further than a *t*-test or Wilcoxon statistic comparing the level of damage between groups. There has been little exploration of the degree of within-person variation in measurements of DNA damage, or of the closely related issue of whether it is in fact more appropriate to make comparisons within the same person before and after exposure, or between groups of exposed and unexposed subjects. Statistical analysis could also usefully adjust for factors such as age and smoking status, if they have not been controlled for in the design. Another important statistical question concerns the choice of exposure 'time window' (Rothman, 1981) to be used in relating external exposure to DNA damage. The most important determinant of adduct levels may be exposure in the immediate past, while chromosomal aberrations may relate to exposure over a longer period.

All of these ideas have parallels in epidemiological studies in which more conventional endpoints, such as cancer incidence, are studied. Their importance is undiminished when DNA damage is the outcome of interest, and they will have to be taken into account before molecular epidemiology can be accepted as a valid part of epidemiology.

Studies of DNA damage as an exposure measurement

Use of DNA damage as a measure of exposure is of great interest to epidemiologists, since it gives them the chance to sharpen their usually clumsy instruments for measuring the true level of risk factors. The hope is that, instead of relying on questionnaires and other indirect sources of exposure information, one could utilize a measurement that reflects directly the immediate consequences of an exposure on the DNA of target cells, or of other cells that may be viewed as surrogates for the target cells. Studies of this kind can define a biological marker of exposure that is a better predictor of cancer risk than the external exposure level itself. If, indeed, for a given level of external dose some individuals suffer more genetic damage than others, this could indicate variation in constitutional susceptibility to the agent and lead to a means of distinguishing individuals who are at higher risk of cancer.

So far, all of the studies in which DNA is used as a measure of exposure have utilized the case-control design, in which patients with cancer are compared with healthy controls with regard to the degree of DNA damage. The comparison has been made between samples of tumour material from cases and normal tissue from controls (Perera *et al.*, 1982) or between peripheral lymphocytes in both groups.

The case-control design offers the advantage that, by using all cases arising in a given population and only a small number of controls per case, the study can give almost as precise results as if all individuals in the population had been studied (Breslow *et al.*, 1983). The design suffers from two major disadvantages, which are particularly pertinent when DNA damage is measured as an indication of exposure. First is the problem of temporal ambiguity. Since the damage is measured in cases after cancer has developed, the malignancy itself may have some effect on the measurement of DNA damage, particularly if tumour material is used. The second problem is that even if the measurement is unaffected by the presence of disease, it may be totally unrelated to exposures that occurred in the past which gave rise to the cancer.

The alternative to the case-control approach is a prospective study, in which DNA damage would be assessed before the onset of disease. While largely circumventing the temporal difficulties of the case-control study, the prospective approach presents huge practical problems. Foremost is the number of study subjects required. Even for a relatively common tumour, such as lung cancer in men, it would be necessary to measure DNA damage in thousands of individuals, and to follow them for decades in order to have sufficient cases in the study population. One alternative would be to study older people, in whom the incidence of cancer is high, but one would then be back in the position of measuring DNA damage that was not necessarily relevant to the subsequent incident cancers.

A further problem with both case-control and the prospective designs arises from the multifactorial nature of most types of cancer. Even a relationship as well established as that between lung cancer and cigarette smoking has not been elucidated to the extent that one could specify which DNA adducts should be measured as predictors of cancer risk.

Using either of the designs, a strict adherence to epidemiological guidelines should be maintained. Just as for the studies of DNA damage as an endpoint, it is essential that studies be large and that subjects be sampled in a way that is clearly defined, and free of bias.

Studies of the correlation between DNA damage and carcinogenicity

A third approach is one in which DNA damage is studied as an endpoint following exposure separately to a number of different agents or to different dose levels of the same agent. The DNA damaging effect of each exposure is then compared with its carcinogenicity, using in each case some appropriate potency index. This approach has been used extensively in animal experiments. Lutz (1986) examined the correlation across chemical compounds between the degree of DNA binding in rat liver and the carcinogenicity of the compound. Parodi *et al.* (1983) carried out similar studies of the relationship between the ability of compounds to induce sister chromatid exchanges and other kinds of DNA damage in rodents, and their carcinogenic potency.

Results of this kind obtained in humans have been far more limited. Bryant *et al.* (this volume) have reported that aromatic amine adducts are formed in haemoglobin *in vitro* more often in smokers of black than in smokers of blond tobacco, in a ratio that is consistent with the relative bladder carcinogenicity of the two tobacco types, but this correlation is limited to two points.

Correlation studies are certainly very useful in exploring relationships between short-term markers of DNA damage and long-term carcinogenicity. Their major drawback is that they often utilize disparate study populations, in whom neither the evaluation of DNA damage nor the assessment of carcinogenicity is carried out in a consistent manner.

Exposure to anticancer drugs as an epidemiological model for studying DNA damage

It is now well established that many drugs and combinations of drugs used in the therapy of malignant disease (and some nonmalignant conditions) greatly increase the risk of acute leukaemia, and possibly other tumours, in survivors (IARC, 1981). Some authors (Pedersen-Bjerregaard & Larsen, 1982) have put the cumulative risk of leukaemia as high as 10% for the ten-year period following the start of cytotoxic therapy for Hodgkin's disease. Although attempts have been made to modify the carcinogenicity of anticancer drug therapy, either by substitution of agents (Bonnadonna & Santero, 1982), modification of intensity, simultaneous administration of neutralizing compounds (Brock *et al.*, 1979) or better targetting of the drug to the tumour rather than to sensitive normal tissues (Eisenbrand *et al.*, 1986), the risk of second malignancy probably remains substantially elevated over the background.

Although it is known that most alkylating cytotoxic drugs are carcinogenic, there remain substantial questions that require quantitative answers. These include the relative carcinogenicity of different agents, the effect of changes in dose rate and of radiotherapy on their carcinogenicity, and the extent to which short-term measurements such as DNA damage can be used to predict carcinogenicity, at both group and individual levels.

While the primary goal of any study of anticancer drug carcinogenicity is to improve therapy, it must be recognized that such studies can also provide a unique window into the mechanisms of leukaemogenesis induced by alkylating agents. Although rather specific, there is perhaps no better human model for quantitative validation of DNA damage as a predictor of carcinogenicity, for several reasons.

Firstly, in addition to being a high-risk exposure, drug therapy is the only known carcinogenic agent that is given to humans at carefully measured doses and for which reliable exposure records are maintained over long periods. Furthermore, cancer patients are followed up rather well and are easily accessible for further clinical or laboratory investigation during the course of treatment. Finally, acute leukaemia induced by alkylating agents is extremely specific, to the extent that it can be characterized by certain karyotypic abnormalities (Rowley *et al.*, 1977). One can therefore be almost certain of the exposure that led to the cancer, and when it occurred. This is in contrast to many other cancers, the etiology of which is either unknown or multifactorial.

A number of investigators have recognized the unique research opportunities presented by this situation. Several studies have been carried out in which DNA damage, as defined by elevated levels of chromosomal aberrations or sister chromatid exchanges in peripheral lymphocytes, was evaluated in patients following treatment with cytotoxic drugs (Lambert *et al.*, 1978; Neistadt *et al.*, 1978; Raposa, 1978; Lambert *et al.*, 1979). For one anticancer agent, cisplatin, it has been possible to demonstrate the presence of adducts in lymphocytes, and to characterize various different types of adducts (Poirier *et al.*, 1982; Fichtinger-Schepman *et al.*, 1986; Reed

Table 1. Studies of SCE frequency in the peripheral lymophocytes of patients following anticancer drug therapy

Drug	Treated patients (SCE/cell)	Average dose (g)	Control group (SCE/cell)	Reference
Busulphan	11.0	0.49	6.9	Misawa (1978)
Cyclophosphamide	17.9	5.0	13.1	Debova *et al.* (1983)
	25.9	1.1	8.9	Raposa (1978)
Chlorambucil	14.5	4.3	8.1	Palmer *et al.* (1984)
Actinomycin D	17.3	2.0	14.0	Lambert *et al.* (1979)

et al., 1986). In the near future, monoclonal antibodies could be available to detect the adducts of many other agents (Tilby *et al.*, 1987).

Unfortunately, it is difficult to draw quantitative conclusions from most of the studies on chromosomal effects following therapy with anticancer agents. One fundamental question is the degree to which the carcinogenicity of anticancer agents correlates with their potency in inducing DNA damage. We have approached this question by using data on sister chromatid exchange (SCE) frequency from the published literature.

By analogy with estimates of potency in inducing SCE calculated from animal experiments (Parodi *et al.*, 1983), one can propose the excess frequency per unit dose as an index for SCE induction in treated patients. This index may be estimated as

$$I = \frac{f_r - f_c}{d},$$

where f_r is the frequency of SCE in treated patients, f_c is the frequency in controls, and d is the dose received by treated patients, in grams.

We conducted a search of publications on SCE, and found for four cytotoxic drugs sufficiently detailed data to allow estimation of the potency index. These data are given in Table 1; in Table 2, we give estimates of SCE potency as well as the corresponding estimates of carcinogenic potency for these drugs in humans, based on studies of

Table 2. Potency estimates for inducing leukaemia and SCE in humans, and all tumours in rats

Drug	Human leukaemia (ten-year incidence/g total dose)	Rat (all tumours, TD50[a])	Human SCE (excess exchanges/ cell per g total dose)
Busulphan	1.3	0.12	8.4
Cyclophosphamide	0.28	2.1	0.96 (Debova)
			15.5 (Raposa)
Chlorambucil	3.0	0.71	1.5
Actinomycin D	—	0.00095	1.6

[a] Daily exposure in mg/kg body weight per day required to halve the tumour-free frequency (Peto *et al.*, 1984)

leukaemia, and in rats, based on long-term experiments (Kaldor *et al.*, 1988). It should be noted that, unlike the human potency index and the SCE index, the potency index for rats is smaller for more potent agents.

A visual examination of the correlation reveals major inconsistencies between the SCE index and the indices of carcinogenic potency. For example, the two estimates for cyclophosphamide differ by over 15-fold. According to one study, it is the strongest agent in inducing SCE, while according to the other it is the weakest. Similarly, actinomycin D is a far more potent animal carcinogen than the other agents, but it is one of the weakest inducers of SCE in humans *in vivo*. These discrepancies could indicate that SCE induction is not a valid predictor of carcinogenicity. However, problems in the calculation of these indices should be taken into consideration in the interpretation of the results.

Firstly, they are based on very few individuals, and no attempt has been made to quantify the precision of each estimate. Another important factor is the time at which evaluation of SCE was made, relative to treatment with the cytotoxic agent. It is known that the SCE frequency is elevated soon after treatment and declines rapidly thereafter (Einhorn *et al.*, 1982), so that a difference among the studies in the time at which cells were sampled could have a significant effect on the observed SCE level. Further differences in the SCE index could be due to the use of total dose in the denominator, when in fact the dose over a restricted period may be more appropriate in the calculation. Finally, in all of the reported studies, the comparisons between exposed and unexposed are not corrected for age or for other factors that may influence SCE level.

This review of data on SCE highlights several of the epidemiological issues that were raised in the earlier part of this article. Obviously, the studies summarized here were not planned as full-scale epidemiological investigations, but it is clear that before real quantitative interpretations can be made, further studies are required that do take these issues into consideration.

A study of DNA damage in relation to carcinogenicity

In parallel to the small-scale investigation of DNA damage in patients following cytotoxic drug therapy have been studies of the long-term cancer risk in these patients. These studies have been principally carried out in survivors of Hodgkin's disease, ovarian cancer and a few other malignancies. We have recently completed evaluations of second cancer risk among over 133 000 survivors of Hodgkin's disease, ovarian cancer and testicular cancer (Kaldor *et al.*, 1987) and are in the process of carrying out detailed case-control studies of specific aspects of therapy in relation to the risk.

The approach we have taken in these studies is to establish as large a cohort as possible of patients treated with one or more index cancers. Treatment details are abstracted only for individuals who develop a second cancer and for several matched controls from the cohort who do not. In this way, a considerable saving is made in resources, and the comparisons between second cancer cases and matched controls with regard to therapy are almost as precise as if the whole cohort had been studied.

Exactly the same approach can be used to study the relationship between DNA damage, whether measured at the molecular or chromosomal level, and the risk of drug-induced second cancer in treated patients. In collaboration with the European Organization for Research and Treatment of Cancer (EORTC), we are currently planning a long-time investigation of this kind.

The study will proceed in several phases. The first is the establishment of a collaborative group of major hospitals, which will vary somewhat according to the index cancer site due to the different degrees of specialization among treatment institutes. A central register will then be established of all patients diagnosed with one of the index cancers at a participating hospital. These patients will then be followed up prospectively for the occurrence of second malignancies, particularly acute non-lymphocytic leukaemia, and, as each second cancer arises, the therapy record for that case and several matched controls will be examined with the goal of signalling carcinogenic therapeutic modalities as rapidly as possible.

In parallel to the registration of index and second cancers, blood samples will be taken from each patient diagnosed with one of the index cancers, immediately following diagnosis and then subsequent to therapy. The red and white blood cells will be separated and stored at −70°C. For patients who develop an acute leukaemia, or possibly other cancers, following therapy for the first cancer, and for matched controls who remain free of a second cancer, the samples will be analysed for DNA adducts and chromosomal aberrations and comparisons made between the second cancer cases and these controls.

This study has two general aims. The first is to find some early marker of second cancer risk, which can be used to identify patients who appear to be placed at higher risk by their therapy. For these individuals, appropriate modifications aimed at reducing risk could be applied. The second goal is to gain information of a fundamental nature on the relationship between short-term measures of DNA damage and the risk of malignancy. The studies carried out to date have revealed that many agents that cause cancer following occupational, iatrogenic or other exposure are also capable of causing various kinds of DNA damage, but it has not yet been possible to establish a link at the level of the individual between DNA damage and cancer risk. Studies of cancer patients would appear to offer the best possibility of obtaining this information, since the carcinogenic effect of exposure to cytotoxic drugs occurs rather rapidly and the fact that patients are already undergoing systematic follow-up facilitates the taking of samples and the recording of the incidence of second cancers.

References

Bonnadonna, G. & Santero, A. (1982) ABVD chemotherapy in the treatment of Hodgkin's disease. *Cancer Treat Rev.*, *9*, 21–35

Breslow, N.E., Lubin, J.H. & Langholz, B. (1983) Multiplicative models and cohort analysis. *J. Am. stat. Assoc.*, *78*, 1–12

Brock, N., Stekar, J., Pohl, J. & Scheef, W. (1979) Antidote against the urotoxic effects of the oxazophosphorine derivatives cyclophosphamide, ifosfamide and trofosfamide. *Naturwissenschaften*, *66*, 60–61

Debova, G.A., Sapacheva, V.A. & Filippova, T.V. (1983) Frequency of sister chromatid exchanges in cancer patients during treatment with cyclophosphane. *Tsitol. Genet.*, *17*, 65–66

Einhorn, N., Eklung, G., Franzén, S., Lambert, B., Lindsten, J. & Söderhäll, S. (1982) Late side effects of chemotherapy in ovarian carcinoma. *Cancer*, *459*, 2234–2241

Eisenbrand, G., Müller, N., Schreiber, J., Stahl, W., Sterzel, W., Berger, M.R., Zeller, W.J. & Fiebig, H. (1986) *Drug design*: *nitrosoureas*. In: Schmähl, D. & Kaldor, J.M., eds, *Carcinogenicity of Alkylating Cytostatic Drugs* (*IARC Scientific Publications No. 78*), Lyon, International Agency for Research on Cancer, pp. 281–294

Everson, R.B., Randerath, E., Santella, R.M., Cefalo, R.C., Avitts, T.A. & Randerath, K. (1986) Detection of smoking-related covalent DNA adducts in human placenta. *Science*, *231*, 55–57

Fichtinger-Schepman, A.M.J., Lohman, P.H.M., Berends, F., Reedijk, J. & van Oosterom,

A.T. (1987) *Interactions of the antitumour drug cisplatin with DNA in vitro and in vivo.* In: Schmähl, D. & Kaldor, J.M., eds, *Carcinogenicity of Alkylating Cytostatic Drugs (IARC Scientific Publications No. 78)*, Lyon, International Agency for Research on Cancer, pp. 83–99

IARC (1981) *IARC Monographs on the Evaluation of the Carcinogenic Risk of Chemicals to Humans,* Vol. 26, *Some Anticancer and Immunosuppressive Drugs,* Lyon

IARC (1985) *IARC Monographs on the Evaluation of the Carcinogenic Risk of Chemicals to Humans,* Vol. 38, *Tobacco Smoking,* Lyon

Kaldor, J.M., Day, N.E. & Hemminki, K. (1988) Quantification of drug carcinogenicity in humans and animals. *Eur. J. Cancer clin. Oncol.* (in press)

Kaldor, J.M., Day, N.E., Band, P., Choi, N.W., Clarke, E.A., Coleman, M.P., Hakama, M., Koch, M., Langmark, F., Neal, F.E., Pettersson, F., Pompe-Kirn, V., Prior, P. & Storm, H.H. (1987) Second malignancies following testicular cancer, ovarian cancer and Hodgkin's disease: an international collaborative study among cancer registries of the long-term effects of therapy. *Int. J. Cancer, 39,* 571–585

Lambert, B., Ringborg, U., Harper, E. & Lindblad, A. (1978) Sister chromatid exchanges in lymphocyte cultures of patients receiving chemotherapy for malignant disorders. *Cancer Treat. Rep., 62,* 1413–1419

Lambert, B., Ringborg, U. & Lindblad, A. (1979) Prolonged increase of sister-chromatid exchanges in lymphocytes of melanoma patients after CCNU treatment. *Mutat. Res., 59,* 295–300

Lutz, W.K. (1986) Quantitative evaluation of DNA binding data for risk estimation and for classification of direct and indirect carcinogens. *J. Cancer Res. clin. Oncol., 112,* 85–91

Misawa, S. (1978) Cytogenetic studies of chronic myelocytic leukaemia. II. Mutagenicity of busulfan detected by the induction of sister chromatid exchanges in human lymphocytes. *J. Kyoto Pref. Univ. Med., 87,* 833–852

Neistadt, E.L., Gershanovich, M.L., Kolygin, B.A., Ogorodnikova, B.N., Fedoreev, G.A., Chekharina, E.Ak. & Filov, V.A. (1978) Effect of chemotherapy on the lymph node and bone marrow cell chromosomes in patients with Hodgkin's disease. *Neoplasia, 25,* 91–94

Palmer, R.G., Doré, C.J. & Denman, A.M. (1984) Chlorambucil-induced chromosome damage to human lymphocytes is dose-dependent and cumulative. *Lancet, i,* 246–249

Parodi, S., Zunino, A., Ottaggio, L., De Ferrari, M. & Santi, L. (1983) Quantitative correlation between carcinogenicity and sister chromatid exchange induction *in vivo* for a group of 11 N-nitroso derivatives. *J. Toxicol. environ. Health, 11,* 337–346

Pedersen-Bjerregaard, J. & Larsen, S.O. (1982) Incidence of acute nonlymphocytic leukemia, preleukemia, and acute myeloproliferative syndrome up to 10 years after treatment of Hodgkin's disease. *New Engl. J. Med., 307,* 965–971

Perera, F.P. (1987) Molecular cancer epidemiology: a new tool in cancer prevention. *J. natl Cancer Inst., 78,* 887–898

Perera, F.P., Poirier, M.C., Yuspa, S.H., Nakayama, J., Jaretzki, A., Curnen, M.M., Knowles, D.M. & Weinstein, I.B. (1982) A pilot project in molecular cancer epidemiology: determination of benzo(*a*)pyrene-DNA adducts in animal and human tissues by immunoassay. *Carcinogenesis, 3,* 1405–1410

Peto, R., Pike, M.C., Bernstein, L., Gold, L.S. & Ames, B.N. (1984) The TD50: a proposed general convention for the numerical description of the carcinogenic potency of chemicals in chronic exposure animal experiments. *Environ. Health Perspectives, 58,* 1–8

Poirier, M.C., Lippard, S.J., Zwelling, L.A., Ushay, H.M., Kerrigan, D., Thill, C.C., Santella, R.M., Grunberger, D. & Yuspa, S.H. (1982) Antibodies elicited against cis-diamminedichloroplatinum(II)-modified DNA are specific for cis-diamminedichloroplatinum(II)-DNA adducts formed *in vivo* and *in vitro*. *Proc. natl Acad. Sci. USA, 79,* 6443–6447

Raposa, T. (1978) Sister chromatid exchange studies for monitoring DNA damage and repair capacity after cytostatics *in vitro* and in lymphocytes of leukaemic patients under cytostatic therapy. *Mutat. Res., 57,* 241–251

Reed, E., Ozols, R.F., Fasy, T., Yuspa, S.H. & Poirier, M.C. (1986) *Biomonitoring of cisplatin-DNA adducts in cancer patients receiving cisplatin chemotherapy.* In: Ramel, C.,

Lambert, B. & Magnusson, J., eds, *Genetic Toxicology of Environmental Chemicals,* Part B, *Genetic Effects and Applied Mutagenesis,* New York, A.R. Liss, pp. 247–252
Rothman, K.J. (1981) Induction and latent periods. *Am. J. Epidemiol., 114,* 253–259
Rowley, J.D., Golomb, H.M. & Vardiman, J. (1977) Nonrandom chromosomal abnormalities in acute nonlymphocytic leukemia in patients treated for Hodgkin's disease and non-Hodgkin lymphomas. *Blood, 50,* 759
Tilby, M.J., Styles, J.M. & Dean, C.J. (1987) Immunological detection of DNA damage caused by melphalan using monoclonal antibodies. *Cancer Res., 47,* 1542–1546

KARYOTYPES OF HUMAN T-LYMPHOCYTE CLONES

B. Lambert[1], K. Holmberg[1], S.-H. He[1] & N. Einhorn[2]

[1]*Department of Clinical Genetics and* [2]*Department of Gynaecological Oncology, Karolinska Hospital, Stockholm, Sweden*

T-Lymphocyte clones from healthy males and females and from melphalan-treated ovarian carcinoma patients were studied with regard to sporadic chromosomal aberrations and clonal karyotype: 85% of the clones showed a normal, diploid karyotype, and sporadic aberrations were found to occur at about the same low frequency as in short-term lymphocyte cultures. An abnormal karyotype was found in 11 of the 72 clones studied. Loss of an X chromosome, which was the most frequent abnormality in female clones, was verified by densitometry of Southern blots of clonal DNA hybridized with a probe for the X-linked *hprt* locus. Abnormal karyotype due to chromosomal rearrangement was found in nine clones, and, in five of these, chromosome 12 was involved in the aberration. About 33% of the clones from melphalan-treated patients had an abnormal karyotype, in comparison with about 10% of clones from healthy control subjects. This difference indicated that melphalan treatment may induce stable chromosomal rearrangements that are compatible with cellular proliferation and clonal expansion.

Improved procedures for long-term culture of human T-lymphocytes have made it possible to study mutations induced *in vivo* in the X-linked locus for hypoxanthine phosphoribosyl transferase (*hprt*) by cloning freshly isolated T-cells in medium containing thioguanine (TG) (Albertini *et al.*, 1982; Morley *et al.*, 1983). The human T-cell cloning technique also provides a new approach for the detection of chromosomal mutations derived *in vivo,* i.e., stable changes in the clonal karyotype which may have important phenotypic effects. We have combined these methods in a project in progress to detect and characterize a wide range of spontaneous and induced mutations in T-cell clones from healthy subjects and patients subjected to cytostatic treatment.

Here, we report on the frequency of sporadic chromosomal aberrations and chromosomal loss in G-banded metaphases from unselected and TG-resistant T-cell clones, as compared to the frequency of aberrations in short-term lymphocyte cultures. A number of clonal chromosomal abnormalities have been identified in the karyotypes of unselected and TG-resistant T-cell clones from healthy males and females, as well as from melphalan-treated ovarian carcinoma patients. In two female clones, the loss of an X chromosome was verified by densitometry of Southern blots of clonal DNA hybridized with labelled *hprt* DNA.

Cell cultivation and T-cell cloning

Cells were obtained from three healthy male and female donors and from three female patients with ovarian carcinoma. The patients had been treated with melphalan 3.5–4 years before the present study with a total dose of 350–420 mg during an 11- to 12-month period.

The procedure for T-cell cloning was adopted from Albertini *et al.* (1982), with minor modifications. Ficoll-isopaque-purified lymphocytes from buffy coats (male donors) or 30–50 ml of peripheral blood (female donors) were resuspended in growth medium (RPMI-1640 containing 1% phytohaemagglutinin, 10% fetal calf serum, 5% human serum, supplemented with 30–40% conditioned medium) and distributed into 96-well microtitre plates. The cell density for unselected clones was five cells/well in 0.2 ml growth medium containing $2–4 \times 10^4$ lethally ultraviolet irradiated lymphocytes; that for TG-selected clones was $3–6 \times 10^4$ cells/well in growth medium containing 2 µg/ml TG. After 10–14 days' incubation, growing clones were resuspended in fresh medium without TG and transferred to new plates for further expansion. Samples for chromosome analysis were taken at several intervals during clonal expansion. The cells were resuspended in fresh medium in 1.5-ml microfuge tubes and colchicine was added for 1–2 h. Short-term (72 h) cultivation of lymphocytes, G-banding and chromosome analysis were as described previously (Lambert *et al.*, 1986). Only cells with 45 or more chromosomes were analysed.

Sporadic loss of a single chromosome

All short-term lymphocyte cultures studied were from female donors. About 13% of the metaphases in these cultures were found to have lost a chromosome, which was usually an X chromosome (Table 1). In addition, chromosomes 19–22 appeared to be lost more often than the larger chromosomes. The three melphalan-treated patients showed a similar pattern of chromosome loss. The predominance of X chromosome loss in these cells speaks against an artefactual cause of the event, and suggests that female lymphocytes have an increased tendency to lose one of the X chromosomes.

The unselected T-cell clones studies so far were obtained from two healthy males. Chromosome loss occurred in 12.5% of the cells, and seemed to affect the chromosomes at random (Table 1).

In TG-resistant T-cell clones, the frequency of metaphases with a missing chromosome was 8–15% in the three groups studied (Table 1). Chromosome loss seemed to be a random event in these TG-resistant clones, although there was a tendency for preferential loss of smaller chromosomes. Sporadic loss of an X chromosome in female cells occurred less often in TG-resistant T-cell clones than in short-term cultures.

Sporadic, structural chromosomal aberrations

The predominating structural aberrations in female short-term cultures were typical, long X fragments. These fragments were not conserved in the T-cell clones, where aberrations seemed to be randomly distributed (Table 1).

As shown in Table 2 and discussed previously (Lambert *et al.*, 1984, 1986), the dominant aberrations in short-term cultures of patients' cells were exchanges and complex aberrations. These types of aberration were less frequent in the T-cell clones than simple breaks and deletions. The total frequency of structural aberrations in T-cell clones from healthy females and patients was 5–7%, as compared to 6% in short-term cultures from female controls (excluding the typical, long X fragments). The relatively high frequency of aberrations in the TG-resistant clones of healthy males (14%) was due partly to an exceptionally high number of sporadic aberrations (10/55 cells) in one subject.

Abnormalities of the clonal karyotype

As shown in Table 3, a total of 72 T-cell clones has been karyotyped. One clonal aberration was found among nine unselected male clones, and one among 15

Table 1. Frequency and distribution of sporadic chromosomal aberrations in metaphases of short-term cultures and T-cell clones from three healthy males and females and from three patients treated with melphalan for ovarian carcinoma

Chromosome	Numbers of missing chromosomes (Mc) and structural aberrations (Ab)											
	Short-term cultures				Unselected T-cell clones		TG-Resistant T-cell clones					
	Females		Patients		Males		Males		Females		Patients	
	Mc	Ab	Mc	Ab	Mc	Ab	Mc	Ab	Mc	Ab	Mc	Ab
1	1	2	0	4	0	0	0	1	0	1	0	2
2	1	2	0	4	0	0	0	4	0	3	1	0
3	0	1	1	1	1	0	0	0	1	0	1	0
4	0	1	0	2	0	0	0	4	1	1	0	1
5	1	1	1	3	0	0	0	1	0	1	0	0
6	0	0	1	2	0	0	1	2	1	2	0	0
7	3	1	0	1	0	1	0	0	1	0	0	1
8	0	0	3	5	1	0	0	1	4	0	0	0
9	0	1	0	4	1	1	1	5	3	1	0	0
10	0	2	3	2	1	2	1	0	1	1	0	1
11	1	1	0	0	1	0	1	3	0	1	0	0
12	1	1	2	0	0	0	1	3	2	1	0	0
13	2	2	0	1	0	0	0	0	3	1	0	0
14	1	0	2	2	1	0	1	0	1	1	2	1
15	0	1	1	0	0	0	3	0	4	0	0	0
16	0	0	1	1	1	0	0	0	0	0	0	0
17	2	0	1	1	2	0	2	1	2	0	0	1
18	1	0	2	1	0	2	0	1	3	2	0	1
19	2	0	6	0	1	0	0	0	5	0	0	0
20	6	0	2	0	2	0	2	0	4	0	2	1
21	0	1	5	1	0	0	1	0	6	0	1	0
22	3	0	4	0	1	0	0	0	2	0	0	0
X	9	12	9	4	0	0	0	0	1	1	1	1
Y	—	—	—	—	0	0	0	0	—	—	—	—
Total no. of Ab:	34	29	44	39	13	6	14	26	45	17	8	10
No. of cells studied:	264		436		105		154		299		99	
No. of clones studied:	—		—		9		19		33		15	

TG-resistant male clones. One of the female patients was found to have a constitutional deletion of the X (F9/86, 46, XXp−).

Loss of an X chromosome (as a single event or in combination with other abnormalities) occurred in one of the clones from healthy females and in two from the melphalan-treated patients, but in none of the male clones. Two other female clones (F17/86-10 and F17/86-18) showed abnormalities of the X chromosome. Thus, loss or rearrangement of the X seems to be a relatively frequent karyotypic abnormality as it was found in 5/48 (10%) of the female, TG-resistant T-cell clones.

Although the data obtained so far are limited, they nevertheless suggest that abnormal karyotypes are relatively infrequent among T lymphocyte clones from healthy subjects. A tendency towards a higher frequency of abnormal karyotypes was observed in clones from patients (5/15, 33%) than in clones from healthy females (4/33, 12%).

Table 2. Types of sporadic chromosomal aberrations in short-term cultures and T-cell clones from healthy males and females and from melphalan-treated ovarian carcinoma patients

Culture	No. of clones/cells	No. and types of aberrations[a]							Total no. of aberrations
		Break	Del	Dic	Exch	Ring	Cx	Add chrom	
Short-term cultures									
Females	—/264	4	2	0	5	0	6	4	17
Patients	—/436	6	6	1	18	0	8	0	39
Unselected T-cell clones									
(males)	9/107	1	0	0	2	0	1	0	4
TG-Resistant T-Cell clones									
Males	19/157	8	2	3	1	1	6	1	22
Females	33/297	9	2	0	1	0	2	1	15
Patients	15/101	4	0	0	1	0	2	0	7

[a] Del, deletion; Dic, dicentric chromosome; Exch, translocations and other exchanges; Cx, cells with complex rearrangements involving several chromosomes; Add chrom, supernumerary chromosome

Table 3. T-Cell clones with abnormal karyotypes from healthy males and females and from melphalan-treated ovarian carcinoma patients

Selection conditions	No. of clones (abnormal/ studied)	Abnormal clones		
		Clone no.	No. of cells (abnormal/ studied)	Karyotype
Unselected clones				
Males	1/9	F3/86-23	(19/19)	46,XY,t(1p;12q)
TG-resistant clones				
Males	1/15	F1/86-21	(7/10)	46,XY,del(2)(p13);del(17)(p12); + mar(11q); + mar(13q?)
			(1/10)	46,XY
			(Two other cells had deletions of 2(p13), 11(q22) or 17p, + unidentified markers)	
Females	4/33	F10/86-8	(12/12)	45,X, − X
		F15/86-6	(9/9)	46,XX,inv(7)(p15q33)
		F17/86-10	(8/16)	47,XX,mar(?Xp−)
			(7/16)	47,XX,mar(?Xp−) + ace
			(1/16)	46,XX
		F17/86-18	(3/3)	46,X, − X, − 12, − 13,del(9)(q22), + mar(Xp;12q), + mar(Xq;13q), + mar(12p;?)
Patients	5/15	F9/86-7	(6/6)	45,Xp − , − X, − 2, − 8, − 8, − 9, + 4mar(?)
		F9/86-8	(4/4)	46,XXp − ,del(12)(p11)
		F9/86-9	(8/8)	47,XXp − , + mar(G?)
		F14/86-8	(5/8)	46,XX,t(5;12)(q31;q13)
			(3/8)	46,XX
		F16/86-16	(10/10)	45,X, − X

No evidence was obtained that any of the 63 TG-resistant clones had acquired the mutant phenotype as a result of gross chromosomal deletion or rearrangement of the distal part of Xq(26–27), where the *hprt* gene is located. In all the clones that had lost an X chromosome, the remaining X seemed normal (except the constitutionally deleted Xp− in F9/86).

Blot hybridization of DNA from TG-resistant T-cell clones

TG-resistant T-cell clones were expanded to $1–3 \times 10^7$ cells. The DNA was extracted, digested with Pst 1-endonuclease and subjected to electrophoresis and Southern blotting according to standard procedures. Blot hybridization was carried out with a full-length *hprt* probe (pHPT-30, kindly provided by C.T. Caskey) labelled by random priming. As shown in Figure 1, the probe recognizes autosomal pseudogenes in two bands of about 7.5 and 8.5 kb, and five X-linked *hprt* bands which range in size from 1.4 to 5.0 kb.

None of the male or female clones in Figure 1 reveals any obvious deviation from the Pst-1 band pattern of normal, unselected clones (not shown). As loss of bands indicating deletions may be difficult to detect in female clones due to the presence of DNA from the inactive X chromosome, the gels were analysed in a scanning densitometer, using the *hprt* pseudogenes as a reference for quantification. These genes are located on the autosomes and show two distinct bands well separated from the X-linked *hprt* bands. The area under the X-linked *hprt* band peaks from the densitometer tracings was divided by the area under the two pseudogene band peaks (Fig. 1). The quotient was found to be 3.37 ± 0.59 for the four male XY clones, and 5.24 ± 0.41 for six of the female clones that had a normal XX karyotype. Two of the female clones that had lost an X chromosome (F9/7 and F16/6, Table 3) had quotients of about 3, i.e., very similar to that of males. All other female clones, including four clones that were not karyotyped, and F9/9 with a clonal abnormality involving a G chromosome (Table 3), showed quotients between 5.2 and 6.0, indicating the presence of a diploid set of X chromosomes. The densitometer tracing did not reveal any other change in the relative size of the X-linked band peaks that could indicate deletions of single bands. Representative densitometer tracings are shown in Figure 1.

Conclusions

Chromosome analysis of T lymphocyte clones appears to be a useful method for detecting genomic damage in human somatic cells. Structural aberration and loss of a chromosome were found to occur at random and with about the same low frequency in the metaphases of T-cell clones as in short-term lymphocyte cultures. No evidence for preferential engagement of specific chromosomes or break points in these sporadic aberrations was obtained.

Out of 72 T-cell clones studied, 85% showed a normal, diploid karyotype. The abnormal karyotypes in females clones often engaged the X chromosome, corroborating previous observations of a high frequency of X-chromosome abnormalities in short-term cultures of female lymphocytes (Lambert *et al.*, 1986). In two female clones, the loss of an X chromosome was verified by densitometry of blot hybridized clonal DNA using a *hprt* probe. The frequency of clonal karyotype abnormalities in TG-resistant clones was 16%, but none of the abnormalities appeared to affect the *hprt* locus. Thus, large chromosomal deletion or rearrangement seems to be a rare cause of the mutation in this locus.

A tendency for a higher frequency of clones with abnormal karyotype was observed in the group of melphalan-treated patients (33%) than in the healthy female controls

Fig. 1. Southern blot hybridization (top left) and densitometer profiles of PstI-digested DNA from TG-resistant T-cell clones hybridized with a full-length *hprt* probe

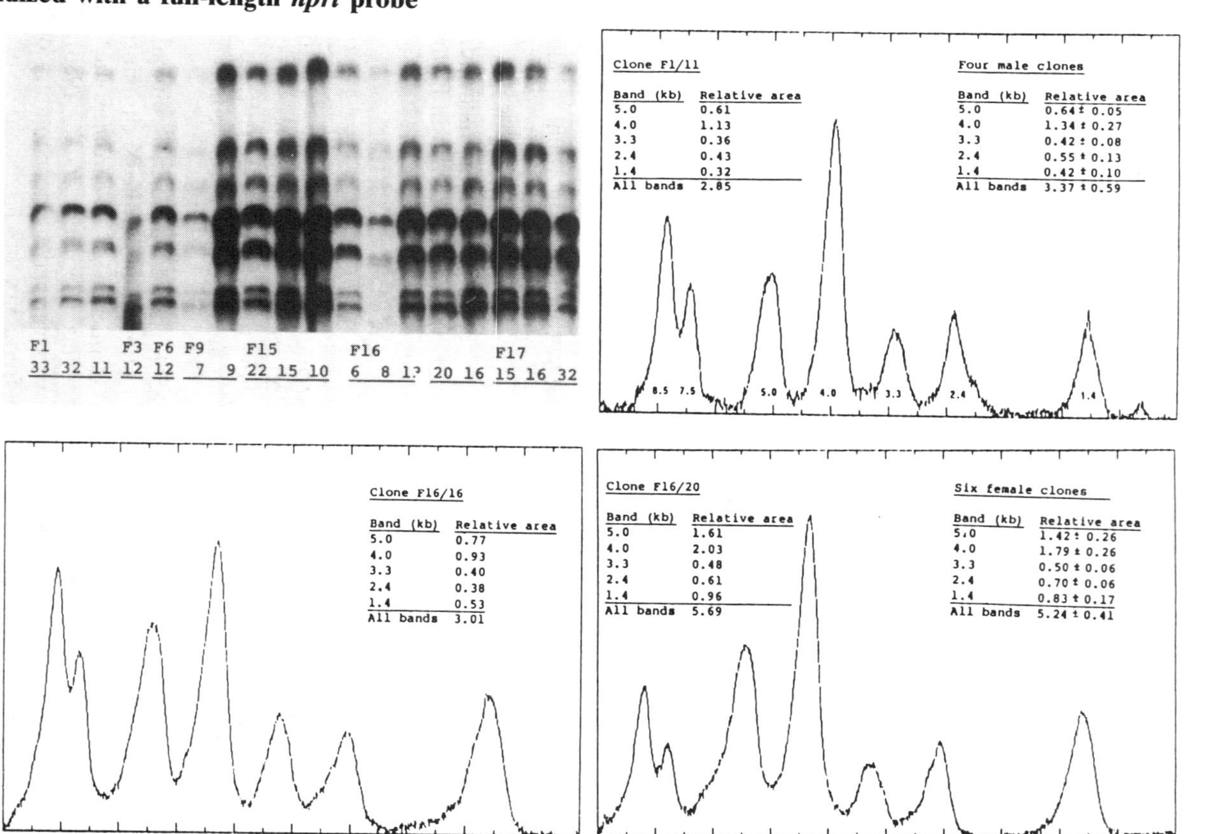

F1/33,32,11, F3/12 and F6/12 are clones from three healthy males. F9/7,9 and F16/6,8,13,20,16 are clones from two melphalan-treated ovarian carcinoma patients. F15/22,15,10 and F17/15,16,32 are clones from two healthy females. The tabular inserts in the densitometer profiles show quotients between the area under X-linked band peaks (1.4, 2.4, 3.3, 4.0 and 5.0 kb) and pseudogene band peaks (7.5 and 8.5 kb). About 10 μg DNA from each of the clones were digested to completion with PstI-restriction enzyme as specified by the deliverer, fractionated on 0.8% agarose gel and transferred to nitrocellulose filter. Prehybridization was carried out overnight at 42°C in 6 × SSC, 10 × Denhardt's, 1% SDS containing 100 μg/ml sonicated, denatured salmon sperm DNA. For hybridization, 25 ng of the ^{32}P-labelled probe was used in 15 ml of 6 × SSC, 1% SDS, 100 μg/ml of salmon sperm DNA, 50% formamide and 5% dextran sulphate. The probe was a PstI-insert of the plasmid pHPT30 (kindly provided by C.T. Caskey) labelled by random priming to a specific activity of 1.3×10^9 cpm/μg. After hybridization, the filter was washed several times in 2 × SSC, 0.1% SDS, then in 0.5 × SSC, 0.1% SDS. The final wash was for 30 min at 65°C. The filter was then exposed overnight at − 70°C, using Fuji XR film backed with intensifying screen. The band patterns of all clones shown in the top left picture (except F9/9 and F3/12) were scanned in a BioRad 1650 densitometer, and the area under the peaks was calculated after base line correction using the Hoefer GS-350 Data system for IBM PC. The direction of scanning was from bottom to top of the blot. The sizes of the bands were determined by comparison with suitable size markers, and they are indicated in the top right profile. The 8.5- and 7.5-kb bands are autosomal pseudogenes, and the others are X-linked (Yang *et al.*, 1984). To calculate the relative amount of X-linked *hprt* DNA, the area under each of the X-linked bands was divided by the area under the pseudogene bands in the same profile. The resulting quotients are shown as inserts over each profile. Also included as inserts in two of the profiles are the mean values for the quotients of six female clones and four male clones showing normal diploid karyotypes. The quotients in the female clones F9/7 (not shown) and F16/16 (bottom left) were similar to male clones, indicating the loss of an X chromosome.

(12%). This may indicate an effect of the cytostatic treatment, but the limited data do not permit a definite conclusion.

Acknowledgements

This work was supported by the Swedish Cancer Society, the Swedish Environment Protection Board, the King Gustav V Jubilee Fund, Stockholm and the Swedish National Board for Animal Care. We thank C.T. Caskey for the gift of pHPT30.

References

Albertini, R.J., Castle, K.L. & Borcherding, W.R. (1982) T-Cell cloning to detect the mutant 6-thioguanine-resistant lymphocytes present in human peripheral blood. *Proc. natl Acad. Sci. USA, 79,* 6617–6621

Lambert, B., Holmberg, K. & Einhorn, N. (1984) Persistence of chromosome rearrangements in peripheral lymphocytes from patients treated with melphalan for ovarian carcinoma. *Human Genet., 67,* 94–98

Lambert, B., Holmberg, K. & Einhorn, N. (1986) *Chromosome damage and second malignancy in patients treated with melphalan.* In: Schmähl, D. & Kaldor, J.M., eds, *Carcinogenicity of Alkylating Cytostatic Drugs* (*IARC Scientific Publications No. 78*), Lyon, International Agency for Research on Cancer, pp. 147–160

Morley, A.A., Trainor, K.J., Seshadri, R. & Ryall, R.G. (1983) Measurement of *in vivo* mutations in human lymphocytes. *Nature, 302,* 155–156

Yang, T.P., Patel, P.I., Chinault, A.C., Stout, J.T., Jackson, L.G., Hildebrand, B.M. & Caskey, C.T. (1984) Molecular evidence for new mutation at the *hprt* locus in Lesch-Nyhan patients. *Nature, 310,* 412–414

ROLE OF ONCOGENES IN CHEMICAL CARCINOGENESIS: EXTRAPOLATION FROM RODENTS TO HUMANS

M.W. Anderson[1], R.R. Maronpot[2] & S.H. Reynolds[1]

[1]*Laboratory of Biochemical Risk Analysis; and* [2]*Chemical Pathology Branch, National Toxicology Program, National Institute of Environmental Health Sciences, Research Triangle Park, NC, USA*

Proto-oncogenes are cellular genes that are expressed during normal growth and development processes. These genes can be activated to cancer-causing oncogenes by point mutations or by gross DNA rearrangement, such as chromosomal translocation or gene amplification. Activated versions of proto-oncogenes have been observed in various human and rodent tumours. Examples will be discussed, as will the activation of proto-oncogenes by chemical carcinogens. Most chemicals are classified as potentially hazardous to humans on the basis of long-term carcinogenesis studies in rodents. Oncogene analysis of tumours of spontaneous origin and from carcinogenesis studies may aid in risk analyses on the basis of rodent carcinogenesis studies.

Increasing evidence suggests that a small set of cellular genes are targets for genetic alterations that contribute to the neoplastic transformation of cells. These genes, termed proto-oncogenes, were initially discovered as the transduced oncogenes of acute transforming retroviruses (Bishop, 1985). Recent studies have established that proto-oncogenes can also be activated as oncogenes by mechanisms independent of retroviral involvement (Varmus, 1984; Weinberg, 1985). These mechanisms include point mutations and gross DNA rearrangements such as translocations and gene amplifications. The activation of proto-oncogenes by genetic alterations results in altered levels of expression of the normal protein product, or in normal or altered levels of expression of an abnormal protein.

Proto-oncogenes are expressed during 'regulated growth' such as embryogenesis, regeneration of damaged liver and stimulation of cell mitosis by growth factors. Proto-oncogenes are highly conserved, being detected in species as divergent as yeast, *Drosophila* and humans. Included are genes which encode for growth factors (*sis*), growth factor receptors (*neu, erb B, fms*), regulatory proteins in signal transduction (*ras* family), nuclear regulatory proteins (*myc, myb, fos*) and tyrosine kinases (*scr, abl*). Thus, the encoded proteins appear to play a crucial role in normal cellular growth and differentiation.

The activation of proto-oncogenes in spontaneous and chemically induced tumours has been studied in great detail over the past several years. The number of proto-oncogenes that must be activated in the multistep process of carcinogenesis is unclear at present, although the concerted expression of at least two oncogenes, *ras* and *myc,* is needed for the transformation of a primary cell *in vitro* (Land *et al.*, 1983). In addition to activation of proto-oncogenes, loss of specific regulatory functions such as tumour suppressor genes may be a distinct step in neoplastic transformation (Barrett *et al.*, 1987). We discuss the detection of activated oncogenes in human tumours and in spontaneous and chemically induced tumours of rodents as well as the role of these oncogenes in the multistep development of neoplasia. The implications of activated oncogenes in rodent tumours are discussed in terms of extrapolation of data on carcinogenesis in rodents to human risk assessment.

Activation of proto-oncogenes by gene amplification and chromosomal translocation

The induction of aberrant expression of proto-oncogenes by gene amplification and chromosomal translocation has been observed in several types of human and rodent tumours (Meltzer *et al.*, 1986; Alitalo *et al.*, 1987; Croce, 1987). Such aberrant expression appears to be an important component in the development of some neoplasias. Aberrant expression consists of elevated levels of the normal protein product or deregulated cell-cycle expression of the gene. It should be emphasized that it is not necessary for the protein to be structurally altered — only abnormal expression of the proto-oncogene. Elevated levels of proto-oncogene expression have been observed when apparently neither gene amplification nor chromosomal rearrangement was involved (Gazdar *et al.*, 1985; Escot *et al.*, 1986). Thus, other, perhaps more subtle, types of mechanism can also induce aberrant expression. For example, deletion of a region upstream from the c-*myb* locus which resulted in increased messenger RNA levels was observed in some leukaemias and lymphomas (Barletta *et al.*, 1987).

The *myc* family of genes and c-*myb* code for nuclear proteins, and these genes respond to mitogenic agents. The *bcl*-2 gene is also expressed in B and T lymphocytes after stimulation of the cells with mitogens (Reed *et al.*, 1987). HER-2/*neu* codes for a protein that is a putative receptor for a growth factor (Schechter *et al.*, 1984). Thus, it is not surprising that aberrant expression of these genes contributes to the development of neoplasia in numerous tumour types (Alitalo *et al.*, 1987; Croce, 1987; Slamon *et al.*, 1987).

Chromosomal translocation can also result in the expression of an altered protein product. The c-*abl* proto-oncogene is recombined with the *bcr* gene in 95% of chronic myelocytic leukaemia. The c-*abl-bcr* protein product is a transforming protein with kinase activity. An altered c-*abl* protein has also been observed in a few cases of acute lymphocytic leukaemia (Croce, 1987).

The development of neoplasia is a multistep process (Land *et al.*, 1983; Weinberg, 1985; Barrett *et al.*, 1987), and activation of a proto-oncogene may represent one step in the tumorigenic process. In general, gene amplification is associated with tumour progression. For example, N-*myc* amplification correlates with the stage classification of neuroblastoma. The degree of gene amplification is a better predictor of survival time than any clinical data (Brodeur *et al.*, 1984; Seeger *et al.*, 1985). Similarly, the degree of HER-2/*neu* amplification is inversely related to survival and time to relapse in women with breast cancer (Slamon *et al.*, 1987). The degree of gene amplification in patients with stage I disease may thus be useful in design of treatment strategy. The high percentage of neoplasms showing c-*myc* translocation in Burkitt's lymphoma, and the prevalence of c-*abl* translocation in chronic myelocytic leukaemia suggest that gene translocation represents one important step in the development of these cancers (Croce, 1987).

Detection of oncogenes by DNA transfection assays

A number of oncogenes present in both human and animal tumours have been detected by the NIH/3T3 transfection assay. This test involves the ability of the NIH/3T3 mouse fibroblast to accept and express genes from donor tumour DNA, resulting in the formation of transformed cells. Shih *et al.* (1979) were the first to show that DNA from carcinogen-transformed cell lines could cause transformation of NIH/3T3 cells after transfection. This transformation was characterized by a change in the morphology of the NIH/3T3 cells and by anchorage-independent growth. Other investigators using this technique were then able to show that dominant transforming genes or oncogenes were present in human tumours and in carcinogen-induced animal tumours. An extension of the NIH/3T3 transfection assay that affords greater sensitivity is the nude mouse tumorigenicity assay (Fasano *et al.*, 1984). This involves cotransfection of NIH/3T3 cells with tumour DNA and a selectable marker gene. The selected cells are then injected subcutaneously into the immuno-compromised mice. The tumours that develop in the nude mice are then analysed to characterize the transforming gene.

Members of the *ras* gene family were the first activated proto-oncogenes detected in the NIH/3T3 assay. H-*ras*, K-*ras* and N-*ras* acquire transforming activity by a point mutation in their coding sequence (Barbacid, 1987). In early studies using the NIH/3T3 assay, *ras* gene activation was detected in a low percentage of human tumours (approximately 10%). However, more recent studies utilizing the nude mouse assay (Bos *et al.*, 1985), gene amplification in conjunction with oligonucleotide hybridization (Bos *et al.*, 1987a) and the RNase A mismatch cleavage method (Forrester *et al.*, 1987) have been used to detect activated *ras* genes at a much higher frequency in some human tumour types (Table 1). For example, Bos *et al.* (1987b) detected activated *ras* genes in 27% of cases of acute myelocytic leukaemia examined; Ananthoswamy *et al.* (1987) detected Ha-*ras* genes in four of six human squamous-cell carcinomas examined; and Bos *et al.* (1987a) and Forrester *et al.* (1987) detected activated K-*ras* in 33–40% of human colon tumours. Other oncogenes have been detected in human tumours, including *lca* (Ochiya *et al.*, 1986), *hst* (Sakamoto *et al.*, 1986), *trk* (Martin-Zanca, 1986) and a transforming gene in human thyroid carcinomas (Fusco *et al.*, 1987). The prevalence of these non-*ras* genes in human tumours is not clear at present.

A variety of animal tumour model systems has also been examined for activated genes using the NIH/3T3 assay (Barbacid, 1987; Sukumar, 1987). These include spontaneous tumours in rats and mice, tumours that arise after single or multiple doses of carcinogen, and tumours that arise after long-term exposure to a carcinogen. Examples of the activated genes in the different tumour model systems are shown in

Table 1. Activated oncogenes detected in human tumours by DNA transfection assay[a]

Tumour	Number positive/ number tested	Oncogene	Reference
Colon	26/66	K-*ras*(26)[b], N-*ras*(2)[c]	Forrester *et al.* (1987)
Colon	11/27	K-*ras*(10), N-*ras*(1)	Bos *et al.* (1987a)
Acute myelocytic leukaemia	12/45	N-*ras*(10), K-*ras*(2)	Bos *et al.* (1987b)
Squamous-cell carcinoma	4/6	H-*ras*(4)	Ananthoswamy *et al.* (1987)
Acute myelocytic leukaemia and preleukaemic cells[d]	3/8	N-*ras*(3)	Hirai *et al.* (1987)
Thyroid papillary carcinoma	5/20	Unknown(5)	Fusco *et al.* (1987)
Hepatocellular carcinoma	2/11	*lca*(2)	Ochiya *et al.* (1986)

[a]The basis for detection of the oncogenes listed in this table is the DNA transfection assay; biochemical methods have enhanced the sensitivity to detect mutations in *ras* genes.
[b]In parentheses, the number of samples with that oncogene
[c]Two tumours contained both K-*ras* and N-*ras*.
[d]Activated N-*ras* was detected in DNA from bone-marrow cells before patients developed acute myelocytic leukaemia.

Table 2. Activated oncogenes detected in rodent tumour models by the DNA transfection assay

Model[a]	Tumour	Number positive/ number tested	Oncogene	Reference
Spontaneous	Mouse liver	17/27	H-*ras*(15)[a], *raf*(1), unknown(1)	Fox & Watanabe (1985); Reynolds *et al.* (1986)
Single dose				
MNU	Rat mammary	61/71	H-*ras*(61)	Sukumar *et al.* (1983)
VC	Mouse liver	10/10	H-*ras*(10)	Wiseman *et al.* (1986)
Continuous dose				
Aflatoxin B_1	Rat liver	10/11	K-*ras*(2), unknown(9)	McMahon *et al.* (1986)
TNM	Rat and mouse lung	18/19, 10/10	K-*ras*(18), K-*ras*(10)	Stowers *et al.* (1987)
DMBA	Mouse skin	4/4	H-*ras*(3), unknown(1)	Bizub *et al.* (1986)
Initiation-promotion				
DMBA + TPA	Mouse skin	33/37	H-*ras*	Balmain & Pragnell (1983)
Transplacental dose				
ENU	Rat neuroblastomas	3/3	*neu*(3)	Schechter *et al.* (1984)
MNU	Rat schwannomas	10/13	*neu*(10)	Sukumar (1987)
Radiation				
Gamma	Mouse skin	4/4	K-*ras*	Guerrero *et al.* (1984)
Ionizing	Rat skin	6/10	H-*ras*	Sawey *et al.* (1987)

[a]MNU, *N*-methyl-*N*-nitrosourea, VC, vinyl carbamate; TNM, tetranitromethane; DMBA, 7,12-dimethylbenz[*a*]anthracene; TPA, 12-*O*-tetradecanoylphorbol 13-acetate; ENU, *N*-ethyl-*N*-nitrosourea
[b]In parentheses, the number of samples with that oncogene

Table 2. Like the human tumours, the majority of the activated oncogenes detected in animal tumours are members of the *ras* gene family. Other oncogenes have been detected in animal tumours using the NIH/3T3 assay, including the activated *neu* oncogene found in nervous-tissue tumours induced in rats by transplacental exposure to *N*-methyl-*N*-nitrosourea (Schechter *et al.*, 1984) or *N*-ethyl-*N*-nitrosourea (Sukumar, 1987). The *neu* proto-oncogene was activated by a single point mutation. *neu* and the *ras* genes are the only examples of oncogenes detected in tumours which are activated by a point mutation.

The non-*ras* genes detected in human and rodent tumours by DNA transfection assays are presented in Table 3. The prevalence of these oncogenes in human and rodent tumours is unclear at present, and, for most of these putative oncogenes, the mechanism of activation of the proto-oncogene is also unknown. The identification of new classes of oncogenes, as well as the detection of novel mutations in the *ras* genes (Reynolds *et al.*, 1987), should be enhanced by extension of the NIH/3T3 transfection method to the nude mouse tumorigenicity assay and by development of transfection assays that include recipient cells other than fibroblasts.

Activation of oncogenes by carcinogens

Studies in animal tumour model systems suggest that chemicals and radiation may play a role in the activation of oncogenes by point mutation. Point mutations resulting in the activation of *ras* proto-oncogenes in several chemically induced rodent tumours are consistent with the known alkylation patterns of the carcinogens. For example, the mutation at the 12th codon of the H-*ras* gene detected in rat mammary tumours induced by *N*-methyl-*N*-nitrosourea (Sukumar *et al.*, 1983) is consistent with the formation of the O^6-methylguanine adduct. The activating mutation in the 61st codon of the H-*ras* gene found in mammary tumours and skin tumours induced by

Table 3. Non-*ras* oncogenes detected in human and rodent tumours by the DNA transfection assay

Tumour[a]	Treatment[b]	Oncogene	Reference
Neuroblastoma (R)	ENU	*neu*	Schechter *et al.* (1984)
Schwannoma (R)	MNU	*neu*	Sukumar (1987)
Stomach carcinoma (H)	—	*hst*	Sakamoto *et al.* (1986)
Colon carcinoma (H)	—	*trk*	Martin-Zanca *et al.* (1986)
Hepatocellular carcinoma (M)	Untreated	*raf*	Reynolds *et al.* (1987)
Hepatocellular carcinoma (H)	—	*lca*	Ochiya *et al.* (1986)
Hepatocellular carcinoma (R)	AFB	?	McMahon *et al.*(1986)
Hepatocellular carcinoma (M)	Untreated	?	Reynolds *et al.* (1987)
Hepatocellular carcinoma (M)	Furfural	?	Reynolds *et al.* (1987)
Pulmonary adenocarcinoma (M)	Untreated	?	U. Candrian and M.W. Anderson, unpublished
Nasal squamous carcinoma (R)	MMS	?	Garte *et al.* (1985)
Skin carcinoma (M)	DMBA	?	Bizub *et al.* (1986)
Skin carcinoma (M)	DBA	?	Bizub *et al.* (1986)
Thyroid carcinoma (H)	—	?	Fusco *et al.* (1987)

[a]In parentheses, species in which tumours occur: R, rat; M, mouse; H, human
[b]ENU, *N*-ethyl-*N*-nitrosourea; MNU, *N*-methyl-*N*-nitrosourea; AFB, aflatoxin B_1; MMS, methylmethane sulphonate; DMBA, 7,12-dimethylbenz[*a*]anthracene; DBA, dibenz[*c,h*]anthracene

7,12-dimethylbenz[*a*]anthracene is consistent with its binding to adenosine residues (Zarbl *et al.*, 1985). 'Hot spots' for activating mutations in *ras* oncogenes have been observed; for example, the G*G*T or G*G*A→G*A*T or G*A*A mutation observed in the 12th codon of H-*ras* oncogenes detected in tumours induced by *N*-methyl-*N*-nitrosourea is always at the second G of this codon, even though a similar mutation at the first G could also produce an activated *ras* oncogene. If the sequence of specificity for the binding of a chemical to DNA corresponds to a known biological 'hot spot' in an oncogene, the chemical is potentially a very potent carcinogen.

Some chemicals may not directly activate the oncogenes detected in tumours. This possibility must be considered when activated oncogenes are observed in spontaneously occurring tumours (Table 2). The chemical may increase the background tumour incidence by a mechanism such as cytotoxicity or receptor-mediated promotion. If the pattern of activated oncogenes in chemically induced tumours is different from that in spontaneously occurring tumours, then the chemical probably causes the mutations, at least in some of the tumours (Reynolds *et al.*, 1987).

Although gene amplification and chromosomal translocation have been observed in several types of human tumours, these activating mechanisms have not been observed or studied extensively in spontaneous or chemically induced rodent tumours. Sawey *et al.* (1987) did observe c-*myc* gene amplification or restriction polymorphisms in addition to activated K-*ras* genes in rat skin tumours induced by ionizing radiation. Quintanilla *et al.* (1986) suggested that amplification of the mutated H-*ras* gene may be involved in the progression of mouse skin papillomas to carcinomas. Further studies are required to determine the possible role of chemicals and radiation in the activation of proto-oncogenes by gene amplification, chromosomal translocation and other mechanisms that can alter gene expression.

Carcinogen-induced rodent tumour models may be useful in determining the temporal activation of oncogenes in tumour development. Evidence from several animal studies suggests that activation of the *ras* proto-oncogene is an early event. The activated *ras* gene has been detected in many benign tumours, including mouse skin papillomas and lung and liver adenomas, and rat basal-cell and clitoral-gland tumours. This implies that activated *ras* was present in the cell that clonally expanded to these benign tumors. In addition, it was shown recently that mouse epidermal cells injected *in vivo* with the viral Ha-*ras* gene can be promoted with 12-*O*-tetradecanoylphorbol 13-acetate to papillomas (Quintanilla *et al.*, 1986). Thus, activation of the *ras* proto-oncogene may be the 'initiation' event in some model systems. Moreover, dormant 'initiated' cells with the activated *ras* gene can survive surrounded by normal cells until stimulated to proliferate by some endogenous or exogenous agent.

Extrapolation from rodents to humans

The transformation of a normal cell into a tumorigenic cell involves the activation and concerted expression of several proto-oncogenes, as well as, perhaps, inactivation of suppressor genes. The activation of *ras* proto-oncogenes represents one step in the multistep process of carcinogenesis for a variety of rodent and human tumours. This activation is probably an early event in tumorigenesis and may be the 'initiation' event in some cases. Thus, a chemical that induces rodent tumours by activation of *ras* proto-oncogenes can potentially invoke one step of the neoplastic process in humans exposed to the chemical. Moreover, dominant transforming oncogenes other than *ras* have been detected in chemical-induced rodent tumours (Table 3). The involvement of these oncogenes in the development of human tumours is unclear at present.

Most chemicals are classified as potentially hazardous to humans on the basis of long-term carcinogenesis studies in rodents. While such studies are often designed to mimic the route of human exposure in the environment or the workplace, the dose given is usually higher than that to which humans are actually exposed. This, coupled with the appearance of species- and strain-specific spontaneously occurring tumours in vehicle-treated rodents, complicates the extrapolation of rodent carcinogenic data to determine human risk. Oncogene analysis of tumours of spontaneous origin and from long-term carcinogenesis studies can be useful in several ways. It allows analysis of the mechanisms of tumour formation at the molecular level. For instance, the finding of activating mutations in different codons of the Ha-*ras* gene in chemical-induced liver tumours, but of activating mutations in only one codon of the Ha-*ras* gene in spontaneous liver tumours, could, in the absence of any cytotoxic effect, indicate that the chemical itself activates the Ha-*ras* proto-oncogene by a genotoxic event. Comparison of patterns of oncogene activation in spontaneous and in chemically induced rodent tumours, together with cytotoxic information, should be helpful in determining whether the chemical is mutagenic, cytotoxic, has a receptor-mediated mechanism of promotion, or has some combination of these (and other) modes of action. This type of analysis might be of particular importance for compounds that are not mutagenic in short-term tests but which are carcinogenic in long-term bioassays (Tennant *et al.*, 1987). Species-to-species extrapolation of risk on the basis of carcinogenesis data may become more reliable with examination of oncogene activation and expression. For example, K-*ras* oncogenes with the same activating lesion in the 12th codon were observed in both rat and mouse lung tumours induced by tetranitromethane (Stowers *et al.*, 1987). Even though nothing is known about the DNA damaging properties of this chemical, these data suggest that it acts in the same manner to induce tumours in both rats and mice. These and similar approaches to explore the mechanisms by which chemicals induce tumours in animal model systems may remove some of the uncertainty in risk analysis on the basis of rodent carcinogenesis data.

References

Alitalo, K., Koskinen, P., Makela, T.P. & Saksela, K. (1987) *myc* oncogenes: activation and amplification. *Biochim. biophys. Acta, 907,* 1–32

Ananthoswamy, H.N., Price, J.E., Goldberg, L.H. & Straka, C. (1987) Simultaneous transfer of tumorigenic and metastatic phenotypes by transfection with genomic DNA from a human cutaneous squamous cell carcinoma. *Proc. Am. Assoc. Cancer Res., 28,* 69

Balmain, A. & Pragnell, I.B. (1983) Mouse skin carcinomas induced *in vivo* by chemical carcinogens having a transforming Harvey-*ras* oncogene. *Nature, 303,* 72–74

Barbacid, M. (1987) *ras* genes. *Ann. Rev. Biochem., 56,* 780–813

Barletta, C., Pelicci, P.-G., Kenyon, L.C., Smith, S.D. & Dalla-Favera, R. (1987) Relationship between the c-*myb* locus and the 6q-chromosomal aberration in leukemias and lymphomas. *Science, 235,* 1064–1067

Barrett, J.C., Oshimura, M. & Koi, M. (1987) *Role of oncogenes and tumour suppressant genes in a multistep model of carcinogenesis.* In: Becker, F., ed., *Symposium on Fundamental Cancer Research,* Vol. 38, pp. 45–56

Bishop, J.M. (1985) Viral oncogenes. *Cell, 42,* 23–38

Bizub, D., Wood, A.W. & Skala, A.M. (1986) Mutagesis of the Ha-*ras* oncogene in mouse skin tumours induced by polycyclic aromatic hydrocarbons. *Proc. natl Acad. Sci. USA, 83,* 6048–6052

Bos, J.L., Toksoz, D, Marshall, C.J., Verlaan-deVries, M., Veeneman, G.H., van der Eb, A.J., van Boom, J.H., Janssen, J.W.G. & Steenvoorden, C.M. (1985) Amino-acid substitutions at codon 13 of the N-*ras* oncogene in human acute myeloid leukaemia. *Nature, 315,* 726–730

Bos, J.L., Fearon, E.R., Hamilton, S.R., Verlaan-deVries, M., van Boom, J.H., van der Eb, A.J. & Vogelstein, B. (1987a) Prevalence of *ras* gene mutations in human colorectal cancers. *Nature, 327,* 293–297

Bos, J.L., Verlaan-de Vries, M., van der Eb, A.J., Janssen, J.W.G., Delwel, R., Lownberg, B. & Colly, L.P. (1987b) Mutations in N-*ras* predominate in acute myeloid leukemia. *Blood, 69,* 1237–1241

Brodeur, G.M., Seeger, R.C., Schwab, M., Varmus, H.E. & Bishop, J.M. (1984) Amplification of N-*myc* in untreated human neuroblastomas correlated with advanced disease stage. *Science, 224,* 1121–1124

Croce, C.M. (1987) Role of chromosome translocations in human neoplasia. *Cell, 49,* 155–156

Escot, C., Theillet, C., Lidereau, R., Spyratos, F., Champema, M.H., Gest, J. & Callahan, R. (1986) Genetic alteration of the c-*myc* protooncogene (MYC) in human primary breast carcinomas. *Proc. natl Acad. Sci. USA, 83,* 4834–4838

Fasano, O., Birnbaum, D., Edlund, L., Fogh, J. & Wigler, M. (1984) New human transforming genes detected by a tumorigenicity assay. *Mol. cell. Biol., 4,* 1695–1705

Forrester, K., Almoguera, C., Han, K., Grizzle, W.E. & Perucho, M. (1987) Detection of high incidence of K-*ras* oncogenes during human colon tumorigenesis. *Nature, 327,* 298–303

Fox, T.R. & Watanabe, P.G. (1985) Detection of a cellular oncogene in spontaneous liver tumours of $B6C3F_1$ mice. *Science, 228,* 596–597

Fusco, A., Grieco, M., Santoro, M., Berlingieri, M.T., Pilotti, S., Pierotti, M.A., Della Porta, G. & Vecchio, G. (1987) A new oncogene in human thyroid papillary carcinomas and their lymph-nodal metastases. *Nature, 328,* 170–172

Garte, S.J., Hood, A.T., Hochwait, A.E., D'Eustachio, P., Snyder, C.A., Segal, A. & Albert, R.E. (1985) Carcinogen specificity in the activation of transforming genes by direct-acting alkylating agents. *Carcinogenesis, 6,* 1709–1712

Gazdar, A.F., Carney, D.N., Nau, M.M. & Minna, J.D. (1985) Characterization of variant subclasses of cell lines derived from small cell lung cancer having distinctive biochemical, morphological, and growth properties. *Cancer Res., 45,* 2924–2930

Guerrero, I., Villasante, A., Corces, V. & Pellicer, A. (1984) Activation of the c-K-*ras* oncogene by somatic mutation in mouse lymphomas induced by gamma radiation. *Science, 225,* 1159–1162

Hirai, H., Kobayashi, Y., Mano, H., Hagiwara, K., Maru, Y., Omine, M., Mizoguchi, H., Nishida, J. & Takaku, F. (1987) A point mutation at codon 13 of the N-*ras* oncogene in myelodysplastic syndrome. *Nature, 327,* 430–432

Land, H., Parada, L. & Weinberg, R.A. (1983) Cellular oncogenes and multistep carcinogenesis. *Science, 222,* 771–778

Martin-Zanca, D., Hughes, S.H. & Barbacid, M. (1986) A human oncogene formed by the fusion of truncated tropomyosin and protein kinase sequences. *Nature, 319*, 743–748

McMahon, G., Hanson, L., Lee, J. & Wogan, G. N. (1986) Identification of an activated c-Ki-*ras* oncogene in rat liver tumors induced by aflatoxin B_1. *Proc. natl Acad. Sci. USA, 83,* 9418–9422

Meltzer, P., Kinzler, K., Vogelstein, B. & Trent, J.M. (1986) Gene amplification in cancer: a molecular cytogenetic approach. *Cancer Genet. Cytogenet., 19,* 93–99

Ochiya, T., Fujiyama, A., Fukushige, S., Hatada, I. & Matsubara, K. (1986) Molecular cloning of an oncogene from a human hepatocellular carcinoma. *Proc. natl Acad. Sci. USA, 83,* 4993–4997

Quintanilla, M., Brown, K., Ramsden, M. & Balmain, A. (1986) Carcinogen-specific amplification of H-*ras* during mouse skin carcinogenesis. *Nature, 322,* 78–80

Reed, J.C., Tsujimoto, Y., Alpers, J.D., Croce, C.M. & Nowell, P.C. (1987) Regulation of *bcl*-2 proto-oncogene expression during normal human lymphocyte proliferation. *Science, 236,* 1295–1299

Reynolds, S.H., Stowers, S.J., Maronpot, R.R., Anderson, M.W. & Aaronson, S.A. (1986) Detection and identification of activated oncogenes in spontaneously occurring benign and malignant hepatocellular tumours of $B6C3F_1$ mouse. *Proc. natl Acad. Sci. USA, 83,* 33–37

Reynolds, S.H., Stowers, S.J., Patterson, R., Maronpot, R.R., Aaronson, S.A. & Anderson, M.W. (1987) Activated oncogenes in $B6C3F_1$ mouse liver tumors: implications for risk assessment. *Science, 237*, 1309–1316

Sakamoto, H., Mori, M., Taira, M., Yoshida, T., Matsukawa, S., Shimizu, K., Sekiguchi, M., Terada, M. & Sugimura, T. (1986) Transforming gene from human stomach cancers and a noncancerous portion of stomach mucosa. *Proc. natl Acad. Sci. USA, 83,* 3997–4001

Sawey, M.J., Hood, A.T., Burns, F.J. & Garte, S.J. (1987) Activation of *myc* and *ras* oncogenes in primary rat tumours induced by ionizing radiation. *Mol. cell. Biol., 7,* 932–935

Schechter, A.L., Stern, D.F., Vaidyanathan, L., Decker, S.J., Drebin, J.A., Greene, M.I. & Weinberg, R.A. (1984) The *neu* oncogene: an *erb*-B-related gene encoding a 185,000-M_r tumour antigen. *Nature, 312,* 513–516

Seeger, R.C., Brodeur, G.M., Sather, H., Dalton, A., Siegel, S.E., Wong, K.Y. & Hammond, D. (1985) Association of multiple copies of the N-*myc* oncogene with rapid progression of neuroblasts. *New Engl. J. Med., 313,* 1111–1116

Shih, C., Shilo, B., Goldfarb, M.P., Dannenberg, A. & Weinberg, R.A. (1979) Passage of phenotypes of chemically transformed cells via transfection of DNA and chromatin. *Proc. natl Acad. Sci. USA, 76,* 5714–5718

Slamon, D.J., Clark, G.M., Wong, S.G., Levin, W.J., Ullrich, A. & McGuire, W.L. (1987) Human breast cancer: correlation of relapse and survival with amplification of the HER-2/*neu* oncogene. *Science, 235,* 177–182

Stowers, S.J., Glover, P.L., Boone, L.R., Maronpot, R.R., Reynolds, S.H. & Anderson, M.W. (1987) Activation of the K-*ras* proto-oncogene in rat and mouse lung tumours induced by chronic exposure to tetranitromethane. *Cancer Res.*, 47, 3212–3219

Sukumar, S. (1987) *Involvement of oncogenes in carcinogenesis.* In: Medina, D., Kidwell, W., Heppnar, G. & Anderson, E., eds, *Cellular and Molecular Biology of Mammary Cancer,* New York, Plenum, pp. 381–398

Sukumar, S., Notario, V., Martin-Zanca, D. & Barbacid, M. (1983) Induction of mammary carcinomas in rats by nitroso-methyl-urea involves malignant activation of H-*ras*-1 locus by single point mutations. *Nature, 306,* 658–661

Tennant, R.W., Margolin, B.H., Shelby, M.D., Zeiger, E., Haseman, J.K., Spalding, J., Caspary, W., Resnick, M., Stasiewicz, S., Anderson, B. & Minor, R. (1987) Prediction of chemical carcinogenicity in rodents from *in vitro* genetic toxicity assays. *Science, 236,* 933–941

Varmus, H.E. (1984) The molecular genetics of cellular oncogenes. *Ann. Rev. Genet., 18,* 553–612

Weinberg, R.A. (1985) The action of oncogenes in the cytoplasm and nucleus. *Science, 230,* 770–776

Wiseman, R.W., Stowers, S.J., Miller, E.C., Anderson, M.W. & Miller, J.A. (1986) Activating mutations of the c-Ha-*ras* proto-oncogene in chemically induced hepatomas of the male B6C3F_1 mouse. *Proc. natl Acad. Sci. USA, 83,* 5285–5289

Zarbl, H., Sukumar, S., Arthur, A.V., Martin-Zanca, D. & Barbacid, M. (1985) Direct mutagenesis of H-*ras*-1 oncogenes by nitroso-methyl-urea during initiation of mammary carcinogenesis in rats. *Nature, 315,* 382–385

LOOKING AHEAD: ALGEBRAIC THINKING ABOUT GENETICS, CELL KINETICS AND CANCER

W.G. Thilly

Center for Environmental Health Sciences, Massachusetts Institute of Technology, Cambridge, MA, USA

The cancer cell phenotype is modelled as a set of necessary genetic changes containing from one to three changes for any particular set of developmentally identical cells (tissue). This analysis suggests that a person in whose tissue an early genetic change occurs during ontogeny will perforce have a markedly higher number of cells in single sectors containing the necessary set of genetic changes. It is hypothesized that this numerical expectation is a dominant factor in determining the probability that a particular person will develop a tumour in a particular tissue.

Lars Ehrenberg started something when he first argued and then sought the relationship between internal dose and cancer in occupationally exposed persons. Today, many of us at this conference honouring Lars are working on ways to measure individual chemicals in human proteins or DNA, as well as in fat depots or excreta. Others are working on techniques to measure specific kinds of genetic change in human cell samples. My own laboratory is engaged in a testing struggle to develop means to obtain mutational spectra from fewer than 10^6 cells.

Personally, I see no reasonable alternative to following Ehrenberg's trail and devising means in analytical chemistry and analytical genetics to study ordinary humans directly. Although I have worked some 20 years in the study and development of human cell systems, I am dissatisfied with this area of research because, while yielding reproducible data, I have never been able to conceive of a means of applying this information in a sensible way to improving our understanding of hazards to human health.

On the basis of the advances in analytical chemistry and genetics, some of which are discussed here, and on what I think are reasonable projections for the next few years, I believe we will be able to obtain knowledge of hundreds of chemical adducts and simultaneously obtain mutational spectra from tissues in several thousand donors over the next ten years.

I for one will not be surprised to see the mutational spectra of particular chemicals in persons experiencing moderate to high exposures. However, how can one foresee what will be found in the vast majority of people who are not exposed to untoward levels of any known carcinogen, save sunlight, and still account for the majority of cancer cases?

Here, I perceive a chasm between geneticists and oncologists that I would like to address: the quantitative relationships between individual cellular genetic changes, growth of mutant cells as colonies (sectors) within tissues and the probability that tumours will appear in these tissues.

$$P(\text{tumour formation}) = f(\text{number of required mutations, mutation rates, sector size, characteristics of phenotypic expression})$$

In introducing the analysis, though incomplete, that follows, I acknowledge my debt to Max Delbrück, whose search for quantitative concepts in biology has benefited us all and also to Lars Ehrenberg, who looked to the actual amounts of chemicals and cancers in people to try to solve one of the more challenging problems of biology.

The case of a single required oncomutation

Let us imagine a tissue which, during gestation, increases from a small number of precursor cells to reach a finite size. For the sake of argument, let this tissue increase from one cell to N cells. During tissue growth to N cells, mutations will of course occur, whether induced by exogenous factors or spontaneously by any of a number of possible processes. Let us consider what would happen if mutation i were to occur at a constant rate r_i over this entire period.

During tissue growth, mutations can occur in every division cycle, and existing mutants will increase each subsequent division by a factor of two. Ignoring turnover of cells during this period, we may write

$$N = 2^g$$

in which g is the number of binary cell divisions necessary to increase a population from one cell to N. For such a population, the expected number of mutants Nm will simply be

$$Nm_i = gr_i\, 2^g = gr_i N.$$

These mutant cells will have arisen in all generations from a smaller number of mutations. The expected number of mutations M_i is just

$$M_i = \ln 2r_i\,(2^g).$$

Let us further posit that no tumour can arise in a tissue in which that mutation i has not occurred. The fraction of all growing tissues in which no mutation has occurred may be estimated by use of Poisson distribution, in which the rare events (mutations) are distributed randomly over all individual cell divisions. Thus, the probability of a tissue at generation g having no mutation of type i is just

$$P(M_i = 0)_g = e^{-\ln 2r_i\,(2^g)}.$$

It follows that the fraction of tissues that have already received at least one mutation is

$$P(M_i \geqslant 1)_g = 1 - e^{-\ln 2r_i\,(2^g)},$$

since

$$P(M_i = 0)_g + P(M_i \geqslant 1)_g = 1.$$

The case of two required oncomutations

With a little paperwork, one may derive similar expressions for the case of two necessary genetic changes (mutations, generically) i and j, having rates of occurrence per cell generation of r_i and r_j, respectively.

Once again, the cell number increases from 1 to $N = 2^g$, but now the number of cells with both mutations is

$$Nm_{ij} = g^2 r_i r_j\, 2^g = g^2 r_i r_j N.$$

When we speak of 'double' mutants, some care must be taken to define the event — mutation — which creates them. A double mutant is defined as containing both mutations i and j, which are independent. The 'mutation' giving rise to the double mutant is here defined as any event which finally yields a double mutant.

Three such events yield double mutants in any generation of population growth:

(1) Simultaneous mutation to i and j in the same cell in the same generation. For any generation, g, the number of such events is

$$r_i r_j\, 2^g$$

(2) Mutation of type i in cells already mutant of type j. The number of mutants of type j is simply $r_j\, g2^g$, so the number of such events is

$$r_i r_j g\, 2^g$$

(3) Mutations of type j in cells already mutant of type i. The number of mutants of type i is $r_i\, g2^g$, so the number of such events is

$$r_i r_j g\, 2^g.$$

Summing all of these events in any generation, we discover them to be

$$\Delta M_{ij} = r_i r_j (2^g + 2g\, 2^g).$$

Integrating this expression over g from $g = 0$ yields an estimate of the number of mutations giving rise to double mutants over the growth of the tissue, given that g is large:

$$M_{ij} = 2.88g\, 2^g r_i r_j.$$

For an expanding tissue, the probability of having no double mutant of type i and type j is again simply the probability that none of these three posited events will yield a double mutant. Using the Poisson expression

$$P(M_{ij} = 0) = \mathrm{e}^{-Mij},$$

where M_{ij} is the average number of any of the three double mutation events in a tissue at generation g,

$$P(M_{ij} = 0) + P(M_{ij} \geqslant 1) = 1;$$

so it follows that the fraction of all tissues containing at least one double mutant is

$$1 - \mathrm{e}^{-Mij} = 1 - \mathrm{e}^{2.88g\, 2^g r_i r_j}.$$

The case of three required oncomutations

By the same reasoning and the back of an infinitely large envelope, we may extend this line of reasoning to the triple mutant M_{ijk}.

The number of triple mutations, i.e., newly arising triple mutants in any generation g is

$$\Delta M_{ijk} = r_i r_j r_k (2^g + 3g\, 2^g + 3g^2\, 2^g).$$

Integrating this expression ΔM_{ijk} over all generations and assuming g is large yields

$$\frac{3g^2\, 2^g}{\ln 2}$$

or

$$M_{ijk} = 4.33g^2\, 2^g.$$

Table 1. Summary of model

Mutation	Rates	Mutants (Nm)	Mutations (M)	$P(M \geqslant 1)$
i	r_i	$g\,2^g r_i$	$0.7\,2^g r_i$	$1 - \exp[-0.7\,2^g r_i]$
$i+j$	$r_i r_j$	$g^2\,2^g r_i r_j$	$2.9g\,2^g r_i r_j$	$1 - \exp[-2.9g\,2^g r_i r_j]$
$i+j+k$	$r_i r_j r_k$	$g^3\,2^g r_i r_j r_k$	$4.3g^2\,2^g r_i r_j r_k$	$1 - \exp[-4.3g^2\,2^g r_i r_j r_k]$

As above, we may predict the fraction of tissues at generation g which do not contain any triple mutants as

$$P(M_{ijk} = 0) = \mathrm{e}^{-4.33g^2 2^g},$$

and the fraction which contains one or more such mutants as

$$P(M_{ijk} \geqslant 1) = 1 - \mathrm{e}^{-4.33g^2 2^g}\, r_i r_j r_k.$$

These formulae are summarized in Table 1.

Oncomutation during organ growth

Figure 1 is a diagram of the relationship between the fraction of persons who would have any particular one of the posited mutations necessary for subsequent tumour formation and cell generations during organ growth. The rate of mutation chosen for the example is close to the average rate at which we think single base-pair substitutions occur in human cells (about 10^{-9} mutations/cell generation).

In the example of Figure 1, we note that 2.2% of the population would have had this particular first necessary mutation by the 25th generation, while 50% of the population would not have it until the 30th generation. For an organ of 2^{37} generations' growth (~10^{11} cells), the mutants in the 2.2% of the population would have formed sectors or colonies of mutant cells at least 2^{37-25}, or 4000 cells. In an average (50th percentile) organ, this sector would be only 2^{37-30}, or 128 cells. Comparison of the experience of 2.2% of the population with the model, or 50% experience, is not done

Fig. 1. Relationship between fraction of persons with a mutation necessary for tumour formation and cell generations during organ growth

$P(M \geq 1) = 1 - e^{-\ln 2\, r_i 2^g}$

GENERATIONS

PROBABILITY OF ≥1 MUTATION

$r_i = 10^{-9}$

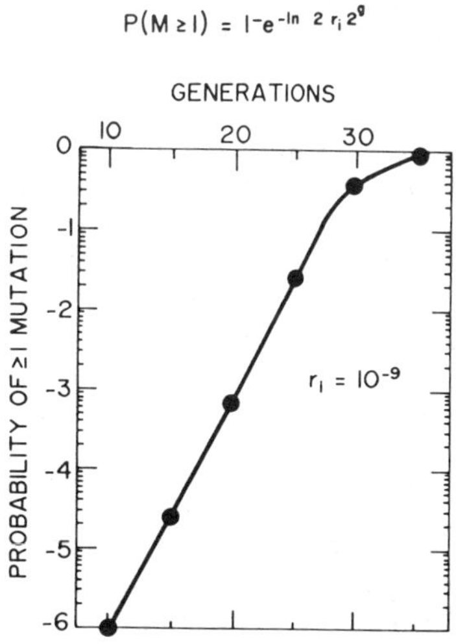

Fig. 2. Age distribution of liver and biliary cancer incidence and relationship between fraction of persons with a mutation necessary for tumour formation and cell generations during organ growth

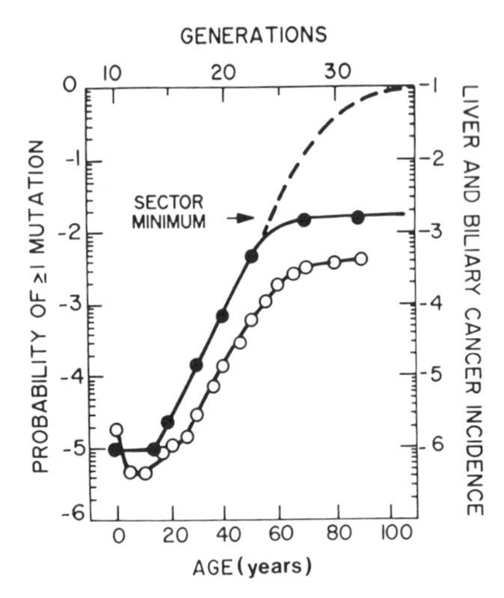

A background hereditary mutant fraction of 10^{-5} is assumed and sector minimum of 4000 mutant cells in a 10^{11} cell organ is shown as an illustration.

without purpose. The integrated frequency of the 15 most common cancers in the US population is about 2%.

This, of course, may just be another random product of academic wool gathering, but I am not so sure. In preparing for this conference, I made reference to the *Atlas of Cancer Mortality for US Counties, 1950–1969 (US HEW Publication No. (NIH) 75–780)* and was struck by the overall shape of the time dependence of actual cancer frequencies with age. In Figure 2, I have plotted the age distribution of liver and biliary cancer appearance (average of males and females). I have also estimated that some fraction of our population will actually inherit the particular necessary mutation and, assuming a generational mutation rate of 10^{-10} and 10^5 generations of humans (2×10^6 years), then we would expect about 10^{-5} of our population to inherit this variant gene.

The shapes of both curves begin with a constant portion to account for inherited oncomutants distributed over the human population. Both curves reach an apparent maximum plateau. Both constant regions are joined by considerable portions which are clearly log-linear with either age or cell generation. The magnitude of the increase for liver cancer frequency is about $4 \times 10^{-4}/5 \times 10^{-7}$, or 800. The maximal magnitude of the increase in probability of obtaining a first oncomutation is $1/10^{-5}$. However, if organ growth and phenotypic expression were such that only the early (2.2 percentile) oncomutations could be expressed, then the oncomutation magnitude (total expressed/inherited) would be about 2200 ($2.2 \times 10^{-2}/10^{-5}$).

What this similarity suggests is a physiological transformation between the generation of jackpots of early oncomutants in a developing organ and the probability of observing a tumour in that organ as a function of age.

Early jackpot → early tumour
Late jackpot → late tumour
No jackpot → no tumour

Furthermore, it appears that 65 years of life are necessary and sufficient to transform the smallest jackpots experienced in human liver into liver tumours and that this age-dependent function has a doubling age of 4.8, or approximately five years.

This reflection on the shape of the human cancer incidence *versus* age function is relatively recent, and is admittedly not yet well thought out. Some ideas that have arisen, however, may be worth noting.

Sector size and parenchymal growth control

Sector size may be important in escaping from normal growth control. This could arise by shielding cells in the sector's interior from a controlling electrochemical network. In three-dimensional tissue such as liver, such control *via* diffusion might depend on the cube root of sector cell number in tissues arrayed in two dimensions, as the square root of sector cell number.

The passing of time, however, after an organ reaches maturity should not permit increase in mutant sector size, *unless the mutation itself confers the ability to escape the cell death phase of normal cell death and division in an organ of constant size.* If such were the case, then oncomutant sectors would expand with a doubling time equal to the turnover time in the normal tissue, or somewhat greater if protection from 'death phase' was not complete. In the liver cancer example, the doubling time would be about five years to transform single mutation jackpot sectors producing a tumour in a particular year of life.

Sector size and the occurrence of necessary second oncomutations

Second mutations are similarly dependent on sector size *via* cell division. In this hypothetical case, the second mutation has not yet occurred in the first oncomutant jackpot sector, but, as the sector grows, the probability of the necessary second change increases to 1 and a tumour appears. This possibility also clearly depends on whether the first oncomutation confers on the sector the ability to avoid fully or in part the death phase of cell turnover in a tissue. Interestingly, a phenotype associated only with the putative first mutation fixes its temporal order in the oncological process. A necessary oncomutation that would not confer the deathless phenotype would not increase in sector size during adult life.

Second mutations could also occur during steady-state turnover, but these would be expected to accumulate linearly, not exponentially, with time.

Dilution of stable wild-type gene products

One perfectly horrible experience I had in my early years on the faculty of the Massachusetts Institute of Technology was discovering that the human B cells in which we were trying to induce mutation to 6-thioguanine resistance had a phenotypic lag of 14 days. This traumatic discovery caused me to suspect the existence of long-lived, or even stainless-steel, gene products that can be removed from genotypically mutant cells only by subsequent cell division and consequent dilution.

If the wild-type gene product of the gene mutated in the jackpot sector were, in fact, infinitely long-lived, then it would be diluted when cells of the sector divided, as in normal cell turnover. I cannot lightly discard this notion because it has the characteristics necessary to accomplish transformation of the jackpot accumulation function into the human cancer incidence function. It requires only a finite jackpot sector size and the passage of time in a normal physiological mode of cell turnover.

One further minor point may be added as a codicil. It may also be possible that the wild-type gene product has a half-life of several years. In such a case, cell turnover would not even be necessary. Eventually, all oncomutant sectors of sufficient size in an organ would be expressed as tumours, and, in advanced human age, the annual incidence would be expected to drop significantly. In this regard, one notes that tumours of the brain and nervous system in which cell turnover, while occurring, does not generally involve the greater mass of the organ, are in fact greatly reduced in frequency after age 65.

Conclusion

This model essentially predicts that jackpot mutations early in organ development are dominant factors in human tumour formation. The model might be tested by treating fetal animals with radiation or directly-acting chemicals as a function of cell number in the developing organ and examining the time to appearance of tumours later in life.

We are now considering examining the mutant sector concept directly in human autopsy samples using our new technique of high-fidelity DNA amplification and denaturing gel electrophoresis. These have the resolving power to observe single base-pair mutations directly in the DNA contained in as few as several dozen cells and, I believe, may put the central hypothesis presented here to the test.

Some may wonder if this hypothesis ignores the role of postnatal exposure to carcinogenic stimuli, and it is important to note that it does not. The model does, however, place such subsequent events within the context of inducing expansion and/or subsequent necessary mutation in an already existing and possibly expanding proto-oncoclone within an organ at risk.

Max Delbrück first worked out the arithmetics of mutational jackpots. John Calvin promoted the idea that we are born predestined to be doomed or blessed. Lars Ehrenberg has kept a focus on the events that occur in real people, and I wish to do the same.

Acknowledgements

I wish to acknowledge support for this work by a US Department of Energy Grant No. DE-FG02-86ER60448 and by the National Institute of Environmental Health Sciences through Grant Numbers NIH-5-P01-ES02109, NIH-5-P01-ES01640 and NIH-5-P01-ES03926.

LIST OF PARTICIPANTS

J. ALEXANDER
Department of Toxicology, National Institute of Public Health, Geitmyrsveien 75, N-0462 Oslo 4, Norway

A. ALHONEN
Institute of Occupational Health, Topeliuksenkatu 41 a A, SF-00250 Helsinki, Finland

B.N. AMES
Department of Biochemistry, University of California, Berkeley, CA 94720, USA

M.W. ANDERSON
Laboratory of Biochemical Risk Analysis, National Institute of Environmental Health Services/National Institutes of Health, PO Box 12233 A3-02, Research Triangle Park, NC 27709, USA

L. ARINGER
Department of Occupational Medicine, National Board of Occupational Safety and Health, S-171 84 Solna, Sweden

H. AUTRUP
Laboratory of Environmental Carcinogenesis, The Fibinger Institute, Danish Cancer Society, Ndr Frihavnsgade 70, DK-2100 Copenhagen, Denmark

R.A. BAAN
Department of Genetic Toxicology, TNO Medical Biological Laboratory, PO Box 45, 2280 AA Rijswijk, The Netherlands

H. BARTSCH
International Agency for Research on Cancer, 150 cours Albert Thomas, F-69372 Lyon Cedex 08, France

F.A. BELAND
National Center for Toxicological Research, HFT 110, Jefferson, AR 72079, USA

E. BERGMARK
Department of Radiobiology, Stockholm University, Svante Arrhenius väg 16–18, S-106 91 Stockholm, Sweden

N. BOMAN
Swedish Work Environment Fund, Box 1122, S-111 81 Stockholm, Sweden

R.P. BOS
Department of Toxicology, Faculty of Medicine, University of Nijmegen, PO Box 9101, 6500 HB Nijmegen, The Netherlands

R. CARTWRIGHT
Imperial Cancer Research Fund, 3K Springfield HS, Hyde Terrace, Leeds LS2 9LU, UK

R. CREBELLI
National Institute of Health, Viale Regina Elena 299, I-00161 Rome, Italy

G.E. DIGGLE
Department of Health and Social Security, Hannibal House, Elephant and Castle, London SE1 6TE, UK

L. EHRENBERG
Department of Radiobiology, Stockholm University, Svante Arrhenius väg 16–18, S-106 91 Stockholm, Sweden

E. ELOVAARA
Institute of Occupational Health, Topeliuksenkatu 41 a A, SF-00250 Helsinki, Finland

P.B. FARMER
Medical Research Council Toxicology Unit, Medical Research Council Laboratories, Woodmansterne Road, Carshalton, Surrey SM5 4EF, UK

A.M.J. FICHTINGER-SCHEPMAN
TNO Medical Biological Laboratory, PO Box 45, 2280 AA Rijswijk, The Netherlands

D. FORMAN
Imperial Cancer Research Fund, Cancer Epidemiology and Clinical Trials Unit, Gibson Building, Radcliffe Infirmary, Oxford OX2 6HE, UK

M. FRIESEN
International Agency for Research on Cancer, 150 cours Albert Thomas, F-69372 Lyon Cedex 08, France

A. FÖRSTI
Institute of Occupational Health, Topeliuksenkatu 41 a A, SF-00250 Helsinki, Finland

R.C. GARNER
Cancer Research Unit, University of York, Heslington, York YO1 5DD, UK

M. GÉRIN
Department of Occupational Medicine and Environmental Hygiene, University of Montréal, CP 6128 Succ. A, Montréal, Québec H3C 3J7, Canada

R.W. GIESE
Department of Medicinal Chemistry in the College of Pharmacy and Allied Health Professions, Northeastern University, 360 Huntington Avenue, Boston, MA 02115, USA

C. GIUNTINI
2nd Medical Clinic, University of Pisa, Via Roma 67, I-56100 Pisa, Italy

N. GORELICK
Massachusetts Institute of Technology, 50 Ames Street, Room E18–572, Cambridge, MA 02139, USA

T. GORSKI
Regulation Sanitary-Epidemiological Station, Ul. Wodna 40, 90–046 Łodz, Poland

Y. GRANJARD
International Agency for Research on Cancer, 150 cours Albert Thomas, F-69372 Lyon Cedex 08, France

C. GREENSTOCK
Atomic Energy of Canada Ltd, Whiteshell Nuclear Research Establishment, Pinawa, Manitoba R0E 1L0, Canada

H.-G. GRIMM
Volkswagen AG, Central Health Unit, D-3180 Wolfsburg 1, Federal Republic of Germany

J.D. GROOPMAN
Environmental Health Section, Boston University School of Public Health, 80 East Concord Street, Boston, MA 02118, USA

C.C. HARRIS
Laboratory of Human Carcinogenesis, Division of Cancer Etiology, National Cancer Institute, National Institutes of Health, Building 37, Room 2C07, Bethesda, MD 20892, USA

H. HAYATSU
Faculty of Pharmaceutical Sciences, Okayama University, Tsushima, Okayama 700, Japan

S.S. HECHT
Division of Chemical Carcinogenesis, American Health Foundation, 1 Dana Road, Valhalla, NY 10595, USA

K. HEMMINKI
Institute of Occupational Health, Topeliuksenkatu 41 a A, SF-00250 Helsinki, Finland

E. HESELTINE
Lajarthe, St Léon-sur-Vézère, 24290 Montignac, France

L.C. HOGSTEDT
Department of Occupational Medicine, National Board of Occupational Safety and Health, S-171 84 Solna, Sweden

B. HOLMBERG
Research Department, National Board of Occupational Safety and Health, S-171 84 Solna, Sweden

N.-H. HUH
Department of Cancer Cell Research, Institute of Medical Science, University of Tokyo, Shirokanedai Minato-ku, Tokyo 108, Japan

K. HUSGAFVEL-PURSIAINEN
Institute of Occupational Health, Topeliuksenkatu 41 a A, SF-00250 Helsinki, Finland

A.M. JEFFREY
Institute of Cancer Research and Division of Environmental Sciences, Columbia University, 701 W 168 Street, New York, NY 10032, USA

D. JENSSEN
Wallenberg Laboratory, Stockholm University, S-106 91 Stockholm, Sweden

G. DE JONG
Shell International Petroleum, Maatschappij BV, PO Box 162, 2501 AN The Hague, The Netherlands

F.F. KADLUBAR
Division of Biochemical Toxicology, National Center for Toxicological Research, Jefferson, AR 72079, USA

J. KALDOR
International Agency for Research on Cancer, 150 cours Albert Thomas, F-69372 Lyon Cedex 08, France

R. KALEJA
Hoechst AG, Postfach 800320, D-6230 Frankfurt am Main 80, Federal Republic of Germany

A. KAUTIAINEN
Department of Radiobiology, Stockholm University, Svante Arrhenius väg 16–18, S-106 91 Stockholm, Sweden

P. KEOHAVONG
Applied Biological Sciences, Massachusetts Institute of Technology, 400 Main Street, Room E18–666, Cambridge, MA 02139, USA

M. KLAUDE
Wenner-Gren Institute, Stockholm University, Biology F3, S-106 91 Stockholm, Sweden

A. KOLMAN
Department of Radiobiology, Stockholm University, Svante Arrhenius väg 16–18, S-106 91 Stockholm, Sweden

A. DE KONINGH
Occupational Health, Shell Rotterdam, PO Box 7000, 3000 HA Rotterdam, The Netherlands

E. KRIEK
Division of Chemical Carcinogenesis, The Netherlands Cancer Institute (Antoni van Leeuwenhoek Huis), Plesmanlaan 121, 1066 CX Amsterdam, The Netherlands

B. LAMBERT
Department of Clinical Genetics, Karolinska Hospital, S-104 01 Stockholm, Sweden

P. LEANDERSON
Department of Occupational Medicine, Faculty of Health Sciences, University Hospital, S-581 85 Linköping, Sweden

M. LE MEUR
Cerchar, BP 2, F-60550 Verneuil en Halatte, France

J. LEWALTER
Medical Department, Bayer AG, D-5090 Leverkusen 1, Federal Republic of Germany

D.W. LINDSAY
Rothmans International Services Ltd, Nevendon Road, Basildon, Essex SS13 1BT, UK

E. LINKOLA
Institute of Occupational Health, Topeliuksenkatu 41 a A, SF-00250 Helsinki, Finland

P.H.M. LOHMAN
Radiation and Chemical Mutagenesis, State University of Leiden, PO Box 9503, 2300 RS Leiden, The Netherlands

W. LUTZ
Institute of Toxicology, University of Zurich, CH-8603 Schwerzenbach, Switzerland

H. MALKER
National Board of Occupational Safety and Health, S-171 84 Solna, Sweden

E. MARAFANTE
Joint Research Centre, Commission of the European Communities, I-21020 Ispra, Italy

G. MARU
International Agency for Research on Cancer, 150 cours Albert Thomas, F-69372 Lyon Cedex 08, France

E. MASSEY
Research and Development Centre, British-American Tobacco Ltd, Regent's Park Road, Southampton SO9 1PE, UK

R. MONTESANO
International Agency for Research on Cancer, 150 cours Albert Thomas, F-69372 Lyon Cedex 08, France

T. MÜLLER
Inbifo Institute for Biological Research, D-5000 Cologne, Federal Republic of Germany

R. MUSTONEN
Institute of Occupational Health, Topeliuksenkatu 41 a A, SF-00250 Helsinki, Finland

K. MÄKELÄ
Kemira Oy, PO Box 330, SF-00101 Helsinki, Finland

H.-G. NEUMANN
Institute of Pharmacology and Toxicology, University of Würzburg, Versbacherstrasse 9, D-8700 Würzburg, Federal Republic of Germany

E. NIEBOER
Department of Biochemistry, McMaster University, 1200 Main Street West, Hamilton, Ontario L8N 3Z5, Canada

C. NYATHI
Preclinical Veterinary Studies, University of Zimbabwe, Box MP 167, Mount Pleasant, Harare, Zimbabwe

N. NÄSLUND
Department of Radiobiology, Stockholm University, Svante Arrhenius väg 16–18, S-106 91 Stockholm, Sweden

H. OHSHIMA
International Agency for Research on Cancer, 150 cours Albert Thomas, F-69372 Lyon Cedex 08, France

I.K. O'NEILL
International Agency for Research on Cancer, 150 cours Albert Thomas, F-69372 Lyon Cedex 08, France

S. OSTERMAN-GOLKAR
Department of Radiobiology, Stockholm University, Svante Arrhenius väg 16–18, S-106 91 Stockholm, Sweden

H. OTT
Environment and Raw Materials Project, Commission of the European Communities, 200 rue de la Loi, B-1049 Brussels, Belgium

R. PAHLMAN
National Public Health Institute, Mannerheimintie 166, SF-00280 Helsinki, Finland

F.P. PERERA
Division of Environmental Sciences, School of Public Health, Columbia University, 60 Haven Avenue B109, New York, NY 10032, USA

S. PETRUZZELLI
2nd Medical Clinic, University of Pisa, via Roma 67, I-56100 Pisa, Italy

D.H. PHILLIPS
Institute of Cancer Research, Chester Beatty Laboratories, Fulham Road, London SW3 6JB, UK

M.C. POIRIER
Laboratory of Cellular Carcinogenesis, National Cancer Institute, National Institutes of Health, Building 37, Room 3B21, Bethesda, MD 20892, USA

K. RANDERATH
Department of Pharmacology, Baylor College of Medicine, Houston, TX 77030, USA

A. RANNUG
Department of Toxicology, National Board of Occupational Safety and Health, S-171 84 Solna, Sweden

J. RANTANEN
Institute of Occupational Health, Topeliuksenkatu 41 a A, SF-00250 Helsinki, Finland

S.M. RAPPAPORT
School of Public Health, University of California, Berkeley, CA 94720, USA

C. REUTERWALL
Department of Occupational Medicine, Institute of Occupational Health, S-171 84 Solna, Sweden

L. ROMERT
Wallenberg Laboratory, Stockholm University, S-106 91 Stockholm, Sweden

G. SABBIONI
Donnerbuhlweg 37, CH-3012 Bern, Switzerland

R. SAFFHILL
Paterson Institute for Cancer Research, Christie Hospital and Holt Radium Institute, Manchester M20 9BX, UK

R.M. SANTELLA
Comprehensive Cancer Center and Division of Environmental Sciences, School of Public Health, Columbia University, 650 W 168 Street, New York, NY 10032, USA

K. SAVELA
Institute of Occupational Health, Topeliuksenkatu 41 a A, SF-00250 Helsinki, Finland

P.-O. SCHULTZ
Företagshälsan Nässjö, Mellangatan 2, S-571 00 Nässjö, Sweden

D. SEGERBÄCK
Department of Radiobiology, Stockholm University, Svante Arrhenius väg 16–18, S-106 91 Stockholm, Sweden

D.E.G. SHUKER
International Agency for Research on Cancer, 150 cours Albert Thomas, F-69372 Lyon Cedex 08, France

L. SHUKER
International Agency for Research on Cancer, 150 cours Albert Thomas, F-69372 Lyon Cedex 08, France

N.J. VAN SITTERT
Health, Safety and Environment Division, Shell International Petroleum Maatschappij BV, PB 162, 2501 AN The Hague, The Netherlands

A.I. SORS
Environment and Raw Materials Research Project, Commission of the European Communities, 200 rue de la Loi, B-1049 Brussels, Belgium

M. SORSA
Institute of Occupational Health, Topeliuksenkatu 41 a A, SF-00250 Helsinki, Finland

H.F. STICH
Environmental Carcinogenesis Unit, British Columbia Cancer Research Centre, 601 West 10th Avenue, Vancouver, BC V5Z 1L3, Canada

K. SVENSSON
Department of Radiobiology, Stockholm University, Svante Arrhenius väg 16–18, S-106 91 Stockholm, Sweden

P. SÖDERKVIST
Department of Occupational Health, University Hospital of Linköping, S-581 85 Linköping, Sweden

G. TALASKA
John L. McClellan Memorial Veterans Administration Medical Center, Little Rock, AR 72205, USA

S.R. TANNENBAUM
Department of Applied Biological Sciences, Massachusetts Institute of Technology, 77 Massachusetts Avenue 56–311, Cambridge, MA 02139, USA

W.G. THILLY
Center for Environmental Health Sciences, Massachusetts Institute of Technology, 400 Main Street, Room E18–666, Cambridge, MA 02139, USA

M. THOMPSON
Public Health Laboratory Service, CAMR, BMRL, Porton Down, Salisbury, Wilts SP4 0JG, UK

L. TOMATIS
International Agency for Research on Cancer, 150 cours Albert Thomas, F-69372 Lyon Cedex 08, France

M. TÖRNQVIST
Biology Laboratory, Department of Radiobiology, Stockholm University, Svante Arrhenius väg 16–18, S-106 91 Stockholm, Sweden

H. VAINIO
Institute of Occupational Health, Topeliuksenkatu 41 a A, SF-00250 Helsinki, Finland

J. VAN BENTHEM
Division of Chemical Carcinogenesis, The Netherlands Cancer Institute (Antoni van Leeuwenhoek Huis), Plesmanlaan 121, 1066 CX Amsterdam, The Netherlands

F.J. VAN SCHOOTEN
The Netherlands Cancer Institute (Antoni van Leeuwenhoek Huis), Plesmanlaan 121, 1066 CX Amsterdam, The Netherlands

P. VODICKA
Institute of Occupational Health, Topeliuksenkatu 41 a A, SF-00250 Helsinki, Finland

K. VÄHÄKANGAS
Department of Pharmacology, University of Oulu, Kajaanintie 52 D, SF-90220 Oulu, Finland

S.A.S. WALLES
Unit of Occupational Toxicology, National Board of Occupational Safety and Health, S-171 84 Solna, Sweden

H. WALLIN
Department of Toxicology, National Institute of Public Health, Geitmyrsveien 75, N-0462 Oslo 4, Norway

W.P. WATSON
Shell Research Ltd, Sittingbourne Research Centre, Sittingbourne, Kent ME9 8AG, UK

A. WESTON
Laboratory of Human Carcinogenesis, Division of Cancer Etiology, National Cancer Institute, National Institutes of Health, Building 37, Room 2C20, Bethesda, MD 20892, USA

C.P. WILD
International Agency for Research on Cancer, 150 cours Albert Thomas, F-69372 Lyon Cedex 08, France

J.C. WILLEY
Laboratory of Human Carcinogenesis, Division of Cancer Etiology, National Cancer Institute, National Institutes of Health, Building 37, Room 2C20, Bethesda, MD 20892, USA

J.W.G.M. WILMER
CIVO-TNO Toxicology and Nutrition Institute, Zeist, The Netherlands

G.N. WOGAN
Department of Applied Biological Sciences, Massachusetts Institute of Technology, Room 16–333, Cambridge, MA 02139, USA

A.S. WRIGHT
Shell Research Ltd, Sittingbourne Research Centre, Sittingbourne, Kent ME9 8AG, UK

R. ZITO
Regina Elena Cancer Institute, viale Regina Elena 291, I-00161 Rome, Italy

A. ÖNFELT
Department of Occupational Health, Institute of Occupational Health, S-171 84 Solna, Sweden

INDEX OF AUTHORS

Alexander, J., 113
Alhonen, A., 329
Ames, B.N., 407
Anderson, M.W., 477
Andersson, B., 232
Annan, R., 356
Arimoto, S., 401
Aringer, L., 232, 396
Autrup, H., 63
Axelson, O., 422

Baan, R.A., 146
Badawi, A.F., 301
Bailey, E., 279
Bartsch, H., 83, 417
Bedford, P., 321
Beland, F.A., 175
Benson, R.W., 166
Berends, F., 321
Béréziat, J.-C., 107
van den Berg, P.T.M., 146
Bingham, S., 107
Bodell, W.J., 217
Bos, R.P., 389
Bosch, F.X., 427
Bradshaw, T.K., 237
Brenner, D., 451
Brésil, H., 75
Brodeur, J., 275
Brooks, A.G.F., 279
Brooks, B.R., 181
Brouet, I., 107
Brunius, G., 396
Bryant, M.S., 133
Bussachini, V., 417

Carmella, S.G., 121
Carnicelli, N., 97
Chapot, B., 67
Choi, J.-S., 181
Cooper, R.S., 372
Creasey, J.S., 341

Day, N.E., 460
Degan, P., 75
DeLeo, V.A., 333
Den Engelse, L., 67, 102, 286

Dijt, F.J., 321
Doskocil, G., 372
Dunn, B.P., 137

Eadsforth, C.V., 271
Egorin, M.J., 313
Ehrenberg, L., 23
Einhorn, N., 469
Ellul, A., 107

Farmer, P.B., 92, 279, 347
Fichtinger-Schepman, A.M.J., 313, 321
Fisher, D., 356
Floyd, R.A., 417
Foiles, P.G., 121
Forman, D., 97
Friesen, M., 417
Frosini, G., 97
Fullerton, N.F., 175

Garner, R.C., 196
Gasparro, F.P., 330
Gérin, M., 275
Giese, R.W., 356
Groopman, J.D., 55
Grover, P.L., 368

Hall, C.N., 301
Harris, C.C., 181, 208, 439
Haugen, A., 439
Hayatsu, H., 401
Hayatsu, T., 401
He, S.-H., 469
Hecht, S.S., 121
Hemminki, K., 190, 306, 329
Hesso, A., 306
Hewer, A., 368
Hietanen, P., 329
Hill, B.T., 321
Hillebrand, M.J.X., 201
Hoffmann, D., 121
Hogstedt, L.C., 21, 265
Holmberg, B., 227
Holmberg, K., 469
Huckle, K.R., 384
Huh, N.-H., 292
Husgafvel-Pursianien, K., 129

Jankowiak, R., 372
Jass, J.R., 368
Jeffrey, A.M., 372
Jongeneelen, F.J., 389

Kadlubar, F.F., 166
Kaldor, J., 460
Kiilunen, M., 329
Kinouchi, T., 175
Knight, T.M., 97
Kolman, A., 258
Kresbach, G., 356
Kriek, E., 201
Krontiris, T., 439
Kuroki, T., 292

Lamb, J., 347
Lambert, B., 469
Lang, N.P., 166
Lawley, P.D., 347
Leanderson, P., 422
Leach, S., 97
Levin, J.-O., 232
Lindblom, K., 227
Linkola, E., 306
Liu, S.F., 217
Lohman, P.H.M., 13
Lorenzini, L., 97
Löf, A., 232

Manchester, D.K., 181
Mann, D.L., 181
Marini, M., 97
Maronpot, R.R., 477
Martel-Planche, G., 75
Maru, G., 417
McDowell, E., 439
Miller, R.H., 361
Minacci, C., 97
Minnetian, O., 356
Mittal, D., 361
Montesano, R., 67, 75
Muñoz, N., 427
Mustonen, R., 329
Mäki-Paakkanen, J., 223

Nair, J., 417
Nair, U., 417
Neumann, H.-G., 157
Newman, M.J., 181
Norppa, H., 223
Norström, Å., 232
Näslund, P., 232

Ohara, Y., 401
Ohshima, H., 83
Olsson, M., 396
O'Neill, I.K., 107
van Oosterom, A.T., 321
Osterman-Golkar, S., 223, 227, 249, 258

Packer, P., 97
Passingham, B.J., 279
Pelkonen, O., 208
Perera, F.P., 190, 451
Phillips, D.H., 190, 196, 368
Poirier, M.C., 175, 181, 313
Pongracz, K., 217
Povey, A.C., 107

Randerath, E., 361
Randerath, K., 190, 361
Rannug, A., 396
Rappaport, S.M., 217
Reddy, M.V., 190
Reed, E., 313
Resau, J., 439
Reynolds, S.H., 477
Roberts, D.W., 166
Rogers, E., 356

Saffhill, R., 301
Saha, M., 356
Santella, R.M., 190, 333, 451
Satoh, M.S., 292
Savela, K., 306
Scherer, E., 67, 102, 286
Segerbäck, D., 258
Selvin, S., 213
Serres, M., 75
Shiga, J., 292
Shuker, D.E.G., 92, 296
Sigvardsson, K., 227
van Sittert, N.J., 271
Skipper, P.L., 133
Small, G.J., 372
Smith, R.J., 384
Sorsa, M., 129, 213
Spratt, T.E., 121
Steenwinkel, M.-J.S.T., 146
Stich, H.F., 137
Strickland, P.T., 341
Svensson, K., 227
Söderqvist, P., 422

Tagesson, C., 422
Talaska, G., 166
Tannenbaum, S.R., 133
Tardif, R., 275
Terheggen, P.M.A.B., 286
Thilly, W.G., 486

Tierney, B., 196
Tosi, P., 97
Trivers, G.E, 181
Trump, B., 439
Trushin, N., 121
Törnqvist, M., 271, 378

Van Benthem, J., 102, 286
Van Schooten, F.J., 201
Vermeulen, E., 102, 286
Vindigni, C., 97
Vineis, P., 133
Vouros, P., 356
Vähäkangas, K., 208

Wakhisi, J., 63
Wallèn, M., 232
Walles, S.A.S., 223, 227
Wallin, H., 113
Watson, W.P., 237, 271, 384
Weinstein, I.B., 451
Welling, M.C., 201
Weston, A., 181, 439
Wild, C.P., 67, 75, 102
Willey, J.C., 181, 439
Wilson, V.L., 181
Winterwerp, H.H.K., 102, 286
Wogan, G.N., 9, 32
Wraith, M.J., 271
Wright, A.S., 237, 271, 384
van der Wulp, C.J.M., 146

Yager, J.W., 213
Yang, X.Y., 333
Yates, D.W., 279
Young, T.-L., 451
Yuspa, S.H., 313

Zamzow, D., 372
Zheng, Q.L., 401

SUBJECT INDEX

A

2-Acetylaminofluorene
  adducts in mouse liver and urinary bladder after feeding, 175–79
  and macromolecular binding in liver, kidney and blood, 162
  antibodies against DNA adducts with, 201–206
  erythrocyte dose and DNA alkylation, 38, 48
  visualization of guanine modified by, 290
trans-4-Acetylaminostilbene, binding to haemoglobin, 162–63
Acetylator phenotype, and cancer, 166–68
Acrylamide, adduct with haemoglobin, 350–52
Acrylonitrile, excretion as measure of exposure, 275–78
Adducts (*see also* DNA adducts; Haemoglobin adducts; Protein, adducts)
  and cancer, 47–48
  and dose, 208–11
  'background', 12, 16, 19, 40–42, 46, 79, 242, 262–63, 274, 284, 379–81
  formation
    effect of antioxidants on, 55, 56–58
    interindividual differences in, 16, 43
  removal of, 36
  stability of, 16, 47, 201, 205–206
  unknown, 12, 16, 28, 45–47, 351–54, 378–82
Aflatoxin
  activation of, 57
  adducts
    in liver cells after exposure, 16, 70, 72, 428, 435
    in lymphocytes, 16, 68, 70
    in white blood cells, 70
    to albumin, 22, 348, 428
    to haemoglobin, 68, 70, 73
  and activated oncogenes, 480
  and hepatitis B virus, 68
  antibodies to, 55–56, 58–62
  assessment of exposure to, 40, 58–62, 63–65, 67–73
  B_1
    in breast milk, 71–72
    in plasma protein, 70
    in serum, 56, 428
    in urine, 9, 10, 22, 39, 55–62, 64, 67, 290, 428
  B_1-formamidopyrimidine, 56, 58
  B_1-7-guanine in urine, 9, 22, 39, 40, 55–62, 63–65, 290
  in diet, 55
  in hepatocellular carcinoma, 67–73, 427, 428–32, 435–36
  risk of, 55–62
Age, relation of
  to micronuclei formation, 213–16
  to oxidative DNA damage, 19, 407–14
Albumin (*see also* Plasma protein)
  adducts
    with aflatoxin, 22, 68, 70, 73, 348, 428
    with 4-aminobiphenyl, 169, 348
  as dose monitor, 37
Alcohol
  in hepatocellular carcinoma, 427, 429, 432–36
  in oesophageal cancer, 80
Algebraic model, of genetics, cell kinetics and cancer, 486–92
Alkylating agent (*see also individual agents*)
  and human cancer, 304, 460, 463–67
  bifunctional, 26, 27
  characterization of, 25–28, 250, 252–55
  detection of adducts with, 10, 40, 347–54
  marker of damage by, 255
  monofunctional, 26–29
  reactivity of, 23–24, 26
  therapy, 460, 463–67
Alkylation (*see also* Alkylating agent; 2-Hydroxyethylation; Methylation; 4-(3-Pyridyl)-4-oxybutylation; *and individual agents*)
  and dose, 38
  and rad-equivalence, 26–29
  as premutagenic event, 26
  'background', 242, 304
  by 4-(3-pyridyl)-4-oxobutylation, 124–27
  detection of, 75–81
  exposure to, 237–309
  of DNA (*see* DNA)
  of DNA bases, 23, 229
  of haemoglobin (*see* Haemoglobin)
  relation to DNA adduct level, 176, 208–11
7-Alkylguanine (*see also individual compounds*)
  detection of, 306–308, 352–54
  stability of, 16
Aluminium manufacturing plants, adducts in workers in, 9, 43, 210
4-Aminobiphenyl
  adducts
    with albumin, 169, 348
    with DNA in lung, 10, 166–72
    with DNA in urinary bladder, 10, 166, 169–72
    with haemoglobin in smokers and non-smokers, 9, 10, 41–42, 133–36, 454–55, 457
  dose-response relationship with, 38
  metabolism of, 168–69
  susceptibility to, 166–72

2-Aminofluorene, 175–76, 201–206
Antibody (*see also* Monoclonal antibody)
  affinity, 201–206, 317, 454
  binding efficiency, 201–206
  characterization of, 12
  to aflatoxin B_1, 64–65, 67–73
  to DNA adducts, 9, 10, 11, 42–43, 77–78, 146–53, 181–87, 190–94, 196–99, 201–206, 296–99, 329–31, 372, 374, 451, 454
  to alkylated DNA, 92–96, 103, 296, 298
Antimutagenic factor, in urine, 401–403
Antioxidant
  defences against, 409–10
  effect of on adduct formation, 55, 56–58
Areca nut, 138, 417–21
Aromatic amine (*see also individual compounds*)
  DNA adducts with, 169–72, 201–206
  exposure to, 133–36, 157–63, 166–72
  haemoglobin adducts with, 133–36
  metabolism of, 166, 168
  risk of, 157–63
  subtoxic level of, 13
  susceptibility to, 166–72
Asphalt exposure, 392, 394
Atomic absorption spectroscopy
  comparison with ELISA, 313, 317–18
  for cisplatin in blood compartments, 329–30
  for platinum products, 322
  sensitivity of, 318

B

Benzidine, haemoglobin adducts with, 159–60
Benzo[*a*]pyrene
  activation of, 209–10
  determination of exposure to, 42–45, 146–53, 208–11
  effects of fibre and meat on, 107–11
  levels in foundries, 191
  metabolism of, 107–11, 210
  metabolites of, 385–86
  visualization of guanine modified by, 290
Benzo[*a*]pyrene diol epoxide–DNA adduct
  antibodies to, 9, 42–43, 146-53, 182–83, 190–94, 196–99, 201–206, 451, 454, 456
  in coke-oven workers, 9, 10, 43, 181, 183, 184–85, 201, 204–205, 210
  in cultured human skin, 384–87
  in liver, 211
  in lung tissue, 42–43, 146–53, 196–99, 211, 451, 454
  in lymphocytes, 209–10
  in microsomes, 209–10
  in mouse skin, 211, 384–87
  in placenta of smokers, 45–47, 208–10
  in white blood cells
    of cancer patients, 183, 184, 456
    of foundry workers, 9, 10, 22, 43, 190–94, 454–55
    of roofers, 9, 43, 182–83
Benzo[*a*]pyrene diol epoxide–haemoglobin adduct, 181–87
Betel quid
  and formation of reactive oxygen species *in vitro,* 417–21
  chewing
    and adducts in oral mucosa, 46
    and micronuclei formation in oral mucosa, 10, 137, 138, 141
    and nitrosation in oral cavity, 87
Binding efficiency, of antibodies to DNA adducts, 201–206
Biochemical epidemiology, 73, 381–82
Bladder (*see* Urinary bladder)
Blood (*see also* Haemoglobin; Lymphocytes; Plasma protein; Serum; White blood cells)
  2-acetylaminofluorene adducts in, 162
  cancers of (*see also* Leukaemia) and exposure to ethylene oxide, 265, 267, 268
  compartments of cancer patients, cisplatin in, 329–31
Blot hybridization, 445, 457–58, 473–75
Blue cotton, 401
Blue rayon, 401–402
Bovine papilloma virus (plasmid pd BPV-1), 139–40
Breast milk, aflatoxin B_1 in, 10, 56, 68, 71–72, 428
Bronchial cells, DNA adducts in, 137, 139, 140–43, 146–53
Bronchioli, O^6- and 7-methylguanine in, 104–105
Bronchogenic carcinoma (*see also* Lung)
  analysis of by restriction fragment length polymorphism, 439–47
  and cigarette smoking, 439, 453, 456–57
  susceptibility to, 439–42, 447
Burkitt's lymphoma, oncogene translocation in, 34, 35, 478

C

C18 Sep-Pak
  for clean up
    of breast milk, 71–72
    of urine, 60, 63, 69, 232
  for recovery of mutagens from urine, 396–99
Cancer
  and acetylator phenotype, 166–68
  and alkylating agents, 304
  and ethylene oxide, 265–69
  and free radicals, 407, 412–14
  and nitrosation, 76, 83–88

Cancer (contd)
  and oxidative DNA damage, 407–14
  etiology, 32–48, 427–94
  genetics and cell kinetics, 486–92
  initiators, 258–59
  patients
    aromatic DNA adducts in, 368–70
    benzo[*a*]pyrene diol epoxide-DNA adducts in, 183, 184, 456
    carboplatin-DNA adducts in, 313–18
    chromosome damage in, 453
    cisplatin in blood compartments of, 329–31
    8-methoxypsoralen-DNA adducts in, 333, 338–39
  prevention, 32–48
  risk (*see* Risk)
  second, 466–67
Carboplatin, DNA adducts with in cancer patients, 313–18
*N*2-Carboxymethyl-7-methylguanine, 296–98
*N*6-Carboxymethyl-3-methyladenine, 92–93
Carcinogenesis, chemical
  and carcinogen-DNA adduct formation, 181
  detection of agents, 237–38
  initial processes of, 286–90
  oncogenes in, 477–82
  stages of, 240–41, 246
Carcinogenicity
  and DNA damage, 56, 175–79, 460–67
  and lipid peroxidation, 407–409
  and mutation, 238–39, 245–46, 445–47, 486–92
  and rad-equivalence values, 245–46
Catechu, as source of reactive oxygen species, 417–21
Cell kinetics, 486–92
Cervix, cancer of, and carboplatin-DNA adducts, 316–17
Chemiluminescence, 10, 417
Chemotherapy
  and chromosomal aberrations, 469–76
  and DNA adducts, 286–90, 313–18, 321–24
  and sister chromatid exchange frequency, 34, 464–66
  dosimetry of response to, 137–44
  optimizing, 23
Chromosome
  aberrations
    and chemotherapy, 469–76
    in T-lymphocyte clones, 469–73
  damage
    in coke-oven workers, 456
    induced by chemicals, 24
    induced by ethylene oxide, 25, 250, 268
    in lung cancer patients, 453
    in tumour initiation, 18
  detection of damage to, 11, 33–35
  instability, and mutation, 442–44, 446–47
  translocation, and oncogene activation, 477–78, 481
Chronic atrophic gastritis, *N*-nitrosoproline in urine of patients, 98–99
Cigarette smoke (*see also* Smoker; Smoking; Tobacco)
  and 4-aminobiphenyl-haemoglobin adducts, 41, 454–55
  and benzo[*a*]pyrene diol epoxide-DNA adducts, 9, 45–47, 208–10, 454–55
  and bronchogenic carcinoma, 439, 453, 456–57
  and cotinine, 129–31, 282–83, 454–55, 457
  and DNA adducts in tissues, 46, 140–43, 361–62, 364
  and micronuclei in bronchi, 141
  and nitrosation, 39
  and sister chromatid exchange in lymphocytes, 34, 454–55
  and urine mutagenicity, 401, 403
  ethene in, 42
Cisplatin
  adducts in white blood cells of patients, 43–45
  adducts with DNA
    and disease response, 315–18
    antibodies to, 321–27, 329–31
    detection of, 44
    in cultured cell lines, 321, 324–26
    induction of *in vivo* and *in vitro,* 321–27
    in tissues of treated patients, 313–18
  in blood compartments of cancer patients, 329–31
  immunostaining of cells treated with *in vivo,* 290
  interindividual differences in adduct formation from, 16, 43
  levels in white blood cells, 329–30
Coal-tar, monitoring of exposure to, 10, 389–95
Coke-oven workers
  benzo[*a*]pyrene diol epoxide-DNA adducts in, 9, 10, 43, 181, 183, 184–85, 201, 204–205, 210
  benzo[*a*]pyrene diol epoxide-haemoglobin adducts in, 181, 184
  chromosomal aberrations in, 456
  1-hydroxypyrene in urine of, 392–94
  polycyclic aromatic hydrocarbon-DNA adducts in, 456
  sister chromatid exchanges in, 456
Collision ion spectrum, 351–53
Colon, cancer of
  activated proto-oncogenes in, 479, 481
  and diet, 368–70
  aromatic DNA adducts in mucosa of patients with, 10, 368–70
Colorectal cavity
  monitoring in, 107–11
  susceptibility of to aromatic amines, 166–67

Complex mixture
analysis of, 181–82, 351–54, 361–64
of polycyclic aromatic hydrocarbons, analysis of, 372
Cotinine, as marker of exposure to tobacco smoke, 129–31, 282–83, 454–55, 457
Covalent binding
index, 35
mechanism of, 37
to DNA of food carcinogens, 113–16
to haemoglobin, 38
to protein by food carcinogens, 113–16
Creosote, monitoring exposure to, 391–92
Cross-linking agent, endogenous, 107–11
Cross-reactivity
of anti-benzo[*a*]pyrene-DNA antibody, 182–83, 185, 186, 198–99
of anti-Guo-2-acetylaminofluorene antibody, 202–204
of anti-hydroxyethylated peptide, 272
of bulky adducts, 296
of purines, 297–98
Cutaneous T-cell lymphoma, 8-methoxypsoralen-DNA adducts in treated patients, 333, 338–39
Cyclobutadithymidine dimers, 10, 341–44
Cyclophosphamide, 213–216, 464
Cytochrome P450, 57, 113, 114, 166, 168
Cytogenetic effect
assays of, 33–34
of ethylene oxide, 268
Cytokinesis block, 213–16
Cytostatic agent (*see also individual agents*)
exposure to, 22, 213–16

D

Dehydroretronecine-DNA adduct, 349
Diet
aflatoxins in, 55
and urine mutagenicity, 397–99, 403
as interfering factor, 48
effects of on benzo[*a*]pyrene metabolism and endogenous cross-linking agents, 107–11
exposure *via,* 55–117
in colorectal cancer, 368–70
DNA
detection of photoproducts in, 341–44
hydrolysates, 10, 293
interaction with tobacco-specific nitrosamines, 121–27
modification, degree of, and effect on antibody affinity, 201–206, 317, 454
rearrangement and oncogene activation, 477–78
surrogates, 107–11
transfection assays, 478–81
DNA adducts
and chemotherapy, 286-90, 313–18, 321–24
and cigarette smoke, 46, 140–43, 361–62, 364
aromatic
in colorectal cancer patients, 368–70
in white blood cells of workers, 22, 36, 45–46, 142, 190–94
as intermediate endpoint, 137–44
general methods for detection of, 14, 15, 17, 25, 36–46, 201–206, 238
in bronchial cells, 137, 139, 140–43, 146–53
in single cells, 286–90
in tumour formation, 18
in urine in relation to formation in liver, 58–59
localization of, 102–106, 142–43, 238
relation to carcinogenicity, 56, 175–79
relation to haemoglobin adducts, 38, 48, 227, 251, 258
DNA alkylation
and erythrocyte dose of 2-acetylaminofluorene, 38, 48
determination of, 10, 92–96, 296–99, 356–59
DNA binding
of food carcinogens to, 113–16
relation to haemoglobin binding, 38, 158, 159, 162
DNA damage (*see also* DNA adduct; Single-strand break)
and carcinogenicity, 460–67
and free radicals, 413–14
and passive smoking, 10
and tobacco use, 121–54
as endpoint, 460–61
as exposure measure, 461–62
by oestrogens, 361, 362–63, 364
detection of, 32–48
epidemiology of, 460–67
in mouse embryo fibroblasts, 374
in tumour formation, 18
oxidative, 9, 407–24
assay for, 40
by man-made mineral fibres, 422–23
relation to cancer and ageing, 19, 407–14
sensitivity of methods for, 14–15
Dose (*see also* Dosimetry; Target dose)
estimation of, 249–55
molecular, 258
monitoring, 23–29, 36, 37, 38, 157–63, 241–42, 258, 261–62
relationship to alkylation, 38
relationship to DNA adduct level, 176, 208–11
Dosimetry
for carcinogens, 451–52
for exposure
to aflatoxins, 55–62
to ethylene oxide, 249–55
to polycyclic aromatic hydrocarbons, 181–87

Dosimetry (contd)
for exposure (contd)
to tobacco-specific nitrosamines, 121–27
molecular, 19, 24–29, 36–38, 271–74
of response to chemotherapy, 137–44

E

Edman degradation, modified, 242, 251, 280, 350, 379, 381
Electrophile, endogenous
metabolism to, 23
persistence of, 27
reaction of, 28, 35, 37, 238, 242, 378–79
Electrophore detection, 356–59
Enzyme-linked immunosorbent assay (ELISA)
comparison with atomic absorbance spectroscopy, 313, 317–18
for aflatoxin B_1 in body fluids, 10, 64, 67–73
for alkylated DNA, 10, 92–96, 296–99
for 4-aminobiphenyl-DNA adducts in cells and tissues, 10, 171–72
for benzo[*a*]pyrene diol epoxide-DNA adducts
in lung tissue, 196–99, 451, 454, 456
in lung tumour and nontumour DNA, 42–43
in white blood cells, 9, 43, 182–83
for cisplatin adducts in blood and tissues, 43–45, 313–18
for cyclobutadithymidine dimers in DNA of cells exposed to ultraviolet light, 10, 341–44
for DNA photoproducts, 341–44
for 8-methoxypsoralen-DNA adducts in cells and tissues, 10, 333–36, 338–39
for nitrosamine exposure, 77
for polycyclic aromatic hydrocarbon-DNA adducts in DNA, 182–83, 190–94, 201–206
validation of, 201, 204–205, 313–14, 317–18
Epidemiology (*see also* Biochemical epidemiology; Molecular epidemiology)
application of methods to, 21–22, 47–48, 83–88
of DNA damage, 460–67
of ethylene oxide, 265–69
of hepatocellular carcinoma, 427–36
Ethene
as source of ethylene oxide, 249, 250–52, 258, 263
as source of hydroxyethylated adducts, 380–81
in cigarette smoke, 42
Ethoxyquin, 55, 56–58
O^4-Ethyl-2′-deoxythymidine, in liver DNA, 292–94
Ethylene oxide
and 'background' adducts, 379–81
and cancer, 25, 237, 246, 252, 262–63, 265–69
and chromosome damage, 25, 250, 268
and haemoglobin adducts, 9, 48, 242
and hydroxyethylvaline, 40–42, 252, 262, 271, 273, 379–80
and liver DNA alkylation, 38
and micronuclei formation, 260
and sister chromatid exchange frequency, 34, 260–61, 268
as model compound for rad-equivalence, 28–29, 37, 271–74
cytogenicity of, 268
dosimetry of, 249–55
effect on lymphocytes, 25
epidemiology of, 265–69
exposure monitoring, 271–74, 275–78
genotoxicity of, 258, 260–61
health hazards of, 24–25
occupational exposure to, 10, 40–41, 262
risk of, 250–52
O^4-Ethylthymine, detection of, 10, 292–94, 356–57
Exposure
assessment, 21–22, 32–48, 58–62, 63–65, 67–73, 75–81, 129–31, 166–72, 181–87, 249, 430–31, 435, 460, 461–62
control, 157–63
dietary, 55–117
medicinal, 313–44
monitoring, 10, 37–38, 48, 59–62, 190–94, 271–74, 275–78, 361–64, 389–95
occupational, 14, 40–41, 157–234, 262, 265–69, 391–94
to alkylation, 237–309

F

Fanconi's anaemia, 440, 443
cell line, cisplatin-DNA adducts in, 324–36
Fast-atom bombardment
for adducts with alkylating agents, 347–50
for benzo[*a*]pyrene diol epoxide-DNA adducts, 147
for haemoglobin adducts, 11
Fibre, effect on benzo[*a*]pyrene and cross-linking agents, 107–11
Fibre, man-made mineral (*see* Man-made mineral fibres)
Fibroblasts
cultured human, DNA adducts in, 146–50
human, cisplatin-DNA adducts in, 321, 324–26
mouse embryo
DNA damage in, 374
transformation of, 258, 260–61
Flow cytometry, for 8-methoxypsoralen-DNA adducts, 333, 336–38
Fluorescence line narrowing spectrophotometry, for carcinogen-DNA adducts, 10, 372–76
Foundry workers
aromatic DNA adducts in, 190–94

Foundry workers (contd)
benzo[*a*]pyrene diol epoxide-DNA adducts in white blood cells of, 9, 10, 22, 43, 190–94
benzo[*a*]pyrene levels in work place, 191
polycyclic aromatic hydrocarbon-DNA adducts in, 455–56, 457
unknown adducts in, 45, 46
Free radicals (*see also* DNA damage; Reactive oxygen species)
and DNA damage, 413–14
contribution of to cancer, 407, 412–14
defences against, 409–10
sources of *in vivo,* 407–409
theory of ageing, 407, 412–14

G

Gas chromatography-mass spectrometry
comparison with HPLC-ELISA, 92, 94–95
comparison with radioimmunoassay, 271, 273–74
for adducts with alkylating agents, 10, 40, 347, 350–51
for haemoglobin adducts, 41, 77, 227, 228, 242, 279–84
for mercapturic acids, 232–33
for *N*-nitrosoproline, 39
sensitivity of, 15
with multiple ion detection, 233–34
with negative-ion chemical ionization, 41, 127, 350
with single-ion monitoring, 39, 40
Gas chromatography-thermal energy analysis, for *N*-nitrosamino acids in urine, 39
Gene amplification, and oncogene activation, 477–78, 481
Genotoxicity
and cancer risk, 258–63
concept of, 25
detection and assessment of, 237–46
markers of, 9, 11, 175
methods for measuring, 17
γ-Glutamyl transpeptidase-positive foci, effect of antioxidants on, 57–58
7-Guanine
alkylation of, 229
postlabelling of, 11, 219
visualization of, 290
O^5-Guanine
alkylation of, 23
visualization of, 290

H

Haemoglobin
alkylation, 37, 38, 48, 227–30
as dose monitor, 157–63, 241–42, 258, 261–62
as exposure monitor, 37–38, 48, 252
binding
as indicator of carcinogenic risk, 160–62
covalent, 38
index, 157, 159–63
of aflatoxin, 70, 73
of *trans*-4-acetylaminostilbene, 162–63
relation to DNA binding, 38, 158, 159, 162
cisplatin in, 329–30
hydroxyethylation of, 9, 279–84, 378–80
interaction of tobacco-specific nitrosamines with, 121–27
Haemoglobin adducts, 9, 22, 27, 40–42, 48, 251
and tobacco smoke, 133–36, 380
detection of, 11, 14, 41, 77, 227, 228, 242, 279–84
relation to DNA adducts, 38, 48, 227, 251, 258
with acrylamide, 350–52
with aflatoxins, 68
with 4-aminobiphenyl, 10, 169
with aromatic amines, 133–36
with benzidine, 159–60
with benzo[*a*]pyrene diol epoxide, 181–87
with ethylene oxide, 9, 48, 242
Heart, adduct levels in mouse, 362–63
Hepatitis B virus
and liver cancer, 65, 427, 428, 430–36
markers, 22
surface antigen and liver cancer, 22
transmission of, 72
Hepatocellular carcinoma (*see also* Liver)
activated oncogenes in, 479, 481
and alcohol, 427, 429, 432–36
and exposure to aflatoxins, 67–73, 427, 428–32, 435–36
and oral contraceptives, 427, 434–36
and tobacco smoke, 427, 432–36
diagnosis of, 64
epidemiological studies of, 427–36
prevalence of, 67
risk for, 428–36
High-performance liquid chromatography (HPLC)
electrophore detection, 359
ELISA for urinary 3-methyladenine, 92, 94–95
for aflatoxin B_1-7-guanine in urine, 39, 40, 56, 59–62, 63
for 4-aminobiphenyl-DNA adducts, 170–71
for benzo[*a*]pyrene metabolites, 108, 385–86
for 8-hydroxydeoxyguanosine, 417, 420
for hydroxymethyldeoxyuridine in urine, 412
for 5-hydroxymethyluracil in urine, 412
for mercapturic acids in urine, 232–33
for separation of mercapturic acid derivatives, 275–76
for thymine glycol in urine, 39, 40, 412

High-performance liquid chromatography (HPLC) (contd)
mass spectroscopy for adducts, 347, 354
sensitivity of, 15
synchronous scanning fluorescence spectrophotometry, 39, 40, 182
with infra-red analysis, 39
Histidine, removal of for mutagenicity testing, 396–99
hprt locus, analysis of, 469–76
Hydrogen peroxide
as reactive oxygen species, 407, 409, 410, 411
formation *in vitro* from betel-quid ingredients, 16, 417–20
1-Hydroxybenzo[*a*]pyrene, in urine after exposure to coal-tar, 10
8-Hydroxy-2′-deoxyguanosine
formation in DNA *in vitro* with betel-quid ingredients, 417–21
formation in DNA *in vitro* with man-made mineral fibres, 422–23
in urine, as assay for oxidized DNA, 412
2-Hydroxyethylation, 40–42, 251–52, 262–63, 271–74, 279–84, 378–81
S-(2-Hydroxyethyl)cysteine, 271, 279, 280
N^{τ}-Hydroxyethylhistidine, 262, 271
N-3-(2-Hydroxyethyl)histidine, 38, 40–42, 278
Hydroxyethylvaline, as monitor of exposure
to ethylene oxide, 9, 40–42, 252, 262, 271, 273, 379–80
to hydroxyethylating agents, 279–84
8-Hydroxyguanine, 10, 12, 410
3-Hydroxyhistidine, 9, 12
5-Hydroxymethyluracil, 358–59, 410–12
1-Hydroxypyrene, in urine, 205–206, 389–95
4-Hydroxy-1-(3-pyridyl)-1-butanone, adduct after exposure to tobacco smoke, 121, 126–27

I

I compound, 361, 363–64
Image processing
for DNA adducts in fibroblasts, 146, 148–49
for immunostaining intensity, 289
for micronuclei, 137, 138
Immunoassay (*see also* ELISA, USERIA)
comparison with postlabelling, 184, 196–99
for alkylation products, 75, 77–79, 296–99
for benzo[*a*]pyrene diol epoxide-DNA adducts, 9, 42–45, 146–53, 181–85, 190–94, 196–99, 201–206, 451, 454, 456
for cisplatin-DNA adducts, 9, 329–31
for cyclobutadithymidine dimers in DNA, 341–44
for DNA photoproducts, 341–44
for ethylene oxide, 271–74
Immunochemical method
competitive, 15
direct, 15
evaluation of, 17
for 4-aminobiphenyl, 169–72
for cisplatin-DNA adducts, 321–27, 329–31
for 3-methyladenine, 92–96, 296, 298
interlaboratory variation in, 16
sensitivity of, 14–15
single-cell, 11, 15, 286–90
Immunocytochemistry, for DNA adducts
in single cells, 11, 286–90
in tissues, 70–71, 102–106
Immunofluorescence microscopy, 14, 143
for benzo[*a*]pyrene diol epoxide-DNA adducts in bronchial cells of smokers, 146–53
for 8-methoxypsoralen-DNA adducts, 333–39
sensitivity of, 15
Immunoperoxidase staining, 102–106
Immunostaining, 289–290
Infra-red analysis–high performance liquid chromatography, 39
Intermediate endpoints, in intervention trials, 137–44
IQ (2-Amino-3,8-dimethylimidazo[4,5-*f*]quinoline), 113–16, 168

K

Karyotype
of human T-lymphocyte clones, 469–76
of tumours, 444–47
Keratinocyte, cultured, 8-methoxypsoralen-DNA adducts in, 333–38
Keyhole limpet haemocyanin (KLH), for antisera
to 4-aminobiphenyl, 171
to 3-methyladenine, 92, 93–94
Kidney, DNA adducts in, 162, 362–63

L

Leucocyte (*see* White blood cells)
Leukaemia
acute, and alkylating agents, 460, 463–67
and exposure to ethylene oxide, 25, 237, 246, 252, 262–63, 265–69
oncogene translocation in, 34, 35, 478–79
T-cell, 35
Leukoplakia, oral, as intermediate endpoint, 137–44
Lipid hydroperoxide, measurement of, 410
Lipid peroxidation, and mutagenicity and carcinogenicity, 407–409
Liver
adducts in after exposure
to 2-acetylaminofluorene, 162, 175–79

Liver (contd)
adducts in after exposure (contd)
to aflatoxin B_1, 16, 70, 72, 428, 435
to *N*-nitroso compounds, 78, 79, 86, 102, 104
alkylation in after exposure to ethylene oxide, 38
benzo[*a*]pyrene diol epoxide-DNA adducts in, 211
cancer
and aflatoxin B_1-7-guanine in urine, 22, 40, 55, 63–65
and hepatitis B virus, 65, 427, 428, 430–36
and risk markers, 22
O^4-ethyl-2′-deoxythymidine in DNA of, 292–94
O^4-ethylthymine in DNA of, 292–94
formation of adducts in, 58–59
microsomes
and 4-aminobiphenyl metabolism, 168–69
and benzo[*a*]pyrene metabolism, 210
single-strand breaks in DNA of mice after exposure to vinyl chloride, 227–31
tumours and 2-acetylaminofluorene-DNA adduct levels in mice, 175–79
Lung (*see also* Bronchial cells; Bronchioli; Bronchogenic carcinoma)
adduct levels after exposure to tobacco smoke, 362–63
4-aminobiphenyl-DNA adduct levels in, 166, 171–72
benzo[*a*]pyrene diol epoxide-DNA adducts in, 42, 43, 146–53, 196–99, 211, 451, 454, 456, 457
cancer
known causes of, 453–54
markers of, 451–58
patients, benzo[*a*]pyrene diol epoxide-DNA adducts in, 183, 184
immunocytochemical analysis
of aflatoxin B_1 in, 71
of alkylated guanine in, 102, 104–106
tumours in mice, and sister chromatid exchange, 34
Lymphatic tumour (*see also* Burkitt's lymphoma)
and ethylene oxide, 265, 267, 268, 269
Lymphocyte (*see also* White blood cells)
activation of benzo[*a*]pyrene in, 209–10
adducts in after exposure
to aflatoxin B_1, 16, 68, 70
to *N*-nitrosodimethylamine, 78
benzo[*a*]pyrene diol epoxide adducts in, 209–10
chromosomal aberrations in, 34
cytokinesis-blocked, micronuclei in, 11, 213–16
effects of ethylene oxide in, 25
8-methoxypsoralen-DNA adducts in *in vitro* and *in vivo*, 333–34, 338–39
single-strand breaks in after exposure to styrene, 223–26
sister chromatid exchange in, 34, 454–55, 463–65
styrene oxide-DNA adducts in, 217–221
T-, karyotypes of clones of in patients, 469–76

M

Mandelic acid, excretion of after exposure to styrene, 223–26
Man-made mineral fibres, oxidative damage by, 422–23
Marker
biological, 11
in lung cancer, 451–59
of genotoxic exposure, 9, 11, 14, 36, 39–40, 129–31, 175, 242, 252, 255, 378
of hepatitis B virus, 22
of intake of tobacco smoke, 129–31, 137, 139, 140–43, 252
of oesophageal cancer, 9, 39
significance of, 12
urinary, 9, 10, 14, 36, 39–40
Mass spectrometry
for identification of haemoglobin adducts, 11
for studies of adducts, 347–54
Meat
cooked, mutagens from, 10, 401–403
effect on benzo[*a*]pyrene and cross-linking agents, 107–11
MeIQ (2-Amino-3,4-dimethylimidazo[4,5-*f*]quinoline), 113–16
MeIQx (2-Amino-3,8-dimethylimidazo[4,5-*f*]quinoxaline), 113–16
Melphalan, 349, 469–76
Mercapturic acid, excretion of as measure of exposure
to acrylonitrile, 275–78
to ethylene oxide, 275–78
to toluene, 232–34
to xylene, 232–34
8-Methoxypsoralen, DNA adducts with, 10, 333–39
3-Methyladenine
antibodies to, 92–96, 296, 298
as indicator of exposure to methylating agents, 9, 39, 40, 92–96
'background', 12
site of alkylation, 77
Methylation (*see also* Alkylation; 2-Hydroxyethylation), 92–96, 102–106
S-Methylcysteine in haemoglobin after exposure, 38, 77
5-Methylcytosine, determination of, 358–59
O^5-Methyldeoxyguanosine
immunoassays for, 43–44, 75, 77–79, 301–304
immunocytochemical localization of, 102–105

O^5-Methyldeoxyguanosine (contd)
in oesophageal and gastric tissue, 43, 44, 302
7-Methyldeoxyguanosine
immunoassays for, 75, 77–79
immunocytochemical localization of, 102–105
O^6-Methylguanine
in blood of people at high risk of oesophageal cancer, 9
in bronchioli, 104–105
in human DNA, detection of, 301–304
in rat oesophagus after exposure to *N*-nitroso compounds, 289
in target tissues for tobacco-specific nitrosamines, 124–25
7-Methylguanine
antibodies to, 296–99
detection of
by immunoassay, 10, 296–99
by ^{32}P-postlabelling, 306–308
in bronchioli, 104–105
natural occurrence in urine, 352–53
Methylmethane sulphonate, 38, 138–39, 242
O^4-Methylthymidine, immunoassay for, 75, 77
O^4-Methylthymine, 10
Microdensitometry, for quantification of DNA adducts *in situ,* 11, 286, 289–90
Micronuclei
and cigarette smoke, 141
as intermediate endpoint in intervention trials, 137–44
effect of age on formation of, 213–16
in cytokinesis-blocked lymphocytes, 11, 213–16
induced by ethylene oxide, 260
induced by methylmethane sulphonate, 138–39
in oral mucosal cells
of betel chewers, 10, 137, 138, 141–44
of tobacco chewers, 138
in urinary bladder, 141
Microsomes
benzo[*a*]pyrene diol epoxide-DNA adduct formation in, 209–10
liver, and 4-aminobiphenyl metabolism, 168–69
RNA adducts in, 114–15
Mitomycin C-DNA adduct, 348–49
Molecular epidemiology
analysis of DNA adducts in, 286–90
applications of methods in, 427–92
definitions of, 21, 185–86, 461
intermediate endpoints in, 144
scope of, 33
Monitoring
biological, 32–33, 389–95, 401–403
dose, 23–29, 36, 37, 38, 157–63, 241–42, 258, 261–62
exposure, 10, 59–62, 190–94, 271–74, 275–78, 361–64, 389–95
use of microencapsulated DNA surrogates in, 107–11
Monoclonal antibody
against aflatoxins, 55–56, 58–62
against benzo[*a*]pyrene diol epoxide-DNA adducts, 146–47
against cyclobutadithymidine dimers, 341–43
against O^5-methyldeoxyguanosine, 77, 301
against 8-methoxypsoralen-DNA adduct, 333–34, 337–38
Multiple-ion detection, 233–34
Mutagenicity
and lipid peroxidation, 407–409
of exposure to coal-tar, 389, 391–95
problems in monitoring, 401–403
testing, 396–99
urinary (*see* Urine, mutagenicity of)
Mutation
analysis of induction *in vivo,* 19, 486–92
and chromosomal instability, 442–44, 446–47
and oncogene activation, 477, 480–82
and rad-equivalence values, 244–45, 258–60, 262
mechanisms of in plants, 23–24
point, 24
role in carcinogenicity, 238–39, 245–46, 445–47, 486–92

N

Nasal cavity, alkylated bases in tissue of, 102, 104
Nasopharynx, cancer of and exposure to *N*-nitrosamines, 76
Negative-ion chemical ionization, 41, 127, 350, 454
Nitrogen mustard–DNA adduct, 349
N-Nitrosamines
and human cancer, 76, 80, 86, 87
as source of 'background' alkylation, 304
exposure determination from adducts, 75–81
in tobacco smoke, 283–84
tobacco-specific, 11, 76, 121–27
N-Nitrosamino acid, in urine, 39, 83–88
Nitrosation
and cigarette smoke, 39
and human cancer, 9, 83–88
in oral cavity, 87
N-Nitroso compounds
and urinary bladder cancer, 76
index of exposure to, 9, 78, 79, 83–88, 102, 104
N-Nitrosodiethylamine, as alkylating agent, 293–94
N-Nitrosodimethylamine
as alkylating agent, 37, 77, 78, 279
dose-response relationship of, 38
4-(*N*-Nitrosomethylamino)-1-(3-pyridyl)-1-butanone (NNK), as alkylating agent, 121–27

N-Nitrosomethylbenzylamine
DNA adduct after exposure to, 102–106
immunocytochemical staining of rat oesophagus after exposure to, 289
N'-Nitrosonornicotine, as alkylating agent, 121–24, 126–27
N-Nitrosoproline (NPRO)
as index of endogenous nitrosation, 9, 83–88
excretion by patients with gastric lesions, 97–100
interindividual differences in excretion of, 39
relationship to gastric cancer risk, 10
in urine of smokers, 39
Nonsmoker
4-aminobiphenyl-haemoglobin adducts in, 9, 10, 41–42, 133–36, 454–55, 457
'background' adducts in, 40–42, 46
bronchogenic carcinoma in, 439
cotinine in plasma of, 454–55
exposure markers in, 129–31, 137, 139, 140–43
hydroxyethylated adducts in, 281–83, 380
1-hydroxypyrene in urine of, 390–94
nitrosation in, 39
sister chromatid exchange in lymphocytes of, 454–55
urine mutagenicity in, 396–99
Nucleic acid (*see also* DNA; RNA)
covalent binding to, 114–16
detection of adducts with, 347–54
methylation of, 92–96
Nucleophile, reaction with electrophile, 37

O

Oesophagus
alkylated bases in, 9, 43, 44, 102–106, 289, 302
cancer of
and alcohol consumption, 80
and chewing of tobacco and of betel quid with tobacco, 417
and exposure to *N*-nitrosamines, 76, 80, 86, 87
and oncogenes, 81
areas of high and low risk, markers in inhabitants of, 9, 39
Oestrogen, DNA damage by, 361, 362–63, 364
Oncogene (*see also* Proto-oncogene)
activated
and DNA rearrangment, 477–78
and mutation, 477, 480–82
and stomach cancer, 481
by carcinogens, 480–82
detection of, 11, 35
in hepatocellular carcinoma, 479, 481
in lung cancer patients, 456–59
and genetic predisposition, 441
c-myc, 34, 456–58, 477, 481
c-ras, 437, 441–42, 444, 446, 457–58, 477, 479–82
in chemical carcinogenesis, 477–82
translocation of
in Burkitt's lymphoma, 34, 35, 478
in leukaemia, 34, 35, 478–79
Oncomutation, 487–92
Oral cavity
cancer of, 76, 417, 420
immunostaining of cells after cisplatin treatment, 290
mucosal cells
adducts in, 9, 16, 45, 46, 150, 152
micronuclei in, 10, 137, 138, 141–44
nitrosation in, 87
Oral contraceptive, in hepatocellular carcinoma, 427, 434–36
Ovary, cancer of, patients
receiving cisplatin, adducts in, 9, 43, 44, 314–16, 318
receiving melphalan, T-lymphocyte clones in, 469–76
Oxygen radical, 422–23

P

Pentafluorobenzylation, for detection of alkylated DNA adducts, 10, 356–59
Peroxidase-antiperoxidase staining, for DNA adducts *in situ*, 11, 286–90
Photoproduct, of DNA, 10, 341–44
Physicochemical methods
for DNA adducts, 9, 10, 17, 36
for haemoglobin adducts, 451
Placenta, adducts in, 9, 45–48, 183, 208–10
Plasma protein
aflatoxin B_1 in, 70
cisplatin in, 329–30
Polycyclic aromatic hydrocarbons (*see also* Benzo[*a*]pyrene)
adducts with DNA in white blood cells of foundry workers, 455–56, 457
antibodies to DNA adducts with, 11, 153, 185, 201–206
aromatic adducts with DNA after exposure to, 22, 190–94
DNA adducts with, 10, 181–87, 196–99, 201–206, 372, 374, 451, 453–57
dosimeters of exposure to, 181–87
metabolites in urine, 43, 205–206, 389–95
Polyethyleneimine, as DNA surrogate, 107–11
Polymorphism, genetic
detection of, 439–47
and cancer of urinary bladder and rectum, 166–68
in polycyclic aromatic hydrocarbon-DNA adduct levels, 186

^{32}P-Postlabelling
  comparison with immunoassays, 193–94, 196–99
  evaluation of, 17
  for adducts
    in bone marrow, 45
    in bronchial mucosa, 137, 139, 140–43
    in colonic mucosa, 10, 368–70
    in oral mucosa, 9, 45
    in placenta, 9, 45
    in skin, 384–87
    in white blood cells, 9, 45
  for analysing DNA of exposed individuals, 9, 10, 19, 45–47
  for 'background' DNA alkylation, 242
  for benzo[*a*]pyrene diol epoxide–DNA adducts, 146–53
  for exposure to 4-aminobiphenyl, 169–72
  for 7-methylguanine, 306–308
  for polycyclic aromatic hydrocarbon–DNA adducts, 10, 183–84, 190–94
  for styrene oxide–DNA adducts, 10, 217–221
  for unknown adducts, 16
  lack of correlation with immunoassays, 184, 196–99
  sensitivity of, 14–15
  of 7-guanine, 11, 219
  ultrasensitive, for unidentified genotoxicants, 361–64
  use of ^{3}H-acetic anhydride in, 11, 306–308
Preparative affinity columns, 55–56, 58–62, 68–69, 196, 197–99
Protein
  adducts
    detection of, 4, 14–15, 17, 347–54
    formation of, 37–38
  binding of food carcinogens to, 113–16
Proto-oncogene
  activation of, 439, 444, 456–58, 477–82
  in restriction fragment length polymorphism, 441
Psoriasis
  8-methoxypsoralen–DNA adducts in patients, 10, 333, 338–39
  skin cancer in, 339
4-(3-Pyridyl)-4-oxybutylation, 11, 124–27

Q

Quality factor, for comparison of chemical and radiation genotoxicity, 252, 258–62

R

Rad-equivalence
  and mutation, 244–45, 258–60, 262
  for estimation of risk, 18, 26–29, 37, 237, 243–46, 258–63, 379
  to relate target dose to effect, 16, 251–52
  validation of, 244–45
Radiation (*see also* Ultraviolet radiation)
  γ, 24, 25, 26, 244, 252, 258–62, 480
  ionizing, 23, 26–28, 407, 410, 480
  linear energy transfer (LET), 24, 27, 243, 251
  solar, simulated, 341–44
  X, 24, 26
Radioimmunoassay
  comparison with gas chromatography-mass spectrometry, 271, 273–74
  for 2-acetylaminofluorene adduct levels, 175–77
  for aflatoxin B_1, 59
  for cotinine in plasma, 454–55
  for O^4-ethyl-2′-deoxythymidine, 293
  for ethylene oxide, 271–74
  for O^6-methyldeoxyguanosine, 43–44, 77, 301–304
  for photoproducts in DNA, 341–44
  validation of, 271, 273–74
Reaction-kinetic characterization
  at critical sites in DNA, 258
  of alkylating agents, 25–28, 250, 252–55
  of 7-methylguanine and styrene-7,8-oxide-7-guanine acetylation, 306–307
Reactive oxygen species, 407, 409, 410, 411
  and oral cavity cancer, 417, 420
  formation of from betel-quid ingredients, 16, 417–21
  in induction of adducts, 238, 422–23
Restriction fragment length polymorphism
  analysis
    for changes in DNA sequence, 35
    of human bronchogenic carcinoma, 439–47, 458
  at *c-mos* in oesophageal cancer, 81
  detection of, 11
Retinoblastoma, 440–41, 443–45
Risk, cancer
  acceptable, 239, 243, 244, 251, 379
  and genotoxicity, 258–63
  and haemoglobin binding, 160–62
  and inherited disorders, 439–40
  and excretion of *N*-nitrosoproline, 10
  assessment of, 23–29, 32, 237–41, 244–46
    for aflatoxin B_1, 55–62
    for aromatic amines, 157–63
    for ethylene oxide, 250–52
    from studies in rodents, 477–82
  estimation of, 17–19
    by rad-equivalence, 18, 26–29, 37, 237, 243–46, 258–63, 379
    quantitative, 13
  factor, identification of, 379–82, 427–36, 452

Risk, cancer (contd)
for hepatocellular carcinoma, 428–36
for second cancer, 465–66
identification of, 28–29
RNA
adducts
as exposure indicators, 35
in microsomes, 114–15
in urine, 36, 59
7-methylguanine from, 352
Roofers, benzo[*a*]pyrene diol epoxide-DNA adduct in white blood cells of, 9, 43, 182–83

S

Salmonella/mammalian microsome test
for exposure to coal-tar products, 389–95
insensitivity for measuring passive exposure to tobacco smoke, 129–31
problems in, 401–403
standardization of, 396–99
Samples, storage of for later analysis, 21–22, 48, 435, 460, 466
Selective-ion monitoring, 11, 280, 282, 347, 350–52
Single-ion monitoring, 39, 40
Single-strand break, in DNA
of lymphocytes of styrene-exposed workers, 223–26
of mouse liver after inhalation of vinyl chloride, 227–31
Sister chromatid exchange
and lung tumours in mice, 34
as indication of chromosomal damage, 11, 33–35
as indicator of DNA damage, 10
frequency, and carcinogenic potency, 34, 462–65
in coke-oven workers, 456
induced by ethylene oxide, 34, 260–61, 268
in lung cancer patients, 453, 456
insensitivity for measuring passive exposure to tobacco smoke, 129, 130–31
in smokers and nonsmokers, 34, 130, 454–55
Skin
benzo[*a*]pyrene diol epoxide-DNA adducts, 211, 384–87
cancer in psoriasis patients, 339
DNA photoproducts in, 341, 343–44
8-methoxypsoralen–DNA adducts in patients', 333, 335–39
Smoker (*see also* Cigarette smoke; Smoking; Tobacco)
adducts in
bronchial cells of, 137, 139, 140–43, 146–53
bronchus and larynx of, 45, 46
oral mucosa of, 9, 16, 46
placenta of, 45–47, 208–10
4-aminobiphenyl–haemoglobin adducts in, 9, 10, 41–42, 133–36, 454–55,
benzo[*a*]pyrene diol epoxide–DNA adducts in, 45, 183, 208–10, 454–55, 458
benzo[*a*]pyrene diol epoxide–haemoglobin adducts in, 184
cotinine in plasma of, 454–55
haemoglobin adducts in, 9, 133–36, 252, 279, 281–84
hydroxyethylation in, 380
1-hydroxypyrene in urine of, 390–94
micronuclei in urinary bladder of, 141
sister chromatid exchange in, 34, 130, 454–55
urine mutagenicity in, 130, 396–99, 401, 402
Smoking (*see also* Cigarette smoke; Smoker; Tobacco)
and benzo[*a*]pyrene diol epoxide–DNA adducts
in lung tissue, 198–99
in lymphocytes, 210
in placenta, 45–47, 208–11
and DNA damage, 10
and nitrosation, 39
inverted, 10, 46
urinary markers of, 10
Soft ionization, 350
Stomach
cancer
and exposure to *N*-nitrosamines, 76, 80, 86, 87
and exposure to ethylene oxide, 265–69
and oncogene activation, 481
risk and *N*-nitrosoproline excretion, 10
lesions and *N*-nitrosoproline excretion, 97–100
tissue, O^6-methyldeoxyguanine lesions in, 43, 44, 302
Styrene, measurment of exposure to, 10, 217–21, 223–26, 306–308
Styrene glycol, in blood of styrene-exposed workers, 223–26
Styrene oxide–DNA adducts in styrene-exposed workers, 10, 217–21, 306–308
Supercritical fluid chromatography, 347
Superoxide ion, formation *in vivo* with betel-quid ingredients, 417–20
Susceptibility
assessment of, for aromatic amines, 166–72
measures of, 47–48, 208
to 4-aminobiphenyl, 166–72
to aromatic amines, 166–72
to bronchogenic carcinoma, 439–42, 447
s-value, 250, 254, 379
Synchronous scanning fluorescence spectrophotometry
for aflatoxin B_1-7-guanine in urine, 39, 63
for benzo[*a*]pyrene diol epoxide–DNA adducts, 9, 10, 208–11

Synchronous scanning fluorescence spectrophotometry (contd)
for polycyclic aromatic hydrocarbon–macromolecular adducts, 181, 184
sensitivity of, 15, 372

T

Tandem mass spectrometry, 11, 14–15, 347, 351–54
Target dose, 16, 237, 241–43, 249, 251–52, 255, 258–59
N-Terminal valine
as basis for immunoassay for ethylene oxide, 271–74
as marker
of damage by alkylating agents, 255
of haemoglobin of smokers, 252
of haemoglobin reaction with electrophiles, 242, 378
from exposure to 2-hydroxyethylating agents, 279–84
Testis
alkylation in DNA of, after exposure to ethylene oxide, 38
cisplatin–DNA adducts
in cell line of, 321, 326–27
in cancer patients, 314–15, 316
Theophylline, as interfering compound in urine, 95
Thiocyanate, as marker of intake of tobacco smoke, 129–31
Thioether, as marker of exposure
compared with other methods, 388, 390–95
to acrylonitrile, 275–78
to ethylene oxide, 275–78
to toluene, 232–34
to xylene, 232–34
Thioguanine resistance, for detecting *hprt* mutations, 469–76
Thymidine glycol, as measure of background levels of radical-induced DNA damage, 39, 40, 407, 412–13
Thymine glycol, as measure of background levels of radical-induced DNA damage, 39, 40, 407, 410–13
T-lymphocyte (*see also* Lymphocyte)
aberrations in clones of, 469–73
karyotypes of clones of, 469–76
Tobacco
and oesophageal cancer, 80, 417
chewing
and adducts in oral mucosa, 46
and micronuclei in oral mucosa, 138
and oesophageal and oral cancer, 417
DNA damage due to use of, 121–54
smoke
and adducts in lung, 362–63
and cotinine, 129–31, 282–83, 454–55, 457
and haemoglobin adducts, 133–36, 380
and hepatocellular carcinoma, 427, 432–36
and 4-hydroxy-1-(3-pyridyl)-1-butanone, 121, 126–27
and thiocyanate, 129–31
and urinary bladder cancer, 133, 135–36
passive and transplacental exposure to, 129–31
specific nitrosamines (*see N*-Nitrosamines)
Toluene, measurement of exposure to, 232–34
Toluene mercapturic acid, 232–34
Tumour
initiation, 18, 19
karyotypes of, 444–47
models, 479–82
second, 22
suppressor gene, 439, 442–45, 446–47, 477, 482
'Two-hit' carcinogenesis model, 443–45

U

Ultrasensitive enzyme radioimmunoassay (USERIA), for benzo[*a*]pyrene diol epoxide–DNA adducts, 9, 43, 182
Ultraviolet radiation
as analytical method, 15
induction of DNA photoproducts by, 10, 340–43
subtoxic dose of, 13
with 8-methoxypsoralen, and DNA adducts, 333–39
Urinary bladder
2-acetylaminofluorene adducts in mouse, 175–79
4-aminobiphenyl-DNA adducts in, 10, 166, 169–72
cancer
and exposure to *N*-nitroso compounds, 76
and genetic polymorphism, 166–68
and tobacco smoking, 133, 135–36
cisplatin-DNA adducts in cultured human cell line, 321, 326–27
micronuclei in, 141
susceptibility of, to aromatic amines, 166
tumours, and 2-acetylaminofluorene-DNA adduct levels, 175–79
Urine
aflatoxin B_1 in, 9, 10, 22, 39, 55, 56–62, 63, 67–70, 72–73
aflatoxin B_1-7-guanine in, 9, 10, 22, 39, 40, 55–62, 63–65, 290, 428
antimutagenic factor in, 401–403
8-hydroxy-2′-deoxyguanosine in, 412
1-hydroxypyrene in, 392–94
markers of genotoxic exposure in, 9, 10, 14, 36, 39–40
mercapturic acids in, 232–34

Urine (contd)
  3-methyladenine in, 92–96
  7-methylguanine in, 352–53
  mutagenicity of
    and diet, 397–99, 403
    comparison with other methods, 389, 391–93
    in nonsmokers, 396–99
    in smokers, 130, 396–99, 401, 402
    problems in measuring, 401–403
  N-nitrosamino acids in, 39, 83–88
  N-nitrosoproline in, 39, 97–100
  RNA adducts in, 36, 59
  theophylline in, 95
  thioethers in, 275–78

V

Validation
  of ELISA, 201, 204–205, 313–14, 317–18
  of existing methods, 12, 47–48
  of rad-equivalence, 244–45
  of radioimmunoassay, 271, 273–74
Variable tandem repeat, 437, 441–42, 445, 447
Vinyl chloride, and single-strand breaks in mice, 227–31

W

White blood cells
  aflatoxin B_1-DNA adducts in, 70
  aromatic DNA adducts in, 22, 45–46, 190–94
  as surrogate for target tissue, 47
  benzo[*a*]pyrene diol epoxide-DNA adducts in, 9, 10, 22, 43, 182–83, 184, 190–94, 454–56
  cisplatin-DNA adducts in, 44, 314–17, 322–24
  cisplatin levels in, 329–30
  measurement of DNA adducts in, 36, 142
  polycyclic aromatic hydrocarbon-DNA adducts in, 11, 454–55
Wood smoke, and adducts in placenta, 45–47

X

XAD-2, 396–97, 399
Xeroderma pigmentosum, 324–26, 440, 442
Xylene, measurement of exposure to, 10, 232–34
Xylene mercapturic acid, 232–34

PUBLICATIONS OF THE INTERNATIONAL AGENCY FOR RESEARCH ON CANCER

SCIENTIFIC PUBLICATIONS SERIES

(Available from Oxford University Press)
through local bookshops

No. 1 LIVER CANCER
1971; 176 pages; out of print

No. 2 ONCOGENESIS AND HERPESVIRUSES
Edited by P.M. Biggs, G. de-Thé & L.N. Payne
1972; 515 pages; out of print

No. 3 *N*-NITROSO COMPOUNDS: ANALYSIS AND FORMATION
Edited by P. Bogovski, R. Preussmann & E. A. Walker
1972; 140 pages; out of print

No. 4 TRANSPLACENTAL CARCINOGENESIS
Edited by L. Tomatis & U. Mohr
1973; 181 pages; out of print

*No. 5 PATHOLOGY OF TUMOURS IN LABORATORY ANIMALS. VOLUME 1. TUMOURS OF THE RAT. PART 1
Editor-in-Chief V.S. Turusov
1973; 214 pages

*No. 6 PATHOLOGY OF TUMOURS IN LABORATORY ANIMALS. VOLUME 1. TUMOURS OF THE RAT. PART 2
Editor-in-Chief V.S. Turusov
1976; 319 pages
*reprinted in one volume, Price £50.00

No. 7 HOST ENVIRONMENT INTERACTIONS IN THE ETIOLOGY OF CANCER IN MAN
Edited by R. Doll & I. Vodopija
1973; 464 pages; £32.50

No. 8 BIOLOGICAL EFFECTS OF ASBESTOS
Edited by P. Bogovski, J.C. Gilson, V. Timbrell & J.C. Wagner
1973; 346 pages; out of print

No. 9 *N*-NITROSO COMPOUNDS IN THE ENVIRONMENT
Edited by P. Bogovski & E. A. Walker
1974; 243 pages; £16.50

No. 10 CHEMICAL CARCINOGENESIS ESSAYS
Edited by R. Montesano & L. Tomatis
1974; 230 pages; out of print

No. 11 ONCOGENESIS AND HERPESVIRUSES II
Edited by G. de-Thé, M.A. Epstein & H. zur Hausen
1975; Part 1, 511 pages; Part 2, 403 pages; £65.-

No. 12 SCREENING TESTS IN CHEMICAL CARCINOGENESIS
Edited by R. Montesano, H. Bartsch & L. Tomatis
1976; 666 pages; £12.-

No. 13 ENVIRONMENTAL POLLUTION AND CARCINOGENIC RISKS
Edited by C. Rosenfeld & W. Davis
1976; 454 pages; out of print

No. 14 ENVIRONMENTAL *N*-NITROSO COMPOUNDS: ANALYSIS AND FORMATION
Edited by E.A. Walker, P. Bogovski & L. Griciute
1976; 512 pages; £37.50

No. 15 CANCER INCIDENCE IN FIVE CONTINENTS. VOLUME III
Edited by J. Waterhouse, C. Muir, P. Correa & J. Powell
1976; 584 pages; £35.-

No. 16 AIR POLLUTION AND CANCER IN MAN
Edited by U. Mohr, D. Schmähl & L. Tomatis
1977; 311 pages; out of print

No. 17 DIRECTORY OF ON-GOING RESEARCH IN CANCER EPIDEMIOLOGY 1977
Edited by C.S. Muir & G. Wagner
1977; 599 pages; out of print

No. 18 ENVIRONMENTAL CARCINOGENS: SELECTED METHODS OF ANALYSIS
Edited-in-Chief H. Egan
VOLUME 1. ANALYSIS OF VOLATILE NITROSAMINES IN FOOD
Edited by R. Preussmann, M. Castegnaro, E.A. Walker & A.E. Wassermann
1978; 212 pages; out of print

No. 19 ENVIRONMENTAL ASPECTS OF *N*-NITROSO COMPOUNDS
Edited by E.A. Walker, M. Castegnaro, L. Griciute & R.E. Lyle
1978; 566 pages; out of print

No. 20 NASOPHARYNGEAL CARCINOMA: ETIOLOGY AND CONTROL
Edited by G. de-Thé & Y. Ito
1978; 610 pages; out of print

No. 21 CANCER REGISTRATION AND ITS TECHNIQUES
Edited by R. MacLennan, C. Muir, R. Steinitz & A. Winkler
1978; 235 pages; £35.-

Prices, valid for October 1987, are subject to change without notice

SCIENTIFIC PUBLICATIONS SERIES

No. 22 ENVIRONMENTAL CARCINOGENS: SELECTED METHODS OF ANALYSIS
Editor-in-Chief H. Egan
VOLUME 2. METHODS FOR THE MEASUREMENT OF VINYL CHLORIDE IN POLY(VINYL CHLORIDE), AIR, WATER AND FOODSTUFFS
Edited by D.C.M. Squirrell & W. Thain
1978; 142 pages; out of print

No. 23 PATHOLOGY OF TUMOURS IN LABORATORY ANIMALS. VOLUME II. TUMOURS OF THE MOUSE
Editor-in-Chief V.S. Turusov
1979; 669 pages; £37.50

No. 24 ONCOGENESIS AND HERPESVIRUSES III
Edited by G. de-Thé, W. Henle & F. Rapp
1978; Part 1, 580 pages; Part 2, 522 pages; out of print

No. 25 CARCINOGENIC RISKS: STRATEGIES FOR INTERVENTION
Edited by W. Davis & C. Rosenfeld
1979; 283 pages; out of print

No. 26 DIRECTORY OF ON-GOING RESEARCH IN CANCER EPIDEMIOLOGY 1978
Edited by C.S. Muir & G. Wagner,
1978; 550 pages; out of print

No. 27 MOLECULAR AND CELLULAR ASPECTS OF CARCINOGEN SCREENING TESTS
Edited by R. Montesano, H. Bartsch & L. Tomatis
1980; 371 pages; £22.50

No. 28 DIRECTORY OF ON-GOING RESEARCH IN CANCER EPIDEMIOLOGY 1979
Edited by C.S. Muir & G. Wagner
1979; 672 pages; out of print

No. 29 ENVIRONMENTAL CARCINOGENS: SELECTED METHODS OF ANALYSIS
Editor-in-Chief H. Egan
VOLUME 3. ANALYSIS OF POLYCYCLIC AROMATIC HYDROCARBONS IN ENVIRONMENTAL SAMPLES
Edited by M. Castegnaro, P. Bogovski, H. Kunte & E.A. Walker
1979; 240 pages; out of print

No. 30 BIOLOGICAL EFFECTS OF MINERAL FIBRES
Editor-in-Chief J.C. Wagner
1980; Volume 1, 494 pages; Volume 2, 513 pages; £55.-

No. 31 *N*-NITROSO COMPOUNDS: ANALYSIS, FORMATION AND OCCURRENCE
Edited by E.A. Walker, L. Griciute, M. Castegnaro & M. Börzsönyi
1980; 841 pages; out of print

No. 32 STATISTICAL METHODS IN CANCER RESEARCH. VOLUME 1. THE ANALYSIS OF CASE-CONTROL STUDIES
By N.E. Breslow & N.E. Day
1980; 338 pages; £20.-

No. 33 HANDLING CHEMICAL CARCINOGENS IN THE LABORATORY: PROBLEMS OF SAFETY
Edited by R. Montesano, H. Bartsch, E. Boyland, G. Della Porta, L. Fishbein, R.A. Griesemer, A.B. Swan & L. Tomatis
1979; 32 pages; out of print

No. 34 PATHOLOGY OF TUMOURS IN LABORATORY ANIMALS. VOLUME III. TUMOURS OF THE HAMSTER
Editor-in-Chief V.S. Turusov
1982; 461 pages; £32.50

No. 35 DIRECTORY OF ON-GOING RESEARCH IN CANCER EPIDEMIOLOGY 1980
Edited by C.S. Muir & G. Wagner
1980; 660 pages; out of print

No. 36 CANCER MORTALITY BY OCCUPATION AND SOCIAL CLASS 1851-1971
By W.P.D. Logan
1982; 253 pages; £22.50

No. 37 LABORATORY DECONTAMINATION AND DESTRUCTION OF AFLATOXINS B_1, B_2, G_1, G_2 IN LABORATORY WASTES
Edited by M. Castegnaro, D.C. Hunt, E.B. Sansone, P.L. Schuller, M.G. Siriwardana, G.M. Telling, H.P. Van Egmond & E.A. Walker
1980; 59 pages; £6.50

No. 38 DIRECTORY OF ON-GOING RESEARCH IN CANCER EPIDEMIOLOGY 1981
Edited by C.S. Muir & G. Wagner
1981; 696 pages; out of print

No. 39 HOST FACTORS IN HUMAN CARCINOGENESIS
Edited by H. Bartsch & B. Armstrong
1982; 583 pages; £37.50

No. 40 ENVIRONMENTAL CARCINOGENS: SELECTED METHODS OF ANALYSIS
Edited-in-Chief H. Egan
VOLUME 4. SOME AROMATIC AMINES AND AZO DYES IN THE GENERAL AND INDUSTRIAL ENVIRONMENT
Edited by L. Fishbein, M. Castegnaro, I.K. O'Neill & H. Bartsch
1981; 347 pages; £22.50

No. 41 *N*-NITROSO COMPOUNDS: OCCURRENCE AND BIOLOGICAL EFFECTS
Edited by H. Bartsch, I.K. O'Neill, M. Castegnaro & M. Okada
1982; 755 pages; £37.50

No. 42 CANCER INCIDENCE IN FIVE CONTINENTS. VOLUME IV
Edited by J. Waterhouse, C. Muir, K. Shanmugaratnam & J. Powell
1982; 811 pages; £37.50

SCIENTIFIC PUBLICATIONS SERIES

No. 43 LABORATORY DECONTAMINATION AND DESTRUCTION OF CARCINOGENS IN LABORATORY WASTES: SOME *N*-NITROSAMINES
Edited by M. Castegnaro, G. Eisenbrand, G. Ellen, L. Keefer, D. Klein, E.B. Sansone, D. Spincer, G. Telling & K. Webb
1982; 73 pages; £7.50

No. 44 ENVIRONMENTAL CARCINOGENS:. SELECTED METHODS OF ANALYSIS
Editor-in-Chief H. Egan
VOLUME 5. SOME MYCOTOXINS
Edited by L. Stoloff, M. Castegnaro, P. Scott, I.K. O'Neill & H. Bartsch
1983; 455 pages; £22.50

No. 45 ENVIRONMENTAL CARCINOGENS: SELECTED METHODS OF ANALYSIS
Editor-in-Chief H. Egan
VOLUME 6. *N*-NITROSO COMPOUNDS
Edited by R. Preussmann, I.K. O'Neill, G. Eisenbrand, B. Spiegelhalder & H. Bartsch
1983; 508 pages; £22.50

No. 46 DIRECTORY OF ON-GOING RESEARCH IN CANCER EPIDEMIOLOGY 1982
Edited by C.S. Muir & G. Wagner
1982; 722 pages; out of print

No. 47 CANCER INCIDENCE IN SINGAPORE 1968-1977
Edited by K. Shanmugaratnam, H.P. Lee & N.E. Day
1982; 171 pages; out of print

No. 48 CANCER INCIDENCE IN THE USSR
Second Revised Edition
Edited by N.P. Napalkov, G.F. Tserkovny, V.M. Merabishvili, D.M. Parkin, M. Smans & C.S. Muir,
1983; 75 pages; £12.-

No. 49 LABORATORY DECONTAMINATION AND DESTRUCTION OF CARCINOGENS IN LABORATORY WASTES: SOME POLYCYCLIC AROMATIC HYDROCARBONS
Edited by M. Castegnaro, G. Grimmer, O. Hutzinger, W. Karcher, H. Kunte, M. Lafontaine, E.B. Sansone, G. Telling & S.P. Tucker
1983; 81 pages; £9.-

No. 50 DIRECTORY OF ON-GOING RESEARCH IN CANCER EPIDEMIOLOGY 1983
Edited by C.S. Muir & G. Wagner
1983; 740 pages; out of print

No. 51 MODULATORS OF EXPERIMENTAL CARCINOGENESIS
Edited by V. Turusov & R. Montesano
1983; 307 pages; £22.50

No. 52 SECOND CANCER IN RELATION TO RADIATION TREATMENT FOR CERVICAL CANCER
Edited by N.E. Day & J.D. Boice, Jr
1984; 207 pages; £20.-

No. 53 NICKEL IN THE HUMAN ENVIRONMENT
Editor-in-Chief F.W. Sunderman, Jr
1984: 530 pages; £32.50

No. 54 LABORATORY DECONTAMINATION AND DESTRUCTION OF CARCINOGENS IN LABORATORY WASTES: SOME HYDRAZINES
Edited by M. Castegnaro, G. Ellen, M. Lafontaine, H.C. van der Plas, E.B. Sansone & S.P. Tucker
1983; 87 pages; £9.-

No. 55 LABORATORY DECONTAMINATION AND DESTRUCTION OF CARCINOGENS IN LABORATORY WASTES: SOME *N*-NITROSAMIDES
Edited by M. Castegnaro, M. Benard, L.W. van Broekhoven, D. Fine, R. Massey, E.B. Sansone, P.L.R. Smith, B. Spiegelhalder, A. Stacchini, G. Telling & J.J. Vallon
1984; 65 pages; £7.50

No. 56 MODELS, MECHANISMS AND ETIOLOGY OF TUMOUR PROMOTION
Edited by M. Börszönyi, N.E. Day, K. Lapis & H. Yamasaki
1984; 532 pages; £32.50

No. 57 *N*-NITROSO COMPOUNDS: OCCURRENCE, BIOLOGICAL EFFECTS AND RELEVANCE TO HUMAN CANCER
Edited by I.K. O'Neill, R.C. von Borstel, C.T. Miller, J. Long & H. Bartsch
1984; 1011 pages; £80.-

No. 58 AGE-RELATED FACTORS IN CARCINOGENESIS
Edited by A. Likhachev, V. Anisimov & R. Montesano
1985; 288 pages; £20.-

No. 59 MONITORING HUMAN EXPOSURE TO CARCINOGENIC AND MUTAGENIC AGENTS
Edited by A. Berlin, M. Draper, K. Hemminki & H. Vainio
1984; 457 pages; £27.50

No. 60 BURKITT'S LYMPHOMA: A HUMAN CANCER MODEL
Edited by G. Lenoir, G. O'Conor & C.L.M. Olweny
1985; 484 pages; £22.50

No. 61 LABORATORY DECONTAMINATION AND DESTRUCTION OF CARCINOGENS IN LABORATORY WASTES: SOME HALOETHERS
Edited by M. Castegnaro, M. Alvarez, M. Iovu, E.B. Sansone, G.M. Telling & D.T. Williams
1984; 53 pages; £7.50

No. 62 DIRECTORY OF ON-GOING RESEARCH IN CANCER EPIDEMIOLOGY 1984
Edited by C.S. Muir & G.Wagner
1984; 728 pages; £26.-

No. 63 VIRUS-ASSOCIATED CANCERS IN AFRICA
Edited by A.O. Williams, G.T. O'Conor, G.B. de-Thé & C.A. Johnson
1984; 774 pages; £22.-

SCIENTIFIC PUBLICATIONS SERIES

No. 64 LABORATORY DECONTAMINATION AND DESTRUCTION OF CARCINOGENS IN LABORATORY WASTES: SOME AROMATIC AMINES AND 4-NITROBIPHENYL
Edited by M. Castegnaro, J. Barek, J. Dennis, G. Ellen, M. Klibanov, M. Lafontaine, R. Mitchum, P. Van Roosmalen, E.B. Sansone, L.A. Sternson & M. Vahl
1985; 85 pages; £6.95

No. 65 INTERPRETATION OF NEGATIVE EPIDEMIOLOGICAL EVIDENCE FOR CARCINOGENICITY
Edited by N.J. Wald & R. Doll
1985; 232 pages; £20.-

No. 66 THE ROLE OF THE REGISTRY IN CANCER CONTROL
Edited by D.M. Parkin, G. Wagner & C. Muir
1985; 155 pages; £10.-

No. 67 TRANSFORMATION ASSAY OF ESTABLISHED CELL LINES: MECHANISMS AND APPLICATION
Edited by T. Kakunaga & H. Yamasaki
1985; 225 pages; £20.-

No. 68 ENVIRONMENTAL CARCINOGENS: SELECTED METHODS OF ANALYSIS VOLUME 7. SOME VOLATILE HALOGENATED HYDROCARBONS
Edited by L. Fishbein & I.K. O'Neill
1985; 479 pages; £20.-

No. 69 DIRECTORY OF ON-GOING RESEARCH IN CANCER EPIDEMIOLOGY 1985
Edited by C.S. Muir & G. Wagner
1985; 756 pages; £22.

No. 70 THE ROLE OF CYCLIC NUCLEIC ACID ADDUCTS IN CARCINOGENESIS AND MUTAGENESIS
Edited by B. Singer & H. Bartsch
1986; 467 pages; £40.-

No. 71 ENVIRONMENTAL CARCINOGENS: SELECTED METHODS OF ANALYSIS VOLUME 8. SOME METALS: As, Be, Cd, Cr, Ni, Pb, Se, Zn
Edited by I.K. O'Neill, P. Schuller & L. Fishbein
1986; 485 pages; £20.

No. 72 ATLAS OF CANCER IN SCOTLAND 1975-1980: INCIDENCE AND EPIDEMIOLOGICAL PERSPECTIVE
Edited by I. Kemp, P. Boyle, M. Smans & C. Muir
1985; 282 pages; £35.-

No. 73 LABORATORY DECONTAMINATION AND DESTRUCTION OF CARCINOGENS IN LABORATORY WASTES: SOME ANTINEOPLASTIC AGENTS
Edited by M. Castegnaro, J. Adams, M. Armour, J. Barek, J. Benvenuto, C. Confalonieri, U. Goff, S. Ludeman, D. Reed, E.B. Sansone & G. Telling
1985; 163 pages; £10.-

No. 74 TOBACCO: A MAJOR INTERNATIONAL HEALTH HAZARD
Edited by D. Zaridze & R. Peto
1986; 324 pages; £20.-

No. 75 CANCER OCCURRENCE IN DEVELOPING COUNTRIES
Edited by D.M. Parkin
1986; 339 pages; £20.-

No. 76 SCREENING FOR CANCER OF THE UTERINE CERVIX
Edited by M. Hakama, A.B. Miller & N.E. Day
1986; 315 pages; £25.-

No. 77 HEXACHLOROBENZENE: PROCEEDINGS OF AN INTERNATIONAL SYMPOSIUM
Edited by C.R. Morris & J.R.P. Cabral
1986; 668 pages; £50.-

No. 78 CARCINOGENICITY OF ALKYLATING CYTOSTATIC DRUGS
Edited by D. Schmähl & J. M. Kaldor
1986; 338 pages; £25.-

No. 79 STATISTICAL METHODS IN CANCER RESEARCH. VOLUME III. THE DESIGN AND ANALYSIS OF LONG-TERM ANIMAL EXPERIMENTS
By J.J. Gart, D. Krewski, P.N. Lee, R.E. Tarone & J. Wahrendorf
1986; 219 pages; £20.-

No. 80 DIRECTORY OF ON-GOING RESEARCH IN CANCER EPIDEMIOLOGY 1986
Edited by C.S. Muir & G. Wagner
1986; 805 pages; £22.-

No. 81 ENVIRONMENTAL CARCINOGENS: METHODS OF ANALYSIS AND EXPOSURE MEASUREMENT. VOLUME 9. PASSIVE SMOKING
Edited by I.K. O'Neill, K.D. Brunnemann, B. Dodet & D. Hoffmann
1987; 379 pages; £30.-

No. 82 STATISTICAL METHODS IN CANCER RESEARCH. VOLUME II. THE DESIGN AND ANALYSIS OF COHORT STUDIES
By N.E. Breslow & N.E. Day
1987; 404 pages; £30.-

No. 83 LONG-TERM AND SHORT-TERM ASSAYS FOR CARCINOGENS: A CRITICAL APPRAISAL
Edited by R. Montesano, H. Bartsch, H. Vainio, J. Wilbourn & H. Yamasaki
1986; 575 pages; £32.50

No. 84 THE RELEVANCE OF *N*-NITROSO COMPOUNDS TO HUMAN CANCER: EXPOSURES AND MECHANISMS
Edited by H. Bartsch, I.K. O'Neill & R. Schulte-Hermann
1987; 671 pages; £50.-

SCIENTIFIC PUBLICATIONS SERIES

No. 85 ENVIRONMENTAL CARCINOGENS: METHODS OF ANALYSIS AND EXPOSURE MEASUREMENT. VOLUME 10. BENZENE AND ALKYLATED BENZENES
Edited by L. Fishbein & I.K. O'Neill
(in press) 1988; 318 pages; £35.-

No. 86 DIRECTORY OF ON-GOING RESEARCH IN CANCER EPIDEMIOLOGY 1987
Edited by D.M. Parkin & J. Wahrendorf
1987; 685 pages; £22.-

No. 87 INTERNATIONAL INCIDENCE OF CHILDHOOD CANCER
Edited by D.M. Parkin, C.A. Stiller, G.J. Draper, C.A. Bieber, B. Terracini & J.L. Young
(in press) 1988; approx. 400 pages; £35.-

No. 88 CANCER INCIDENCE IN FIVE CONTINENTS. VOLUME V
Edited by C. Muir, J. Waterhouse, T. Mack, J. Powell & S. Whelan
1988; 1004 pages; £50.-

No. 89 METHODS FOR DETECTING DNA DAMAGING AGENTS IN HUMANS: APPLICATIONS IN CANCER EPIDEMIOLOGY AND PREVENTION
Edited by H. Bartsch, K. Hemminki & I.K. O'Neill
1988; approx. 520 pages; £45.-

No. 90 NON-OCCUPATIONAL EXPOSURE TO MINERAL FIBRES
Edited by J. Bignon, J. Peto & R. Saracci
(in press) 1988; approx. 500 pages; £45.-

No. 91 TRENDS IN CANCER INCIDENCE IN SINGAPORE 1968-1982
Edited by H.P. Lee, N.E. Day & K. Shanmugaratnam
1988; approx. 160 pages; £25.-

IARC MONOGRAPHS ON THE EVALUATION OF THE CARCINOGENIC RISK OF CHEMICALS TO HUMANS

(English editions only)

(Available from booksellers through the network of WHO Sales Agents*)

Volume 1
Some inorganic substances, chlorinated hydrocarbons, aromatic amines, N-*nitroso compounds, and natural products*
1972; 184 pages; out of print

Volume 2
Some inorganic and organometallic compounds
1973; 181 pages; out of print

Volume 3
Certain polycyclic aromatic hydrocarbons and heterocyclic compounds
1973; 271 pages; out of print

Volume 4
Some aromatic amines, hydrazine and related substances, N-*nitroso compounds and miscellaneous alkylating agents*
1974; 286 pages;
Sw. fr. 18.-

Volume 5
Some organochlorine pesticides
1974; 241 pages; out of print

Volume 6
Sex hormones
1974; 243 pages;
out of print

Volume 7
Some anti-thyroid and related substances, nitrofurans and industrial chemicals
1974; 326 pages; out of print

Volume 8
Some aromatic azo compounds
1975; 357 pages; Sw.fr. 36.-

Volume 9
Some aziridines, N-, S- *and* O-*mustards and selenium*
1975; 268 pages; Sw. fr. 27.-

Volume 10
Some naturally occurring substances
1976; 353 pages; out of print

Volume 11
Cadmium, nickel, some epoxides, miscellaneous industrial chemicals and general considerations on volatile anaesthetics
1976; 306 pages; out of print

Volume 12
Some carbamates, thiocarbamates and carbazides
1976; 282 pages; Sw. fr. 34.-

Volume 13
Some miscellaneous pharmaceutical substances
1977; 255 pages; Sw. fr. 30.-

Volume 14
Asbestos
1977; 106 pages; out of print

Volume 15
Some fumigants, the herbicides 2,4-D and 2,4,5-T, chlorinated dibenzodioxins and miscellaneous industrial chemicals
1977; 354 pages; Sw. fr. 50.-

Volume 16
Some aromatic amines and related nitro compounds — hair dyes, colouring agents and miscellaneous industrial chemicals
1978; 400 pages; Sw. fr. 50.-

Volume 17
Some N-*nitroso compounds*
1978; 365 pages; Sw. fr. 50.

Volume 18
Polychlorinated biphenyls and polybrominated biphenyls
1978; 140 pages; Sw. fr. 20.-

Volume 19
Some monomers, plastics and synthetic elastomers, and acrolein
1979; 513 pages; Sw. fr. 60.-

Volume 20
Some halogenated hydrocarbons
1979; 609 pages; Sw. fr. 60.-

Volume 21
Sex hormones (II)
1979; 583 pages; Sw. fr. 60.-

Volume 22
Some non-nutritive sweetening agents
1980; 208 pages; Sw. fr. 25.-

Volume 23
Some metals and metallic compounds
1980; 438 pages; Sw. fr. 50.-

Volume 24
Some pharmaceutical drugs
1980; 337 pages; Sw. fr. 40.-

Volume 25
Wood, leather and some associated industries
1981; 412 pages; Sw. fr. 60.-

Volume 26
Some antineoplastic and immunosuppressive agents
1981; 411 pages; Sw. fr. 62.-

*A list of these Agents may be obtained by writing to the World Health Organization, Distribution and Sales Service, 1211 Geneva 27, Switzerland

IARC MONOGRAPHS SERIES

Volume 27
Some aromatic amines, anthraquinones and nitroso compounds, and inorganic fluorides used in drinking-water and dental preparations
1982; 341 pages; Sw. fr. 40.-

Volume 28
The rubber industry
1982; 486 pages; Sw. fr. 70.-

Volume 29
Some industrial chemicals and dyestuffs
1982; 416 pages; Sw. fr. 60.-

Volume 30
Miscellaneous pesticides
1983; 424 pages; Sw. fr. 60.-

Volume 31
Some food additives, feed additives and naturally occurring substances
1983; 14 pages; Sw. fr. 60.-

Volume 32
Polynuclear aromatic compounds, Part 1, Chemical, environmental and experimental data
1984; 477 pages; Sw. fr. 60.-

Volume 33
Polynuclear aromatic compounds, Part 2, Carbon blacks, mineral oils and some nitroarenes
1984; 245 pages; Sw. fr. 50.-

Volume 34
Polynuclear aromatic compounds, Part 3, Industrial exposures in aluminium production, coal gasification, coke production, and iron and steel founding
1984; 219 pages; Sw. fr. 48.-

Volume 35
Polynuclear aromatic compounds, Part 4, Bitumens, coal-tars and derived products, shale-oils and soots
1985; 271 pages; Sw. fr.70.-

Volume 36
Allyl compounds, aldehydes, epoxides and peroxides
1985; 369 pages; Sw. fr. 70.-

Volume 37
Tobacco habits other than smoking; betel-quid and areca-nut chewing; and some related nitrosamines
1985; 291 pages; Sw. fr. 70.-

Volume 38
Tobacco smoking
1986; 421 pages; Sw. fr. 75.-

Volume 39
Some chemicals used in plastics and elastomers
1986; 403 pages; Sw. fr. 60.-

Volume 40
Some naturally occurring and synthetic food components, furocoumarins and ultraviolet radiation
1986; 444 pages; Sw. fr. 65.-

Volume 41
Some halogenated hydrocarbons and pesticide exposures
1986; 434 pages; Sw. fr. 65.-

Volume 42
Silica and some silicates
1987; 289 pages; Sw. fr. 65.-

Volume 43
Man-made mineral fibres and radon
(in press)

Volume 44
Alcohol and alcoholic beverages
(in preparation)

Supplement No. 1
Chemicals and industrial processes associated with cancer in humans (IARC Monographs, Volumes 1 to 20)
1979; 71 pages; out of print

Supplement No. 2
Long-term and short-term screening assays for carcinogens: a critical appraisal
1980; 426 pages; Sw. fr. 40.-

Supplement No. 3
Cross index of synonyms and trade names in Volumes 1 to 26
1982; 199 pages; Sw. fr. 60.-

Supplement No. 4
Chemicals, industrial processes and industries associated with cancer in humans (IARC Monographs, Volumes 1 to 29)
1982; 292 pages; Sw. fr. 60.-

Supplement No. 5
Cross index of synonyms and trade names in Volumes 1 to 36
1985; 259 pages; Sw. fr. 60.-

Supplement No. 6
Genetic and related effects: An updating of selected IARC Monographs from Volumes 1-42
1987; 750 pages; Sw. fr. 80.-

Supplement No. 7
Overall evaluations of carcinogenicity: An updating of IARC Monographs Volumes 1-42
1987; 440 pages; Sw. fr. 65.-

*From Volume 43 onwards, the series title has been changed to IARC MONOGRAPHS ON THE EVALUATION OF CARCINOGENIC RISKS TO HUMANS

INFORMATION BULLETINS ON THE SURVEY OF CHEMICALS BEING TESTED FOR CARCINOGENICITY

(Available from IARC and WHO Sales Agents)

No. 8 (1979)
Edited by M.-J. Ghess, H. Bartsch
& L. Tomatis
604 pages; Sw. fr. 40.-

No. 9 (1981)
Edited by M.-J. Ghess, J.D. Wilbourn,
H. Bartsch & L. Tomatis
294 pages; Sw. fr. 41.-

No. 10 (1982)
Edited by M.-J. Ghess, J.D. Wilbourn
& H. Bartsch
362 pages; Sw. fr. 42.-

No. 11 (1984)
Edited by M.-J. Ghess, J.D. Wilbourn,
H. Vainio & H. Bartsch
362 pages; Sw. fr. 50.-

No. 12 (1986)
Edited by M.-J. Ghess, J.D. Wilbourn,
A. Tossavainen & H. Vainio
385 pages; Sw. fr. 50.-

NON-SERIAL PUBLICATIONS

(Available from IARC)

ALCOOL ET CANCER
By A. Tuyns (in French only)
1978; 42 pages; Fr. fr. 35.-

CANCER MORBIDITY AND CAUSES OF DEATH AMONG DANISH BREWERY WORKERS
By O.M. Jensen
1980; 143 pages; Fr. fr. 75.-

DIRECTORY OF COMPUTER SYSTEMS USED IN CANCER REGISTRIES
By H.R. Menck & D.M. Parkin
1986; 236 pages; Fr. fr. 50.-